# Resorcinol

## Chemistry, Technology and Applications

Raj B. Durairaj

# Resorcinol

## Chemistry, Technology and Applications

With 740 Figures

Springer

**Raj B. Durairaj**
Indspec
Chemical Corporation
William Pitt Way 1010
15238 Pittsburgh, USA

Library of Congress Control Number: 2005921447

**ISBN-10 3-540-25142-1 Springer Berlin Heidelberg New York**
**ISBN-13 978-3-540-25142-2 Springer Berlin Heidelberg New York**
**DOI 10.1007/b982897**



**Springer is a part of Springer Science+Business Media**
springeronline.com

Printed in Germany



Cover design: *design&production* GmbH, Heidelberg
Typesetting and production: LE-TEX Jelonek, Schmidt & Vöckler GbR, Leipzig, Germany
Printed on acid-free paper 2/3141/YL - 5 4 3 2 1 0

*This book is dedicated to my father, Sri. R. Bojayan (deceased), and mother, Mrs. B. Chinthamani, for their inspirational wish to see their son succeed in life, and also Lord Sri. Venkateswara of Pittsburgh for his blessings, which gave me the strength, motivation and good excellent health for the successful completion of this book.*

# Preface

In my view, resorcinol is a "wonder chemical" and the chemistry associated with resorcinol is fascinating, challenging and rewarding. Ever since this chemical was synthesized about 150 years ago, the tremendous efforts made on the investigation of resorcinol chemistry produced unique and novel products for various applications. I was quite fortunate to have the opportunity to work on resorcinol chemistry in 1986. During the past 18 years of my R&D experience with resorcinol chemistry, I have admired and been fascinated with this chemistry. Because of this, I had been thinking of writing a true chemistry book on resorcinol so that the scientific community around the world could learn about the uniqueness of this chemical. For a very long time I hesitated to write this book mainly because I doubted whether I could do it.

In April 2003, one of my friends strongly suggested and convinced me that I could write this resorcinol chemistry book. Since then, I have taken this work as a special project for myself. The complete book manuscript planning, formatting and writing was done during the evening and nighttime hours at home, spending on average about six to eight hours per day. I never believed myself that I could finish this monumental task in less than 15 months time, particularly when this was done after my regular office hours.

Resorcinol chemical has a unique molecular structure. It provides the opportunity for chemists, scientists and technologists to explore and develop advanced technologies in the fields of organic, polymer, agricultural and pharmaceutical chemistries with this chemical. In the modern scientific world, a breakthrough technology does not come very often. But with resorcinol chemistry, novel technologies have been developed and reported. Some of the notable technologies developed using the resorcinol chemical are RFL adhesives, Lexan SLX, Zylon fibers, PEN (cyanoarylene ether polymer), UV absorbers, RDP Flame retardant, Penacolite resins (for radial tires), Intal (asthma drug), Osten (osteoporosis), Mikado (herbicide) and Goal (herbicide).

In the scientific community, though remarkable progress has been made in the past with the resorcinol chemistry, a lot of potential still exists for the exploration and development of advanced chemistries and technologies with this chemistry. This reminds me of a quote from the great Tamil Poetess

Avvaiyar who lived thousands of years ago in India: "*What we have learned is like a handful of earth. What we have yet to learn is like the whole world.*" Therefore, by acquiring knowledge in resorcinol chemistry, scientists could expand their skills and explore further in their advanced technologies for humankind. With this in mind, my intention in this book is to outline the advantages of resorcinol chemistry and technology in various applications. The unique feature in this book is that each chapter includes a brief critical review on the subject topic.

Chapter 1 outlines the general structure and physical properties of resorcinol. Chapter 2 gives an account of the current resorcinol manufacturing technologies and also suggests a need for an alternate economical process for the production of resorcinol. In Chap. 3, a general overview of various resorcinolic derivatives is presented. Special compounds based on resorcinol chemical and their industrial applications are briefly detailed in Chap. 4. The RFL technology was developed 50 years ago. Still this chemistry and technology dominates in the adhesion development of various synthetic fibers to rubber compounds. A detailed outline of the chemistry and mechanisms of fibers to rubbers is presented in Chap. 5. For the first time, based on the experimental conditions, a mechanism has been proposed for the bonding of polyester to rubber compound. Resorcinolic resins play a vital role in the bonding of steel tire cords to rubber compounds, which results in the production of high-performance steel-belted radial tires. A bonding mechanism explaining the advantages of resorcinolic resins in the steel cords adhesion improvements is presented in Chap. 6.

Due to meta-substitution, resorcinol provides unexpected and also outstanding properties to polymers synthesized using this chemical. For example, the weatherable thermoplastic polyestercarbonate material developed and sold under the trade names Sollx and Lexan SLX by the General Electric Company showed an outstanding performance of retaining its gloss and color after exposure to sunlight for 10 to 15 years. Similarly, a highly crystalline super-thermoplastic poly (cyanoarylene ether) material was synthesized from resorcinol and sold under the trade name PEN. A resorcinol derivative, namely 4,6-diaminoresorcinol, was used in the production of world's strongest fiber Zylon. Various polymeric materials produced from resorcinol chemistry and their applications are presented in Chap. 7.

A very detailed chemistry and an overview of various resorcinol based UV-absorber compounds and their industrial applications are discussed in Chap. 8. The advantages of resorcinol-derived phosphate ester compounds as flame retardants in the advanced thermoplastic polymers such as the PC, PPO, PC/ABS and PPO/HIPS are also outlined in Chap. 8. In pharmaceutical applications, resorcinol chemistry offers numerous possibilities for developing drug compounds to treat and cure various human illnesses and diseases. In Chap. 9, current activities on resorcinol-based drug chemistries and their developments are discussed. For photoresist applications, resorcinol-based sensitizer compounds offer several advantages and are discussed in Chap. 10.

Throughout the book, wherever possible, procedures for the preparation of compounds and derivatives from resorcinol have been provided. I would like to point out that this book contains information on the chemistry pertaining only to the reactions of resorcinol at the 2, 4 and 6 positions and the two hydroxyl groups. I have written this book with the hope that it may provide enough and suitable information to scientists and professional communities to obtain more knowledge and generate new ideas about this chemistry. I strongly believe that my book will be of value to chemists, researchers, students, professors and other scientific people working in the universities, colleges, industrial and scientific R&D labs around the world.

In the preparation of this manuscript, I made every effort to avoid any mistakes. But, for some reason, if there are any errors or incorrect citations in this book, I accept sole responsibility and personally apologize.

*Raj B. Durairaj, Ph.D.*

**Acknowledgement** The first and most important person I have to thank and express my gratitude is my wife Latha Durairaj. Her patience, understanding, support and encouragement at all stages of preparation of this manuscript in the evening and nighttime hours at home for 15 months were invaluable. Without my wife's support, I could not have achieved my success in this personal book project. I also have to express my deepest thanks to my children, Mohanraj (son) and Umadevi (daughter) for their enthusiastic support, tolerance and love in spite of my preoccupation with the manuscript preparation at home. It is also equally important to acknowledge and thank my friend Dr. G. Rathinakumar and his family members for their extraordinary help giving me the peace of mind necessary for the planning and completion of this book.

My special thanks to Dr. Lalit Chordia, CEO and president, Thar Technologies, who was instrumental in first convincing me that I could write this chemistry book. I am very deeply indebted and grateful to my supervisor Mr. Thomas K. Volek, vice president of Technology, for his outstanding support and for reviewing the complete manuscript and providing constructive and positive comments. I appreciate and acknowledge the help of Ms. Nancy Manion and Prof. C.T. Vijayakumar, Ph.D for their partial review of this manuscript.

It is always nice to recognize people who are instrumental and responsible for a person's success. In this respect, I would like to thank Prof. Frank D. Blum (University of Missouri-Rolla), Prof. M.T. Shaw (University of Connecticut), Prof. G. Ramamurthy (Retd), and Prof. T.S. Venkatraman (Drexel University) for their motivational advice, help and guidance at different stages in my life.

My special and sincere thanks to Mrs. Dhanalakshmi Rathinakumar, who reviewed the entire manuscript for spelling, typos and grammar during her vacation in Pittsburgh, USA. Also, I would like to acknowledge the moral support and encouragement received from my family members and friends in USA and India during the preparation of this manuscript.

Finally, it is my obligation and duty to sincerely thank all of the Research and Development, and Analytical Department scientists and staff at Indspec Chemical Corporation's Harmarville R&D Center for their appreciation, kindness and wonderful support of my "dream come true" endeavor of publishing this resorcinol chemistry book for the scientific community around the world.

# Contents

# 1 Resorcinol Structure and Physical Properties

## 1.1 Introduction

Resorcinol has been widely known as a versatile chemical compound utilized extensively in the development of advanced chemistries and technologies to benefit humans and other living things. Resorcinol is one of the most remarkable chemicals known in organic chemistry, helping scientists, technologists and researchers around the world to research and develop modern technologies in the diversified fields of chemistry. Due to its unique structure, resorcinol is providing fascinating chemistries and technologies that not many chemicals known in organic chemistry can do. A tremendous amount of exciting and most rewarding chemistries have been developed using resorcinol over the past 125 to 150 years and this trend will be expected to continue in the future.

Resorcinol is a white crystalline dihydric phenol having a faint, characteristic aromatic odor, and with a sweetish bitter taste. This compound is also known as resorcin, meta-dihydroxybenzene, 1, 3-dihdroxybenzene, 1,3-benzenediol and 3-hydroxyphenol. Structurally, the chemical resorcinol can be represented as follows (Figure 1.1).

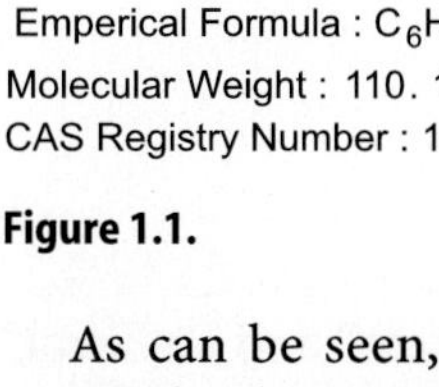


Emperical Formula : $C_6H_6O_2$
Molecular Weight : 110. 11
CAS Registry Number : 108 - 46 - 3

**Figure 1.1.**

As can be seen, the resorcinol molecule has two hydroxyl groups in the aromatic ring structure, and they are located at their meta-positions with respect to each hydroxyl group. The high reactivity of resorcinol is primarily associated with the location of these two hydroxyl groups in the benzene ring.

As far as the reactivity of resorcinol is concerned, the hydrogen atoms adjacent to the hydroxyl groups, namely at carbon atoms 2,4 and 6, are particularly reactive. The hydrogen atom located at the 5-position of the resorcinol molecule is basically non-reactive, and therefore does not take part in any chemical reactions under normal reaction conditions. Also, the carbon atoms present at the 4 and 6 positions may be designated as "beta or $\beta$", the carbon atom at the 2-position may be designated as "gamma or $\gamma$", and the 5-position carbon atom as "alpha or $\alpha$".

## 1.2 Crystalline Structure and Polymorphism

Resorcinol exists in two crystalline forms, the $\alpha$-form and the $\beta$-form, as evidenced from X-ray diffraction measurements [1–3]. $\alpha$-Resorcinol can be obtained through recrystallization from alcohol or benzene or spontaneous transformation from $\beta$-resorcinol below 71 °C. The $\alpha$-form ordinarily obtained by recrystallization is stable up to about 71 °C under normal pressure conditions. This $\alpha$-resorcinol is converted to the $\beta$-form at about 74 °C, and $\beta$-resorcinol is stable up to the melting point.

$\beta$-Resorcinol could be obtained through recrystallization of the melt, through sublimation or by heating above the transformation point. The heat of transition from $\alpha$- to $\beta$-resorcinol was determined to be (220±5) cal/mol [2].

The melting point for $\alpha$-resorcinol is 108 °C, and for $\beta$-resorcinol is 110.5 °C. Resorcinol in the $\beta$-form is somewhat more dense (density = 1.327) than the $\alpha$-form (density = 1.28), which suggests hydrogen bonding between oriented molecules in the crystal lattice [4]. Both $\alpha$- and $\beta$-resorcinol crystallize to produce orthorhombic hemimorphic crystals.

Infrared and Raman spectroscopic studies have been carried out on $\alpha$-resorcinol crystals at high pressure, and also at low temperature conditions [5,6]. The $\alpha$-resorcinol crystals behaved like hydrogen bonded phenolic polymer, exhibiting pyroelectric and piezoelectric properties [2,3,5]. The equilibrium $\alpha\beta$ transition temperature, $T_{\alpha\beta}$, between the $\alpha$-crystalline phase and $\beta$-crystalline phase has been determined [7]. From this study, it was concluded that the $\alpha\beta$ transition would be induced by drastic breakdown of the hydrogen bonds accompanied by redistribution of protons between the covalent structure and ionic structure associated with the transition.

## 1.3 Tautomeric Form of Resorcinol

Resorcinol can exist in two tautomeric forms, namely a "keto" and an "enol" form (Figure 1.2).

Organic compounds exhibiting the "keto" and "enol" structure can show reactions characteristics of these two structures. Resorcinol shows some of the

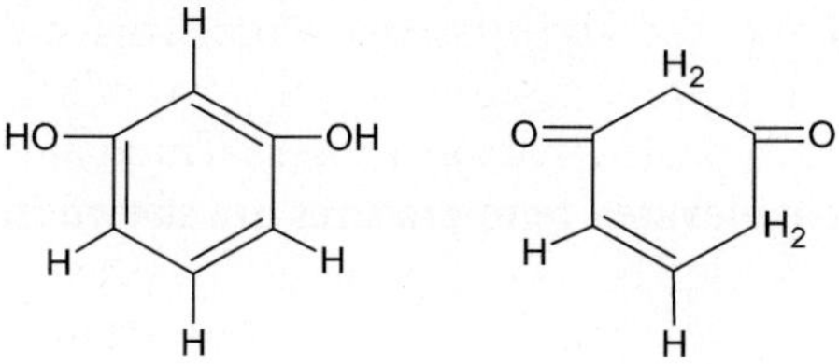


**Figure 1.2.**

reactions of the ketonic form, but the equilibrium seems to be more in favor of the completely enolic form [8].

## 1.4 Physical Properties of Resorcinol

### 1.4.1 Vapor Pressure Data

Resorcinol melts at 108.8 °C and tends to volatilize. This compound has the tendency to sublime at its melting point. Vapor pressure data obtained for resorcinol are presented in Table 1.1 [4,9,10].

**Table 1.1.** Vapor pressure data for resorcinol

| Vapor pressure (mm Hg) | Temperature (°C) |
|---|---|
| 1 | 108.4 |
| 3 | 130 |
| 5 | 138 |
| 8.5 | 150 |
| 10 | 152.3 |
| 20 | 168 |
| 23.5 | 170 |
| 40 | 185.3 |
| 53 | 190 |
| 60 | 195.8 |
| 100 | 209.8 |
| 200 | 230.8 |
| 400 | 253.4 |
| 760 | 276.5 |

Data from [4,9,10]

This data can be used to separate resorcinol from a mixture of organic compounds, and also to purify resorcinol.

### 1.4.2
### Melt Viscosity Data

Molten resorcinol viscosities measured at elevated temperatures are shown in Figure 1.3 [11].

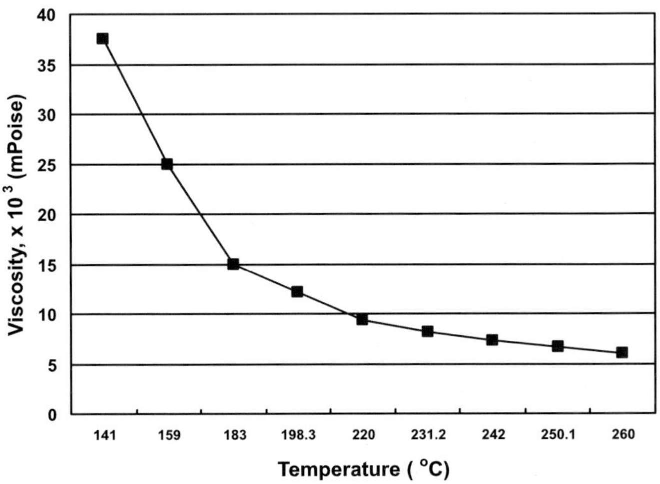


**Figure 1.3.** Viscosity of resorcinol as a function of temperature

### 1.4.3
### Solubility of Resorcinol in Solvents

Resorcinol is highly soluble in water. The density of water solutions of resorcinol appears to increase approximately proportionally to the resorcinol concentration [12]. The solubility of resorcinol in water at various temperatures is plotted and shown in Figure 1.4 [13].

Resorcinol solubility in solvents may vary widely due to the presence of $\alpha$- and $\beta$-phases [12]. Solubility data of resorcinol in different organic solvents is presented in Table 1.2 [4, 13–15].

### 1.4.4
### Weight Loss of Resorcinol

The percent weight loss temperature of resorcinol determined from thermogravimetric analysis (TGA) is shown in Table 1.3 [4, 8].

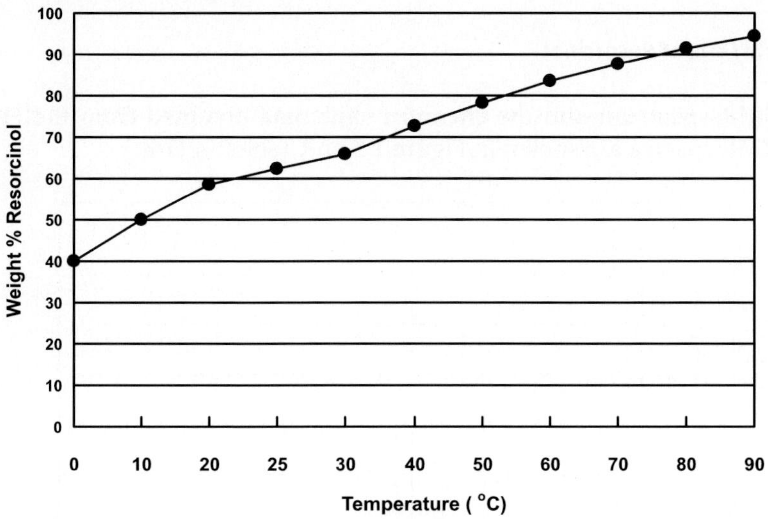


**Figure 1.4.** Solubility of resorcinol in water at various temperatures

**Table 1.2.** Solubility of resorcinol in various solvents

| Solvent | Solubility (Weight percent) at 20 °C | 60 °C |
|---|---|---|
| Water | 58.4 | 83.5 |
| Ethanol | 61 | 73.2 |
| Acetone | 66.9 | 75.1 |
| Benzene | 2.2 | 14.1 |
| Chloroform | NA | 1.2 |
| Carbon tetrachloride | NA | 0.3 |
| Nitrobenzene | 6 | 6.8 |

Resorcinol is also soluble in liquid ammonia, liquid sulfur dioxide, liquid hydrogen sulfide and pyridine
Data from [4, 13–15]

**Table 1.3.** Resorcinol volatility determination by TGA methods

| Temperature (°C) | Weight loss (%) |
|---|---|
| 50 | < 0 |
| 100 | < 0. 1 |
| 150 | 5 |
| 200 | 55 |
| 250 | 100 |

Heating rate at 20 °C/min under nitrogen
Data from [4, 9]

## 1.4.5
## Spectral Data of Resorcinol

$^{13}$C-NMR spectrum and the chemical shift data obtained from the $^{1}$H and $^{13}$C-NMR spectra are shown in Figure 1.5 and Table 1.4 [16].

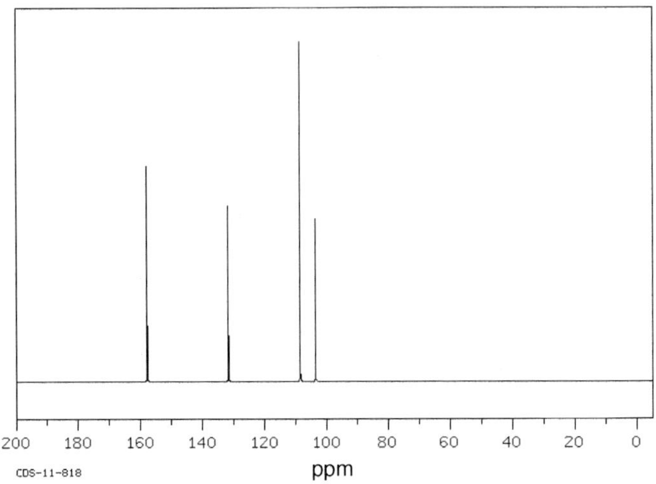


**Figure 1.5.** $^{13}$C-NMR spectrum of resorcinol recorded in $D_2O$ [16]

**Table 1.4.** $^{1}$H and $^{13}$C NMR chemical shift data for resorcinol

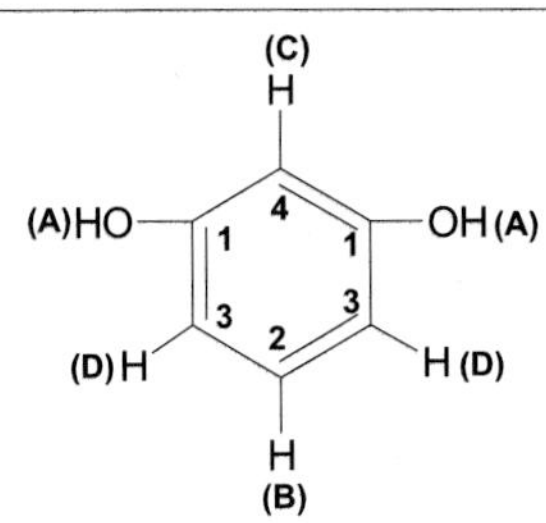


| $^{1}$NMR (1) | | $^{13}$C NMR (2) | |
|---|---|---|---|
| Assign | Shift (ppm) | Assign | Shift (ppm) |
| A | 9.15 | 1 | 157.73 |
| B | 6.929 | 2 | 131.65 |
| C | 6.22 | 3 | 108.52 |
| D | 6.21 | 4 | 103.55 |

(1) 400 MHz in DMSO-$D_6$
(2) 25.16 MHz in $D_2O$
Data from [16]

## 1.5
## General Properties of Resorcinol

Resorcinol is triboluminescent, and therefore its crystal can emit light when struck or ground together [4, 17]. The needle-like crystals of resorcinol can become pink on exposure to light and air, or by contact with iron. Resorcinol is hygroscopic, and also has the tendency to pick up moisture when exposed to humid atmosphere. The presence of moisture in resorcinol can reduce the

**Table 1.5.** General physical properties of resorcinol

| Property | Value |
|---|---|
| Freezing point (°C) | 110 |
| Melting point (°C) | 108.8 |
| Boiling point (°C, at 760 mm Hg) | 281.4 |
| Density (g/cm$^3$) | |
| Solid at 20 °C | 1.292 |
| Molten at 150 °C | 1.151 |
| $\alpha$-phase at 20 °C | 1.278 ** |
| $\beta$-phase at 20 °C | 1.327 ** |
| Flash point (°C) | |
| Tag open cup | 168 |
| Tag closed cup | 127 |
| Specific gravity (15 °C/4 °C) | 1.272 |
| Dipole moment (benzene) | 207 D |
| Auto ignition temperature (°C) | 608 |
| Deflagration index (Kst, Bar. m/s) | 134 |
| Heat capacity at 298 °K (J/mol) | 131 |
| Heat of combustion (cal/g) | 6200 |
| Heat of crystallization (kcal/mol) | 5.09 |
| pH (10% aqueous) | 4.5 |
| Dissociation constants (water) | |
| Ionization constant (25 °C) | $K_1 = 1.55 \times 10^{-10}$ |
| Activity constants (30 °C) | $K_1 = 7.11 \times 10^{-10}$ |
| | $K_2 = 4.78 \times 10^{-12}$ |
| Heat of sublimation (kcal/mol) | |
| Vapor temperature = 56.5 °C | 22.8 |
| Vapor temperature = 10 – 50°C | 21.7 |
| Distribution between solvents at 20 °C | |
| K [benzene/water] | 0.0073 |
| K [dichloroethane/water] | 0.0318 |
| K [ether/water] | 4.67 |
| K [ether/dichloroethane] | 146 |

** Data from [1–3]
Data from [4, 9]

freezing point, increase the tendency for caking and may cause discoloration. On exposure to light, air and chemical vapors, resorcinol may form color bodies. When exposed to ammonia vapors, resorcinol develops a blue color. Contact of resorcinol with metals other than aluminum or stainless steel may also cause darkening of the product [4,9].

Some of the other physical properties of resorcinol are shown in Table 1.5 [4,9].

## 1.6 Resorcinol Specifications

Resorcinol as a chemical is available in technical grade and USP grade for use in different applications. The properties of these two grade of resorcinol are shown in Table 1.6.

**Table 1.6.** Resorcinol specifications

| Property | Technical grade | USP grade* |
|---|---|---|
| Physical state | Flake or granules | Crystals or powder |
| Color, solid | Light colored | White |
| Color, molten (NPA standards, max.) | 2.5 | |
| Purity, % resorcinol, min. (dry basis) | 99.6 | 99.5 |
| Freezing point, °C, min. (dry basis) | 109.4 | 109.6 |
| Bulk density (kg/m$^3$) | 721 (Flakes) | 854 (Crystals) |
| Phenol content, %, max. | 1 | ** |
| Catechol content, %, Max. | 0.1 | ** |
| Ash, %, max. | 0.005 | 0.05 |
| Water insoluble, %, max. | 0.005 | ** |
| Moisture, %, max. | 0.2 | 1 |

* Meets all requirements of the U.S Pharmacopeia XVI
** Meets USP XVI requirement
Note: (1) USP grade for medical use.
(2) Technical grade for the manufacture of most chemicals and resins
Data from [4,9]

## References

1. J. Robertson, Proc. R. Soc. (London), Vol. A157, p 79 (1936).
2. J. Robertson and A. Ubbelohde, Proc. R. Soc. (London), Vol. A167, p 122 (1938).
3. J. Robertson, Trans. Faraday Soc., Vol. 36, p 913 (1940).
4. "Resorcinol", Technical Bulletin No. CD-2-424, Koppers Company, Inc., Pittsburgh, PA, USA.
5. G. Tripathi, J. Chem. Phys., Vol. 74, p 250 (1981).
6. R. Rao, T. Sakuntala and B. Godwal, Phys. Rev., Vol. B65, p 054108 (2002).

7. M. Yoshino, K. Takahashi, Y. Okuda, T. Yoshizawa, N. Fukushima and M. Naoki, J. Phys. Chem., Vol. 103, p 2775 (1999).
8. G. Wheland, "Resonance in Organic Chemistry", John Wiley & Sons, Inc., New York, p 403 (1955).
9. "The Resorcinol Handbook", Indspec Chemical Corporation, Pittsburgh, PA, USA, March (1998).
10. D. Stull, Ind. Eng. Chem., Vol. 39, p 517 (1947).
11. J. Friend and W. Hargreaves, Phil. Mag., Vol. 37, p 120 (1946).
12. K. Schmiedel and D. Decker, "Resorcinol", Ullmann's Encyl. Ind. Chem., 5th edn, Wiley-VCH, Vol. A23, p 111 (1999).
13. A. Seidell, "Solubilities of Organic Compounds", 3rd Ed., Vol. 11, D. Van Nostrand Company, Inc., Princeton, NJ, USA (1941).
14. A. Mortimer, J. Am. Chem. Soc., Vol. 45, p 635 (1923).
15. W. Walker et al., J. Phys. Chem., Vol. 35, p 3259 (1931).
16. SDBSWeb:http://www.aist.go.jp/RIODB/SDBS, Dated July 8, 2004.
17. M. Trautz, Z. Phys. Chem., Vol. 53, p 55 (1905).

# 2 Manufacturing Processes for Resorcinol

Resorcinol was first prepared in the laboratory in 1864 [1–3]. Soon after the first preparation, enormous potential was anticipated for the use of resorcinol in industrial as well as in scientific applications. Therefore, considerable effort was made by the scientific community to investigate and develop commercial processes to manufacture resorcinol in an economical way. The earliest description of a process for the manufacture of resorcinol appeared in 1878, and involved the disulfonation of benzene and fusion of the disodium salt of benzenedisulfonic acid [4–6]. Over the past 130 years there have been enormous efforts made to improve the benzene disulfonation process to increase the yield and improve the purity of resorcinol. Resorcinol has been produced industrially for more than 100 years. In the USA, Indspec Chemical Corporation, the largest producer of resorcinol as of now, still uses the benzene disulfonation process to manufacture resorcinol in a cost-effective way.

Though the disulfonation of benzene is an economical process, alternative synthetic routes have been constantly sought for the manufacture of resorcinol. In this respect, the following processes have been developed.

- Hydrolysis of *meta*-phenylenediamine (MPDA).
- Hydroperoxidation of *meta*-diisopropylbenzene (*m*-DIPB).
- Keto-acid cyclization and dehydrogenation [7–10].
- Cyclohexanone oxidation [11].
- Hydrolysis of haloaromatics [12,13].

In spite of various process developments, only two commercial processes are currently used to manufacture resorcinol, namely the disulfonation of benzene and hydroperoxidation of *meta*-diisopropylbenzene. There are two major resorcinol producers in the world, Indspec Chemical Corporation in USA and Sumitomo Chemical Company in Japan. The disulfonation process is used by Indspec Chemical Corporation and few other small plants in China and India. In addition to Sumitomo Chemical Company, Mitsui Petrochemical is also producing resorcinol using the hydroperoxidation process.

In this chapter, the disulfonation of benzene, hydrolysis of *meta*-phenylenediamine and hydroperoxidation of *meta*-diisopropylbenzene processes for the manufacture of resorcinol have been discussed.

## 2.1 Resorcinol by the Benzenedisulfonation Process

### 2.1.1 Introduction

The manufacture of resorcinol by benzenedisulfonation is the oldest process known. Resorcinol has been produced industrially by the disulfonation process for more than 100 years [2, 3, 6]. The sulfonation of benzene to the disulfonic acids has been exhaustively investigated and documented, both with reference to the reaction conditions and the quantities of various isomers produced [4, 5, 14, 15]. The first process for the manufacture of resorcinol by the disulfonation of benzene process appeared in 1878 [4–6]. After this, extensive effort has been made in the development of an improved process for the production of resorcinol by the widely known benzene "disulfonation–fusion" process [2, 16–22].

Constant and consistent effort has been made in the cost-effective process improvements in the disulfonation process. With these kinds of improvements, Indspec Chemical Corporation in USA is still using disulfonation technology to manufacture resorcinol.

The manufacture of resorcinol by the disulfonation of benzene involves the following processing steps.

- Disulfonation of benzene.
- Neutralization and sodium sulfate recovery.
- Alkali fusion of benzenedisulfonic acid sodium salt.
- Acidification of fusion mixture.
- Extraction and distillation.

In this section, a brief description of the chemistry and process parameters involved in each of the processing steps obtained from the published literature are discussed.

### 2.1.2 Disulfonation of Benzene

In the manufacture of resorcinol by this process, the first step is the production of a highly purified *meta*-benzenedisulfonic acid (called "diacid") from the disulfonation reaction of benzene with sulfuric acid or oleum ($H_2SO_4 + SO_3$). Sulfur trioxide ($SO_3$) is the preferred sulfonating agent in the disulfonation of benzene because it can react in substantially stoichiometric quantities. Sulfur trioxide is also an aggressive electrophilic reagent that can react rapidly with any organic compound such as benzene. Sulfonation reaction with sulfur trioxide is rapid and highly exothermic. A precise control of the molar ratio of sulfur trioxide to benzene is essential to maximize the formation of *meta*-benzenedisulfonic acid, and also to avoid by-products formation.

While sulfur trioxide is an effective sulfonating agent, it is highly reactive and its reactions are extremely exothermic which can lead to the charring of reaction mixtures. In order to control the sulfur trioxide reactivity, mixing or complexing the sulfur trioxide to moderate the rate of reaction has been done. For this reason, sulfur trioxide is often mixed with sulfuric acid and used as oleum in the benzene disulfonation process [22–25]. The reaction of benzene with oleum is carried out under carefully controlled conditions to ensure good yield of *meta*-benzenedisulfonic acid. To achieve an optimum yield of the diacid, benzene has been preferentially reacted with 65% oleum in the presence of anhydrous sodium sulfate [25].

Improvements in the production of *meta*-diacid were achieved with the use of catalyst mixtures containing sodium sulfate and vanadium salts [22, 26]. During the sulfonation of benzene, *p*-benzenedisulfonic acid (*para*-diacid), sulfonated diphenylsulfone and disulfonated diphenylsulfone compounds are also produced [27]. The addition of sodium sulfate is the most effective way of reducing sulfone formation during the sulfonation reaction with oleum [28].

This disulfonation reaction of benzene with sulfur trioxide or oleum can be done either continuously or batch wise. The chemical reaction of benzene with oleum can be illustrated by the following scheme [6] (Figure 2.1).

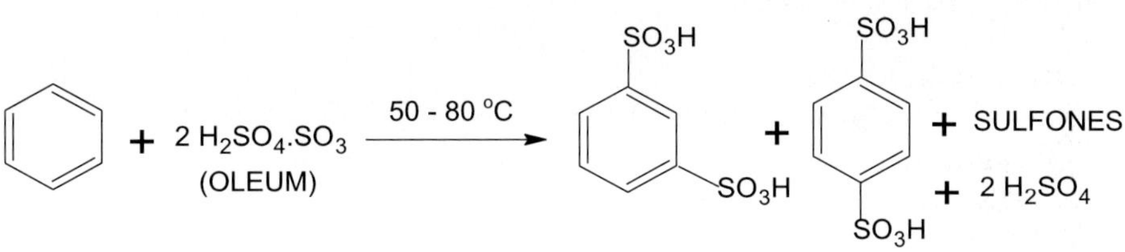


**Figure 2.1.**

## 2.1.3
## Neutralization and Sodium Sulfate Recovery

The disulfonation mixture contains sulfuric acid. Therefore, it has been necessary to remove the sulfuric acid prior to fusion in order to obtain higher yields of resorcinol. The sulfuric acid is commonly removed by adding an alkaline compound that will form a disulfate salt with sulfuric acid. Since sodium sulfite is obtained as a by-product from this process, this compound can be conveniently used to neutralize both the *meta*-benzenedisulfonic acid and the sulfuric acid simultaneously (Figure 2.2).

For this, the *meta*-diacid produced in the disulfonation process step is dissolved in water and neutralized with sodium sulfite to produce the disodium salt of *m*-benzenedisulfonic acid (called “*meta*-disalt”). The sodium sulfate

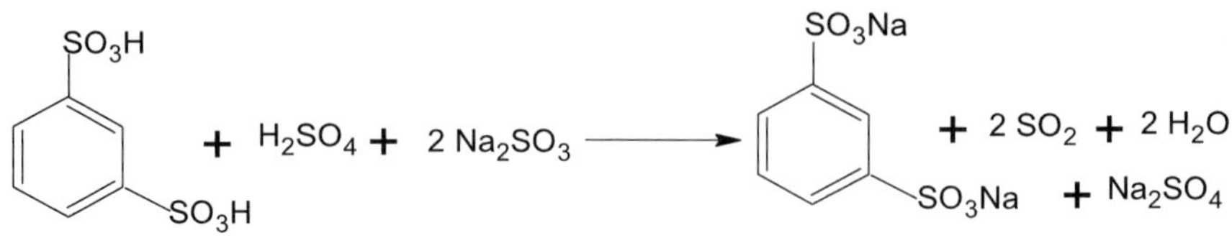


**Figure 2.2.**

formed can be removed, and the aqueous solution containing the disalt is dried to obtain the disalt for the fusion operation [2].

## 2.1.4 Alkali Fusion of Disalt

The disalt thus produced is then fused with molten sodium hydroxide, which could be obtained by the concentration of commercial aqueous sodium hydroxide solutions. Fusion temperature used in this operation could be in the range of 300 to 360 °C [29, 30]. In this processing step, an excess of caustic may be necessary to maintain the fluidity of the fusion mass (Figure 2.3).

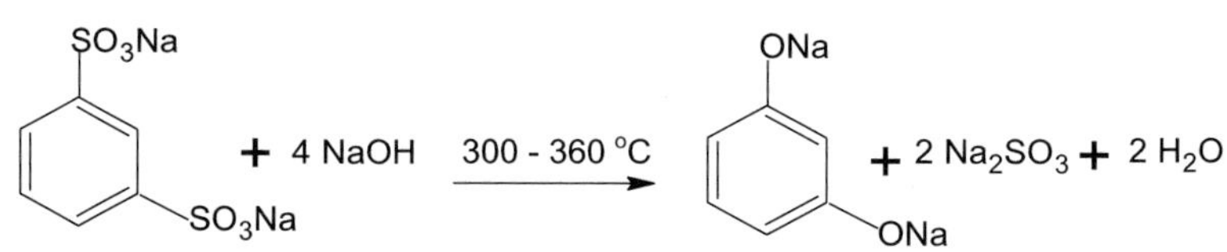


**Figure 2.3.**

From the fusion step, the disalt of resorcinol and sodium sulfite are produced. After this fusion, the mass is quenched in water. In this way the solid fusion product is solubilized for easier handling and further processing.

## 2.1.5 Acidification of Fusion Mixture

The aqueous solution containing the disodium salt of resorcinol can be acidified with sulfuric acid or other acidic substance, adjusting the pH to about 4 [31]. In this acidification step, sodium bisulfite is the preferred acidifying agent, since the by-product sodium sulfite formed from the reaction can be conveniently separated and sold in the commercial markets [2]. Resorcinol is generated from the disodium salt in this step of the process (Figure 2.4).

The sodium sulfite thus produced is removed by centrifugation and dried.

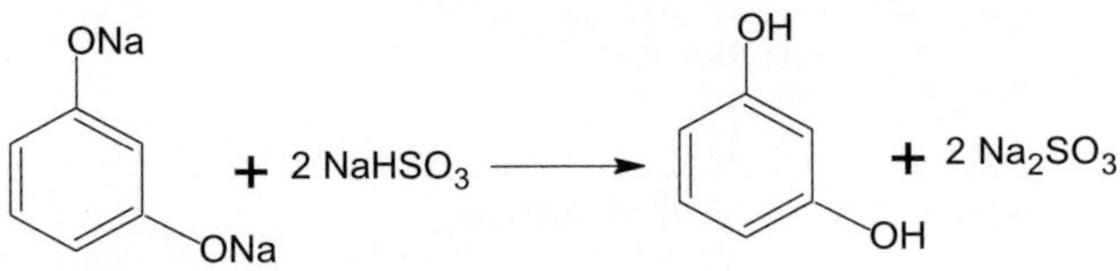


**Figure 2.4.**

### 2.1.6 Extraction and Distillation

The dissolved resorcinol is now extracted with an organic solvent immiscible with water, and also has the high extraction efficiency for resorcinol. Suitable organic solvents for the resorcinol extraction are diisopropyl ether, diethyl ether, benzene, methyl isobutyl ketone (MIBK) and iso-butanol [17,19,25,32].

After the evaporation of extraction solvent, the resorcinol is purified by means of a distillation in vacuo. The yield of resorcinol from the sulfonation process is about 80% [6].

The flow diagram outlining the unit operations of the benzene disulfonation process for the manufacture of resorcinol has been shown in Figure 2.5.

Since Indspec Chemical Corporation has been developing and implementing process modification in the disulfonation process, its process is the most economical process for the production of resorcinol.

## 2.2 Hydrolysis of *meta*-Phenylenediamine (MPDA) Process

### 2.2.1 Introduction

Another method developed to produce resorcinol is the hydrolysis of *m*-phenylenediamine. The first attempt to synthesize resorcinol by an acid hydrolysis of *m*-phenylenediamine with dilute hydrochloric acid was made in 1897 [33]. This process has several advantages over the other resorcinol processes. *m*-Phenylenediamine (MPDA) is readily available, and could be easily made from the catalytic hydrogenation of *m*-dinitrobenzene. Therefore, in a single processing step resorcinol can be obtained in greater than 95% yields. Since MPDA is commercially available, in this section only the process on the hydrolysis of MPDA to produce resorcinol has been detailed.

### 2.2.2 MPDA Hydrolysis Processes Developed

For the hydrolysis of MPDA, strong mineral acids such as hydrochloric, sulfuric and phosphoric acid have been employed [34–38]. When phosphoric acid

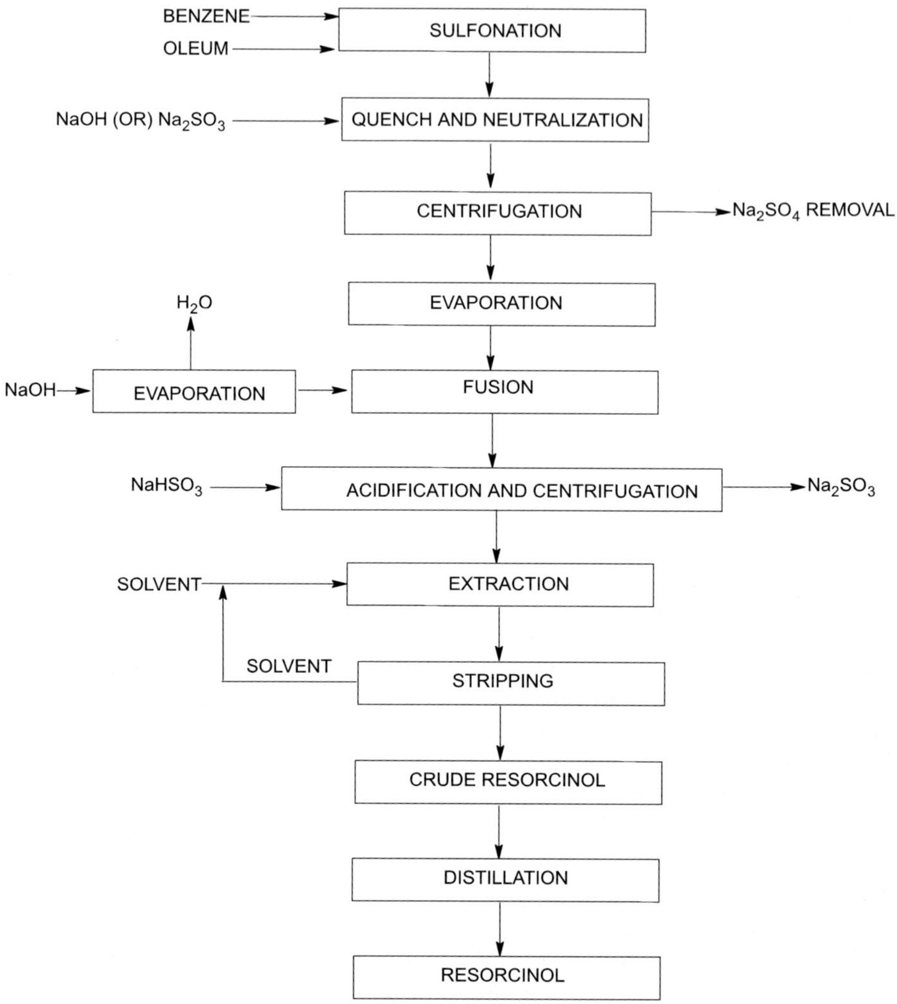


**Figure 2.5.** Flow diagram for the manufacture of resorcinol by sulfonation-fusion process

was used, resorcinol was obtained in about 86 to 96% yields after the ether extraction. The ammonium phosphate by-product from this process could be used as a fertilizer. Though an autoclave fitted with Monel liner was used to carry out the reaction, a resinous type product was formed during the hydrolysis process [34,35].

A resorcinol yield of 67.5% was obtained with sulfuric acid using $H_2SO_4$/MPDA molar ratio of 1.6 to 1.8. A Tantalum tubular reactor was employed in this hydrolysis process [36].

Ammonium bisulfate was also employed as the acid for the hydrolysis of MPDA [37] (Figure 2.6).

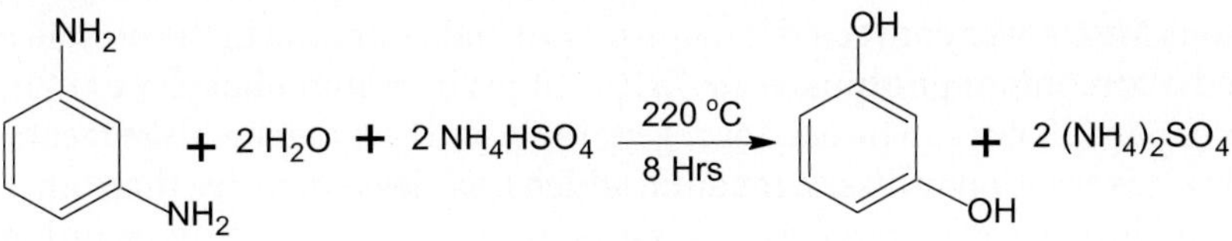


**Figure 2.6.**

The ammonium sulfate formed in the hydrolysis process was decomposed to obtain the acid ammonium bisulfate, which could be reused effectively (Figure 2.7).

$$(NH_4)_2SO_4 \xrightarrow[12\ \text{Min}]{312\ ^\circ\text{C}} NH_4HSO_4 + NH_3$$

**Figure 2.7.**

## 2.2.3
## Major Problems in MPDA Hydrolysis Process

The MPDA hydrolysis process requires high temperature and strong acids. Employing strong acids in the presence of water brings the pH of the solution to a very low value. The combination of low pH and high temperature makes the hydrolysis process very dangerous. Therefore, this process requires the use of non-corrosive materials to withstand these processing conditions. Reactor vessels made of zirconium, zirconium alloy, Hastelloy alloy, molybdenum, platinum, titanium and tungsten were corrosion resistant, and found to be suitable for the MPDA hydrolysis process [38].

Resorcinol produced from the hydrolysis process contained both water soluble and insoluble resinous products and tars. These substances are formed as a result of the reaction of resorcinol with other materials present in the reaction mixture, which could include the starting material (MPDA) or reaction intermediates such as *m*-aminophenol. The insoluble materials can build up in the reactor and can block process lines, resulting in a potentially hazardous situation.

## 2.2.4
## Newer Process for MPDA Hydrolysis

Newer processes have been developed to overcome the problems of corrosive environments of the MPDA hydrolysis process. One such process reported in the literature was the use of Zeolite catalysts. Resorcinol was produced in higher yields from the hydrolysis of MPDA in the presence of alumina-silicates such as Zeolite H-USY, H-Beta and amorphous silica-alumina [39].

Aqueous MPDA was converted into resorcinol and ammonia in the presence of layered $\alpha$-zirconium phosphates ($\alpha$-ZrP) and $\gamma$-zirconium phosphates [40]. Zirconium phosphates can be used as regenerable and heterogeneous reagents. The hydrolysis reaction utilizes a medium which is far less corrosive than other reaction media generally employed. This process offers an alternative solution to the extremely corrosive properties of mineral acids at higher temperatures experienced in the MPDA hydrolysis process.

## 2.2.5 MPDA Hydrolysis Using Sulfuric Acid

Indspec Chemical Corporation has developed a new MPDA hydrolysis process for the production of resorcinol using sulfuric acid [41]. In this process, a zirconium reactor was used as the corrosion resistant vessel for the MPDA hydrolysis reaction. The formation of tar or other resinous by-products during the hydrolysis reaction was either minimized or eliminated. This was achieved by the proper selection of MPDA/$H_2SO_4$ molar ratios and hydrolysis reaction conditions.

**Formation of Salt from MPDA and $H_2SO_4$** When MPDA is mixed with sulfuric acid, two kinds of salts can be expected depending upon the molar ratios of these two chemicals in the mixture. For example, if 1 mol of MPDA is mixed with 1 mol of sulfuric acid, a salt having the "sulfate ion" structure can be formed. Similarly, mixing 2 mol of sulfuric acid with 1 mol of MPDA, a "bisulfate ion" structure containing compound can be produced (Figure 2.8).

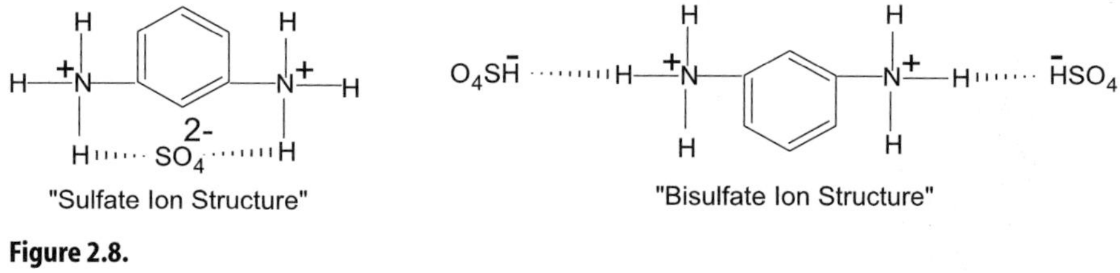


**Figure 2.8.**

These two salts on hydrolysis with water can produce resorcinol, *m*-aminophenol and ammonium salts according to the following reaction schemes. (Figures 2.9 and 2.10)

The formation of *m*-aminophenol in the reaction mixture during the hydrolysis process could be responsible for the production of "tarry" material. FT-IR and FT-NMR analysis techniques were utilized to characterize the nature of tar materials present in the crude resorcinol product. The analysis data suggested the following structure for the tar (Figure 2.11).

This compound can be produced from the reaction of *m*-aminophenol and resorcinol in the presence of strong acids such as sulfuric acid and ammonium bisulfate (Figure 2.12).

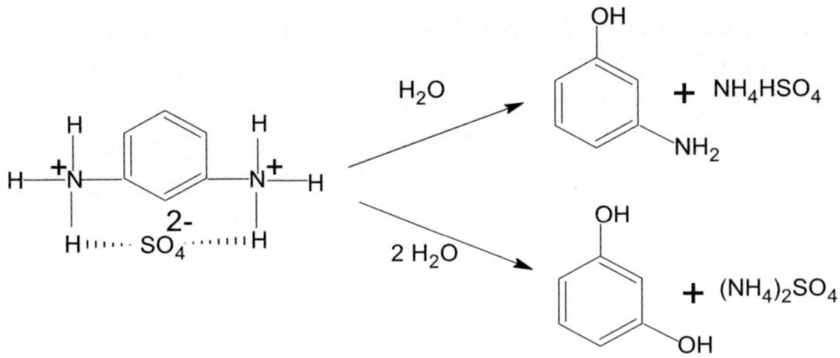


**Figure 2.9.**

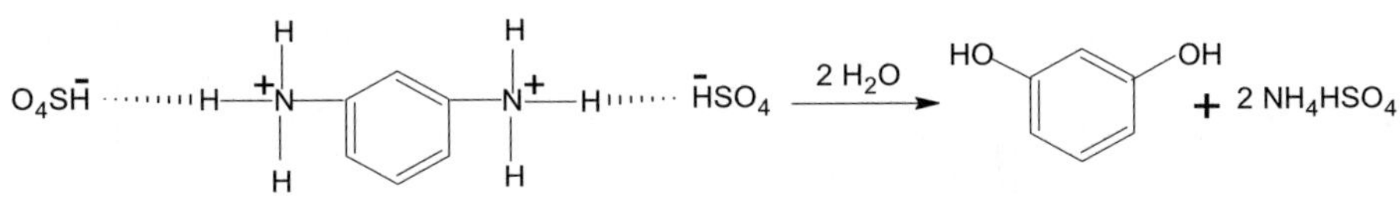


**Figure 2.10.**

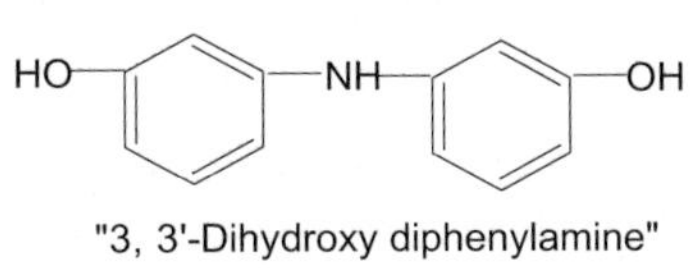


**Figure 2.11.**

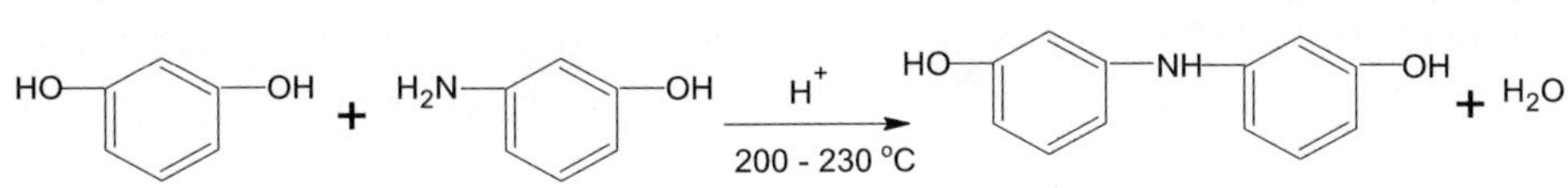


**Figure 2.12.**

Resorcinol and ammonium bisulfate are the products expected from the hydrolysis of the salt having the "bisulfate ion" structure. Therefore, with the use of 2 mol of sulfuric acid per mole of MPDA in the hydrolysis process, the formation of *m*-aminophenol can either be avoided or minimized. This may result in the hydrolysis process without much tarry substance in the reactor.

## 2.2.6 Single Pass MPDA Hydrolysis Process [41]

MPDA, sulfuric acid and water were charged into a zirconium reactor and heated at 215–230°C for 3–6h. After cooling, the hydrolysis mixture was extracted with diethyl ether. The ether solution was stripped to recover ether for recycle, and the residue was distilled to obtain pure resorcinol in a single process operation. A series of single-pass experiments were run, and the results are presented in Table 2.1 [41].

**Table 2.1.** Resorcinol production from *m*-phenylenediamine hydrolysis process (single pass process and zirconium reactor)

| Experiment | 1 | 2 | 3 | 4 | 5 |
|---|---|---|---|---|---|
| Reactor feed (mol) | | | | | |
| *m*-Phenylenediamine | 1 | 1 | 1 | 1 | 1 |
| Sulfuric acid | 1.8 | 2 | 1.8 | 2 | 2 |
| Water | 80 | 60 | 60 | 80 | 60 |
| Reactor conditions | | | | | |
| Temperature (°C) | 215 | 215 | 230 | 230 | 240 |
| Time (h) | 2 | 6 | 6 | 6 | 3 |
| Crude resorcinol | | | | | |
| Yield (wt%) | 44.8 | 87.4 | 90.7 | 96.3 | 95.3 |
| Resorcinol content (wt% by NMR) | 94.4 | 98.8 | 98.5 | 99.8 | 99.7 |
| Soluble tar content (wt% by NMR) | 5.5 | 1.2 | 1.5 | 0.2 | 0.3 |
| Aqueous phase (wt% by NMR) | | | | | |
| *m*-Phenylenediamine | 0.7 | 0.1 | 0.1 | ND | 0.1 |
| *m*-Aminophenol | 2.1 | 0.8 | 0.1 | 0.2 | 0.2 |
| Soluble tar | 0.7 | 0.3 | 0.3 | 0.2 | 0.1 |
| Inorganic material (wt%) | | | | | |
| Ammonium bisulfate | 16.4 | 17.9 | 10.3 | 14.8 | 18.2 |
| Ammoniun sulfate | 1.5 | 5.5 | 2 | 3.2 | 4.4 |

ND = None Detected
Data from [41]

The zirconium reactor used in the hydrolysis process was weighed before and after the reaction for any weight loss due to corrosion. The weight remained unchanged indicating no corrosion to the zirconium reactor suggesting that this reactor might be well suited for the MPDA hydrolysis reactions.

A very high resorcinol yield of 96.3% was achieved in a single-pass MPDA hydrolysis reaction.

### 2.2.7 Double Pass MPDA Hydrolysis Process [41]

A schematic diagram illustrating the two-pass MPDA hydrolysis process is shown in Figure 2.13 [41].

Following the conditions outlined in Figure 2.13, a series of two-pass MPDA hydrolysis reactions were performed employing slight changes in the $H_2SO_4$/MPDA ratios. The details are summarized in Table 2.2.

The total resorcinol yield from pass-1 and pass-2 MPDA hydrolysis process for $H_2SO_4$/MPDA $= 2 : 1$ was 96.5%.

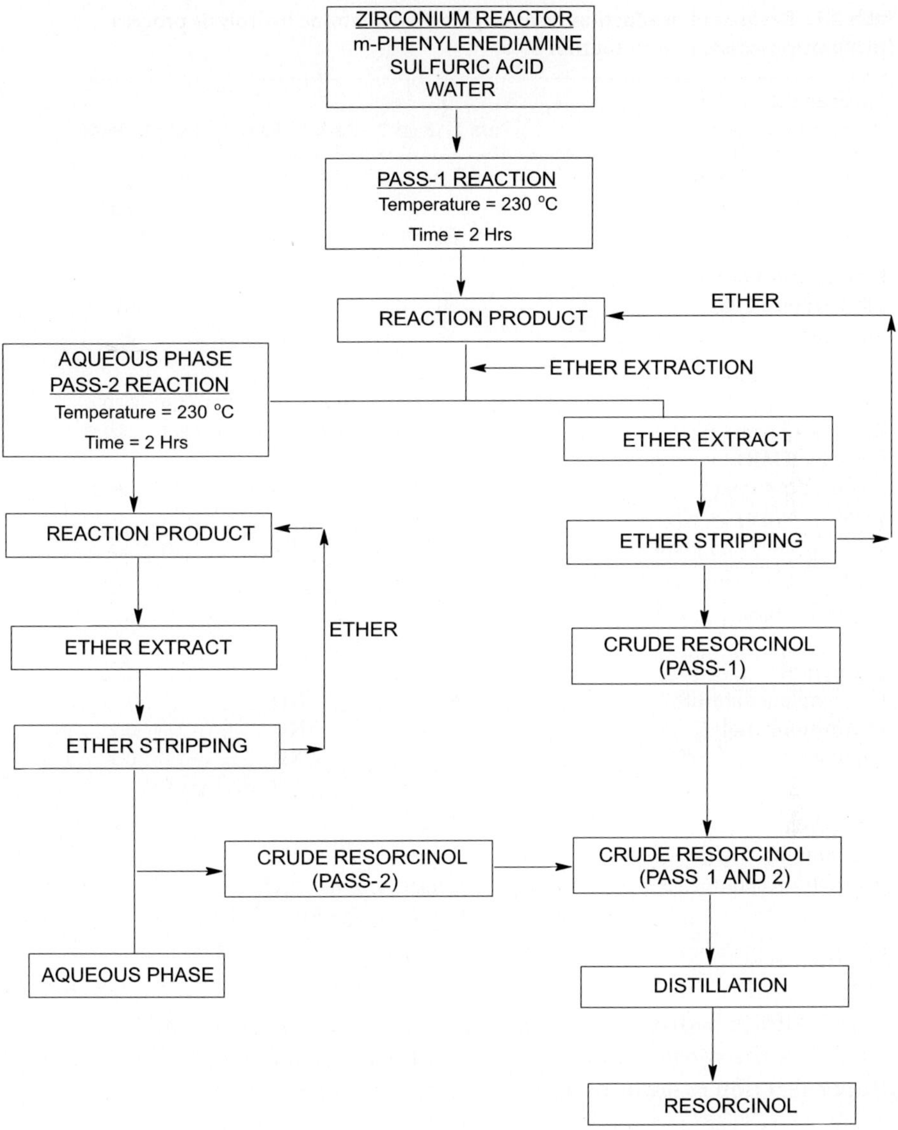


**Figure 2.13.** Two pass *m*-phenylenediamine hydrolysis process

**Table 2.2.** Resorcinol production from *m*-phenylenediamine hydrolysis process (two pass process and zirconium reactor)

| Experiment | 1 | | 2 | | 3 | |
|---|---|---|---|---|---|---|
| Reactor Feed (Mol) | Pass 1 | Pass 2 | Pass 1 | Pass 2 | Pass 1 | Pass 2 |
| *m*-Phenylenediamine | 1 | | 1 | | 1 | |
| Sulfuric acid | 1.9 | | 2 | | 2.1 | |
| Water | 60 | | 60 | | 60 | |
| Reactor conditions | | | | | | |
| Temperature (°C) | 230 | | 230 | | 230 | |
| Time (h) | 2 | | 2 | | 2 | |
| Crude resorcinol | | | | | | |
| Yield (wt%) | 82.1 | 12.6 | 79.6 | 16.9 | 74.9 | 20.4 |
| Resorcinol content (wt% by NMR) | 97 | 99.9 | 98.2 | > 99.9 | 98.8 | 99.9 |
| Soluble tar content (wt% by NMR) | 2.9 | ND | 1.8 | ND | 1.2 | ND |
| Overall resorcinol yield (pass 1 and 2) | 94.7 | | 96.5 | | 95.3 | |
| Aqueous phase (wt% by NMR) | | | | | | |
| Resorcinol | ND | | ND | | ND | |
| *m*-Phenylenediamine | ND | | ND | | ND | |
| *m*-Aminophenol | 0.1 | | ND | | 0.2 | |
| Soluble tar | 0.1 | | ND | | 0.1 | |
| Inorganic material (wt%) | | | | | | |
| Iron (ppm) | 2.4 | | 8 | | 2.9 | |
| Ammonium bisulfate | 16.6 | | 18.2 | | 19.3 | |
| Ammoniun sulfate | 4.3 | | 3.8 | | 4.3 | |

ND = none detected
Data from [41]

The MPDA hydrolysis process could be a viable process and can be considered as the most economical process for the manufacture of resorcinol if proper reaction conditions and reactor vessels are employed.

## 2.3 Hydroperoxidation of *m*-Diisopropylbenzene Process

### 2.3.1 Introduction

One of the unique technologies developed in the 1950s for the commercial production of resorcinol utilizes the hydroperoxidation of 1,3- or *meta*-diisopropylbenzene (*m*-DIPB). Unlike the benzene disulfonation or MPDA hydrolysis process, both processes produce inorganic salts as the by-products, the hydroperoxidation process producing acetone as the co-product. This makes the hydroperoxidation process more attractive and valuable for the construction of new resorcinol plants. Also, this process appeared to be the most economical and efficient way of producing resorcinol on a larger scale. Therefore, the two Japanese companies, Sumitomo Chemical and Mitsui Petrochemical, have built their resorcinol plants based on hydroperoxidation technologies. Indspec Chemical Corporation in USA has also developed, as an alternate to the benzene disulfonation process, an economical, capital and energy intensive hydroperoxidation technology, and is ready for commercialization.

### 2.3.2 Hydroperoxidation Process

Review articles and large number of patents covering the technologies of resorcinol via the hydroperoxidation process have been published [42–46]. In this chapter, a comprehensive overview of the hydroperoxidation process has been discussed.

In the hydroperoxidation method for the manufacture of resorcinol, the following process steps are involved.

- Preparation of high purity *m*-diisopropylbenzene (*m*-DIPB).
- Oxidation of *m*-diisopropylbenzene.
- Caustic extraction of *m*-DIPB oxidation product.
- Cold MIBK extraction to separate hydroxyhydroperoxide (HHP).
- Hot MIBK extraction to produce pure dihydroperoxide (DHP).
- Cleavage of DHP to produce resorcinol.
- Distillation and purification of resorcinol and acetone.
- Decomposition of HHP to produce dicarbinol (DCL).

### 2.3.3 Preparation of *m*-Diisopropylbenzene (*m*-DIPB)

The synthesis of resorcinol by the hydroperoxidation process requires a high purity *m*-DIPB feed or raw material. Commercial production of *m*-DIPB can be achieved by the Friedel-Crafts alkylation of benzene with propylene, and is based on the following reaction [47–49] (Figure 2.14).

The commercial DIPB process normally produces all three isomers, namely *ortho*-, *meta*-, and *para*-diisopropylbenzenes. Boiling points of these DIPB isomers are as follows: *ortho*-, 203.8 °C; *meta*-, 203.2 °C; and *para*-, 210.4 °C [43]. Due to very close boiling points between the ortho and meta isomers, it may not be practically possible to separate these two isomers completely. Since *ortho*-DIPB cannot be oxidized during the oxidation of *m*-DIPB, it will accumulate in the unreacted DIPB stream and tend to slow the oxidation rate

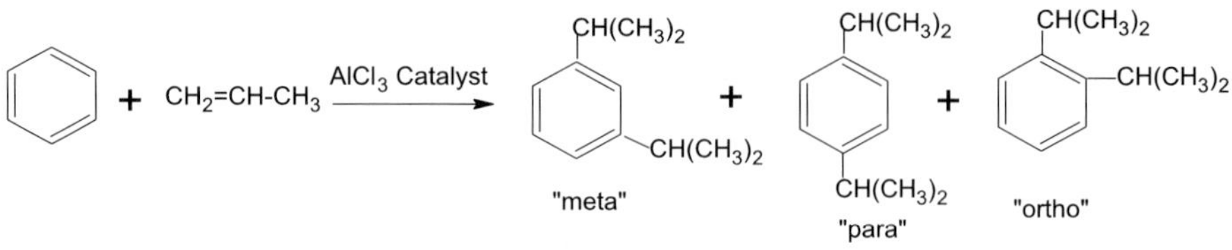


**Figure 2.14.**

of *m*-DIPB. Therefore, it is desirable to use *m*-DIPB feed material containing the least amount of *o*-DIPB ($< 0.5$ wt.%) to prevent the build-up of *o*-DIPB in the recycle, and also the gradual deterioration of hydroperoxidation with increasing recycle.

In order to develop *m*-DIPB having higher purity for the hydroperoxidation process, solid [50, 51], Nafion-H [52], and Zeolite [53–56] catalysts have been employed in the alkylation reaction of benzene with propylene. The commercial *m*-DIPB material normally contains less than 1% *o*-DIPB and less than 0.5% trimethylindane impurities, and may be used in the hydroperoxidation process.

## 2.3.4 Oxidation of *m*-Diisopropylbenzene

The first step in the production of resorcinol by hydroperoxidation technology is the oxidation of *m*-DIPB to produce *m*-di-(2-hydroperoxy-2-propyl)-benzene (known as DHP). This oxidation is generally carried out in a pressurized reactor with either air or oxygen in the presence of sufficient dilute caustic catalyst to maintain the pH of reaction medium between 7 and 9 [57–65]. The main initial product from the *m*-DIPB oxidation is *m*-isopropyl-$\alpha\alpha$-dimethylbenzyl-hydroperoxide (called MHP), from which all other oxidation products are produced. The following products are formed during the *m*-DIPB oxidation.

- *m*-Di-(2-hydroperoxy-2-propyl)-benzene (DHP).
- *m*-2-Hydroxy-2-propyl-$\alpha\alpha$-dimethylbenzyl hydroperoxide (HHP).
- *m*-Isopropylphenyl dimethyl carbinol (MCL).
- *m*-Isopropyl acetophenone (MKT).
- *m*-Acetyl-$\alpha\alpha$-dimethylbenzyl hydroperoxide (KHP).
- *m*-Di-(2-hydroxy-2-propyl)-benzene (DCL).

For the production of resorcinol, DHP is the product required from the *m*-DIPB oxidation. However, the formation of other oxidation products during the oxidation cannot be avoided. HHP can be formed by the thermal decomposition of DHP, and the HHP may further decompose to give the DCL compound.

The main reactions occurring during the oxidation of pure *m*-DIPB are outlined in the following scheme (Figure 2.15).

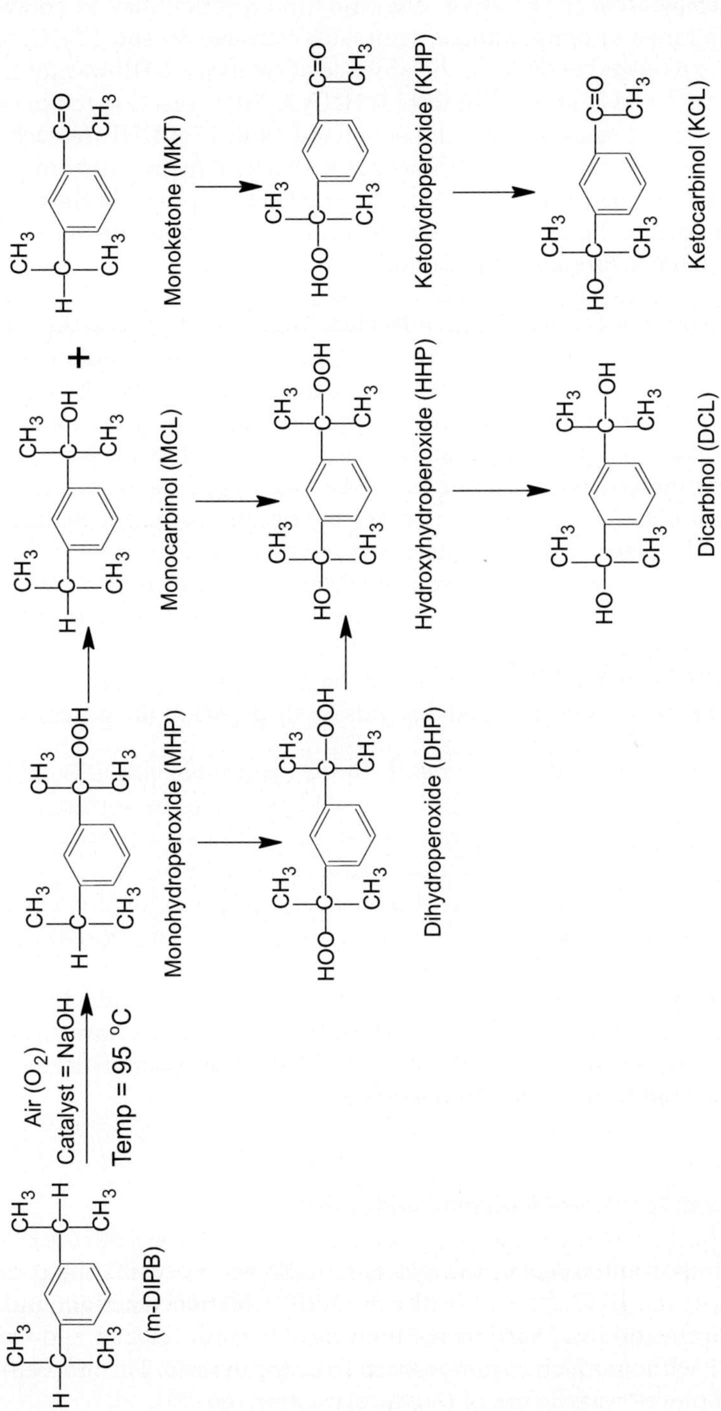


**Figure 2.15.** *m*-Diisopropylbenzene oxidation chemistry

**Oxidation Temperature and Pressure** The oxidation reaction may be conducted over a wide range of temperatures, preferably between 80 and 120 °C. When *m*-DIPB was oxidized at 80 °C in the absence of catalysts, DHP was formed at 8 to 9.5 $gl^{-1}hr^{-1}$ and contained 10 to 11% HHP. At 90 °C reaction temperature, DHP formed at 18 to 20 $gl^{-1}hr^{-1}$ and contained 14 to 15% HHP. In each case, the oxidation reactor was run continuously, with a total hydroperoxide content of about 70% calculated as MHP [42]. For practical purposes, in the presence of caustic solution, the oxidation reaction may preferably be run at $(90\pm5)$ °C and 0.14 to 0.6 MPa pressure conditions.

**Catalyst and pH of Oxidation Reaction Medium** The oxidation reaction is normally done in the presence of a base catalyst, preferably either sodium hydroxide or sodium carbonate. These basic substances can increase the oxidation reaction efficiency by retarding the development of excessive acidity associated with the formation of carboxylic acids during the *m*-DIPB oxidation. The preferred pH for the oxidation reaction ranges between 6 and 10. At high pH values (greater than pH = 10), some alkali catalyzed decomposition of hydroperoxides could take place, leading to increased production of carbinols relative to hydroperoxides. In order to achieve a consistent pH of 6 to 10 in a continuous oxidation process, it may be necessary to treat the recycled oxidation material (known as the oxidate) to remove the entrained alkali.

The *m*-DIPB oxidation reaction can be performed either continuously [59,63] or batch-wise [64], and depends on the needs of the process.

**Composition of *m*-DIPB Oxidation Product** In the continuous *m*-DIPB oxidation process, the concentration of DHP in the oxidate should be kept steady, and can be obtained approximately from the determination of total hydroperoxide (calculated as MHP). The total hydroperoxide level in the oxidation mixture is a function of residence time when other factors, such as extraction efficiency, are kept constant. An approximate equilibrium composition of *m*-DIPB oxidation product is shown in Table 2.3 [45,46].

Since the oxidation mixture contained different reaction products, in order to separate the desired DHP, caustic and MIBK extractions have been performed. In this way, a comparatively pure DHP for the production of resorcinol has been isolated from the *m*-DIPB oxidate.

### 2.3.5 Separation and Recovery of Dihydroperoxide (DHP)

The most important aspect of the hydroperoxidation process is the complete separation of the DHP, formed in the *m*-DIPB oxidation reaction, and uses it to produce resorcinol. Various methods have been developed and tried to isolate DHP without much decomposition from the oxidate. The most efficient method employed was the use of caustic extraction [66–75].

**Table 2.3.** *m*-Diisopropylbenzene oxidation material composition[a] NMR analysis results (weight ratios)

| Component | Oxidation Product |
|---|---|
| *m*-Diisopropylbenzene (*m*-DIPB) | 26.6 |
| Monohydroperoxide (MHP) | 38.8 |
| Dihydroperoxide (DHP) | 18.2 |
| Hydroxyhydroperoxide (HHP) | 4.3 |
| Monocarbinol (MCL) | 3.5 |
| Dicarbinol (DCL) | 0.6 |
| Monoketone (MKT) | 0.4 |
| Ketohydroperoxide (KHP) | 1 |
| Ketocarbinol (KCL) | 0.2 |
| Organic peroxides (OP) | 6 |

[a]Oxidation reaction was performed at 90 °C temperature and in the presence of NaOH catalyst
Data from references [45, 46]

**Caustic Extraction of Oxidate** Isolation of DHP from the oxidation mixture has been the objective of several processes developed. In practice, DHP in the oxidate material has been effectively separated by means of 4 to 8% caustic solution [59, 75]. Though DHP is produced in the *m*-DIPB oxidation step, complete recovery of DHP from the oxidate and using it in the cleavage step of hydroperoxidation process could increase the production of resorcinol. During the caustic extraction process, it has been realized that not only DHP but also the other impurities present in the oxidate, such as HHP, KHP, MHP, etc., are also extracted. For example, when 8% caustic was used, about 90 to 95% of KHP and HHP present in the oxidation mixture were known to be extracted along with 1 to 2% MHP.

Depending upon the extraction procedures used, the extraction efficiency of caustic solutions was varied. Literature information showed that DHP was extracted from the oxidation mixture using either a mixer-settler [77] or Karr column [45,46] type operations. When Karr column was used, the DHP present in the oxidate was easily and effectively extracted into the caustic solution. For this operation, to minimize the DHP decomposition, the extraction temperature should be kept preferably between 20 and 40 °C.

**DIPB Wash of Caustic Extract** The MHP extracted into the caustic, unless recovered, will end up in the cleavage operation, and may result in the overall yield loss in the hydroperoxidation process. MHP is the product which produces DHP on further oxidation. Therefore, the recovery of MHP material from the caustic extract should be regarded as very important. In order to recover MHP, the caustic extract has been back extracted with DIPB solvent. In the Karr column operation, this extraction step was very effective in the recovery of MHP. This DIPB extraction did not have much effect on removing other impurities such as HHP and KHP from the caustic extract solutions. The oxidate recycle generated from the caustic extraction and DIPB wash operation step, containing the MHP, MCL and other oxidation products, was fed back into the oxidation reactor for further oxidation.

Using a single Karr column operation, it was demonstrated that both the oxidation recycle (for further oxidation) and caustic DHP extract (for isolation of DHP) were obtained simultaneously [45,46].

**Separation of DHP and HHP from the Caustic Extract** The DHP and HHP extracted into the caustic solutions exist in the form of their sodium salts. In order to isolate these two compounds, several methods have been developed. Of these methods, the most preferred method is the extraction of DHP and HHP into an organic solvent, and preferably MIBK [72,73,75–77]. Using MIBK solvent, a high proportion of DHP was extracted at 70 to 80 °C temperature conditions with negligible losses by decomposition. To improve the extraction efficiency, about 1% ammonia was added into the 8% caustic solution containing 12.3% DHP. MIBK solvent was more effective even at 60 °C for the extraction of DHP through three counter-current contact stages [68].

Four stage mixer-settler type and Karr column counter current extraction methods were used to extract DHP and HHP materials from the caustic solutions. For the effective utilization of HHP in the production of resorcinol, the compositions containing the HHP were treated with hydrogen peroxide in the presence of an acid catalyst. In this way, the HHP and DCL by-products were converted into DHP and used in the cleavage step of the hydroperoxidation process [78–80].

In another approach, the DHP and HHP materials were selectively and successively separated from the caustic extract, and used to manufacture resorcinol and dicarbinol (DCL) respectively [45,46]. This operation was done more effectively by carrying out the MIBK extraction at low and also at high temperature conditions using Karr columns.

**Cold MIBK Extraction** If the HHP has to be removed selectively from the caustic solutions, then the MIBK solvent extraction should be performed at low temperatures. Using Karr column and an extraction temperature of 10 to 30 °C, the MIBK extraction performed on the caustic extract containing the DHP/HHP material selectively removed the HHP. Analysis of the cold MIBK extract showed that all or nearly all of the KHP, KCL, MHP and other oxidation impurities present in the caustic solutions were extracted. The caustic solution after this extraction mainly contained DHP product [45,46].

**Hot MIBK Extraction** The DHP present in the caustic solution is in the form of its disodium salt, and can undergo decomposition at elevated temperatures, and also on prolonged storage. This DHP rich caustic solution was then fed into another Karr column and extracted with MIBK solvent at 40 to 70 °C temperature conditions. After this extraction, the concentration of DHP in the

MIBK extract was typically between 6 to 12 weight percent, and the amount water was about 1 to 3%.

The extraction chemistry of the *m*-DIPB oxidation material is outlined below (Figure 2.16).

The composition of oxidate, oxidation recycle, cold MIBK and hot MIBK extract is presented in Table 2.4.

### 2.3.6 Production of Resorcinol from DHP Cleavage

Before using the hot MIBK solution for cleavage operation, it must be concentrated in order to reduce its water content and also to increase the DHP concentration. By means of this operation, the DHP content was increased to between 20 and 40%, and water content was reduced to less than 0.3%. The DHP cleavage reaction can be conveniently carried out using strong acidic catalysts such as sulfuric acid, sulfur trioxide, phosphoric acid, hydrochloric aid, boron trifluoride, *p*-toluene sulfonic acid, acid clays and ion-exchange resins [80–88].

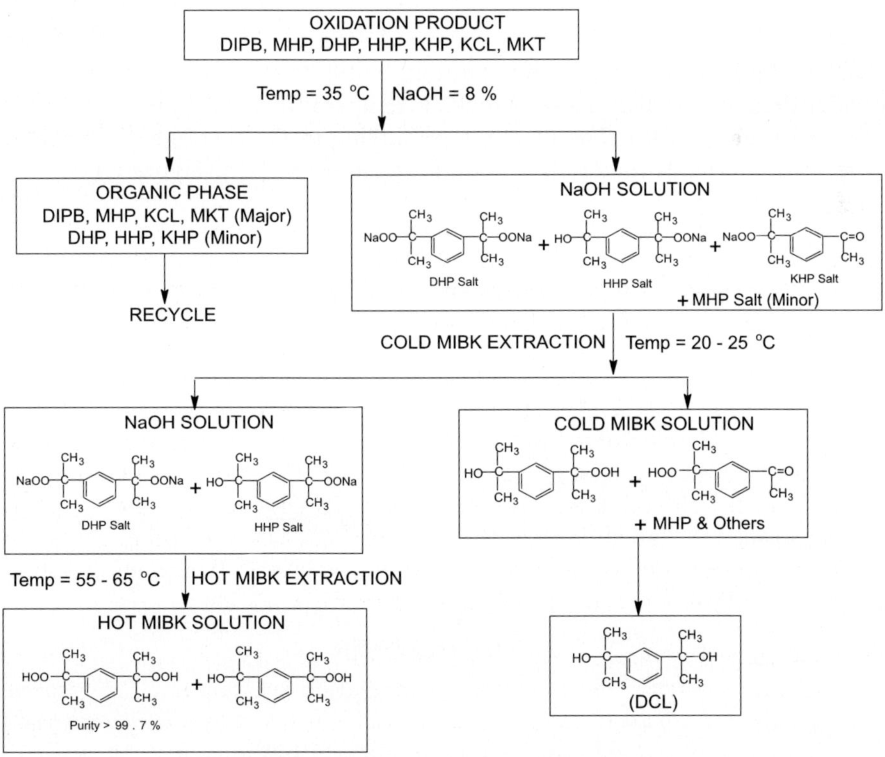


**Figure 2.16.** Extraction chemistry of *m*-DIPB oxidation products

**Table 2.4.** Extraction and simultaneous separation of DHP and HHP from DIPB oxidation mixture[a]; Composition of different streams; NMR analysis results (weight ratios)

| Component | *m*-DIPB oxidation material | Oxidation recycle | Cold MIBK extract (solvent free) | Hot MIBK extract (solvent free) |
|---|---|---|---|---|
| *m*-Diisopropylbenzene (*m*-DIPB) | 24.9 | 44 | | |
| Monohydroperoxide (MHP) | 38.2 | 42.7 | 8.7 | |
| Dihydroperoxide (DHP) | 18.7 | 0 | 1.4 | 99.7 |
| Hydroxyhydroperoxide (HHP) | 4.3 | 2.7 | 76.2 | 0.3 |
| Monocarbinol (MCL) | 4 | 6.1 | | |
| Dicarbinol (DCL) | 0.6 | 0.8 | 0.4 | |
| Monoketone (MKT) | 0.4 | 0.1 | | |
| Ketohydroperoxide (KHP) | 1 | 0 | 12.4 | |
| Ketocarbinol (KCL) | 0.2 | 0 | 0.9 | |
| Organic peroxides (OP) | 5.9 | 3.6 | | |

[a]Oxidation reaction was carried out at 90 °C temperature and in the presence of NaOH catalyst Extraction was performed using Karr Columns
Data from [45, 46]

Sulfur trioxide was found to be the most effective cleavage catalyst. When used in the form of an acetone solution, sulfur trioxide catalyst produced greater than 99.5% cleavage at 60 to 80 °C temperatures with a residence time of 5 to 10 min [45, 46]. The reactions occurring in the cleavage of DHP and other hydroperoxide materials are summarized below [17] (Figure 2.17).

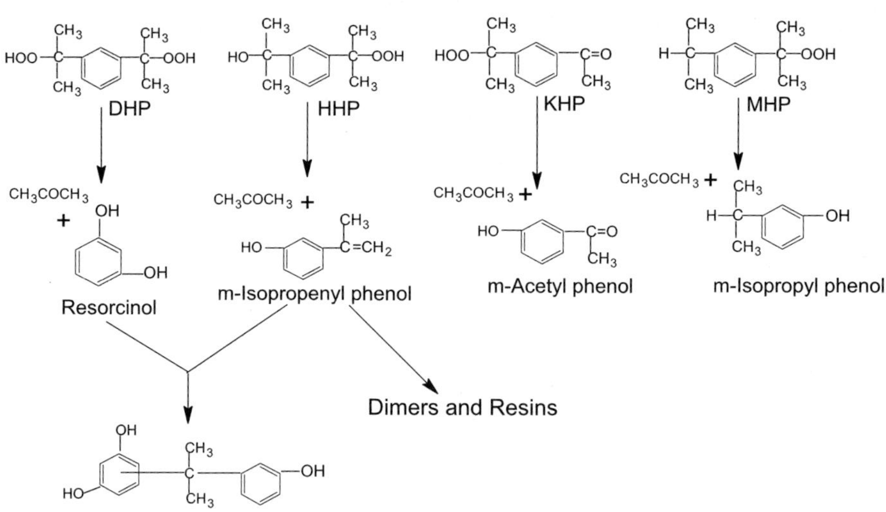


**Figure 2.17.** Cleavage chemistry of *m*-DIPB hydroperoxide materials

As can be seen in the cleavage reaction, DHP cleaves to produce resorcinol and acetone. The presence of HHP in the DHP cleaves to *m*-isopropenyl phenol and acetone. In the presence of acidic conditions, resorcinol can react with acetone to produce resins having isopropylidene structures. Also, the reaction of resorcinol with *m*-isopropenyl phenol can produce higher boiling point or higher molecular weight products. Isopropenyl phenol can also dimerize to give solid and liquid dimer compounds. MHP and KHP, on the other hand, can cleave to produce *m*-isopropyl phenol and *m*-acetyl phenol derivatives, and acetone.

Once the DHP has been cleaved to resorcinol and acetone, the cleavage reaction product should be neutralized to eliminate the acid, otherwise resorcinol can react with acetone to produce resinous type products.

### 2.3.7 Distillation and Recovery

The recovery of resorcinol and acetone from the cleavage reaction mixture can be achieved by distillation, extraction and crystallization techniques. The reported resorcinol yield from the DHP cleavage reaction was about 95%. The resorcinol purity after distillation was greater than 99.7% [45,46].

### 2.3.8 Manufacture of Dicarbinol (DCL) from HHP

The HHP raw material obtained in the cold MIBK extraction operation can be used for the production of dicarbinol. DCL has been used in the manufacture of various organic and polymeric materials for various applications [89–92]. DCL was obtained by the reduction reaction of HHP in the presence of hydrogen and palladium–alumina catalyst [93,94]. Using an aqueous sodium hydroxide or

**Table 2.5.** Decomposition of hydroxyhydroperoxide (HHP) to dicarbinol (DCL) using aqueous NaOH and $Na_2SO_3$ solutions NMR analysis of components (weight ratios, solvent free)

| Component | Cold MIBK Extract | Decomposition Using NaOH | $Na_2SO_3$ |
|---|---|---|---|
| Hydroxyhydroperoxide (HHP) | 77.9 | 0 | 0 |
| Ketohydroperoxide (KHP) | 10.6 | 0 | 0 |
| Monohydroperoxide (MHP) | 9.6 | 0 | 0 |
| Ketocarbinol (KCL) | 0.9 | 2.6 | 1.9 |
| Dicarbinol (DCL) | 0.7 | 94.6 | 97.6 |
| Monocarbinol (MCL) | 0 | 2.8 | 0.4 |
| MIBK (wt%) | 61 | 0 | 0.1 |

Data from [45,46]

sodium sulfite solution, the HHP and other hydroperoxides present in the cold MIBK extract were completely decomposed to their corresponding carbinols (Table 2.5) [45,46].

The DCL product obtained can be easily filtered and purified for commercial markets.

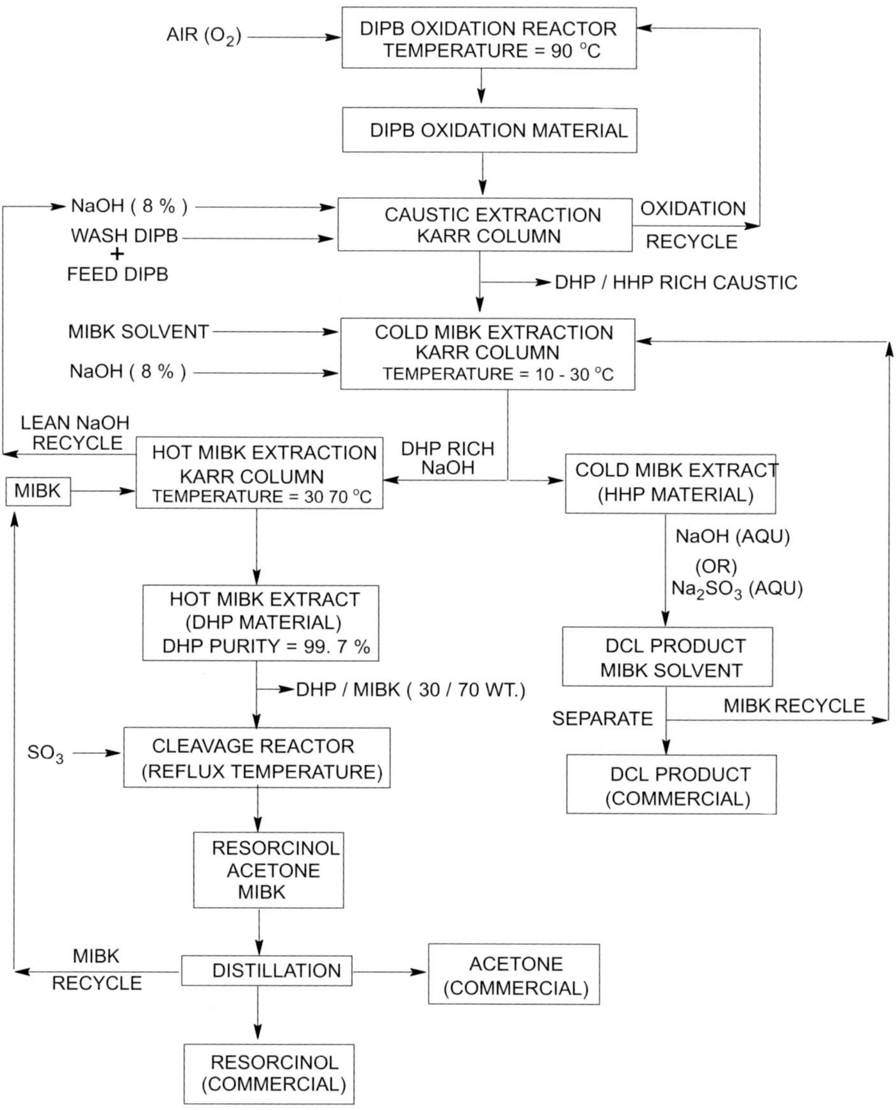


**Figure 2.18.** Indspec Chemical Corporation hydroperoxidation process

A schematic flow diagram outlining Indspec Chemical Corporation's hydroperoxidation technology for the production of resorcinol, acetone and dicarbinol materials, as published in the patent literature, is presented below [45,46] (Figure 2.18).

The hydroperoxidation technology developed by Indspec Chemical was also scaled up, and can be implemented if there is a need for another resorcinol plant in future.

### 2.3.9 Summary and Future Outlook

Benzene disulfonation and hydroperoxidation of *m*-DIPB processes have been used to produce resorcinol. With the proper selection of reactor vessels, acid catalysts and control of MPDA/acid ratios, the hydrolysis of MPDA could also be the potential cost-effective process for the manufacture of resorcinol. The hydroperoxidation process developed by Indspec Chemical Corporation could also be practiced to build new resorcinol plant, if the demand for resorcinol increases in future.

## References


1. H. Hlasiwetz and L. Bath, Ann. Chem. Pharm., Vol. 130, p 354 (1864).
2. H. Scheeline, "Dihydroxybenzenes", SRI Process Economics Program, Report No. 79, September 1972.
3. K. Schmiedel and D. Decker, "Resorcinol", Ullmanns' Encyclopedia of Industrial Chemistry, Vol. A23, p 111 (1993).
4. L. Binschedlar and Y. Busch, Mon. Scientif., p 1168 (1878).
5. L. Binschedlar and Y. Busch, Chem. Ind., I, p 370 (1878).
6. S. Holter and H. Dressler, Paper Presented at the American Institute of Chemical Engineers, Denver, CO, August 21–24 (1988).
7. W. Mueller, US Patent 4,377,709 (1983).
8. W. Mueller, K. Mack and H. Hey, US Patent 4,250,336 (1981).
9. N. Greco, US Patent 4,399,310 (1983).
10. N. Greco, US Patent 4,431,848 (1984).
11. J. Fellman and R. Saxton, Ger. Offen., 3, 708, 851 (1987).
12. K. Bursack and H. Moltzav, US Patent 3,413,341 (1968).
13. J. Kamlet, US Patent 2,835,708 (1958).
14. C. Suter, "The Organic Chemistry of Sulfur", John Wiley & Sons, Inc., New York (1944).
15. J. Garrick, Z. Chem., p 550 (1869).
16. J. Carr and M. Dahlen, US Patent 1,999,955 (1935).
17. Staatsmijnen Company, British Patent 672,511 (1952).
18. C. Noller, Textbook of Organic Chemistry, 2nd edn, Saunders Publishers, Philadelphia, p 394 (1966).
19. Mitsui Toatsu Chemical Company, British Patent 1,237,349 (1971).
20. G. Leston, US Patent 4,009,212 (1977).
21. F. Allen, Chem. Eng., Sep. 25, p 118 (1967).
22. W. Cotton, US Patent 1,915,925 (1933).

23. Monsanto Chemical Company, British Patent 679,827 (1949).
24. W. Cake, US Patent 2,856,437 (1958).
25. Koppers Company, Inc., British Patent 1,146,408 (1969).
26. A. Sensemann, Ind. Eng. Chem., Vol. 13, p 1124 (1921).
27. A. Shestov, et al., J. Gen. Chem., (USSR) (Engl. Transl.), Vol. 29, p 591 (1959).
28. E. Gilbert, "Sulfonation and Related Reactions", Interscience, New York (1965).
29. G. Webb, US Patent 2,736,754 (1956).
30. Mitsui Toatsu Chemical Company, British Patent 1,214,118 (1970).
31. P. Oppenheimer and W. Cake, US Patent 2,579,122 (1951).
32. S. Burroughs, US Patent 2,043,232 (1936).
33. R. Meyer, Ber., Vol. 30, p 2568 (1897).
34. N. Greco, Canadian Patent 783,607 (1968).
35. N. Greco, US Patent 3,462,497 (1969).
36. N. Greco, US Patent 3,862,245 (1975).
37. N. Greco, US Patent 3,862,246 (1975).
38. K. Terada and T. Kamiyama, Jap. Pat. Disclosure, S52-153919 (1977).
39. B. Brack, D. Gammon and E. Steen, J. Molecular Catalysis A, Vol. 154, p 73 (2000).
40. B. Brack, D. Gammon and E. Steen, Microporous and Macroporous Materials, Vol. 41, p 149 (2000).
41. R. Durairaj and G. Jesionowski, US Patent 6,531,637 (2003).
42. A. Graham, World Petrol. Congress, Proceedings, 7th, Vol. 5, p 29 (1967).
43. Dihydroxybenzenes, SRI Process Economics Program, Report No. 79, Chapter 7, Menlo Park, CA, 1972.
44. I. Pokrovskaya, et al., Int. Chem. Eng., Vol. 18, p 100 (1978).
45. R. Durairaj, V. Romell, M. Tackie and T. Volek, PCT, WO 99/43637 (1999).
46. R. Durairaj, V. Romell, M. Tackie and T. Volek, US Patent 6,350,921 (2002).
47. H. Enos, Canadian Patent 570,340 (1959).
48. R. Landau and R. Egbert, Canadian Patent 571,083 (1959).
49. A. Saffer, Canadian Patent 571,084 (1959).
50. Bataafsche Petroleum, British Patent 809,908 (1959).
51. Bataafsche Petroleum, British Patent 809,909 (1959).
52. G. Olah, US Patent 4,339,613 (1982).
53. W. Weber, C. Smith, F. Bryan, S. Brown and J. Cheng, US Patent Appln., 2003/0,018,227 (2003).
54. W. Weber, C. Smith, F. Bryan, S. Brown and J. Cheng, US Patent Appln., 2003/0,125,591 (2003).
55. C. Chi and S. Chen, US Patent Appln., 2004/ 0,030,211 (2004).
56. J. Cheng, W. Weber and F. Bryan, US Patent Appln., 2004/ 0,097,777 (2004).
57. E. Hawkins, D. Quin and F. Salt, British Patent 641,250 (1950).
58. E. Hawkins, D. Quin and F. Salt, British Patent 646,102 (1950).
59. W. Weber, British Patent 727,498 (1955).
60. J. Connor and V. Verplank, US Patent 2,856,432 (1958).
61. Y. Ogata and M. Haba, J. Org. Chem., Vol. 38, p 2779 (1973).
62. H. Suda, et al., US Patent 3,933,921 (1976).
63. H. Suda, et al., US Patent 3,953,521 (1976).
64. H. Nambu, I. Hashimoto and I. Imai, US Patent 4,237,319 (1980).
65. I. Imai, I. Hashimoto, K. Suzuki and H. Nakagawa, US Patent 4,267,387 (1981).
66. R. Hall and W. Weber, British Patent 785,607 (1957).
67. Bataafsche Petroleum, British Patent 816,200 (1959).
68. H. Suda, et al., US Patent 3,932,528 (1976).

69. H. Suda, et al., US Patent 3,950,431 (1976).
70. K. Kobayashi, I. Dohgane and Y. Nakao, US Patent 3,993,696 (1976).
71. H. Hosaka, K. Tanimoto, H. Okabe, K. Tanaka and I. Dohgane, US Patent 4,059,637 (1977).
72. S. Tokumasu, H. Ohki and T. Ohmae, US Patent 5,962,747 (1999).
73. T. Ohmae, S. Tokumasu and H. Ohki, US Patent 5,959,155 (1999).
74. T. Noritake, S. Goto and K. Itoh, US Patent 6,369,278 (2002).
75. J. Conner and V. Verplank, US Patent 2,856,432 (1958).
76. A. Graham, US Patent 2,789,142 (1957).
77. B. Thompson, British Patent 746,388 (1956).
78. H. Nakagawa, N. Hirowatari and T. Kakamura, US Patent 4,283,570 (1981).
79. H. Miki, S. Araki and M. Masatoshi, US Patent 4,835,325 (1989).
80. T. Ogino, S. Arita, M. Takeda and M. Kaya, US Patent 4,933,506 (1990).
81. K. Berneis, British Patent 754,864 (1956).
82. Bataafsche Petroleum, British Patent 819,450 (1959).
83. Distillers Company, British Patent 982,514 (1965).
84. H. Suda, et al., US Patent 3,923,908 (1975).
85. T. Taguchi, T. Ago and I. Hashimoto, US Patent 4,229,597 (1980).
86. S. Tokumasu, US Patent 6,423,876 (2002).
87. T. Noritake, US Patent 6,433,232 (2002).
88. J. Vodnar, J. Farkas and S. Bekassy, Appl. Catal. A: General, Vol. 185, p 165 (1999).
89. G. Gregory, US Patent 3,919,326 (1975).
90. H. Buysch, H. Oertel and E. Roos, US Patent 4,116,932 (1978).
91. W. Beijleveld and W. Endstra, US Patent 5,091,479 (1992).
92. R. Kazmierezak and R. MacLeay, US Patent 5,539,061 (1996).
93. N. Toshio, I. Masaru and T. Shigefumi, Jap. Patent JP 9,143,112 (1997).
94. N. Toshio and I. Masaru, Jap. Patent JP 9,194,411 (1997).

# 3 Resorcinol Derivatives

Resorcinol is a phenolic compound having two hydroxyl groups in the benzene structure. These two hydroxyl groups are located at the 1 and 3 positions of the benzene molecule. Therefore, resorcinol can undergo electrophilic substitution reactions similar to phenol. The two phenolic hydroxyl groups present in the resorcinol molecule could strongly activate the *ortho-* and *para-* positions of the benzene ring structure (Figure 3.1).

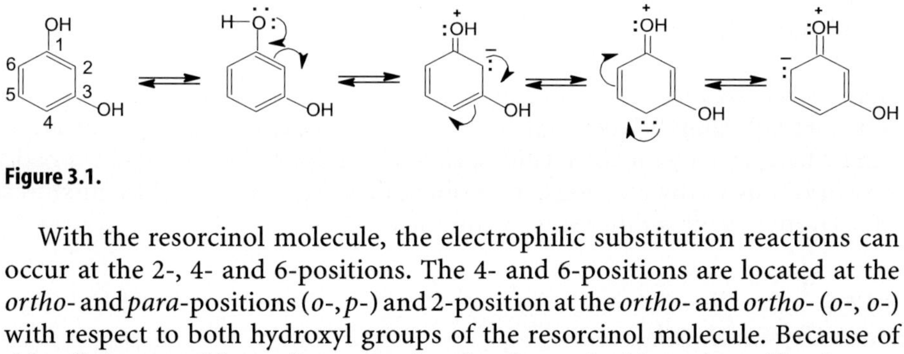

**Figure 3.1.**

With the resorcinol molecule, the electrophilic substitution reactions can occur at the 2-, 4- and 6-positions. The 4- and 6-positions are located at the *ortho-* and *para-*positions (*o-*, *p-*) and 2-position at the *ortho-* and *ortho-* (*o-*, *o-*) with respect to both hydroxyl groups of the resorcinol molecule. Because of this, all three positions of resorcinol molecule are doubly activated by the two hydroxyl groups.

Since the substitution in phenols occurs first at the most reactive *para-*position rather than the *ortho-*, electrophilic reactions with resorcinol are expected to take place readily at the 4- and 6-positions. Even though the 2-position of resorcinol is sterically hindered by the two adjacent *ortho-*hydroxyls and is expected to be less reactive when compared to the 4- and 6-positions, reactions still occur at the 2-position at a relatively faster rate. Due to stronger activation of the resorcinol molecule, substitution reactions can take place under milder conditions than phenol.

Electrophilic substitution reactions such as alkylation, acylation, halogenation, nitration and sulfonation can occur predominantly at the 2-, 4- and 6-positions of resorcinol. The 5-position of resorcinol is considered to be a non-reactive position in the resorcinol, and therefore under normal reaction conditions, electrophilic substitution reaction may not take place at this position. Because of the high reactivity of resorcinol, it may be difficult to obtain exclusively the mono-substituted product. Polysubstitution with resorcinol can often present a problem of separating different derivatives from the mixture. When the 4- and 6-positions of resorcinol are occupied, the substitution reaction occurs at the 2-position. However, the bulkier groups such as *tert*-butyl group cannot be substituted at the 2-position because of steric hindrance.

The reactivity, coupled with the three reactive positions for the electrophilic substitution reactions and two hydroxyl groups for etherification and esterification reactions, offers enormous possibilities to develop various derivatives for numerous applications. In this chapter, different kinds of resorcinolic derivatives that were developed from the resorcinol molecule and their chemistries have been briefly discussed.

## 3.1 Alkyl and Aralkyl Resorcinols

### 3.1.1 Alkyl Resorcinols

Alkyl derivatives of resorcinol can be prepared from the alkylation reaction of resorcinol using Friedel-Craft's as well as Lewis acid catalysts. Alkylation can be performed with alkyl halides, alkyl alcohols and unsaturated aliphatic hydrocarbons in the reactions. Very often, these reactions produce mixtures of derivatives having different substitutions.

#### 3.1.1.1 Mono-Alkyl Resorcinols

Mono-alkyl substituted resorcinol compounds have shown excellent antiseptic, anthelmintic, bactericidal and germicidal properties, and are used in various therapeutic, agricultural and organic synthesis applications. These derivatives were generally prepared by condensing aliphatic carboxylic acids with resorcinol to produce the acyl derivatives first, and then reducing the resulting ketones with zinc amalgam [1–5] (Figure 3.2).

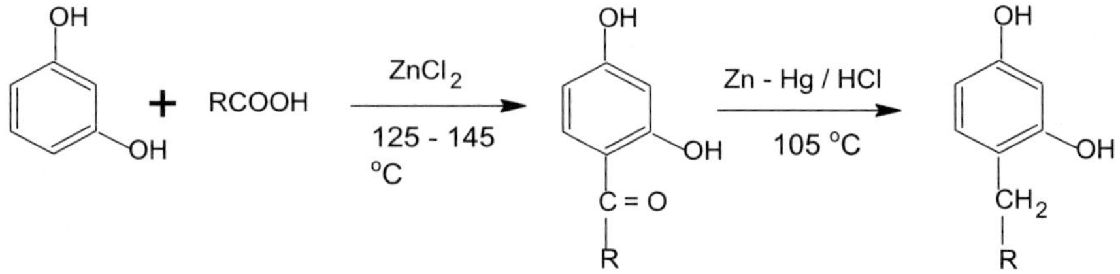


**Figure 3.2.**

In addition to Zn-Hg/HCl reduction, 4-acyl resorcinol was almost quantitatively reduced under the hydrogenation conditions using a copper catalyst [6] (Figure 3.3).

The boiling points and melting points of mono-alkyl resorcinol derivatives are presented in Table 3.1.

**Table 3.1.** Physical properties of 4-alkyl substituted resorcinols

| Substitution | Group structure | Melting point (°C) | Boiling point (°C/mm) |
|---|---|---|---|
| Methyl[a] | $CH_3$ | 104–105 | 267–70 |
| Ethyl | $CH_3(CH_2)$ | 95–98 | NA |
| Propyl | $CH_3(CH_2)_2$ | 80–81.5 | 158–159/6–7 |
| Butyl | $CH_3(CH_2)_3$ | 47–48 | 164–166/6–7 |
| Amyl | $CH_3(CH_2)_4$ | 71.5–73 | 168–170/6–7 |
| Hexyl | $CH_3(CH_2)_5$ | 57 | 178–179/6–7 |
| Heptyl | $CH_3(CH_2)_6$ | 73–74.5 | 186–188/6–7 |
| Octyl | $CH_3(CH_2)_7$ | 74–75 | 199–201/6–7 |
| Decyl[b] | $CH_3(CH_2)_9$ | 73–74 | 235–240 /11 |
| Dodecyl | $CH_3(CH_2)_{11}$ | 80–81.5 | 224–226/6–7 |
| iso-Propyl[c] | $(CH_3)_2$-CH | 98 | 114/0. 2 |
| iso-Butyl | $(CH_3)_2$-$CHCH_2$ | 62–63.5 | 166–168 |
| iso-Amyl | $(CH_3)_2$-$CH(CH_2)_2$ | 61–62.5 | 177–178/6–7 |
| iso-Hexyl | $(CH_3)_2$-$CH(CH_2)_3$ | 70–71.5 | 182–183/6–7 |

Data from [1]
[a]Data from [4,5]
[b]Data from [2]
[c]Data from [3]

4-Hexylresorcinol has several pharmaceutical applications. In the cosmetic applications, this chemical has been shown to improve the storage and oxidative stabilities of compositions containing terpenoids [7,8]. Aliphatic ether and ester compounds were synthesized from 4-hexylresorcinol [9] (Figure 3.4).

Dihydroresorcinols, useful as intermediates for the preparation of pharmaceutical and other organic compounds, were synthesized by the catalytic hydrogenation process from their corresponding 4-alkylresorcinol derivatives [10] (Figure 3.5).

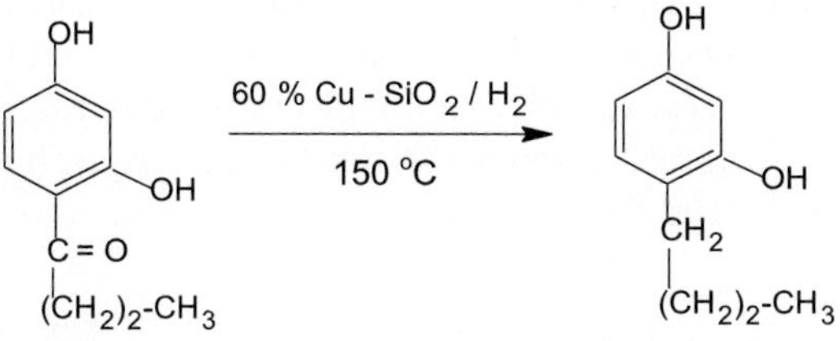


**Figure 3.3.**

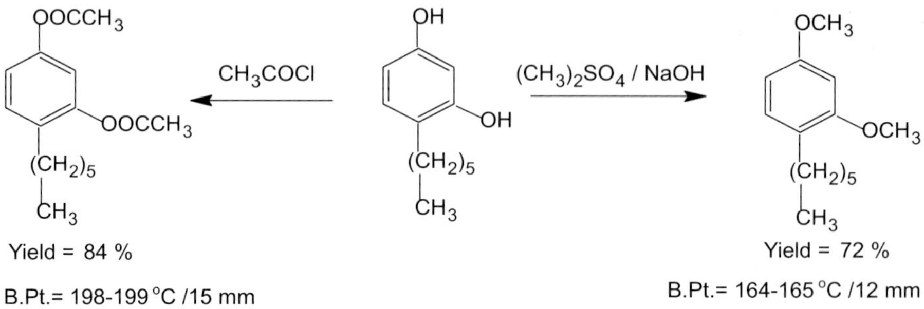


**Figure 3.4.**

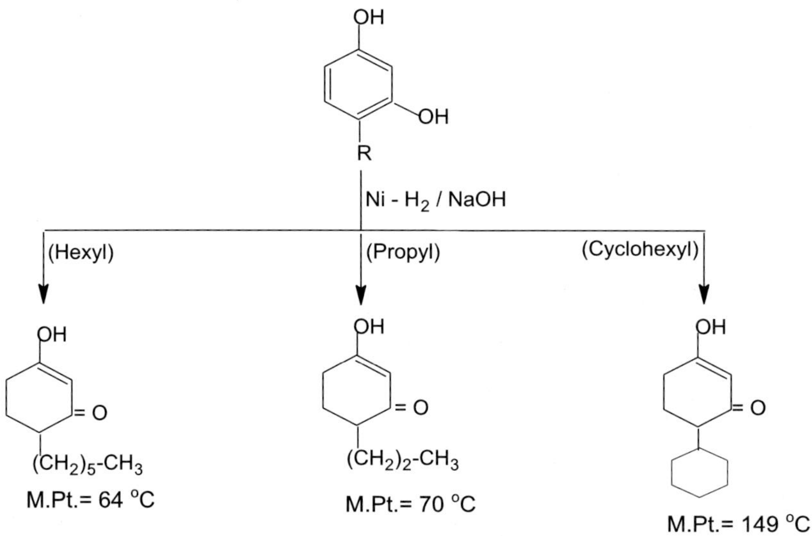


**Figure 3.5.**

### 3.1.1.2
### 2-Alkyl Substituted Resorcinols

The 2-substituted resorcinols have been used as raw materials in the synthesis of various pharmaceutical compounds, agricultural chemicals, dyes, pigments and photographic reagents. Under the Friedel-Craft's alkylation conditions, substitution normally takes place preferentially at the 4 and 6 positions, and therefore the desired 2-substituted alkylresorcinols can not be obtained in high yields. The 2-substituted resorcinols can be conveniently synthesized from the following reaction scheme [11–13] (Figure 3.6).

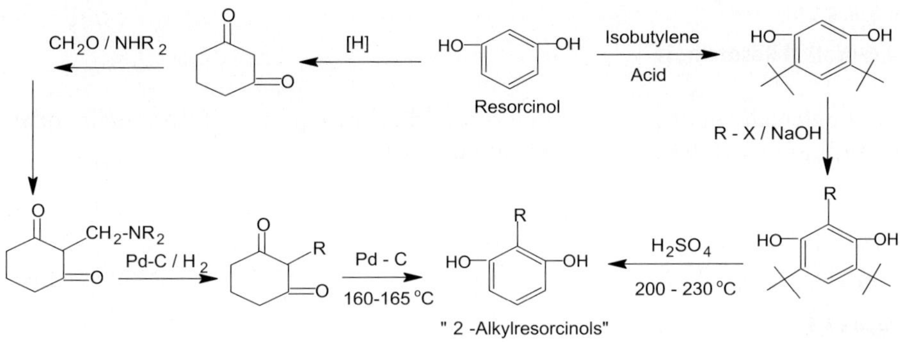


**Figure 3.6.**

Table 3.2 presents the melting points and boiling points of 2-alklyresorcinols synthesized from their corresponding 4,6-di-*tert*-butylresorcinol derivatives, which were basically synthesized using 4,6-di-*tert*-butylresorcinol.

**Table 3.2.** Melting and boiling points of 2-substituted 4,6-di-*tert*-butylresorcinols and resorcinols

| 2-Substitution | Group structure | 4,6-Di-*tert*-butylresorcinol M.Pt. (°C) | Resorcinol M.Pt. (°C) | B.Pt. (°C/mm) |
|---|---|---|---|---|
| Methyl | $CH_3$ | 117–119 | 118–121 | 138/10 |
| Ethyl | $CH_3(CH_2)$ | 112–114 | 98–100 | 145/10 |
| Propyl | $CH_3(CH_2)_2$ | 67–70 | NA | NA |
| Isopropyl | $(CH_3)_2$-CH | 103–106 | 79–82 | 144–150/10 |
| Allyl | $CH_2$=CH-$CH_2$ | 59–60 | 48–50[a] | NA |
| Benzyl | $C_6H_5$-$CH_2$ | 120–121 | 80–82 | NA |
| Dodecyl | | NA | 75–77 | NA |

Data from references [11, 12]
[a]Data from reference [37]

2-Methylresorcinol has a wide range of industrial applications, particularly in the hair dyeing and treatment compositions [14–16]. Psoralen compound useful in these applications was synthesized from the reaction of 2-methylresorcinol with ethylpropiolate [17] (Figure 3.7).

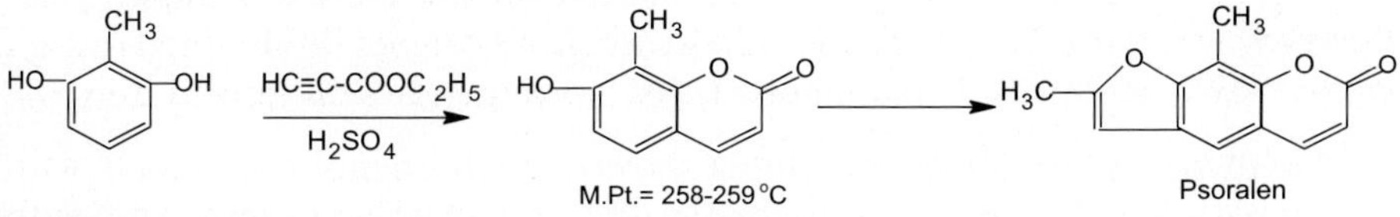


**Figure 3.7.**

### 3.1.1.3
### 4,6-Dialkyl Resorcinols

Symmetrically dialkyl substituted resorcinols were prepared from their corresponding diacyl derivatives [18] (Figure 3.8).

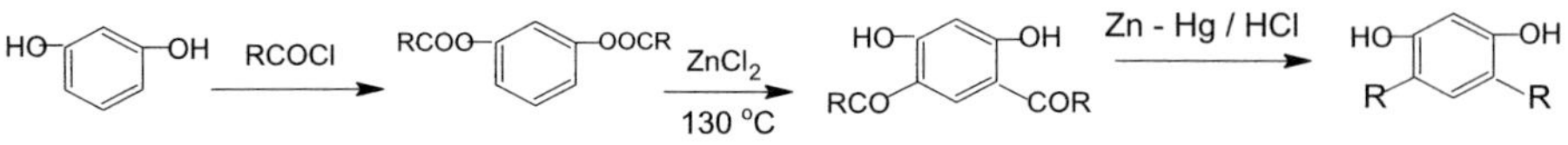


**Figure 3.8.**

Table 3.3 presents the melting and boiling points of some of the dialkylresorcinol derivatives [19,20].

**Table 3.3.** Symmetrically substituted 4,6-dialkylresorcinols (melting and boiling points data)

| 4,6-Substitution | Group structure | Melting point (°C) | Boiling point (°C/mm) |
|---|---|---|---|
| Methyl | $CH_3$ | 124.5–125 | 279[a] |
| Ethyl | $CH_3(CH_2)$ | 71 | NA |
| Propyl | $CH_3(CH_2)_2$ | 59 | NA |
| Butyl | $CH_3(CH_2)_3$ | NA | 194–196/14 |
| Hexyl | $CH_3(CH_2)_5$ | NA | 205/7 |
| Decyl | $CH_3(CH_2)_9$ | 65–66 | NA |

Data from [18–20]
[a]Data from [20]

Unsymmetrically substituted dialkylresorcinols possessing the germicidal and antiseptic properties were prepared according to the following scheme (Figure 3.9). The boiling point data of these compounds are presented in Table 3.4 [21].

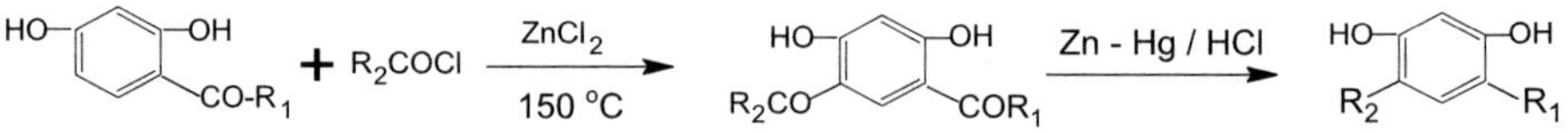


**Figure 3.9.**

The di-secondary alkyl resorcinols showing high phenol coefficients with potent germicidal properties were prepared by the reaction of resorcinol with primary aliphatic alcohols [22] (Figure 3.10).

The boiling points of some of these derivatives are presented in Table 3.5.

**Table 3.4.** Unsymmetrically substituted 4,6-dialkylresorcinols (boiling points data)

| 4,6-Substitution $R_1$ Group | $R_2$ Group | Boiling point (°C) | (mm) |
|---|---|---|---|
| $CH_3$ | $CH_3(CH_2)_3$ | 160–170 | 6 to 8 |
| | $CH_3(CH_2)_4$ | 170–178 | 6 |
| | $CH_3(CH_2)_5$ | 170–180 | 5 |
| | $CH_3(CH_2)_6$ | 172–183 | 4 to 5 |
| $CH_3(CH_2)$ | $CH_3(CH_2)_2$ | 140 | 3 |
| | $CH_3(CH_2)_3$ | 140–150 | 3 |
| | $CH_3(CH_2)_4$ | 170–175 | 3 |
| | $CH_3(CH_2)_5$ | 172–182 | 2 |
| | $CH_3(CH_2)_6$ | 175–185 | 3 |
| $CH_3(CH_2)_2$ | $CH_3(CH_2)_3$ | 165–170 | 3 |
| | $CH_3(CH_2)_4$ | 170–180 | 4 |
| | $CH_3(CH_2)_5$ | 180–190 | 2 |
| | $CH_3(CH_2)_6$ | 185–200 | 4 |
| $CH_3(CH_2)_3$ | $CH_3(CH_2)_4$ | 175–190 | 5 |
| | $CH_3(CH_2)_5$ | 195–205 | 5 |
| | $CH_3(CH_2)_6$ | 170–190 | 2 |

Data from [21]

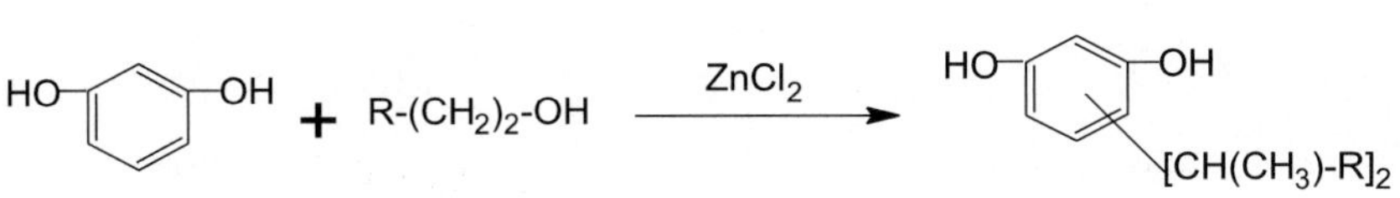


**Figure 3.10.**

**Table 3.5.** Di-sec-alkyl substituted resorcinols from alkylalcohols

| Alcohol Used | R Group | Boiling Point (°C/mm) |
|---|---|---|
| Ethyl | H | 135–137/4–5 |
| Propyl | $CH_3$ | 156–158/6–7 |
| Amyl | $CH_3(CH_2)_2$ | 168–175/6–7 |
| Hexyl | $CH_3(CH_2)_3$ | 178–182/6–7 |
| Heptyl | $CH_3(CH_2)_4$ | 165–175/2 |

Data from reference [22]

### 3.1.1.4
### 2,4-Dimethyl Resorcinol

The synthesis process employed in the preparation of 2,4-dimethylresorcinol from 2-methylresorcinol can be schematically shown in the following reaction [23] (Figure 3.11).

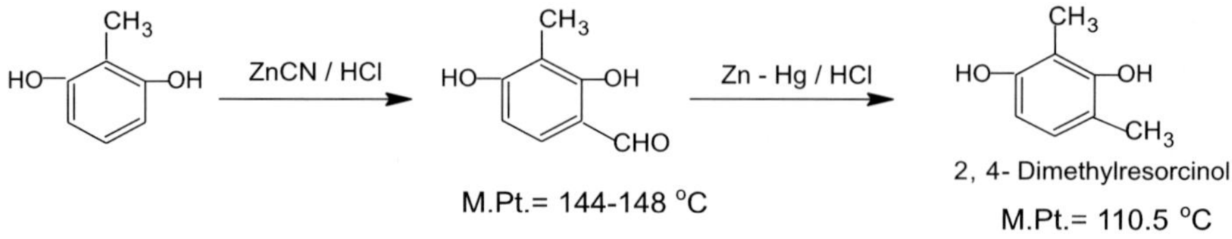


**Figure 3.11.**

This resorcinol derivative was used in the synthesis of pharmaceutically and biologically important organic compounds (Figure 3.12) [24].

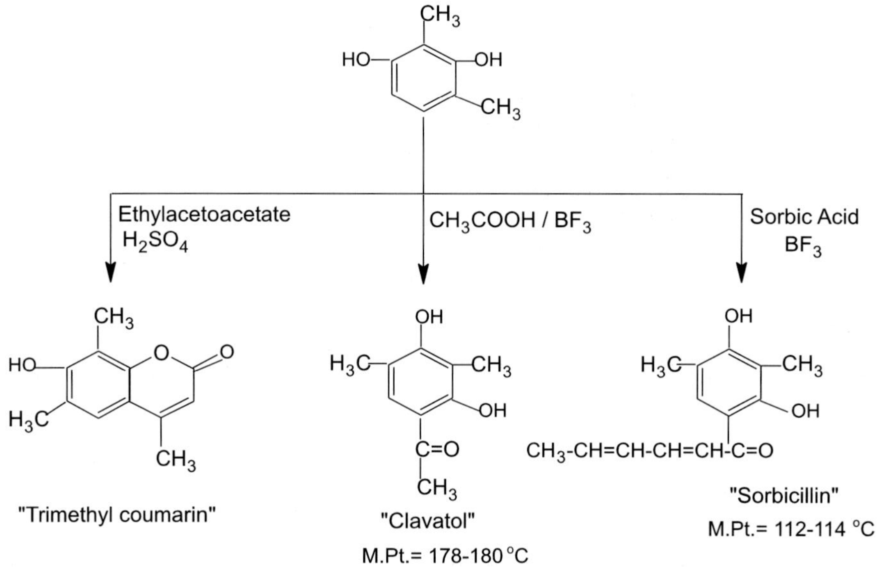


**Figure 3.12.**

### 3.1.1.5
### 2,4,6-Trimethyl Resorcinol

This derivative was obtained when resorcinol was reacted with methanol in the presence of sodium methoxide catalyst [25] (Figure 3.13).

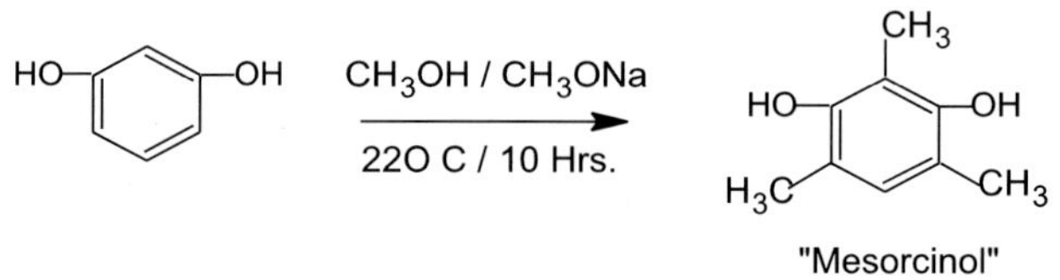


**Figure 3.13.**

### 3.1.1.6
### Tertiary-Alkyl Resorcinols

In the presence of an acid catalyst, resorcinol reacts with either an isobutylene or *tert*-butylalcohol compound to produce 4,6-di-*tert*-butylresorcinol [26–28] (Figure 3.14).

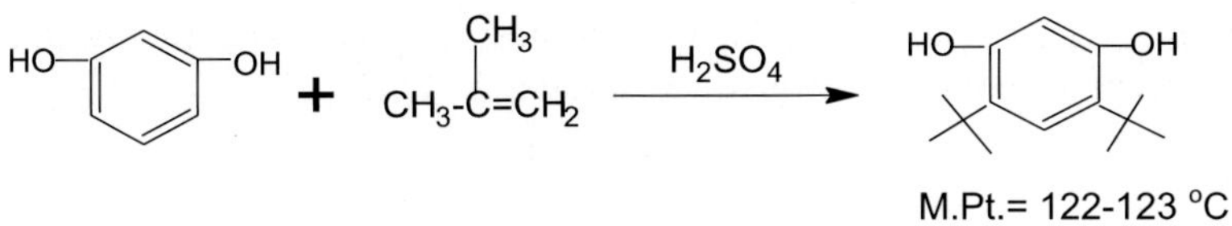


**Figure 3.14.**

Due to steric hindrance, the *tert*-butyl group does not enter into the 2-position of the resorcinol molecule. This compound can be useful to synthesize 2-substituted resorcinols by reacting with non-bulkier organic compounds (Figure 3.15) [29, 30].

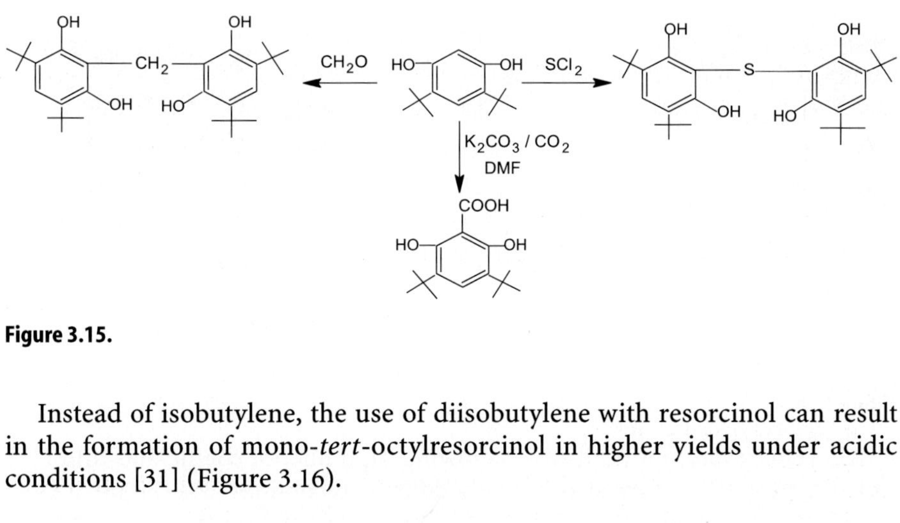


**Figure 3.15.**

Instead of isobutylene, the use of diisobutylene with resorcinol can result in the formation of mono-*tert*-octylresorcinol in higher yields under acidic conditions [31] (Figure 3.16).

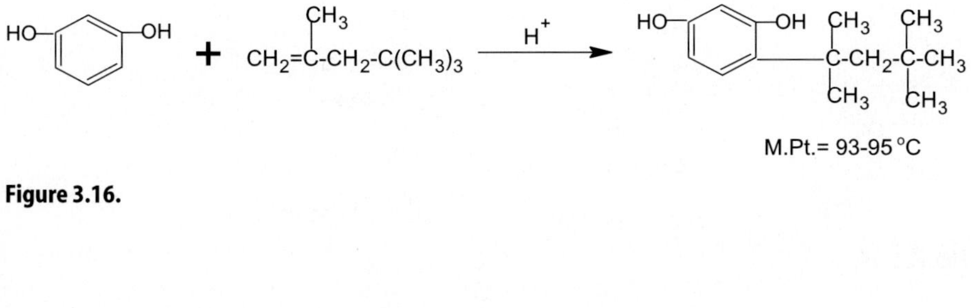


**Figure 3.16.**

### 3.1.1.7
### Cycloaliphatic Substituted Resorcinols

Cyclohexyl and cyclopentyl resorcinols can be obtained by condensing resorcinol with cyclohexanol or cyclopentanol in the presence of a suitable dehydrating or condensing agent [32] (Figure 3.17).

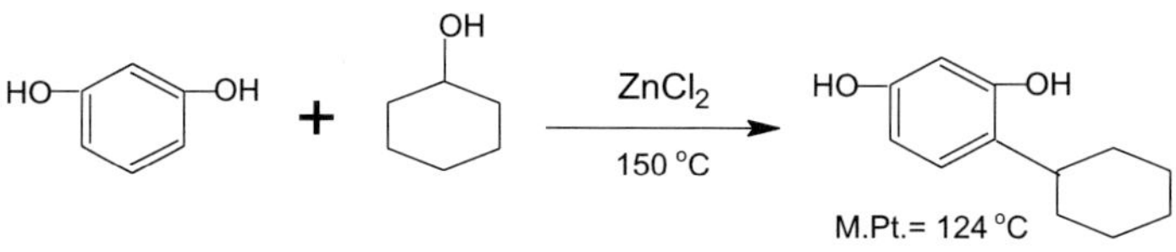


**Figure 3.17.**

The cyclohexylresorcinol can be further derivatized to synthesize pharmaceutically important compounds [33,34] (Figure 3.18).

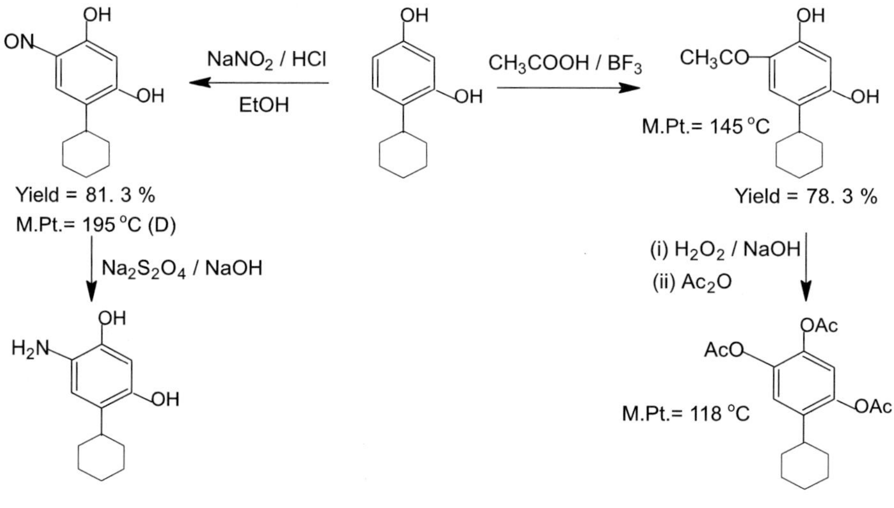


**Figure 3.18.**

Alicyclic resorcinol derivatives showing high phenol coefficients and bactericidal properties were also synthesized from resorcinol [35] (Figure 3.19).

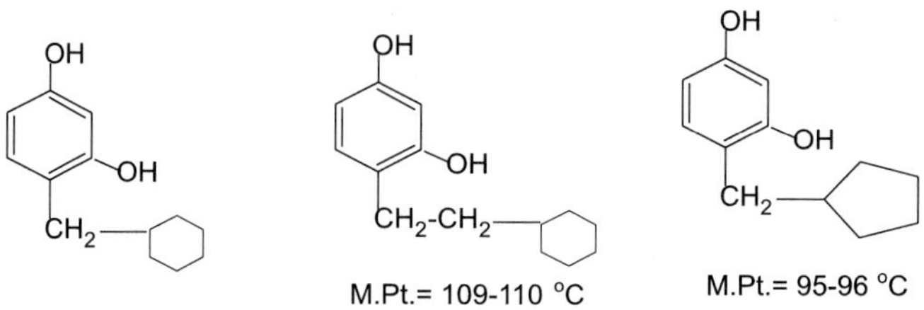


**Figure 3.19.**

### 3.1.1.8
### Allyl and Vinyl Resorcinols

Vinyl substituted resorcinol was synthesized using paraldehyde as the reactant [36] (Figure 3.20).

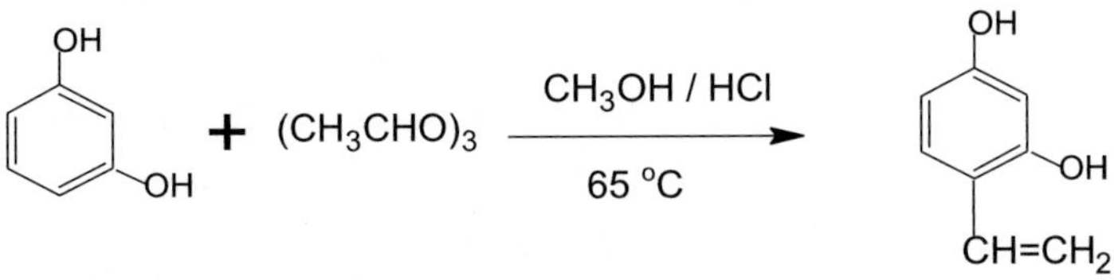


**Figure 3.20.**

Mono and diallylresorcinol compounds obtained from the allylhalide reactions were known to be useful in various industrial and pharmaceutical applications. Direct allylation of resorcinol with allylchloride was achieved by the use of cuprous chloride or sodium hydrogen phosphate in the reaction [37–39] (Figure 3.21).

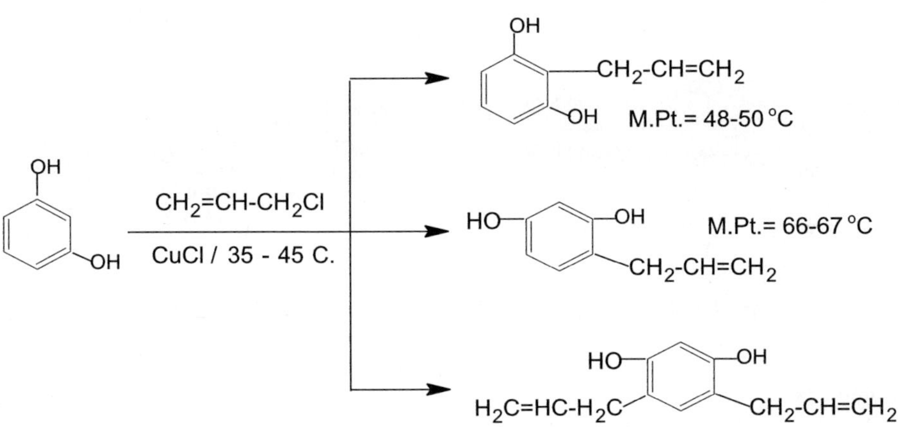


**Figure 3.21.**

Di-*n*-propyl resorcinol obtained from the diallylresorcinol was used in the development of drug compounds for the treatment of diabetes and lipid disorders [40] (Figure 3.22).

### 3.1.1.9
### Prenylated Resorcinols

Prenylated phenols play an important role in mediating many biological processes, are also known to exhibit anti-fungal, anti-tumor, anti-HIV and anti-Alzheimer's properties, and could function as potent anti-bacterial agents [41].

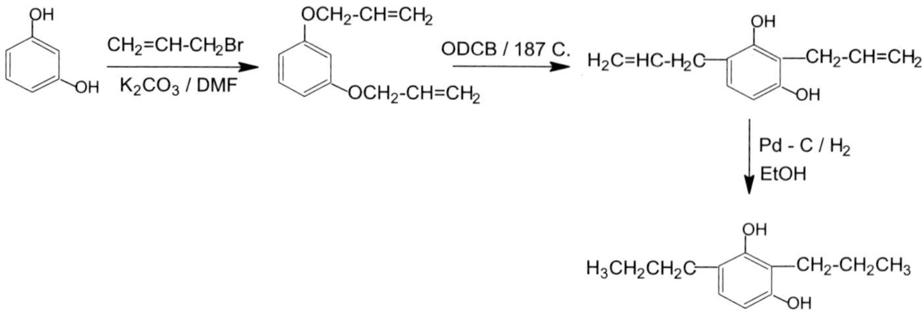


**Figure 3.22.**

The 2 and 4-prenyl substituted resorcinol compounds were synthesized according to the following reaction schemes (Figure 3.23).

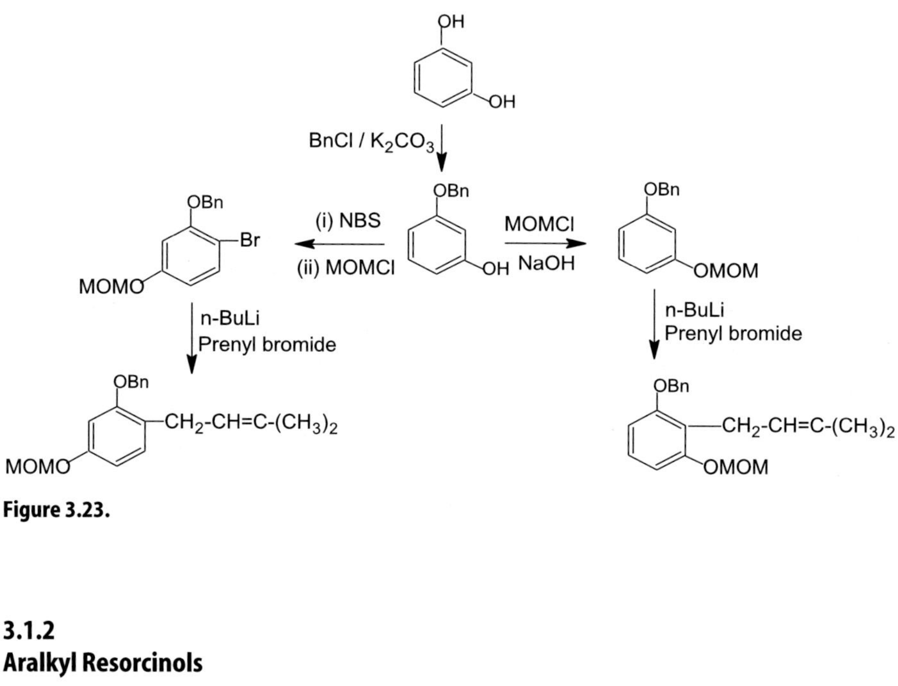


**Figure 3.23.**

## 3.1.2 Aralkyl Resorcinols

Aralkyl substituted resorcinol derivatives have been utilized as intermediates in the preparation of various organic compounds, polymers, and also used as polymer stabilizers and reactive additives to cure the epoxy resins. These compounds were normally synthesized by reacting resorcinol with aromatic vinylic or allylic compounds. Using styrene monomer, mono, di- and tri-$\alpha$-methyl benzyl substituted resorcinol derivatives were prepared in the presence of an acid catalyst [42, 43] (Figure 3.24).

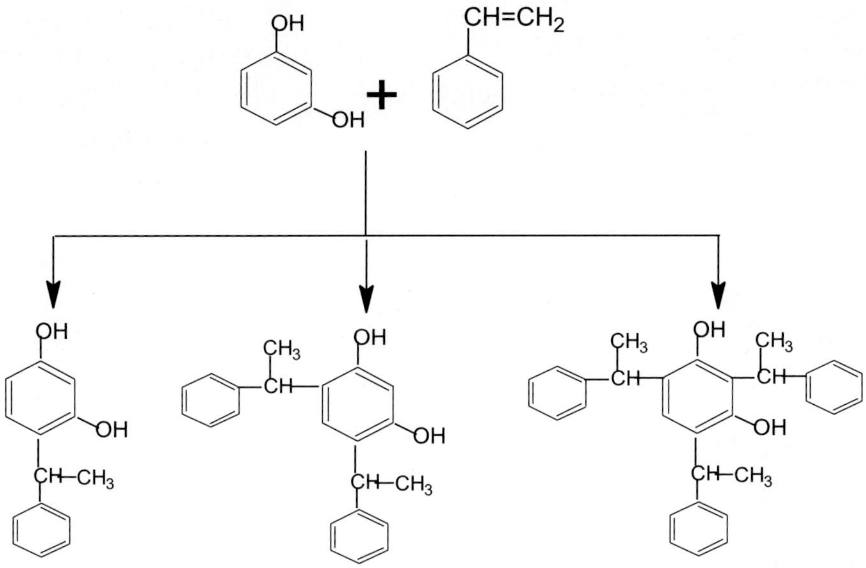

M.Pt.= 79-80 °C

M.Pt.= 88-89 °C

**Figure 3.24.**

In addition to styrene monomer, mono, di and tri-isopropenyl benzene monomers were also employed to obtain the aralkyl resorcinols. Unlike the styrene monomer, the isopropenyl benzene does not react with resorcinol at the 2-position due to the formation of the bulkier $\alpha,\alpha$-dimethylbenzyl group [44,45] (Figure 3.25).

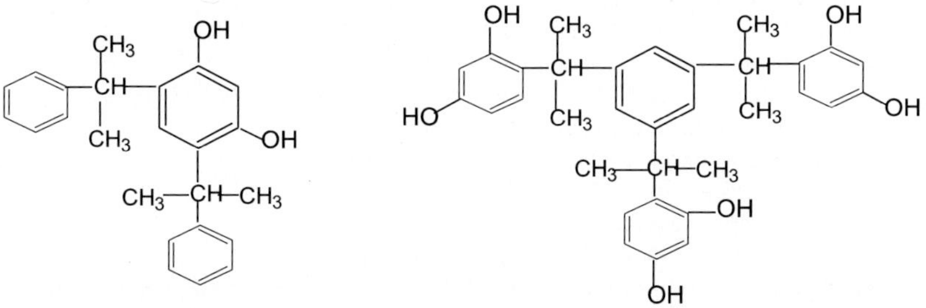

M.Pt.= 135.9 °C

M.Pt.= 147 °C

**Figure 3.25.**

Isopropylidene bridged resorcinols were also prepared using phenolic compounds containing the isopropenyl functional group [46] (Figure 3.26).

In the presence of an acid, acetone produced 4,4′-isopropylidene-bis-(2-methylresorcinol) from the 2-methylresorcinol derivative [47] (Figure 3.27).

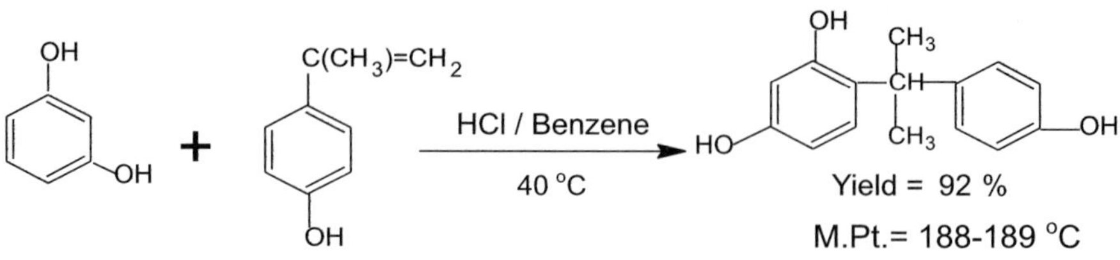


**Figure 3.26.**

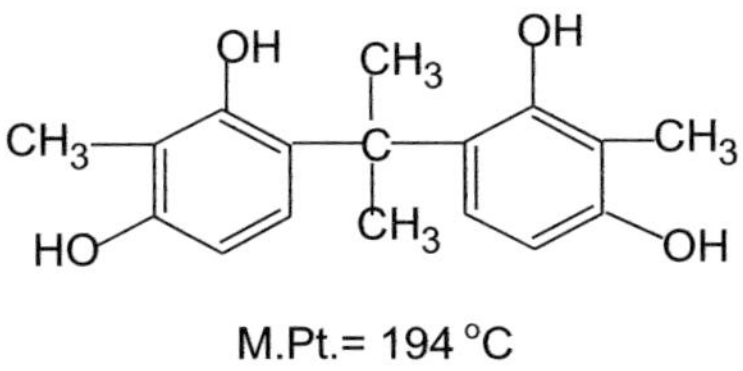


**Figure 3.27.**

## References

1. A. Dohme, E. Cox and E. Miller, J. Am. Chem. Soc., Vol. 48, p 1688 (1926).
2. E. Bousquet, US Patent, 2,146,007 (1939).
3. J. Brown, D. Johnson, A. Robertson and W. Whalley, J. Chem. Soc., p 2019 (1951).
4. D. Cram, J. Am. Chem. Soc., Vol. 70, p 4240 (1948).
5. Dictionary of organic compounds, 4th edn, Eyre & Spottiswoode Publishers, London, Vol. 2, p 1112 (1965).
6. A. Steinmetz, US Patent, 4,419,529 (1983).
7. J. Kerschner, A. Love and M. Barratt, US Pat. Appln., 2003/0,185,867 (2003).
8. A. Love, J. Kerschner, M. Barratt and Y. Zhou, US Pat. Appln., 2003/0,180,234 (2003).
9. D. Twiss, J. Am. Chem. Soc., Vol. 48, p 2206 (1926).
10. Hoffmann-La Roche, British Patent, 427,297 (1935).
11. Y. Takisawa and S. Hasegawa, US Patent, 4,337,370 (1982).
12. Y. Takisawa and S. Hasegawa, US Patent, 4,398,047 (1983).
13. T. Nakamatsu, Y. Nishida and N. Kometani, Euro. Pat. Appln., 0,529,517 (1993).
14. J. Jachowicz and L. Wolfram, US Patent, 4,588,760 (1986).
15. J. Cotteret and M. Audousset, US Patent, 5,599,353 (1997).
16. J. Cotteret and M. Audousset, US Patent, 5,964,898 (1999).
17. K. Kaufman, US Patent, 3,201,421 (1965).
18. E. Klarmann, J. Am. Chem. Soc., Vol. 48, p 2358 (1926).
19. Dictionary of Organic Compounds, Fourth Edition, Eyre & Spottiswoode Publishers, London, Vol. 2, p 1115 (1965).
20. U. Lille and H. Kundel, J. Chromatogr., Vol. 116, p 1 (1976).
21. W. Austin, US Patent, 2,006,039 (1935).
22. A. Rawlins and H. Hamilton, US Patent, 2,107,307 (1938).
23. W. Baker, H. Bondy, J. McOmie and H. Tunnicliff, J. Chem. Soc., p 2834 (1949).
24. J. McOmie and M. Tute, J. Chem. Soc., p 3226 (1958).
25. J. Cornforth and R. Robinson, J Chem. Soc., p 684 (1942).
26. T. Evans and K. Edlund, US Patent, 2,051,473 (1936).
27. F. Bentley, US Patent, 3,055,862 (1962).

28. S. Narayanan and K. Murthy, Appl Catal, Vol. 213, p 273 (2001).
29. T. Yamano, K. Jono and S. Sato, US Patent, 3,784,699 (1974).
30. C. Fuchsman, US Patent, 3,726,928 (1973).
31. F. Nishio and R. Oi, US Patent, 3,373,210 (1968).
32. M. Phillips, US Patent, 1,650,036 (1927).
33. W. McLamore, J. Am. Chem. Soc., Vol. 73, p 2225 (1951).
34. L. Hu, US Patent, 6,132,740 (2000).
35. R. Talbot and R. Adams, J. Am. Chem. Soc., Vol. 49, p 2040 (1927).
36. J. Kindall et al., WADD Technical Report 60-791, Part-I, p 126 (1961).
37. S. Nikiforova and S. Melkanovitskaya, J. Org. Chem. USSR, Vol. 2, p 1240 (1966).
38. C. Hurd, H. Greengard and F. Pilgrim, J. Am. Chem. Soc., Vol. 52, p 1700 (1930).
39. K. Moore, US Patent, 2,459,835 (1949).
40. K. Liu, L. Xu and A. Brian, US Patent, 6,569,879 (2003).
41. C. Hoarau and R. Pettus, Synlett., No. 1, p 127 (2003).
42. Y. Kozlikovskii et al., Zhur. Organ. Khim. (Translated), Vol. 22, p 606 (1986).
43. U. Neumann, US Patent, 6,277,944 (2001).
44. H. Dressler, US Patent, 4,599,464 (1986).
45. N. Sunao and K. Hisaharu, Jap. Patent, 2002-060,361 (2002).
46. I. Mizuo and I. Shigeru, Jap. Patent, 05-201,903 (1993).
47. I. Shigeki, M. Shigeru and N. Toshikatsu, Jap. Patent, 07-285,901 (1995).

## 3.2 Resorcinol Ethers

Resorcinol ether compounds can be classified into aliphatic and aromatic ethers. They are valuable in the synthesis of various organic derivatives useful for pharmaceutical and agricultural applications. In this chapter, some of the important ether derivatives of resorcinol have been discussed.

### 3.2.1 Alkyl Ethers of Resorcinol

Alkyl ether derivatives of resorcinol were prepared by treating the resorcinol mono or di-resorcinate, obtained from the reaction of resorcinol with alkali metal hydroxides or carbonates, with alkyl halides or sulfates [1–3] (Figure 3.28).

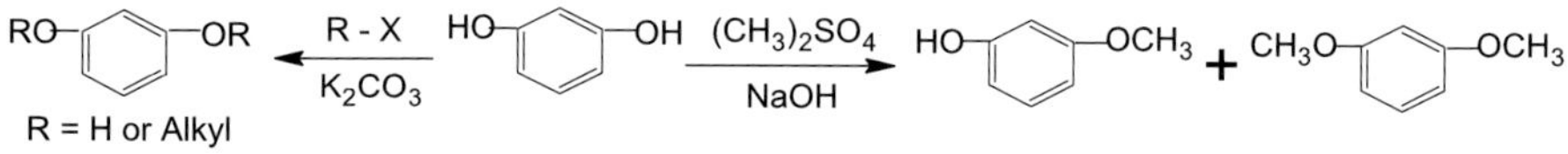


**Figure 3.28.**

In these preparations, alkyl sulfates are excellent alkylating agents because of the ease of displacement of the alkyl sulfate anion. Depending upon the reaction conditions and the amount of reactants used in the reaction, both the mono and dialkyl ether derivatives can be obtained.

#### 3.2.1.1 Monoalkyl Ethers

The first member in this derivative series is the mono-methyl ether of resorcinol. This compound was prepared from the reaction of resorcinol with methyl chloride [4], dimethyl sulfate [5], methanol in the presence of metal oxide catalyst [6], or dimethyl carbonate [7]. Resorcinol mono-methyl ether was also obtained in higher yields from the dihydroresorcinol reaction with methanol [8] (Figure 3.29).

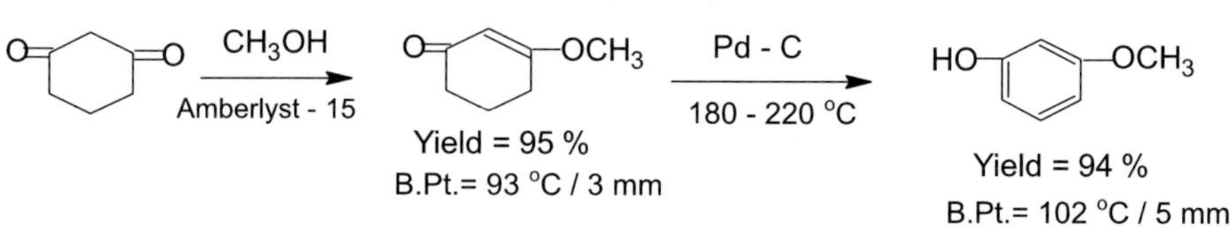


**Figure 3.29.**

In addition to methanol, 2-propanol, *n*-butanol, allyl alcohol and 1,4-butanediol was also employed in the reaction with dihydroresorcinol to obtain their corresponding mono-alkyl ether derivatives.

More often, the yields of mono-alkyl ether compounds from the alkyl halide reaction were low. In order to produce the 3-alkoxy resorcinols in excellent yields (90–99%), resorcinol monobenzoate was employed in the reaction [9] (Figure 3.30).

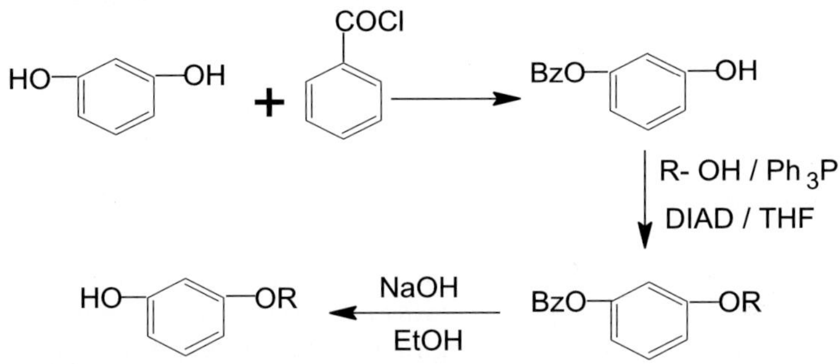


**Figure 3.30.**

High purity mono-propyl, butyl and pentyl ethers of resorcinol were prepared by reacting resorcinol with alkyl bromide compounds in the presence of sodium ethoxide [2] (Figure 3.31).

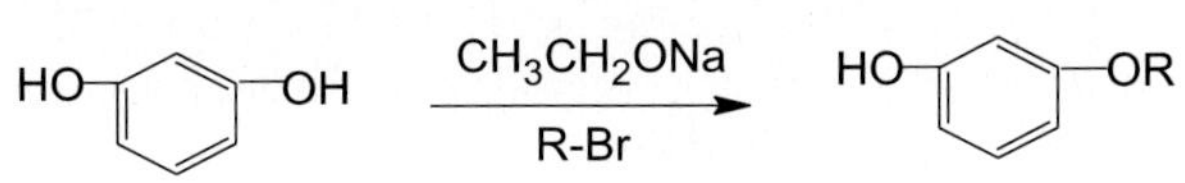


R = Methyl (B.Pt. = 244 °C; 121-122 °C / 10 mm)

= Propyl (B.Pt. = 118-120 °C / 6 mm)

= Butyl (B.Pt. = 148 °C / 12 mm)

= Pentyl (B.Pt. = 128 -131 °C / 5 mm)

**Figure 3.31.**

Mono-alkyl ether compounds of resorcinol can undergo nitration and bromination reactions similar to resorcinol to produce mono-, di- and tri-substituted nitro or bromo derivatives. These ether compounds could be used as intermediates in the synthesis of pharmaceuticals, dyes and other specialty resorcinol derivatives.

When the resorcinol mono-methyl ether was reacted with 1, 2-dichloro-4-trifluoromethyl benzene, a phenyl ether compound useful as an intermediate for the synthesis of herbicides was obtained [10] (Figure 3.32).

Mono-alkyl ether of resorcinol was used in the development of pharmaceutically important compounds useful as medicaments in local anesthetics [2,11] (Figure 3.33).

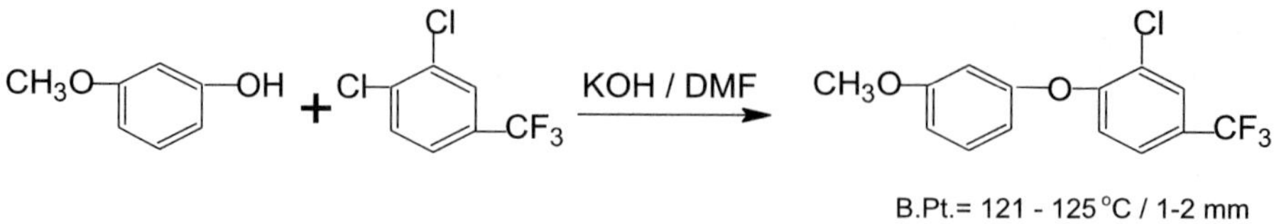


**Figure 3.32.**

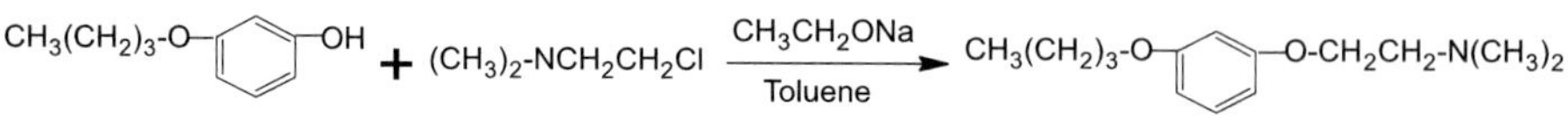


**Figure 3.33.**

### 3.2.1.2 Resorcinol Dimethyl Ether

This derivative is an important intermediate in the synthesis of pharmaceuticals, 2-substituted resorcinols, dye intermediates and photographic chemicals. Resorcinol dimethyl ether has two activating methoxy groups that could increase the electron density at the *ortho*- and *para*-positions of the resorcinol ring. Therefore, many reactions typical of resorcinol can be expected with this compound.

By employing *n*-butyl lithium (*n*-BuLi) in the reaction, resorcinol dimethyl ether can be used to synthesize various 2-substituted resorcinol compounds [12,13] (Figure 3.34).

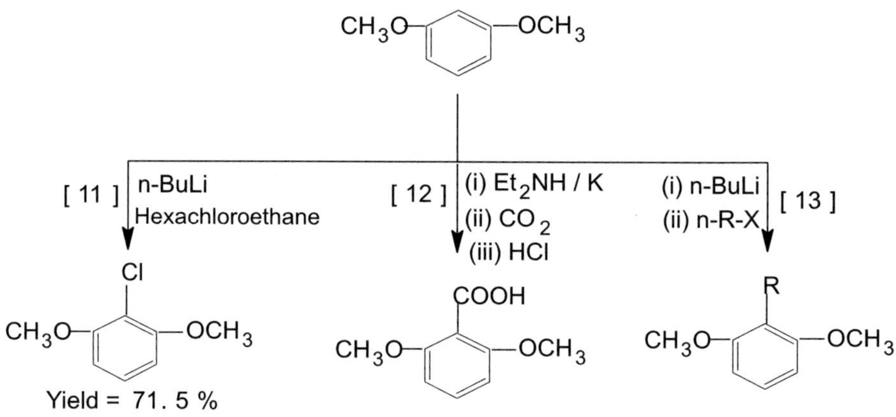


**Figure 3.34.**

The melting and boiling points of some 2-substituted dimethylether compounds are presented in Table 3.6.

After the substitution, the 2-substituted dimethyl ether derivatives can be demethylated using hydrobromic (HBr) solutions to obtain the 2-substituted resorcinols.

**Table 3.6.** 2-Substituted resorcinol dimethylether derivatives[a] melting and boiling points data

| 2-Substitution | | Melting point (°C) | Boiling point (°C/mm) |
|---|---|---|---|
| Methyl | $CH_3$ | | 120–122/30 |
| Ethyl | $CH_2CH_3$ | 54–55 | |
| Butyl | $(CH_2)_3CH_3$ | | 116/15 |
| Chloro | Cl | 71–73[b] | |
| Bromo | Br | 91–93 | |
| Iodo | I | 102–103 | |
| Carboxylic acid | COOH | 192–194 | |
| Methyl carboxylate | $COOCH_3$ | 93–94 | |
| Aldehyde | CHO | 96–98 | |

[a] Data from [14]
[b] Data from [11]

4-Substituted compounds were also obtained directly from the resorcinol dimethyl ether reaction with various reactants [13, 15–17] (Figure 3.35).

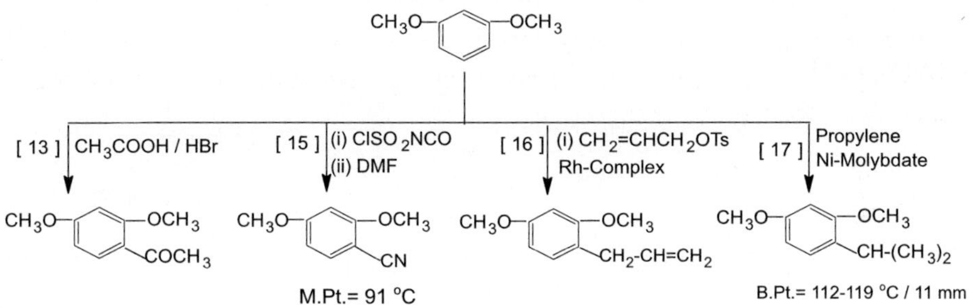


**Figure 3.35.**

These dimethoxy derivatives were used further in various organic reactions to obtain synthetically as well as pharmaceutically important chemical compounds. For example, 2,4-dimethoxybenzaldehyde was further reacted with the 2-butenolide compound to synthesize 4-(2,4-dimethoxybenzyliden)-2-butenolide [18] (Figure 3.36).

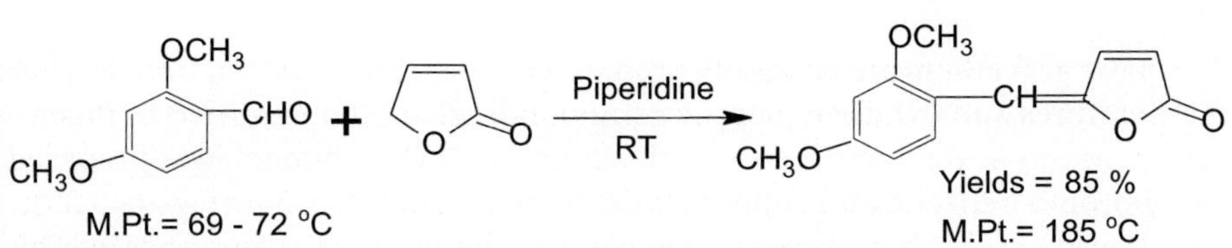


**Figure 3.36.**

### 3.2.1.3
### Allyl Ether Compounds

Mono- and di-allyl resorcinol derivatives were synthesized using allyl bromide or allyl carbonate in this reaction [19,20] (Figure 3.37).

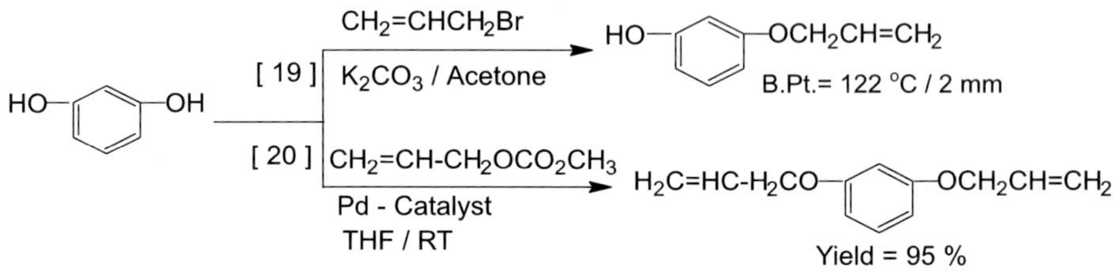


**Figure 3.37.**

### 3.2.1.4
### Hydroxyethyl Ethers of Resorcinol

Ethylene chlorohydrin and ethylene carbonates were employed in the synthesis of mono-hydroxyethyl and bis-(hydroxyethyl) ether derivatives [21,22] (Figure 3.38).

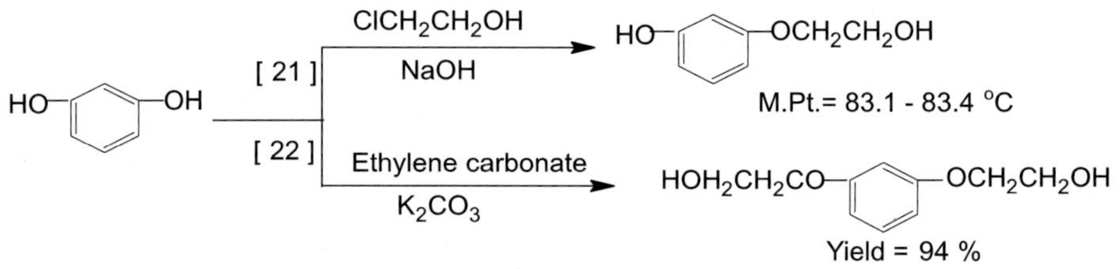


**Figure 3.38.**

Bis-(hydroxyethyl) ether of resorcinol (HER®) has been used in the manufacture of cast and thermoplastic polyurethanes.

### 3.2.2
### Aromatic Ethers of Resorcinol

Aromatic aryl ethers are generally prepared by the Ullmann's reaction. Higher temperatures and the use of copper compounds were often required in this reaction. Mono-aryl ether of resorcinol, namely *m*-aryloxyphenol, was prepared using bromo benzene and copper oxide in the reaction [23,24] (Figure 3.39).

It is well known that certain halogenated diphenyl ether compounds exhibit excellent antimicrobial activity. The non-halogenated aryl ether compounds

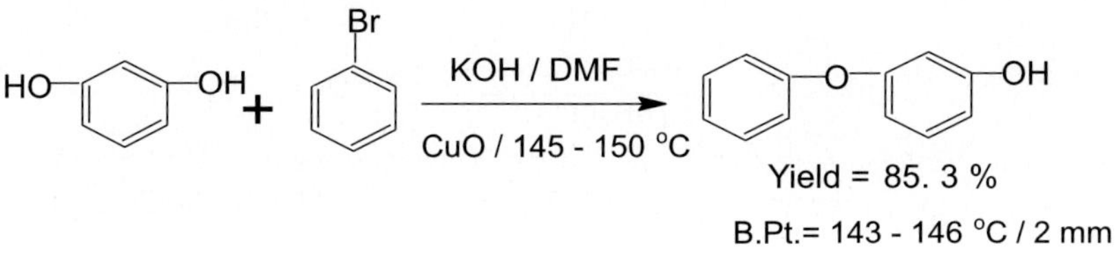


**Figure 3.39.**

derived from the *m*-phenoxyphenol were found to be highly effective as antimicrobial agents, and were found to be suitable in applications such as textiles, paper, plastics, detergent additives, soaps and dental hygiene products [25] (Figure 3.40).

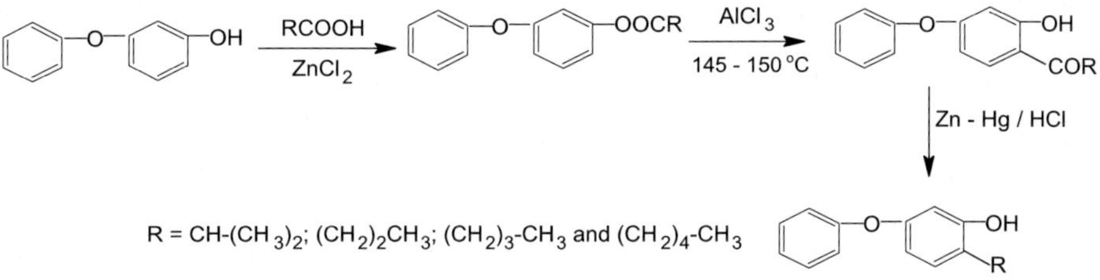


**Figure 3.40.**

Resorcinol diphenyl ether compound was synthesized by using pyridine and soluble copper salts in the reaction of resorcinol with bromo benzene, and also in the absence of oxygen and moisture during this reaction [26, 27] (Figure 3.41).

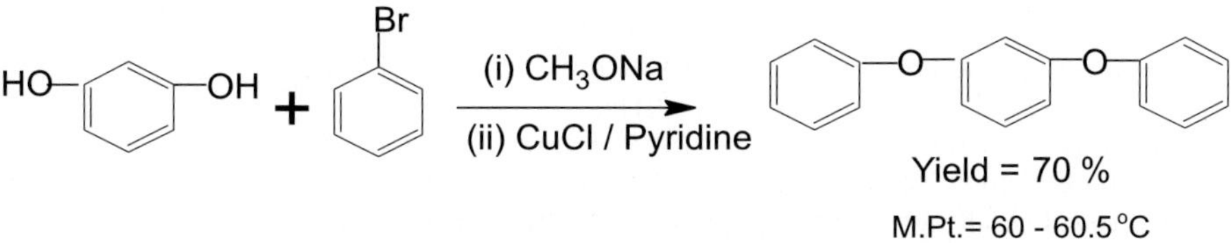


**Figure 3.41.**

Diphenoxy benzaldehyde compounds, important precursors for pharmaceuticals and organic syntheses in general, were obtained via two different synthetic methods [28] (Figure 3.42).

Goal® is a diphenyl ether compound synthesized using resorcinol as the main reactant, and has been extensively used as a selective pre-emergence and post-emergence herbicide [29–32] (Figure 3.43).

Goal® herbicide has been used for the selective control of broad leaf weeds and some grasses in fallow conditions, tree and nut crops, brassica vegetables, onions, tobacco, coffee and pyrethrum.

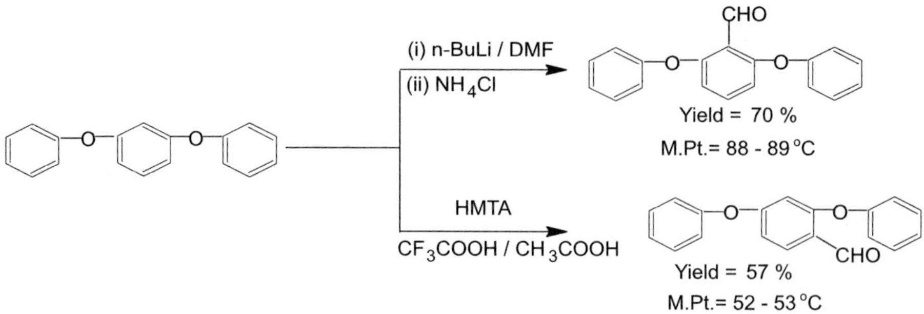


**Figure 3.42.**

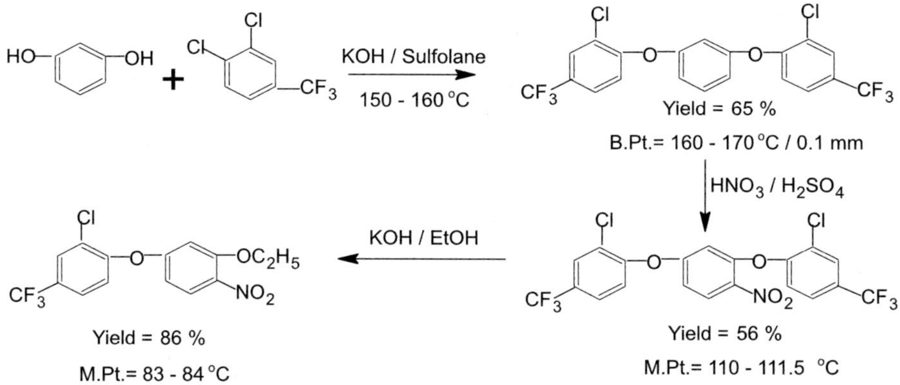


**Figure 3.43.**

## References

1. W. Perkin, J. Ray and R. Robinson, J. Chem. Soc., p 941 (1926).
2. R. Sandberg, US Patent, 6,310,252 (2001).
3. R. Mirrington and G. Feutrill, Organic Syntheses, Coll. Vol. 6, Wiley, p 859 (1988).
4. P. Gradeff and C. Bertrand, British Patent, 1,277,186 (1972).
5. Universal Oil Products, British Patent, 1,000,932 (1965).
6. K. Ariyoshi, Y. Satoh and N. Saio, US Patent, 5,817,886 (1998).
7. F. Merger, F. Towae and L. Schroff, US Patent, 4,192,949 (1980).
8. W. Mueller, Canadian Patent, 1,060,478 (1979).
9. J. Boxhall et al., Synlett., No. 7, p 997 (2003).
10. H. Forster and U. Priesnitz, US Patent, 4,450,308 (1984).
11. G. Shutske, T. Lee and G. Jobin, US Patent, 4,579,981 (1986).
12. L. Yu, US Patent, 4,845,276 (1989).
13. W. Marshall, S. Sigmund and C. Whitesitt, US Patent, 4,777,299 (1988).
14. U. Azzena et al., J. Org. Chem., Vol. 55, p 5386 (1990).
15. G. Lohaus, Organic Syntheses, Coll. Vol. 6, Wiley, p 465 (1988).
16. N. Tsukada, Y. Yagura, T. Sato and Y. Inoue, Synlett., No. 10, p 1431 (2003).
17. F. Fawcett and B. Howk, US Patent, 2,572,019 (1951).
18. L. Sorotskaya et al., Molbank, No. 4, M108 (1999).

19. C. Hurd, H. Greenhard and F. Pilgrim, J. Am. Chem. Soc., Vol. 52, p 1700 (1930).
20. R. Kolodziuk et al., Synth. Commn., Vol. 30, p 3955 (2000).
21. L. Kyrides, US Patent, 2,015,115 (1935).
22. C. Summer, B. Hitch and B. Bernard, US Patent, 5,068,460 (1991).
23. J. Livak and E. Britton, US Patent, 3,294,849 (1966).
24. G. Irick and G. Hawkins, US Patent, 3,487,114 (1969).
25. W. Holzl et al., US Pat. Appln., 2003 / 0,162,836 (2003).
26. R. Bridger, R. Kinney and A. Williams, US Patent, 3,651,151 (1972).
27. A. Williams, R. Kinney and R. Bridger, J. Org. Chem., Vol. 32, p 2501 (1967).
28. A. Katritzky, H. He, Q. Long and A. Wilcox, Arkivoc, p 3 (2001).
29. H. Bayer, C. Swithenbank and R. Yih, US Patent, 4,093,446 (1978).
30. H. Bayer, C. Swithenbank and R. Yih, US Patent, 4,185,995 (1980).
31. H. Bayer and C. Swithenbank, US Patent, 4,772,744 (1988).
32. H. Bayer and C. Swithenbank, US Patent, 4,795,825 (1989).

## 3.3
## Resorcinol Esters

Alkyl and aryl esters of resorcinol have been synthesized from the reactions of resorcinolic hydroxyl groups with aliphatic and aromatic carboxylic acids, acyl halides and acid anhydrides. In the esterification processes, the two hydroxyl groups of resorcinol can take part in the reactions. Depending upon the reaction conditions and catalysts used, mono and diesters of resorcinol could be produced (Figure 3.44).

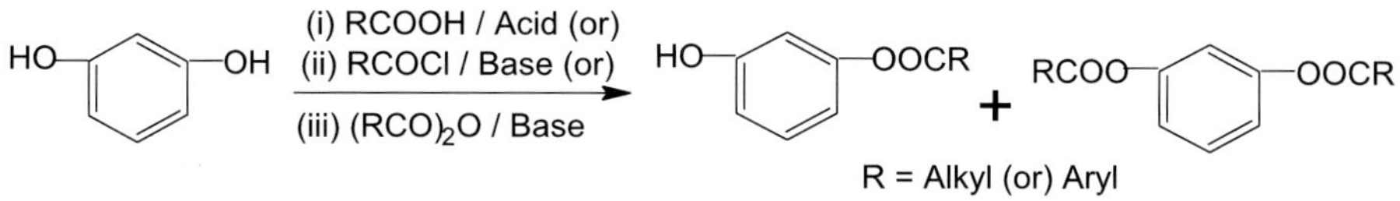


**Figure 3.44.**

Resorcinol mono and diesters were found to be useful in various industrial applications, and also used as raw materials in the synthesis of different derivatives.

### 3.3.1
### Aliphatic Monoesters

The first member in this derivative series is the monoacetate. This compound was prepared from the reaction of disodium resorcinate with acetic anhydride [1] (Figure 3.45).

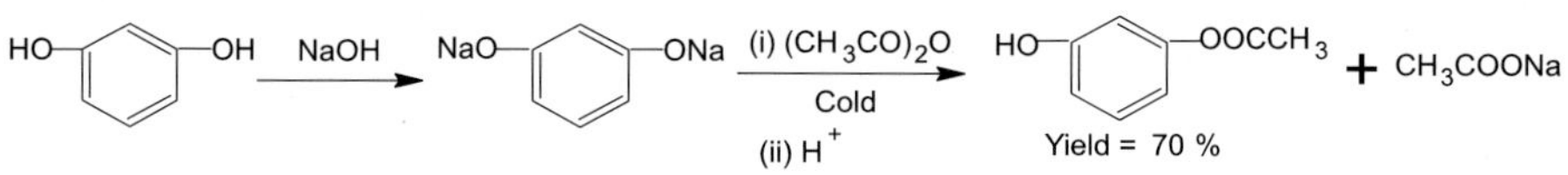


**Figure 3.45.**

In addition to the above process, the monoacetate was synthesized in about 83% yield when the resorcinol diacetate was treated with disodium resorcinate [1] (Figure 3.46).

Hexyl resorcinol has been used as an anthelmintic and antiseptic compound in various pharmaceutical compositions. The monoacetate derivative of hexyl resorcinol was determined to be hydrolyzable and non-toxic [2] (Figure 3.47).

This compound could undergo a gradual hydrolysis in the intestine due to the pH change of the environment, and also in the presence of lipases. Because of this, a gradual release of hexyl resorcinol over a considerable period of time can take place.

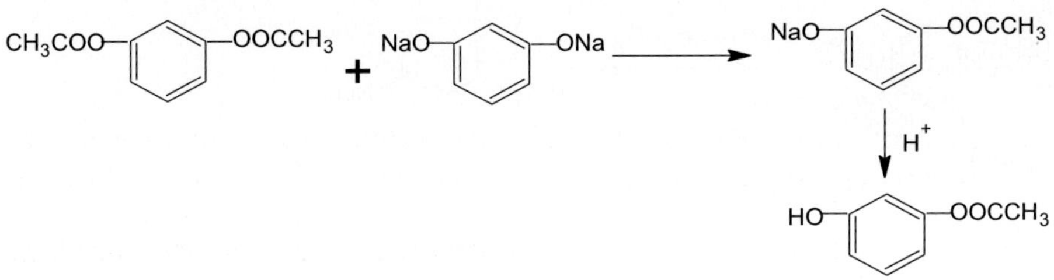


**Figure 3.46.**

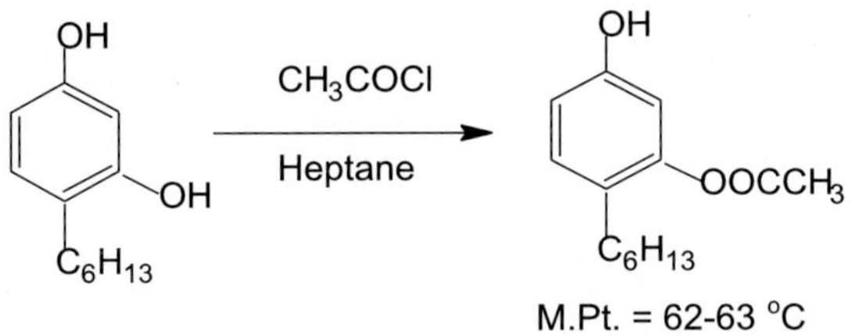


**Figure 3.47.**

Monooctanoate derivative, obtained from the resorcinol and octanoic anhydride reaction, was used in the peroxygen bleaching compound formulations to bleach fabrics, textiles and other materials [3].

### 3.3.2 Aliphatic Diesters

Resorcinol diacetate was synthesized in quantitative yields by using pyridine as the catalyst in the esterification reaction with acetic anhydride [4] (Figure 3.48).

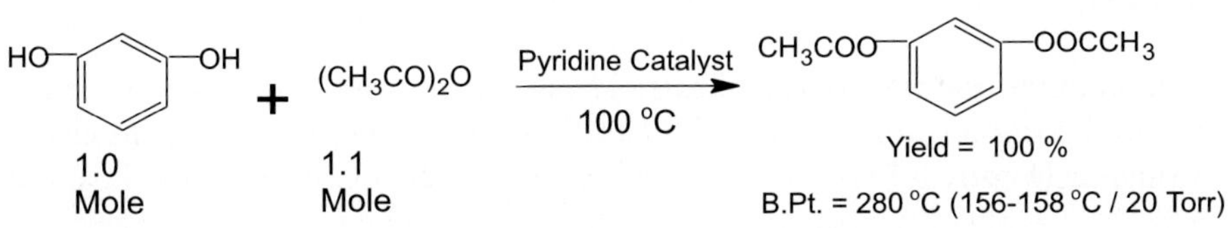


**Figure 3.48.**

In addition to pyridine, Montmorillonite K-10 [5,6] and sulfamic acid [7] catalysts were also employed in the acetylation reaction.

Resorcinol diacetate was used as a monomer in the synthesis of aromatic polyester materials, and also employed as the raw material in the preparation of 4,6-dinitroresorcinol intermediate to synthesize 4,6-diaminoresorcinol monomer for poly-(*p*-phenylene benzobisoxazole), (PBO), polymer (Figure 3.49).

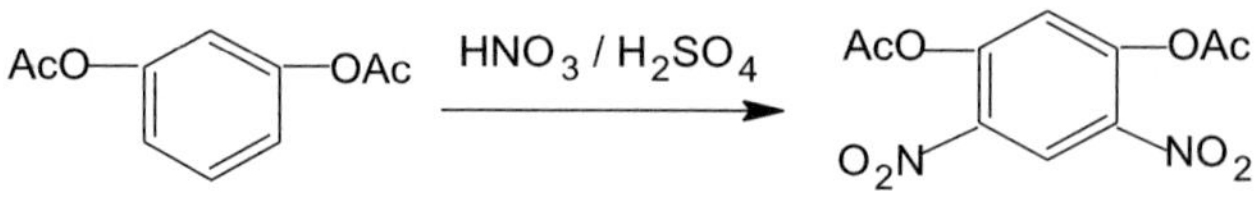


**Figure 3.49.**

The acetylenic ester of resorcinol was prepared using propiolyl chloride in the reaction [8] (Figure 3.50).

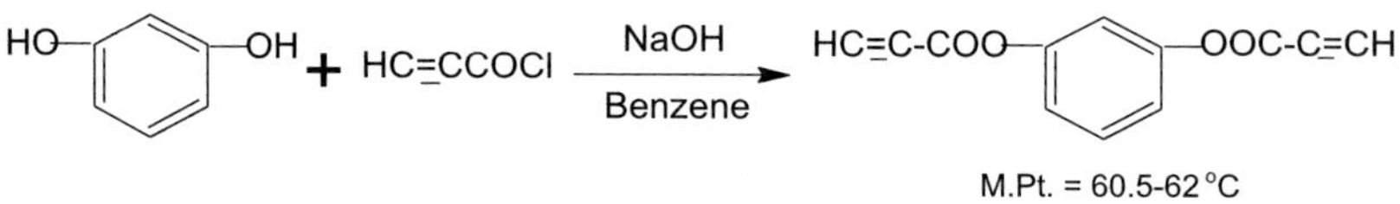


**Figure 3.50.**

Compounds having acetylenic ester groups could serve as the essential effective ingredients of fungicidal and bactericidal compositions.

### 3.3.3 Monoaryl Esters

Resorcinol monobenzoate was prepared in greater than 90% yield, when resorcinol was reacted with benzoyl chloride in the presence of sodium hydroxide at 19 to 21 °C and 8 to 8.3 pH conditions [9] (Figure 3.51).

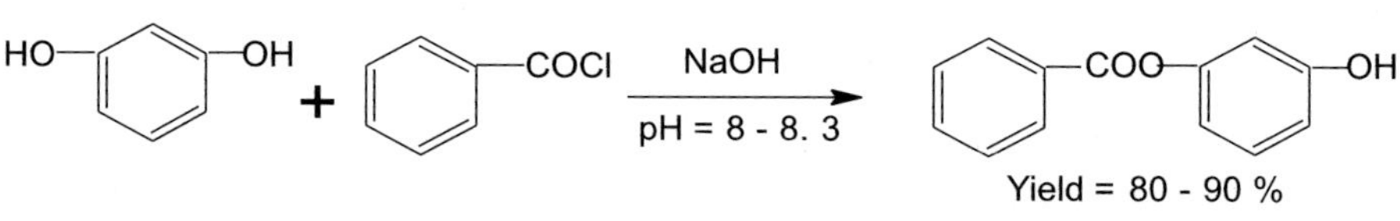


**Figure 3.51.**

This compound was also synthesized from the resorcinol dibenzoate derivative employing cesium carbonate [10] and sodium methoxide [11] as the cleavage agents (Figure 3.52).

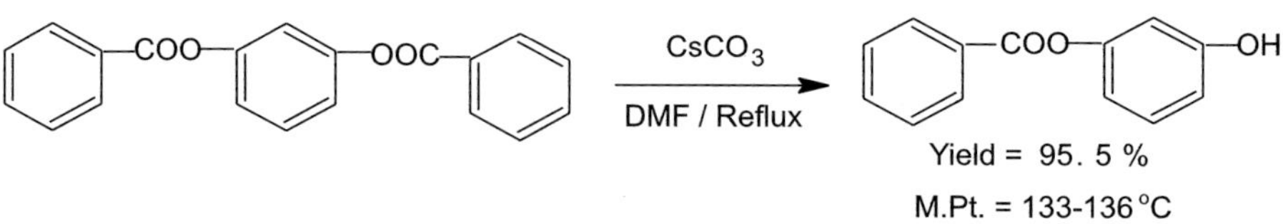


**Figure 3.52.**

Resorcinol mono-*m*-hydroxybenzoate [12] and mono-piperonylate [13] compounds were synthesized from resorcinol and the respective carboxylic acid chlorides generated in-situ in the presence of phosphorus oxychloride (Figure 3.53).

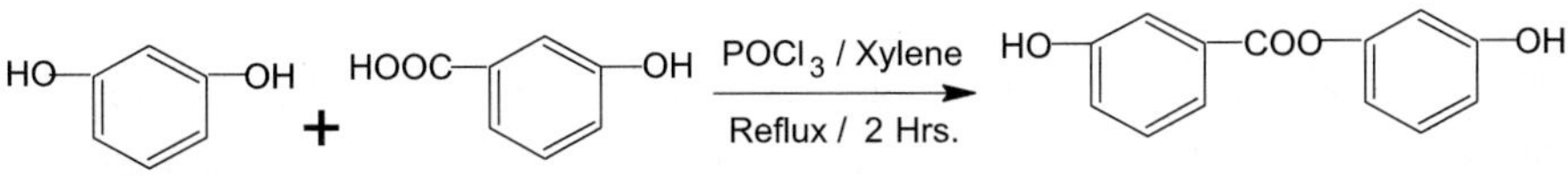


**Figure 3.53.**

Resorcinol monoaryl derivatives have been primarily used as ultraviolet (UV) stabilizers for poly (olefins) and poly (vinyl chloride). Alkyl ether derivatives obtained from the resorcinol monobenzoate were found to be effective UV stabilizers for various polymers (Figure 3.54) [14].

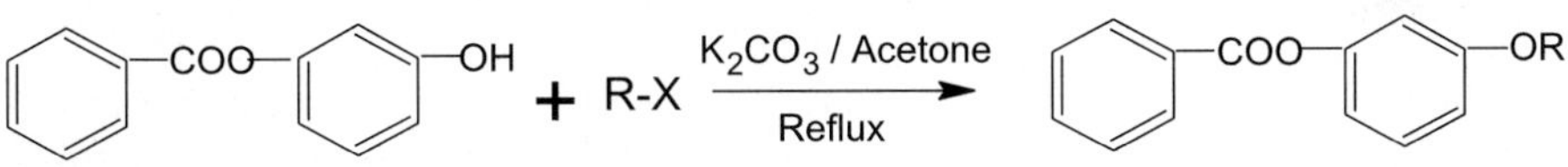


**Figure 3.54.**

The boiling and melting points of these alkyl ether compounds are presented in Table 3.7.

**Table 3.7.** Melting and boiling points of alkylether substituted phenylbenzoates

| Alkyl group (R) | Boiling point (°C/mm) | Melting point (°C) |
|---|---|---|
| $CH_3$ | 152–155/1 | |
| $C_2H_5$ | 155–158/1 | 25–27 |
| $n$-$C_3H_7$ | 157–163/1 | |
| $n$-$C_4H_9$ | 165–168/1 | 26–28 |
| $n$-$C_5H_{11}$ | 178–182/1 | |
| $n$-$C_6H_{13}$ | 170–173/0.5 | 28–30 |
| $n$-$C_7H_{15}$ | 180–184/0.5 | 27–29 |
| $n$-$C_8H_{17}$ | 88–92/0.5 | 28–30 |
| $n$-$C_9H_{19}$ | 99–101/0.5 | |
| $n$-$C_{10}H_{21}$ | | 36–38 |
| $n$-$C_{12}H_{25}$ | | 43–44 |
| $n$-$C_{14}H_{29}$ | | 48–51 |
| $n$-$C_{16}H_{33}$ | | 57–59 |
| $n$-$C_{18}H_{37}$ | | 65–67 |

Data from [14]

Using resorcinol monobenzoate as the starting material, several brominated derivatives of resorcinol were synthesized [15] (Figure 3.55).

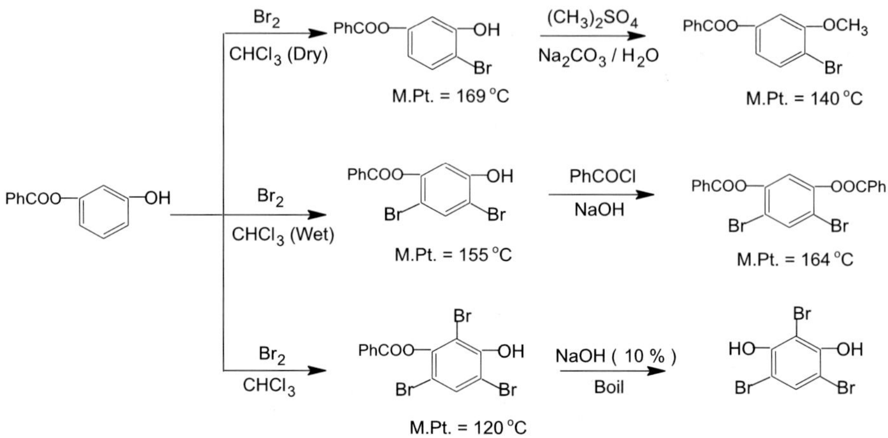


**Figure 3.55.**

The 3-benzoyloxyphenyl-2-methacryloxy ethyl carbonate and other mathacryl derivatives obtained from the resorcinol monobenzoate were useful in the surface coatings of polycarbonate articles to protect them against UV radiation [16–18] (Figure 3.56).

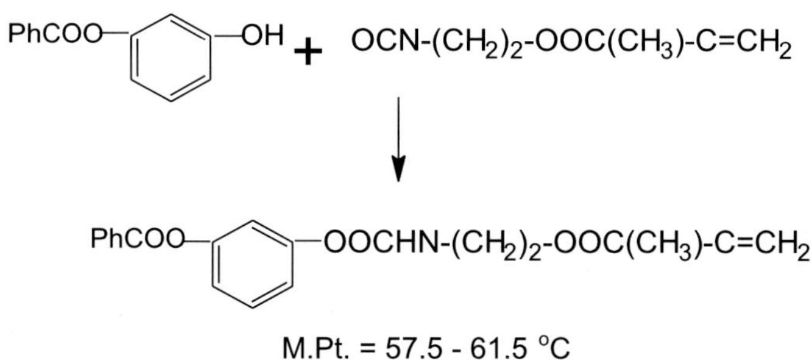


**Figure 3.56.**

*m*-(2-Propynyloxy) phenyl benzoate obtained from the propargyl chloride and resorcinol monobenzoate reaction was used as a parasiticide for the control of various insect, worm, bacterial, fungal and mite organisms [19].

## 3.3.4
## Dibenzoate Esters

Resorcinol dibenzoate derivatives can be obtained from the reactions of resorcinol with various aromatic substituted and un-substituted acid halide compounds [20,21] (Figure 3.57).

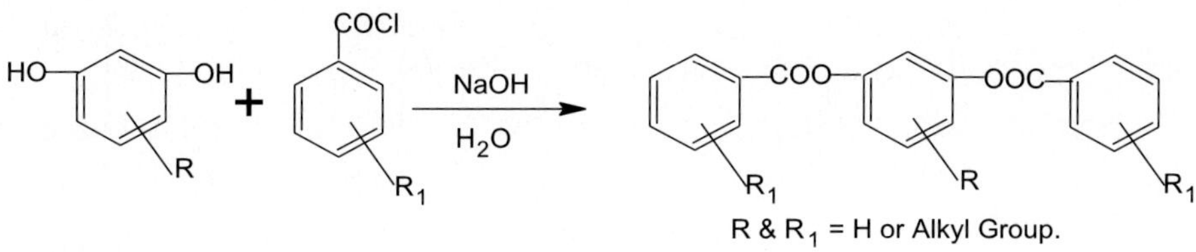


**Figure 3.57.**

Resorcinol dibenzoate has been used primarily as a latent UV stabilizer in the coating formulations containing PMMA. This compound was also used in the synthesis of monobenzoate and benzoyl resorcinol derivatives.

The first achiral bent core banana-shaped mesogen consisted of thermally and hydrolytically stable dibenzoate compounds synthesized from resorcinol exhibited liquid crystalline properties. The textural and electro-optic switching characteristics of these novel compounds could be valuable in potential photonic switching applications [22–24] (Figure 3.58).

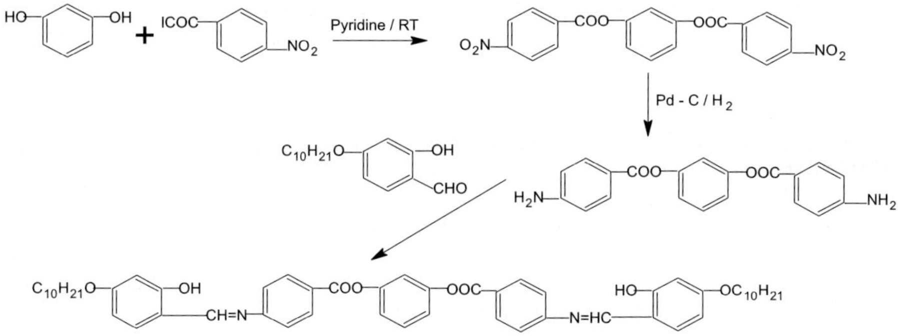


**Figure 3.58.**

## 3.3.5 Carbonate Esters

With phosgene, resorcinol produced carbonate esters having different molecular weights and melting points [25] (Figure 3.59).

The free hydroxyl groups present in these molecules can be etherified, esterified or urethanized.

Allylic carbonate esters of resorcinol having the following structure were synthesized and utilized in the preparation of polymers [26,27] (Figure 3.60).

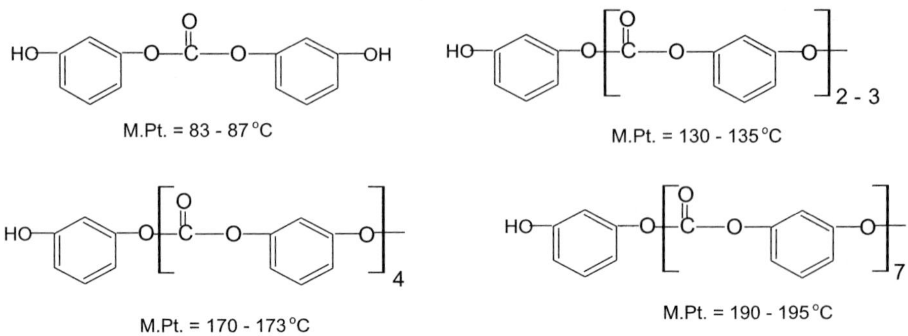


**Figure 3.59.**

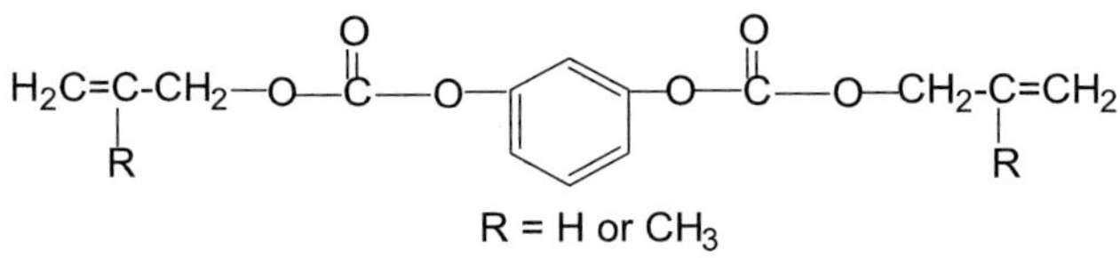


**Figure 3.60.**

## References


1. S. Israelstam and J. Simpson, J. South African Chem. Inst., Vol. 9, p 92 (1956).
2. J. Dereska, US Patent, 2,752,388 (1956).
3. R. Fong and R. Rowland, US Patent, 4,814,110 (1989).
4. I. Nitta and K. Asai, US Patent, 5,488,138 (1996).
5. A. Li, T. Li and T. Ding, Chem. Comm. p 1389 (1997).
6. T. Li and A. Li, J. Chem. Soc., Perkin Trans. 1, p 1913 (1998).
7. T. Jin, Y. Ma, Z. Zhang and T. Li, Synth. Commun., Vol. 28, p 3173 (1998).
8. L. Miller and J. Buttler, US Patent, 3,082,242 (1963).
9. W. Gearhart, R. Hill and M. Broyles, US Patent, 2,571,703 (1951).
10. H. Zaugg, J. Org. Chem., Vol. 41, p 3419 (1976).
11. W. Gearhart and R. Pugh, US Patent, 2,588,978 (1952).
12. L. Meyer and M. Broyles, US Patent, 2,811,461 (1957).
13. L. Meyer and M. Broyles, US Patent, 2,708,638 (1955).
14. G. Newland and G. Lippin, US Patent, 3,519,599 (1970).
15. H. Hodgson and R. Dyson, J. Chem. Soc., p 946 (1935).
16. T. O'Donnell and D. Olson, US Patent, 4,520,074 (1985).
17. M. DeBergalis, US Patent, 4,522,971 (1985).
18. M. DeBergalis, US Patent, 4,636,431 (1987).
19. C. Pawloski, US Patent, 3,314,984 (1964).
20. J. Hosler, US Patent, 3,391,108 (1968).
21. T. Ching, Euro. Pat. Appln., 0,097,841 A1 (1983).
22. C. Yelamaggard, U. Hiremath, S. Nagamani, D. Shankar Rao and S. Krishna Prasad, J. Mater. Chem., Vol. 11, p 1818 (2001).

23. D. Walba, E. Korblova, R. Shao and N. Clark, J. Mater. Chem., Vol. 11, p 2743 (2001)
24. W. Weissflog, H. Nadasi, U. Dunemann, G. Pelzl, S. Diele, A. Eremin and H. Kresse, J. Mater. Chem., Vol. 11, p 2748 (2001).
25. R. Schubart and K. Albrecht, US Patent, 4,125,512 (1978).
26. Wingfoot Corporation, British Patent, 585,775 (1947).
27. I. Muskat and F. Strain, US Patent, 2,385,932 (1945).

## 3.4 Halo Resorcinols

Resorcinol can undergo electrophilic aromatic substitution reaction with active halogen compounds to produce halogenated resorcinols. With phenols in general, halogenation normally occurs at the *ortho*- and *para*-positions relative to the hydroxyl group. Due to higher activation by the two hydroxyl groups of resorcinol, the halogenation occurs readily at the 2, 4 and 6 positions of resorcinol, and thus tri-halo resorcinol compounds could be easily produced.

Halogenated resorcinols have a wide range of industrial, pharmaceutical and agricultural applications.

### 3.4.1 Fluorinated Resorcinols

#### 3.4.1.1 Monofluoro Resorcinols

The 2 and 4-fluoro resorcinol compounds were synthesized by reacting resorcinol with cesium fluoroxy sulfate in the presence of boron trifluoride catalyst [1] (Figure 3.61).

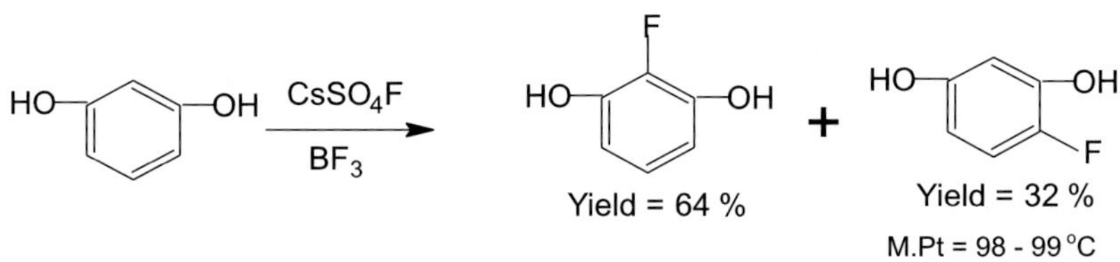


**Figure 3.61.**

Fluorinated resorcinols have been used in the preparation of fluorinated fluorescein and rhodol dyes [2].

#### 3.4.1.2 Fluoracyl Resorcinols

The fluoracyl resorcinolic compounds were useful as active ingredients in pharmaceutical, cosmetic and pesticidal compositions. These derivatives were synthesized from the reaction of resorcinol with trifluoro acetic anhydride [3] (Figure 3.62).

A series of alkyl, aralkyl and halo substituted fluoracyl resorcinols obtained from the above reaction scheme is presented in Table 3.8.

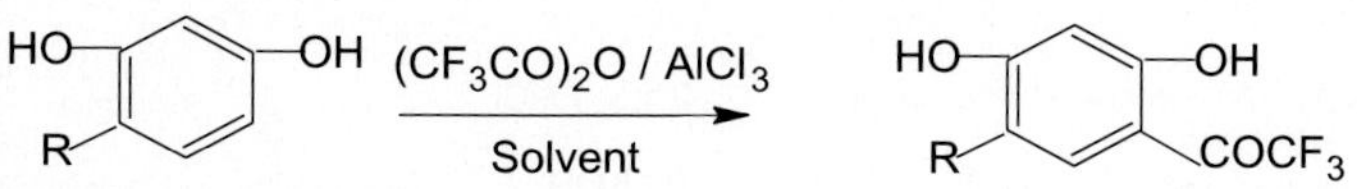


**Figure 3.62.**

**Table 3.8.** Melting points of fluoracyl resorcinols

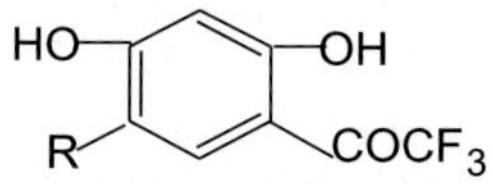


| R group | Melting point (°C) | R group | Melting point (°C) |
|---|---|---|---|
| *n*-Propyl | 95 | *n*-Hepyl | 85 |
| iso-Propyl | 97 | *n*-Octyl | 87 |
| *n*-Butyl | 96 | *n*-Nonyl | 87 |
| iso-Butyl | 90 | Dodecyl | 92 |
| *n*-Pentyl | 97 | Benzyl | 114 |
| Cyclopentyl | 94 | Chloro | 110 |
| Cyclohexyl | 80 | Bromo | 81 |
| *n*-Hexyl | 90 | Methylthio | 57 |

Data from [3]

### 3.4.1.3 Fluoro Resorcinol Ethers

2,4-Dimethoxyfluoro benzene compound obtained from the 2,4-dimethoxy aniline was based on the following reaction [4] (Figure 3.63).

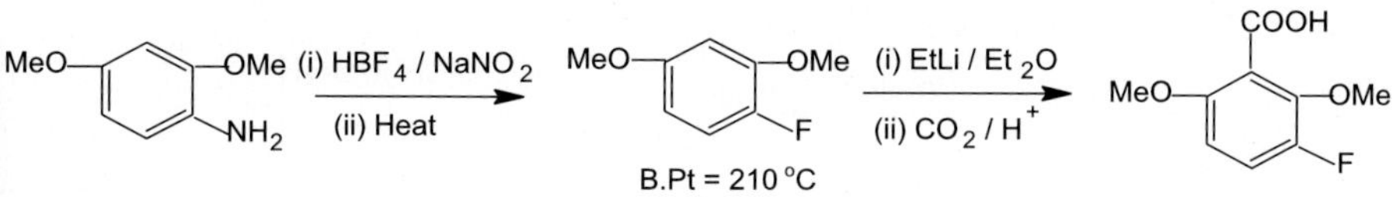


**Figure 3.63.**

## 3.4.2 Chlorinated Resorcinols

### 3.4.2.1 2-Chloro-Substituted Resorcinols

2-Chlororesorcinol was prepared by sulfonating the resorcinol first, and then halogenating the disulfonic acid compound with potassium chlorate [5] (Figure 3.64).

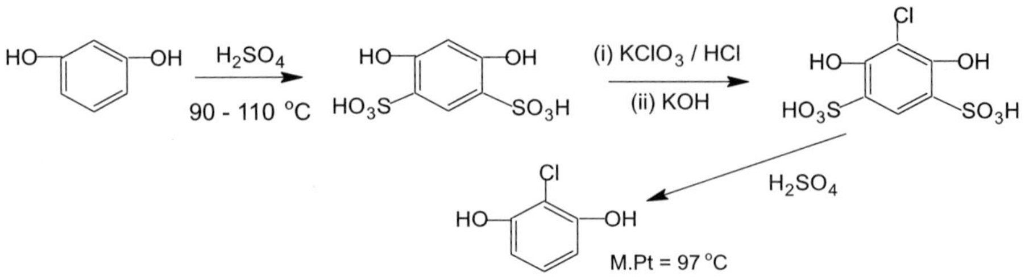


**Figure 3.64.**

This derivative can be used to synthesize coumarins containing halogens in the 8-position, and could be useful for medicinal applications.

Using *N*-chlorosuccinimide as the chlorinating agent, 2-chloro-1,3-dimethoxybenzene was synthesized from 1,3-dimethoxybenzene [6] (Figure 3.65).

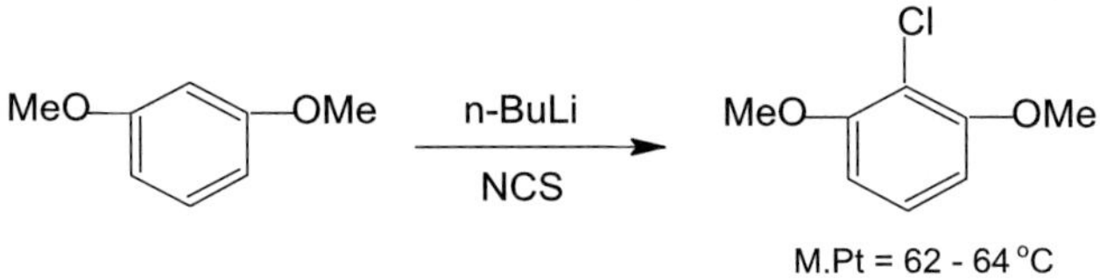


**Figure 3.65.**

This 2-chloro derivative was used as the starting material in the synthesis of pharmacologically valuable diuretic and antihypertensive 1,2-benzisoxazoloxy acetic acids [6] (Figure 3.66).

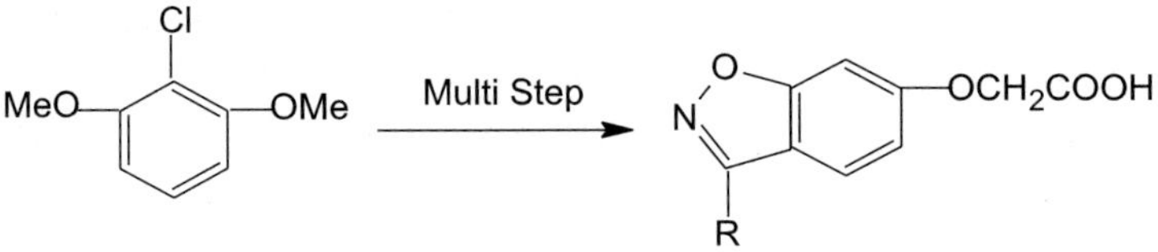


**Figure 3.66.**

Compositions containing benzisoxazole carboxylic acid were found to be useful in the treatment of hypertension, cardiac failure and edema. These pharmaceutically important compounds were synthesized using 2-chloro-1,3-dimethoxybenzenze [7] (Figure 3.67).

### 3.4.2.2
### 4-Chloro-Substituted Resorcinols

Chlorination of resorcinol with sulfuryl chloride gave 4-chlororesorcinol in high yields [8] (Figure 3.68).

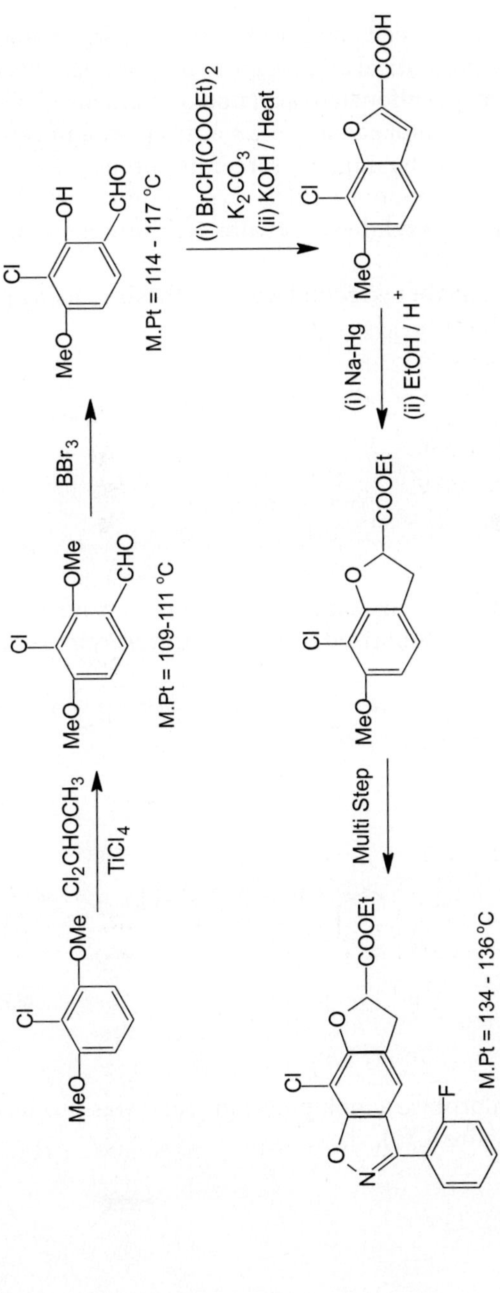
MeO
Cl
OMe
$Cl_2CHOCH_3$
$TiCl_4$
CHO
M.Pt = 109-111 °C
$BBr_3$
OH
M.Pt = 114 - 117 °C
(i) $BrCH(COOEt)_2$
$K_2CO_3$
(ii) KOH / Heat
O
COOH
(i) Na-Hg
(ii) EtOH / $H^+$
COOEt
Multi Step
N
F
M.Pt = 134 - 136 °C

**Figure 3.67.**

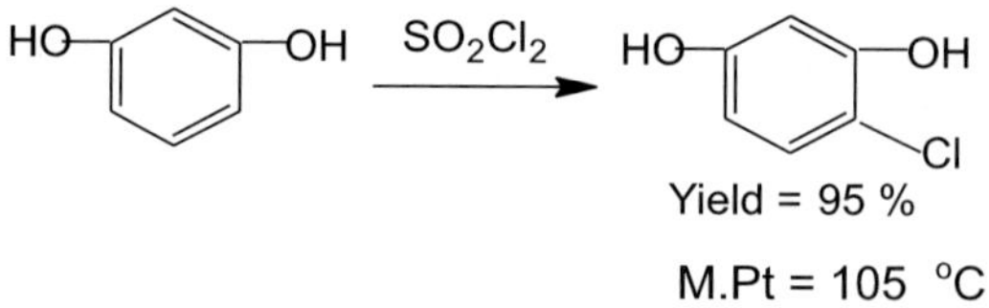


**Figure 3.68.**

4-Chlororesorcinol has been employed as the starting material for the synthesis of various pharmaceutical and organic compounds. This compound can be halogenated, nitrated, sulfonated, acylated or condensed with beta ketonic esters. Monochlororesorcinol can act as an antiseptic, and can be used in the preparation of coumarins having UV screening properties. With diazonium compounds, 4-chlororesorcinol produces dyes of different colors.

Some of the derivatives synthesized from 4-chlororesorcinol are discussed below.

Imino-thio carbonates of resorcinol were synthesized using potassium thiocyanate [9] (Figure 3.69).

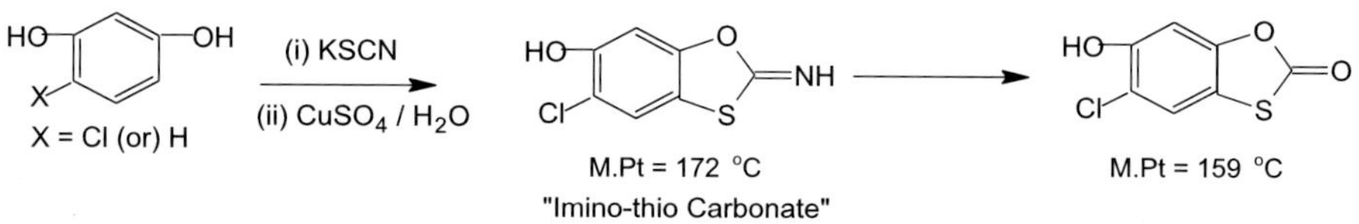


**Figure 3.69.**

5-Chloro-6-methoxy-2-benzoxazolinone has been found to possess hypotensive properties, and was synthesized from 4-chlororesorcinol [10] (Figure 3.70).

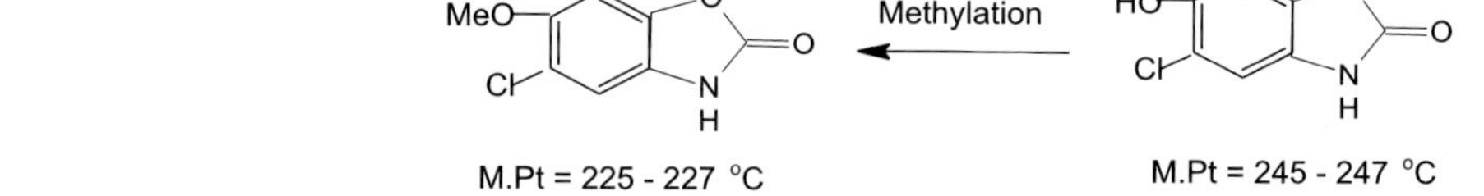


**Figure 3.70.**

Starting with 4-chlororesorcinol, phloroglucinol was produced in about 50 to 70% yield [11] (Figure 3.71).

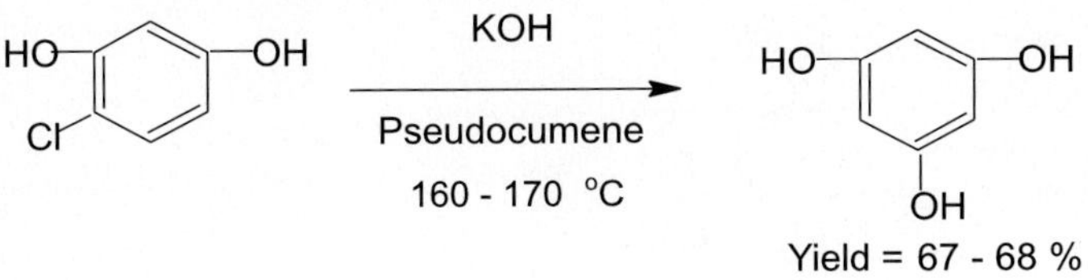


**Figure 3.71.**

For the preparation of photoactive compounds useful in the applications of photolithography, the following derivative was prepared using 4-chlororesorcinol [12] (Figure 3.72).

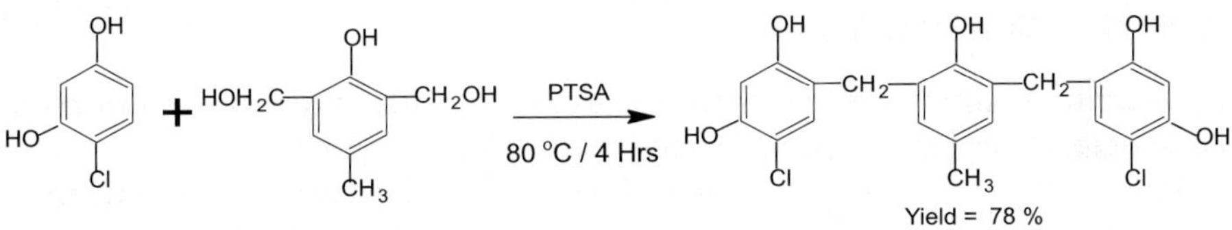


**Figure 3.72.**

Compounds exhibiting good herbicidal activities have also been synthesized [13] (Figure 3.73).

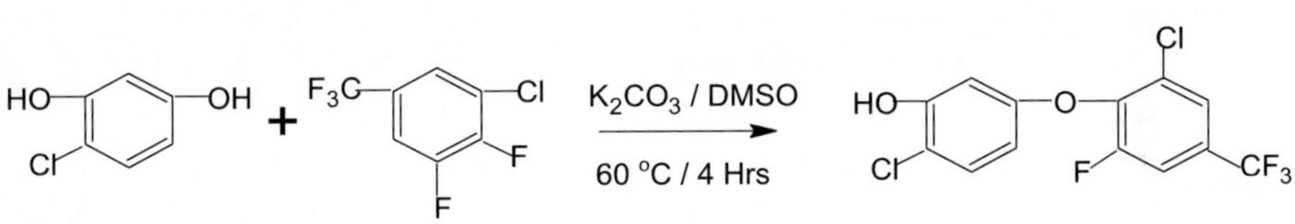


**Figure 3.73.**

The *t*-butyl derivative of 4-chlororesorcinol exhibiting germicidal and antiseptic activities coupled with low toxicity was prepared [14] (Figure 3.74).

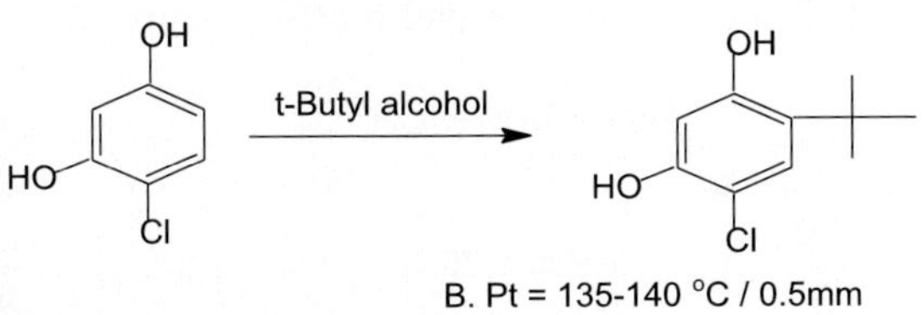


**Figure 3.74.**

Monoalkyl ether derivatives of 4-chlororesorcinol possessing germicidal and antiseptic properties were synthesized using alkyl halides in the reaction [15] (Figure 3.75).

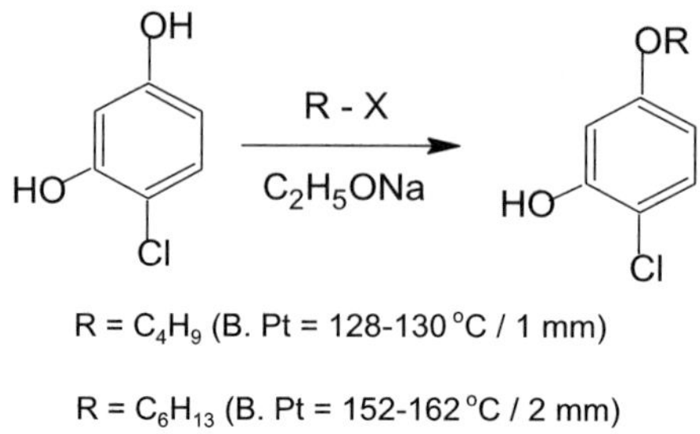


**Figure 3.75.**

### 3.4.2.3
### Monochloro Alkyl Resorcinols

By reacting 4-alkyl resorcinols with sulfuryl chloride, 4-alkyl-6-chlororesorcinol derivatives have been prepared [16, 17] (Figure 3.76).

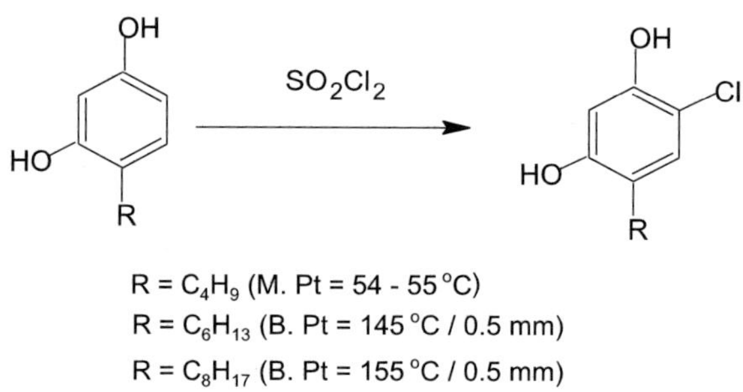


**Figure 3.76.**

### 3.4.2.4
### 4,6-Dichlororesorcinol

When resorcinol was reacted with sulfuryl chloride in the presence of an acetate solvent, 4,6-dichlororesorcinol was obtained in higher yields [18] (Figure 3.77).

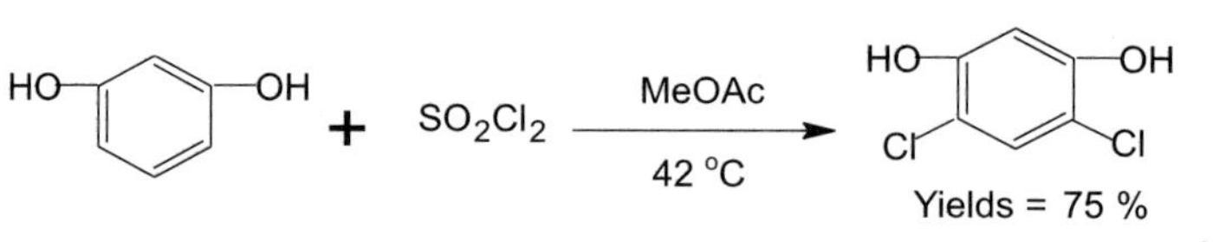


**Figure 3.77.**

Dimethylether and phosphorthioate derivatives of 4,6-dichlororesorcinol were developed and used in organic synthesis and new pesticidal compositions respectively [19] (Figure 3.78).

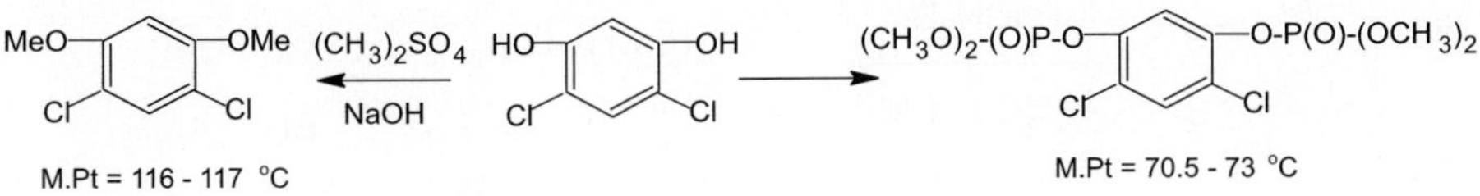


**Figure 3.78.**

### 3.4.2.5
### 2,4,6-Trichlororesorcinol

This compound was synthesized from the reaction of resorcinol with sulfuryl chloride in acetic acid solvent [20] (Figure 3.79).

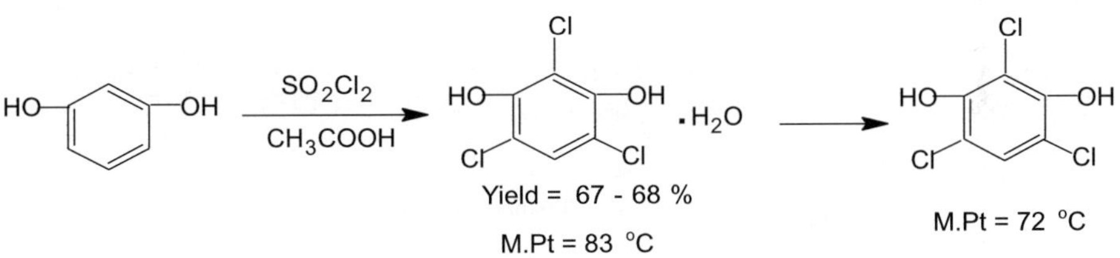


**Figure 3.79.**

### 3.4.3
### Bromoresorcinols

### 3.4.3.1
### 4-Bromoresorcinol

This derivative was synthesized from $\beta$-resorcylic acid [21] (Figure 3.80).

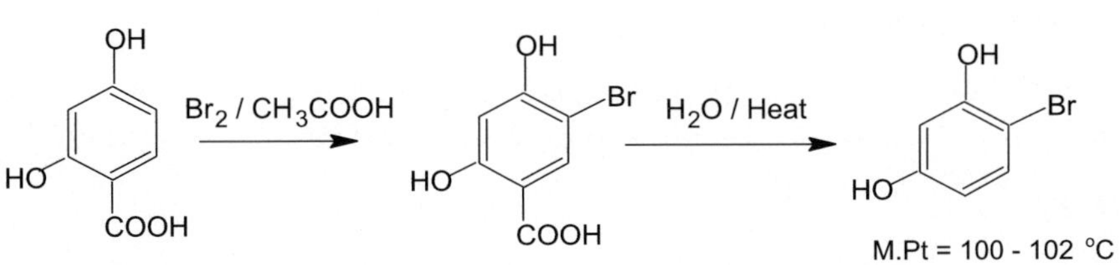


**Figure 3.80.**

4-Bromoresorcinol has been utilized in the preparation of pharmaceutically important compounds and derivatives [22–24] (Figure 3.81).

A phenol compound intermediate for the total synthesis of *seco*-lateriflorone was synthesized using the 4-bromoresorcinol derivative [25] (Figure 3.82).

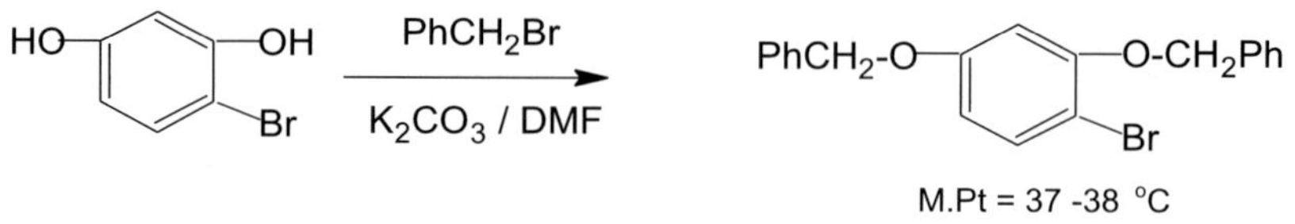


**Figure 3.81.**

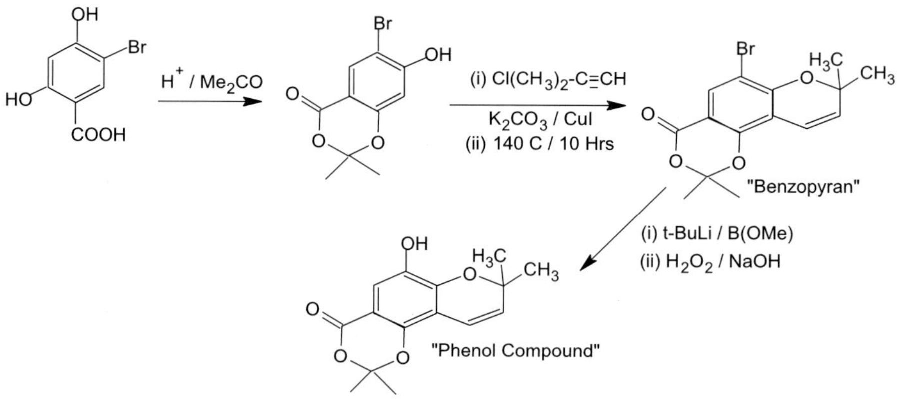


**Figure 3.82.**

### 3.4.3.2 4,6-Dibromoresorcinol

Employing chloroform as the solvent, the addition of bromine solution into resorcinol produced 4,6-dibromoresorcinol in high yields [26] (Figure 3.83).

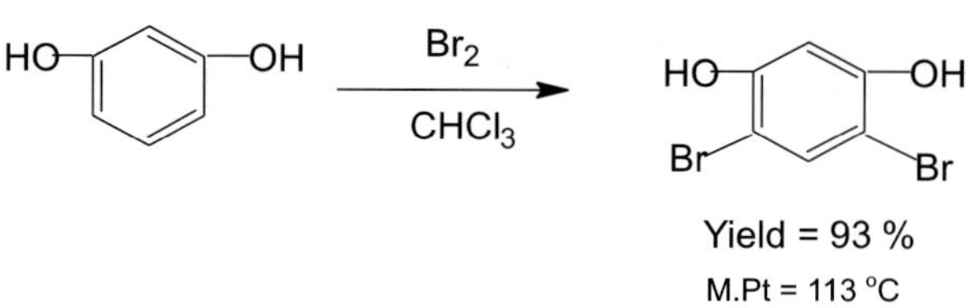


**Figure 3.83.**

This compound was further derivatized according to the following scheme [27] (Figure 3.84).

### 3.4.3.3 2,4,6-Tribromoresorcinol

The reaction of resorcinol with bromine produced tribromoresorcinol, and was used in different applications [2] (Figure 3.85).

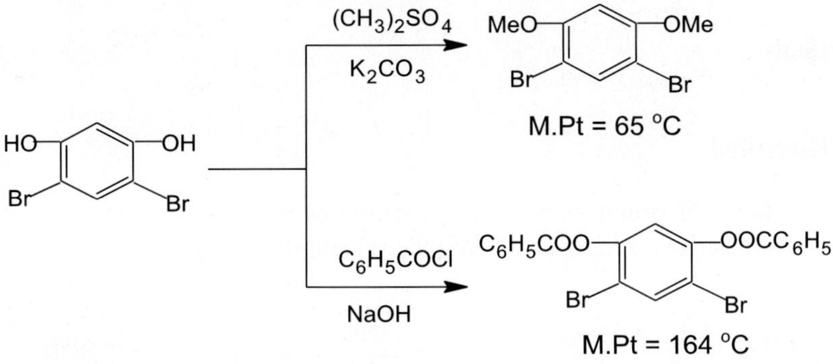


**Figure 3.84.**

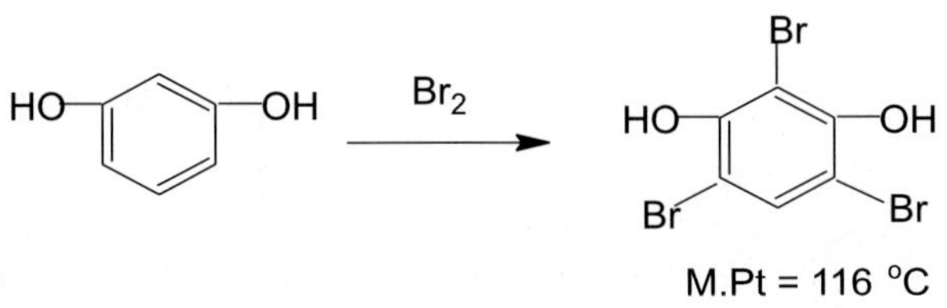


**Figure 3.85.**

### 3.4.3.4
### Dibromo Derivatives of Resorcinol Ethers

Brominated resorcinol compounds useful as flame retardant materials in the applications of polyurethanes and other thermoplastic polymers have been synthesized [28, 29] (Figure 3.86).

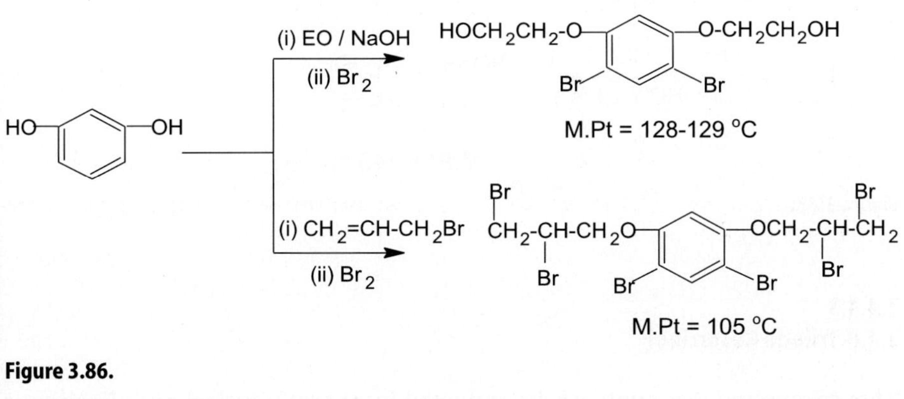


**Figure 3.86.**

### 3.4.4
### Iodo Resorcinols

#### 3.4.4.1
#### Monoiodo Resorcinol

A Suzuki synthesis of precursors to furanoenolates was involved the use of 2-iodoresorcinol as the starting material [30, 31] (Figure 3.87).

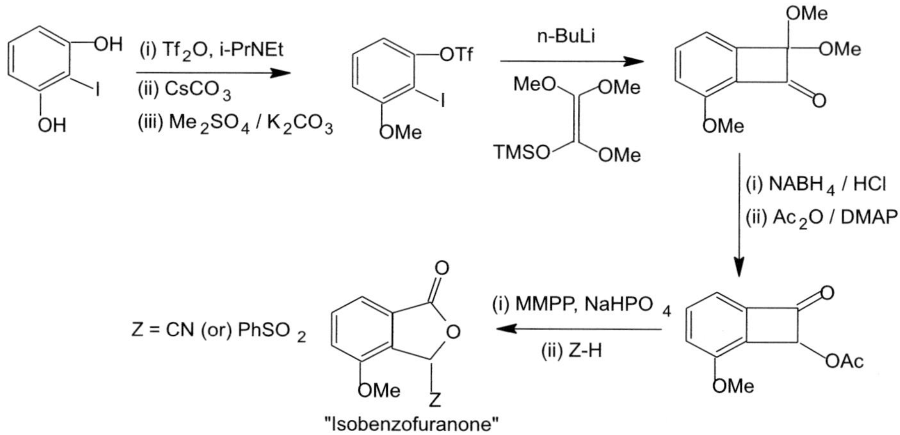


**Figure 3.87.**

#### 3.4.4.2
#### 4,6-Diiodo Resorcinol

By reacting resorcinol with iodine monochloride, 4,6-diiodoresorcinol was obtained [32] (Figure 3.88).

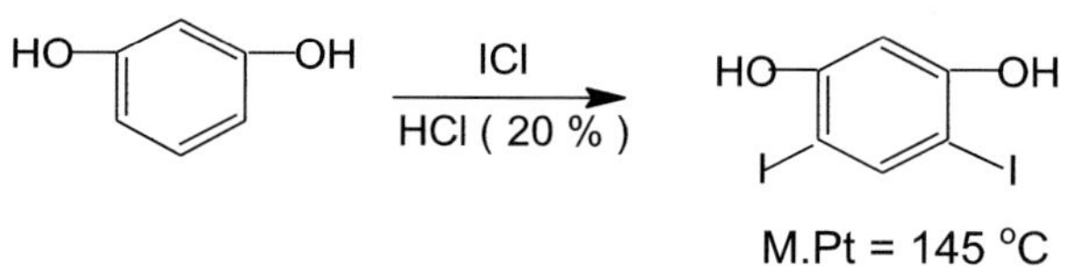


**Figure 3.88.**

#### 3.4.4.3
#### 2,4,6-Triiodo Resorcinol

This compound was synthesized and used in pharmaceutical and diagnostic radiology applications [33, 34] (Figure 3.89).

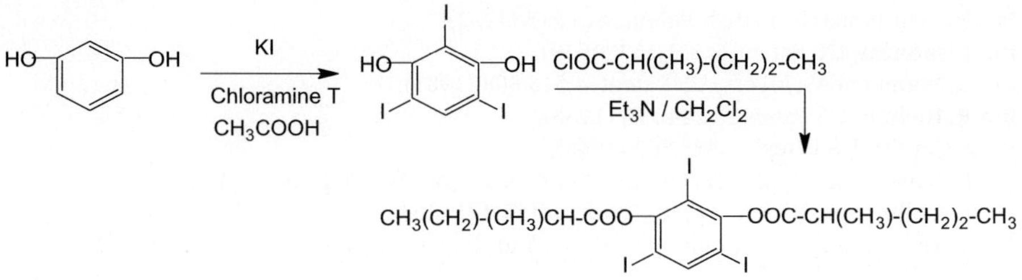


**Figure 3.89.**

## 3.4.5 Properties of Halo Resorcinols

The melting points of halogenated resorcinols obtained from different methods have been presented in Table 3.9.

**Table 3.9.** Melting points (°C) of halogenated resorcinols

| Halo substitution: position | Fluoro | Chloro | Bromo | Iodo |
|---|---|---|---|---|
| 2-Mono | 114–116 [2] | 97 [35] | 102 [35] | 103 [35] |
| 4-Mono | 94–96 [2] | 105 [8] | 100–102 [21] | 67 [32] |
| 2,4-Di | 100–101 [2] | | | |
| 4,6-Di | | 112–113 [8] | 110 [36] | 145 [32] |
| 2,4,6-Tri | | 72 [20] | 116 [35] | 154 [33] |

The references are shown in parentheses

## References

1. T. Patrick and D. Darling, J. Org. Chem., Vol. 51, p 3242 (1986).
2. D. Klaubert and K. Gee, US Patent, 6,299,055 (2001).
3. R. Brick, H. Eberhardt, K. Apel, U. Lechner and W. Merk, US Patent, 4,225,619 (1980).
4. A. Durrani and J. Tyman, J. Chem. Soc., Perkin I, p 1658 (1980).
5. F. Valle and Romeo, US Patent, 4,704,469 (1987).
6. G. Shutske, T. Lee and G. Jobin, Euro. Pat. Appln., 0,067,342 (1982).
7. J. Platter and A. Fung, US Patent, 4,456,612 (1984).
8. M. Moore, A. Day and C. Suter, J. Am. Chem. Soc., Vol. 56, p 2456 (1934).
9. G. Werner, US Patent, 2,332,418 (1943).
10. McNeil Laboratories, British Patent, 1,015,713 (1966).
11. N. deHeij and A. Hendrick, US Patent, 3,959,388 (1976).
12. A. Jefferies, A. Blankeney and M. Toukhy, US Patent, 4,992,596 (1991).
13. D. Cartwright, US Patent, 5,120,869 (1992).
14. W. Austin, US Patent, 2,023,160 (1935).
15. R. Reed, US Patent, 2,036,826 (1936).

16. Farbenindustrie, British Patent, 415,715 (1932).
17. E. Moness, US Patent, 2,151,137 (1939).
18. R. Wang and G. Myers, US Patent, 4,518,808 (1985).
19. R. Richter, US Patent, 3,244,775 (1966).
20. J. Cavelti, US Patent, 2,485,562 (1946).
21. R. Sandin and R. McKee, Organic Syntheses, Coll. Vol. 2, p 100 (1943).
22. E. Collington, US Pat. Appln., 2002 / 0,155,075 (2002).
23. A. Browning, W. Collington, M. Proctor and J. Geden, US Pat. Appln., 2002 / 0,161,041 (2002).
24. T. Althius, C. Harbert, M. Johnson and L. Melvin, US Patent, 4,284,829 (1981).
25. E. Tisdale, B. Vong, H. Li, S. Kim, C. Chowdhary and A. Theodorakis, Tetrahedron, Vol. 59, p 6873 (2003).
26. N. Lorie, Technical Report, AFML-TR-65-377, p 29 (1965).
27. H. Hodgson and R. Dyson, J. Chem. Soc., p 946 (1935).
28. M. Kulka, M. Borr and K. Macphee, Canadian Patent, 681,805 (1964).
29. K. Masashi, Jap. Kokai., JP 08,337,681 (1996).
30. J. Wang, L. Pettus and T. Pettus, Tetrahedron Lett., Vol. 45, p 1793 (2004).
31. T. Hamura et al., Helv. Chim. Acta, Vol. 85, p 3589 (2002).
32. B. Nicolet and J. Sampey, J. Am. Chem. Soc., Vol. 49, p 1796 (1927).
33. E. Roberts, J. Chem. Soc., p 2707 (1923).
34. E. Bacon et al., US Patent, 5,318,769 (1994).
35. Chemistry of carbon compounds, E. Rodd, Ed., Elsevier, New York, Vol. III, Part A, p 470 (1954).
36. K. Rosen and W. Kuhnhenn, Ber. Vol. 56, p 1262 (1923).

## 3.5 Aldehyde and Ketone Compounds

### 3.5.1 Resorcinol Aldehydes

The preparation of resorcinol aldehydes can be accomplished by either the Gattermann or the Reimer-Tiemann or the Vilsmeier reaction. In the Gattermann process, resorcinol or certain resorcinol derivatives can be treated with hydrocyanic acid and hydrogen chloride in ether solution to yield resorcinol aldehydes by hydrolysis of the ketonimine chloride intermediates formed. In the case of Reimer-Tiemann synthesis, chloroform and dilute aqueous sodium hydroxide are used to obtain the aldehydes [1] (Figure 3.90).

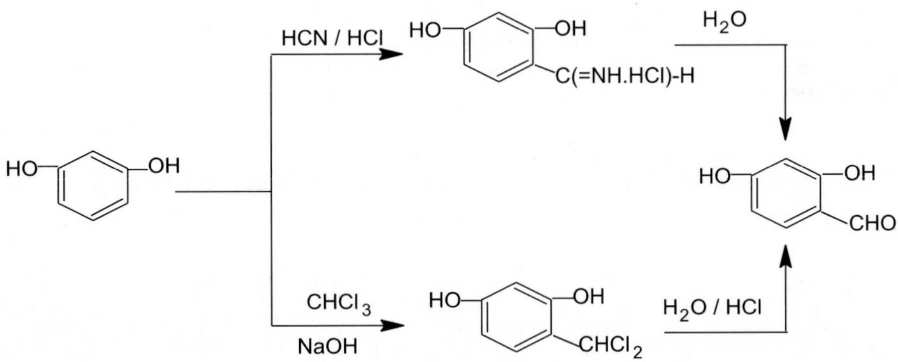


**Figure 3.90.**

#### 3.5.1.1 2,4-Dihydroxybenzaldehyde

Vilsmeier formylation reaction of resorcinol with either phosphorus oxychloride ($POCl_3$)/dimethyl formamide (DMF) or oxalyl chloride – DMF in acetonitrile at lower temperature can be used to produce 2,4-dihydroxybenzaldehyde in good yields and high purity [2,3] (Figure 3.91).

2,4-Dihydroxybenzaldehyde has been utilized in the synthesis of various organic compounds and intermediates by the pharmaceutical and dye industries.

The 2,4-di-(2′-vinyloxy ethoxy)-benzylidene melononitrile having highly polarizable $\pi$-conjugated dipolar system was prepared from 2,4-dihydroxybenzaldehyde [4] (Figure 3.92).

Using cationic initiators, this monomer was polymerized to obtain cross-linked poly (vinyl ether) containing oxybenzylidene melononitrile groups. The resultant polymer was a potential effective non-linear optical (NLO) chromophore for second-order NLO applications.

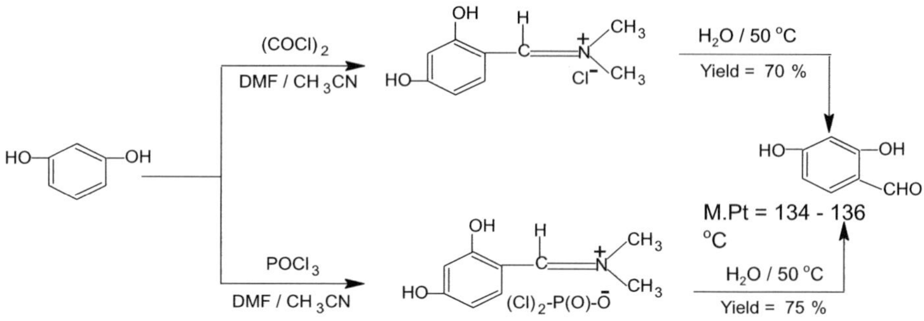


**Figure 3.91.**

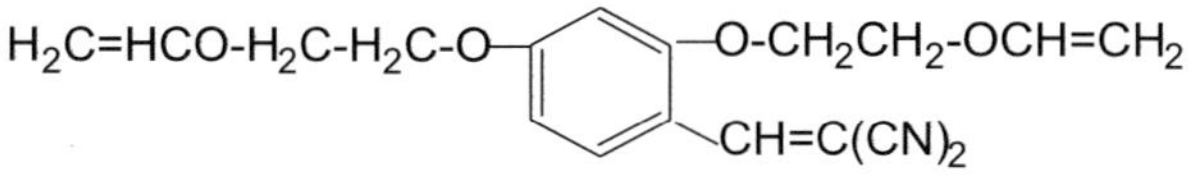


**Figure 3.92.**

In the preparation of an $\alpha$-(2,4-dihydroxy phenyl)-$N$-phenyl nitrone compound, 2,4-dihydroxybenzaldehyde was used [5] (Figure 3.93).

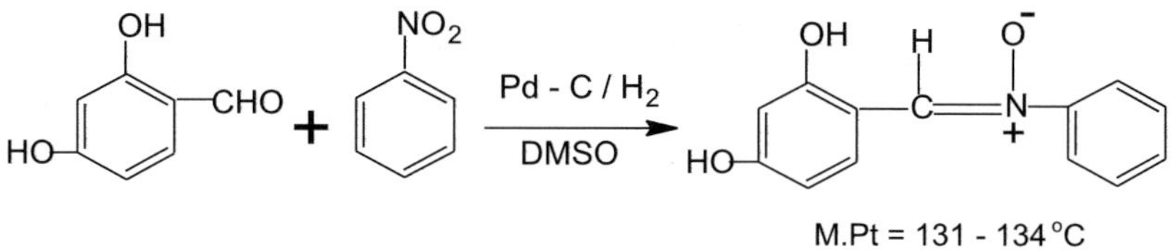


**Figure 3.93.**

A series of diene-containing polymers was modified with the phenyl nitrone compound, and Strebler adhesion measurements were made to determine the adhesion performance of the rubber formulation to itself. A large increase in Strebler adhesion was observed in *cis*-polybutadiene and styrene butadiene rubbers modified with this phenyl nitrone derivative.

### 3.5.1.2
### 2,6-Dihydroxybenzaldehyde

Based on the following reaction scheme, 2,6-dihydroxybenzaldehyde was prepared from the use of 3-methoxyphenol [6] (Figure 3.94).

Tucaresol is an immuno stimulant compound, and was synthesized from 2,6-dihydroxybenzaldehyde [6] (Figure 3.95).

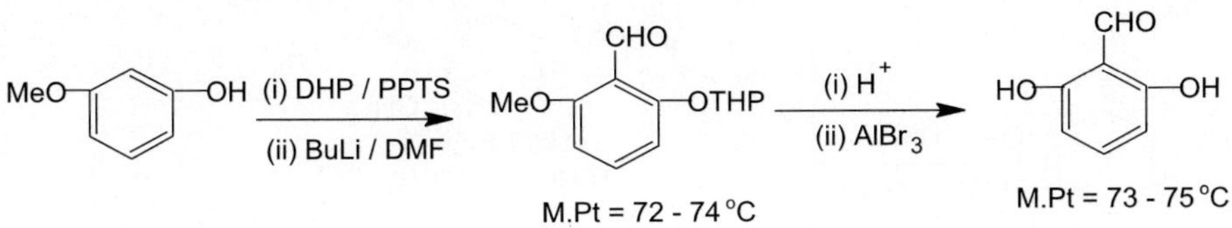


**Figure 3.94.**

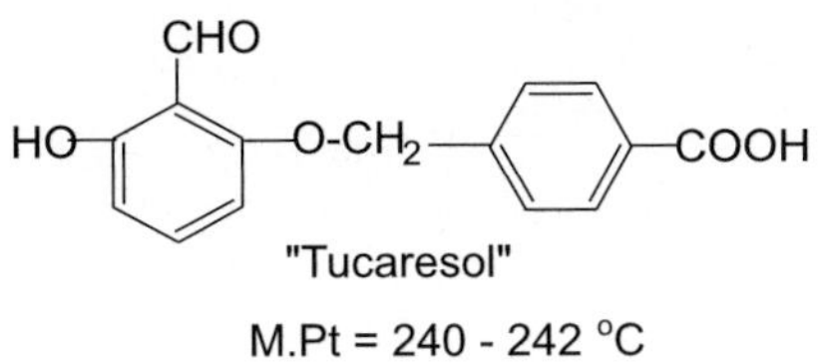


**Figure 3.95.**

### 3.5.1.3
### 2,4-Dialkoxybenzaldehydes

Aromatic aldehyde compounds possessing additional substituents or functional groups are important as versatile intermediates in the pharmaceutical industry and for organic syntheses. In the preparation of 2,4-alkoxybenzaldehydes, two general methods have been reported. With 2,4-dihydroxybenzaldehyde as the starting material, the alkyl ether groups can be formed by the alkylation reaction using alkyl halides [7–9] (Figure 3.96).

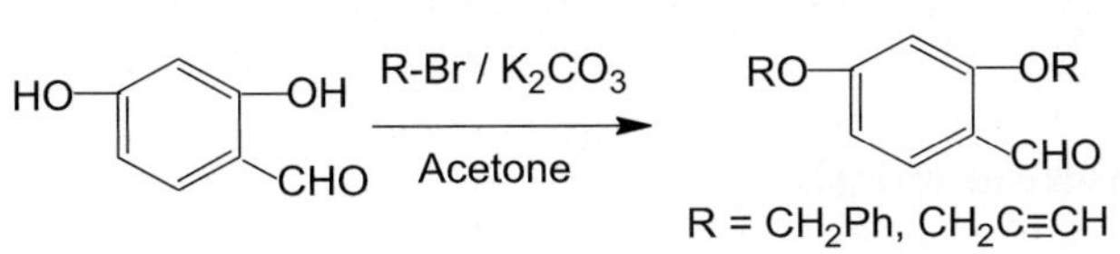


**Figure 3.96.**

Direct formylation of 1,3-dialkoxybenzene could also be achieved with the use of hexamethylenetetramine (HMTA) and an acid system. Suzuki modified the direct formylation reactions with a stronger acid, such as methane sulfonic acid, trifluoroacetic acid or polyphosphoric acid as a solvent, and successfully introduced the formyl group onto electron deficient aromatic rings [10] (Figure 3.97).

2,4-Dimethoxybenzaldehyde was employed in the preparation of 4-methylene-1,3-dioxolane (4-MDO) derivative having dimethoxyphenyl group on the 2-position of 1,3-dioxolane ring [11] (Figure 3.98).

The polymerization of 2-(dimethoxyphenyl)-4-MDO monomer using boron trifluoride as a cationic initiator underwent polymerization with ring opening as well as cyclization reaction to afford a mixture of ring opened polymer and dihydrofuranone derivatives.

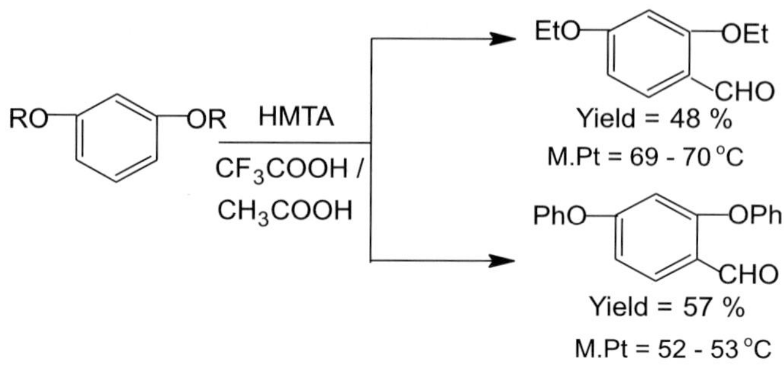


**Figure 3.97.**

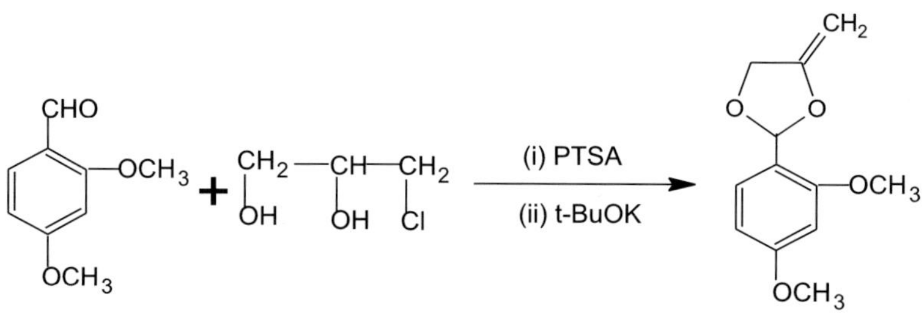


**Figure 3.98.**

2,4-Dimethoxybenzaldehyde was also used to prepare 4-(2,4-dimethoxy benzylidene)-2-butenolide [12, 13].

### 3.5.1.4 Aldehyde Compounds from Alkylresorcinols

2-Methylresorcinol was used to synthesize 2,4-dihydroxy-3-methyl-benzaldehyde [14] (Figure 3.99).

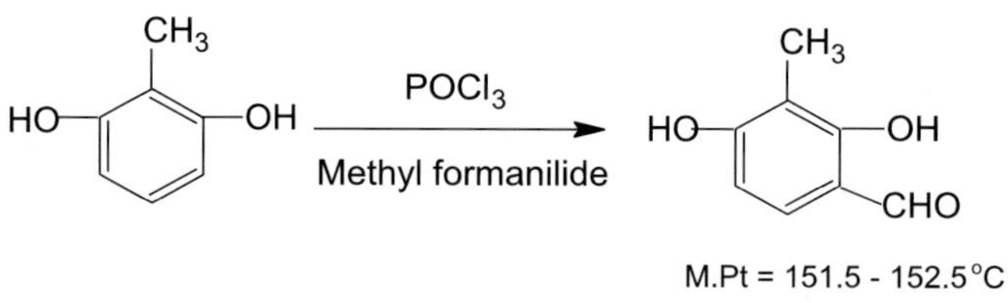


**Figure 3.99.**

This compound could replace partially or completely the "Oak Moss" extract in perfume compositions.

An aldehyde compound prepared from 2-propylresorcinol has been found to be useful in pharmaceutical applications [15] (Figure 3.100).

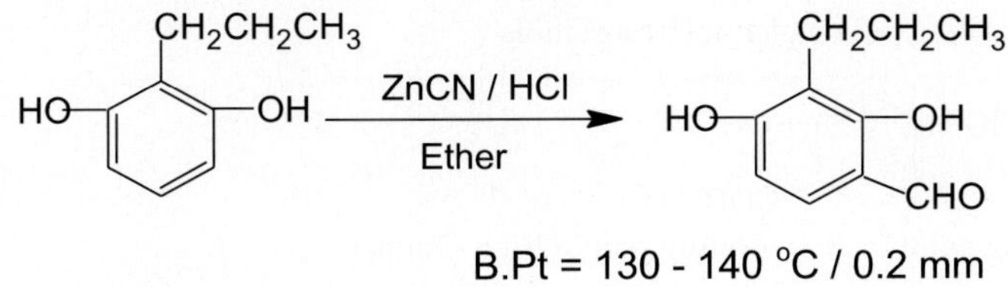


**Figure 3.100.**

## 3.5.2 Resorcinol Ketones

Resorcinol ketone compounds may be prepared by rearrangement of the esters under Fries reaction conditions. Catalysts such as zinc chloride, anhydrous aluminum chloride, boron trifluoride, and strong mineral acids are effective in this rearrangement. Hoesch first discovered that nitriles, when reacted with phenols such as resorcinol in the presence of anhydrous zinc chloride and dry hydrogen chloride, produced ketone derivatives [16] (Figure 3.101).

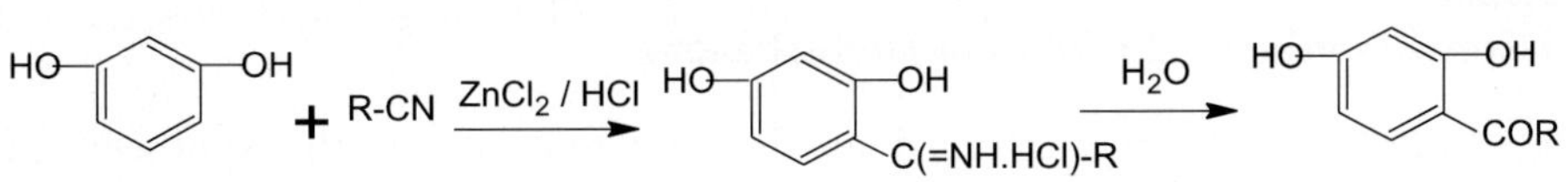


**Figure 3.101.**

Depending upon the nature of the nitrile compounds, aromatic or aliphatic, acyl or benzoyl-type derivatives could be obtained from Hoesch's reaction. The benzoyl derivatives of resorcinol have been discussed in greater detail in the UV absorber section of this book, and therefore only the acyl derivatives of resorcinol are discussed in this section.

### 3.5.2.1 Acylresorcinols

Acylresorcinols were prepared directly from resorcinol by acylating with acyl halides, anhydrides or acids in the presence of zinc chloride at elevated temperatures [17] (Figure 3.102).

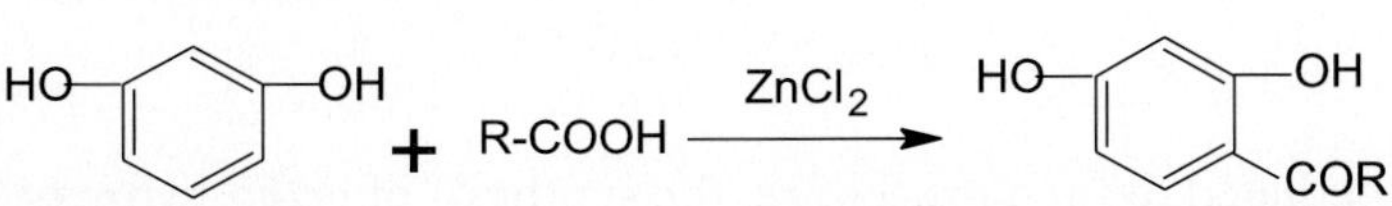


**Figure 3.102.**

A series of derivatives has been synthesized using the above reaction scheme and the melting point and boiling point data are presented in Table 3.10.

**Table 3.10.** Melting and boiling points of 4-acylresorcinols

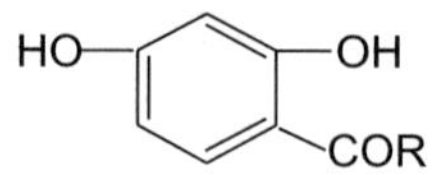


| Acyl group | Melting point (°C) | Boiling point (°C, 6–7 mm) |
|---|---|---|
| Propionyl | 95–96 | 176–178 |
| Butyryl | 68–69.5 | 186–188 |
| Valeryl | 58.5–60 | 190–192 |
| Caproyl | 54.6–56 | 196–197 |
| Heptylyl | 48–50 | 204–206 |
| Caprylyl | 62.5–64 | 214–216 |
| Dodecylyl | 84–85.5 | 237–239 |
| Isobutyryl | 67–68.5 | 173–175 |
| Iso-valeryl | 108–110 | 183–185 |
| Iso-hexylyl | 76–77.5 | 192–194 |

Data from [17]

### 3.5.2.2 2-Acetylresorcinol or 2,6-Dihydroxyacetophenone

1,3-Cyclohexanedione was used as the raw material in the synthesis of 2-acetylresorcinol [18] (Figure 3.103).

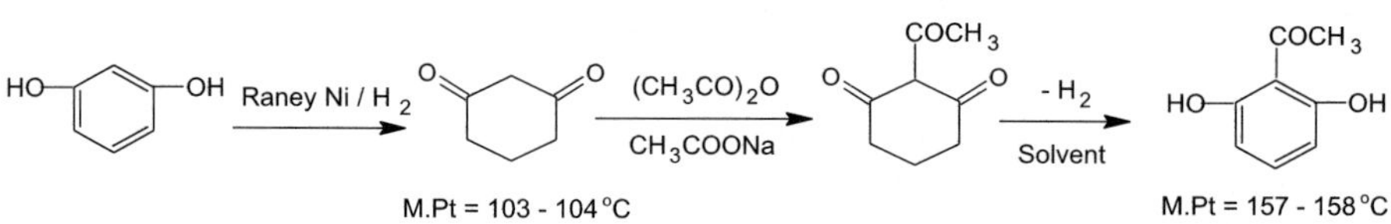


**Figure 3.103.**

This compound was also synthesized from 7-hydroxycoumarin [19] (Figure 3.104).

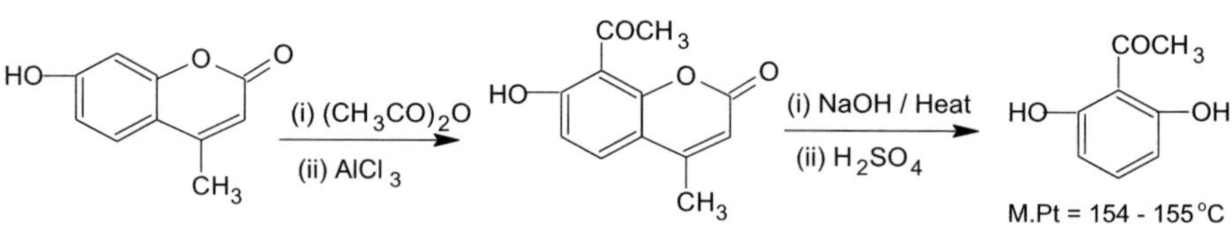


**Figure 3.104.**

By the iodine mediated oxygen procedure, the synthesis of *ortho*-hydroxyacetophenone derivative was achieved by the use of 2-acyl-cyclohexane-1,3-dione. The reaction of cyclohexane-1,3-diones with electron-withdrawing groups at the 2-position with iodine in methanol produced the corresponding monomethoxy resorcinol [20] (Figure 3.105).

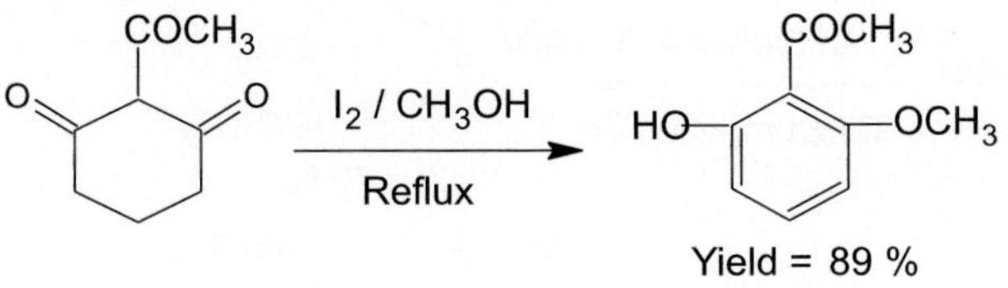


**Figure 3.105.**

Polymerizable monomers that incorporate moieties derived from energy-absorbing molecules were synthesized from 2,6-dihydroxyacetophenone [21] (Figure 3.106).

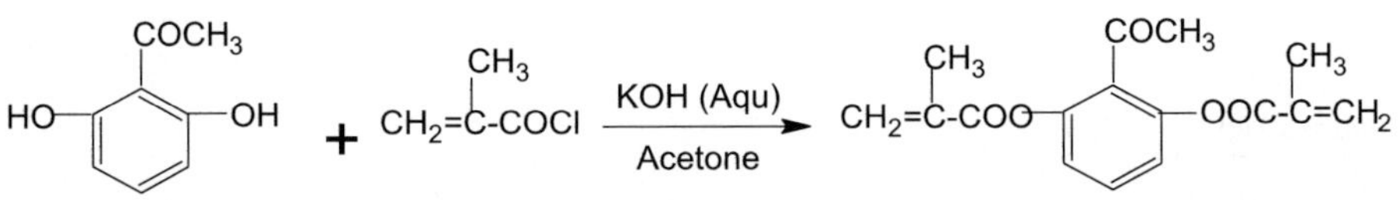


**Figure 3.106.**

Polymers obtained from this type of monomer possess unique properties that make them ideally suited for use in diverse analyses, which includes dispersion/ionization mass spectroscopy of analytes.

2-Acetylresorcinol was converted into 2,2-dimthyl-8-acetyl-7-hydroxychroman in a single step reaction with isoprene in the presence of Amberlyst® 15 catalyst [22] (Figure 3.107).

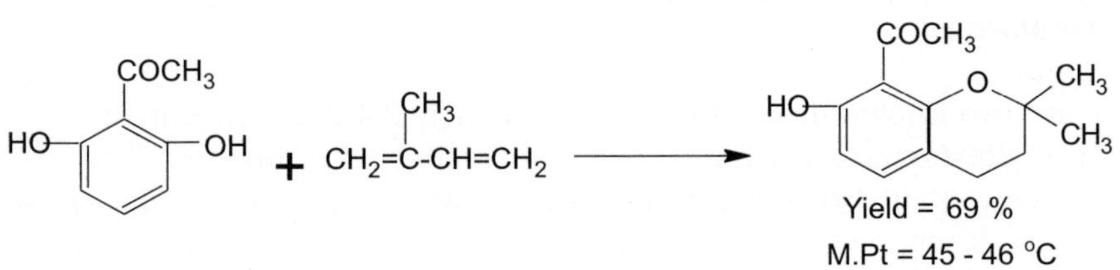


**Figure 3.107.**

### 3.5.2.3
### 4-Acetylresorcinol or 2,4-Dihydroxyacetophenone

When resorcinol and acetic acid are reacted together, 2,4-dihydroxyacetophenone is produced. In this reaction, several strong acidic catalysts have been employed [23–26] (Figure 3.108).

From 4-acetylresorcinol, 3-propyl-2,4-dihydroxyacetophenone was prepared and used in pharmaceutical applications for the treatment of asthma and allergic diseases [27–29] (Figure 3.109).

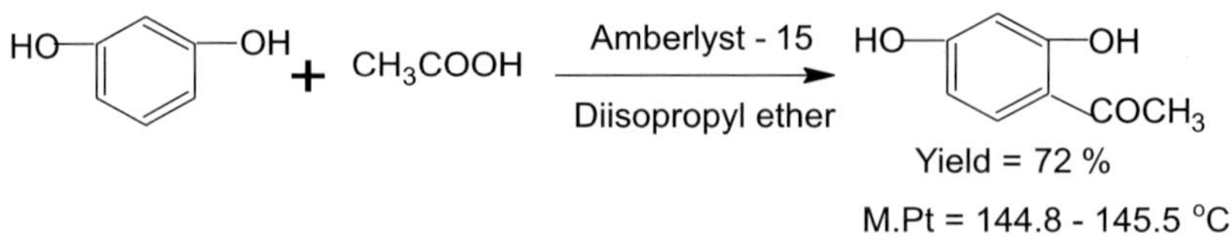


**Figure 3.108.**

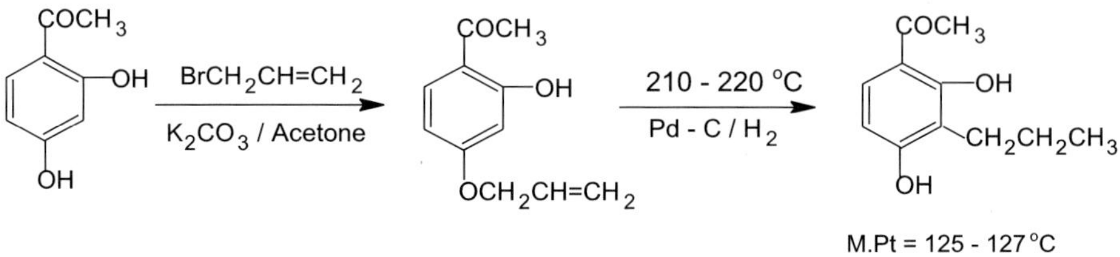


**Figure 3.109.**

2,4-Dihydroxyacetophenone has been employed in a wide range of applications, including use as intermediates in the manufacture of pharmaceuticals and perfumes [30], cosmetics [31], fine chemicals [32, 33], and dyes [34–36]. In addition, this compound was used in the spectroscopic determination of iron [37], and as an intermediate for the preparation of therapeutic compounds [38] and food preservatives [39].

### 3.5.2.4 Diacetylresorcinols

There are two known diacetylresorcinols, namely 2,4-diacetylresorcinol and 4,6-diacetylresorcinol, which have been prepared from resorcinol. These two derivatives contain two nuclear substituted acetyl groups each adjacent to a hydroxyl group, and therefore both compounds may resemble each other closely in their physical and chemical properties [40] (Figure 3.110).

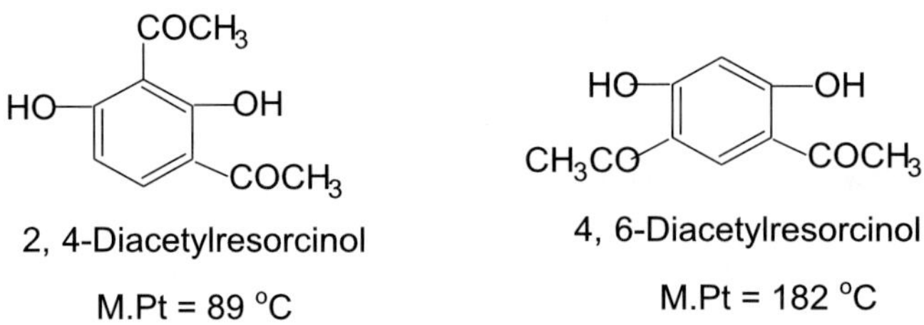


**Figure 3.110.**

4,6-Diacetylresorcinol was prepared from resorcinol and acetic anhydride in the presence of zinc chloride [41] (Figure 3.111).

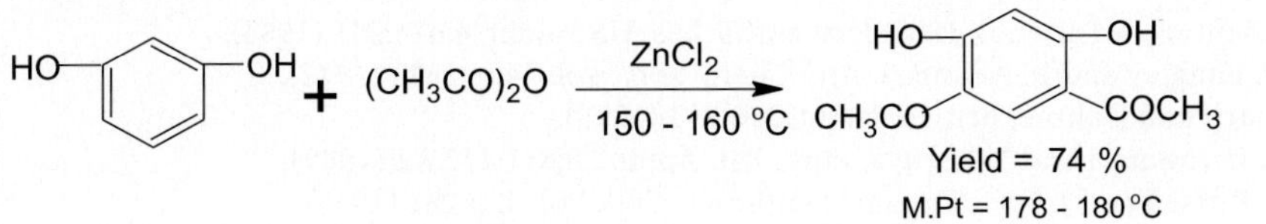


**Figure 3.111.**

2-Allyl and 2-alkyl derivatives of 4,6-diacetylresorcinol have been synthesized and reported [42] (Figure 3.112).

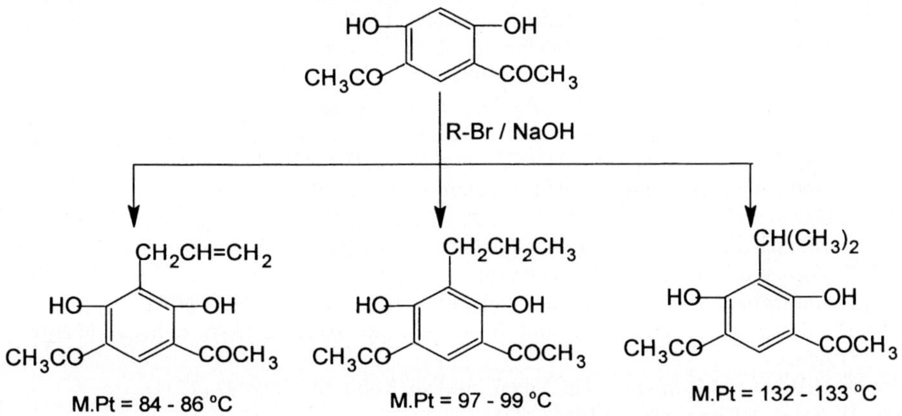


**Figure 3.112.**

## References


1. Technical Bulletin CD-2-424 on "Resorcinol" by Koppers Company Inc., Pittsburgh, PA, USA.
2. W. Mendelson et al., Synth. Commun., Vol. 26, p 603 (1996).
3. W. Mendelson, US Patent, 5,599,988 (1997).
4. J. Lee, J. Kim and M. Kim, Bull. Korean Chem. Soc., Vol. 20, p 307 (1999).
5. J. Cantillo, R. Strozier and W. Francik, US Patent, 5,091,449 (1992).
6. B. Zacharie, G. Attardo, N. Barriault and C. Penny, J. Chem. Soc., Perkin Trans. 1, p 2925 (1997).
7. A. Katritsky, H. He, Q. Long, X. Cui, J. Level and A. Wilcox, Arkat-USA J, Vol. 1, Part 3 (2003).
8. T. Kimachi, T. Tanaka and F. Yoneda, J. Heterocylic Chem., Vol. 28, p 439 (1991).
9. M. Thompson et al., J. Am. Chem. Soc., Vol. 101, p 2483 (1979).
10. Y. Suzuki and H. Takahashi, Chem. Pharm. Bull., Vol. 31, p 1751 (1983).
11. J. Kim and M. Gong, Bull. Korean Chem. Soc., Vol. 20, p 663 (1999).
12. L. Sorotskaya, T. Kaklyugina and L. Bodovskaya, Molecules, Vol. 4, M108 (1999).
13. L. Sorotskaya et al., J. Org. Chem. (Russia), Vol. 25, p 175 (1989).
14. K. Kulka and T. Zazula, US Patent, 3,975,309 (1976).

15. J. Bantick, J. Fuher, D. Harddern and T. Lee, US Patent, 4,424,231 (1984).
16. W. Langley and R. Adams, J. Am. Chem. Soc., Vol. 44, p 2320 (1922).
17. Sharp and Dohme, British Patent, 250,893 (1925).
18. N. Hirowatari and Y. Furuya, Euro. Pat. Appln., 0,331,442 A2 (1989).
19. A. Russell and J. Frye, Organic Syntheses, Coll. Vol. 3, p 281 (1955).
20. J. Kim, K. Lee and J. Kim, Bull. Korean Chem. Soc., Vol. 24, p 1057 (2003).
21. N. Nitagawa, US Pat. Appln., 2003 / 0,207,460 (2003).
22. G. Kalena, A. Jain and A. Banerji, Molecules, Vol. 2, p 100 (1997).
23. S. Cooper, Organic Syntheses, Coll. Vol. 3, p 761 (1955).
24. R. Brown, R. Glasser, C. Liotta and C. Eckert, Chem. Commun., p 1295 (2000).
25. G. Yadav and A. Joshi, Clean Tech. Environ. Policy, Vol. 4, p 157 (2002).
26. T. Fukumoto, K. Ujita, T. Mori, K. Nakao and Y. Tamai, US Patent, 5,621,146 (1997).
27. W. Baker and O. Lothian, J. Chem. Soc., p 628 (1935).
28. R. Rokach, P. Hamel and R. Hirschmann, US Patent, 4,252,818 (1981).
29. N. Miki, Japanese Patent, 5,017,389 (1993).
30. T. Konechika, S. Osada and N. Ando, Jap. Kokai. Tokkyo, Koho, JP 02011536 (1990).
31. T. Satoh, H. Matsumoto and T. Miki, PCT Int. Appln., WO 9,205,137 (1992).
32. K. Tanaka and K. Goto, Jap. Kokai. Tokkyo, Koho, JP 61176548 (1986).
33. United Kingdom Chemicals, British Patent, 714,772 (1953).
34. Y. Uyeno and M. Terao, Japanese Patent, 178,046 (1949).
35. H. Morschel and K. Wedemeyer, US Patent, 2 726,238 (1955).
36. R. Botros, US Patent, 4,066,388 (1978).
37. G. Chowdhary and N. Appalaraju, Current Science, Vol. 44, p 343 (1975).
38. T. Fukumoto, J. Sato, T. Mori and T. Oonishi, Jap. Kokai. Tokkyo, Koho, JP 07291889 (1995).
39. A. Nishina and Y. Sakurai, Jap. Kokai. Tokkyo, Koho, JP 06217747 (1994).
40. W. Baker, J. Chem. Soc., p 1684 (1934).
41. J. Kawachi, H. Matsubara, Y. Nakahara and Y. Watanabe, US Patent, 5,892,118 (1999).
42. D. Ramey and M. Fanthorpe, US Patent, 4,400, 542 (1983).

## 3.6 Carboxylic Acid Derivatives

Resorcinol and its derivatives can undergo, similar to phenols, carboxylation reactions in the presence of carbon dioxide atmosphere under alkaline conditions, to produce carboxylic acid compounds. With an un-substituted resorcinol, the carboxylation could lead to mono and dicarboxylated resorcinol. In the case of mono carboxylation of resorcinol, the predominant products are either the 2,4-dihydroxybenzoic acid ($\beta$-resorcylic acid or BRA) or 2,6-dihydroxybenzoic acid ($\gamma$-resorcylic acid or GRA) or a mixture of these two carboxylic acids. Similarly, the dicarboxylic acid derivatives that could be produced from the Kolbe-Schmitt type reaction are the 2,4- and 4,6-dicarboxylic acids.

In this section, the synthesis of carboxylic acid derivatives pertaining to the carboxylation reaction of resorcinol at the 2, 4 and 6 positions, and some of their potential applications are discussed.

### 3.6.1 $\beta$-Resorcylic Acid (2,4-Dihydroxybenzoic Acid)

This compound can be prepared by the reaction of resorcinol with an alkali metal carbonate in the presence of carbon dioxide [1–3] (Figure 3.113).

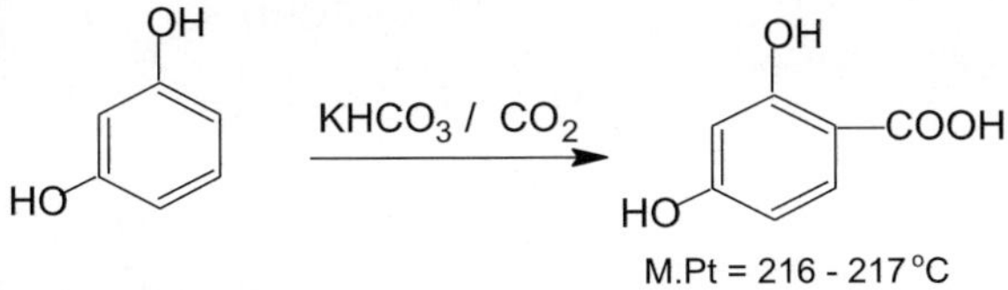


**Figure 3.113.**

A wide range of industrial, pharmaceutical and agricultural applications have been explored for $\beta$-resorcylic acid. This derivative was utilized in the development of commercially important derivatives and drug compounds.

The bis-3,5-[4-(2′-methoxyphenyl)-piperazinylmethyl]-2,4-dihydroxybenzoic acid prepared from $\beta$-resorcylic acid, O-methoxyphenyl piperazine and formaldehyde showed outstanding analgesic, antipyretic, and anti-inflammatory properties [4] (Figure 3.114).

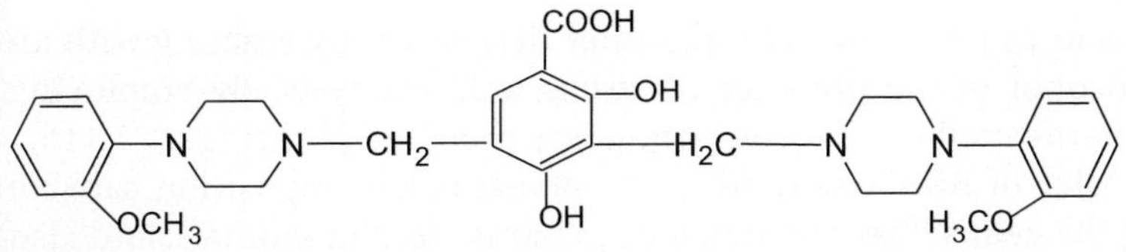


**Figure 3.114.**

This compound exhibited extremely low toxicity ($LD_{50} \Rightarrow 3\,g/kg$), when determined in mice by an oral administration. In humans suffering from chronic polyarthritis, an oral administration decreased the articular-pain and inflammatory symptoms in a few days. The drug was well tolerated and no side effects were observed [4].

$\beta$-Resorcylic acid was used in the preparation of cyclic calixarene compounds, which were found to be suitable as thermal stabilizing additives in aviation fuel compositions, and also in lubricating oil compositions used in four-stroke engines [5, 6]. When these additives were used in B99/1II aviation base fuel, less deposits, such as gums, sediments or granular materials, had been formed from the decomposition of fuel at elevated temperatures. Because of this, the performance of engines or oil compositions could be enhanced.

$\beta$-Resorcylic acid was used to improve the sweetness delivery profile of the sweetener sucralose [7].

Azo compounds and dyes have been developed using $\beta$-resorcylic acid (BRA), and applied to dye shaped articles of polypropylene, polyester, polyacrylics, polyamides and polyolefin materials. The BRA-based dyes did not exhibit the bathocromic coloration phenomenon in polyesters, and also showed good dyeing and fastness properties [8–10] (Figure 3.115).

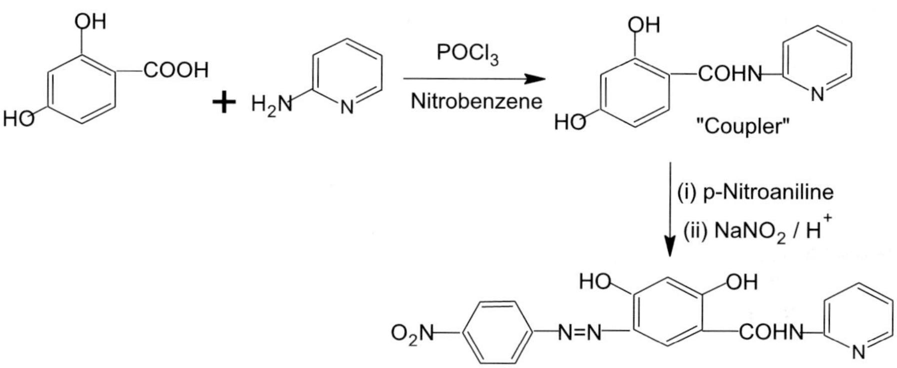


**Figure 3.115.**

### 3.6.1.1 Esters of $\beta$-Resorcylic Acid

$\beta$-Resorcylic acid can be converted into ester derivatives by reacting with an appropriate alcohol in the presence of strong acid catalysts. By employing methyl alcohol, the methyl ester compound was prepared [11] (Figure 3.116).

The butyl ester of BRA was used in the nematicide composition capable of controlling the genus *Toxocara* nematodes parasitized in animals, and also effective for preventing larvae migrans caused by the parasitism in men [12].

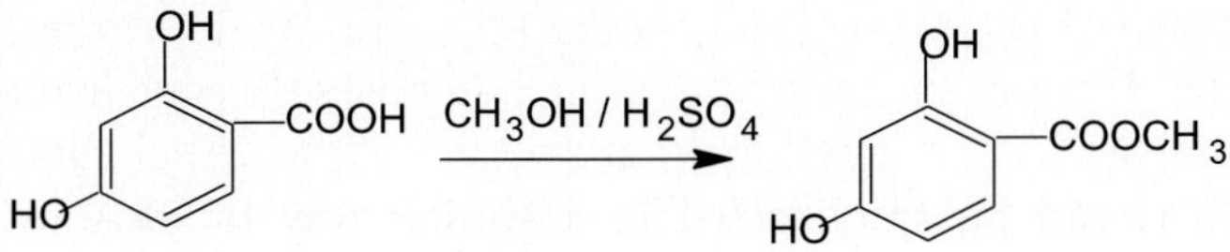


**Figure 3.116.**

Heat sensitive recording paper has been useful in wide fields such as medical or industrial recording instruments, terminals of computers or communications and facsimile equipment. The docosyl ester of BRA, when used as a color developing agent for heat sensitive recording paper, gave high color density and was free from spontaneous color development before heat treatment [13].

Methyl-$\beta$-resorcylate was used to prepare methyl-2-hydroxy-4-(2,2,3,3,3-pentafluoropropoxy)-benzoate. In pharmaceutical compositions, this compound was found to be useful for the treatment of psoriasis and other immune diseases [14] (Figure 3.117).

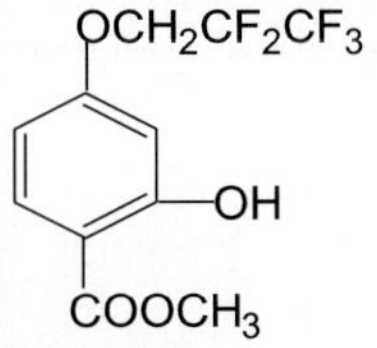


**Figure 3.117.**

### 3.6.1.2 Ether Derivatives of BRA

The 2,4-dimethoxybenzoic acid was prepared according to the following reactions [15] (Figure 3.118).

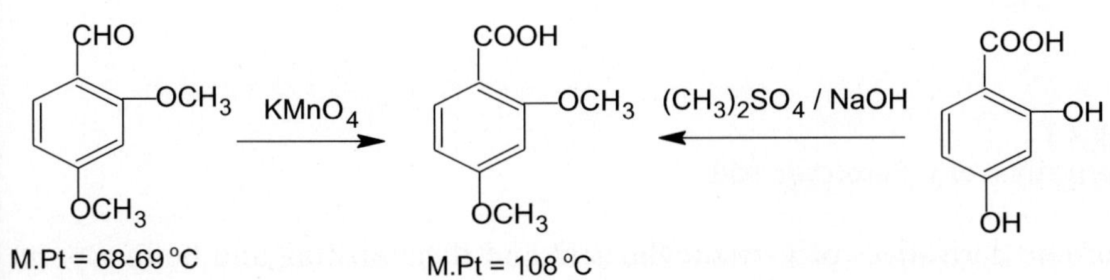


**Figure 3.118.**

## 3.6.2
## γ-Resorcylic Acid

γ-Resorcylic acid (GRA) can be synthesized in a similar way to BRA by the Kolbe-Schmitt process, but it requires more anhydrous conditions. GRA was prepared from the reaction of resorcinol with potassium carbonate in the presence of carbon dioxide and dimethyl formamide solvent [16–18] (Figure 3.119).

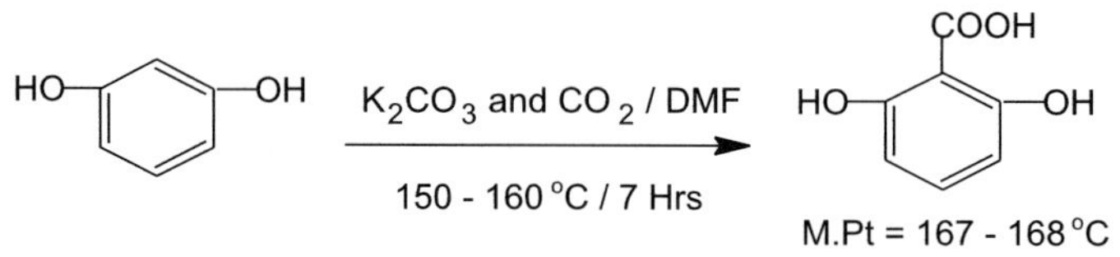


**Figure 3.119.**

A mixture containing GRA and BRA was prepared from resorcinol, selectively decomposed the BRA present in the reaction product, and then separated the GRA in a highly purified form [19] (Figure 3.120).

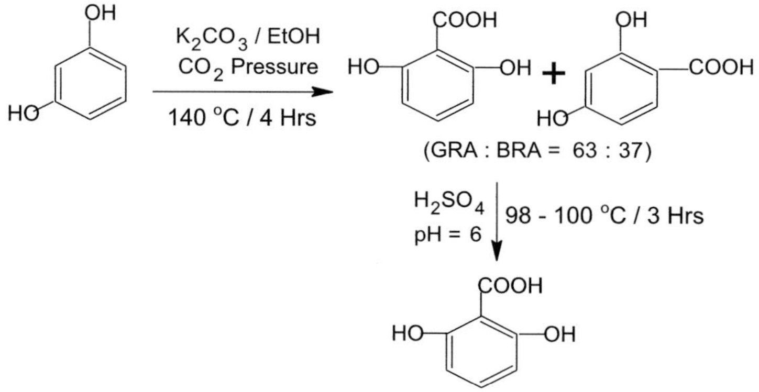


**Figure 3.120.**

### 3.6.2.1
### Derivatives of γ-Resorcylic Acid

Various derivatives of γ-resorcylic acid and their melting and boiling point data are presented in Table 3.11 [20,21].

2,6-Dimethoxybenzoic acid was prepared by acidifying the reaction mixture obtained from the reaction of 1,3-dimethoxybenzene with potassium dialkylamine in the presence of triethylamine and carbon dioxide [22,23] (Figure 3.121).

**Table 3.11.** Melting points of $\gamma$-resorcylic acid derivatives

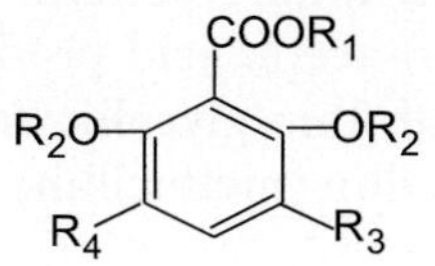


| $R_1$ | $R_2$ | $R_3$ | $R_4$ | Melting point (°C) |
|---|---|---|---|---|
| H | H | H | H | 160 – 161 (Decomp)[a] |
| H | H | H | H | 166 – 168 (Hydrate)[a] |
| $CH_3$ | H | H | H | 69 – 71[a] |
| $CH_3$ | $CH_3$ | H | H | 88 – 90[a] |
| H | $(CH_2)_2CH_3$ | H | H | 54 – 56[a] |
| H | $(CH_2)_3CH_3$ | H | H | 82 – 84[a] |
| H | $CH_2$-Ph | H | H | 124 – 126[a] |
| H | $CH_2$-CH=$CH_2$ | H | H | 71 – 72[a] |
| H | H | Cl | Cl | 216 – 218[b] |
| $CH_3$ | $CH_3$ | Cl | Cl | B.Pt. = 168/16 mm [b] |
| H | $CH_3$ | Cl | Cl | 104 – 106[b] |
| H | H | Cl | H | 189[b] |
| H | $CH_3$ | Cl | H | 133[b] |
| H | $CH_3$ | Br | Br | 109 – 111[b] |
| H | $CH_3$ | I | I | 136 – 138[b] |

[a] Data from [20]
[b] Data from [21]

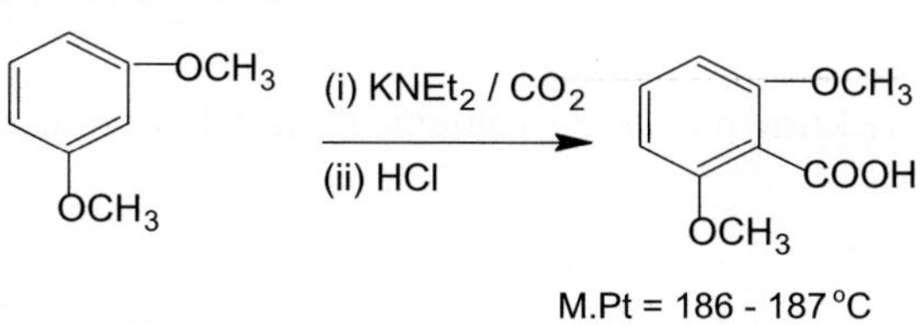


**Figure 3.121.**

1,2,3- and 1,2,4-trihydroxybenzene compounds have been prepared in a one-step process by reacting substituted resorcinols, including the GRA, with hydrogen peroxide [24] (Figure 3.122).

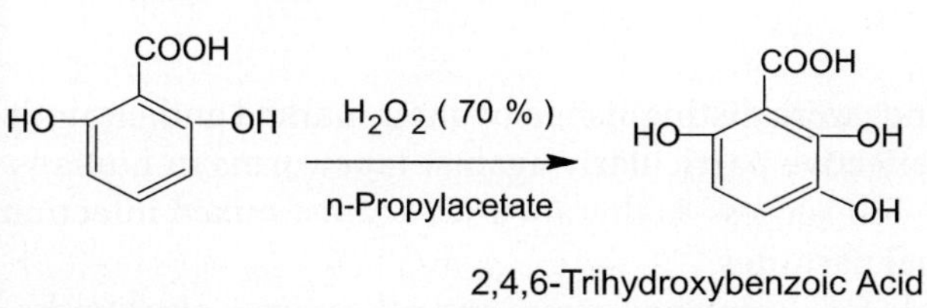


**Figure 3.122.**

This trihydroxybenzoic acid was obtained in 75% yield, and was based on the amount of hydrogen peroxide used in the reaction.

The 2-(2,6-dimethoxybenzoyl thio)-acetic acid produced by the reaction of 2,6-dimethoxybenzoyl chloride with thio-glycollic acid was useful for the production of dimethoxyphenyl penicillin (methicillin) [25] (Figure 3.123).

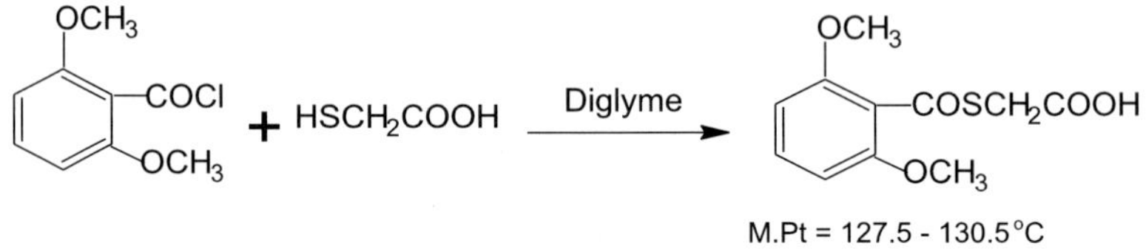


**Figure 3.123.**

Compounds having superior acaricidal and mite ovicidal activity against mites such as Citrus red mite, Desert spider mite and European red mite were prepared from the reaction of 2,6-dimethoxybenzoyl chloride with *O*-alkylhydroxylamines [26] (Figure 3.124).

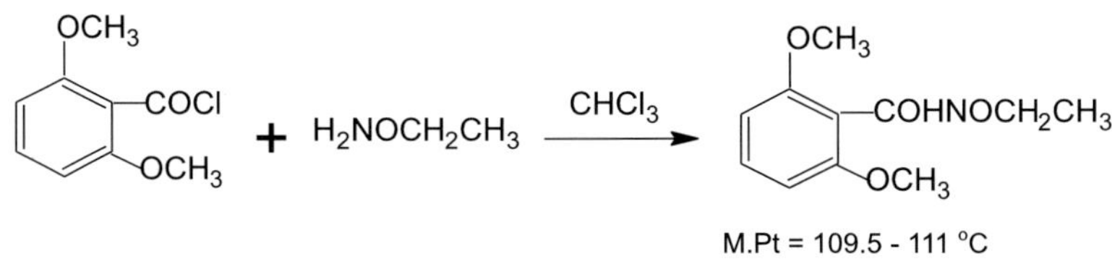


**Figure 3.124.**

$\gamma$-Resorcylic acid anilides were known for their valuable highly therapeutic properties [27,28] (Figure 3.125).

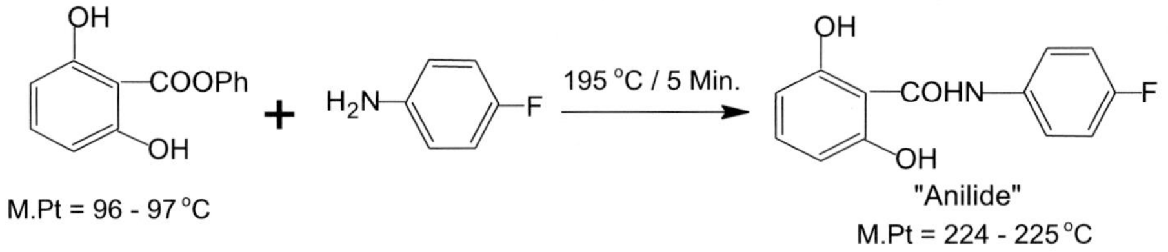


**Figure 3.125.**

These anilide compounds were distinguished by their marked anthelmintic action, and found to be effective particularly against tapeworms in humans. They were administered with success as therapeutics against mixed infection caused by several intestinal parasites.

Azo compounds suitable for dyeing polyesters and others were also synthesized [29,30] (Figure 3.126).

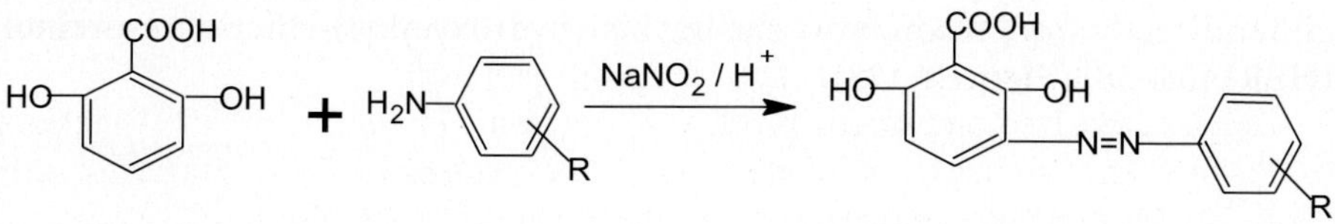


**Figure 3.126.**

Lanthanide complexes of 2,6-dihydroxybenzoic acid were synthesized, and their crystal structures and luminescent properties were studied. Photoluminescence measurements showed an intense photoluminescence both in the solid and in ethanolic solution for the terbium complex [31].

### 3.6.3 Resorcinol Dicarboxylic Acids

The dicarboxylation reaction of resorcinol had been studied and documented. Resorcinol, on carboxylation with potassium carbonate at 210 °C, gave exclusively resorcinol-4,6-dicarboxylic acid (M.Pt. = 312 °C) [32, 33]. Resorcinol-2,4-dicarboxylic acid, on the other hand, was obtained from 4-methyl resorcinol [32] (Figure 3.127).

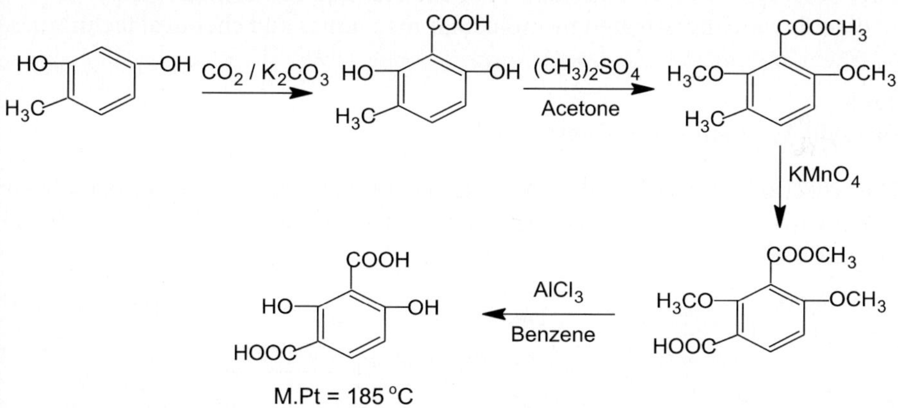


**Figure 3.127.**

### 3.6.4 Resorcinol Dioxydiacetic Acid Compounds

Resorcinol dioxydiacetic acid (RDOA) derivative was used as a monomer in the synthesis of polyester materials to enhance their oxygen ($O_2$) and carbon dioxide ($CO_2$) permeability properties. This compound could be synthesized either directly from resorcinol or using bis-(hydroxyethyl)-ether of resorcinol (HER) [34–38] (Figure 3.128).

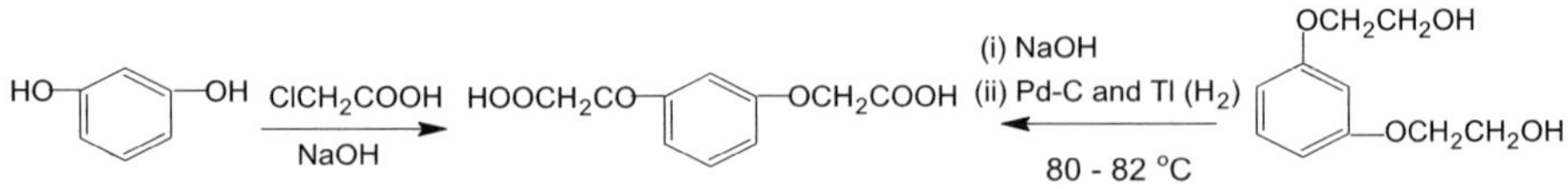


**Figure 3.128.**

A series of dicarboxylic acids having the following general structure was prepared from resorcinol and its derivatives [39] (Figure 3.129).

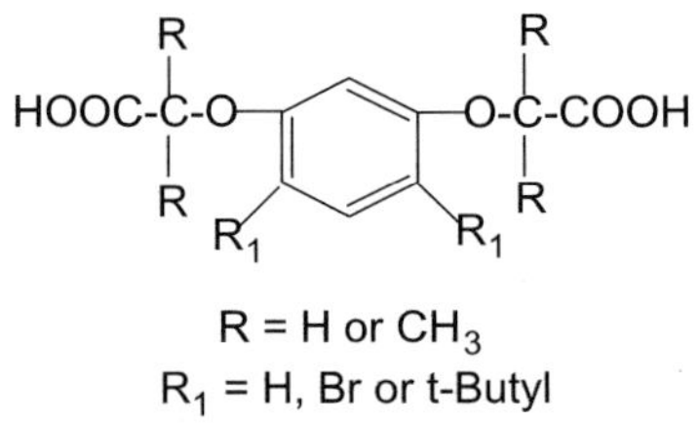


**Figure 3.129.**

Monochelate and bischelate complexes of dioxydicarboxylic acid derivatives have been synthesized. The stability of chelates and conformational properties of diacids were investigated by molecular mechanics and chemical techniques.

## 3.6.5 Glyoxylic Acid Derivatives of Resorcinol

The reaction of resorcinol and its alkyl derivatives with cyanoformate or ethyloxalyl chloride produced resorcinol glyoxylic acid compounds [40, 41] (Figure 3.130).

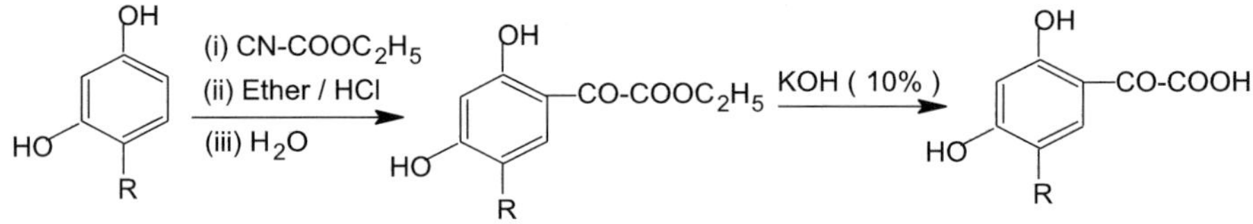


**Figure 3.130.**

The antibacterial activities of alkyl substituted glyoxylic acid derivatives had been compared to that of alkyl resorcinols. For antibacterial activity, these compounds were tested against *S. aureus*, *E. coli* and *B. subtilis* by both the broth serial dilution (BSD) and agar streak plate (ASP) techniques. The sodium salt of 5-*n*-octyl-2,4-dihydroxyphenyl glyoxylic acid was the most active, and completely inhibited the growth of *S. aureus*.

## References


1. M. Nierenstein and D. Clibbens, Organic syntheses, Coll. Vol. 2, p 557 (1943).
2. T. Kenji, M. Hirishi and H. Haruhisa, Jap. Pat. Publn., 59-184146 (1984).
3. P. Neumann and U. Eichenauer, US Patent, 4,996,354 (1991).
4. G. Regnier, R. Canevari, L. Roses and J. Le Douarec, US Patent, 3,280,127 (1966).
5. S. Taylor and M. Wilson, US Pat. Appln., 2002 / 0,077,254 (2002).
6. S. Taylor and M. Wilson, US Pat. Appln., 2002 / 0,107,153 (2002).
7. C. Merkel and M. Lindley, US Pat. Appln., 2003 / 0,059,519 (2003).
8. E. May and A. Fono, US Patent, 3,257,379 (1966).
9. E. May and A. Fono, US Patent, 3,251,822 (1966).
10. R. Botros, US Patent, 4,118,383 (1978).
11. R. Angelo, US Patent, 3,328,351 (1967).
12. T. Yoshinori, Jap. Pat. Publn., 04-217602 (1992).
13. M. Masato and S. Sumi, Jap. Pat. Publn., 63-041444 (1988).
14. J. Bartroliorpi et al., PCT, WO 2004 / 009528 (2004).
15. Robinson and Venkataraman, J. Chem. Soc., p 62 (1929).
16. E. Stove, British Patent, 916,548 (1961).
17. J. Cartwright, J. Jones and D. Marmion, J. Chem. Soc., p 3499 (1952).
18. D. Hake, A. Hawdon, J. Jones and D. Packham, J. Chem. Soc., p 3503 (1952).
19. T. Nakamatsu, Y. Nishida and N. Kometani, US Patent, 5,304,677 (1994).
20. F. Doyle, K. Hardy, J. Nayler, M. Soulal, E. Stove and H. Waddington, J. Chem. Soc., p 1453 (1962).
21. F. Doyle, J. Nayler, H. Waddington, J. Hanson and G. Thomas, J. Chem. Soc., p 497 (1963).
22. L. Yu, US Patent, 4,845,276 (1989).
23. N. Cartwright, J. Jones and D. Marmion, J. Chem. Soc., p 3500 (1952).
24. G. Prescher, G. Ritter and H. Sauerstein, US Patent, 4,762,953 (1988).
25. W. McDowell and W. Schmidt, US Patent, 3,178,466 (1965).
26. Nippon Soda Company, British Patent, 1,221,280 (1971).
27. H. Ruschig, B. Soden, T. Konig, N. Taunus and D. Duwel, US Patent, 3,449,420 (1969).
28. Hoechst, British Patent, 1,315,526 (1970).
29. E. May and A. Fono, US Patent, 3,236,829 (1966).
30. E. May and A. Fono, US Patent, 3,249,599 (1966).
31. P. Soares-Santos et al., Euro. J. Inorg. Chem., p 3609 (2003).
32. T. Gore, P. Inamdar and P. Nagarkar, Ind. J. Chem., Vol. 12, p 946 (197).
33. D. Sikkema and A. Reichwein, US Patent, 6,040,478 (2000).
34. J. Morimoto, T. Kamikawa and H. Ueda, Euro. Pat. Appln., EP 1,026,146 (2000).
35. J. Morimoto, T. Kamikawa and H. Ueda, US Patent, 6,316,666 (2001).
36. M. Komatsu and J. Ishikawa, Euro. Pat. Appln., EP 1,067,108 (2001).
37. C. Sumner, US Patent, 4,804,777 (1989).
38. C. Sumner, US. Patent, 4,935,540 (1990).
39. J. Bickley, R. Bonar-Law, McGrath, N. Singh and A. Steiner, N. J. Chem., Vol. 28, p 425 (2004).
40. I. Hunsberger and E. Amstutz, J. Am. Chem. Soc., Vol. 70, p 671 (1948).
41. R. Sprenger, P. Ruoff and A. Frazer, J. Am. Chem. Soc., Vol. 72, p 2874 (1950).

## 3.7 Sulfur Compounds

Sulfur derivatives of resorcinol have been used in the preparation of new organic compounds for use in pharmaceutical, dyes and photographic applications.

### 3.7.1 Sulfonic Acid Derivatives of Resorcinol

In the preparation of sulfonic acid derivatives, resorcinol has been reacted with sulfuric acid, sulfur trioxide or oleum ($H_2SO_4 + SO_3$) under suitable reaction conditions and solvent medium to obtain either mono-, di- or tri-sulfonic acid compounds [1–3] (Figure 3.131).

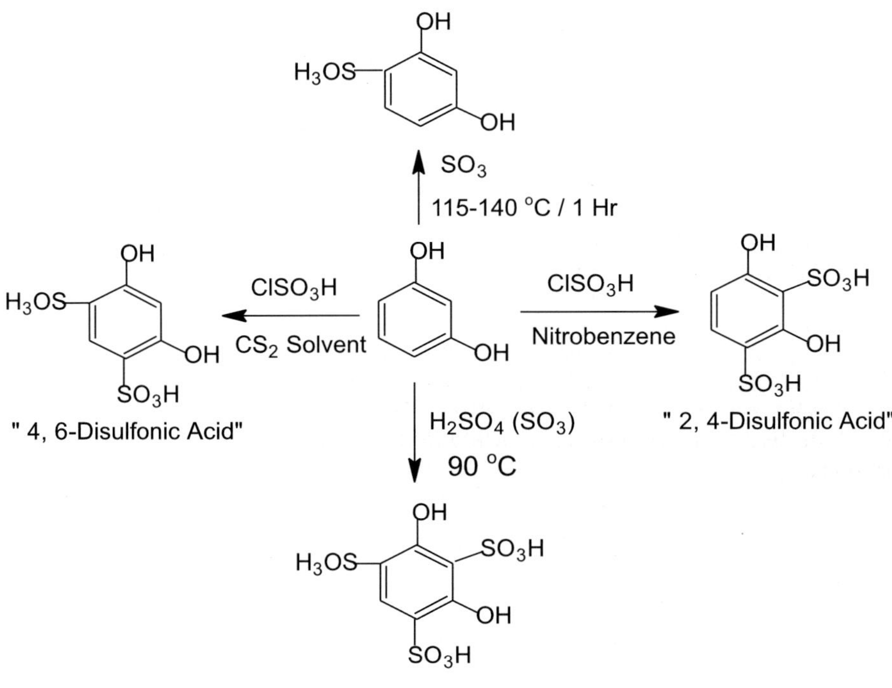


**Figure 3.131.**

The 2,4-dihydroxybenzene sulfonic acid and its derivatives were useful as intermediates in the preparation of dyestuffs.

The hexamethylenetetramine-mono-hexylresorcinol sulfonic acid salt possessed an antiseptic action superior to that of either hexylresorcinol or the sulfonic acid. It was synthesized by sulfonating hexylresorcinol with concentrated sulfuric acid [4] (Figure 3.132).

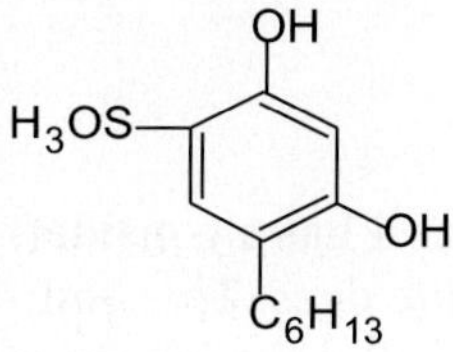


**Figure 3.132.**

### 3.7.2 Thiobisresorcinols

Thiobisresorcinol compounds could be conveniently prepared by reacting resorcinol or its derivatives with sulfur dichloride in the presence of non-interacting solvents at relatively lower temperature conditions [5, 6] (Figure 3.133).

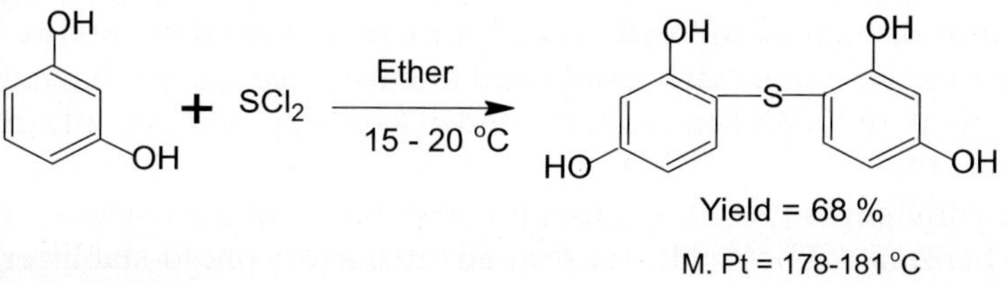


**Figure 3.133.**

The 2,2′-thiobis-(4,6-di-*tert*-butylresorcinol) compound exhibited rodent repellent properties [7]. Similarly, the substituted diresorcyl sulfide derivative obtained from 2-methyl-4-chlororesorcinol was found to be useful as an azo coupling component in light sensitive diazo compositions and diazo type materials [8].

Novel chemical compounds prepared from the diresorcyl sulfide were useful in the treatment, and also prolonging the life, of animals infected with the distemper virus [9]. These compounds were synthesized by the reaction of a resorcinol derivative with dialkylamine and formaldehyde (Figure 3.134).

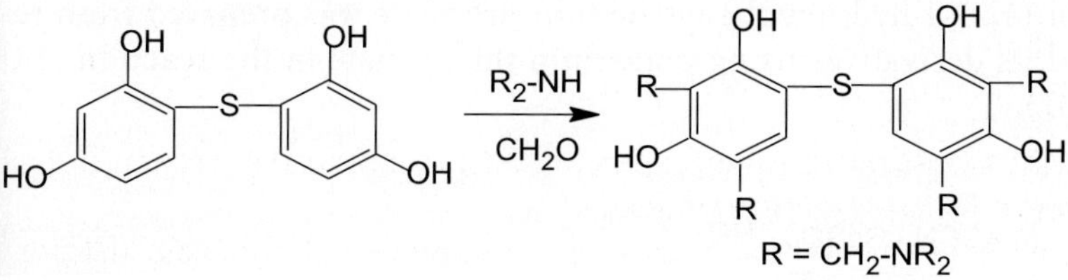


**Figure 3.134.**

### 3.7.3 Diresorcyl Sulfoxide

In the diazo printing process, diresorcyl sulfoxide compound has a considerable value. It shows an excellent shade property for the blue dyes. This compound was readily prepared by the oxidation of diresorcyl sulfide by means of 30% hydrogen peroxide in an aqueous medium [10] (Figure 3.135).

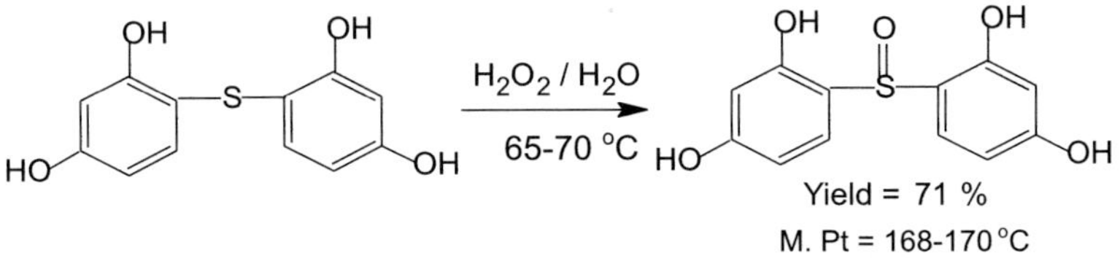


**Figure 3.135.**

A storage stable black line diazo type material can be obtained by replacing the resorcinol with resorcinol sulfoxide [11]. Hair dyes obtained using diresorcyl sulfoxide can give particularly bright and intensive dye finishes having high fastness to light and high heat stability, and also having tones mainly in the brown region [12].

2,4-Dihydroxydiphenyl sulfoxide compound prepared from the reaction of resorcinol with benzene sulfonyl chloride showed satisfactory photo-stabilizer action for polyolefin [13] (Figure 3.136).

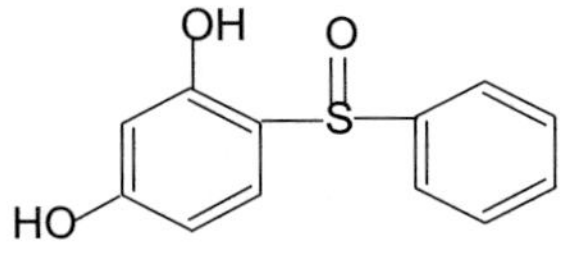


**Figure 3.136.**

### 3.7.4 Tioxolone and Thio Derivatives

Tioxolone or (1,2)-4-hydroxyphenylene thiocarbonate was prepared from resorcinol and its derivatives using potassium thiocyanate in the reaction [14] (Figure 3.137).

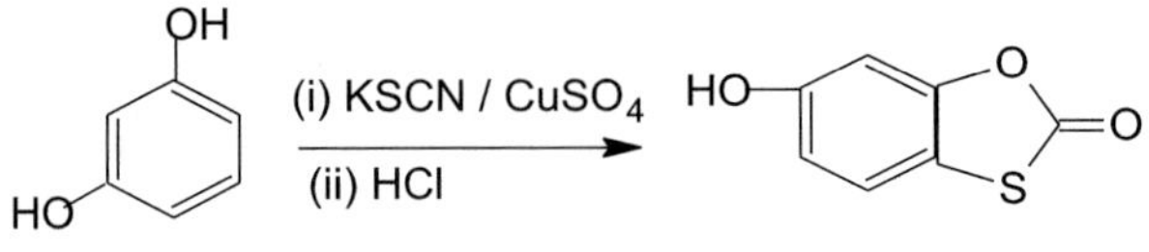


**Figure 3.137.**

Various 4-thio-substituted resorcinol derivatives were prepared using tioxolone as the starting material [14–16] (Figure 3.138).

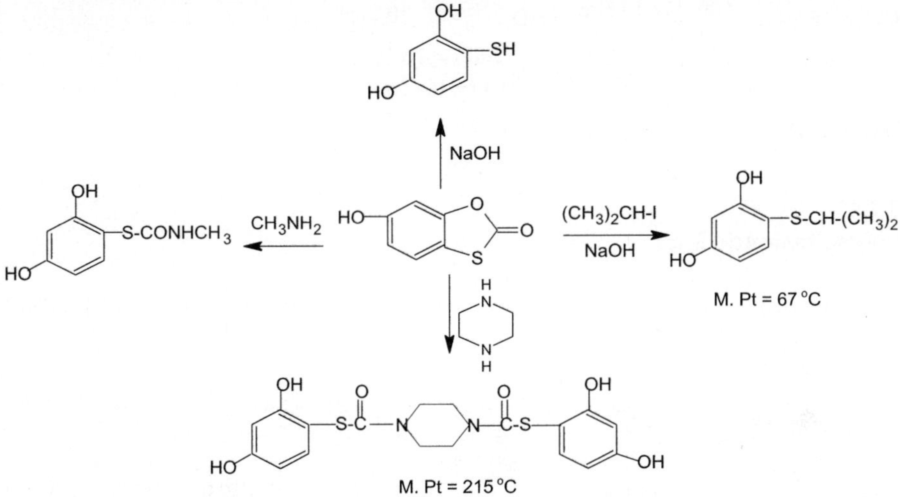


**Figure 3.138.**

A dithiane resorcinol derivative, useful in skin lightening agents, was prepared by reacting 2,4-dihydroxybenzaldehyde with 1,3-dimercapto propane in the presence of an acid catalyst [17] (Figure 3.139).

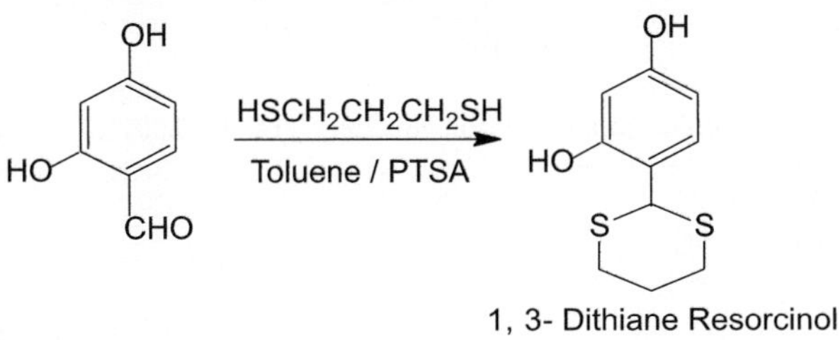


**Figure 3.139.**

In general, the thio resorcinol derivatives were employed in compositions used as skin lightening agents, and they were also used as coupling components in the preparation of azo coloring materials.

### 3.7.5 Bisulfite Compound

3,5-Dihydroxy-2,4-cyclohexadien-1-sulfonic acid was obtained when resorcinol was allowed to react with sodium sulfite under controlled pH conditions [18, 19] (Figure 3.140).

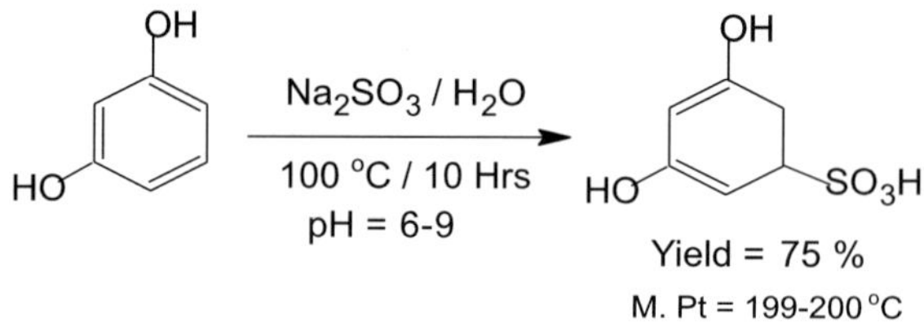


**Figure 3.140.**

## 3.7.6 Vinylsulfonic Acid Ester

The reaction of carbyl sulfate with resorcinol produced resorcinol bis-vinyl sulfonate [20] (Figure 3.141).

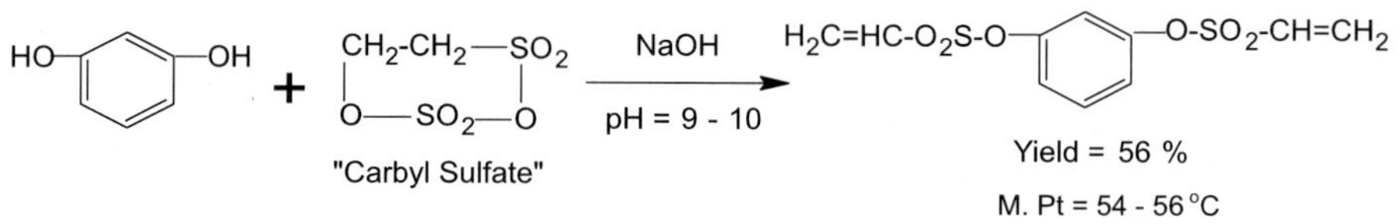


**Figure 3.141.**

## References


1. B. Davidson and L. Byrne, British Patent, 820,659 (1957).
2. W. Peterson, US Patent, 2,487,586 (1949).
3. K. Yukihiro, M. Hideki and K. Masahiko, Jap. Pat. Publ. No. 2001-114748 (2001).
4. H. Legerlotz, British Patent, 492,914 (1938).
5. R. Moore, US Patent, 2,755,306 (1956).
6. R. Moore, US Patent, 2,760,989 (1956).
7. C. Fuchsman, US Patent, 3,726,928 (1973).
8. R. Desjarlais, US Patent, 3,857,896 (1976).
9. C. Geschickter, US Patent, 3,056,801 (1962).
10. A. Comer and R. Moore, US Patent, 2,661,376 (1953).
11. J. Sulich, US Patent, 2,717,832 (1965).
12. N. Maak and E. Lieske, US Patent, 4,552,565 (1985).
13. D. Chaser, US Patent, 4,032,505 (1977).
14. G. Werner, US Patent, 2,332,418 (1943).
15. A. LaGrange and H. Borowiak, US Patent, 5,468,472 (1995).
16. V. Ouperoff-Urne, US Patent, 2,650,925 (1953).
17. B. Harichian, M. Barrett, C. Bosko and M. Tallman, US Pat. Appln., 2004 / 0,115,145 (2004).
18. G. Zagorodnikova, J. Org. Chem. (USSR), Vol. 31, p 904 (1995).
19. W. Lauer and C. Langkammerer, J. Am. Chem. Soc., Vol. 56, p 1628 (1934).
20. Badische Anilin, British Patent, 921,073 (1961).

# 3.8 Nitro and Nitrosoresorcinols

Resorcinol forms both nitro and nitroso type compounds, which are valuable raw materials in the synthesis of pharmaceutically and industrially important organic derivatives. In this section, some of these derivatives have been discussed.

## 3.8.1 Nitroresorcinols

### 3.8.1.1 Mono-Nitroresorcinol

2-Nitroresorcinol and 4-nitroresorcinol (M.Pt. = 117–119°C) were prepared by the reaction of resorcinol with a nitrating mixture containing sulfuric and nitric acids [1,2] (Figure 3.142).

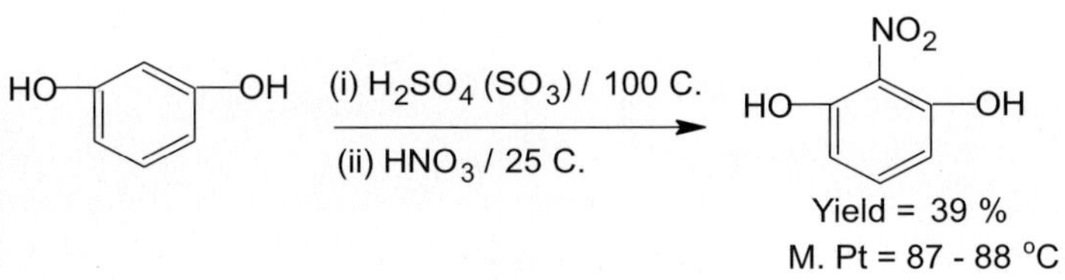


**Figure 3.142.**

From the 2-nitro derivative, the mono-methyl ether and dibromo compounds were developed [1,2] (Figure 3.143).

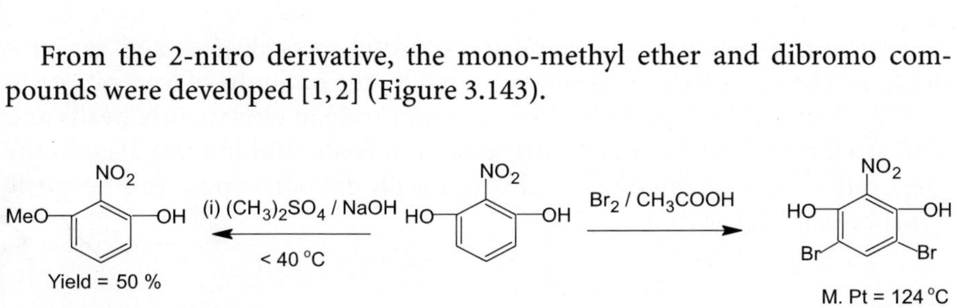


**Figure 3.143.**

A class of enzymatically activated agents, and also the first MRI contrast agent DOPTA-Gd that can conditionally enhance image intensity, was synthesized from 2-nitroresorcinol [3] (Figure 3.144).

An immediate application of this agent was to study the changes in cellular calcium activity during embryogenesis.

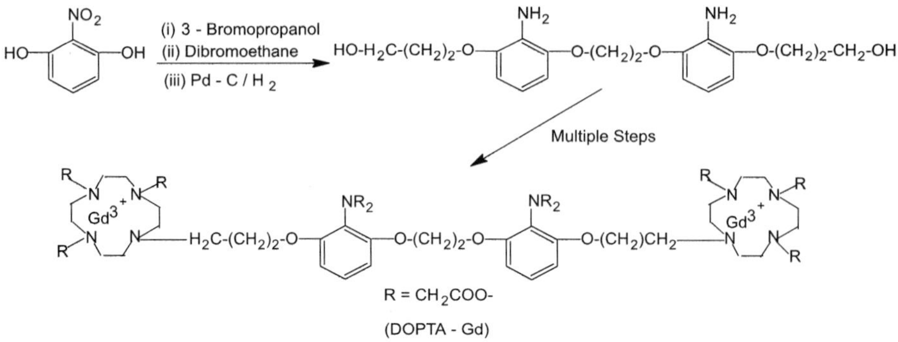


**Figure 3.144.**

## 3.8.1.2 Dinitroresorcinols

2,4-Dinitroresorcinol was prepared from the nitration reaction of resorcinol, [4], whereas 4,6-dinitroresorcinol was selectively synthesized in a white fuming nitric acid reaction with resorcinol diacetate [5] (Figure 3.145).

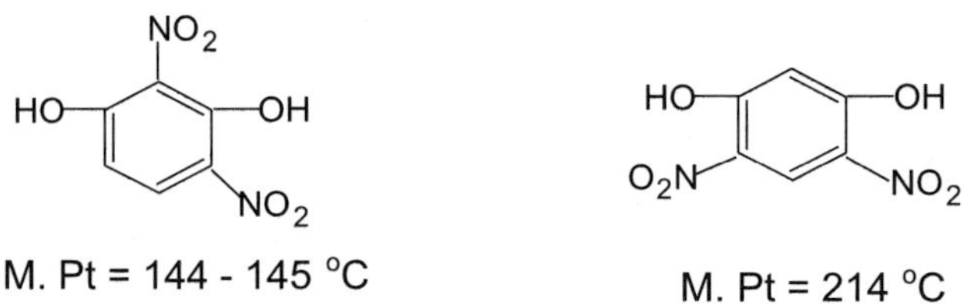


**Figure 3.145.**

2,4-Dinitroresorcinol was found to be useful in explosives and gas production. The monobasic lead salt prepared by the addition of lead nitrate to the disodium salt of 2,4-dinitroresorcinol was used in electric fuse heads and similar electric igniters. 4,6-Dinitroresorcinol is a useful intermediate in the preparation of 4,6-diaminoresorcinol for the production of poly-(*p*-phenylene benzobisoxazole), (PBO).

## 3.8.1.3 Trinitroresorcinol

2,4,6-Trinitroresorcinol, generally known as styphnic acid, has been extremely useful in explosives production, particularly in the form of its normal and basic lead salt [6–9] (Figure 3.146).

This compound is a powerful explosive, and therefore no attempt should be made to produce it, even in small quantities, under normal laboratory conditions.

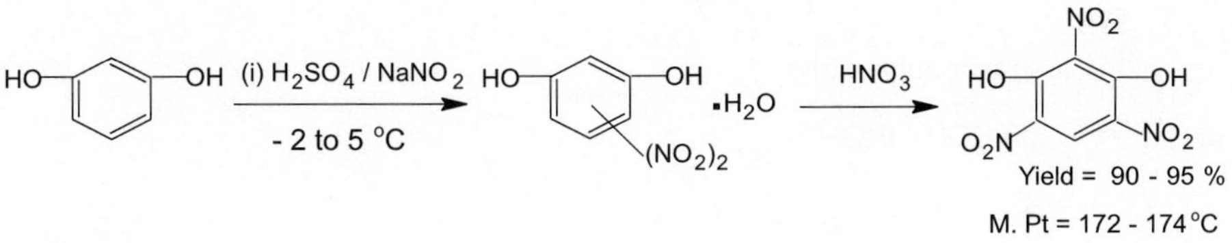


**Figure 3.146.**

Nitroresorcinols can be reduced to their corresponding amino derivatives by hydrogenation reactions.

### 3.8.1.4 Octanitroresorcinol Derivatives

Supersonic aircraft, new types of missiles and equipment for space exploration required explosives possessing better heat resistance than conventional explosives such as TNT. Resorcinol-based octanitro macrocyclic explosives possessed good thermal stability for these applications (Figure 3.147) [10].

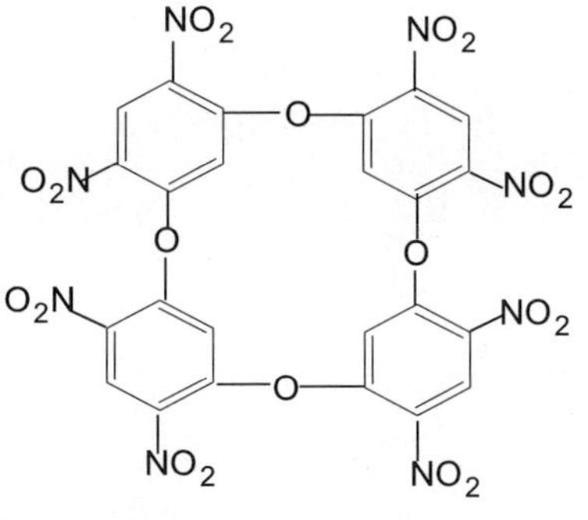


**Figure 3.147.**

## 3.8.2 Nitrosoresorcinols

### 3.8.2.1 4-Nitrosoresorcinol

Mono-nitrosoresorcinol derivatives can be conveniently prepared by the action of amyl nitrite or isopentyl nitrite in an alcoholic potassium hydroxide solution [11] (Figure 3.148).

4-Nitrosoresorcinol can be oxidized in the presence of hydrogen peroxide to obtain 4-nitroresorcinol, and also reduced to 4-aminoresorcinol.

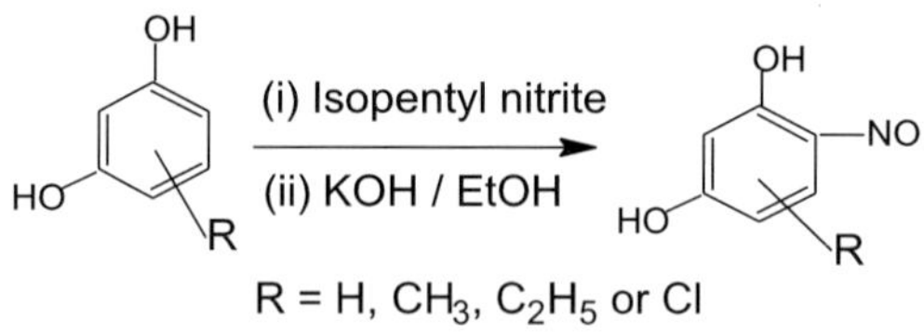


**Figure 3.148.**

4-Nitrosoresorcinol has been used to synthesize resorufin derivatives. *N*-Acyldihydroresorufin-4-carboxylic acid morpholide was synthesized by the reaction of nitroresorcinol with $\gamma$-resorcylic acid [11, 12] (Figure 3.149).

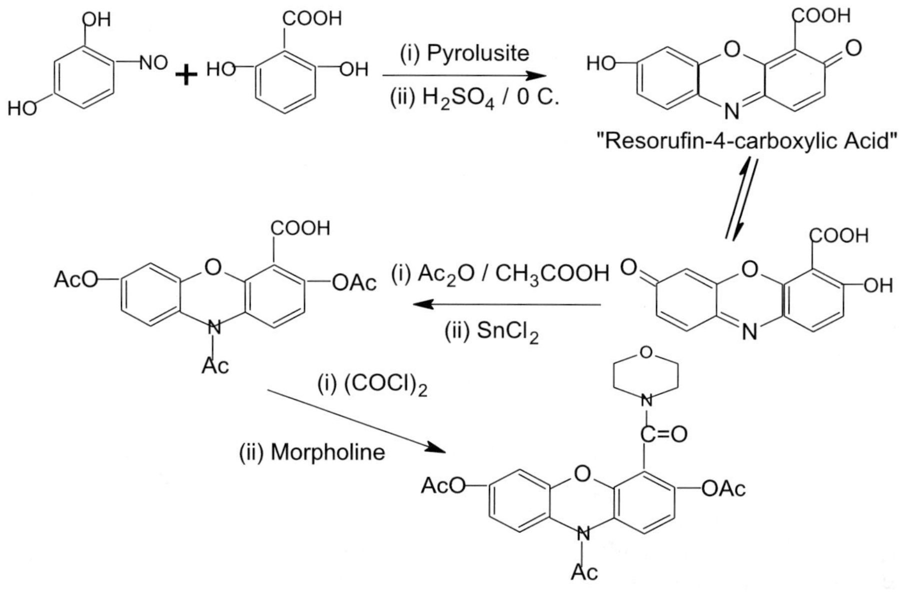


**Figure 3.149.**

*N*-Acyldihydroresorufin derivatives were suitable for the determination of hydrogen peroxide, peroxidase-acting compounds and peroxidase. Resorufin compounds can also be used as florescent labeled low or high molecular weight compounds. They can be useful for the determination of enzymes. Resorufins prepared from the reaction of nitroso derivatives of 2-methyl, 4-ethyl, and 4-chlororesorcinols with 2,6-dihydroxybenzoic acid can be bound to a substrate, which can be split by the enzyme to be determined [11].

Resorufin glycosides that could incorporate a lipophilic group may be useful in detecting cells producing enzymes that hydrolyze a glycoside [12]. The reddish fluorogenic substance produced from the reaction of nitrosoresorcinol with 4-dodecylresorcinol or 4-octadecylresorcinol was used to analyze glucosidic enzyme activity (Figure 3.150).

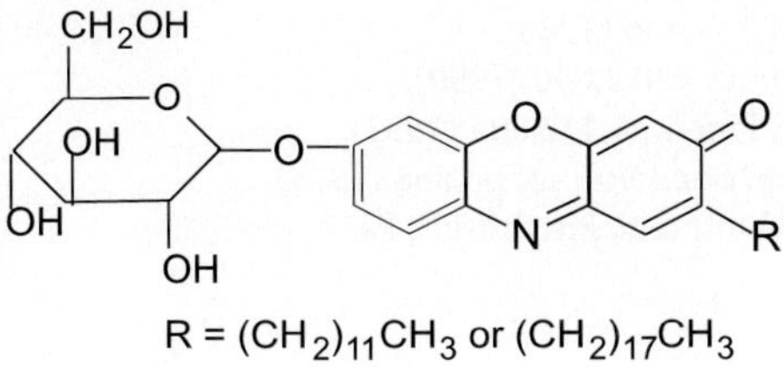


**Figure 3.150.**

### 3.8.2.2 2,4-Dinitrosoresorcinol

This compound was formed by the reaction of nitrous acid on resorcinol in aqueous acidic solution [15] (Figure 3.151).

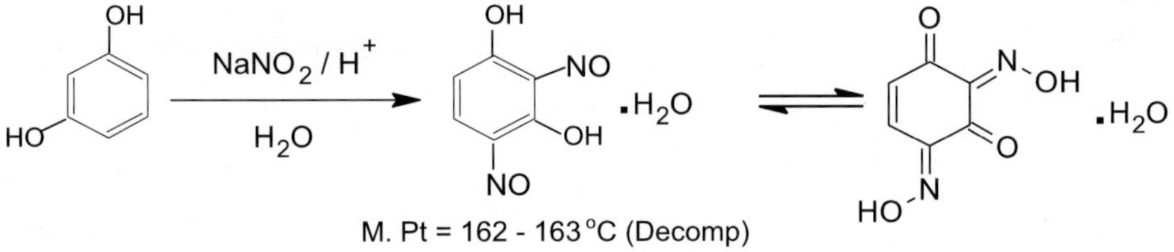


**Figure 3.151.**

Dinitrosoresorcinol formed chelates with heavy metals. These derivatives not only offset the deleterious effects of iron and iron compounds, but actually considerably improved the stability of gasoline. Dinitrosoresorcinol had been used as a coupling agent with diazonium salts for the preparation of organic pigments, and found to be useful in cross-linking unsaturated polymers. By employing mild oxidation and reduction conditions, this compound can be converted to 2,4-dinitroresorcinol and 2,4-diaminoresorcinol derivatives [16].

## References


1. H. Hodgson and R. Dyson, J. Chem. Soc., p 946 (1935).
2. C. Jeager, US Patent, 2,385,282 (1945).
3. W. Li, S. Fraser and T. Meade, J. Am. Chem. Soc., Vol. 121, p 1413 (1999).
4. W. Allan, US Patent, 2,945,890 (1960).
5. R. Schmitt, D. Ross and J. Wolfe, US Patent, 4,745,232 (1988).
6. D. Jones and C. Roberts, US Patent, 2,301,912 (1942).
7. D. Salter and R. Simkins, US Patent, 3,933,926 (1976).
8. G. Taylor and A. Thomas, US Patent, 3,894,068 (1975).
9. K. Holloway, G. Taylor and A. Thomas, US Patent, 3,954,822 (1976).
10. E. Gilbert, US Patent, 3,941,812 (1976).
11. C. Klein, H. Batz and R. Herrmann, US Patent, 5,304,645 (1994).

12. C. Klein and H. Von Der Eltz, US Patent, 4,737,466 (1988).
13. C. Klein, H. Batz and R. Hermann, US Patent, 4,954,630 (1990).
14. J. Naleway, Y. Zhang and R. Haugland, US Patent, 5,242,805 (1993).
15. W. Orndorff and M. Nicholls, J. Am. Chem. Soc., Vol. 45, p 1538 (1923).
16. Technical Bulletin K-3703 from Koppers Company, Pittsburgh, Pa.

# 3.9 Aminoresorcinols

Amino derivatives can be prepared conveniently by reduction reactions of nitro compounds of resorcinol. In this section, some of the well-known aminoresorcinols and their applications have been discussed.

## 3.9.1 Mono Aminoresorcinols

### 3.9.1.1 2-Aminoresorcinol

This compound has been synthesized by the hydrogenation of 2-nitroresorcinol [1] (Figure 3.152).

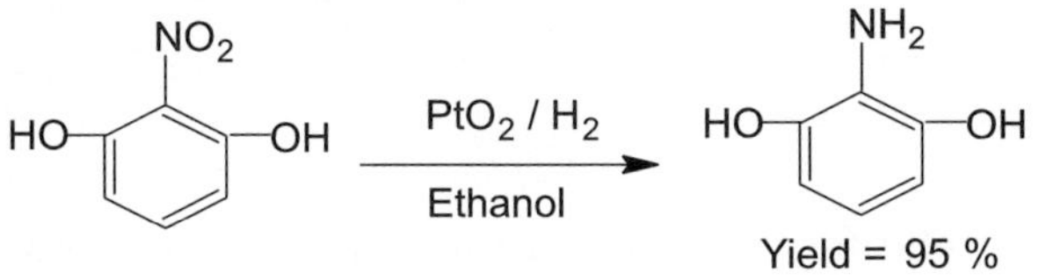


**Figure 3.152.**

5-Hydroxy-3-phenyl-2*H*-1,4-benzoxazin-2-one compound was prepared from 2-aminoresorcinol and ethyl phenylglyoxylate [2] (Figure 3.153).

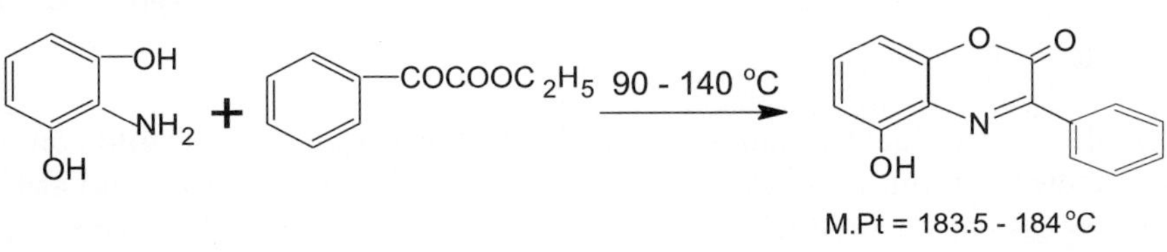


**Figure 3.153.**

These benzoxazinone compounds exhibit tranquilizing and sedative activity, and therefore can be used as central nervous system depressants.

Benzoheterocyclic carbamates are known to be active anti-lipolytic agents. In order to prepare these compounds, 2-aminoresorcinol was allowed to react with triethylorthoformate to produce 4-hydroxybenzoxazole first, and then the carbamate derivatives were synthesized from this compound [3] (Figure 3.154).

The resulting carbamate compounds are capable of inhibiting lipolysis associated with ischemic heart disease, and found to be useful in the treatment of ischemic disease and hypertriglyceridemia.

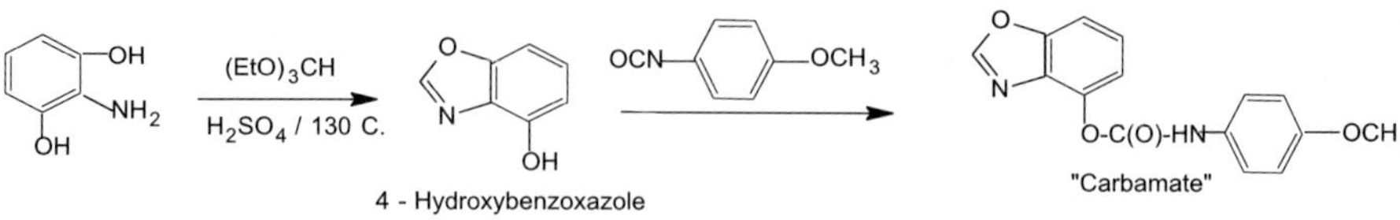


**Figure 3.154.**

## 3.9.1.2 4-Aminoresorcinol

This derivative of resorcinol was synthesized by reducing either 4-nitroresorcinol or 4-nitrosoresorcinol. In addition to the above reactants, 4-phenylazo resorcinol obtained from resorcinol was also used to prepare 4-aminoresorcinol [4] (Figure 3.155).

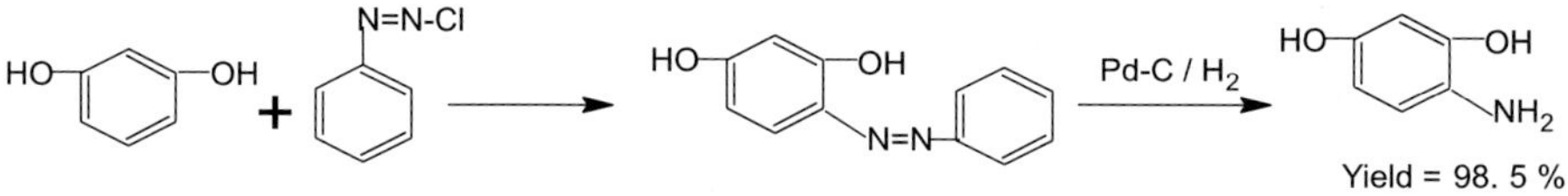


**Figure 3.155.**

Fluorogenic compounds useful for the determination of hydrolyzing enzymes were prepared from 4-aminoresorcinol [5] (Figure 3.156).

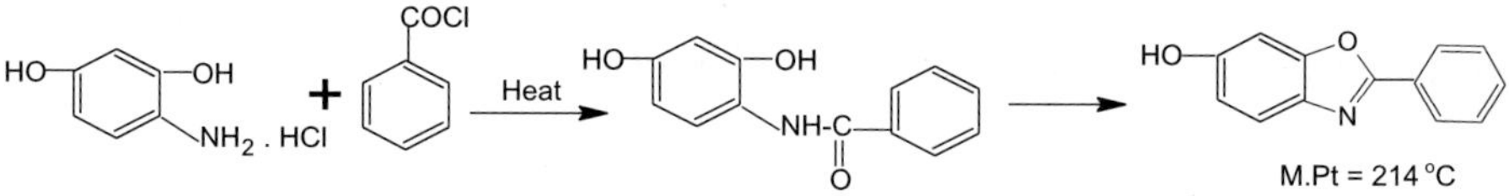


**Figure 3.156.**

The fluorogenic data obtained on 2-phenyl-6-hydroxy-benzoxazole compound showed excitation peaks at 308 nm (acid) and 370 nm (base), and also emission peaks at 480 nm (acid) and 480 nm (base). Derivatives obtained from the 6-hydroxybenzoxazole compound could be useful for the detection of various hydrolytic enzymes.

## 3.9.2 Diaminoresorcinols

2,4-Diaminoresorcinol and 4,6-diaminoresorcinol compounds have been prepared from 2,4-dinitroresorcinol and 4,6-dinitroresorcinol derivatives (Figure 3.157).

The major application for 2,4-diaminoresorcinol is in dyes, and similarly 4,6-diaminoresorcinol has been used as a monomer in the manufacture of Zylon® fibers.

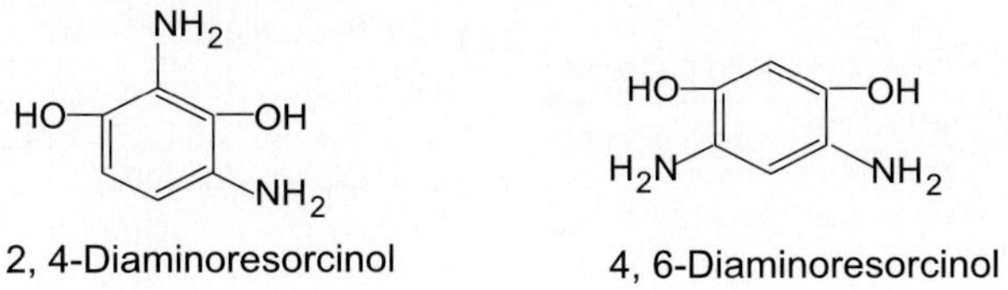


**Figure 3.157.**

### 3.9.3
### *m*-Aminophenol (or) 3-Aminoresorcinol

To prepare *meta*-aminophenol, there are several methods and procedures available in the literature. Resorcinol has also been used as the raw material for the manufacture of *m*-aminophenol. When resorcinol was allowed to react with ammonia in the presence of an acid catalyst at elevated temperature, *m*-aminophenol was obtained in good yields [6–9] (Figure 3.158).

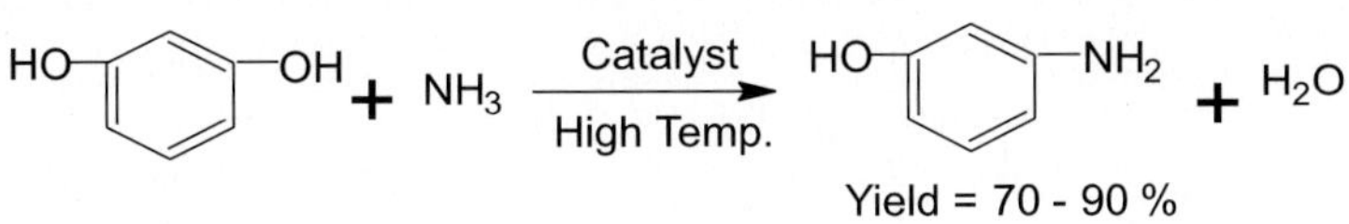


**Figure 3.158.**

3-Aminoresorcinol was also synthesized in higher yields and purity by using dihydroresorcinol in the reaction [10] (Figure 3.159).

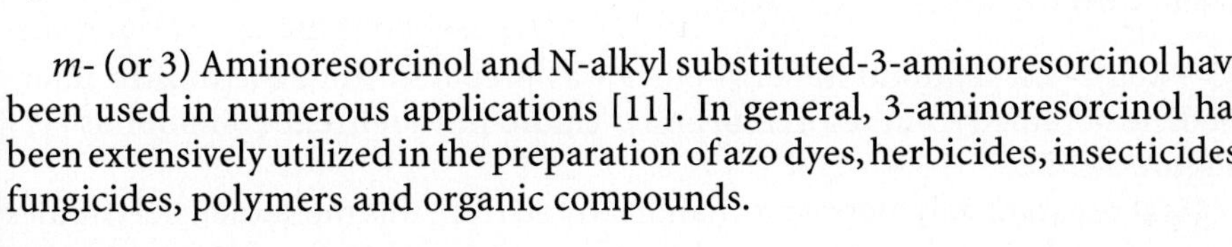


**Figure 3.159.**

*m*- (or 3) Aminoresorcinol and N-alkyl substituted-3-aminoresorcinol have been used in numerous applications [11]. In general, 3-aminoresorcinol has been extensively utilized in the preparation of azo dyes, herbicides, insecticides, fungicides, polymers and organic compounds.

### 3.9.4
### *m*-Hydroxy Diphenylamines

Diphenylamine compounds were synthesized by the condensation reaction of resorcinol with primary aromatic amines in the presence of catalytic amounts of an acid catalyst at elevated temperature conditions [12–14] (Figure 3.160).

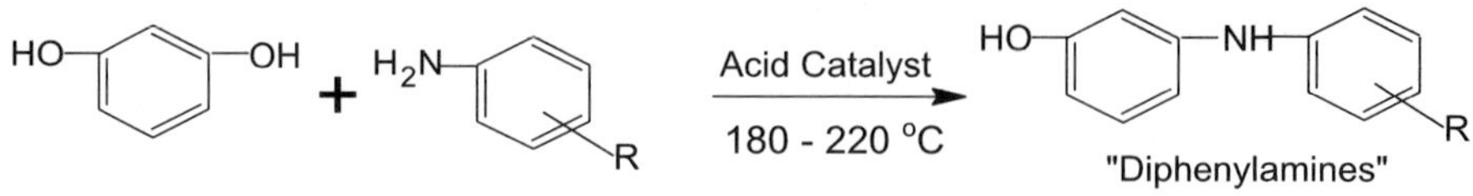


**Figure 3.160.**

3-Hydroxydiphenylamine derivatives were found to be useful as intermediates for the manufacture of crop protection agents, optical brighteners, and dyes, especially of the xanthene, pyronine, rhodamine, oxazine, azo, triphenylmethane and diphenylmethane series.

## 3.9.5 Morpholinomethyl Resorcinols

Morpholinomethyl substituted resorcinolic compounds were prepared from the reaction of resorcinol with morpholine and formaldehyde [15, 16] (Figure 3.161).

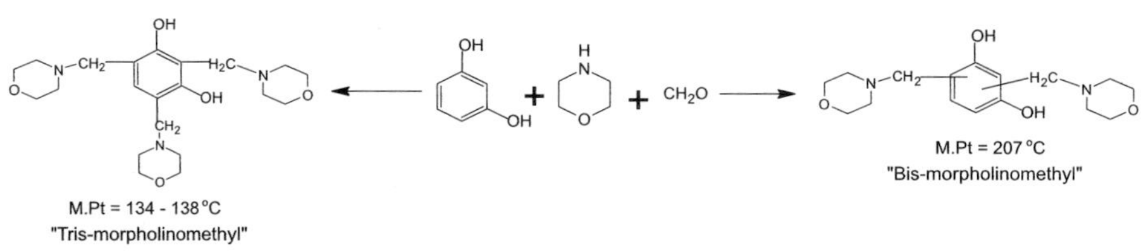


**Figure 3.161.**

These compounds were shown to be useful in the ingredients of bactericidal and insecticidal compounds, as fungicides, anti-oxidants, dyestuff intermediates, and also in the adhesion of rubber compounds to tire cords.

## 3.9.6 Amino Ether Derivatives of Resorcinol

1,3-Phenyl-4,4′-diaminodiphenyl ether was prepared from the dinitro intermediate obtained from resorcinol and *p*-chloro nitrobenzene compounds [17] (Figure 3.162).

Heat resistant polymers were manufactured from this monomer. Polyimides having low dielectric constant and excellent moldability were prepared using 1,3-bis-(3-amino-5-trifluoromethyl phenoxy)-benzene monomer, which was synthesized from resorcinol and 3,5-dinitrobenzo trifluoride compounds [18] (Figure 3.163).

Polyimides, prepared from this diamine monomer, were found to be suitable as insulating materials of a flexible printed circuit board for a highly integrated electronic circuit.

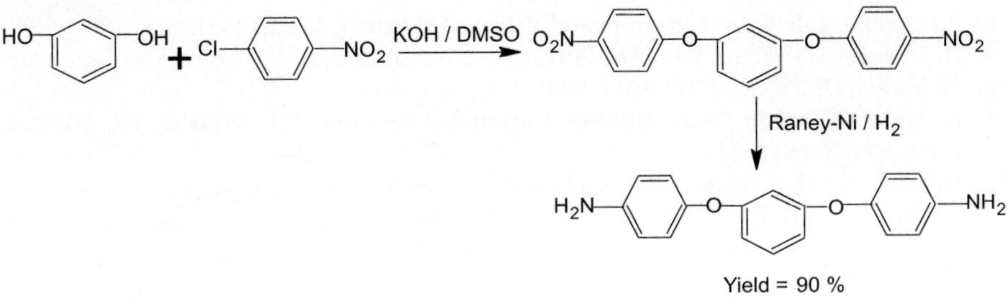


**Figure 3.162.**

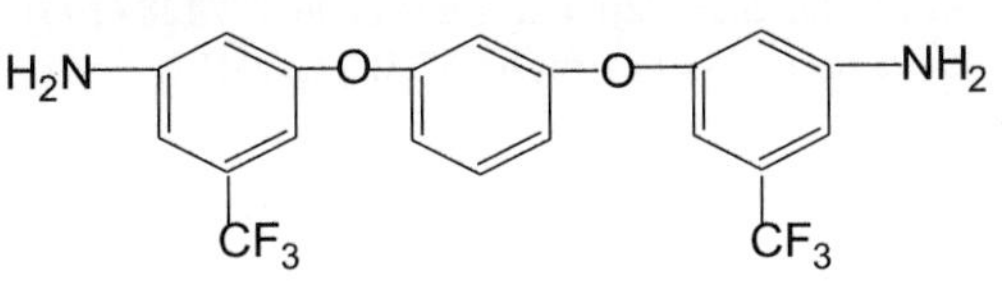


**Figure 3.163.**

Polyimides exhibiting an excellent heat resistance, chemical resistance, mechanical strength, electrical and, also adhesion properties suitable for low-temperature adhesive bonding were prepared from the following diamine compounds synthesized from resorcinol [19] (Figure 3.164).

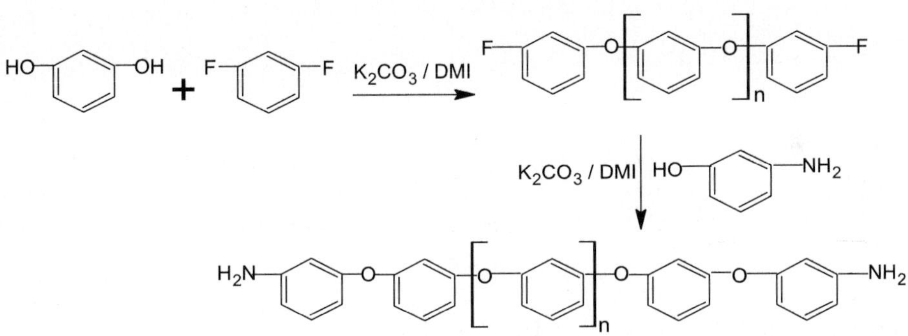


**Figure 3.164.**

## References


1. J. Cavalla, L. Dolby, P. Hofer and M. Chasin, US Patent, 5,665,737 (1997).
2. R. Moffett, US Patent, 3,255,188 (1966).
3. J. Musser and C. Sutherland, US Patent, 4,579,865 (1986).
4. S. Hideo, A. Kazuhiko and T. Kenichi, Jap. Pat. Publn., 11-158124 (1999).
5. K. Habenstein, US Patent, 5,696,001 (1997).
6. F. Bean and T. Donovan, US Patent, 2,376,112 (1945).
7. H. Harada, H. Maki and S. Sasaki, US Patent, 4,585,896 (1986).

8. F. Matsunaga, E. Kato, T. Kimura and Y. Isota, US Patent, 4,675,444 (1987).
9. H. Dressler, US Patent, 5,130,489 (1992).
10. W. Muller, US Patent, 4,212,823 (1980).
11. *m*-Aminophenol in "Resorcinol-its uses and derivatives", H. Dressler, Ed., Plenum Press, New York (1992).
12. H. Hoch and H. Scheuermann, US Patent, 4,067,903 (1978).
13. F. Werner and H. Blank, US Patent, 4,265,833 (1981).
14. A. Bugaut and Estradier, US Patent, 4,494,953 (1983).
15. H. Bruson, US Patent, 2,040,040 (1936).
16. A. Kolka and T. Sai, US Patent, 3,462,382 (1969).
17. M. Minami, Jap. Pat. 73 33736 (1973).
18. I. Tsutomu, M. Mitsuki, Y. Keisaburo and Y. Terihiro, Jap. Pat. Publn., 06-157428 (194).
19. Y. Kodama, M. Mori, N. Ngai and M. Kawaguchi, US Patent, 6,737,503 (2004).

# 4 Special Compounds

## 4.1 Flavans and Spirodichromans

### 4.1.1 Resorcinol-Ketone Reactions

The reaction of resorcinol with ketones is completely different from that of aldehydes. Ketones and aldehydes look similar in most of their properties. By comparing their molecular structures, a hydrogen atom is attached to the carbonyl group in aldehydes (–CHO), and two organic groups (namely, an alkyl, aryl or alkyl and aryl) are attached to the carbonyl group in ketones ($R_1R_2{-}C = O$). Generally, aldehydes are more reactive than ketones towards nucleophilic addition reactions.

The resorcinol molecule can undergo electrophilic substitution reactions. The presence of two hydroxyl groups at the *meta*-positions powerfully activates the benzene ring towards electrophilic substitutions at the *ortho*- and *para*-positions of resorcinol. Therefore, the reaction of resorcinol with various ketones can take place readily, and is expected to be faster when compared to phenol and its derivatives.

Resorcinol-ketone reactions are generally catalyzed by acids. In the presence of an acid, the hydrogen ion becomes attached to the carbonyl oxygen. This lowers the activation energy for the nucleophilic attack at the carbon atom of the carbonyl group. The nucleophilic addition to ketones, in the presence of an acid catalyst, can be represented as below (Figure 4.1).

With resorcinol, the reaction of ketones does not stop at the tertiary alcohol formation stage. The reaction may continue until the reactive positions of

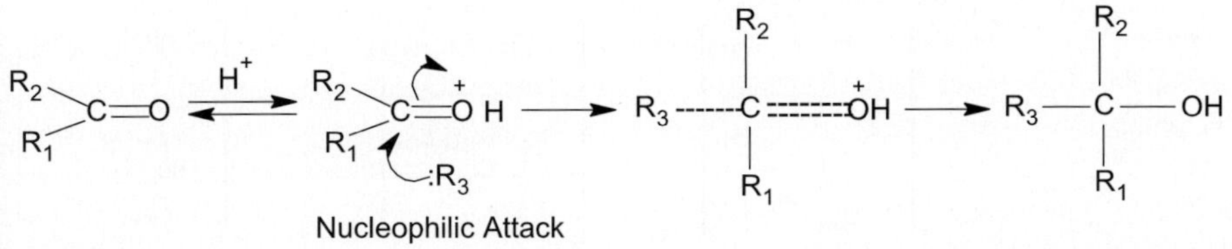


**Figure 4.1.**

resorcinol are consumed or exhausted. On further reaction, the tertiary alcohol condenses with other resorcinol or substituted resorcinol molecules to form the isopropylidene bridged resorcinols.

In this chapter, the reaction of resorcinol with various ketones including aliphatic, cycloaliphatic and unsaturated ketones, has been discussed.

### 4.1.2 Reaction of Resorcinol with Acetone

Resorcinol and acetone are well-known organic compounds, and have been used extensively in the development of various kinds of derivatives for numerous applications. Several studies have shown that, on reaction with resorcinol, acetone has formed products having different chemical structures [1–3]. Depending upon the synthetic procedures employed, acetone can react with resorcinol and produce compounds having different melting points.

The condensation reaction of resorcinol with acetone was primarily carried out in the presence of hydrochloric acid catalyst [4]. In one study, resorcinol and acetone were mixed first in water and then heated to 70 °C before the addition of concentrated hydrochloric acid was achieved [4]. After this, the reaction mixture was stirred for 4 to 8 h. White powdery materials obtained from the resorcinol-acetone reactions were analyzed for their melting points and molecular weights. The details are presented in Table 4.1.

**Table 4.1.** Physical properties of resorcinol-acetone reaction products

| Expt. no. | Resorcinol (mole) | Acetone (mole) | M. point (°C) | Molecular weight |
|---|---|---|---|---|
| 1 | 1 | 1 | 108 – 110 | 1200 |
| 2 | 1 | 2 | 175 – 180 | 1700 |
| 3 | 1 | 3 | 218 – 220 | 3700 |
| 4 | 1 | 4 | 210 – 220 | 3000 |
| 5 | 1 | 8 | 200 – 205 | 3600 |

Data from [4]

A mechanism proposed for the formation of resorcinol-acetone reaction product and its structure is given below (Figure 4.2) [4].

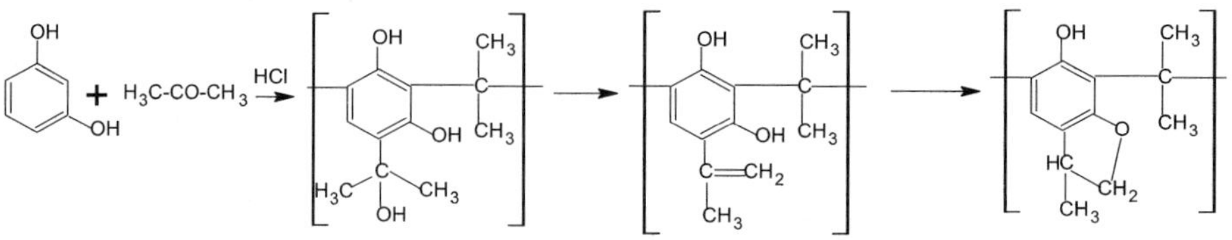


**Figure 4.2.** Proposed structure for the resorcinol-acetone reaction product

From the results in Table 4.1, the melting points and molecular weights were observed to increase when the acetone molar ratios per mole of resorcinol employed were increased, and reached a maximum at 3.0 moles of acetone. The proposed structure for the acetone-resorcinol reaction product indicated the formation of a resinous type and not a crystalline compound from the reaction. The apparent melting points determined for the isolated products were primarily for the polymeric structures and not necessarily for a simple molecular compound.

Later on, a flavans-type structure (Figure 4.3) was proposed for the resorcinol and acetone reaction. This compound was conveniently prepared from the procedure outlined below [5,6].

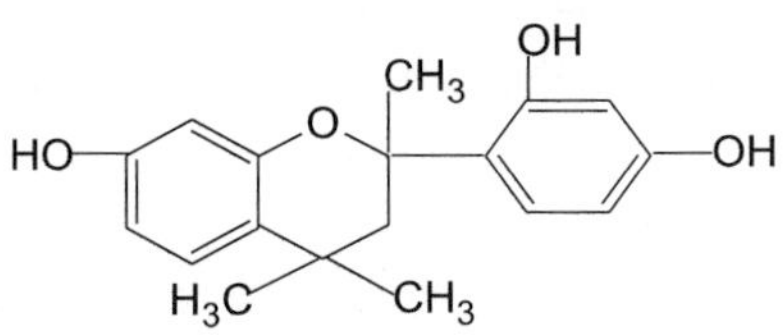


2, 4, 4 - Trimethyl-7, 2', 4'-trihydroxy Flavan

**Figure 4.3.**

**Procedure for the Preparation of 2,4,4-Trimethyl-7,2′,4′-trihydroxy Flavan [5]** By weight, 550 parts of resorcinol were dissolved in 550 parts by weight of water; 145 parts by weight of acetone (molar ratio of acetone to resorcinol = 1 : 2) and 94 parts by weight of 37% hydrochloric acid were added to this solution. The mixture was then allowed to stand at room temperature for 48 h, when crystalline slurry was separated off. The crystals were then filtered off and washed with distilled water and dried in vacuo at 70 °C. The crystalline powder was white and had a melting point of 225 °C. The molecular weights determined were 285 and 290 (calculated for $C_{18}H_{20}O_4 = 300$). The hydroxyl number was found to be 1.02 eq./100 g (theoretical value for 2,4,4-trimethyl-7,2′,4′-trihydroxy flavan = 1.0). The quantity of the product was 345 parts by weight (yield calculated on resorcinol).

In order to determine the structure of the resorcinol and acetone reaction product, a detailed study was made and reported [7]. In this work, resorcinol and acetone were reacted in the presence of hydrochloric acid having different concentrations. A synthetic procedure employed is given below [7].

**Procedure for the Synthesis of 2,4,4-Trimethyl-7,2′,4′-trihydroxy Flavan** Resorcinol (2.063 g, 18.74 mmol), acetone (0.362 g, 6.23 mmole) and 2 ml HCl (10 or 36%) were stirred and refluxed in a mixture of 30 ml of ether and 30 ml of methylene chloride for 12 h. After the removal of solvents, the reaction mixture was treated with water and the precipitate thus formed was collected by filtration, washed thoroughly with water, and dried at 80 °C overnight.

The product mixtures were analyzed by the HPLC on a C18 column using acetonitrile:water at 65 : 35, with UV detection at 280 nm. Structural characterizations were performed using $^1$H NMR and $^{13}$C NMR and X-ray crystal structure analysis methods. Results obtained from these experiments are presented in Table 4.2.

**Table 4.2.** Synthesis of 2,4,4-trimethyl-7,2′,4′-trihydroxy flavan

| Expt. no. | Resorcinol (mole) | Acetone (mole) | HCl (mole) | HCl (conc) (%) | % of Mixtures Str. 1 | Str. 2 | Resorcinol |
|---|---|---|---|---|---|---|---|
| 1 | 6 | 1 | 2 | 10 | 93 | 3.9 | 3.1 |
| 2 | 6 | 1 | 1 | 10 | 99.8 | ND | 0.2 |
| 3 | 3 | 1 | 1 | 10 | 99 | 0.7 | 0.2 |
| 4 | 2 | 1 | 1 | 10 | 99.4 | ND | 0.6 |
| 5 | 6 | 1 | 1 | 36 | 99.6 | ND | 0.4 |

Structure 1 = Trihydroxy flavan. Structure 2 = Spirodichroman
ND = None detected (or) not measured
Data from [8]

Experimental results have confirmed the structure of resorcinol and acetone reaction product as 2,4,4-trimethyl-7,2′,4′-trihydroxy flavan (Figure 4.4).

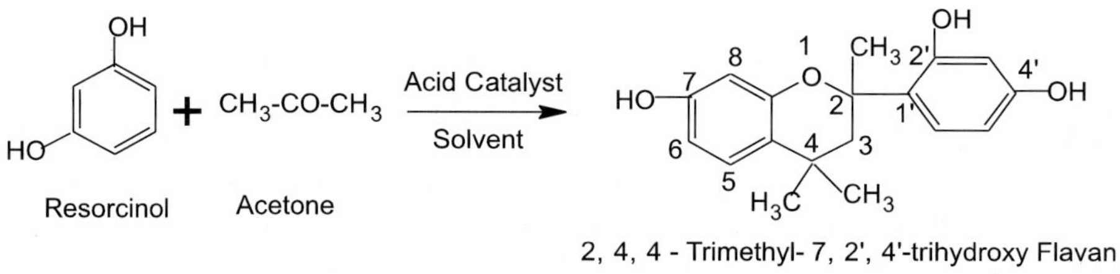


**Figure 4.4.**

High purity flavan was obtained when the resorcinol molar ratio used in the reaction was greater than two per mole of acetone. The formation of spirodichroman was minimal at higher resorcinol concentration in the reaction mixture.

The production of spirodichroman and other higher molecular weight products was increased when the acetone molar ratio was increased as shown in Table 4.3.

From earlier study results, no mechanism was proposed for the formation of flavan and spirodichroman structures from the resorcinol and acetone reaction. It was believed that these structures might have been formed from the intermediate structures of mesityl oxide or phorone that could be formed from acetone.

Studies have shown that aliphatic ketones, such as acetone, can be self condensed in the presence of an acid catalyst to form di- or tri-nuclear con-

**Table 4.3.** Synthesis of flavans-effect of acetone concentrations

| Expt. no. | Resorcinol (mole) | Acetone (mole) | HCl (mole) | HCl (conc) (%) | % of Mixtures Str. 1 | Str. 2 | Resor-cinol | Higher m. wts. |
|---|---|---|---|---|---|---|---|---|
| 1 | 1 | 1 | 1 | 10 | 49.8 | 29.5 | 10.8 | 9.9 |
| 2 | 1 | 2 | 1 | 10 | 6.9 | 58.7 | 26.2 | 7.3 |
| 3 | 1 | 3 | 1 | 10 | 17.4 | 48.6 | 21.2 | 12.8 |
| 4 | 1 | 6 | 1 | 36 | 22.8 | 13.5 | 3.8 | 60.2 |

Structure 1 = Trihydroxy flavan. Structure 2 = Spirodichroman
Data from [7]

densates. With acetone, for example, this can be condensed to form mesityl oxide or phorone based on the following reactions (Figure 4.5).

$$2\ CH_3\text{-}CO\text{-}CH_3 \xrightarrow{H^+} CH_3\text{-}CO\text{-}CH{=}C\text{-}(CH_3)_2$$

Acetone — Mesityl Oxide

$$3\ CH_3\text{-}CO\text{-}CH_3 \xrightarrow{H^+} (CH_3)\text{-}C{=}CH\text{-}CO\text{-}CH{=}C\text{-}(CH_3)_2$$

Phorone

**Figure 4.5.**

Two acetone molecules can condense to produce the mesityl oxide compound having a carbonyl and unsaturated carbon groups in the molecule. Similarly, the phorone structure could be produced by the condensation of three molecules of acetone, and contains two double bonds and a keto group. The double bonds of the unsaturated ketones are activated by the adjacent keto group and therefore, a reaction between resorcinol and the double could be faster. The reaction of resorcinol with unsaturated aliphatic compounds is well known in the literature. The formation of flavans and spirodichromans was reported from the reactions of mesityl oxide and phorone compounds with resorcinol [8,9].

### 4.1.3 Resorcinol-Unsaturated Ketones Reaction

Unsaturated aliphatic ketones having an adjacent carbonyl and carbon-carbon double bond are activated by each other, and therefore they are considered as highly reactive molecules. Due to their high reactivity, they can be expected to react readily with reactive molecule like resorcinol. With mono-unsaturated ketones, in the presence of an acid catalyst, resorcinol produces chemical compounds predominantly having the flavan structure (Figure 4.6) [10].

Flavan compounds prepared using different mono-unsaturated ketone derivatives and their physical properties are presented in Table 4.4.

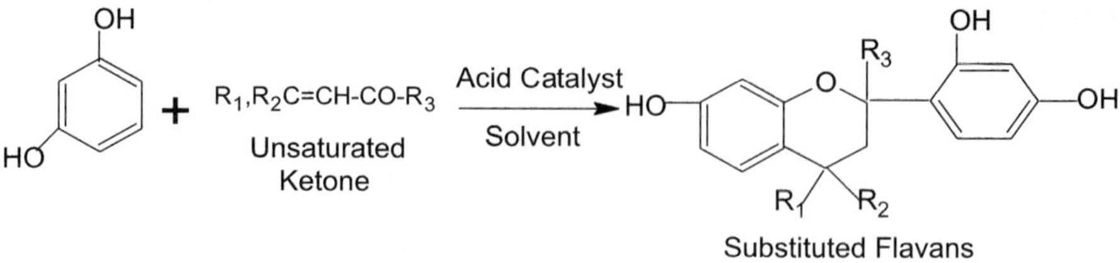


**Figure 4.6.**

**Table 4.4.** Synthesis of flavans from resorcinol and mono-unsaturated ketones

| Ketone used | $R_1$ | $R_2$ | $R_3$ | Yield (%) | Melting point (°C) |
|---|---|---|---|---|---|
| 1. Mesityl oxide | $CH_3$ | $CH_3$ | $CH_3$ | 91 | 231 – 232 |
| 2. Methyl vinylketone | H | H | $CH_3$ | 78 | 174 – 76 |
| 3. Ethyl vinylketone | H | H | $CH_2CH_3$ | 85 | 162 – 168 |
| 4. 4-Phenyl-3-butene-2-one | H | Ph | $CH_3$ | 90 | ND |

ND = Not determined, Ph = Phenyl group
Data from [10]

Di-unsaturated ketones, such as phorone and dibenzylidene acetone, produced compounds having spirodichroman-type structures. These compounds can be formed from the following reaction scheme (Figure 4.7) [11].

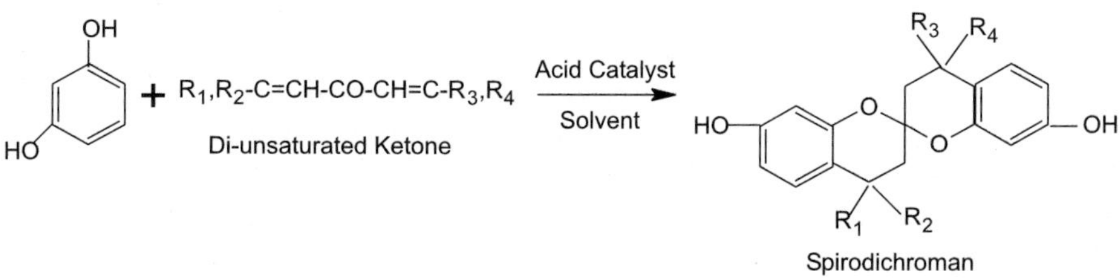


**Figure 4.7.**

In general the yields were higher. The spirodichroman compounds exhibited higher melting points (Table 4.5).

## 4.1.4 Resorcinol-Cyclic Ketones Reaction

Cyclic ketones are known to react with resorcinol, and the reactions are similar to those of aliphatic ketones. Both saturated and unsaturated cyclic ketones have been known to react with resorcinol and produce compounds having the flavan-type structures [10–12].

**Table 4.5.** Synthesis of spirodichromans from di-unsaturated ketones and resorcinol

| Expt. no. | Ketone used | $R_1$ | $R_2$ | $R_3$ | $R_4$ | Yield (%) | Melting point (°C) |
|---|---|---|---|---|---|---|---|
| 1 | Phorone | $CH_3$ | $CH_3$ | $CH_3$ | $CH_3$ | 85 | 199–200 |
| 2 | Dibenzylidene acetone | H | Ph | H | Ph | 76 | 270–283 |

Ph = Phenyl Group
Data from [10]

The reaction of resorcinol with cyclopentanone produced a flavan as shown in Figure 4.8. This derivative was very difficult to purify. Therefore, a triacetate derivative was synthesized from this flavan compound, and confirmed the molecular structure of resorcinol-cyclopentanone reaction product (Figure 4.8).

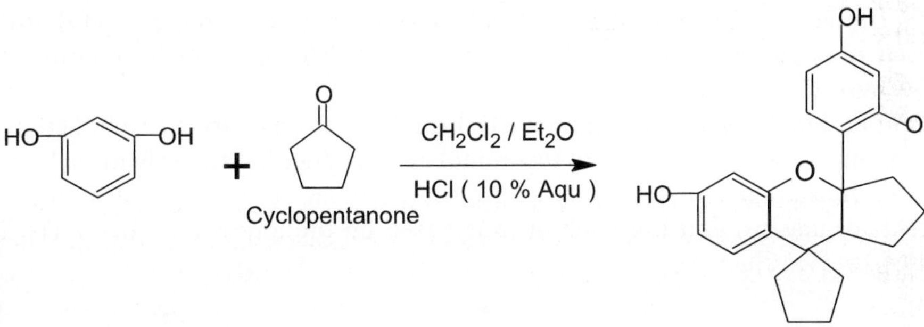


**Figure 4.8.**

Cyclohexanone also produced a flavan-type reaction product with resorcinol. A synthetic scheme describing the preparation of this compound is outlined in Figure 4.9.

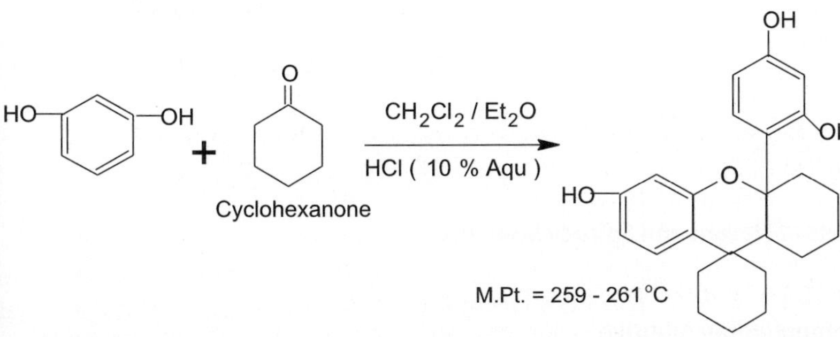


**Figure 4.9.**

The preparative procedure used for the synthesis of trihydroxy flavan is given below [11].

**Preparation of Trihydroxy Flavan from Resorcinol and Cyclohexanone [11]** In a 500 ml four-necked flask, resorcinol (88 g), water (241.4 g) and 36% hydrochloric acid (22.6 g) were charged and heated to 50 to 55 °C to obtain a homogeneous solution. To the solution, cyclohexanone (19.6 g) was added dropwise over 60 min, and the mixture was stirred at 50 to 55 °C for 6 h. After cooling to room temperature, a precipitated cake was recovered by filtration and washed with water (2 l). The wet cake was dissolved in ethyl acetate (500 g) and washed with water (500 g) till an aqueous phase became neutral.

After separation, ethyl acetate was evaporated off under reduced pressure from the aqueous phase to obtain a residue (75.2 g). To the residue, toluene (75 g) was added and heated to 80 °C to form a homogeneous solution. After cooling to 25 °C., the mixture was cooled with ice and kept at a temperature not higher than 5 °C for 1 h, followed by filtration.

A wet cake was washed with toluene (30 g). The washed cake was added to toluene (48 g) and heated to 80 °C followed by stirring for 1 h. Then the mixture was cooled to 25 °C and filtered. A cake was washed with toluene (30 g) and dried to obtain a white crystal compound (27.6 g) having a melting point of 259–261 °C.

Similar to the reaction of monocyclic ketones, $\alpha,\beta$-unsaturated cyclic ketone compounds can also react with resorcinol and produce flavan derivatives [10]. 2-Cyclohexenone and 4,4-dimethyl-2-cyclohexenone compounds gave higher yields of flavans, and the reaction was based on the following scheme (Figure 4.10) [7, 10].

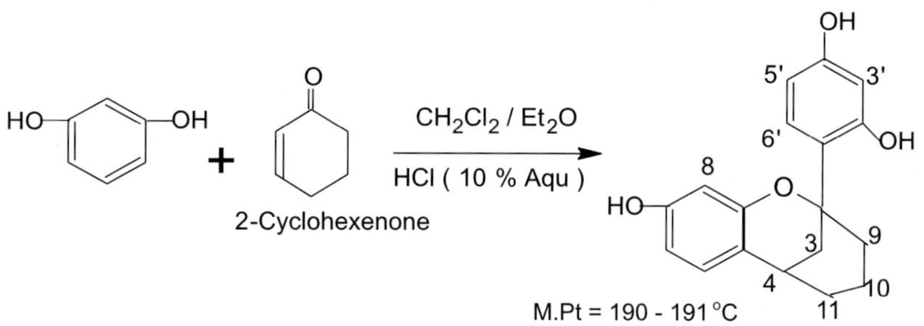


**Figure 4.10.**

## 4.1.5 Applications of Flavans and Spirodichromans

### 4.1.5.1 Rubber Compounding Additive

The trihydroxy-flavan compound made from the resorcinol-acetone reaction was used as the methylene acceptor in rubber compositions. When used with a methylene donor and a cobalt salt, the flavan derivative has shown to improve the vulcanized rubber compound properties such as the hardness, processing and provided good adhesion to reinforcements. This compound was determined to be a non-fuming material [13, 14].

### 4.1.5.2
### Curing Agents for Epoxy Resins

Trihydroxy flavans were partially modified with acetic anhydride and benzoyl chloride to obtain esterified compounds as shown below (Figure 4.11) [15–17].

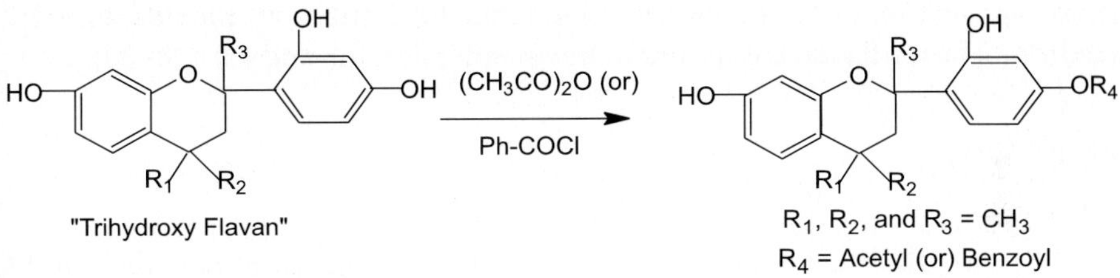


**Figure 4.11.**

When these esterified flavans were used as the curing agents, the cured products exhibited low dielectric constant and low dielectric loss tangent properties. Due to excellent dielectric properties, copper clad laminates prepared from the esterified flavans and epoxy resin systems were found to be suitable for the multi-layered printed wiring boards for high speed operations, especially in the high frequency region.

A partially allylated flavan compound was also prepared by employing an allyl chloride in the following reaction (Figure 4.12).

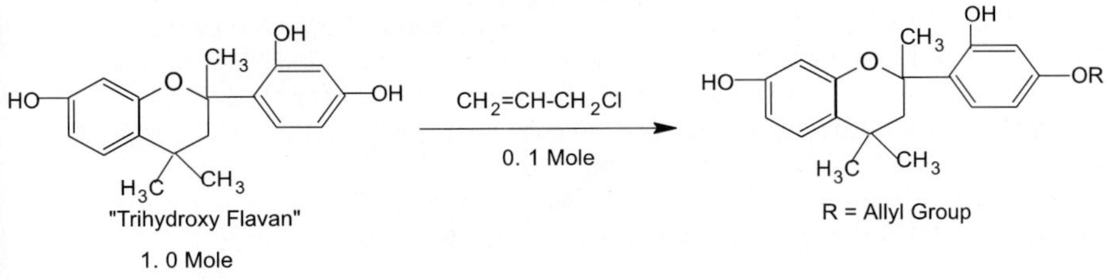


**Figure 4.12.**

This allylated compound was blended with an epoxy resin, silica and other additives, and then transfer molded and post cured. The cured product showed a glass transition temperature (Tg) of 143 °C and water absorption of 0.24% after 72 h at 85 °C and 85% RH conditions [21].

### 4.1.5.3 Photoresist Material

In the manufacture of integrated circuits, positive resist compositions containing a novolak resin and quinone diazide compounds have been widely used because of their excellent resolution. The positive resists exhibiting higher resolution and excellent patterning are required to form sub-micron patterns for the advanced integrated circuits.

Positive resists with improved profiles and undiminished sensitivity and resolution properties were prepared with the use of sensitizer compounds prepared from the reaction of 1,2-naphthoquinone diazide-5-sulfonyl chloride with the resorcinol based flavan compounds shown in Figures 4.3 and 4.9 [18–20].

## References

1. R.N. Shen and M.J. Quadrat-I-Khuda, J. Indian Chem. Soc, Vol. 7, p 167 (1930).
2. R.N. Shen, N.C. Chattopadhya and S.C. Sen-Gupta, J. Indian Chem. Soc, Vol. 7, p 997 (1930).
3. H.C. Malhotra and S. Banerjee, J. Indian Chem. Soc, Vol. 67, p 117 (1990).
4. C.L. Segal and J.B. Rust, Ind. Eng. Chem, Vol. 52, p 324 (1960).
5. British Patent, 822,659 (1959).
6. P. Bruin and A. Klootwijk, US Patent, 2,947,760 (1960).
7. P. Livant, T.R. Webb and W. Xu, J. Org. Chem, Vol. 62, p 737 (1997).
8. K. Hirotoshi, N. Sunao and K. Tomoyoki, Jap. Patent, 2002193970 (2002).
9. H. Haruhisa et al., Jap. Patent, 62, 111988 (1987).
10. P. Livant and W. Xu, J. Org. Chem, Vol. 63, p 636 (1998).
11. Y. Uetani, H. Nakanishi and Y. Doi, US Patent, 5,283,155 (1994).
12. Y. Uetani and H. Nakanishi, US Patent, 5,290,657 (1994).
13. N. Inui, H. Iyama, K. Tsuta and H.H. Nagasaki, US Patent, 5,665,799 (1997).
14. N. Inui, H. Iyama, K. Tsuta and H.H. Nagasaki, US Patent, 5,688,871 (1997).
15. Y. Ueda, Y. Endo, M. Shibata and K. Yamasaki, US Patent, 5,916,683 (1999).
16. Y. Ueda, Y. Endo, M. Shibata and K. Yamasaki, US Patent, 5,726,257 (1998).
17. Y. Ueda, Y. Endo, M. Shibata and K. Yamasaki, US Patent, 6,469,109 (2002).
18. Y. Uetani, H. Naganishi and Y. Doi, US Patent, 5,283,155 (1994).
19. Y. Uetani and H. Naganishi, US Patent, 5,290,657 (1994).
20. Y. Uetani, J. Tomioka and S. Lee, US Patent, 6,383,709 (2002).
21. Y. Endo and T. Hayashi, Japanese Patent, 1,149,766 (1999).

## 4.2 Coumarins and Chromanones

### 4.2.1 Introduction

Resorcinol can react with $\alpha$, $\beta$-unsaturated acids and esters and produce compounds with two distinctive chemical structures. With $\alpha$, $\beta$-unsaturated acids, chromanone-type products are formed. On the other hand, $\alpha$, $\beta$-unsaturated esters can produce coumarin-type compounds with resorcinol. Similarly, compounds such as the $\beta$-keto esters can undergo Pechmann-type reactions with resorcinol under acidic conditions and thereby, forming coumarins of various structures.

Coumarins and chromanones are an important group of organic compounds that have been used widely in various food, pharmaceutical and agricultural applications [1–3]. The reactions of resorcinol with different unsaturated and ketonic acids and esters, and the formation of different chromanone and coumarin-type products from these reactions are discussed in this chapter. With the use of resorcinol-based derivatives and compounds, a large number of coumarin and chromanone compounds were developed. These compounds have been used as additives or active ingredients in food and cosmetics, optical brightening agents, dispersed fluorescent and laser dyes, medical drugs and insecticides.

### 4.2.2 Reaction of Resorcinol with $\alpha$, $\beta$-Unsaturated Carboxylic Acids

Unsaturated carboxylic acids, such as 3,3′-dimethyl acrylic acid and 3-phenyl acrylic acid can react with resorcinol under strong acidic conditions to produce chromanone-type compounds based on the following reactions (Figures 4.13 and 4.14) [4–7].

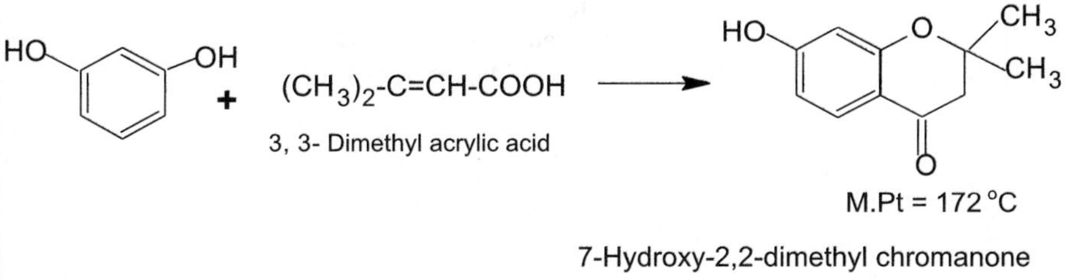


**Figure 4.13.**

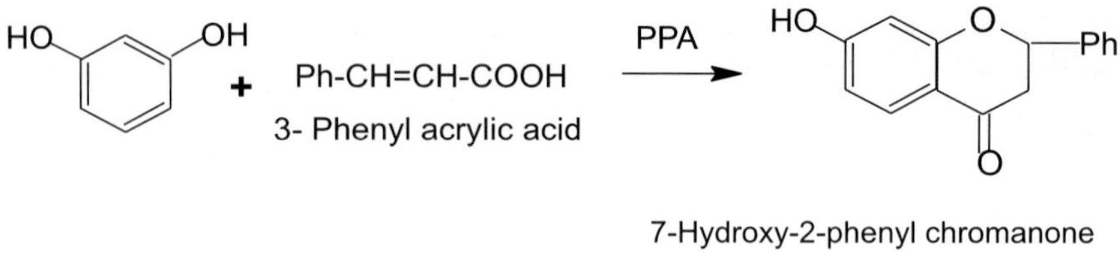


**Figure 4.14.**

The formation of these compounds might have resulted from the reaction of carboxylic acids with resorcinol to form keto substituted products followed by a cyclization reaction. The mechanism of chromanone formation can be illustrated in the scheme shown below (Figure 4.15).

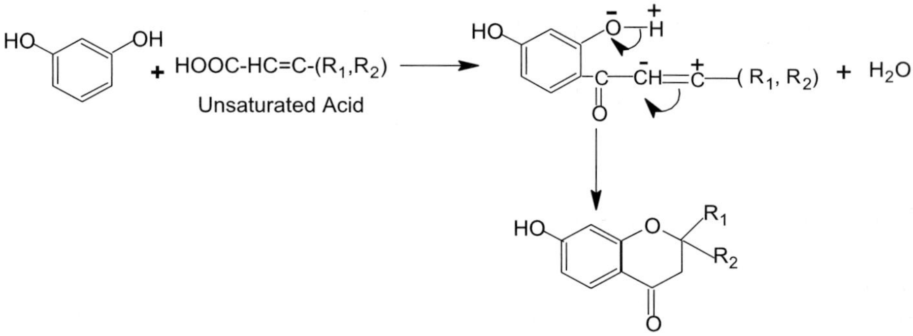


**Figure 4.15.**

Chromanone derivatives have exhibited pharmaceutical activities and therefore are more useful in the manufacture of drug compounds.

### 4.2.3 Reaction with $\alpha,\beta$-Unsaturated Carboxylic Acid Esters

With $\alpha,\beta$-unsaturated esters, resorcinol reacts and produces chromone-type compounds in the presence of strong acid catalysts (Figure 4.16) [8].

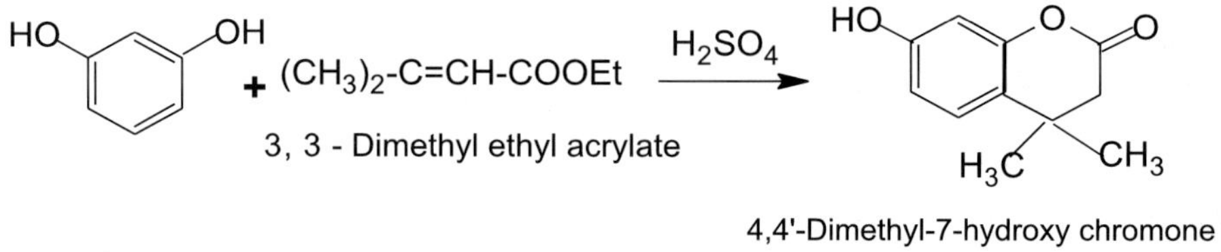


**Figure 4.16.**

From the above reaction, the unsaturated group of the ester molecule reacts first with the resorcinol to produce the alkyl substituted compound before the esterification and cyclization reaction can take place. Chemical derivatives having chromone structures could be conveniently obtained with the use of $\alpha,\beta$-unsaturated esters in the reactions.

### 4.2.4 Hydroxycoumarins from Resorcinol

Reactions of resorcinol with keto acids and hydroxy acids to form umbelliferones or coumarins have been well known for a long time. The derivatives of coumarins are commonly present as secondary metabolites in seed, roots and leaves of many plant species. Their functions in the plant are thought to be as plant growth regulators, fungistats and bacteristats [9]. Coumarins are known to possess antibacterial and antifungal properties [10, 11]. Coumarins synthesized from resorcinol and its derivatives have been utilized in numerous applications.

### 4.2.5 Reaction of Resorcinol with Malic Acid

Umbelliferone or 7-hydroxy coumarin has been synthesized by the reaction of resorcinol with malic acid in the presence of sulfuric acid. This process has been long known as the Pechmann reaction (Figure 4.17).

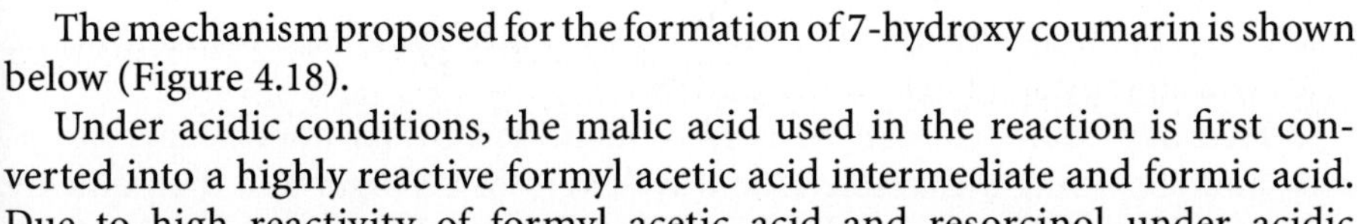


**Figure 4.17.**

The mechanism proposed for the formation of 7-hydroxy coumarin is shown below (Figure 4.18).

Under acidic conditions, the malic acid used in the reaction is first converted into a highly reactive formyl acetic acid intermediate and formic acid. Due to high reactivity of formyl acetic acid and resorcinol under acidic medium, both could react faster and condense to form the second ring of the 7-hydroxy coumarin. A preparative procedure employed to synthesize 7-hydroxy coumarin from resorcinol is given below.

**Preparation of 7-Hydroxycoumarin [12]** To a 200-gallon reactor, there was fed 635 lb (6.3 mol) of 98% sulfuric acid to an agitated heel of 17.0 lb (1.55 mol) of resorcinol, 230 lb (1.72 mol) malic acid and 86 lb (0.70 mol) nitrobenzene

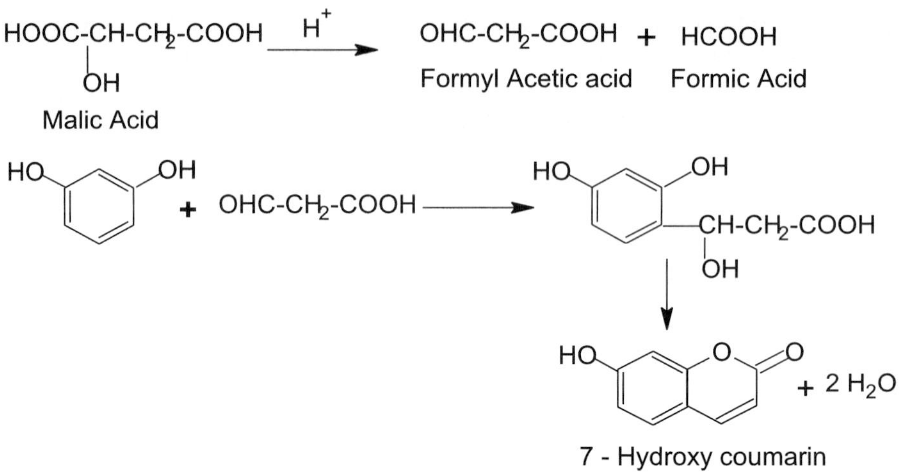


**Figure 4.18.**

at about 100 °C. The acid was added over 5 min, the temperature rising to 106 °C. After 12 min from the start of acid addition, the temperature peaked at 118 °C and steam was applied to the reactor jacket to hold the temperature at 115–118 °C. After 40 min, from the start of acid addition, the reaction was cooled to 30 °C. The homogeneous reaction mixture (80 gal) was poured into 200 gallons of ice water at a rate to keep the drowned liquor below 30 °C. The 7-hydroxycoumarin crystals settled and the spent acid was decanted. Fresh water (160 gal) was added to the settled crystals and decantation repeated. The 7-hydroxycoumarin crystals were heated to 100 °C in the presence of water to remove nitrobenzene by steam distillation, filtered and vacuum dried at 80 °C and 120 torr to give 112 lb of 7-hydroxycoumarin (45% yield based on resorcinol).

Umbelliferone is now available from various commercial sources. This derivative appeared as a white or pale yellow crystalline powder with a melting point in the range between 226–230 °C, which depends upon the purity. This compound is soluble in organic solvents such as alcohol, chloroform and acetic acid, and slightly soluble in water (0.02%).

7-Hydroxy coumarin can be readily identified by its fluorescence property, and absorbing a characteristic wavelength of UV light and then quickly emitting light of longer wavelength. This compound is a pH-sensitive fluorescent indicator. It is highly fluorescent in an alkaline solution (emission wavelength = 455 nm, pKa = 7.7) and the fluorescent intensity is seven times stronger than that of erythrodine [13]. Umbelliferone is a common component of sun-screens and suntan lotions.

### 4.2.6 Reaction with Ethyl Acetoacetate

The most industrially important coumarin compound, namely 7-hydroxy-4-methyl coumarin, is prepared by the Pechmann reaction of resorcinol with ethyl acetoacetate, and is based on the following reaction scheme (Figure 4.19).

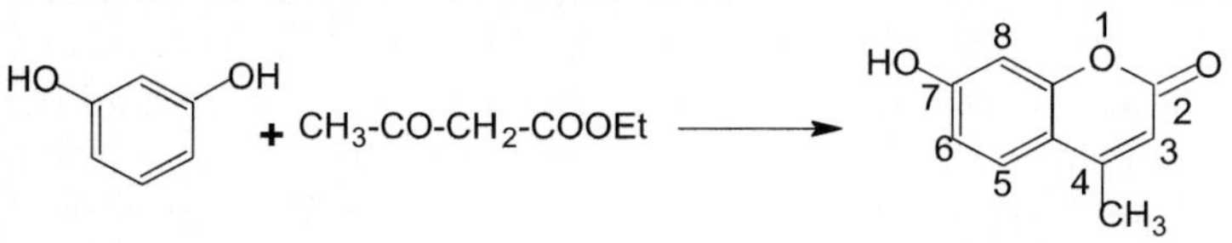

7-Hydroxy -4- methyl Coumarin

**Figure 4.19.**

A mechanism proposed for the formation of 7-hydroxy-4-methyl coumarin (also known as 4-methyl umbelliferone) can be illustrated in the following scheme (Figure 4.20).

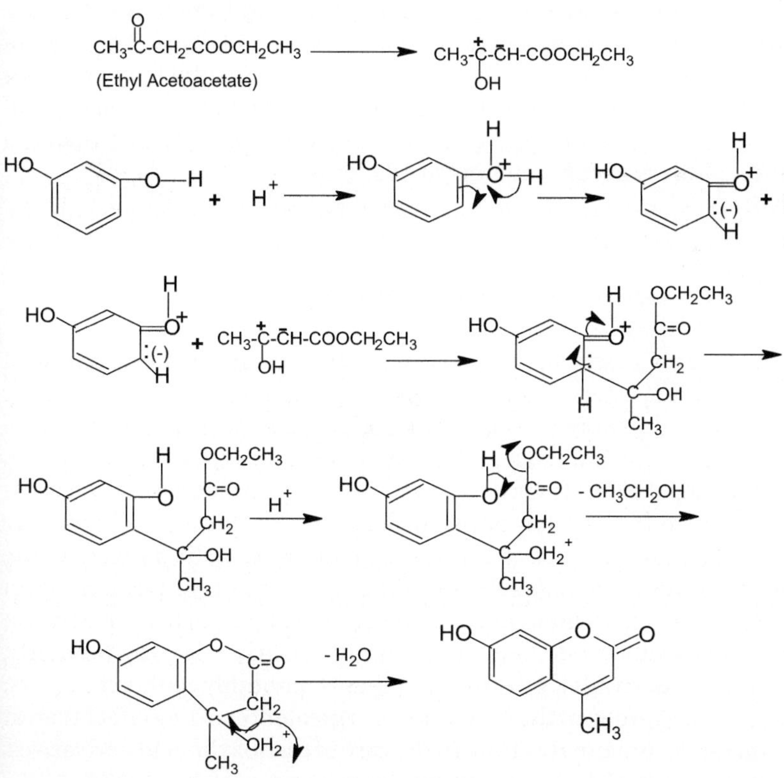

**Figure 4.20.** Mechanism of 7-hydroxy-4-methylcoumarin formation from resorcinol

A published synthesis procedure for the preparation of 4-methyl umbelliferone is given below [15].

**Synthesis of 4-Methyl-7-hydroxycoumarin from Resorcinol [15]** Into a 5-l three necked flask, fitted with a mechanical stirrer, a thermometer reaching to the bottom, and a dropping funnel, is placed 2 l of concentrated sulfuric acid (Sp. Gr. 1.84). The flask is surrounded by an ice bath, and when the temperature falls below 10 °C a solution of 220 g (2 mol) of resorcinol in 260 g (2 mol) of freshly distilled ethyl acetoacetate is added dropwise. The mixture is stirred, and the temperature is kept below 10 °C by means of ice and salt. After all the solution has been added (about 2 h) the reaction mixture is set aside for 12–24 h without further cooling. The reaction mixture is now poured with vigorous stirring into a mixture of 4 kg of ice and 6 l of water. The precipitate is collected on a filter and washed with three 50-ml portions of cold water. The crude product is then dissolved in 3 l of 5% aqueous sodium hydroxide solution, the solution is filtered, and the substituted coumarin is re-precipitated from the filtrate by the slow addition of dilute (1 : 10) sulfuric acid until the solution is acid to litmus. About 1.1 l of dilute sulfuric acid is required. During the neutralization, the reaction mixture must be stirred well. The product is collected on a Buchner funnel, washed with four 50-ml portions of cold water and dried. The yield of 4-methyl-7-hydroxycoumarin is 230–320 g (82–90%). It is sufficiently pure for use in the next step but may be purified by recrystallization from 95% ethanol using about 15 ml of ethanol for 5 g of product. The recrystallized material forms almost colorless needles melting at 185 °C.

## 4.2.7 Substituent Effect of Resorcinol on Pechmann Reaction [14]

From the Pechmann reaction, coumarins are formed from the reaction of resorcinol with $\beta$-keto esters. Their rate of formation depends on the reactivity of resorcinol and its derivatives. Different substituents present in the resorcinol molecule have different influences on the reactivity towards the keto esters and also on the rate of formation of coumarins. Unsubstituted resorcinol is the most reactive compound and may be due to the presence of two equally reactive positions (4 and 6) for the reactions. 4-Substituted resorcinolic derivatives are less reactive than resorcinol. A substitution at the 2-position has less effect on the reactivity of resorcinol. Alkyl substituted resorcinols (2 or 4) react readily with ethyl acetoacetate in the presence of sulfuric acid [16]. Similarly, 4-chloro or bromo derivative of resorcinol reacts smoothly with keto esters. The presence of nitro group at the 4-position has greater retarding effect than at the 2-position for Pechmann reaction. In the case of carboxylic acid derivatives, the ester group, compared to the acid group, at the 4-position enhances the reactivity.

### 4.2.8
### Applications of Hydroxycoumarin and Chromanone Derivatives

Coumarins and chromanones have been utilized in the development and manufacture of various industrially important compounds for different applications. In the following section, some of their chemistries are discussed. In each of these applications, one example of the coumarin derivative, out of various derivatives reported in the respective references, is outlined and discussed.

Geiparvarin compounds having 3(2H)-furanone moiety possess significant tumor-inhibiting properties. Geiparvarin analogue containing fluorinated segment in the structure joining the 3(2H)-furanone and coumarin skeletons has been synthesized from 7-hydroxycoumarin and studied [17] (Figure 4.21).

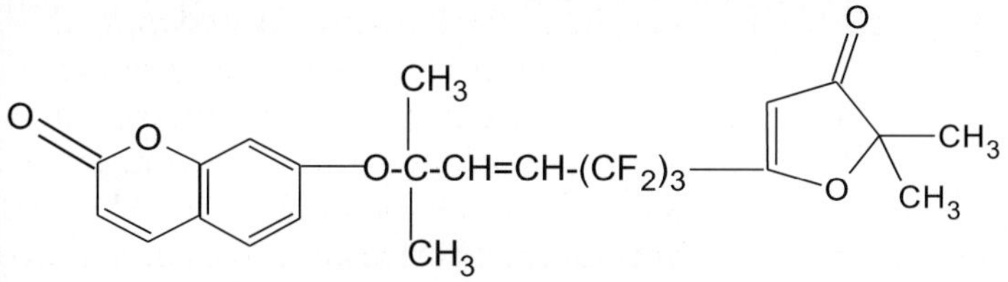


**Figure 4.21.**

The fluorescence property of coumarin compounds depends on the nature of substituents present in the molecules. The strong fluorescence of 7-hydroxycoumarin in aqueous media is completely lost with 7-aryloxycoumarins. The aryloxycoumarins prepared from 7-hydroxycoumarin may be employed as specific fluorescence probes for the hydroxyl radicals in an aqueous environment [18].

Novel derivatives of coumarin useful as medicaments, notably as analgesics, were prepared from 4-methyl-7-hydroxycoumarin. An example of one such compound prepared is shown below (Figure 4.22) [19].

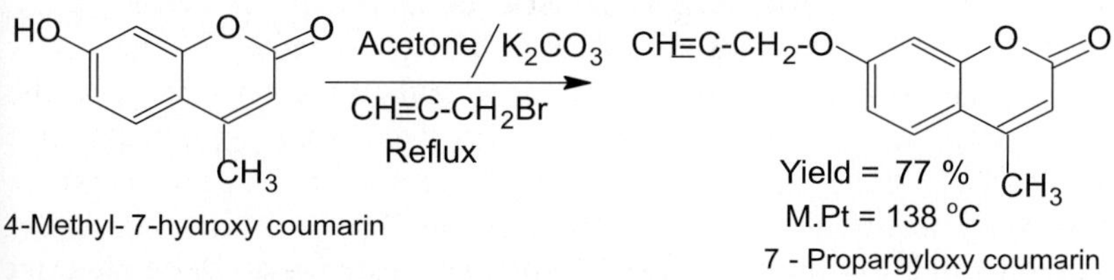


**Figure 4.22.**

Compounds containing the propargyloxy group showed valuable analgesic activity, making them utilizable in the treatment of chronic and acute pain, traumatic algias, visceral and neurologic, dental pains and diverge algias such as headaches and cancerous pains.

The butynoxychromene compounds synthesized from hydroxychromanone were utilized as insect control agents, either as insecticides or anti-juvenile hormones (Figure 4.23) [20].

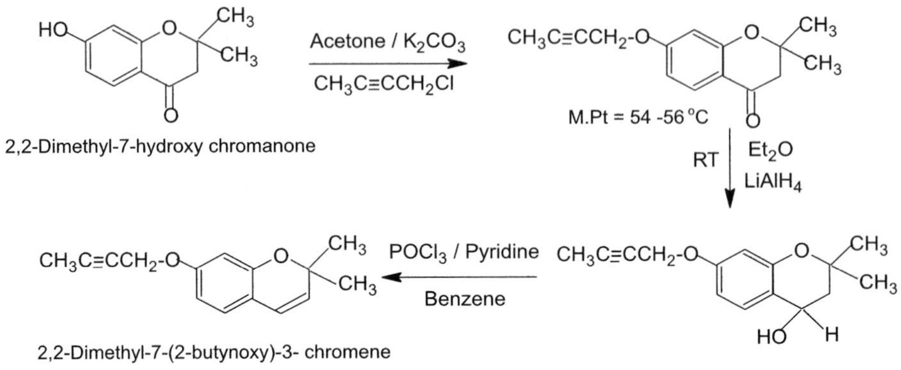


**Figure 4.23.**

These compounds were tested for anti-juvenile hormone activity and insecticidal activity against the milkweed bug, *Oncopeltus fasciatus* (Dallas).

2-Alkyl substituted resorcinol derivatives were used in the synthesis of pyranochromo derivatives as shown below (Figure 4.24) [21].

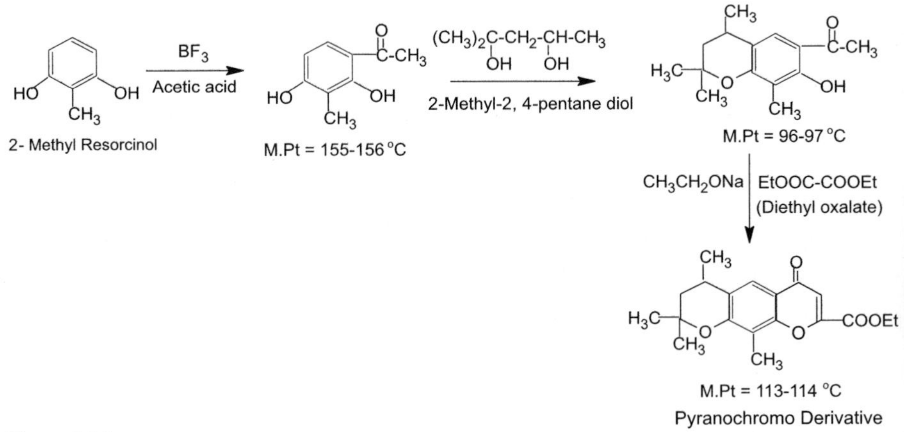


**Figure 4.24.**

Novel medicaments containing these derivatives were found to be effective for the treatment of allergic diseases. They were believed to show the anti-allergic effect in accordance with the same function mechanism as that of cromoglycate.

Monohalogenated resorcinolic compounds were employed in the development of 7-hydroxycoumarin derivatives based on the following scheme (Figure 4.25) [22,23].

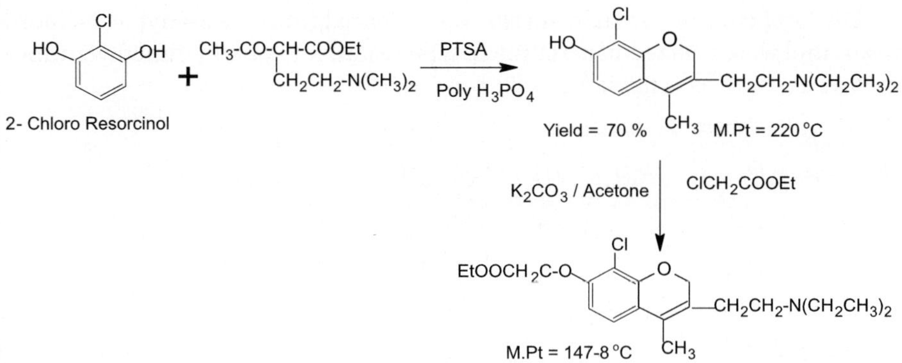


**Figure 4.25.**

The resulting coumarins showed valuable specific coronary vasodilating activity, and in particular when the halo substituent is present at the 8 position of these compounds. These derivatives could be used to exercise a vasodilatory action effectively to coronary vessels in animals and humans.

7-Hyroxycoumarin glucuronide compound, an important metabolite, was synthesized using 7-hydroxycoumarin (Figure 4.26) [24].

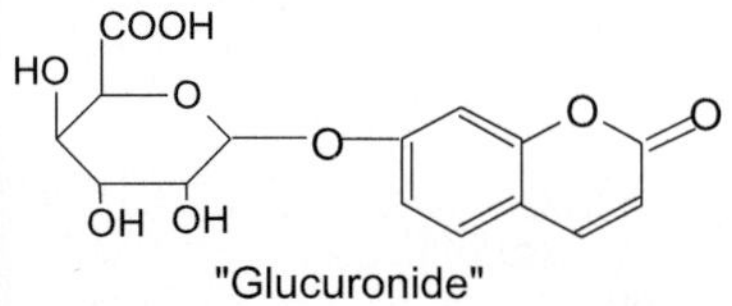


**Figure 4.26.**

3,4-Dimethyl hydroxycoumarin was synthesized from the reaction of resorcinol with 2-methyl ethylacetoacetate and subsequently used in the preparation of sulfonates of hydroxycoumarins (Figure 4.27) [25].

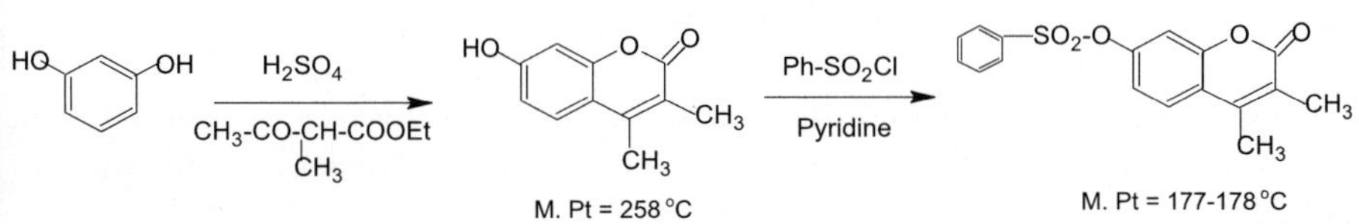


**Figure 4.27.**

Pharmaceutical formulations containing these sulfonate compounds as the active ingredients were found to be useful in the treatment of psychological disturbances, in particular depression and allergies. On the basis of pharmacological findings, these novel substances in appropriate pharmaceutical formulations were useful for the pharmacotherapy of psychological disturbances.

Allyl and chlorocoumarin derivatives prepared from the 2-allyl or 2-chloro resorcinol were utilized in a Mannich-type reaction to obtain dimethylaminomethyl derivatives (Figure 4.28) [26].

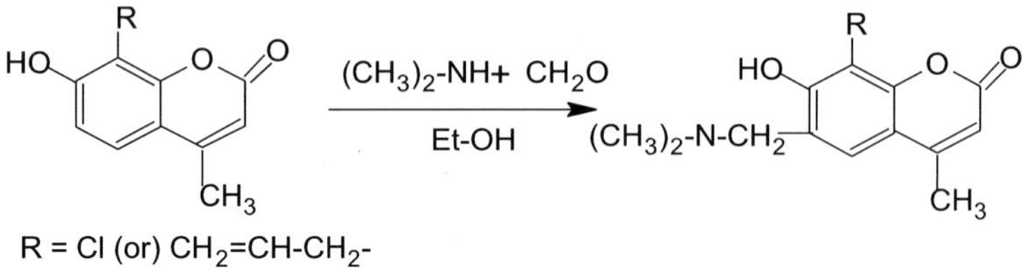


**Figure 4.28.**

These compounds showed antiaggregating properties and were active in inhibiting platelet aggregation. Mannich-type coumarin compounds prepared were proposed for the experimental, diagnostic or therapeutic purposes in veterinary or human medicines and especially in an antithrombotic drug.

Compounds having utility as fluorophores were synthesized from an ether derivative of coumarin based on the following reaction (Figure 4.29) [27].

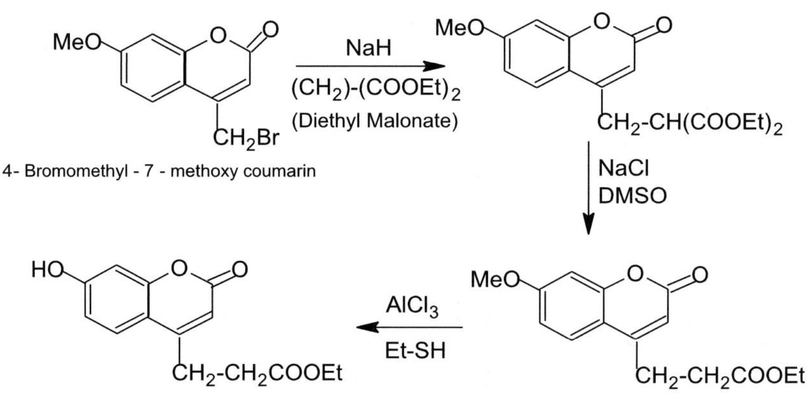


**Figure 4.29.**

The 7-hydoxy-4-methylcoumarins are known to fluoresce. By conjugating the compounds to a biological molecule, the presence or absence of the macromolecule can be quantified. These coumarin compounds may be conjugated to antibodies for the determination of immunoassay. They may also be conjugated to oligonucleotides or other nucleic acid hybridization assays.

Fluorine substituted fluorescent dyes typically possess greater photostability and lower pH sensitivity in the physiological pH range. Fluorescent dyes suitable for biological applications should possess high sensitivity for the detection. Hydroxycoumarins containing the fluorine substitutions at the 6 and 8 positions were used in the preparation of fluorescent dye-conjugates and fluorescent-labeled enzyme substrates (Figure 4.30) [28].

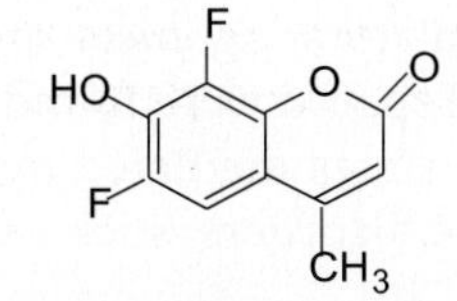


6, 8 - Difluoro Coumarin

**Figure 4.30.**

Benzo dipyranamine compounds were synthesized starting from resorcinol in a multi-step procedure as shown below (Figure 4.31) [29].

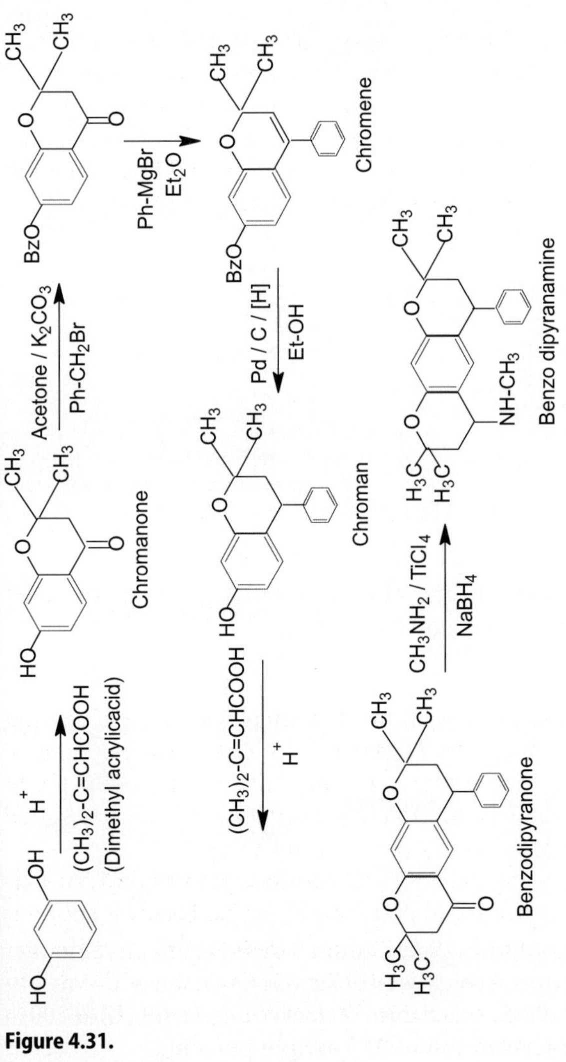


**Figure 4.31.**

Substituted benzo dipyranamines were used as modulators, agonists or antagonists, of the CCR5 receptor [30]. The compounds prepared were found to be utilized in the treatment and prevention of disease states mediated by CCR5 receptor mechanism including the asthma and topic disorders, such as topic dermatitis and allergies.

Chromone derivatives are important in the pharmaceutical and chemical industry for various applications due to their biological activity. Biologically interesting group of aldehydes containing chromone derivatives were developed from resorcinol and acetyl resorcinol (Figures 4.32 and 4.33) [31].

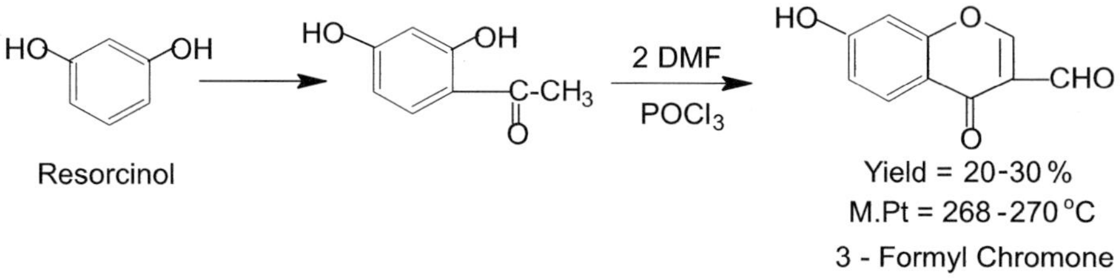


**Figure 4.32.**

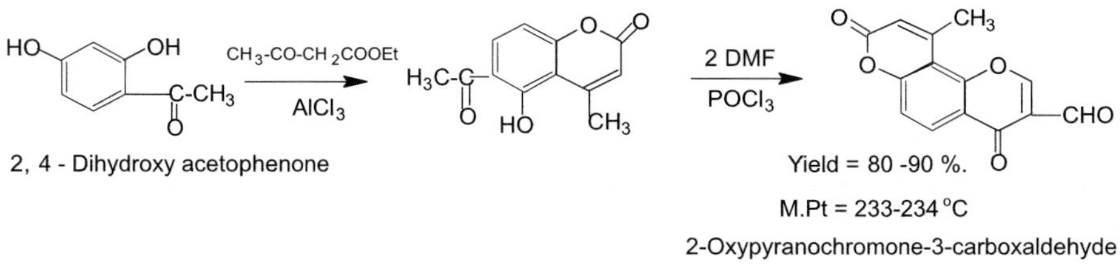


**Figure 4.33.**

Dialkyl phosphinothioates having insecticidal activity were synthesized from 4-methyl-7-hydroxy coumarin (Figure 4.34) [32].

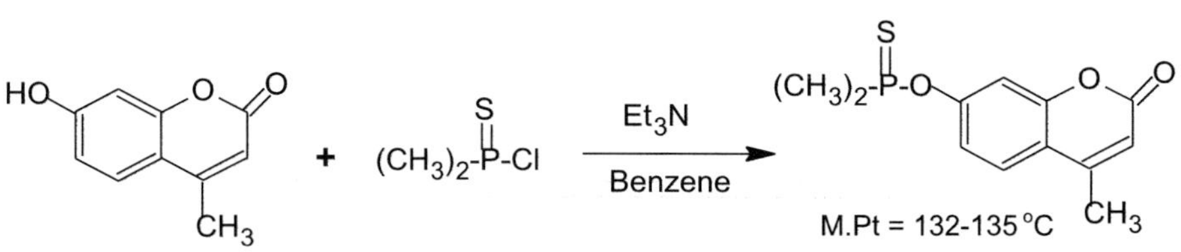


**Figure 4.34.**

Insecticidal compositions containing these coumarins were effective against a variety of insect pests. Systemic activity was observed against a wide variety of insects, for example two-spotted spider mite, *Tetranychustelarius* (L), a 100% kill was observed even at a concentration of 0.01 weight percent.

Introduction of a closed ring methylene chain between the 3 and 4 positions of coumarins resulted in compounds of high insecticidal activity coupled with low toxicity towards warm blooded animals (Figure 4.35) [33].

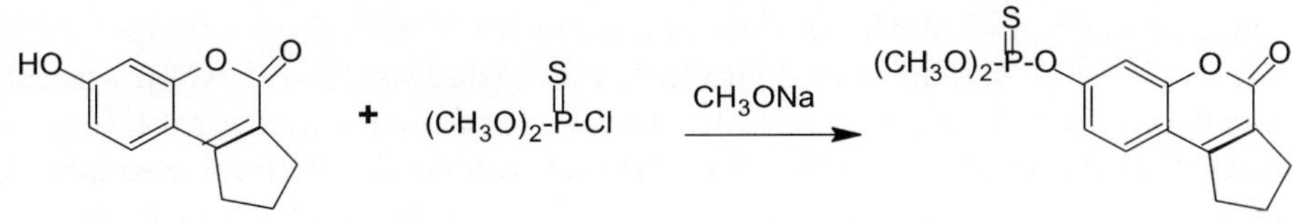


3,4-Trimethylene-7-hydroxycoumarin

**Figure 4.35.**

Compounds prepared were found to be useful in the control of houseflies.

Organophosphorus compounds have been widely used in agriculture as pesticides for the control of insect pests (insecticides), acarids (acaricides), nematodes (nematocides), plant diseases (fungicides) and weeds (herbicides). These pesticides are used to control vermin of fruit trees (apples, pear, plum), and vegetable crops. They are also used to protect tree nurseries and to destroy flies and parasites of domestic animals. Some organophosphorus pesticides have strong biological activity not only for insects and acarids but also warm-blooded animals and humans.

A well-known coumarin-based organophosphorus compound, sold under the trade names Coumaphos and Co-ral, has been used as an insecticide and acaricide. This compound was synthesized in a two-step procedure and the starting material for the synthesis was resorcinol. A synthetic scheme outlining the preparation of this compound is shown below (Figure 4.36) [34, 35].

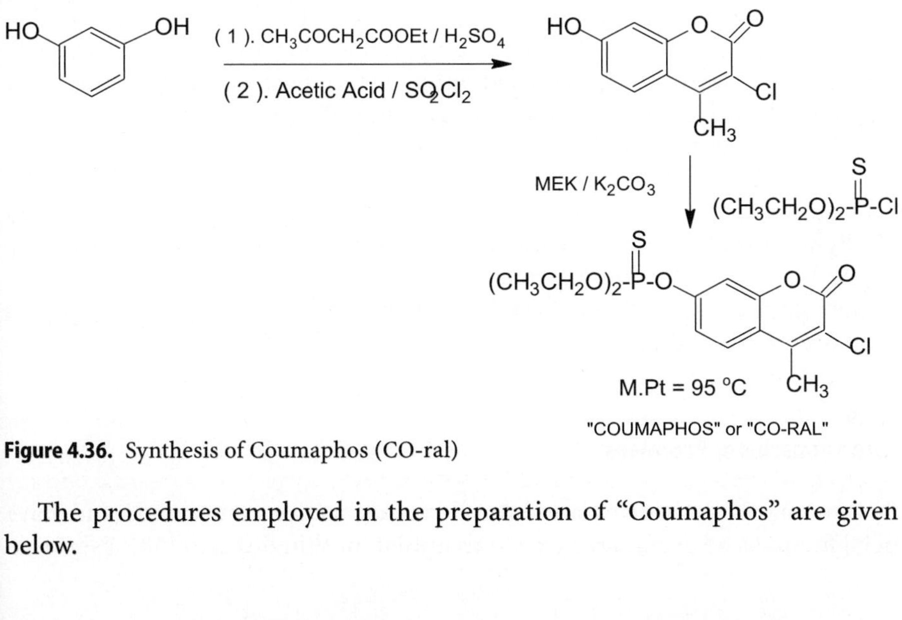


**Figure 4.36.** Synthesis of Coumaphos (CO-ral)

The procedures employed in the preparation of "Coumaphos" are given below.

**Synthesis of 3-Chloro-7-hydroxy-4-methylcoumarin Raw Material for "Coumaphos" Manufacture [34]** A mixture of resorcinol (11.0 g; 0.1 mol), methyl acetoacetate (13.9 g; 0.12 mol), sulfuric acid (0.1 g) and octane (40 ml) was heated to reflux for 5 min at about 100 °C (Pot) and 75 °C (Head). After 18.5 ml of distillate were removed slowly and continuously over a period of 20 min, the pot temperature rose to 120 – 125 °C. The reaction mixture was heated at 120 – 125 °C for 1 h and then cooled to 60 °C. Acetic acid (200 ml) was added and the mixture was heated to 75 °C. Sulfuryl chloride (15 g; 0.11 mol) was added slowly to the mixture at below 85 °C over a period of 11 min. The resulting mixture was heated at 75 – 80 °C for 2 h and then acetic acid (160 to 180 ml) was distilled off under reduced pressure at a temperature below 100 °C. The viscous reaction mixture was cooled to room temperature and 200 ml water was added with vigorous stirring. After cooling to 10 – 15 °C the reaction mixture was filtered and the product was washed with water and dried. The yield of crude product (20.7 g) was 98% based on the resorcinol used. The product assayed 88% 3-chloro-7-hydroxy-4-methylcoumarin by G. C. analysis.

**Synthesis of Coumaphos from 3-Chloro-7-hydroxy-4-methylcoumarin [35]** 3-Chloro-4-methyl-7-hydroxycoumarin (84 g) was suspended in 400 ml of methyl ethyl ketone and 42 g of pulverized and sifted potassium carbonate and 2 g of copper powder were added thereto. Diethyl chlorothionophosphate (72 g) was run into the mixture at 75 °C with stirring. The temperature was maintained at 75 – 85 °C for a further 15 min while stirring was continued. The mixture was then cooled, the salts were separated, the solvent was evaporated, and the residue was recrystallized from ethanol. As a result, 105 g of diethyl thiophosphoric acid ester of 4-methyl-3-chloro-7-hydroxycoumarin with a melting point of 95 °C were obtained.

Coumaphos compounds showed excellent insecticidal effect, especially against fly larvae, cotton pests such as pink boll worm and boll weevil and also low toxicity towards warm blooded animals. Compositions containing these coumarin-based derivatives were effective in killing moth larvae, bugs, lice, aphids, roach and Colorado beetles and their larvae.

7-(8-Trimethoxysilyloctyloxy)-coumarin can self-assemble into optically active monolayers on substrates. When irradiated with linearly polarized UV light, the self-assembled coumarin monolayer can induce a homogeneous alignment of nematic liquid crystals [36].

Various 4-alkyl substituted coumarin derivatives and their melting points have been reported [37].

### 4.2.9
### Furocoumarins or Psoralens

Furocoumarin derivatives are natural photosensitive compounds and have been found to be useful for medical treatment of skin diseases [38]. Psoralens are the linear furocoumarins and have pyrone and furan rings in their molecular structures.

Dihydrofuro-[3,2-*g*]-coumarin-6-one compound was synthesized from resorcinol. Furocoumarins (Psoralens) are used clinically for PUVA (psoralen + UV-A radiation) treatment of psoriasis and some other skin disorders (Figure 4.37) [39].

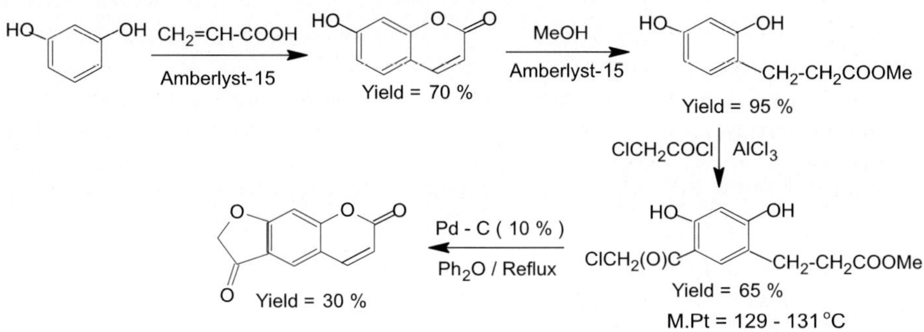


**Figure 4.37.**

Psoralens have been used in the photo-chemotherapy or psoriasis. In such treatments, psoralens are administered orally or topically to a patient.

When 7-chloroacetoxy coumarins were subjected to Fries rearrangement, dihydrofuro-[2,3-*h*]-coumarin-9-one or didydrofuro- [2,3-*g*]-coumarin-6-one were formed (Figure 4.38) [40,41].

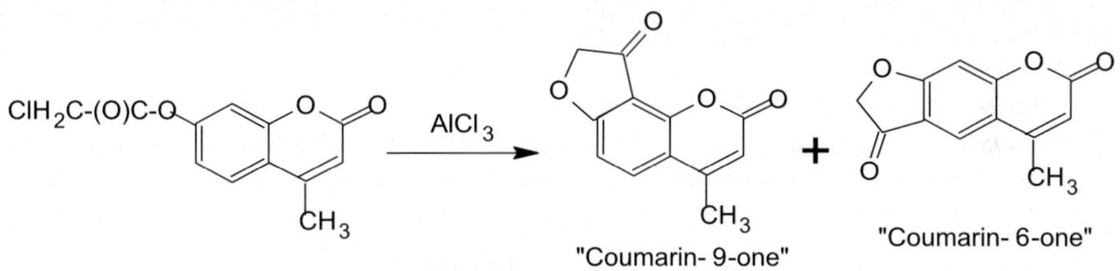


**Figure 4.38.**

Coumarin-9-one was formed when the rearrangement temperature was 115–120 °C. If a substituent is present at the 8-position of 7-chloroacetoxy coumarins, then coumarin-6-one compounds were the predominant products.

Tetracyclic psoralen derivatives were prepared from 2-methyl resorcinol. In these derivatives, the fourth ring is constituted by cyclopentane, cyclohexane and benzene fused to tricyclic furocoumarins moiety. The photo antiproliferative activity of these tetracyclic compounds tested on two human tumor cell lines (HeLa and HL-60), appeared from 8 to 22 times higher than that of the well known photochemotherapeutic drug 8-methoxypsoralen in HeLa, and slightly higher in HL-60 [42].

## References


1. R. O'Kennedy and R.D. Thornes, Coumarins: biology, applications and mode of action, Wiley, Chichester, 1997.
2. M. Maeda, Laser Dyes, Academic Press, New York, 1994.
3. M. Zahradnik, The production and application of fluorescent brightening agents, Wiley, 1992.
4. H.A. Offe and W. Barkon, Chem. Ber, Vol 80, p 458 (1947).
5. C.C.R. Mercier, Acad. Sci. Ser. C, Vol. 276, p 1053 (1973).
6. B. Talapatra, T. Deb and S. Talapatra, Ind. J. Chem, Vol. 25 B, p 1122 (1990).
7. Benger Lab, British Patent, 1,077,066 (1967).
8. J. Cologne, E. LeSech and R. Marey, Bull. Soc. Chim. Fr, Ser. Vol. 5, p 776 (1957).
9. R.D.H. Murray, J. Mendez and S.A. Brown, The natural coumarins, Wiley, New York (1982).
10. Kumari Rao and N.V. Subba Rao, Proc. Ind. Acad. Sci, Vol. 77, p 149 (1973).
11. D.P. Chakraborty, Mira Sen and P.K. Bose, Trans. Bose. Research. Inst. Vol. 24, p 31 (1961).
12. H. Dressler and K.G. Reabe, US Patent, 3,503,996 (1970).
13. D.W. Fink and W.R. Koehler, Anal. Chem, Vol. 42, p 990 (1970).
14. S. Sethna and R. Phadke, Organic reactions, Vol. 7, p 1 (1953).
15. A. Russel and J.R. Frye, Org. Syn. Collect. Vol III, p 281 (1955).
16. Chudgar and Shah, J. Univ. Bombay, Vol. 13, p 18 (1944).
17. C. Amato and P. Calas, Synthesis, No. 17, p 2709 (2003).
18. K. Setsukinai et al., J. Chem. Soc., Perkin Trans. 2, p 2453 (2000).
19. F.M. J. Vallet, US Patent, 4,151,291 (1979).
20. A. Mihailovski, US Patent, 4,162,326 (1979).
21. N. Minami, S. Kijima, S. Katayama and H. Shionoya, US Patent, 4,208,426 (1980).
22. F.D. Valle, US Patent, 4,296,039 (1981).
23. F.D. Valle, US Patent, 4,452,811 (1984).
24. R. Brown, F. Scheinmann and A. Stachulski, J. Chem. Res. (S), p 370 (1997).
25. R. Schlecker et al., US Patent, 4,618,622 (1986).
26. F.D. Valle and A. Romeo, US Patent, 4,737,517 (1988).
27. J.E. Celebuski, US Patent, 5,247,099 (1993).
28. K.R. Gee, R.P. Haugland and W.C. Shen, US Patent, 5,830,912 (1998).
29. F.E. Blaney, F.E. Bondinell and J.A. Chan, US Patent, 6,506,790 (2003).
30. Nature Medicine, Vol 2, p 1174 (1996).
31. M. Lacova, D. Loos, M. Furdik, M. Matulova and H. El-Shaaer, Molecules, Vol. 3, p 149 (1998).
32. P.E. Newallis, J.W. Baker and J.P. Chupp, US Patent, 3,061,613 (1962).
33. R. Fusco, G. Losco and C. Augusto, US Patent, 2,860,085 (1958).
34. R.H.S. Wang, US Patent, 4,788,298 (1988).
35. Bayer, British Patent, 713,142 (1954).
36. J. Fang, C. Whitaker, B. Weslowski, M. Chen, J. Naciri and R. Shashidar, J. Mater. Chem, Vol. 11, p 2992 (2001).
37. M. Afri, H. Gottlieb and A. Frimer, Free radical biology & medicine, Vol. 32, p 605 (2002).
38. R. Murray, The natural coumarins, occurrence, chemistry and biochemistry, Wiley Interscience, New York, 1982.
39. J. Gonzalez-Gomez, L. Santana and E. Uriarte, Synthesis, p 27 (2003).
40. V. Traven, Arkivoc, Vol. 1, p 523 (2000).
41. V. Traven, Molecules, Vol. 9, p 50 (2004).
42. L. Via, E. Uriarte, L. Santana, S. Magno and O. Gia, Arkivoc, Vol. 5, p 131 (2004).

## 4.3 Resorcinarenes

### 4.3.1 Introduction

Resorcinol, in general, reacts with various aliphatic aldehydes to produce novolak-type resin structures in the presence of acid catalysts. The molar ratios of resorcinol and aldehydes used to obtain novolak resins are, normally, less than one mole of aldehyde per mole of resorcinol. However, a new class of materials, known as calixarenes, was obtained from the equimolar ratios of resorcinol to various aldehydes in the reaction [1]. From the molecular weight determinations, it was found out that resorcinol reacted with the aldehydes at a ratio of 4 : 4 to form these novel chemical compounds. A cyclic structure was proposed initially and then confirmed later on by a single crystal X-ray analysis performed on these materials [2, 3]. Different nomenclature was used to name this new class of materials, which included resorcinol based calixarenes and calix-[4]-resorcinarenes [4]. Later on, for the cyclic oligomeric products from the reaction of resorcinol and aldehydes, they were named and called resorcinarenes [5].

A large number of articles appeared on the chemistry of resorcinarenes, including the synthesis, structural and conformational characterization, complexational properties and their conversion into cavitands and carcerands [4, 6–8].

### 4.3.2 Synthesis of Resorcinarenes

Resorcinarenes are conveniently synthesized in high yields from the one step acid catalyzed reaction of resorcinol or resorcinol ethers with various aldehydes [9, 10]. A reaction scheme outlining the preparation of resorcinarenes is shown in Figure 4.39.

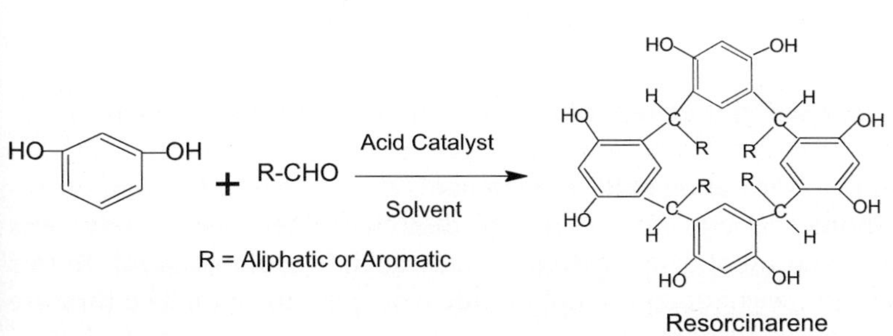


**Figure 4.39.**

The reaction of resorcinol with an aldehyde is a condensation reaction and can be carried out in the presence of a solvent and an acid catalyst. Suitable solvents used in the preparation of resorcinarenes are generally alcohols, such as the methanol and ethanol. Reaction mixtures are refluxed for 30 min to several hours and the products formed are crystallized out from the solution on cooling [9, 11].

Resorcinarenes obtained from the resorcinol-aliphatic or aromatic aldehydes reactions (Figure 4.40) are presented in Table 4.6.

**Table 4.6.** Resorcinarenes from resorcinolic derivatives

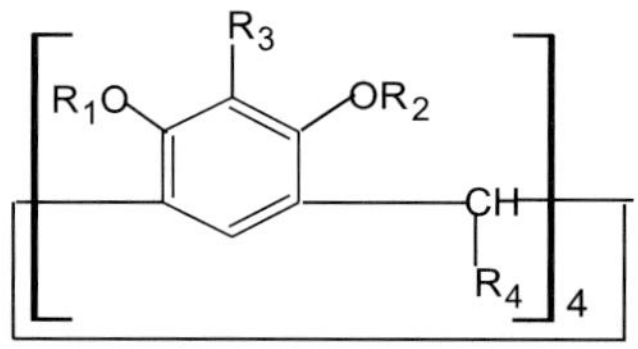


| Expt. no. | $R_1$ | $R_2$ | $R_3$ | $R_4$ | Yield (%) |
|---|---|---|---|---|---|
| 1 | H | H | H | $CH_3$ | 88 |
| 2 | H | $CH_3$ | H | $CH_3$ | 2 |
| 3 | $CH_3$ | $CH_3$ | H | $CH_3$ | 0.2 |
| 4 | H | H | H | Ph | 90 |
| 5 | H | H | H | 2-OH-Ph | 78 |
| 6 | H | H | H | 4-OH-Ph | 91 |
| 7 | H | H | H | 4-$CH_3$-Ph | 94 |

Note: Ph = Phenyl group
Data from [12]

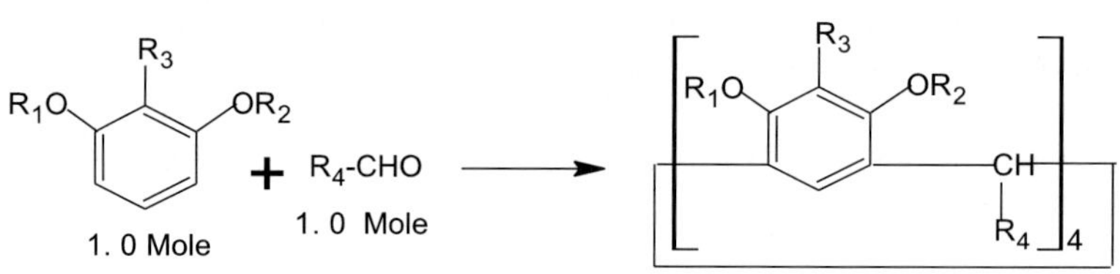


**Figure 4.40.**

A procedure used in the preparation of resorcinarenes is given below [12].

**Procedure for the Preparation of Resorcinarenes [12]** Resorcinol (0.1 mol; or resorcinol monomethyl ether or resorcinol dimethyl ether, respectively) was dissolved in a solution of dried hydrogen chloride in 95 g of methanol. To this refluxing solution was added 0.1 mol of aldehyde over 30 min. The mixture was stirred for an additional 30 min; in the case of using acetaldehyde, half of the solvent was distilled off. The reaction mixture was cooled to room temperature, and the precipitated macrocycles were collected, washed acid-free with water and dried at 80 °C.

From the results in Table 4.6 it was shown that, when the hydroxyl groups of resorcinol were substituted with methyl ether groups, then the resorcinarenes formation was restricted. This indicates that activation of benzene ring positions in the resorcinol molecule by the hydroxyl groups is important for the reaction of aldehydes to form the cyclic structures. With the unsubstituted resorcinol, resorcinarenes were produced in good yields with aliphatic and aromatic aldehydes.

Resorcinarene compounds were found to be higher melting solids. In order to determine the effect of alkyl chain length of an aliphatic aldehyde on the physical properties of resorcinarenes, compounds were made using different aldehydes. The results are presented in Table 4.7 [9, 11].

**Table 4.7.** Resorcinarenes: effect of varying alkyl chain length of aldehydes

| Expt. no. | R-CHO; R = | Yield (%) | Melting point (°C) |
|---|---|---|---|
| 1 | $CH_3$ | 60 | > 360 |
| 2 | $CH_2$-$CH_3$ | 88 | > 360 |
| 3 | $(CH_2)_2CH_3$ | 92 | > 360 |
| 4 | $(CH_2)_3CH_3$ | 89 | 344–345 |
| 5 | $(CH_2)_4CH_3$ | 77 | 329–330 |
| 6 | $(CH_2)_{10}CH_3$ | 68 | 285 |

Data from [9, 11]

From this study (Table 4.7) it was determined that, when the aldehyde alkyl chain length was increased, the resorcinarenes produced were observed to be more soluble in hexane, benzene and chloroform solvents. In addition, the melting points of resorcinarene compounds were found to be decreased with increase in chain length of the aldehydes.

### 4.3.3 Mechanism of Resorcinarene Formation

A mechanism proposed on the formation of resorcinarene from the hydrochloric acid catalyzed condensation of resorcinol with acetaldehyde in the presence of methanol solvent is shown below (Figure 4.41) [12].

$^1$H NMR spectroscopy was used to study the mechanism of formation of resorcinarene. Condensation reaction appeared to proceed in a step-wise manner to reach the linear tetramer molecule. At this stage, the linear tetramer reacts (cyclizes) faster to produce the cyclic tetramer molecule. The fast cyclization of linear tetramer was due to the formation of hydrogen bonds between the hydroxyl groups of adjacent resorcinolic molecules in the folded or cyclized chain structure.

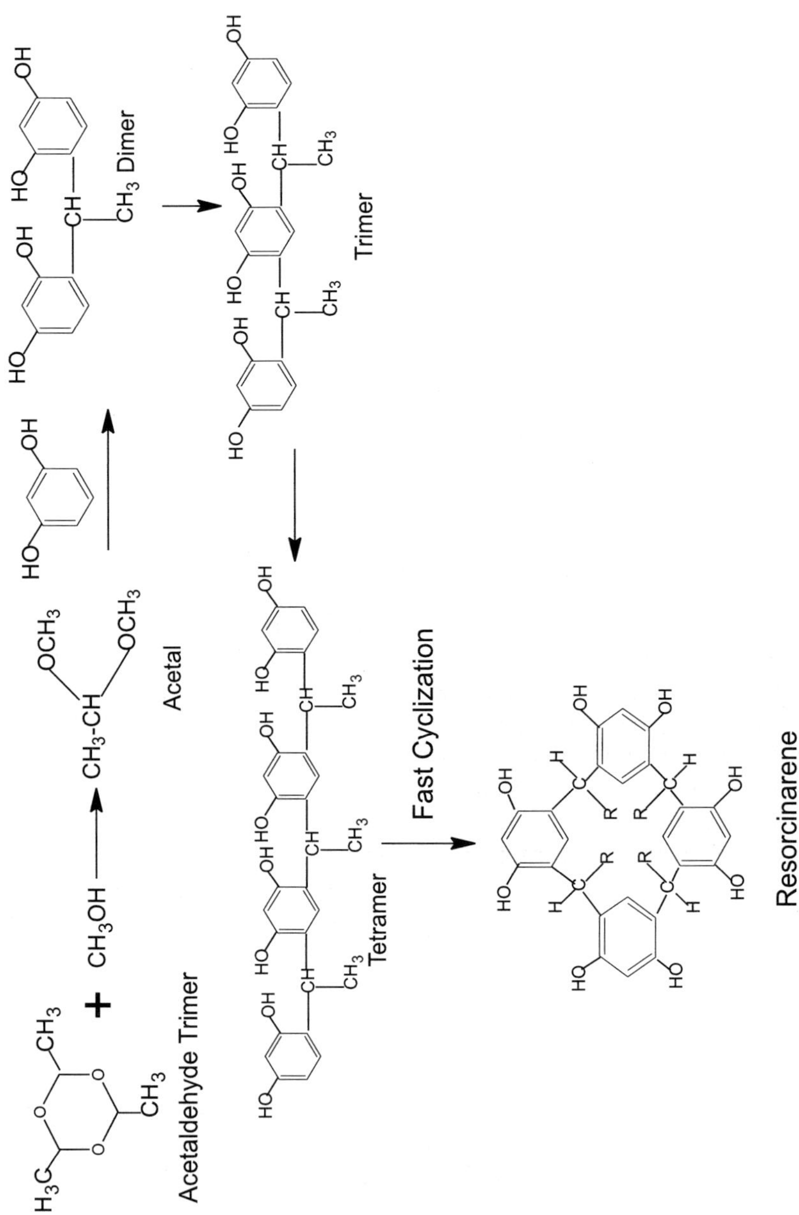


**Figure 4.41.** Mechanism of resorcinol formation from resorcinol and acetaldehyde

### 4.3.4
### Resorcinarene Containing Methylene Bridges

Methylene bridged resorcinarenes were difficult to obtain from the reaction of resorcinol with formaldehyde because of the formation of highly crosslinked resin-type structures from the equimolar quantities of these two reactants. A novel synthetic route used to synthesize a methylene bridged resorcinarene was based on the following scheme (Figure 4.42) [13].

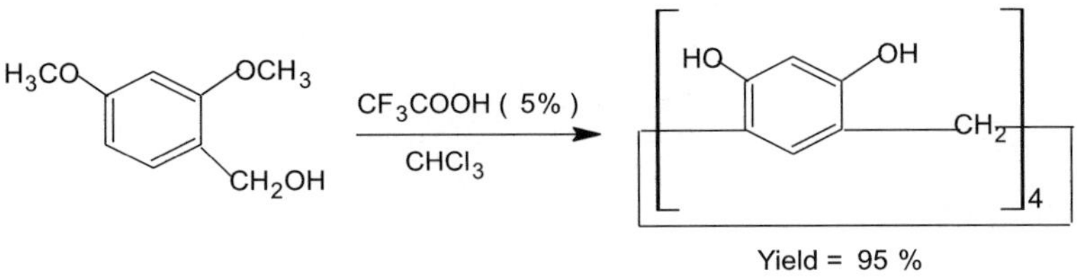


**Figure 4.42.**

Using hydroxymethyl substituted dimethyl ether of resorcinol, this resorcinarene was made in high yields employing a trifluoro acetic acid (catalyst) and chloroform (solvent) system. Since there is no alkyl chain in the methylene bridge of this molecule, conformational flexibility can be expected in this structure.

### 4.3.5
### Properties of Resorcinarenes

Resorcinarenes are cyclic oligomers consisting of a benzene ring with polar cavities. The cavities and their walls contain active sites which can serve for substrate binding. These macrocyclic molecules are capable of complexing with all the inorganic cations. Therefore, phenolate and coordination complexes can be formed with a variety of metal compounds. Resorcinarenes can bind metal ions by means of ionic interactions (resorcinate ions), coordination and arene interactions.

In an alkaline medium, half of the hydroxyl groups are deprotonated and the resulting bowl-shaped tetra-anion has a great affinity for other cations such as alkyl ammonium cations. These materials are known to bind polar organic molecules like D-ribose and dicarboxylic acids through hydrogen bonding [6].

### 4.3.6
### Resorcinarene Reactions

The electron releasing nature of hydroxyl groups present in the resorcinarenes activates the benzene ring for electrophilic substitution reactions. Therefore, electrophilic substituents such as bromo or diazo groups can readily be introduced at the *ortho*-position of the benzene ring.

#### 4.3.6.1
#### Bromination Reaction

Bromination of resorcinarenes using *N*-bromo succinimide (NBS) at room temperature gave bromo substituted derivative in about 80% yields (Figure 4.43) [14].

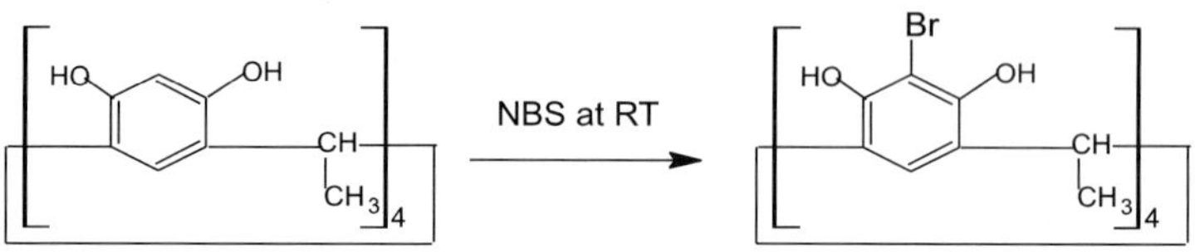


**Figure 4.43.**

#### 4.3.6.2
#### Aminomethylation Reaction

Employing secondary amine and formaldehyde as the reactants, aminomethylated resorcinarenes were obtained based on the following reaction (Figure 4.44) [15,16].

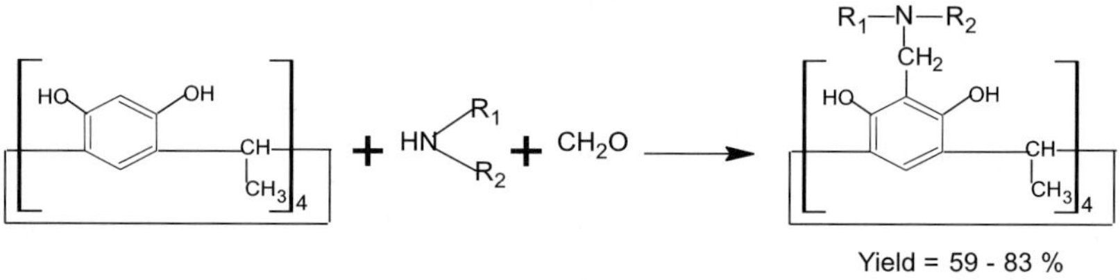


**Figure 4.44.**

The hydroxy groups of resorcinarenes can also be functionalized in several ways. By the reaction of hydroxyl groups with various chemical compounds, functional groups such as ester, ether, alcohol, silicon, phosphorus and bipyridyl may be introduced into the molecule. In addition, a selective modification of four of the hydroxyl groups that resulted in acetyl derivatives was also reported.

### 4.3.7
### Applications of Resorcinarenes

The presence of eight hydroxyl groups in resorcinarene molecules enables them to complex with the organic molecules containing polar functional groups. The complexing possibilities are not limited to neutral molecules but also to other metals and ammonium cations. Resorcinarene complexes may serve as enzyme mettalosite models, redox sensors, selective extractants for radioactive metal ions and molecular magnetic materials. A number of significant industrial applications for resorcinarenes have been reported [4]. Current interests in the possible applications include binding and incorporation of guest compound [17,18].

#### 4.3.7.1
#### Resorcinarenes in Nanotechnology

Nanoparticles are used to develop new generation of stronger materials for applications ranging from the medical to aerospace. These particles are fragile and unstable. When their surfaces are touched, they fuse themselves and lose their properties and structure. Resorcinarene molecules have bowl shaped heads which can enable them to adhere readily to the surface of nanoparticles. Gold nanoparticles having a diameter of 10 to 20 nanometers were successfully encapsulated with resorcinarenes [19,20]. By this process, a polymer cage was created around the surface of gold particles by chemically stitching the resorcinarenes together. The porous coating of gold nanoparticles permitted them to interact with the substances outside the coating but prevented them from touching each other. This process has a significant impact on the physical properties and stability of nanoparticles. Potential use for the coated nanoparticles with resorcinarenes could be in the drugs delivery applications. This method can be employed to stabilize nanoparticles with magnetic properties that may be used in the development of microelectronic devices and magnetic sensors.

### 4.3.8
### Cavitands and Carcerands

Cavitands are synthesized by the covalent linking of neighboring resorcinolic hydroxyl groups in the resorcinarenes as shown below (Figure 4.45) [21].

Cavitands are extremely rigid molecules.

Two cavitands are covalently linked via their upper rims to form the carcerands [22].

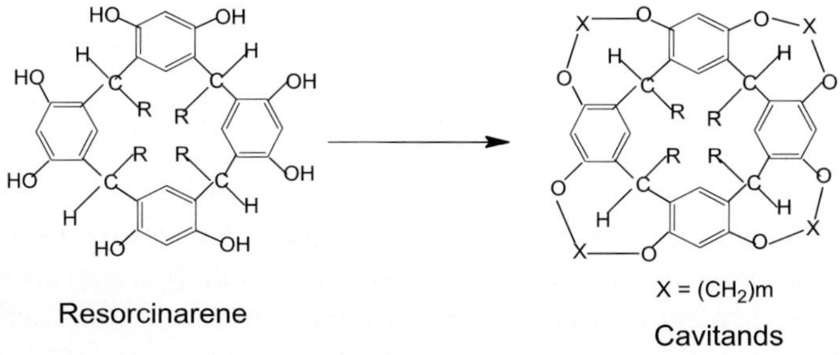


**Figure 4.45.**

## References

1. J.B. Neiderl and H.J. Vogel, J. Am. Chem. Soc, Vol. 62, p 2512 (1940).
2. H. Erdtman, S. Hogberg, S. Abrahansson and B. Nilsson, Tetrahedron Lett, p 1679 (1968).
3. B. Nilsson, Acta. Chem. Scand, Vol. 22, p 732 (1968).
4. C.D. Gutsche, Calixarenes, Monograph in Supermolecular Chemistry, J.F. Stoddart Ed., Royal Society of Chemistry, Cambridge, Vol. 1 (1989).
5. U. Schneider and H.J. Schneider, J. Am. Chem. Soc, Vol. 127, p 2455 (1994).
6. P. Timmerman, W. Verboom and D.N. Reinhoudt, Tetrahedron, Vol. 52, p 2663 (1996).
7. Calixarenes in action, L. Mandolini and R. Ungaro Eds., Imperial College Press, London (2000).
8. Calixarenes 2001, Z. Asfari, V. Bohmer, J. Harrofield and J. Vicens Eds., Kluwer, Dordrecht (2001).
9. L.M. Tunstad, J.A. Tucker, E. Dalcanale, J. Weisser, J.A. Bryant and J.C. Sherman, J. Org. Chem, Vol. 54, p 1305 (1989).
10. E.U. Thodenvan Velzen, J.F.J. Engbersen and J.N. Reinhoudt, J. Am. Chem. Soc, Vol. 116, p 3597 (1994).
11. A.G.S. Hogberg, J. Org. Chem, Vol. 45, p 4498 (1980).
12. F. Weinelt and H.J. Schneider, J. Org. Chem, Vol. 56, p 5527 (1991).
13. O.M. Falana, E. Al-Farhan, P.M. Keehn and R. Stevensen, Tetrahedron Lett, Vol. 35, p 65 (1994).
14. J.R. Morgan, S. Karbach and D.J. Cram, J. Am. Chem. Soc, Vol. 104, p 5826 (1982).
15. Y. Matsushita and T. Matsui, Tetrahedron Lett, Vol. 34, p 7433 (1993).
16. D.A. Leigh, P. Lannane, R.G. Pritchard and G.J. Jackson, J. Chem. Soc, Chem. Commun, p 389 (1994).
17. K. Ichimura et al., Langmuir, Vol. 13, p 7680 (1996).
18. Y. Matsuzawa, T. Seki and K. Ichimura, Chem. Lett, p 411 (1998).
19. R. Balasubramanian, B. Kim, S.L. Tripp, M. Liberman and A. Wei, Langmuir, Vol. 18, p 3676 (2002).
20. Purdue News, October 2000.
21. D.J. Cram, Science, Vol. 219, p 1777 (1983).
22. D.J. Cram and J.M. Cram, Container molecules and their guests, Monographs in Super Molecular Chemistry, J.F. Stoddart Ed., Royal Society of Chemistry, Cambridge, Vol. 4 (1994).

# 4.4 Dihydroresorcinol

## 4.4.1 Introduction

Dihydroresorcinol (DHR) or 1,3-cyclohexanedione (CHD) is a most versatile and valuable raw material for the preparation of derivatives of carbo- or heterocyclic condensation products. The compounds derived from dihydroresorcinol are used in pharmaceutical, agricultural chemicals, dyestuffs, organic intermediates and polymer additives applications.

In this chapter, a general outline on the synthesis of dihydroresorcinol, its properties and chemical reactions that led to the development of important commercial products are discussed.

## 4.4.2 Synthesis of Dihydroresorcinol

Dihydroresorcinol and its derivatives have been synthesized using different synthetic pathways. The hydrogenation of resorcinol and its derivatives in aqueous alkali with Raney® nickel catalyst under various conditions produced dihydroresorcinols or 1,3-cyclohexanediones [1–3]. The synthesis of DHR from resorcinol using nickel catalyst is outlined in a reaction scheme shown below (Figure 4.46).

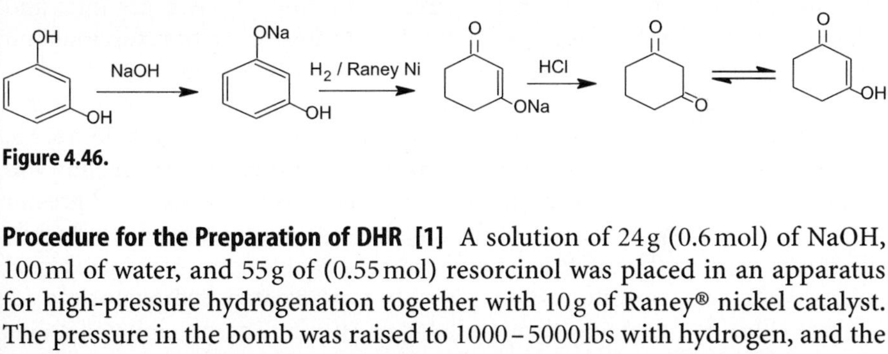


**Figure 4.46.**

**Procedure for the Preparation of DHR [1]** A solution of 24 g (0.6 mol) of NaOH, 100 ml of water, and 55 g of (0.55 mol) resorcinol was placed in an apparatus for high-pressure hydrogenation together with 10 g of Raney® nickel catalyst. The pressure in the bomb was raised to 1000–5000 lbs with hydrogen, and the temperature was adjusted to 50 °C. The bomb was shaken and the reaction was allowed to proceed for 10 to 12 h, during which time 0.5 mol of hydrogen was absorbed.

The apparatus was allowed to cool to room temperature, the pressure was released, and the catalyst was removed by filtration. The filtrate was made acidic to Congo red with concentrated hydrochloric acid, and the solution was cooled to 0 °C in an ice-salt bath and held at that temperature for 30 min before filtration. The DHR which crystallized was separated by filtration and dried, 50–60 g of crude dry product containing sodium chloride was obtained.

The crude DHR was dissolved in 125–150 ml of hot benzene, filtered to remove the sodium chloride, and allowed to crystallize. The solid was separated by filtration and dried overnight in a vacuum desiccator. The product melted at 103–104 °C and weighed 48–53 g (85–95% yield).

Milder conditions and noble metal catalysts other than nickel, such as Rh/$Al_2O_3$ and Pd/carbon were also employed in the hydrogenation of resorcinol in aqueous alkali at 25 °C and low hydrogen pressure (0.35 MPa) and obtained greater than 85% yield of DHR [4–8].

A catalytic transfer hydrogenation method was also employed to produce DHR in higher yields. The advantage of this method was the use of atmospheric condition for the hydrogenation reactions. DHR was made in an excellent yield and high purity by reacting resorcinol and sodium formate in an aqueous medium in the presence of 5% Pd/C catalyst at a pH of 5–11 (Figure 4.47) [9].

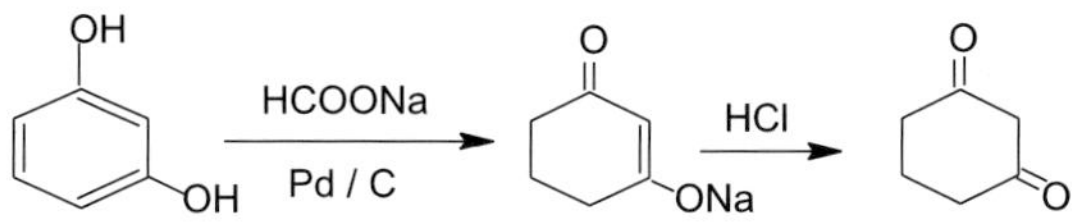


**Figure 4.47.**

The synthesis procedure used in the DHR preparation by the catalytic transfer hydrogenation method is described below.

**Catalytic Transfer Hydrogenation Procedure for DHR Synthesis [9]** A 500 ml four-neck flask fitted with a mechanical stirrer, a thermocouple, a gas inlet and a reflex condenser was charged with 125 ml of water, 55.0 g of resorcinol and 40.8 g of sodium formate. The reaction mixture was heated to 40 °C while stirring and purging the reaction medium with nitrogen gas for 20 min. At this temperature, 2.0 g of 50% wet 5% Pd/C catalyst were added and held for 3 h. The reaction mixture was then heated to 50 °C for another 3 h. The reaction was analyzed by HPLC which indicated greater than 98% conversion and greater than 96% selectivity. At the end of the reaction, the mixture was filtered while still hot to remove the catalyst. The filtrate was cooled to 0–5 °C by external means while adjusting the pH to 3.0 with concentrated hydrochloric acid. Sodium chloride (40 g) was then added over 20 min. The precipitated solids were isolated by filtration and dried to obtain 52 g of 1,3-cyclohexanedione (yield, 91%).

**Synthesis of Cyclohexanediones from Resorcinol Derivatives** Alkyl, cycloalkyl and aralkyl resorcinols were catalytically hydrogenated into their corresponding 1,3-cyclohexanone derivatives in the presence of nickel catalyst (Figure 4.48) [3, 10].

The product yield and melting point obtained on these derivatives are presented in Table 4.8.

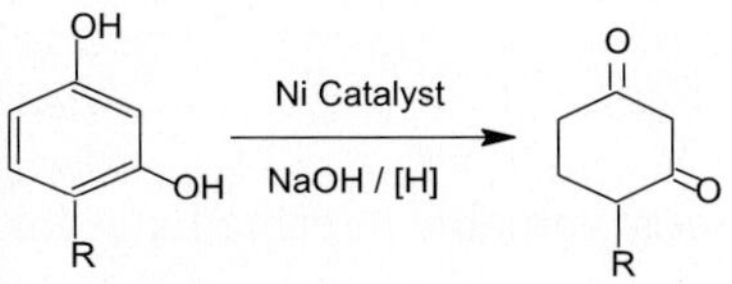


**Figure 4.48.**

**Table 4.8.** Synthesis of 4-substituted dihydroresorcinols

| Substituent | Yield (%) | M. Pt (°C) |
|---|---|---|
| Methyl | ND | 65–70 |
| *n*-Propyl | 90 | 70 |
| *n*-Hexyl | 96 | 64 |
| Cyclohexyl | 0 | 149 |
| 4-Phenyl ethyl | 95 | 126 |

Data from [3,10]

## 4.4.3 Properties of Dihydroresorcinol

Highly pure DHR is a white crystalline solid material with a melting point in the range of 105–108 °C [11,12]. It is soluble in water, chloroform, acetone and boiling benzene and slightly soluble in ether, carbon disulfide and petroleum ether. This compound is non-volatile in steam.

Dihydroresorcinol may exist in three tautomeric forms (Figure 4.49).

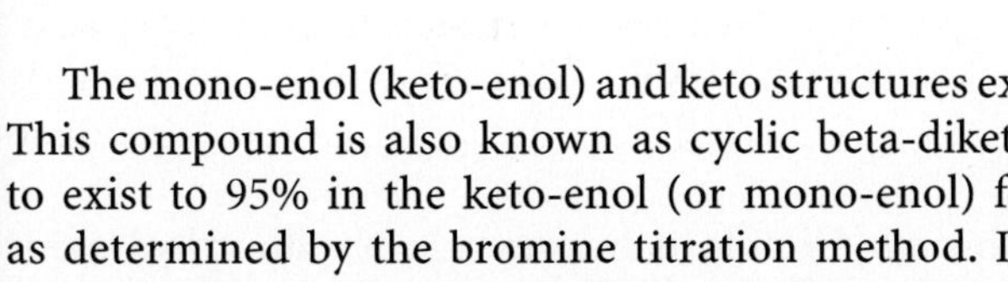


**Figure 4.49.**

The mono-enol (keto-enol) and keto structures exist in the solution of DHR. This compound is also known as cyclic beta-diketone and has been shown to exist to 95% in the keto-enol (or mono-enol) form in aqueous solution, as determined by the bromine titration method. In non-polar solvents, the amount of the enol in the equilibrium mixture is reduced.

The pKa value for DHR is about 4.8 and, therefore, it is considered an acid and almost as strong as acetic acid.

Purified and dried DHR is stable practically without limit. Impure and moist DHR can deteriorate rapidly and undergo extensive decomposition on storage.

## 4.4.4
## Derivatives of 1,3-Cyclohexanedione

The derivatives of DHR have been found to be most valuable intermediates for the synthesis of various organic compounds.

### 4.4.4.1
### Alkyl Substituted Dihydroresorcinol

When DHR was allowed to react with alkyl halide compounds, 2-alkyl substituted DHR derivatives were produced. In the presence of alkali hydroxides, DHR exists in the ionic mesomeric forms (Figure 4.50).

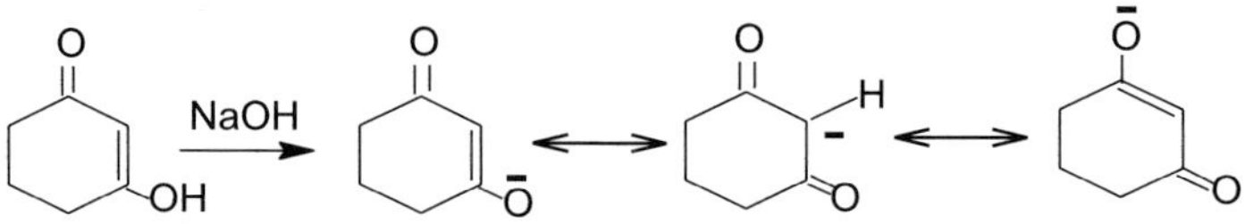


**Figure 4.50.**

Therefore, with alkyl halides, the possibilities on the formation of both enol-ethers and C-alkylated products exist (Figures 4.51 and 4.52).

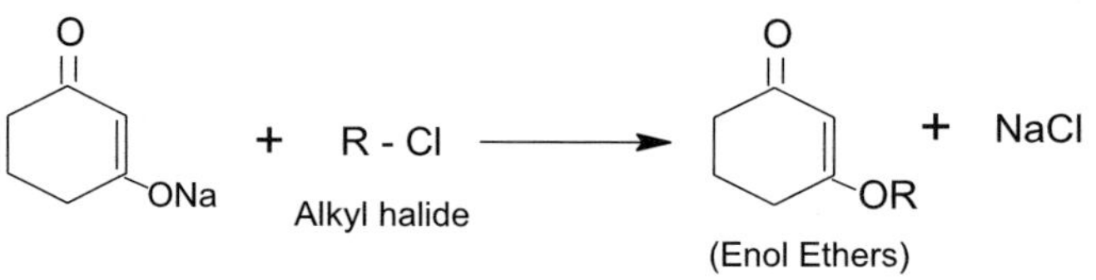


**Figure 4.51.**

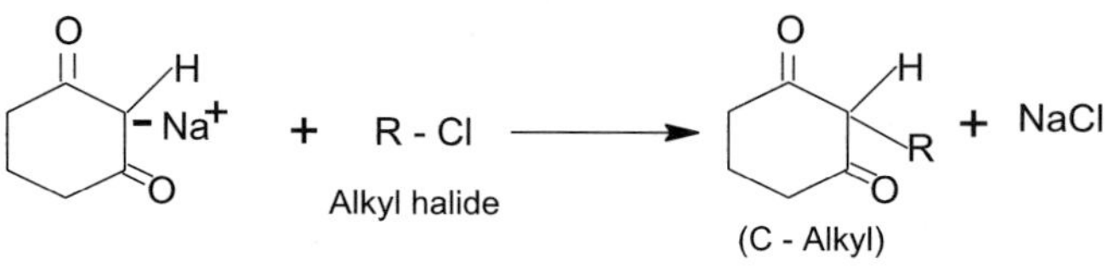


**Figure 4.52.**

Methylation of DHR with methyl iodide in the presence of sodium hydroxide produced 2-methyl substituted (C-alkylated) product (Figure 4.53) [13, 14].

Mono-alkylated DHR still possesses an acidic hydrogen at the 2-position and is therefore capable of forming a salt with sodium hydroxide, which may be soluble in water. On the other hand, the enol-ethers that could be formed in the reaction are insoluble in alkali and therefore can be completely removed from a mixture containing the C-alkylated and enol-ether products.

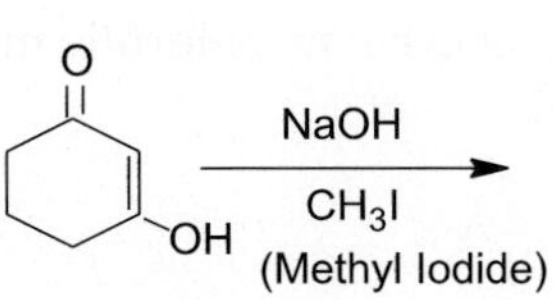


**Figure 4.53.**

By employing two different alkyl halides, di-substituted (at the 2-position) DHR of different alkyl chain lengths were synthesized and reported (Table 4.9) [15–18].

**Table 4.9.** Synthesis of mono-2-alkyl dihydroresorcinols

| Substituent | R-Group | M. Pt (°C) |
|---|---|---|
| Methyl | $CH_3$ | 204 |
| Ethyl | $CH_2CH_3$ | 178 |
| *n*-Propyl | $(CH_2)_2CH_3$ | 137 |
| *n*-Butyl | $(CH_2)_3CH_3$ | 116 |

Data from [15]

Due to the presence of an acidic hydrogen at the 2-position of mono-2-alkyl substituted resorcinol, on further alkylation with an alkyl halide it can form 2,2-dialkyl-1,3-cyclohexanedione products as shown below (Figure 4.54) [19].

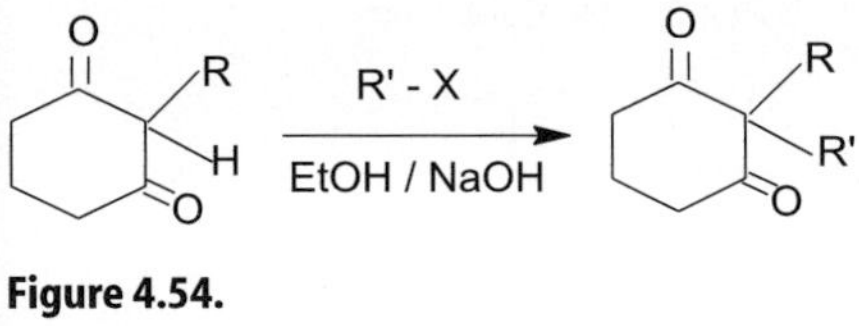


**Figure 4.54.**

#### 4.4.4.2
#### 2-Bromo and 2,2-Dibromo-1,3-Cyclohexanediones

2-Bromo-1,3-cyclohexanedione was prepared by the reaction of bromine in the presence of an aqueous medium (Figure 4.55) [20].

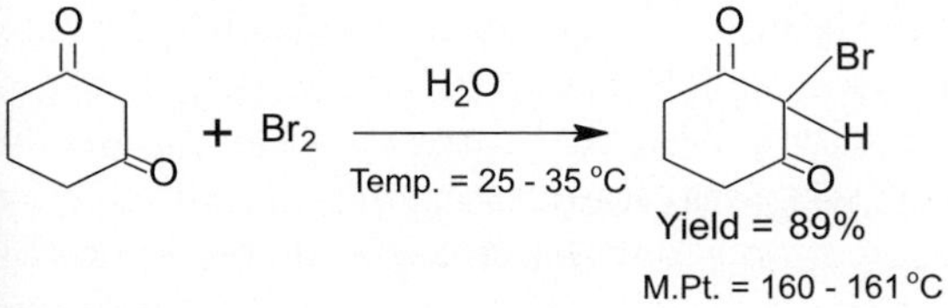


**Figure 4.55.**

The addition of bromine to the 2-bromo DHR dissolved in chloroform solvent produced 2,2′-dibromo DHR (Figure 4.56) [20].

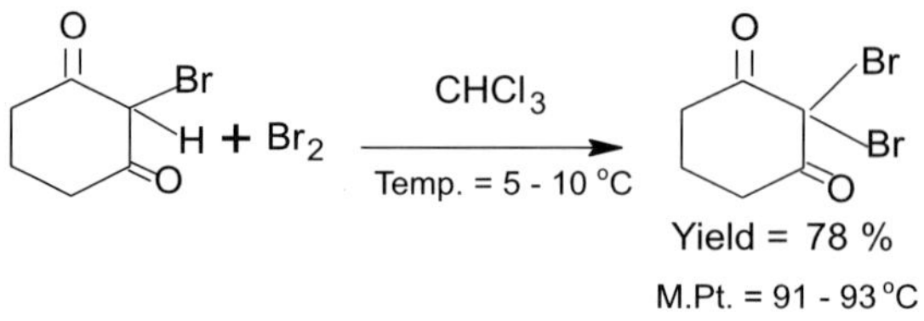


**Figure 4.56.**

### 4.4.4.3
### 2-Acetyl Dihydroresorcinol

DHR was reacted with acetic anhydride in the presence of sodium acetate catalyst to obtain 2-acetyl dihydroresorcinol (Figure 4.57) [21].

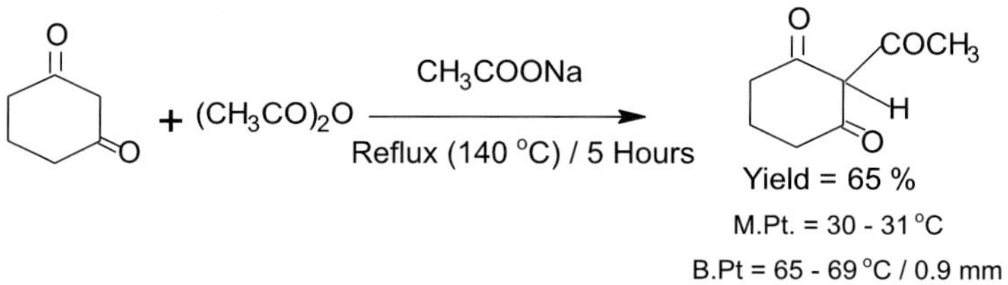


**Figure 4.57.**

### 4.4.4.4
### 2-Chloro and 2,2-Dichloro-1,3-Cyclohexanediones

The preparation of 2,2-dichloro-1,3-cyclohexanedione and its precursor 2-chloro-1,3-cyclohexanedione compound was carried out by passing gaseous chlorine into a solution of DHR dissolved in chloroform (Figures 4.58 and 4.59) [22, 23].

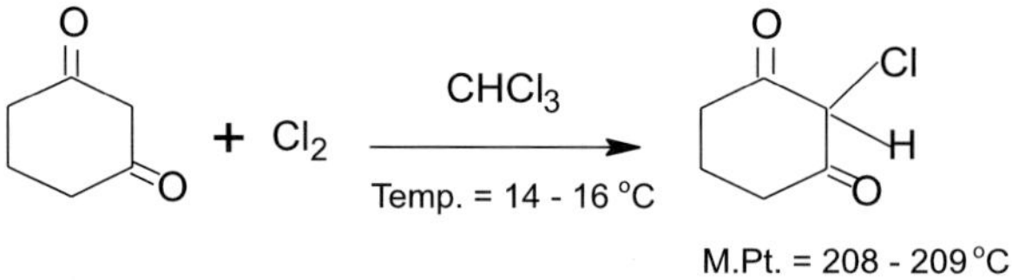


**Figure 4.58.**

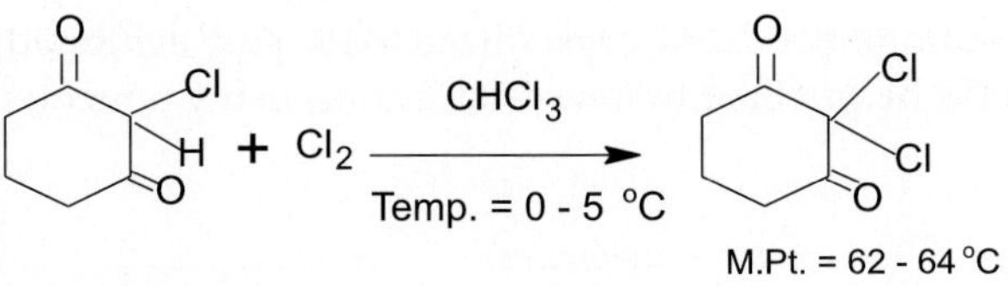


**Figure 4.59.**

### 4.4.4.5 2-Allyl Dihydroresorcinol

Allyl bromide was allowed to react with DHR dissolved in aqueous potassium hydroxide solution to obtain 2-allyl substituted dihydroresorcinol (Figure 4.60) [24].

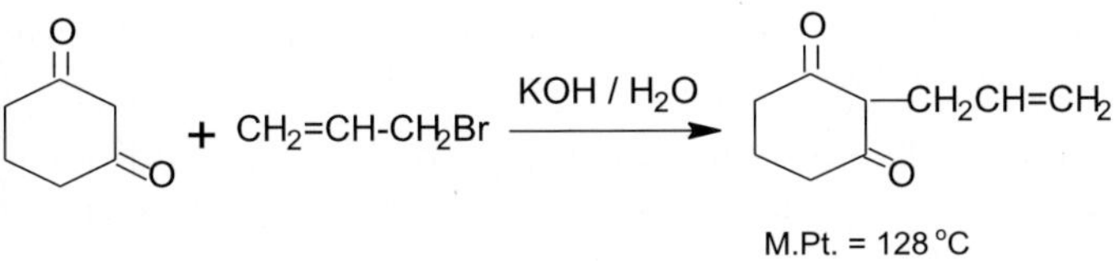


**Figure 4.60.**

## 4.4.5 Reactions of Dihydroresorcinol

### 4.4.5.1 Reaction with Aldehydes

DHR reacts with aldehydes and produces alkylidene-di-1,3-cyclohexanedione derivatives (Figure 4.61).

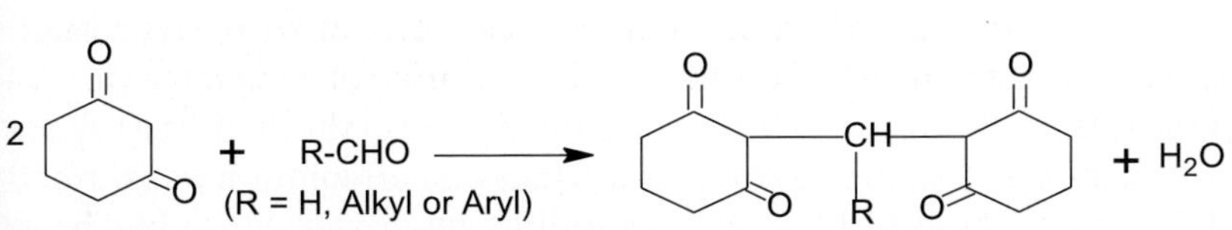


**Figure 4.61.**

The reaction was observed to proceed well without the use of a condensing agent and in aqueous or alcoholic solutions. With the use of formaldehyde, a quantitative yield of methylene-di-1,3-cyclohexanone was obtained [25].

In general, aliphatic aldehydes readily reacted or condensed with DHR in a warm alcoholic solution in the presence of small quantities of piperidine catalyst. The di-cyclohexanedione products crystallized well, and exhibited high melting points and could be purified by recrystallization using aqueous ethanol (Table 4.10) [26].

**Table 4.10.** Synthesis of alkylidene di-(1,3-cyclohexanediones)

| Aldehydes Used | R-CHO (R=) | M.Pt (°C) |
|---|---|---|
| Formaldehyde | H | 134 |
| Acetaldehyde | $CH_3$ | 155 |
| Propionaldehye | $CH_2CH_3$ | 118 – 119 |
| *n*-Butyraldehyde | $CH_2CH_2CH_3$ | 97 – 98 |

Data from [26]

### 4.4.5.2
### Formation of Ethers

DHR can be condensed with alcohols under azeotropic distillation conditions, and in the presence of strong acid catalysts, this leads to monoether derivatives (Figure 4.62) [27–29].

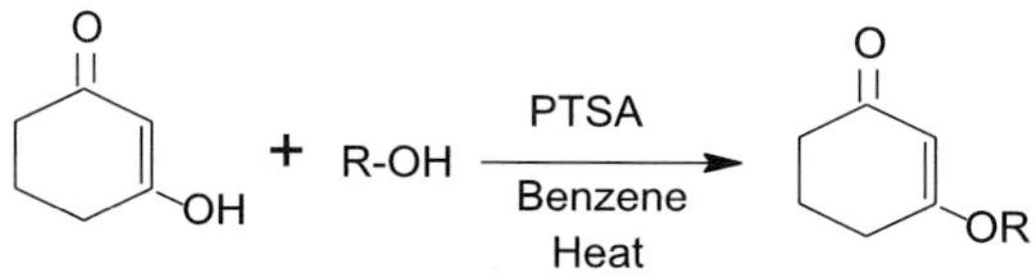


**Figure 4.62.**

**Procedure for the Preparation of 3-Ethoxy-2-cyclohexene-1-one [28]** In a 2-l flask fitted with a total reflux, variable take-off distillation head was placed a solution of 53 g (0.472 mol) of dihydroresorcinol, 2.3 g of *p*-toluene sulfonic acid (PTSA) and 250 ml of absolute alcohol in 900 ml of benzene. The mixture was heated to boiling and the azeotrope composed of benzene, alcohol and water was removed at the rate of 100 ml per h. When the temperature of the distilling vapors reached 78 °C, the distillation was stopped and the residual solution was washed with four 100-ml portions of 10% aqueous sodium hydroxide which had been saturated with sodium chloride. The resulting organic solution was washed with successive 50-ml portions of water until the aqueous washings were neutralized and then concentrated under reduced pressure. The residual liquid was distilled under reduced pressure. The yield of 3-ethoxy-2-cyclohexene-1-one was 46.6 – 49.9 g. (B.Pt. = 66–68.5/0.4 mm Hg; Yield = 70–75%).

The above procedure can be generally used to prepare 3-alkoxy-2-cyclohexenone-type compounds using various aliphatic-type alcohols. A high yield of 3-isobutoxy-2-cyclohexenone was reported from the reaction of DHR and isobutyl alcohol [30].

Acidic ion-exchange resin catalysts could be effectively utilized in the preparation of alkyl ethers from DHR (Figure 4.63) [27].

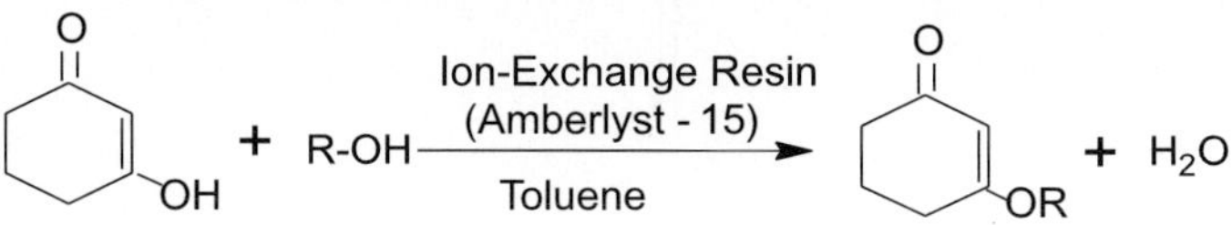


**Figure 4.63.**

With methanol, 3-methoxy-2-cyclohexenone was produced in about 95% yield even in the absence of an azeotroping solvent. Similarly, in the case of 3-*n*-butoxy-2-cyclohexanone preparation, an azeotrope forming agent was not required, since *n*-butanol used in the synthesis formed an azeotrope with water and separated into two phases on cooling. In this way, the *n*-butanol was recycled much more effectively. With an ion-exchange method, the catalyst can be effectively removed by filtration after the reaction and therefore, distillation of the reaction product could be much easier.

### 4.4.5.3 Reaction with Benzoyl Halides

1,3-Cyclohexanedione (CHD) has been reacted with various benzoyl chloride derivatives to produce 2-benzoyl substituted cyclohexanone derivatives (Figure 4.64) [31–37].

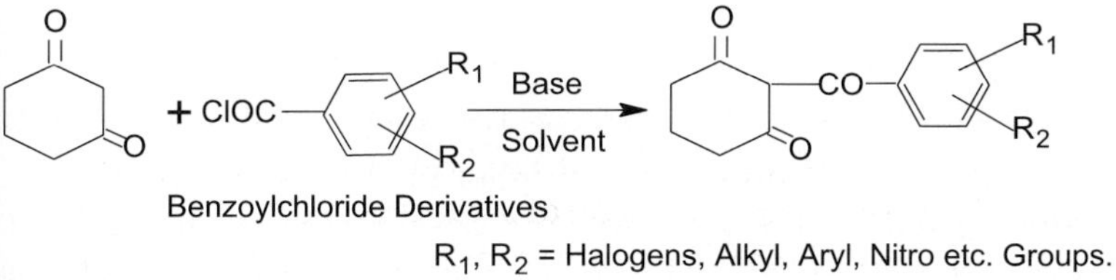


**Figure 4.64.**

2-Benzoyl-cyclohexanedione derivatives could exhibit the following four structural formulae because of tautomerism (Figure 4.65).

The proton (shown in structure) present on each of the four tautomers is reasonably labile. Due to its acidic nature, this proton can be reacted with a base to form a salt having an anion showing four resonance structures (Figure 4.66).

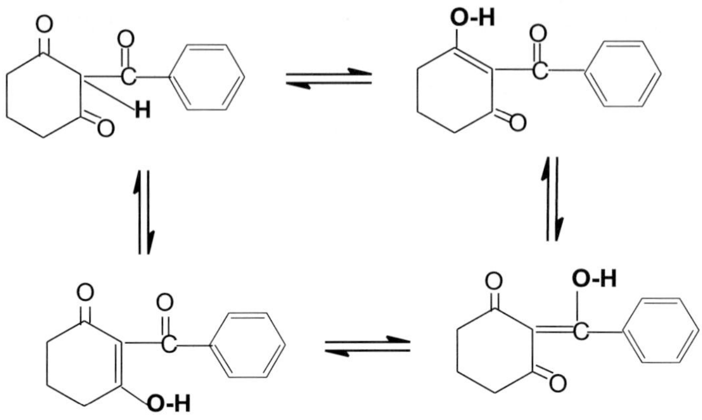

**Figure 4.65.**

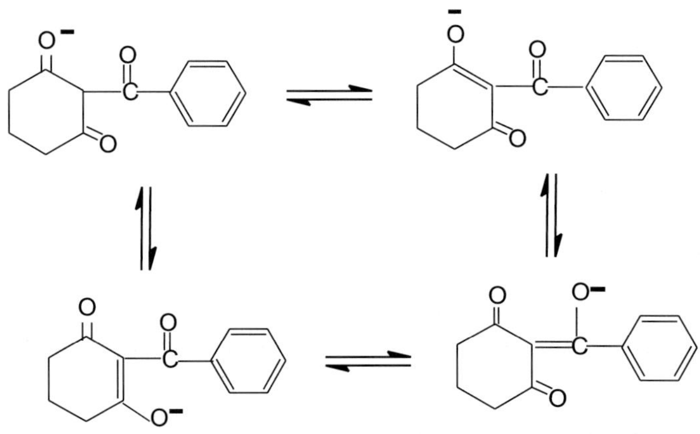

**Figure 4.66.**

The synthesis of benzoyl substituted cyclohexanediones was performed using an organic solvent in the presence of a base having moderate basic strength in a two-step process. In the first step, an enol-ester intermediate was produced. The final keto product was obtained by the rearrangement of the enol-ester structure in the succeeding step (Figure 4.67).

#### 4.4.5.4 Methyl Vinyl Ketone Reaction

$\alpha,\beta$-Unsaturated ketones, carboxylic acid esters and nitriles can undergo Michael addition-type reactions with DHR. Methyl vinyl ketone, for example, reacted with DHR and produced 2-mono and 2,2-disubstituted cyclohexanedione derivatives (Figure 4.68) [38].

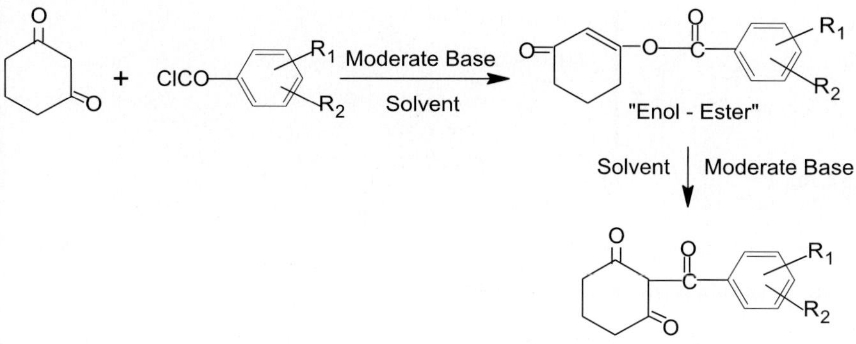


**Figure 4.67.**

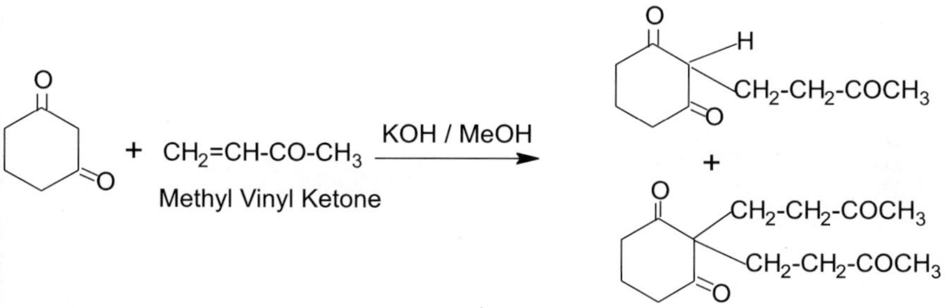


**Figure 4.68.**

### 4.4.5.5
### Reaction with Vinyl Ethers

When subjected to an electrochemical reaction, DHR and ethyl vinyl ether reacted to form a hydrocoumarone derivative (Figure 4.69) [39].

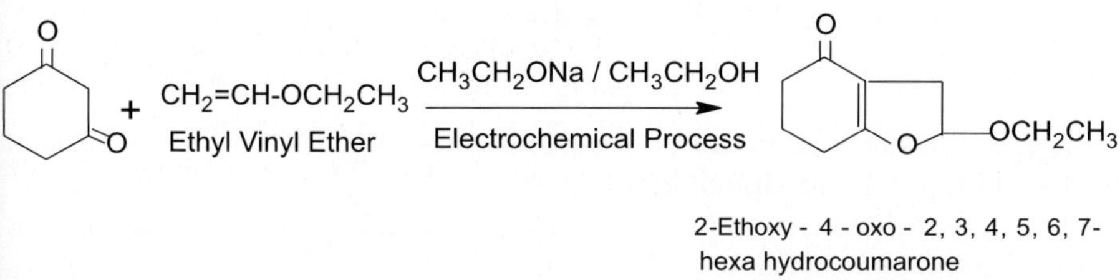


**Figure 4.69.**

The reaction mechanism involved in the electrochemical method had not been elucidated, but a tentative reaction mechanism was proposed for the hydrocoumarone formation (Figure 4.70).

Compounds prepared from this process can be converted into indole derivatives by reacting with ammonia, ammonium salts or primary amines.

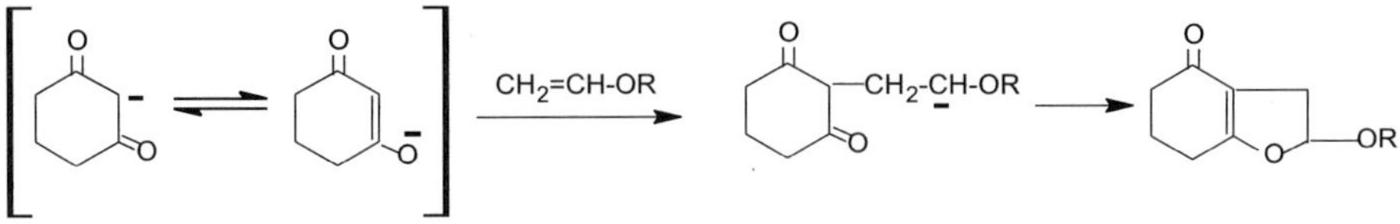


Figure 4.70.

### 4.4.5.6
### Synthesis of Indole Derivatives from DHR

A derivative of 4-oxo-4,5,6,7-tetrahydro benzofuran was prepared by reacting DHR with chloro acetaldehyde in the presence of a base (Figure 4.71) [40].

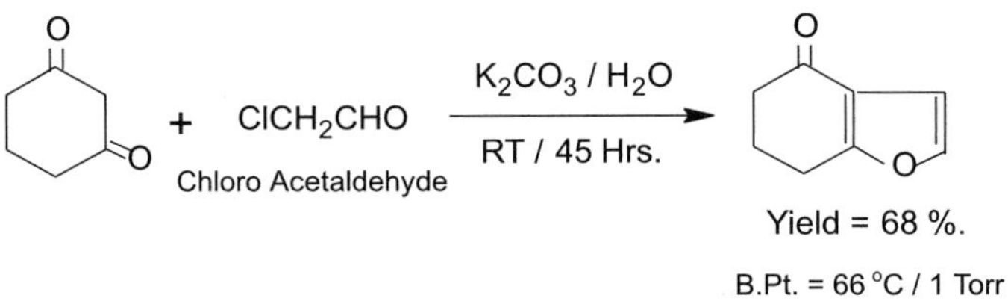


Figure 4.71.

On further reaction with ammonia, the benzofuran compound was converted into a 4-oxo-4,5,6,7-tetrahydroindole derivative (Figure 4.72).

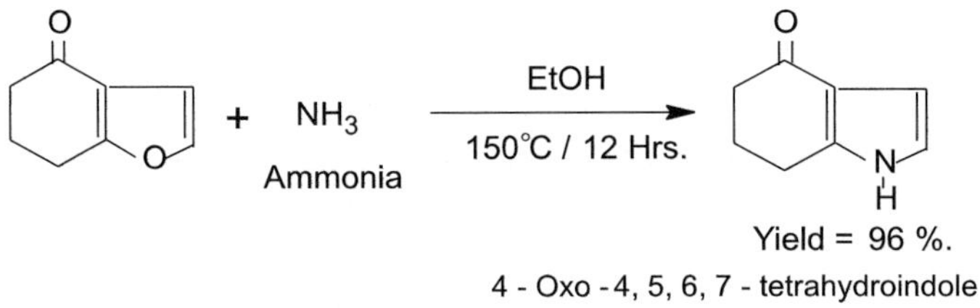


Figure 4.72.

This indole compound can be used as an intermediate for the synthesis of various kinds of 4-substituted indole drugs.

A tetrahydroindole compound was also prepared from the reaction of DHR with oxyiminoglyoxal (Figure 4.73) [41].

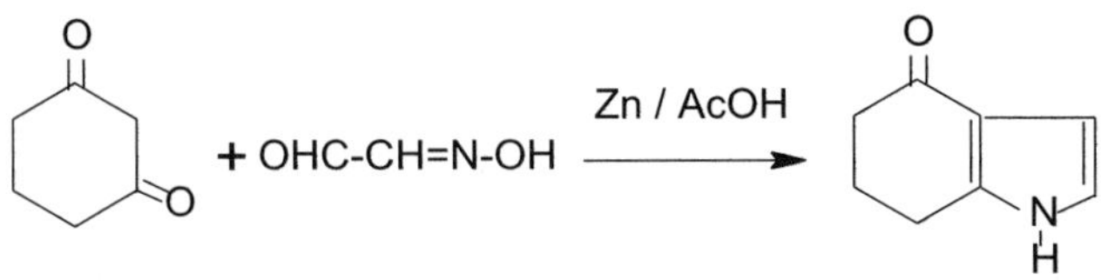


Figure 4.73.

Aminoacetaldehyde dimethyl acetal also produced tetrahydroindole derivative in the presence of *p*-toluene sulfonic acid (PTSA) catalyst (Figure 4.74) [42, 43].

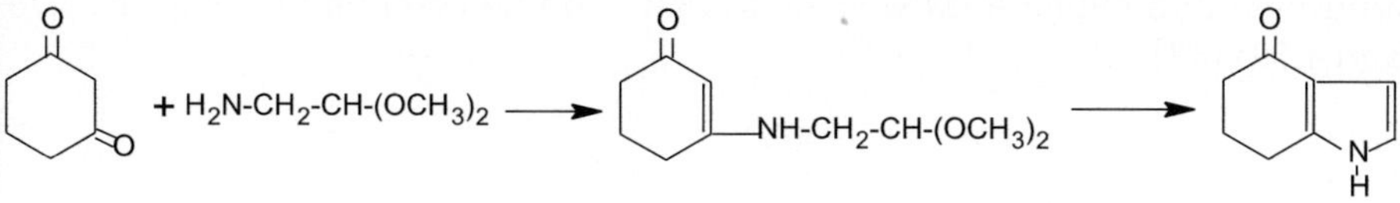


**Figure 4.74.**

Tetrahydroindole was produced in a two-step process from the reaction of an alkyl halopyruvate compound with DHR (Figure 4.75) [44].

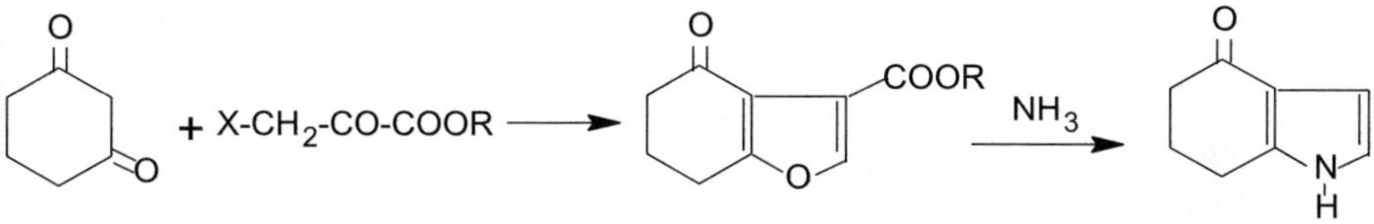


**Figure 4.75.**

By reacting DHR with an ethanolamine compound, *N*-3-oxo-cyclohexylidene ethanol was prepared first and then, in the next step, dehydrated the resulting material to obtain tetrahydroindole derivative (Figure 4.76) [45, 46].

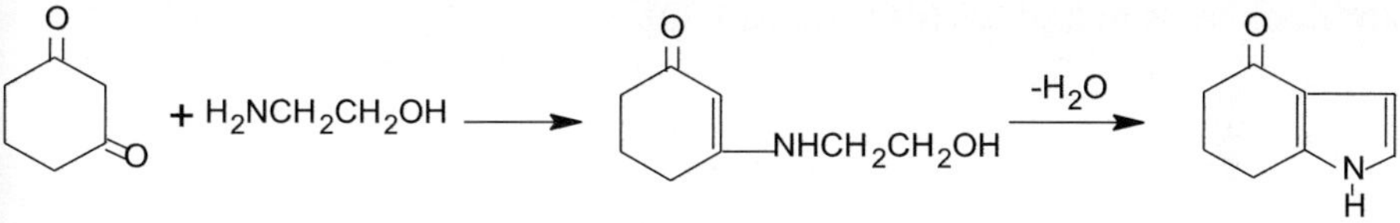


**Figure 4.76.**

Substituted 4,5,6,7-tetrahydro-4-oxindoles, used as psychotherapeutic agents, were prepared based on the following chemical reaction (Figure 4.77) [47].

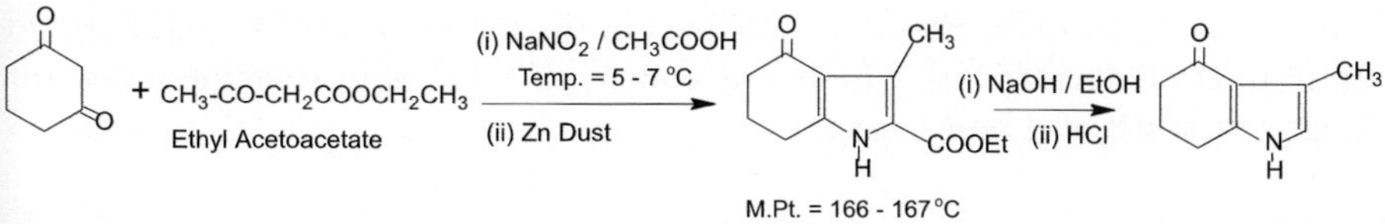


**Figure 4.77.**

Certain indole derivatives exhibited anti-inflammatory and analgesic activities.

### 4.4.5.7
### Reaction with Ammonia and Primary Amines

3-Aminocyclohexenone derivatives were obtained when DHR was reacted with ammonia and aliphatic as well as aromatic amine compounds (Figures 4.78 and 4.79) [48].

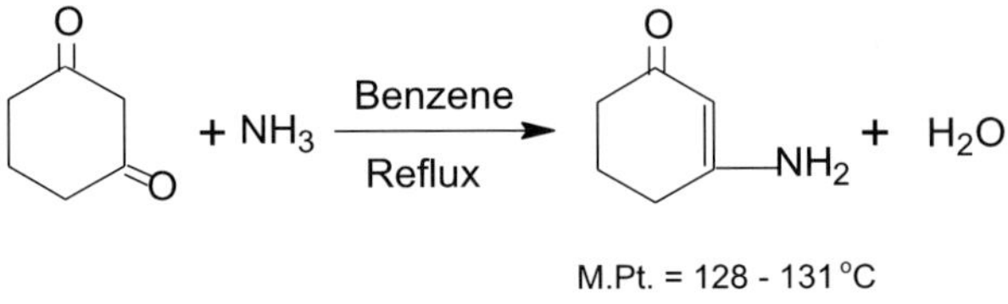


**Figure 4.78.**

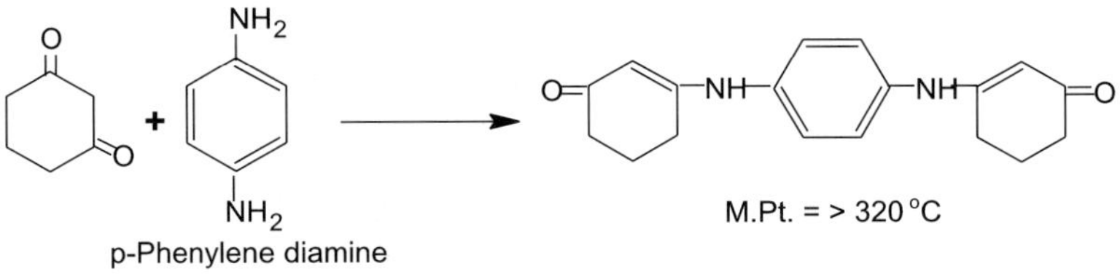


**Figure 4.79.**

Dehydrogenation of 3-aminocyclohexenone derivatives in the presence of Pd/carbon at elevated temperature conditions gave the corresponding aminophenols in high yields (Figure 4.80).

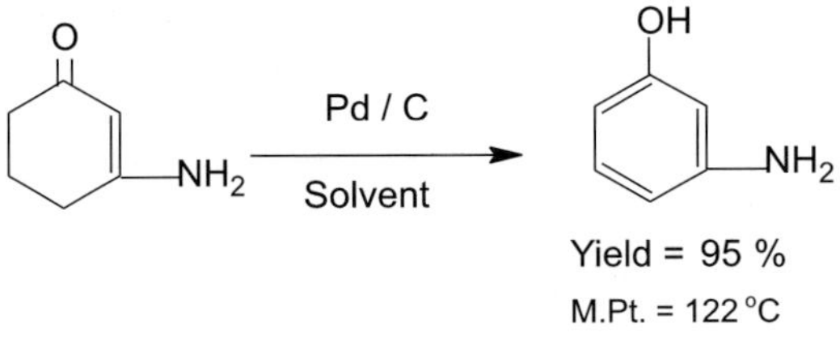


**Figure 4.80.**

This method can be effectively utilized in the manufacture of 3-substituted aminophenol derivatives, which are considered valuable intermediates for drugs, dyestuffs and herbicides.

### 4.4.5.8
### Reaction with $PCl_3$ and $POCl_3$ Compounds

Phosphorus trichloride reacted with DHR and produced 3-chloro-2-cyclohexen-1-one (Figure 4.81) [49].

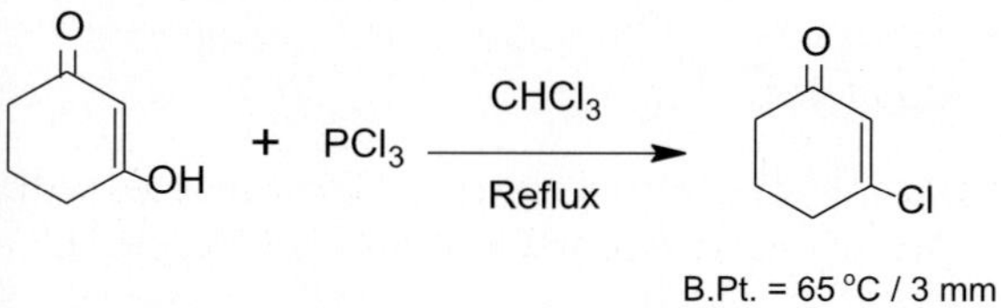


**Figure 4.81.**

2-Azasteroids displaying valuable pharmacological properties, such as antiviral, were manufactured by a total synthesis originated with 3-chloro-cyclohexenone compounds.

The reaction of phosphorus oxychloride with 2-allyl-1,3-cyclohexanedione gave 2-allyl-3-chloro-2-cyclohexen-1-one (Figure 4.82) [24].

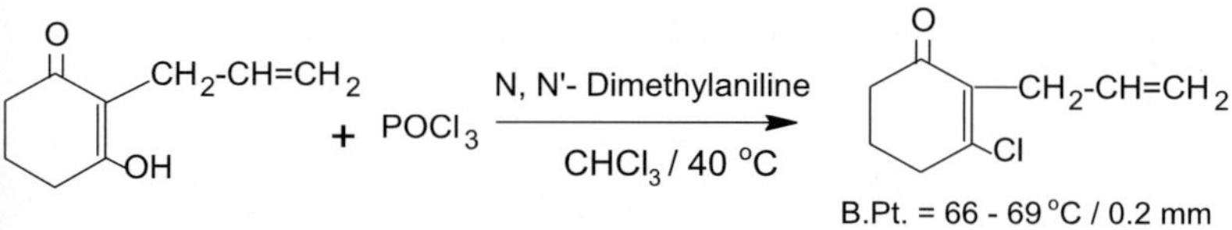


**Figure 4.82.**

### 4.4.5.9 Mannich-Type Compounds

Mannich-type derivatives used in the synthesis of 2-methyl resorcinol were obtained from the reaction of DHR with formaldehyde and dialkylamine compounds (Figure 4.83) [50].

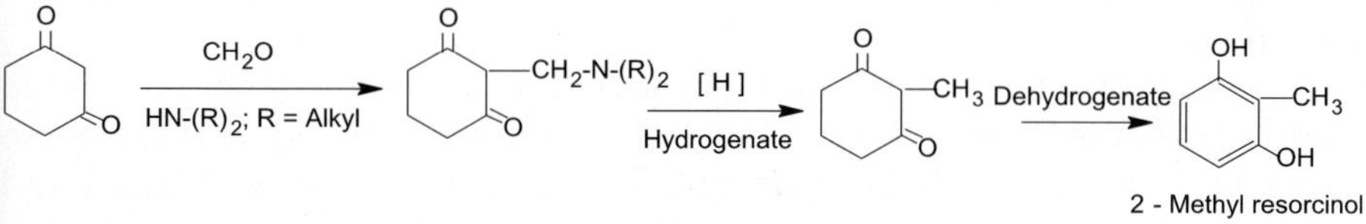


**Figure 4.83.**

### 4.4.5.10 General Reactions of DHR

Novel heterocyclic bis-hydrazones possessing tranquilizing properties were prepared by the reaction of DHR with nicotinic acid hydrazide (Figure 4.84) [51].

Ceric ammonium nitrate (CAN) mediated cycloaddition of 1,3-cyclohexanedione to phenyl vinyl sulfide produced substituted dihydrofuran-type derivative in a simple and efficient synthesis process (Figure 4.85) [52].

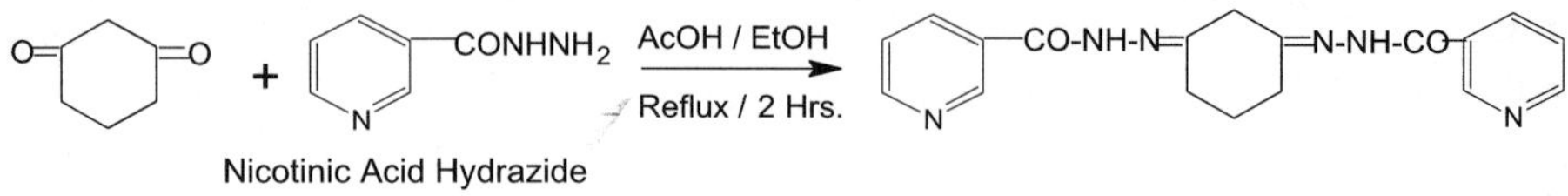


**Figure 4.84.**

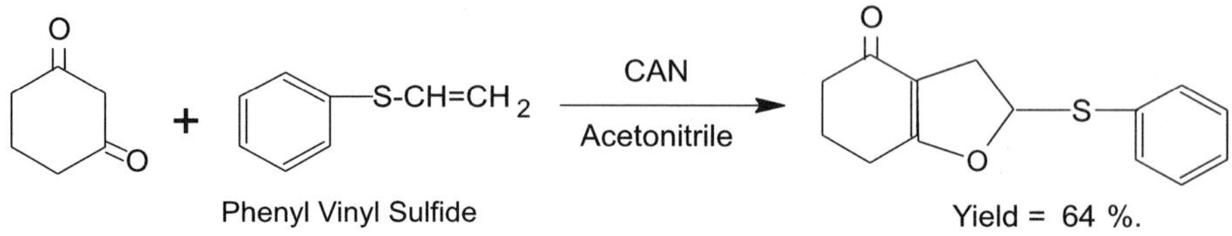


**Figure 4.85.**

Silylether compounds were obtained from DHR derivatives and organo halosilane by effecting the reaction in the presence of an acid acceptor (Figure 4.86) [53, 54].

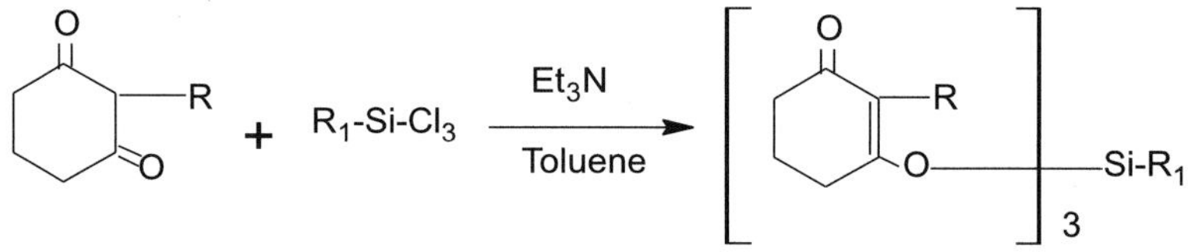


**Figure 4.86.**

The silylether compounds were useful as vulcanizing agents for silanol terminated poly(diorganosilanes). A copper substrate coated with the silylether mixture first and then held at 90% relative humidity conditions was found to be corrosion free.

The reaction of 3-cyanobenzaldehyde with 3-amino-2-cylohexen-1-one gave 9-(3-cyanophenyl)-3,4,6,7,9,10-hexahydro-1,8-(2H,5H)-acridinedione compound having the following structural formula. This compound was found to be useful as potassium channel in cells of mammals such as humans (Figure 4.87) [55].

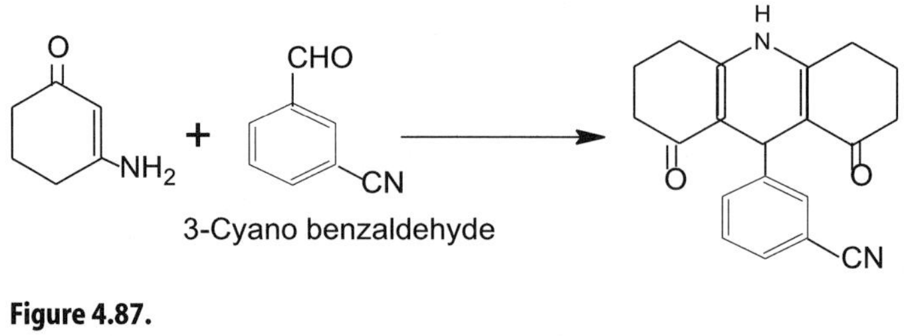


**Figure 4.87.**

## 4.4.6 Applications of 1,3-Cyclohexanedione Derivatives

### 4.4.6.1 Manufacture of Herbicides

Herbicides are used in agriculture to increase crop yields. These herbicides are effective in killing or controlling unwanted weeds, which compete for the soil nutrients with the crop plants. Two categories of herbicides, namely the pre-emergence and post-emergence, of different types have been widely used. The pre-emergence herbicides are normally incorporated into or applied to the soil prior to the emergence of the weed plants from the soil. On the other hand, the post-emergence herbicides are applied to plant surfaces after the emergence of weeds.

Mikado®, also known as Sulcotrione, is a well-known and novel broad spectrum maize herbicide, active against dicotyledonous weeds and barnyard grass (Echinochloacrusgalli). Sulcotrione is a foliar-applied post-emergence herbicide triketone class of compounds. This triketone compound is chemically known as 2-(2-chloro-4-mesylbenzoyl)-cyclohexane-1,3-dione and has been synthesized from 1,3-cyclohexanedione and 2-chloro-4-methanesulfonyl-benzoyl chloride compounds (Figure 4.88) [56].

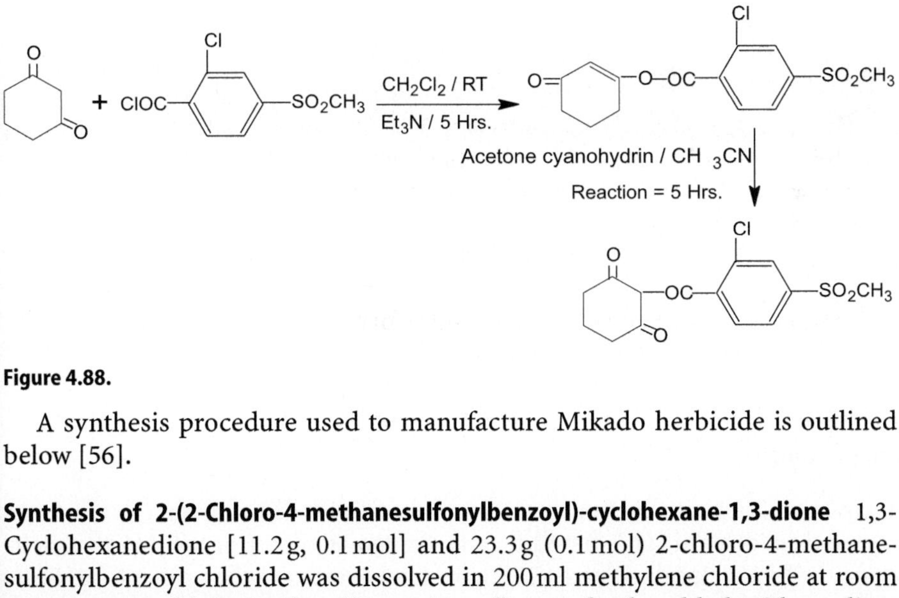


**Figure 4.88.**

A synthesis procedure used to manufacture Mikado herbicide is outlined below [56].

**Synthesis of 2-(2-Chloro-4-methanesulfonylbenzoyl)-cyclohexane-1,3-dione** 1,3-Cyclohexanedione [11.2 g, 0.1 mol] and 23.3 g (0.1 mol) 2-chloro-4-methane-sulfonylbenzoyl chloride was dissolved in 200 ml methylene chloride at room temperature. Triethylamine (11 g, 0.11 mol) was slowly added with cooling. The reaction mixture was stirred at room temperature for 5 h and then poured into hydrochloric acid (2N). The aqueous phase was discarded and the organic phase dried with $MgSO_4$ and then evaporated to yield the intermediate enol-ester 3-(2-chloro-4-methanesulfonylbenzoyloxy)-cyclohex-2-enone. The 3-(2-chloro-4-methanesulfonylbenzoyloxy)-cyclohex-2-enone was dissolved in 200 ml acetonitrile and triethylamine (22 g, 0.22 mol) was added all at once, followed by acetonecyanohydrin (0.8 g, 0.01 mol). The solution was stirred for 5 h and then poured into 2N HCl and extracted twice with ethyl acetate. The organic layer was dried with $MgSO_4$ and the solvent evaporated to yield the product.

Another synthetic method used to produce Sulcotrione® compound is based on the reaction scheme outlined below (Figure 4.89) [57].

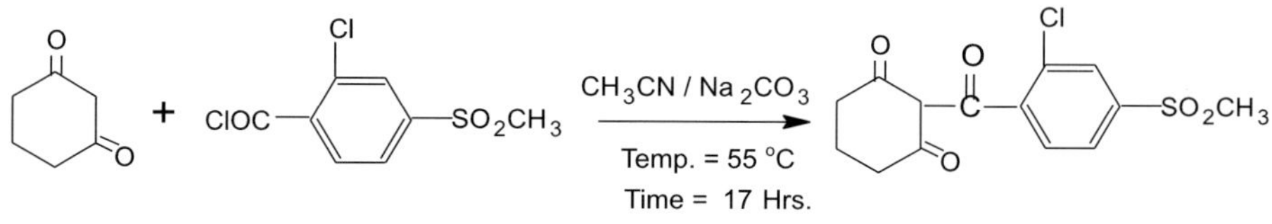


**Figure 4.89.**

With the use of $NaCO_3$ as the base, the enolization and rearrangement reactions were performed in one single step. The product obtained was a sodium salt of the triketone compound.

Mesotrione® is also another herbicide manufactured using DHR to control a wide spectrum of broad-leaf weeds affecting corn. This triketone compound is based on the following chemistry (Figure 4.90) [56].

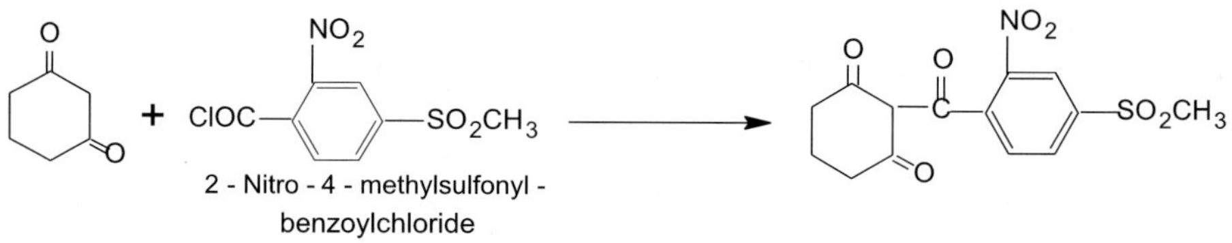


**Figure 4.90.**

Mesotrione® can be synthesized following the procedures outlined for Mikado® herbicide compound [56].

### 4.4.6.2
### Pharmaceuticals

Tyrosinemia-I is a rare but devastating metabolic disorder that presents as severe liver disease. This hereditary tyrosinemia-type I occur due to a deficiency of fumarylacetoacetate (FAH), the final enzyme in the tyrosine catabolic pathway.

This disorder is characterized by progressive liver failure, increased risk of hepatocellular carcinoma, coagulopathy, painful neurologic crises and renal tubular dysfunction resulting in rickets.

Orfadin® capsule contains nitisinone, which is 2-(2-nitro-4-trifluoromethyl benzoyl)-cyclohexane-1,3-dione (NTBC). The compound NTBC was synthesized from DHR and 2-nitro-4-trifluoromethyl benzoyl chloride reaction (Figure 4.91) [58].

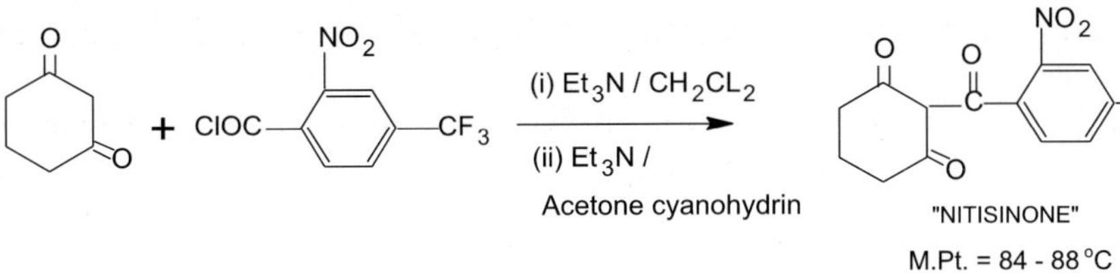


**Figure 4.91.**

A synthesis procedure used in the preparation of "nitisinone" is outlined below [58].

**Procedure for the Preparation of Nitisinone** Triethylamine (3.4 ml, 25 mM) is added dropwise to a stirred solution of 2-nitro-4-trifluromethylbenzoyl chloride (19 mM; itself obtained by reaction of 2-nitro-4-trifluoromethylbenzoic acid with an excess of oxalyl chloride) and cyclohexane-1,3-dione (19 mM) in dichloromethane (100 ml). The mixture is stirred for 1 h at ambient temperature and then further triethylamine (57 mM) and acetone cyanohydrin (0.4 ml) is added. Stirring is continued for a further 2.5 h and then the mixture is washed thoroughly with 2 M hydrochloric acid to remove triethylamine. The organic phase is then extracted thoroughly with potassium carbonate (5% w/v). The combined basic extracts are then acidified with 2 M hydrochloric acid and extracted with ether. The ether extracts are washed with saturated sodium chloride solution, dried (magnesium sulphate) and the volatile material removed by rotary evaporation in vacuo. There may thus be obtained 2-(2-nitro-4-trifluoromethylbenzoyl)-1,3-cyclohexanedione as a solid, M.Pt. = 84–88 °C (recrystallized from ethyl acetate).

Nitisinone (NTBC) is a white to yellowish-white crystalline powder, practically insoluble in water, sparingly soluble in alcohol, soluble in 2 N NaOH and methanol solutions.

NTBC is an inhibitor of the catalytic activity of the enzyme HPPD and may be useful in the treatment of diseases where the products of the action of HPPD are involved, for example tyrosinemia. NTBC was successful in reversing the symptoms of tyrosinemia and renal tissue to normalize.

**Indole Drugs** Pindolol, brand name Visken, is used to treat high blood pressure and has the following structural formula (Figure 4.92).

4-Oxo-4,5,6,7-tetrahydroindole, synthesized from DHR, can be transformed into the drug "Pindolol" via its dehydrogenation product 4-hydroxyindole (Figure 4.93).

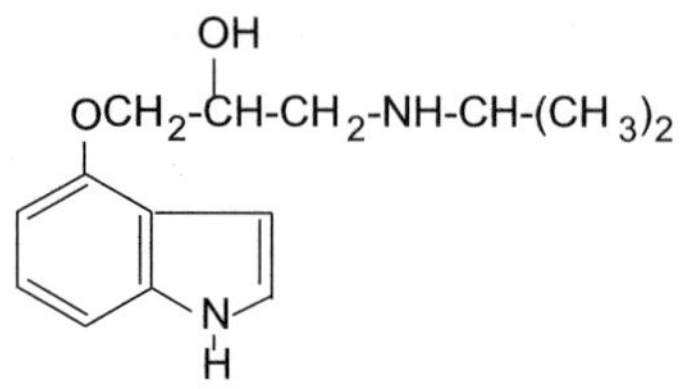


**Figure 4.92.**

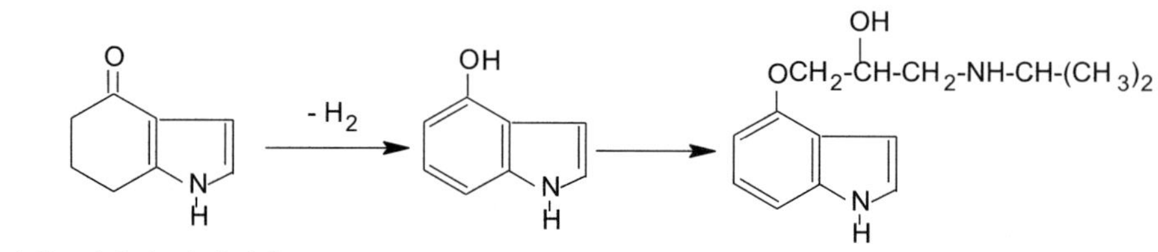


**Figure 4.93.**

## 4.4.6.3
## Building Block for Steroids and Terpenoids

2-Methyl-1,3-cyclohexanedione has been used as a starting material for the synthesis of several polycyclic compounds. This was used in the synthesis of a racemic 8a-methyl-3,4,8,8a-tetrahydro-1,6-(2H, 7H)-naphthalenedione, a well known building block for the steroids and terpenoids, particularly progesterone synthesis (Figure 4.94) [59–61].

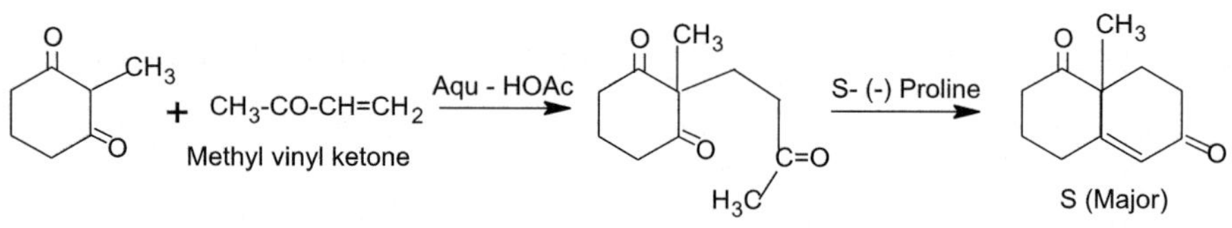


**Figure 4.94.**

2,2-Disubstituted-1,3-cyclohexanedione derivative was obtained by alkylating 2-(3′-butenyl)-1,3-cyclohexanedione with methyl iodide (Figure 4.95) [62]. This compound could be used as an intermediate for the synthesis of steroids.

## 4.4.6.4
## Chemical Intermediates

2-Cyclohexenone was prepared by the reduction of 3-ethoxy-2-cyclohexenone with lithium aluminum hydride ($LiAlH_4$) followed by hydrolysis and dehydration of the reduction product (Figure 4.96) [63].

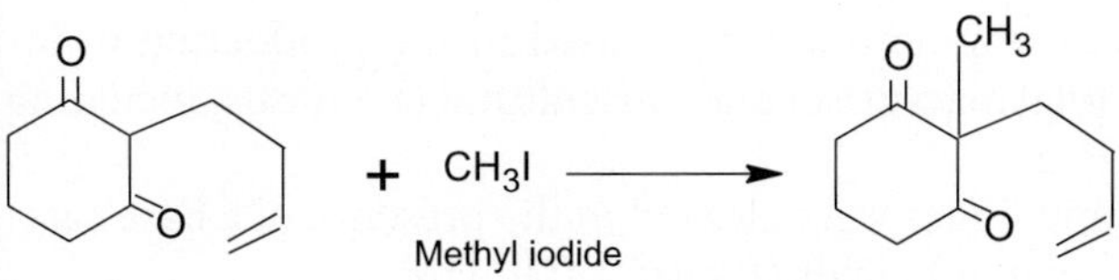


**Figure 4.95.**

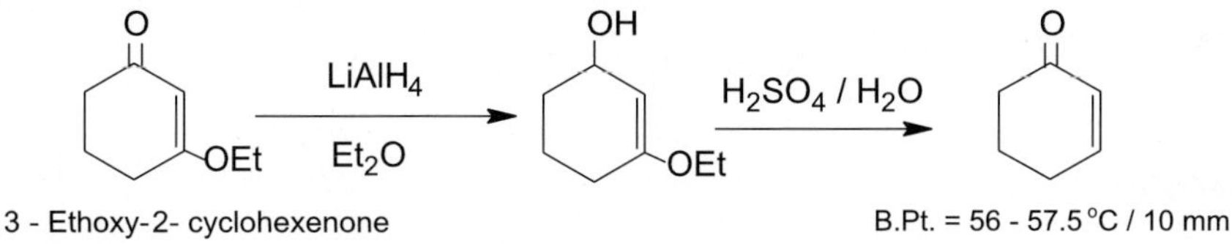


**Figure 4.96.**

Dehydrogenation of 3-alkoxy-cyclohexenone compounds in the presence of Pd/C catalyst at elevated temperatures produced mono-ether derivatives of resorcinol (Figures 4.97 and 4.98) [27].

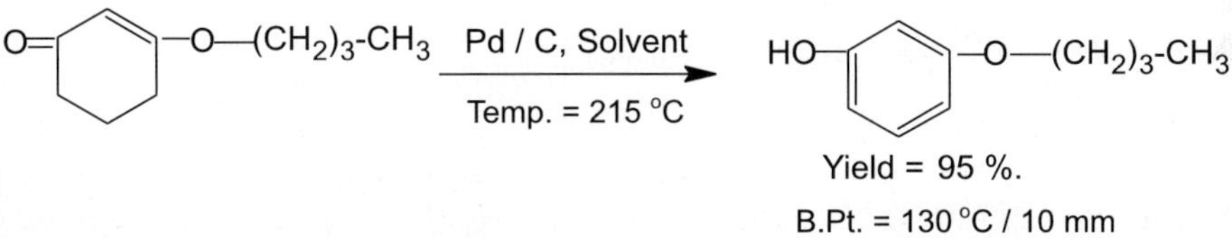


**Figure 4.97.**

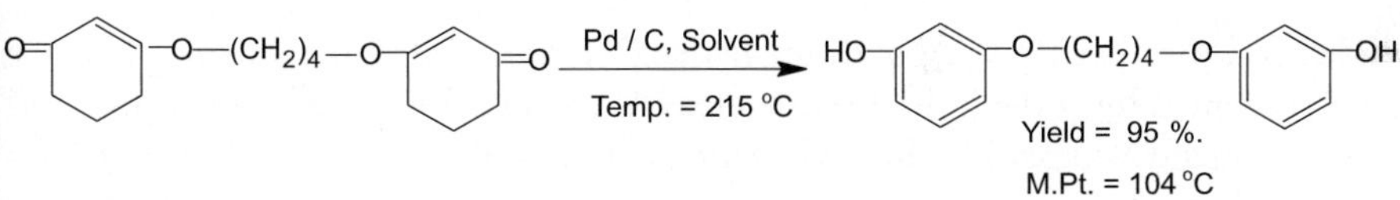


**Figure 4.98.**

2-Acetyl resorcinol was synthesized in a single step dehydrogenation of 2-acetyl-1,3-cyclohexanedione, and obtained in high yields (Figure 4.99). [21,64]

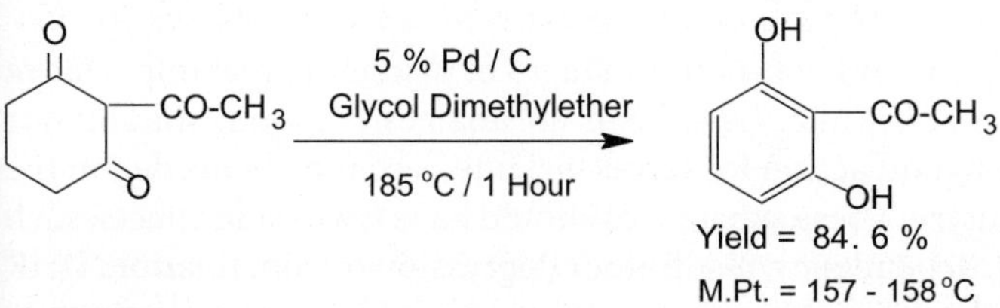


**Figure 4.99.**

This acetyl resorcinol compound has been used in the production of intermediates for various pharmaceutical and agricultural chemicals, including cough and phlegm medicines.

2-Alkyl-1,3-cyclohexanediones were cleaved in the presence of a base catalyst to obtain δ-oxoacids in good yields (Figure 4.100) [16].

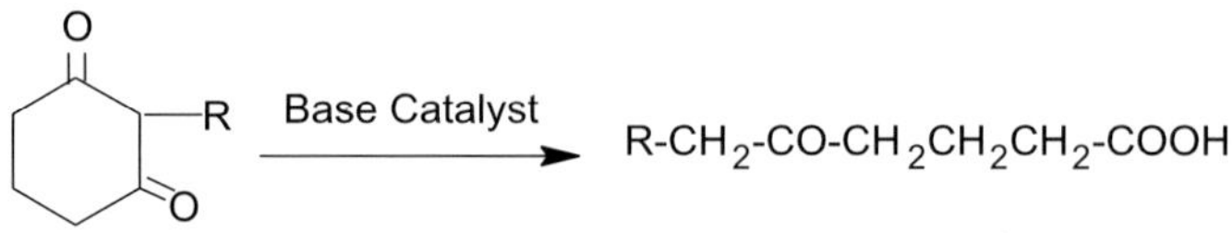


**Figure 4.100.**

Barium hydroxide was reported to be the best catalyst for acid cleavage reactions. The above reaction could also be carried out using sodium or potassium hydroxide catalysts.

Methylene di-(1,3-cyclohexanedione) produced practically a quantitative yield of brassylic acid with the use of reductive acid-cleavage method (Figure 4.101) [25].

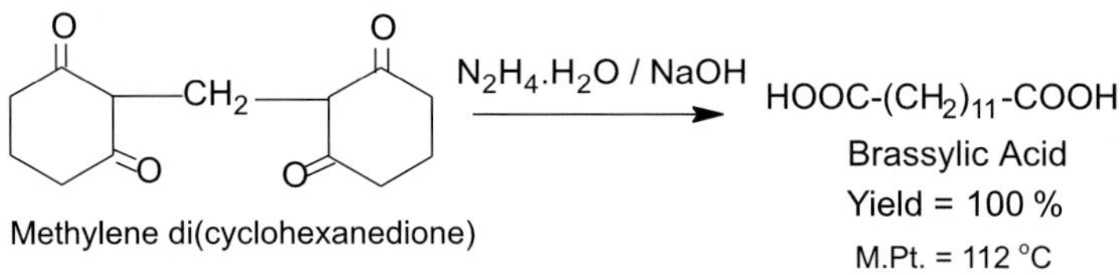


**Figure 4.101.**

In the preparation of brassylic acid, the dicyclohexanedione compound was reduced by refluxing with sodium hydroxide and hydrazine in the presence of a solvent. The sodium hydroxide cleaved the cyclohexanedione compound and produced δ-oxoacid. The hydrazine present in the reaction mixture immediately converted the keto group into hydrazone which subsequently lost the nitrogen in the normal manner to give the reduced dicarboxylic acid.

#### 4.4.6.5 Synthesis of Drugs for Pain Treatment

The treatment of chronic and non-chronic stages of pain is of great importance in medicine. Substances, which have been an analgesic action, suitable for pain treatments with good action for target-oriented are mostly needed in the pharmaceutical industry. These substances should have fewer side effects, such as nausea, vomiting, dependency, respiratory depression or constipation. DHR was used in the synthesis of phenolic derivatives useful in the applications of pain treatments (Figure 4.102) [65].

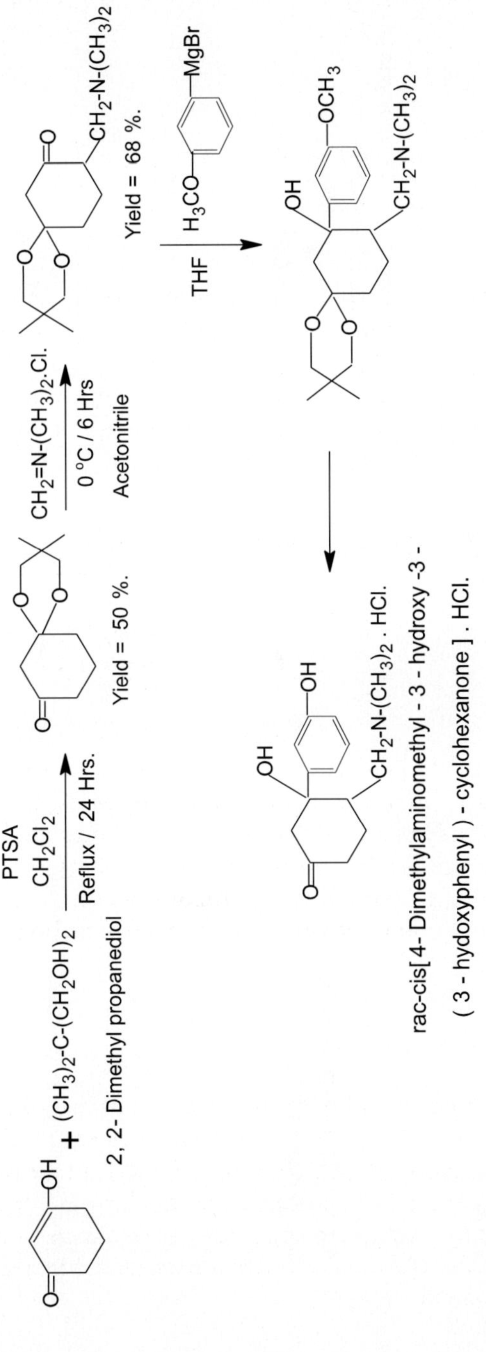
OH
$(CH_3)_2$-C-$(CH_2OH)_2$
2, 2- Dimethyl propanediol
PTSA
$CH_2Cl_2$
Reflux / 24 Hrs.
Yield = 50 %.
$CH_2$=N-$(CH_3)_2$.Cl.
0 °C / 6 Hrs
Acetonitrile
$CH_2$-N-$(CH_3)_2$
Yield = 68 %.
THF
$H_3CO$
MgBr
OH
$OCH_3$
$CH_2$-N-$(CH_3)_2$
OH
OH
$CH_2$-N-$(CH_3)_2$ . HCl.
rac-cis[ 4- Dimethylaminomethyl - 3 - hydroxy -3 -
( 3 - hydoxyphenyl ) - cyclohexanone ] . HCl.

Figure 4.102.

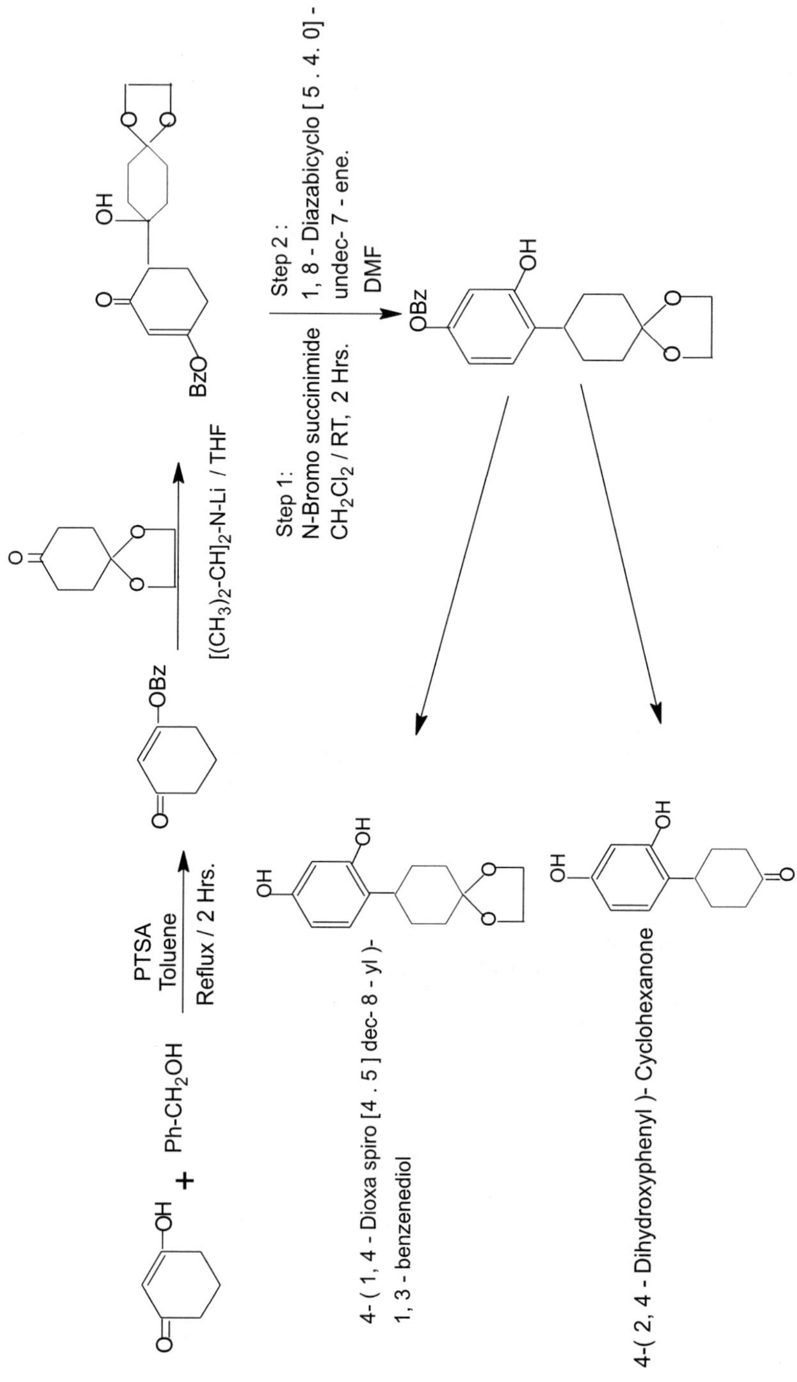
Ph-CH$_2$OH
PTSA
Toluene
Reflux / 2 Hrs.
OH
O
OBz
[(CH$_3$)$_2$-CH]$_2$-N-Li / THF
BzO
OH
Step 1:
N-Bromo succinimide
CH$_2$Cl$_2$ / RT, 2 Hrs.
Step 2 :
1, 8 - Diazabicyclo [ 5 . 4. 0] -
undec- 7 - ene.
DMF
4- ( 1, 4 - Dioxa spiro [4 . 5 ] dec- 8 - yl )-
1, 3 - benzenediol
4-( 2, 4 - Dihydroxyphenyl )- Cyclohexanone

**Figure 4.103.**

**Skin Lightening Agents** Human skin color arises from a series of complex cellular processes carried out within melanocyte cells. These cells are located in the lower part of epidermis and their main function is to synthesize melanin pigments, which protects the body from the damaging effects of UV rays from the sunlight. When the skin is exposed to sunlight, melanocytes increase their melanin synthesis. Melanins produced are first deposited in melanosomes and then carried to the surface of the skin by keratinocytes, which results in the skin exhibiting a brown color, known as a "tan". The darkness of the skin is proportionate to the amount of melanin synthesized. The mechanism of pigment or melanin formation is particularly complex and involves schematically shown below.

Tyrosine → Dopa → Dopaquinone → Dopachrome → Melanin.

The enzyme, which takes part in the sequence of reactions, is essentially tyrosinase. The activity of tyrosinase is promoted by the action of UV rays, eventually having melanins formed as chromatism in the skin.

The depigmenting or skin lightening agent or substance interferes with one of the stages of melanin biosynthesis. The skin lightening agents inhibit or decrease the tyrosinase activity in the melanin synthesis.

Resorcinol derivatives have been found to be very good tyrosinase activity inhibiting property and therefore may act as skin lightening agents. In the pharmaceutical industry, there is a strong demand for chemical compounds or agents, which can enable acquired spots or freckles to be restored to a normal skin color. Resorcinol compounds known to be useful in the cosmetic field as skin lightening agents are primarily 4-substituted derivatives. DHR has been utilized in the synthesis of 4-substituted resorcinolic compounds for the skin lightening agent applications and the chemistry is outlined below (Figure 4.103) [66–70].

## References


1. R.B. Thompson, Org. Synth. Coll. Vol. 3, p 278 (1955).
2. Hoffmann-La Roche, Brit. Patent, 416,892 (1934).
3. Hoffmann-La Roche, Brit. Patent, 427,297 (1935).
4. H.A. Smith and B.L. Stump, J. Am. Chem. Soc, Vol. 83, p 2739 (1961).
5. J.C. Sincar and A.I. Meyers, J. Org. Chem, Vol. 30, p 3206 (1965).
6. K. Masahiro and M. Hisaya, Jap. Pat. Appln, 02-117862 (1990).
7. O. Kyoko et al., Jap. Pat. Appln, 2000-168386 (2000).
8. Mokotoff et al., J. Org. Chem, Vol. 39, p 409 (1974).
9. V. Elango and R. Sakamuri, US Patent, 5,744,648 (1998).
10. J.N. Chatterjea and V.N. Mehrotra, J. Indian Chem. Soc, Vol. 39, p 599 (1962).
11. Dictionary of organic compounds, 4th Edn, Eyre & Spottiswoode Publishers, London, 1965.
12. The Merck index, 10th Edn, Merck & Co, Rahway, NJ, USA, 1983.
13. H. Stetter and M. Coenen, Chem. Ber, Vol. 87, p 992 (1954).
14. A.M. Mekler, S. Ramachandran, S. Swaminathan and M.S. Newman, Org. Synth. Coll. Vol. 5, p 743 ( 1973).

15. H. Stetter, in Newer methods of preparative organic chemistry, Vol. II, W. Foerst Ed, Academic Press, New York, 1963.
16. H. Stetter and W. Dierichs, Chem. Ber, Vol. 85, p 61 (1952).
17. H. Stetter and M. Coenen, Chem. Ber, Vol. 87, p 992 (1954).
18. H. Stetter, C. Buntgen and M. Coenen, Chem. Ber, Vol. 88, p 79 (1955).
19. H. Stetter and E. Klanke, Chem. Ber. Vol. 86, p 513 (1953).
20. I.N. Nazarov and S.I. Zavyalov, Bull. Acad. Sci, USSR (Engl. Transl) p 639 (1959).
21. K. Tanaka, Euro. Pat. Appln, 0219282 (1986).
22. S. Mohrmann, Ber, Vol. 96, p 2257 (1963).
23. N. Schamp, Bull. Soc. Chim. Belg, Vol. 73, p 38 (1964).
24. J.G. Strong, US Patent, 3,852,357 (1974).
25. H. Stetter and W. Dierichs, Chem. Ber, Vol. 85, p 290 (1952).
26. F.E. King and D.G.I. Felton, J. Chem. Soc, p 1371 (1948).
27. W.H. Muller, US Patent, 4,046,816 (1977).
28. W.F. Gannon and H.O. House, Org. Syn. Coll. Vol. 5, p 539 (1973).
29. D.C. Rodrigues, A. Meijere and A.J. Marsaioli, J. Braz. Chem. Soc, Vol. 13, p 664 (2002).
30. J.J. Panouse and C. Sannie, Bull. Soc. Chim. France, p 1272 (1956).
31. W.J. Michaely and G.W. Kratz, US Patent, 4,816,066 (1989).
32. C.G. Carter, US Patent, 4,822,906 (1989).
33. C.G. Carter, D.L. Lee, D.J. Michaely and G.W. Kratz, US Patent, 5,006,158 (1991).
34. I. Nasuno, M. Shibata, M. Sakamoto and K. Koike, US Patent, 5,468,878 (1995).
35. I. Nasuno, M. Shibata, M. Sakamoto and K. Koike, US Patent, 5,480,858 (1996).
36. M.P. Esminger and J.M. Shribbs, US Patent, 5,506,195 (1996).
37. H.B. Schar and J. Chen, 6, 541, 422 (2003).
38. I.N. Nazarov and S.I. Zavyalov, Zhur. Obshchei. Khim, Vol. 23, p 1703 (1953); CA. Vol. 48, 13667 (1954).
39. S. Torii et al., US Patent, 4,336,202 (1982).
40. M. Matsumoto and N. Watanabe, US Patent, 4,558,143 (1985).
41. Japanese Patent, 9904 (1969).
42. J.M. Bobbitt and C.P. Dutta, J. Chem. Soc. Commun, p 1429 (1968).
43. J.M. Bobbitt et al., J. Org. Chem, Vol. 43, p 3541 (1978).
44. Japanese Patent Appln, 19971 (1979).
45. I. Fumito, H. Katsumasa and N. Takashito, Jap. Pat. Publ, 59-144758 (1984).
46. I. Fumito, H. Katsumasa and N. Takashito, Jap. Pat. Publ, 59-144757 (1984).
47. K. Scheon, K. Gardens and I.J. Patcher, US Patent, 3,467,755 (1969).
48. W.H. Miller, US Patent, 4,212,823 (1980).
49. R.J. Chorvat and R. Pappo, US Patent, 4,007,194 (1977).
50. T. Nakamatsu, Y. Nishida and N. Kometani, US Patent, 5,276,198 (1994).
51. M. Zimmermann, US Patent, 3,015,663 (1962).
52. Y.R. Lee, G.J. Lee and K.Y. Kang, Bull. Korean. Chem. Soc, Vol. 23, p 1477 (2002).
53. J.A. Cella, US Patent, 4,210,596 (1980).
54. J.A. Cella, US Patent, 4,257,957 (1981).
55. O. Cyrus, Jap. Pat. Publ, 05-201986 (1993).
56. H.M. Chin, US Patent, 4,781,751 (1988).
57. H. Rawlinson, J.W. Wiffen and S.M. Brown, US Patent, 5,728,889 (1998).
58. M.K. Ellis et al., US Patent, 5,550,165 (1996).
59. P. Buchschacher, A. Furst and J. Gutzwiller, Org. Syn. Coll. Vol. 7, p 368 (1990).
60. S. Dadishefsky, P. Cain and A. Nagel, J. Am. Chem. Soc, Vol. 97, p 380 (1975).
61. T.A. Spencer et al., J. Org. Chem, Vol. 33, p 712 (1968).
62. H. Yoshio and K. Koichi, Jap. Pat. Publ, 07-291888 (1995).

63. W.F. Gannon and H.O. House, Org. Syn. Vol. 40, p 14 (1960).
64. N. Hirowatari and W.K. Yamaguchi, Euro. Pat. Appl, 0331422 (1989).
65. C. Puetz, H. Buschmann and B.Y. Koegel, US Pat. Appl, 2003 / 0096811 (2003).
66. S.E. Bradley et al., US Patent, 6,504,037 (2003).
67. S.E. Bradley et al., US Pat. Appl, 2002 / 0032352 A1 (2002).
68. S.E. Bradley et al., US Pat. Appl, 2002 / 0114770 A1 (2002).
69. S.E. Bradley et al., US Pat. Appl, 2002 / 0137961 A1 (2002).
70. S.E. Bradley et al., US Pat. Appl, 2003 / 0088113 A1 (2003).

# 5 Resorcinol Based Resins and Applications

## 5.1 Resorcinol-Aldehyde Resins

### 5.1.1 Introduction

Resorcinol has been considered to be one of the important raw materials to develop various kinds of synthetic resins and adhesives with outstanding properties. The resorcinolic adhesives developed from the reaction of resorcinol and formaldehyde were primarily used for the fabrication of plywood aircraft bombers and lamination of hardwood helicopter propellers during World War II. These adhesives were found to be excellent in their durability and unique low temperature properties. The adhesive bonds resulting from the use of resorcinolic adhesives had good structural strength. These joints or bonds are known to be resistant to many kinds of conditions including boiling water, high humidity and most types of solvents except highly caustic solutions.

Resorcinol-based adhesives have been widely used in the bonding of wood materials, particularly in the construction of laminated wooden beams and arches. Similarly, these adhesives are known to be useful in the bonding of many other materials such as paper, textiles, rubbers, metals, glass, ceramics and concrete. In the manufacture of synthetic adhesives for specific applications, various raw materials are reacted with resorcinol. The most important raw materials that are often reacted with resorcinol come from the family of aliphatic aldehydes. Starting from the lowest member, namely formaldehyde, to higher members in the aliphatic aldehyde series, various aldehydes including acetaldehyde, propionaldehyde, butyraldehyde and so on, have been used in the synthesis of different resorcinolic resins and adhesives for numerous high performance applications.

When developing resorcinol-aldehyde resins for different applications, it is important to understand the chemistry and mechanism by which these materials are developed. With a full understanding and knowledge of the reactions of resorcinol with aldehydes, one can develop better chemistries and high performance materials for the demanding market needs. Therefore, this chapter mainly focuses on the chemistry of resorcinolic resins from resorcinol and aldehydes reactions.

### 5.1.2 Resorcinol-Formaldehyde (RF) Resins

The reaction of resorcinol (R) with formaldehyde (F) to produce resorcinol-formaldehyde (RF) resins is well known. RF resins have a wide range of industrial applications particularly in the wood and rubber industry. In the wood industry, RF resins have been used as room temperature curable adhesives to bond wood products. In the rubber industry, these resins are widely used to enhance the adhesion between various synthetic fibers, such as steel, polyester, nylon, rayon, aramid and glass, and rubber compounds.

The reaction of formaldehyde with resorcinol can be similar to phenol. However, the reactivity of resorcinol with formaldehyde is completely different, and therefore the products formed from the reaction can be different. With phenol, complete reactivity and mechanistic studies have been made and well documented. On the other hand, the resorcinol-formaldehyde reaction may not be clearly understood and properly documented. Therefore, in the following sections, an outline of the chemistry, reactions and mechanisms of formaldehyde reaction with the resorcinol molecule under various catalytic conditions is presented.

#### 5.1.2.1 Resorcinol Structure and Reactivity for Formaldehyde Reaction

Resorcinol, also known as 1,3-dihydroxy benzene, has two hydroxyl groups on the same benzene ring as shown in the following structure (Figure 5.1).

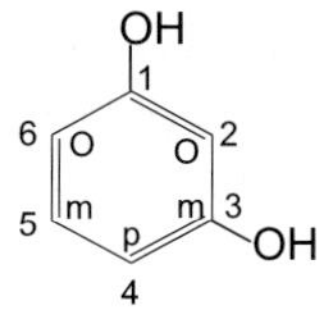


**Figure 5.1.**

These two hydroxyl groups are located at the 1- and 3-positions in the benzene ring. The molecular structure of resorcinol is very unique due to the location of these two hydroxyl groups in the molecule. The reactive sites or positions in the resorcinol molecule for various chemical reactions, including formaldehyde, are 2, 4 and 6. Of these three, the 4- and 6-positions are located at either "ortho" (adjacent) or "para" (opposite) to the two hydroxyl groups. The 2-position of resorcinol is located in between the two hydroxyl groups, and only ortho to the hydroxyl groups. Therefore all three reactive positions of the resorcinol molecule are doubly activated by the two hydroxyl groups. The electron density of $\delta$ (delta) and $\pi$ (pi) electrons of resorcinol molecule contributes to the electrophilic substitution of protons at the 2-, 4- and 6-positions of the benzene ring.

In order to compare the reactivity of resorcinol against various phenolic compounds, studies have been made [1–5]. Table 5.1 summarizes the reaction rate of formaldehyde with different phenolic derivatives including resorcinol. Assuming the reactivity of phenol, with one hydroxyl group and no other substituent in the benzene ring, with formaldehyde is equal to one, the reactivity of other phenolic derivatives were compared.

**Table 5.1.** Relative reaction rates of phenolic compounds with formaldehyde

| Phenolic compound | Comparative rate |
|---|---|
| Phenol | 1 |
| *p*-Methyl phenol | 0.35 |
| *o*-Methyl phenol | 0.26 |
| *m*-Cresol | 2.88 |
| 3,5-Dimethyl phenol | 7.75 |
| Resorcinol | 12.2[a] |

[a] Rate constant quoted at pH 3.5 is minimum for resorcinol
Data from [1–5]

From Table 5.1 results, it can be seen that the methyl group present in the ortho or para position in the benzene ring of a phenolic molecule tends to retard or slow the rate of reaction with formaldehyde. On the other hand, the presence of a methyl group at the meta position appeared to increase the reactivity. At the same time, the introduction of a hydroxyl group in the phenolic nucleus at the meta position increased the rate of reaction more than the alkyl group. With resorcinol this effect was particularly obvious where the two hydroxyl groups are meta to one another. The resonance structures produced by the meta hydroxyl groups activate the *ortho*- and *para*-positions of the benzene nucleus, which accounts for the enhanced reactivity of resorcinol towards the formaldehyde reaction.

Due to the high reactivity of resorcinol when compared to phenol, the reaction conditions required to produce resins from resorcinol and formaldehyde could be different. Resorcinol can undergo all of the typical reactions of phenol, but at a much faster rate because of the enhanced electron density at the 2-, 4- and 6-positions. The ratio of electron density at the 2-position (ortho) to the 4- or 6- (*ortho-para*) position is about 5.8 [6].

Though the electron density is higher at the 2-position, it is sterically hindered by the adjacent hydroxyl groups and therefore the substitutions primarily occur at the 4- and 6-positions [7].

### 5.1.2.2
### Uncatalyzed Reaction of Resorcinol with Formaldehyde

Because of the high electron density at the 2, 4 and 6 reactive positions, resorcinol can react with formaldehyde at room temperature conditions without any catalysts, and at a faster rate. Kinetic studies were performed to determine the reaction velocity of resorcinol with formaldehyde in the absence of a catalyst, meaning under neutral conditions [3]. In the reported study, solutions were prepared first by mixing resorcinol and unstabilized (no alcohol present) aqueous formaldehyde solution at different molar ratios and the pH of the mixture were determined. The results are presented in Table 5.2.

**Table 5.2.** Resorcinol-formaldehyde solution pH

| Resorcinol (mol) | Formaldehyde (mol, 37% aqu.) | pH |
|---|---|---|
| 1 | 0.5 | 3.87 |
| 1 | 1 | 3.93 |
| 1 | 1.5 | 4.09 |

Note: Solution does not contain any alcohol
Data from [3]

As can be seen from Table 5.2, the pH of resorcinol and formaldehyde solutions appeared to vary between 3.87 and 4.09 depending on their molar ratios. Reaction kinetics was also followed by determining the concentrations of unreacted formaldehyde as a function of reaction time on the RF reactions carried out at different temperature conditions. Kinetics data showed that the uncatalyzed RF reaction appeared to be first order with activation energy of 19 kcal/mol for all the temperatures studied.

The reaction velocity constant of uncatalyzed RF reaction was compared with the published values of base catalyzed formaldehyde reactions with various substituted phenolic derivatives. From this study, it was concluded that the reaction velocity constant for the uncatalyzed RF reaction might have been several orders of magnitude greater than that of the uncatalyzed phenol-formaldehyde (PF) reaction. The study results clearly suggested or indicated that RF reactions could be carried out without any additional catalyst, and the reaction could be faster even at the pH conditions of these two mixes (pH approximately 4.0).

Quantitative $^{13}C$ NMR analyses were performed on the uncatalyzed RF reaction carried out at room temperature with the R/F ratio of 1:1 [8]. Aliphatic carbons associated with the methylene bridges of the RF resin or reaction product appeared at 29.93, 24.23 and 18.53 ppm peak positions in the $^{13}C$ NMR spectrum for the 4,4′-, 2,4′- and 2,2′-methylene bridged carbons respectively. From this analysis, it was observed that the uncatalyzed RF reaction at room temperature and a pH of 4.0 produced only methylene bridged substitutions. The absence of methylol groups in the uncatalyzed RF reaction product strongly suggests a very fast reaction occurred between resorcinol and formaldehyde at room temperature conditions.

### 5.1.2.3
### Determination of Resorcinol Reactivity on Varying pH

The reaction of resorcinol with formaldehyde appeared to proceed at a good rate even in the absence of any catalyst. This prompts one to wonder at what conditions, such as temperature, pH and catalyst, the curing reaction of RF condensation can be best controlled. From the literature, it was found that a remarkable discovery was made in 1946 at Aero Research Ltd in Duxford, England [9]. A published report from this company had shown that on heating a fusible RF (novolak type) resin with additional formaldehyde, the gelation time (crosslinking) was dependent on the pH of the curing mixture. On measuring the gelation time at various pH values, these data points were plotted and a bell-shaped curve was obtained (Figure 5.2).

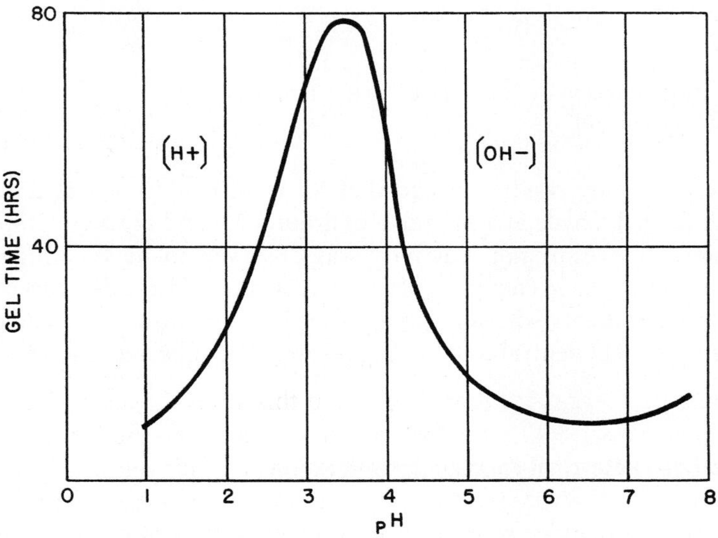


**Figure 5.2.** Resorcinol-aldehyde condensation – gel time versus pH relation

From this curve, the maximum gelation time was determined at a pH of between 3 and 4. This data strongly suggested that with resorcinol or RF resins at pH values less than 3, the reaction may be catalyzed by hydrogen ions $H^+$. When pH values are greater than 3 then the reaction could be catalyzed by the hydroxide ions $OH^-$.

To enhance the knowledge on determining the effect of pH on the reactivity of formaldehyde with the RF resin, another study was made and the results were published in 1947 [10]. In this study, a 50% aqueous RF solution containing no catalyst was made. Then, various resin and catalyst mix solutions having different pH values were made by mixing 100 g of catalyst free RF solution with an acid or base. After adjusting the pH of solutions to the desired values, 25 g of 37% aqueous formaldehyde solution were then added, and the gel times were measured. The results obtained are presented in Table 5.3.

**Table 5.3.** Reactivity of RF resin with formaldehyde at varying solution pH

| Aqueous solution pH (50%) | Gel time at 25 °C | Gel time at 140 °F |
|---|---|---|
| 1.92 | 5 min | |
| 2.96 | >30 h | 20 |
| 5.11 | 7 h | 10 |
| 5.78 | 4.5 h | 8.5 |
| 6.23 | 2.75 h | 6.5 |
| 6.51 | 1 h | 4.5 |
| 6.65 | 50 min | 4 |
| 6.9 | 40 min | 3.5 |
| 7.04 | 24 min | 3 |
| 7.11 | 20 min | 2.25 |

RF resin used for reactivity was made using R/F ratio 1:0.67
Data source: from [10]

These two study results showed that RF resin could react at the slowest rate with formaldehyde at a pH value of around 3.0 and at room temperature conditions. The resorcinol reactivity was observed to be very high in the presence of strong acid catalysts ($pH < 2$). Also, the gel time decreased rapidly when the pH of the RF solution was increased to 7 or above. This data enabled the development of neutral cold-setting resorcinolic adhesives from resorcinol.

### 5.1.2.4
### Base Catalyzed Resorcinol-Formaldehyde Reaction

The reaction velocity of resorcinol with formaldehyde is much greater than phenol. Under basic or alkaline conditions, phenol produces a stable methylol compounds, namely the monomethylol, dimethylol and trimethylol phenols. Phenol, like resorcinol, has three nuclear 2-, 4- and 6-positions for the formation of methylol phenols. Due to low reactivity, the methylol compounds formed from the phenol are stable and can be isolated without much difficulty. On the other hand, the methylol compounds that can be formed from the RF reaction may not be stable enough for isolation because of the high reactivity of resorcinol.

Resorcinol, under basic conditions, can react with formaldehyde and is expected to produce the methylol compounds having the following structures (Figure 5.3).

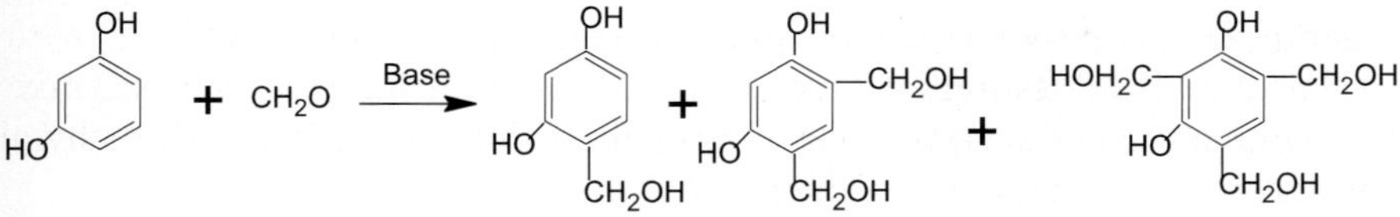


Methylol Resorcinols

**Figure 5.3.** Base catalyzed resorcinol formaldehyde reaction

Based on the uncatalyzed RF reactions, it was determined that if the pH is greater than 3 then the RF reaction is assumed to be catalyzed by the hydroxide ions [3, 10]. Unlike phenol, the reactivity of methylol resorcinols is very high and therefore the isolation of these compounds in the pure form cannot be possible.

**Synthesis and Properties of Monomethylol and Dimethylol Resorcinols** Patent literature detailing the preparation and isolation of methylol resorcinols from the RF reaction was found [11] (Figure 5.4).

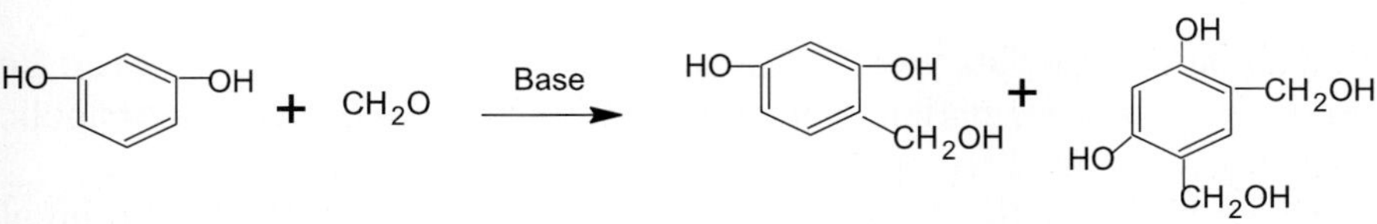


**Figure 5.4.**

Based on the patent information, well controlled conditions and lots of precautions were employed to avoid resin formation on the synthesis of 2,4-dihydroxy-hydroxymethyl benzene (4-methylol resorcinol) and 2,4-dihydroxy-1,5-(dihydroxymethyl) benzene.

Regarding 4,6-dimethylol resorcinol compounds, in the preparation of these derivatives, resorcinol was dissolved first in an aqueous formaldehyde solution and then sodium hydroxide solution was added in 10 min into RF solutions in such a way that the temperature did not exceed 40–42 °C. Immediately after the NaOH addition, it was cooled to 18 °C. The pH of the solution was never allowed to fall below 7.0 during this preparation. The resulting solution was cooled to about – 5 °C in the presence of a carbon dioxide-free nitrogen atmosphere. During this time, crystals started to appear and were complete in about 1 h. The crystals were filtered and rapidly washed with water and acetone and finally dried in a vacuum desiccator.

In the preparation of these compounds, it was required to maintain the pH of RF solutions above 7.0. The pH might be falling due to the absorption of carbon dioxide ($CO_2$) from the surrounding atmosphere. Similarly, metal impurities, especially iron, were excluded because they can catalyze RF condensation.

The monomethylol and dimethylol resorcinol derivatives obtained appeared as colorless crystals of generally acicular form, which were frequently grouped together in clusters with an appearance resembling that of echinoid.

The dimethylol resorcinol appeared to melt over 200 °C and a technical mixture containing about two-thirds of monomethylol and one-third dimethylol showed an instantaneous melting point of the order of 148 to 155 °C.

The solubility of this technical mixture in water at 20 °C was 30 to 40 g/l, while it reached about 300 g in pyridine. Contrary to all expectations, the dimethylol resorcinol was less soluble in water than the monomethylol resorcinol.

These methylol derivatives appeared to be stable if kept away from $CO_2$ and trace amounts of catalysts. In the dry state, they undergo very fast polycondensation reactions under heat and also in the presence of $CO_2$ at room temperature conditions. In an aqueous solution, spontaneous polycondensation reaction occurred even at room temperature.

Though this work reported the synthesis and isolation of methylol resorcinols, practical problems exist with the storage and handling of these materials in the stable form. These derivatives continue to react even at room temperature and under uncatalyzed conditions to form higher molecular weight polymers having branched as well as linear configurations of greater complexity.

**Mechanism of Base Catalyzed RF Reaction** The following mechanism can be proposed for the formation of methylol and methylene bridged resorcinolic structures from the base catalyzed RF reaction (Figure 5.5).

Under basic condition, the first step in the RF reaction is the formation of a resorcinol anion from the resorcinol and hydroxide ion. This resorcinate ion now activates the benzene ring, and the electron density at the 4 (or 6) increases for the formaldehyde attack and formation of methylol group (as shown in Figure 5.5). The introduction of a methylol group into the ring increases the activity of other nuclear positions [1]. The dimethylol resorcinol could be formed from the reaction of a monomethylol resorcinol with formaldehyde. This reaction might have occurred from the delocalization of electrons by the resorcinate anion.

Simultaneously, due to the presence of a methylol group and resorcinate anion in the same benzene ring, the formation of quinone methide structure (an intermediate) can be expected by the elimination of a hydroxyl group from the molecule. A compound having a quinone methide structure is very unstable and highly reactive, and therefore it can react fast with another resorcinol molecule or reactive sites present in the resorcinol to produce a methylene bridged structure. The formation of quinone methide structure in the presence of basic catalysts, and the presence of high electron densities at the 2-, 4- and 6-positions of resorcinol ring might be the reasons for the enhanced reactivity of RF reactions. Due to these facts, in RF reactions, there is a greater tendency

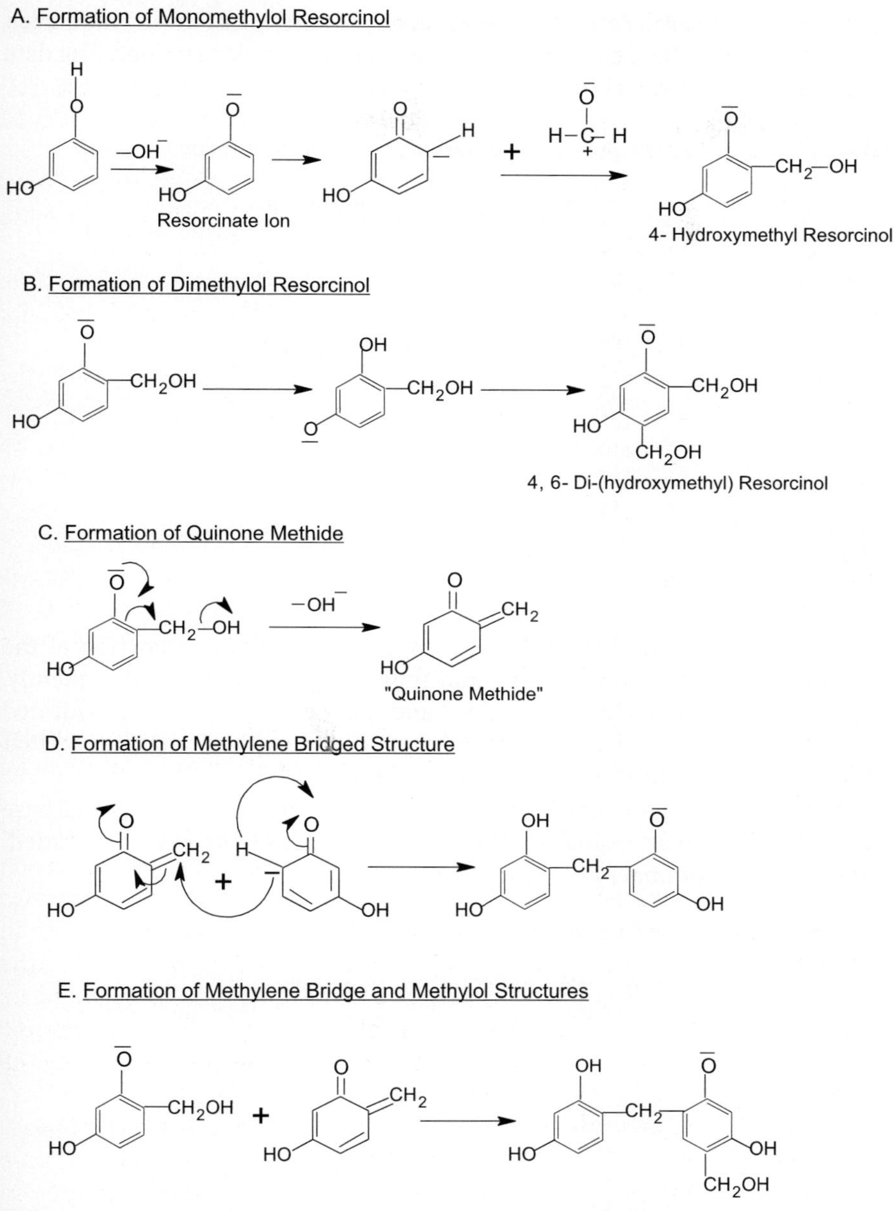


**Figure 5.5.** Mechanisms of base catalyzed resorcinol formaldehyde reaction

for the methylol group or formaldehyde to continue to react as long as there are active sites in the resorcinol molecule or RF resins.

Therefore, the RF resins produced by the base catalyzed RF reactions predominantly contain methylene bridged novolak structures.

**Base Catalyzed Novolak Resins from RF Reaction** A series of RF resins have been prepared from the base catalyzed resorcinol-formaldehyde reactions. The data is presented in Table 5.4 [10, 12].

**Table 5.4.** RF novolak resin prepared with varying catalyst concentration

| Sodium hydroxide used (weight) | Sticking point (or) softening point (°C) | pH (50% aqu. solution) |
|---|---|---|
| None | 106 | 2.96 |
| 0.2 | 103 | 5.11 |
| 0.4 | 103 | 5.78 |
| 0.6 | 104 | 6.23 |
| 0.8 | 105 | 6.51 |
| 1 | 107 | 6.65 |
| 1.2 | 110 | 6.9 |
| 1.4 | 112 | 7.04 |
| 1.6 | 114 | 7.11 |

RF resin used for reactivity was made using R/F ratio 1:0.67
Data source: from [10]

The resins presented in Table 5.4 were made by mixing a portion of the formaldehyde with resorcinol first and then heating to reflux. Subsequently, the rest of the formaldehyde was added and then the catalyst. The dehydrated resin appeared brittle. The sticking point or softening point data showed that these properties were affected by the pH of resins.

In order to study the effect of mol ratio of R/F on the properties of base catalyzed RF novolak resins, another series of experiments were performed. The results are summarized in Table 5.5.

**Table 5.5.** Preparation of RF novolak resins in the presence of a base catalyst

| Resorcinol/formaldehyde (mol) | Sticking point (or) softening point (°C) | pH (50% aqu. solution) |
|---|---|---|
| 1/0.5 | 69 | 6.62 |
| 1/0.55 | 75 | 6.63 |
| 1/0.60 | 82 | 6.53 |
| 1/0.65 | 104 | 6.69 |
| 1/0.70 | 111 | 6.56 |
| 1/0.75 | 131 | 6.59 |

Catalyst: sodium hydroxide at 1.0 wt. % level
Data source: from [10]

The pH of the final resin materials was kept constant by using a known amount of catalyst in all the resin synthesis. From the results in Table 5.5, it was very apparent that the softening point of the resins was directly related to the molar ratios of R/F used. Since the softening points are related to the molecular weights of the final resin products, higher F/R ratios produced higher molecular weight resins, which can be seen from the softening point data.

**Resorcinolic Resoles from the Base Catalyzed RF Reaction** The base catalyzed RF reactions are important in improving the adhesion of resorcinol-formaldehyde-latex (RFL) treated synthetic fibers, such as nylon, polyester, rayon and aramid, with rubber compounds. In order to develop and enhance the adhesion, the synthetic fibers are treated first with RFL adhesive in which the major chemistry involved is the resorcinolic resoles or methylols. The formation and stability of methylol derivatives of resorcinol and RF resins are important to develop and strengthen the bond between the fibers and RFL adhesives. Therefore, studying the chemistry on the formation of methylol resorcinols and determining their reactivity could provide valuable information on the RF formulations and their usefulness in various bonding applications.

In general, the formation of resorcinolic resins from the RF reactions can take place in three stages as shown in Figure 5.6.

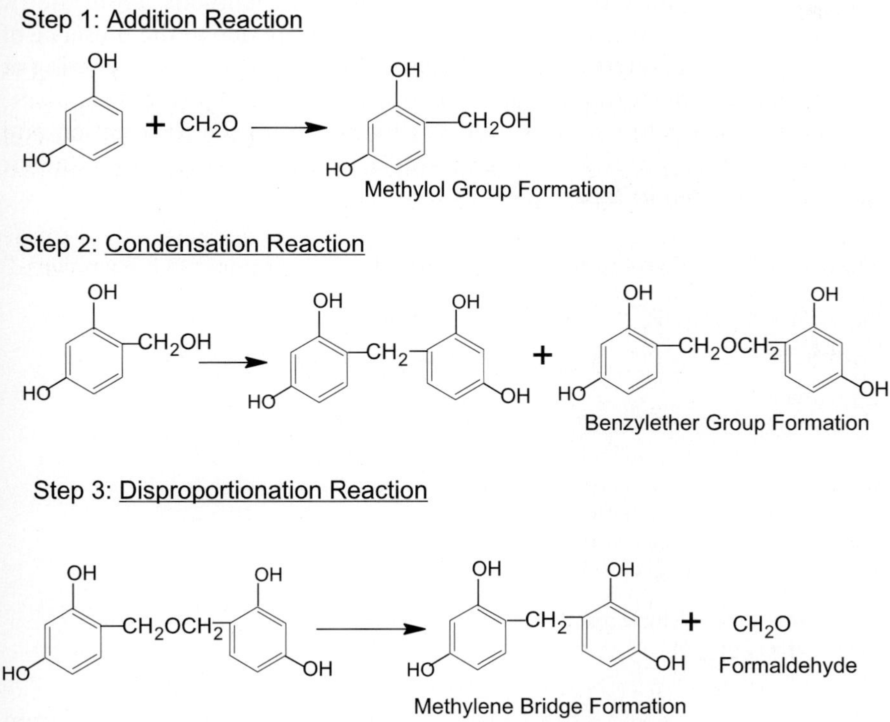


**Figure 5.6.** RF resin formation from the resorcinol-formaldehyde reaction

The first step of the RF reaction is called an addition reaction in which resorcinol combines with formaldehyde to form hydroxymethyl (or methylol) resorcinol. In the second stage, the condensation of methylol groups takes place, forming the methylene and methylene ether (or benzylether) bridged structures. Finally, in the disproportionation reaction the benzylether structures are decomposed to produce the methylene bridged structures and formaldehyde.

The addition rate of formaldehyde to resorcinol depends on the concentration of the resorcinate ion [13]. It was observed that the addition reaction of RF was fast even at room temperature. Differential Scanning Calorimetry (DSC), High Pressure Liquid Chromatography (HPLC) and $^{13}C$ NMR techniques were used to study the reaction of sodium hydroxide catalyzed RF reactions. In this study, RF solutions were prepared using R/F molar ratios of 1:1, 1:2 and 1:3 and containing 0.001 mol NaOH solutions.

DSC analysis showed two peaks for the base catalyzed RF reaction. By combining DSC with the $^{1}H$ and $^{13}C$ NMR analyses results, it was proposed that the first peak appearing in the DSC was due to the addition reaction and the second peak was associated with a condensation reaction.

The reaction mixtures obtained from RF reactions carried out at 40 °C and after 30 min was separated by the HPLC technique, and the individual components were then analyzed by FT-IR and NMR methods. From the IR spectra, the peaks appearing at 765 and 740 $cm^{-1}$ were due to the presence of 2-hydroxymethyl resorcinol in the solution. Similarly, the peak appearing at 830 $cm^{-1}$ showed the formation of 4- and 6-hydroxymethyl resorcinols.

From this study, the various products formed from the RF reaction and identified by the combined HPLC and NMR methods for different F/R molar ratios are presented in Table 5.6.

**Table 5.6.** Concentration of various products formed in the base catalyzed R-F reaction[a]

| Mol ratio of R/F/NaOH: Component | 1/1/0.001 | 1/2/0.001 | 1/2/0.001 |
|---|---|---|---|
| 1. Resorcinol | 31.4 | 20.4 | 9.1 |
| 2. 2-Hydroxymethyl resorcinol | 5 | 4.5 | 7.8 |
| 3. 4-Hydroxymethyl resorcinol | 39.3 | 33.4 | 30.1 |
| 4. 2,4-Di-(hydroxymethyl) resorcinol | 2.1 | 5.7 | 6.3 |
| 5. 4,6-Di-(hydroxymethyl) resorcinol | 1.5 | 2.4 | 3.2 |
| 6. 2,4,6-Tri-(hydroxymethyl) resorcinol | 15.4 | 26.7 | 30.5 |
| 7. High molecular weight product | 5.3 | 6.9 | 13 |

[a]Resorcinol-formaldehyde reaction at 40 °C for 30 min
Data source: from [6]

The ratio of 2 to 4 and 6 substituted resorcinolic derivatives produced from the RF was found at 1:7. At low R/F molar ratio, the most favored positions in the resorcinol molecule for the formaldehyde to react are 4 and 6. Though the 2-position is sterically hindered, it is still reactive enough to produce 2-methylol resorcinol.

**Resoles and Hemiformals from the RF Reaction** Based on published information, it was understood that resorcinolic resoles can exist only under dilute solution conditions and their stability was limited. The methylol groups formed by the reaction of formaldehyde with resorcinol can react with an additional formaldehyde molecule to form the hemiformal structure as shown below (Figure 5.7).

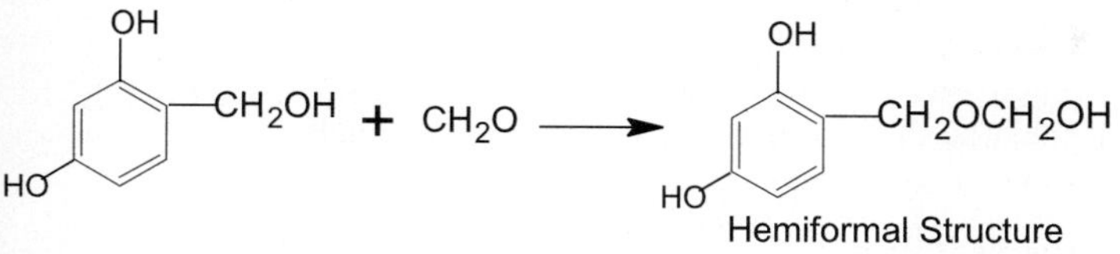


**Figure 5.7.**

Similarly, the benzyl ether group can also be formed from the reaction of methylol groups (Figure 5.8).

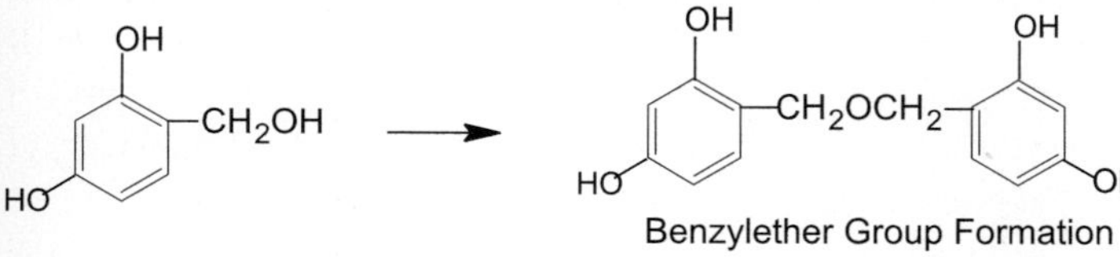


**Figure 5.8.**

Since the formation of hemiformal and benzyl ether structures should come from the methylol groups, they were considered very unstable structures and their formation was questionable. This is because when the methylol groups are formed in the RF reaction, due to instability and high reactivity they are immediately converted into carbonium ion (with acid catalysts) and quinone methide (with base catalysts) intermediates [2, 14]. These reactive intermediates could react very quickly with the reactive positions in the resorcinol molecule to produce methylene bridged structures.

The hemiformals and benzyl ethers of resorcinol form and disappear at a faster rate at elevated temperatures, tying in with their instability and the difficulty in isolating them from the RF reaction mixture. Though the benzyl ether structures formed from the phenol and formaldehyde can be stable up to 130 °C, with resorcinol the formation and decomposition of such structures can happen even at room temperature conditions.

$^{13}$C NMR spectroscopy was used to monitor the NaOH catalyzed RF reaction under dilute solution conditions at 30 °C [8, 15, 16]. From the study results it was determined that the methylol (-$CH_2OH$) groups were primarily attached to the 4-position of resorcinol. The hemiformals were observed to be formed in the earlier stages of RF reaction and then disappeared quickly as the reaction progressed. The first methylene bridge formation occurred after 20 min and

2,4′-methylene bridge formed after 2.5 h of the reaction started. After 1.7 h of reaction time, about 95% of the formaldehyde present in the reaction mixture reacted with resorcinol. Among the methylene species formed in the first few minutes of RF reaction, the methylol groups dominated. The ratio of 4-methylol to 2-methylol groups formed was 6:1.

The $^{13}C$ NMR chemical shift data for 4-hydroxymethyl resorcinol and also for various other resole and hemiformal structures formed in the RF reaction are presented in Tables 5.7 and 5.8.

**Table 5.7.** $^{13}C$ NMR chemical shift positions for 4-methylol resorcinol[a] (in dilute NaOH solution, pH approximately 9.0)

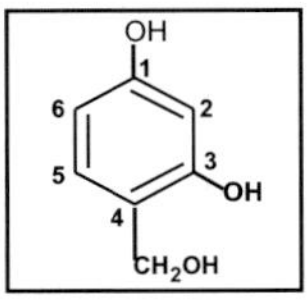


| Carbon position in the molecule | Chemical shift, ppm |
|---|---|
| C-1[b] | 158.7 |
| C-2 | 159 |
| C-3 | 133.8 |
| C-4 | 121.2 |
| C-5 | 109.9 |
| C-6 | 106.2 |
| $CH_2$ | 62 |

[a] Data from [15]
[b] Data from [8]

The methylene bridged carbons showed $^{13}C$ NMR chemical shifts in the region of between 30 and 38 ppm for 4,4′-, 22 and 30 ppm for 2,4′-, and between 18 and 20 ppm for 2,2′-methylene bridges.

Table 5.9 presents the data obtained on the $^{13}C$ NMR analysis of RF resins or reaction products obtained after 5 h of mixing of resorcinol and formaldehyde at two different molar ratios in the presence of NaOH catalyst at room temperature conditions [8].

The results shown in Table 5.9 suggest that the ring positions of resorcinol were highly substituted when the formaldehyde concentration was higher. More methylene bridged structures were formed at higher formaldehyde levels. This study gave valuable information on the nature of reaction products that were formed during the RF reaction and provided an extensive data base for the future solid state NMR studies.

**Stabilized Resorcinol-Formaldehyde Resoles** A monomeric stabilized resorcinol-formaldehyde resole type product containing polyoxymethylene groups attached to resorcinolic molecule was reported [32].

**Table 5.8.** $^{13}C$-NMR analysis of resorcinolic resoles

| Resole structure | Chemical shift, ppm |
|---|---|
| 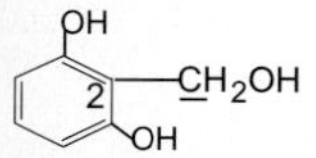 | 57.8–58.8 |
| 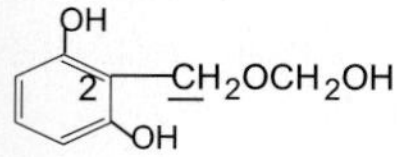 | 63.50 |
| 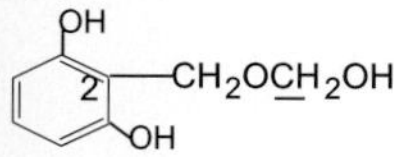 | 89.90 |
| 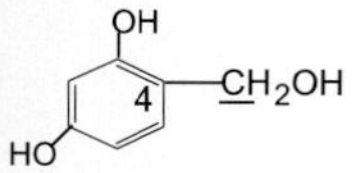 | 61.49 |
| 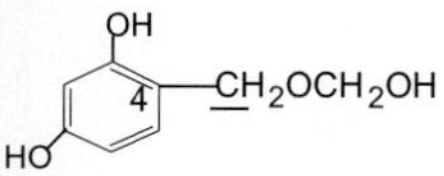 | 66.50 |
| 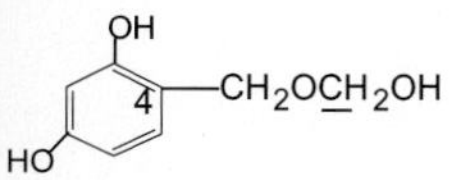 | 89.27 |
| 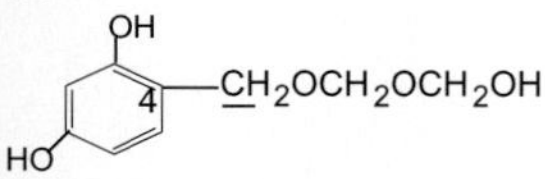 | 67.11 |

Data from [8]

The synthesis of this resole was based on the following reaction (Figure 5.9).

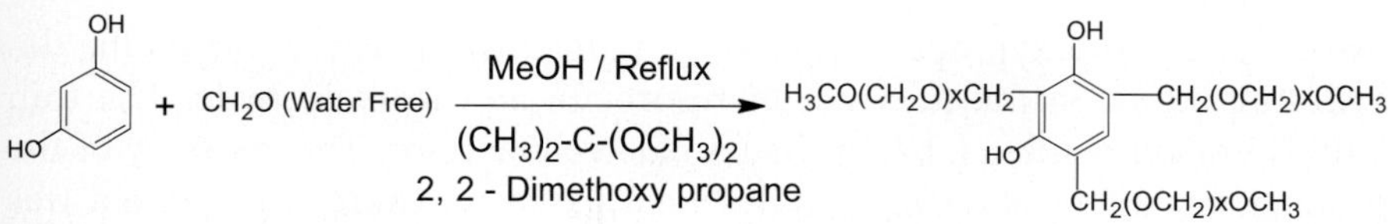


**Figure 5.9.**

The synthesis procedure used to produce this stabilized resorcinolic resole is outlined below [32].

**Procedure for the Preparation of Resorcinolic Resole [32]** To a three-necked flask equipped with stirrer, thermometer and reflux condenser, was added dried methanolic formaldehyde (58.3 wt. % formaldehyde, 216 g) and 67.6 g of 2,2-dimethoxypropane. The mixture was heated with stirring to 80 °C and then 22 g (0.2 mol) of solid resorcinol was added over 6 min. A clear solution formed which was then refluxed (82–88 °C) for 6.3 h. The cooled solution was distilled to remove volatiles at a temperature of 39 °C at 28 mm of Hg vacuum. The volatile distillate contained acetone, a product of reaction be-

**Table 5.9.** $^{13}C$ NMR data on the base catalyzed resorcinol-formaldehyde resins[a]

| Resorcinol/formaldehyde (mol) | 1.0:1.0 | 1.0:2.0 |
|---|---|---|
| Unreacted resorcinol | 35.2 | 19 |
| Total resorcinol ring positions | Mol% | Mol% |
| 1. Unreacted | 77.6 | 46.1 |
| 2. Reacted | 22.4 | 53.9 |
| (a) Methylol (Ph-$CH_2OH$) | 8.2 | 7.3 |
| (b) Hemiformal (Ph-$OCH_2OCH_2OH$) | 2.3 | 7.8 |
| (c) Methylene bridge (Ph-$CH_2$-Ph) | 11.9 | 38.8 |
| Resorcinol 4-positions | | |
| 1. Unreacted | 70.5 | 30.1 |
| 2. Reacted | 29.5 | 69.9 |
| (a) Methylol (Ph-$CH_2OH$) | 10.2 | 9 |
| (b) Hemiformal (Ph-$OCH_2OCH_2OH$) | 3 | 9.6 |
| (c) Methylene bridge (Ph-$CH_2$-Ph) | 16.3 | 51.3 |
| Resorcinol 2-positions | | |
| 1. Unreacted | 91.8 | 78 |
| 2. Reacted | 8.2 | 22 |
| (a) Methylol (Ph-$CH_2OH$) | 4.3 | 4 |
| (b) Hemiformal (Ph-$OCH_2OCH_2OH$) | 1 | 4.3 |
| (c) Methylene bridge (Ph-$CH_2$-Ph) | 2.9 | 13.7 |
| Ratio of reaction at 4- and 2-positions (4/2) | 3.6 | 3.18 |

[a]Data obtained after 5 h of reaction at room temperature
Data source: from [8]

tween 2,2-dimethoxypropane and water, indicating an active role by the dehydrating agent in the presence of resorcinol and formaldehyde. The thin liquid product residue (164.5 g) had a pale yellow color. The viscosity of the product was 90 cps at 24 °C, and the yield based on charged resorcinol was 80%.

A series of experiments were performed and reported by employing higher molar ratios of formaldehyde per mol of resorcinol, and the final products were analyzed for their structural characterization, viscosity and room temperature stabilities. The details are presented in Table 5.10.

Structural characterization by IR and NMR analyses showed that the products obtained from the resorcinol-formaldehyde reaction were tri-substituted monomeric resorcinolic compounds. In these molecules, the hydroxy groups of resorcinol were not substituted. The substituted groups were polyoxymethylene chains terminated by methyl groups and no methylene bridges were present in the molecules.

In order to obtain this resorcinolic resole, anhydrous conditions were essential and the reaction mixture was kept free from water. The water formed

**Table 5.10.** Stabilized resorcinol-formaldehyde resoles

| Expt. # | R/F (mol) used | R/F (mol) reacted | Resorcinol converted (%) | Product initial viscosity (cps, 23 °C) | No. of days at RT without viscosity increase | Days to gel at RT |
|---|---|---|---|---|---|---|
| 1 | 1.0 / 4.0 | 1.0 / 1.0 | 50 | 100–150 | 23 | 36 |
| 2 | 1.0 / 8.0 | 1.0 / 2.7 | 60 | 100–150 | 37 | 40 |
| 3 | 1.0 / 12.0 | 1.0 / 4.6 | 55 | 100–150 | 37 | 42 |
| 4 | 1.0 / 20.0 | 1.0 / 8.8 | 45 | 100–150 | 75 | 77 |
| 5 | 1.0 / 20.0 | 1.0 / 16.7 | 55 | 100–150 | 56 | 59 |

Data from [32]

from the resorcinol and formaldehyde reaction was immediately removed by the reaction of water with the dehydrating solvent used in the reaction.

The monomeric resorcinolic resoles obtained were activated by mild heating to form cross-linked resin products with resin forming materials and also were self reacting above 100 °C. These products were found to be reactive with urea, melamine and resorcinol and its resins, and formed resinous products when heated.

**Acid Catalyzed Resorcinol-Formaldehyde Reaction** The condensation reaction of resorcinol with formaldehyde can produce methylene bridged resorcinolic resin structures in the presence of an acid catalyst. Resinous compounds having methylene bridged structures are often called novolak materials. These resins have no reactive methylol groups in their structures and therefore cannot be self cured like the phenolic resoles. They are thermoplastic in nature and completely fusible compounds. To promote further resinification or crosslinking with these materials, additional formaldehyde or formaldehyde donors should be added.

Since the RF reaction is similar to the phenol-formaldehyde reaction, the initial attack of formaldehyde can take place at the 2-, 4- or 6-position of resorcinol. Acid catalyzed RF reaction can produce only the methylene bridged structures and therefore the initial reaction of RF may produce the simplest novolaks having the structures shown in Figure 5.10.

The reactivity at the 4- or 6-position of resorcinol is about five times faster than that at the 2-position [14]. For this reason, the predominant product expected from the RF reaction may contain 4,4′-methylene bridged resorcinol. Once produced, the reactivity of this compound with additional formaldehyde may be low due to the presence of sterically hindered 2-positions in the molecule. Similarly, though the initial formation of 2,2′-methylene bridged resorcinolic compounds is difficult, their formation can enhance the reactivity or cure behavior with the methylene donor compounds. Under neutral and uncatalyzed conditions, the novolak resins predominantly contain 4,4′-

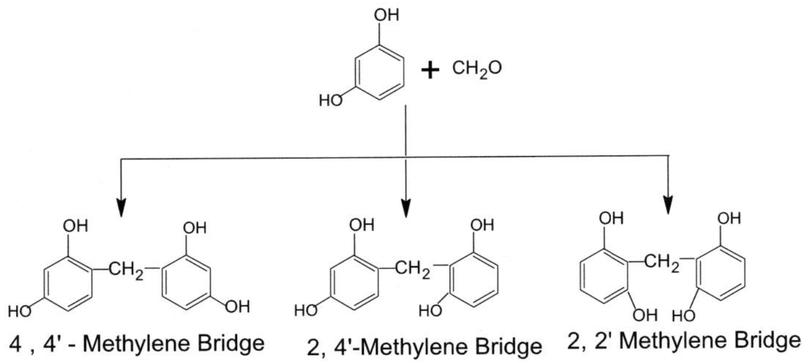


**Figure 5.10.**

bridged structures. But, in the presence of an acid catalyst this might not be the case.

The mechanism by which the methylene bridged structures formed from the RF reaction can be seen in Figure 5.11.

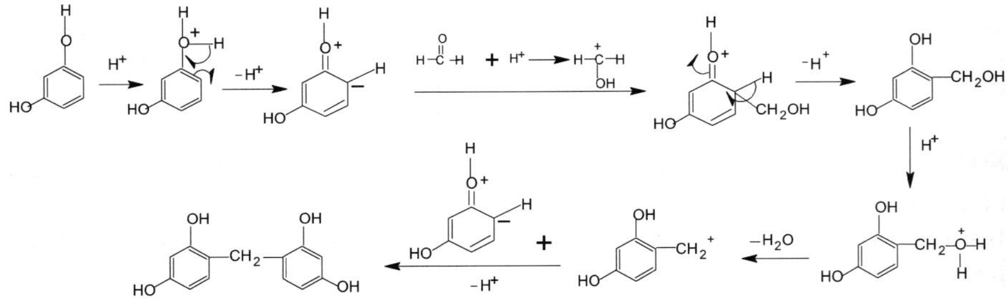

**Figure 5.11.** Mechanism of acid catalyzed resorcinol formaldehyde reaction

In the presence of an acid catalyst the formaldehyde is activated by the protonation reaction. This protonated formaldehyde reacts with resorcinol to produce an unstable methylol or hydroxymethyl intermediate at the resorcinolic reactive sites. On further protonation of the methylol group, it forms a benzyl carbonium ion as the reactive intermediate, which immediately reacts with another resorcinol molecule to form the methylene bridged resorcinolic structure.

**Novolak Resins from the RF Reaction** Due to high reactivity of resorcinol, the RF reaction produces a mixture of methylene bridged structures containing dimers, trimers, tetramers and so on, under acid as well as base catalyzed conditions. With the combination of Mass and NMR Spectroscopic Techniques, unambiguous interpretation of the novolak structures of resorcinol was achieved [17]. The reaction scheme outlining the formation of various novolak structures are shown in Figure 5.12.

The identification of structures "a", "b" and "e" (shown in Figure 5.12) revealed that the reaction of formaldehyde occurred at the 4- and 6-positions

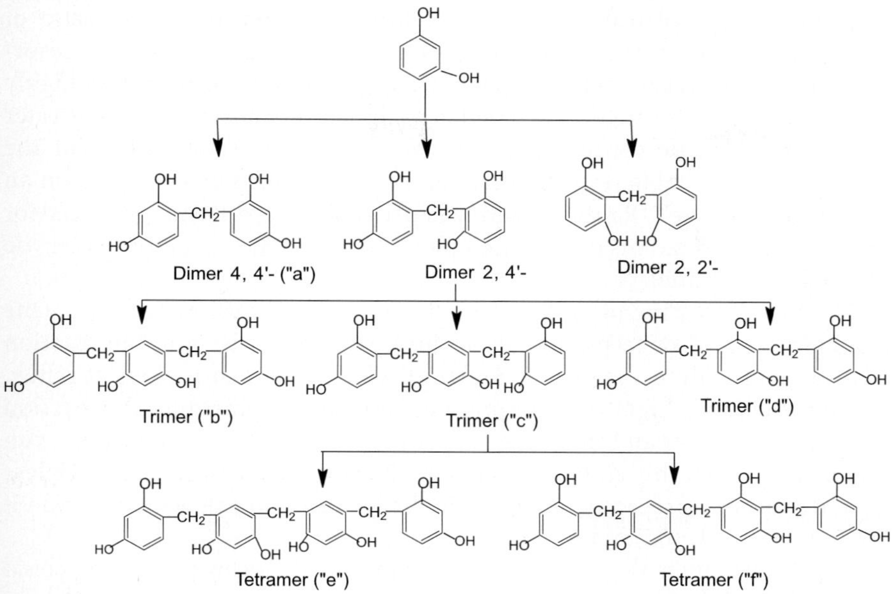


**Figure 5.12.**

of the resorcinol molecule. Mass and NMR determinations identified the trimer structures "c" and "d" which clearly showed the high reactivity of the sterically hindered 2-position for the formaldehyde reaction. The symmetrical tetramer structure ("e") might be produced from the symmetrical trimer ("b") molecule. Similarly, the unsymmetrical tetramer ("f") molecule might have resulted from either the symmetrical or unsymmetrical trimers. Based on the structural analysis of the RF reaction products, it was observed on a molar basis that the proportion of 4- and 6-methylene bridges relative to the 2-methylene bridge is of the order of 10.5:1.

The absence of methylol groups in the isolated compounds showed the high reactivity of resorcinol towards formaldehyde under acidic conditions. The kinetics work also showed, on an equal molar and identical reaction conditions basis, that the RF reaction proceeded at a rate of approximately 10–15 times faster than the phenol-formaldehyde reaction [18].

**Synthesis of RF Novolak Resins** RF novolak resins have been synthesized by acid catalyzed reactions and produced in the form of aqueous resin solutions and brittle solid materials for use as adhesives in the wood and tire industries. A fusible novolak resin can normally be obtained by employing the conventional molar ratios of R/F (1:<0.75 mol) in the reaction. The uniformity of RF resin and the rate of RF reaction are dependent upon the R/F ratios, the concentration of reaction solutions, reaction temperatures, pH and the presence of various catalysts.

Theoretically, RF resins can become infusible when the molar ratio of formaldehyde in a resin synthesis is equal to or greater than that of resorcinol. In the RF resin solution preparation, the stability of resin was highly influenced by the formaldehyde content. Gelation was observed to be greater if the formaldehyde concentration reached beyond a certain level. For the preparation of a stable resin solution, and also a fusible novolak resin, on an industrial scale, the F/R ratio should be kept below 0.7 mol. Resins produced for commercial applications normally employed 0.57 to 0.65 mol of formaldehyde per mol of resorcinol.

The manufacturing of an RF resin can be done in a variety of ways. One common method or technique of producing this resin is from the reaction of resorcinol with formaldehyde under alkaline or acid conditions at reflux. The RF reaction is highly exothermic and very rapid. Sometimes the reaction may be very violent and therefore proper precautions should be taken to run this reaction smoothly and also to achieve the desired resin properties. Unless a good synthetic procedure is followed, the RF reaction can gel easily even if the R/F ratio is 1:<0.7 mol.

Depending upon the method of preparation, RF resin properties could vary considerably. These properties depend mainly on the distribution of resin components having different molecular weights. Molecular size distribution in the novolak resin depends primarily on the rate of an initial reaction between R and F. By careful control of RF reaction, avoiding exothermic reaction, a novolak resin having uniform molecular weights can be made.

The uncatalyzed RF reaction product was found to be unstable and precipitated during the reaction. This might be due to the formation of RF novolaks having linear structures, which may be highly crystalline in nature. The resins could show predominantly 4,4′-methylene bridged structures from the uncatalyzed RF reaction. From the resin synthesis, it was observed that when the reaction temperature was lowered below the reflux and the solids content of the reaction mixture was below 65 wt. % precipitation of the reaction product occurred. Even at 5 to 10% solids concentration, the uncatalyzed RF reaction could produce as much as 30% crosslinked and infusible material during the reaction.

Various synthetic methodologies were developed to obtain a fusible RF resin that can be solubilized in water to obtain a stable solution for commercial applications. In a well known method, resorcinol was first mixed with water and an oxalic acid catalyst [19]. The mixture was heated under reflux and then the formaldehyde was added dropwise. By this method the formaldehyde reacted instantly with resorcinol. Finally, the RF reaction product was dehydrated to obtain a brittle resin solid. A preparative procedure outlining the synthesis of a RF is given below.

**Synthesis of Resorcinol-Formaldehyde (RF) Novolak Resin (R/F = 1:0.68 mol) [19]**
Resorcinol (400 g, 3.64 mol) was mixed with 400 g of water to make a mixture

and then about 8 g of oxalic acid was added to the mixture. Then the contents were heated in a closed container which has reflux condenser. Since the dilution medium was water, the mixture was heated to about 100 °C, the boiling point of water. After the mixture had been brought to the boiling point, about 200 g of 37% formaldehyde (2.47 mol) were carefully added. It usually took about 15 to 20 min to add the formaldehyde solution to the resorcinol solution, such time being sufficient for all the resorcinol to react with the formaldehyde. After the addition of formaldehyde solution had ceased, the mixture was refluxed for a suitable period of time in order to ensure complete combination of the formaldehyde with the resorcinol, and an adjustment of the reaction products to their final end point. In order to remove the water of condensation, the liquid was evaporated under vacuum for a suitable length of time, such as for example 55 min, at a temperature of 85 °C. After the water of condensation had been removed, the temperature was raised somewhat to about 116 °C. At this stage, the reaction mass was still a hot viscous liquid. Upon cooling, the mass solidified to a clear amber-like resinous product which was brittle and grindable. At this stage, the cooled product was thermoplastic and would not set on heating to 150 °C.

In another method, a portion of formaldehyde was first mixed with resorcinol. Heat was applied to produce a homogeneous solution [20]. Then the rest of the formaldehyde was added. As soon as the formaldehyde addition was completed, the oxalic acid catalyst was added to drive the reaction to completion at a pH of 1.5 to 3.0. The final solution pH was adjusted with alkali to 7 to 9. With this method, a stable RF solution could be made with an R/F ratio of 1:<0.7 mol but was unsuccessful in making a stable solution with formaldehyde concentration in the range of 0.8 to 0.9 mol.

A novel synthetic technique was developed to produce a stable and infinitely dilutable RF resin solution with > 0.8 mol of formaldehyde per mol of resorcinol in the reaction. A brittle resin could be produced by the dehydration of this resin solution [21]. The synthesis procedure used in the preparation of RF resin is described below.

**Synthesis of Resorcinol-Formaldehyde (RF) Novolak Resin (R/F = 1:0.84 mol) [21]**
Resorcinol, 264 parts, and 48 parts of a 37% aqueous formaldehyde solution were charged to a 100 gallon kettle to provide a mixture having a resorcinol to formaldehyde ratio of 1 mol resorcinol to 0.246 mol of formaldehyde. Heat was applied and agitation begun. The reaction mixture was heated for 15 min at reflux (100 °C) and then 1.6 parts of oxalic acid dihydrate was added thereto. At this point the measured pH was 0. After 1 h, 74.4 parts of the formaldehyde solution was slowly added over a period of 30 min with cooling being applied to the kettle jacket to give a reaction mixture containing 1 mol of resorcinol and 0.628 mol of formaldehyde and having a pH of 0.4–0.6. Heat was then applied and 200 parts of water was slowly added to the mixture over a 15-min period to give a solution having solids content about 53.5%. The solution was heated to

reflux and 41.6 parts of formaldehyde solution was added thereto with cooling to bring the mol ratio of resorcinol to formaldehyde to 1 mol of resorcinol to 0.84 mol of formaldehyde. At this point, the solids content was 52% and the pH ranged between 1.3 and 1.7. The third formaldehyde addition required about 10 min and the reaction mixture was thereafter heated to reflux. After 20 min at reflux, the viscosity had reached 60 million cps at 73 °F. The solution was cooled and the pH thereof was adjusted to 6.5–7.0 with 12 parts of 50% caustic solution.

Water (148 parts) was then added to the solution thus prepared to give a clear solution having a viscosity measured by Gardner–Holdt viscometer of 36.200 – 46.300 cps at 73 °F. This solution was completely stable and the viscosity did not increase even with additional heating. The solids content at this point measured 36.7% by weight. The solution was found to be infinitely dilutable and capable of being dehydrated to a brittle solid without difficulty to give solid resin having a ring and ball softening point of 116 °C (ASTM-E 28-51 T).

As a general practice, in the synthesis of an RF resin, the resorcinol should be solubilized first in water and then mixed thoroughly with the catalyst. Good catalysts that may be employed for RF resin preparation are oxalic acid, *para*-toluene sulfonic acid, benzene sulfonic acid, sulfuric acid and phosphoric acid. The formaldehyde addition should be carried out slowly and possibly under reflux conditions. In this way the heat generated from the RF reaction can be dissipated very effectively. Since the RF reaction is instantaneous under reflux conditions, there may not be any excess formaldehyde present at any given time during the reaction. If excess formaldehyde is present, this can lead to various side reactions. This could result in the formation of different kinds of reaction products. The presence of excess formaldehyde not only affects the final properties of RF reaction products but also leads to highly exothermic conditions. As a result, the reaction could be potentially explosive and dangerous.

**RF Novolak Resin Properties – Softening Points** The melting characteristics of novolak resins are generally determined by their softening point measurements. The softening points of novolak resins depend on their molecular weights and molecular weight distributions. Depending upon the R/F ratios used in the resin synthesis, the softening points can vary. High R/F ratios could produce high molecular weight materials and therefore, high softening points can be expected from these resins. For a given R/F ratio, the softening point also depends on the amount of unreacted resorcinol present in the resin. Therefore, a combination of R/F ratio and the amount of unreacted resorcinol determines the softening point of a particular novolak resin.

Figures 5.13 and 5.14 show the effect of R/F and R/aldehyde ratios and the free resorcinol content on the softening points of resorcinolic novolak resins.

The softening points were observed to be increased when the aldehyde levels were increased in the resorcinol based novolak resins synthesis. As the

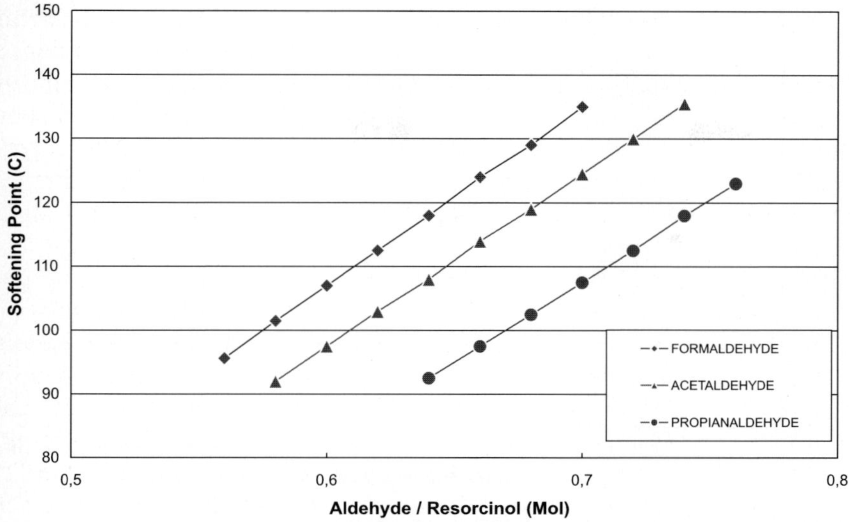


**Figure 5.13.** Softening point versus aldehyde/resorcinol ratio

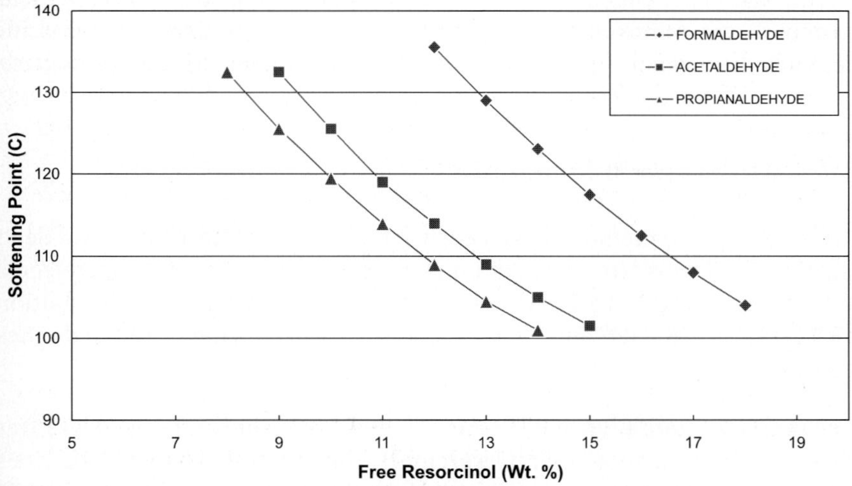


**Figure 5.14.** Effect of free resorcinol on softening points novolak resins

aldehyde contents were increased, the unreacted resorcinol amount in the final resin products was reduced.

In order to observe the effect of removing the free resorcinol content of a known RF resin, a novolak resin having a softening point of 113 °C and free resorcinol content of 16.5 wt. % was subjected to heating just above the sublimation point of resorcinol. After the continued removal of resorcinol from this resin, the softening points and free resorcinol determinations were made. Figure 5.15 shows the results of this experiment.

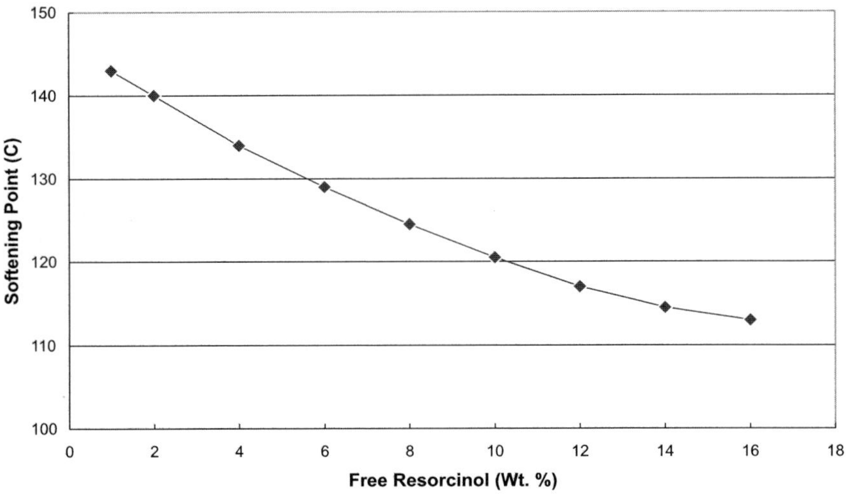


**Figure 5.15.** Effect on the removal of free resorcinol from the RF novolak resin

From the experimental results, softening point increased when unreacted resorcinol was removed from the resin. This suggests that free resorcinol present in the novolak resin reduces its softening point by acting as a plasticizer.

### 5.1.2.5
### Analytical Methods for RF Resins Analysis

Various analytical methods have been used in the characterization and determination of various structural parameters such as the molar ratios, molecular weights distribution, average chain length and methylene bridge distributions of RF resins. NMR, GPC and FT-IR analyses have been widely used to do these determinations.

**Structural Characterization of RF Resins** RF novolak resin synthesized was fractionated to obtain monomer (resorcinol), dimers and trimers [22]. These dimers and trimers were novolak type compounds obtained from the RF reaction. Various molecular weight fractions and the oligomeric products isolated from the RF resin were analyzed by the $^{13}C$ NMR analysis. The analysis showed that the methylene bridges formed during the RF reaction primarily occurred between 4–4′ and 2–4′ carbons of resorcinol and minor amounts between 2–2′ carbons. $^{13}C$ NMR chemical shift assignments obtained for the monomer, dimers and trimers are presented in Figure 5.16.

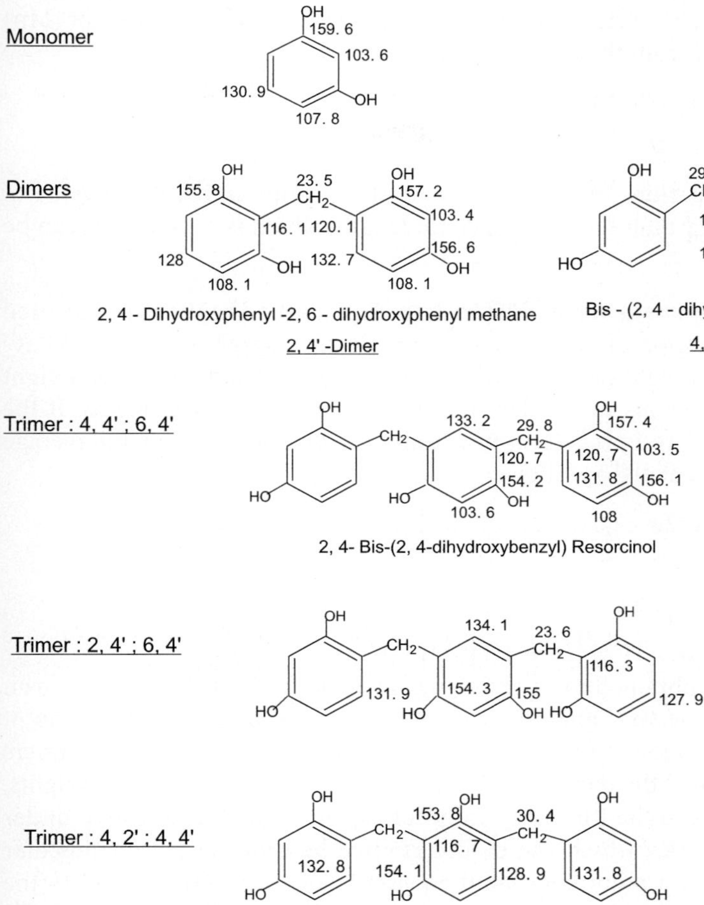


**Figure 5.16.** $^{13}$C NMR chemical shift assignments for monomer, dimers and trimers of RF reaction products [22]

**Average Chain Length of RF Resins** A fusible RF novolak resin can be represented by the following chemical structure (Figure 5.17).

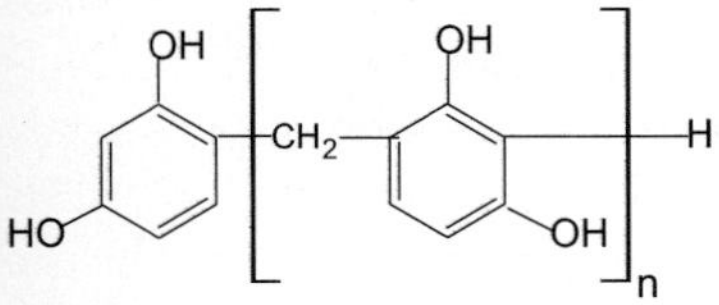


**Figure 5.17.**

Based on the above structure, the number average molecular weight (Mn) can be calculated from the following equation:

$$\mathrm{Mn} = 110 + \mathrm{n}(14 + 108)$$
$$\mathrm{Mn} = 110 + 122\mathrm{n}$$

In the above equation, "n" is the number of repeating units. If the value of "n" is determined then the average chain length, which is N = n + 1, can be calculated.

**Gel Permeation Chromatography (GPC) Analysis** This method has been widely used for the determination of number average molecular weight (Mn), weight average molecular weight (Mw), polydispersity (Mw/Mn) and molecular weight distribution of various molecular sizes present in the novolak resins. If the value of Mn for RF resin is determined from the GPC analysis, then the average length can be calculated as follows:

$$\mathrm{Mn} = 110 + 122\mathrm{n}$$
$$122\mathrm{n} = \mathrm{Mn} - 110$$
$$\mathrm{n} = (\mathrm{Mn} - 110)/122$$

Therefore, $\mathrm{N} = (\mathrm{n} + 1) = \{(\mathrm{Mn} - 110)/122\} + 1$

A GPC curve obtained for a resorcinol-formaldehyde (RF) resin is shown in Figure 5.18. The GPC analysis can also be used to determine the ratio of distribution of monomer (M, resorcinol), dimers (D), trimers (Tri), tetramers (Tetra) and above – the pentamers (pentamers and higher molecular weights, P + Plus) present in the RF resins. In order to determine this, the area under the individual peaks, which can be determined by drawing a perpendicular line from the lowest point between the peaks (as shown in the figure) to the base line. This line is supposed to be the boundary between the adjoining oligomeric compounds or molecules having different molecular weights. After calculating the individual peak areas, the percent concentration of different molecules or compounds having similar molecular weights can be calculated from the simple formula shown below.

For calculating the monomer concentration (resorcinol):

$$\text{Monomer, M (\%)} = \frac{\text{Area of monomer}}{\text{Area of }\{\text{M+Di+Tri+Tetra+(P+Plus)}\}} \times 100$$

Similarly, for the "dimer" concentration calculation:

$$\text{Dimer, D (\%)} = \frac{\text{Area of dimer}}{\text{Area of }\{\text{M+Di+Tri+Tetra+(P+Plus)}\}} \times 100$$

The area ratios calculated from the GPC curves for the two commercially known RF resins supplied and sold for various applications including the preparation of RFL adhesives are presented in Table 5.11.

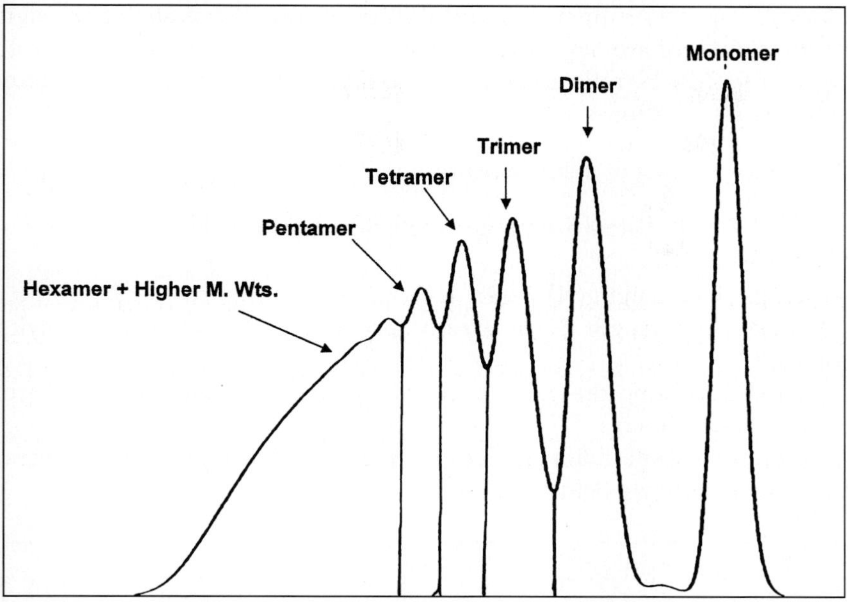


**Figure 5.18.** Method of calculating monomer to over pentamer distributions from gel permeation curve of RF resin

**Table 5.11.** GPC analysis of two commercial RF resins in solution

| RF resins used: components | Penacolite R-50 | Hodogaya |
|---|---|---|
| | Area percent | |
| 1. Monomer (resorcinol, M) | 15.6 | 30.2 |
| 2. Dimers (Di) | 17.1 | 11.7 |
| 3. Trimers (Tri) | 14.2 | 8.6 |
| 4. Tetramers (Tetra) | 11.6 | 7.5 |
| 5. Pentamer + high m.wts. (P + Plus) | 41.5 | 42 |

**Proton and $^{13}$C NMR Analyses** NMR spectroscopy is a powerful tool to characterize and determine some important parameters of novolak resins. This technique can be used to determine the F/R ratio of a resorcinolic novolak resin fairly accurately. With this technique, an average chain length (N) and the methylene bridge distributions can be determined [23].

Using pyridine (or any other solvent) as the solvent, which can mask the signal due to water, a proton NMR was taken to determine the values of N and F/R molar ratio of RF resins. Novolak resin of RF generally contains the aromatic ring, hydroxyl groups and methylene bridges in the structure. In the $^{1}$H NMR spectrum, the methylene bridged protons appear at $\delta = 3{-}4$pm, the aromatic ring protons at $\delta = 6{-}8$ppm and the hydroxyl group protons observed at $\delta = 9-10$ ppm chemical shift positions. By measuring the signal integrals (area) of aromatic and methylene bridged protons from the proton NMR, the F/R molar ratio and the average chain length can be measured from the following equations:

$$\text{F/R (mol)} = \text{area of methylene protons} / \left\{\frac{1}{2}\left[\text{area}\left(\text{aromatic} + \text{methylene protons}\right)\right]\right\}$$

Average chain length (N) = (2n + 4)/2n

Average chain length (N) = area of aromatic protons/area of methylene protons.

After determining the value of n, the chain length was calculated from N = n + 1 equation.

A proton NMR spectrum of RF resin that can be used to calculate the average chain length is shown below (Figure 5.19).

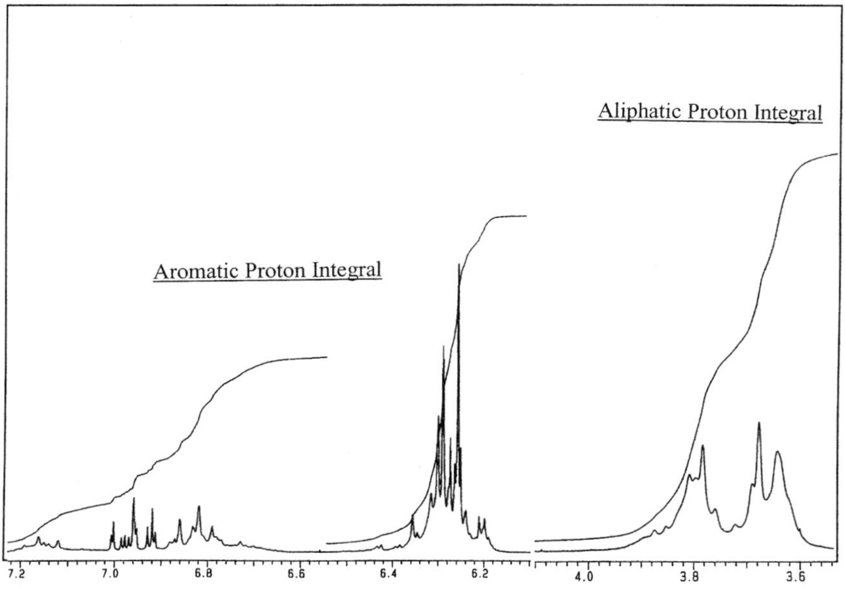


**Figure 5.19.** Proton NMR spectrum of RF resin

$^{13}$C NMR spectrum showed methylene bridged carbons chemical shifts at 29.93, 24.23 and 18.53 respectively for the 4,4′-, 2,4′- and 2,2′-substitutions. Therefore, by measuring the area of signal integrals corresponding to these signal positions, the methylene bridges distribution of a RF novolak could be calculated.

$^{13}$C NMR spectra of RF resin to show the aromatic and aliphatic carbons and an integrated view for the methylene bridged carbons is shown in Figure 5.20.

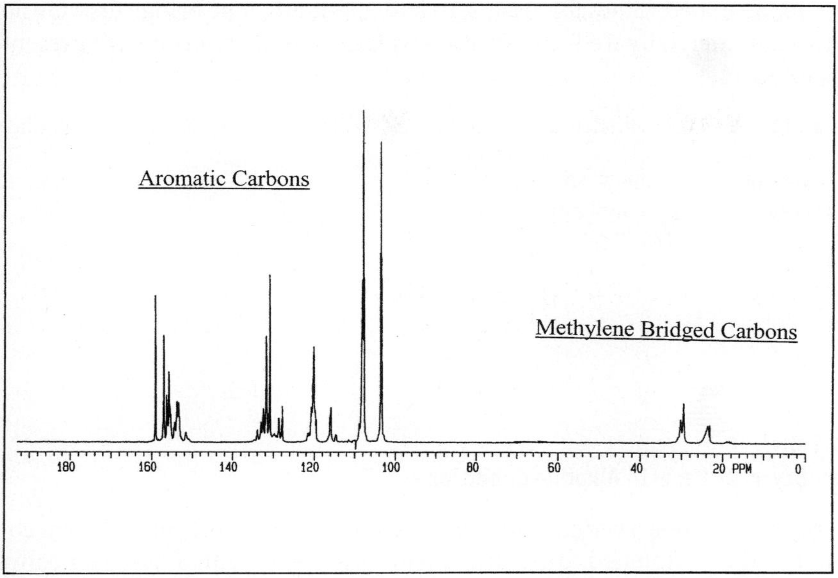


**Figure 5.20.** $^{13}$C-NMR Spectrum of RF Resin

In order to determine the methylene bridge distribution of a commercially well-known RF resin (Penacolite resin B-1A), the sample was dissolved in dioxane-$d_8$ solvent and $^{13}$C NMR spectrum was recorded and the integral areas corresponding to 4,4′-, 2,4′- and 2,2′-methylene bridged carbon atoms were measured (Figure 5.21).

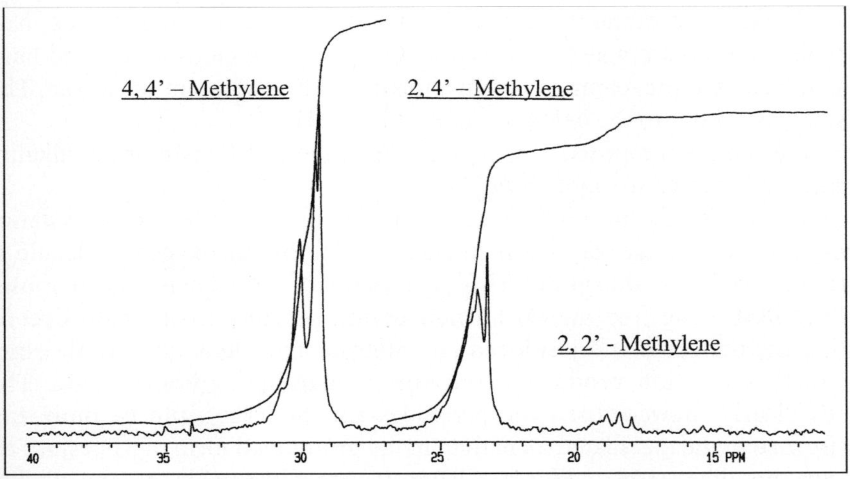


**Figure 5.21.** $^{13}$C-NMR spectrum of RF resin – methylene bridge positions

The calculated values for 4,4′-, 2,4′- and 2,2′-methylene bridge distributions of two commercially well-known RF novolak resins in solutions are presented in Table 5.12.

**Table 5.12.** $^{13}$C NMR methylene distribution analysis of two commercial RF resins in solution

| Bridge type components | Penacolite R-50 | Hodogaya |
|---|---|---|
| | mol ratio | |
| 4,4′- | 57 | 74 |
| 2,4′- | 39 | 25 |
| 2,2′- | 4 | 1 |

### 5.1.2.6 Stability of RF Resin in Alkaline Conditions

RF resins were observed to degrade more rapidly in the metallated form compared to the protonated form. Treatment of RF resin with a base can convert the hydroxyl groups to the corresponding hydroxide ions. The formation of these metallated ions can make the RF resin vulnerable for decomposition. For the determination of this decomposition, a cross linked RF novolak resin was prepared and converted into protonated and metallated forms by the treatment with $HNO_3$ and LiOH respectively [24]. These two modified resins were stirred for three days in distilled water (protonated resin) and 2.0 mol/l LiOH solution (metallated resin) and then analyzed by the solid state $^{13}$C NMR method.

With protonated resin, there was no change observed from the analysis. The LiOH treated resin showed a new resonance above 160 ppm which indicated the formation of quinone, carbonyl and other carbonyl type groups. New resonance peaks were also observed at 40 – 100 ppm regions as a broad band which indicated the formation of an aliphatic carbon adjacent to oxygen. This result was similar to the oxidation of RF resin [25].

A mechanism proposed for the decomposition of RF resin under alkaline condition is shown in Figure 5.22 [24].

According to the proposed mechanism, the first step in the degradation process was an attack at the methylene bridge by an oxygen molecule to form the hydroperoxide group. This hydroperoxide subsequently decomposed into radicals. The free radical formed on the polymer main chain decomposed the molecule and produced an aldehyde and phenolic radical, which on further reaction produced a mixture containing carbonyl group. This study clearly showed that hydroperoxides and radicals could be more easily formed in the metallated form than in the protonated form of RF resin. This means that the protonated form of RF resins are more stable than an alkaline resin.

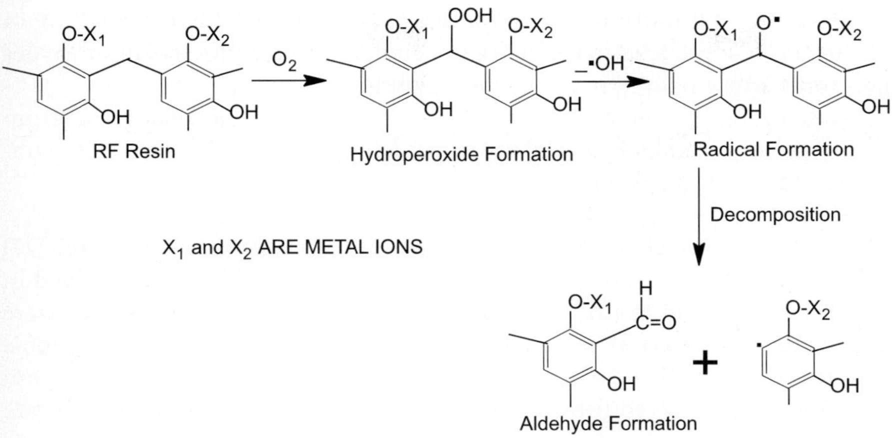


**Figure 5.22.** Proposed degradation mechanism for the RF resin under alkaline condition

## 5.1.2.7 Resorcinolic Novolak Resins from Aliphatic Aldehydes

Novolak resins of resorcinol can also be made using other aliphatic aldehydes, such as acetaldehyde, propionaldehyde, butyraldehyde etc. These higher aldehyde modified novolaks possess certain characteristics which the straight RF resins may not provide. Modified resorcinolic resins confer special solubility or compatibility properties particularly in rubber compounding applications. The alkyl groups are present at the methylene bridges formed between the resorcinol molecules.

Novolak resinous products that can be obtained by the use of different aliphatic aldehydes with resorcinol and its derivatives have the following generalized structure (Figure 5.23) [26].

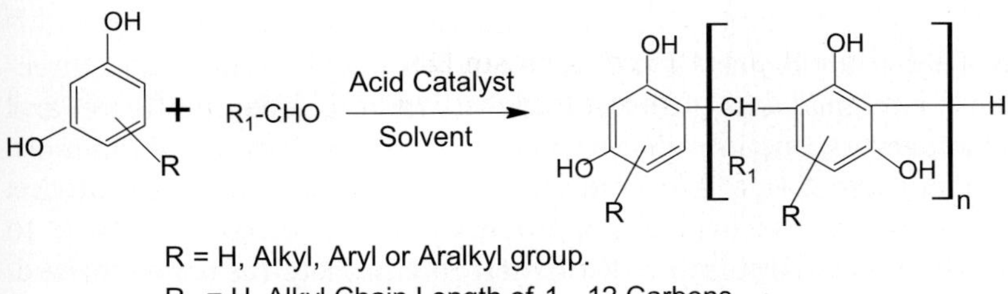


R = H, Alkyl, Aryl or Aralkyl group.
$R_1$ = H, Alkyl Chain Length of 1 - 12 Carbons

**Figure 5.23.**

In the preparation of these resins, suitable solvents such as alcohols, toluene, and xylene should be employed for the proper mixing and smoother reaction of resorcinol with aliphatic aldehydes. Sometimes azeotropic separation of water formed from the reaction can help improve the rate of reaction and also the formation of uniform resin structure. All these depend on the nature of aldehydes, catalyst, reaction temperature, molar ratio of resorcinol to aldehydes and conditions involved in the novolak synthesis.

Some of the procedures employed in the synthesis of novolak resins from resorcinol and acetaldehyde and propionaldehyde, and obtained from the published literature, are given below [27].

**Synthesis of Resorcinol-Acetaldehyde Novolak Resin (R/Aldehyde = 1:0.48 Mol) [27]** Resorcinol (250 g, 2.3 mol) and acetaldehyde (50 g, 1.1 mol) were dissolved in 275 ml of water and to this solution 5 g of oxalic acid in 10 ml of water were added. The solution was refluxed for 2 h, cooled and about 15 g of insoluble material was filtered and discarded. The filtrate was evaporated under vacuum up to 100 °C. A tacky reddish oily condensate of the resorcinol and acetaldehyde was recovered.

**Synthesis of Resorcinol-Propionaldehyde Novolak Resin (R/Aldehyde = 1:0.5 Mol) [27]** Resorcinol (110 g, 1.0 mol), propionaldehyde (30 g, 0.5 mol) and 2.5 g of oxalic acid were dissolved in 200 ml of alcohol. The solution was refluxed for 12 h. The alcohol and water were removed under vacuum up to 100 °C. A tacky reddish oily condensate of the resorcinol and propionaldehyde was recovered.

The novolak resins prepared using 1.0 mol of resorcinol and 0.5 mol of acetaldehyde or propionaldehyde appeared as tacky materials. This might be due to a very low softening point and higher free resorcinol content of the resin. In order to obtain a brittle resin from the acetaldehyde or propionaldehyde and resorcinol, higher molar ratio of aldehyde should be employed in the synthesis. When used in the rubber compounding applications, these resins acted as methylene acceptors and improved the steel cord to rubber compound adhesion.

Another synthesis procedure, obtained from the published source, for the C 10 aldehyde based novolak type resin is outlined below [28].

**Synthesis of Resorcinol-Higher Aliphatic Aldehyde Novolak Resin [28]** Into a three-necked flask equipped with a Dean–Stark separator, a dropping funnel and a stirring apparatus, one introduces while stirring 550 g resorcinol (5.5 mol), 200 g of toluene and 2.8 g of a 65% solution of *para*-toluenesulfonic acid; this is brought to 100 °C and over 2 h one pours in 625 g (4.4 mol) of the C 10 aldehyde while using azeotropic distillation to eliminate the water formed. The C 10 aldehyde is obtained by "oxo" synthesis from propylene trimers; titration shows that it contains 98.5% C 10 aldehyde and is comprised of a large number of branched isomers. The azeotropic distillation is then continued for 2 h; concentration is then carried out at atmospheric pressure at 150 °C; it is put under vacuum and maintained for 30 min at 150 °C at a pressure of 50 mm Hg. The resin is decanted in the form of a red solid with a yield of 1032 g (i.e., a yield of 94% in relation to the raw materials employed, taking into account the purity of the aldehyde). The ball-and-ring melting point is 96 °C and the free-resorcinol content is 8.4%.

Compared to the unmodified RF resin, the C 10 aldehyde modified resin was non-hygroscopic and contained less free resorcinol with an optimum softening point for the rubber compounding applications. This resin was used as a reinforcing material in rubber compounds.

**Resorcinol-Crotonaldehyde Resins** Crotonaldehyde is an unsaturated aliphatic aldehyde compound having both unsaturated carbon atoms and an aldehyde group in the same molecule. As a reactive chemical compound, resorcinol can react with these two functional groups at the same time in the presence of an acid catalyst. A novolak type resin produced from the crotonaldehyde and resorcinol reaction is illustrated in the following synthetic procedure [29].

**Synthesis of Resorcinol-Crotonaldehyde Novolak Resin (R/Aldehyde = 1:0.5 Mol) [29]** A three-necked flask having a thermometer, stirrer and reflux condenser was charged with 330 g of resorcinol (3 mol). Then the catalyst oxalic acid (10 g dissolved in 10 g of water) was added to the flask and then heated to 190 °F temperature. At this point the crotonaldehyde (118 g of 89% solution, 1.5 mol) was added very slowly. A slight exothermic reaction occurs. After the crotonaldehyde addition, the mixture was refluxed for 2 to 4 h at 205–215 °F and then distilled under vacuum until most of the unreacted volatiles were removed and the temperature of the resin had reached 265–280 °F. The resin produced by this procedure was a semi solid material.

The mechanism on the formation of a novolak resin from the resorcinol-crotonaldehyde reaction is outlined in Figure 5.24 [30].

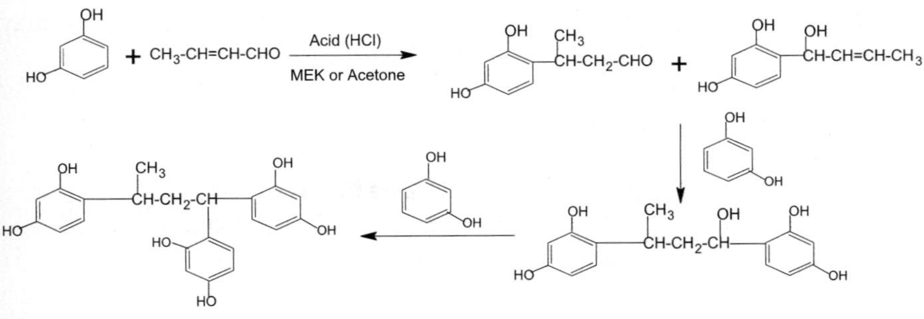


**Figure 5.24.** Mechanism of resorcinol-crotonaldehyde resin formation

**Resorcinol-Glutaraldehyde Resins** Glutaraldehyde contains two aldehyde groups in the aliphatic chain. Therefore, on reaction with resorcinol, the glutaraldehyde reaction product can contain four resorcinol groups attached in the same aliphatic chain as can be seen from the following reaction scheme (Figure 5.25).

A novolak resin developed on the basis of above reaction scheme was used as an accelerator for the thermosetting phenolic resin (PF) [31]. When this

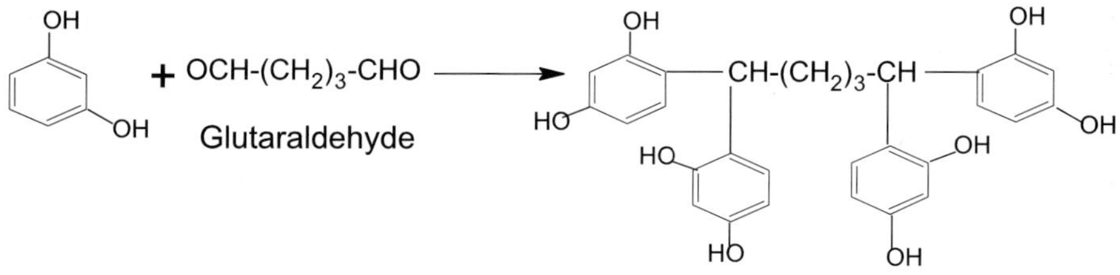


**Figure 5.25.** Synthesis of resorcinol-glutaraldehyde resin

resin was blended with a PF resin, the blend was used in the manufacture of cellulosic board, oriented strand board and plywood.

### 5.1.2.8
### Modified Resorcinolic Novolak Resins with Phenolic Derivatives

When a modification is needed in resorcinolic resins for some special application or to reduce the cost associated with a straight RF resin, phenol and phenolic derivatives are often used in the synthesis [33,34]. In order to prepare resorcinol-phenol-formaldehyde (RPF) resins, the phenol should be reacted with formaldehyde first before resorcinol is introduced for reaction. If resorcinol was added first in RPF resin synthesis, it would lead to gel formation because of much higher reactivity of resorcinol with formaldehyde than phenolic compounds. As a result, most of the phenol used in the resin preparation would remain unreacted [33].

Different synthetic methodologies have been used to develop a fusible RPF resins for various applications. One of the possible reaction schemes that may be used for RPF resin synthesis is given in Figure 5.26.

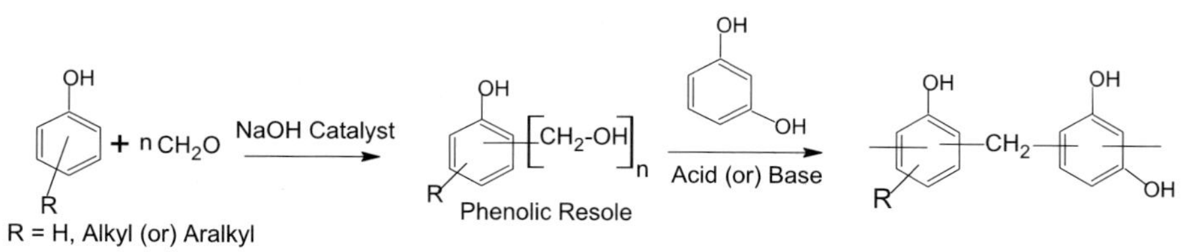


**Figure 5.26.** Resorcinolic resin modified with phenolics

Based on the above scheme, phenol or its derivatives can be reacted first with formaldehyde in the presence of a base catalyst (preferred way) to produce a resole type compound. By this procedure, most of the phenol present in the reaction mixture would react and attach into the final resin molecule. Resorcinol would then be added in sufficient amount to combine with all the methylol groups present in the phenolic molecule. A fusible resinous copolymer of resorcinolic novolak could be produced by following the above reaction scheme. The final novolak resin properties depend on the molar ratios of phenol, resorcinol and formaldehyde and other variables used in the synthesis procedure to develop these resins. The synthesis procedure could be modified in different ways such as varying the catalysts, concentrations, molar ratios of the reactants and condensation conditions to develop a variety of resorcinolic resins.

## References


1. M. M. Sprung, J. Am. Chem. Soc, Vol. 63, p 334 (1941)
2. N. J. L. Megson, J. Soc. Chem. India, Vol. 57, p 189 (1938)
3. R. A. V. Raff and B. H. Silverman, Ind. Eng. Chem, Vol. 43, p 1423 (1951)
4. J. Delmonte, "The tech. of adhesives", Chap. 2, p 26, Reinhold (1947)
5. H. E. Euler, Angew. Chem, Vol. 54, p 458 (1941)
6. A. Sebenik, U. Osredkar and I. Vizovisek, Polymer, Vol. 22, p 804 (1981)
7. J. Varagnat, in Encyclopedia of Chem. Tech., Edited by Mark et al., Wiley, New York, Vol. 13, p 39
8. D. D. Werstler, Polymer, Vol. 27, p 757 (1986)
9. Aero Research Limited, Duxford, England, Sci. and Tech. Memo No. 1/47, December 1946
10. P. H. Rhodes, Modern plastics, August 1947, p 145
11. British Patent, 840,150 (1960)
12. P. H. Rhodes, US Patent, 2,385,372 (1945)
13. A. A. Zavitsas, R. D. Beaulien and J. R. Leblane, J. Polym. Sci., Vol. A-16, p 2541 (1968)
14. M. Horak, A. Pizzi, D. Ferreira and D. G. Roux, J. Appl. Polym. Sci., Vol. 24, p 1571 (1979)
15. A. W. Christiansen, J. Appl. Polym. Sci., Vol. 75, p 1760 (2000)
16. R. Lamartine, Plas. Rubber. Process. Appl. Vol 6, p 313 (1986)
17. A. Pizzi, R. M. Horak, D. Ferreira and D. G. Roux, Cellulose Chem. Tech., Vol. 13, p 753 (1979)
18. D. T. Rossouw, A. Pizzi and G. McGillivray, J. Polym. Sci., Polym. Chem. Edn, Vol. 18, p 3323 (1980)
19. A. J. Norton, US Patent, 2,385,370 (1945)
20. P. H. Rhodes, US Patent, 2,385,374 (1945)
21. W. E. St. Clair and R. H. Moult, US Patent, 3,133,034 ( 1964)
22. M. G. Kim, L. W. Amos and E. E. Barnes, J. Polym. Sci, Polym. Chem, Vol. 31, p 1871 (1993)
23. W. Dankelman and J. de Wit, Die Makromol. Chem, Vol. 62, p 101 (1977)
24. L. M. Pratt, R. Szostak and I. M. Khan, J.M.S. Pure Appl. Chem., Vol. A 34, p 281 (1997)
25. R. L. Ward and R. W. Pekala, Polym. Preprint, Vol. 31, p 167 (1990)
26. R. Durairaj and M. Lawrence, US Patent Appln., 2004 / 0,162,391 (2004)
27. P. T. Paul, US Patent, 3,266,970 (1966)
28. T. Dreyfus and N. L. Bourt, US Patent, 5,412,058 (1995)
29. British Patent, 1,079,909 (1967)
30. M. Zigon, A. Sebenik and U. Osredkar, J. Macromol. Sci. Chem. Vol. A25, p 935 (1988)
31. D. W. T. Shiau, W. D. Detlefsen and E. K. Phillip, US Patent, 5,700,587 (1997)
32. B. T. Larkin, US Patent, 3,367,976 (1968)
33. P. H. Rhodes, US Patent, 2,385,373 (1945)
34. R. Durairaj, M. Walkup and M. Lawrence, US Patent Appln., 2004 / 0,116,592 (2004)

## 5.2 Resorcinolic Resins in Steel Cords Adhesion

### 5.2.1 Introduction

In the manufacturing of radial belted passenger and light truck tires, steel cords are incorporated into the belts and carcass to improve the mechanical stability. The performance of tires depends on the strength and durability of cord to rubber bonds that comprise these materials. A common mode of failure is the separation between the steel belt components. In tires, the belt area has the highest amount of stress and weakest adhesion potential because of the difficulty in bonding rubber to metal. Due to problem with the bonding of bright steel to rubber compounds, steel is often coated with brass to improve the adhesion. The adhesion between the brass coated steel and rubber is important for the durability of tires. In this adhesion, the structure and composition of adhesion interface determines the strength and stability of brass plated steel adhesion with rubber compounds.

### 5.2.2 Brass Plated Steel Cord to Rubber Adhesion

Steel cords are normally coated with a rubber stock containing various components including the sulfur for vulcanization. For optimum adhesion, the copper (Cu) content of brass plating should be in the range of 67 to 72 weight percent. During the vulcanization process, according to various published reports and data, copper present in the brass coating forms a non-stoichiometric copper sulfide ($Cu_xS$) layer at the brass-rubber interface [1]. This copper sulfide layer is dendritic in nature. For optimum adhesion, a critical level of $Cu_xS$ formation is needed. This copper sulfide layer interlocks with rubber compounds. For the efficient interaction of $Cu_xS$ and the sulfur-containing rubber layer, sulfidation ($Cu_xS$ formation) and vulcanization (Sx-Rubber bond formation) should be synchronized. High bond strength can be achieved primarily by a light, physical interlocking between the $Cu_xS$ layer and vulcanized rubber.

#### 5.2.2.1 Factors Affecting the Brass Plated Steel Cord Adhesion

Achieving higher levels of adhesion between the brass plated steel and rubber compounds and maintaining this adhesion under various aged conditions during the entire service life of the tire are important. In general, adhesion is affected by various manufacturing processes, stress, heat and humidity aging conditions.

During the life of tires, heat buildup occurs due to stresses and strains. This causes the rubber to degrade which in turn will result in the deterioration of adhesion layer. In addition, the wear and tear of tires promote the corrosion of steel wires because of the water or moisture due to penetration.

Under heat aging, copper present in the brass migrates through the ZnS and $Cu_xS$ layers formed on the brass surface during the vulcanization process, and thickens the existing $Cu_xS$ layer. Heat can also cause the polysulfidic rubber to break down to become monosulfidic with the release of additional sulfur for further reaction. Therefore, with the availability of additional sulfur, more sulfidation can occur and transform the $Cu_xS.ZnS$ to $Cu_2S.ZnS$ layer. This phenomenon of copper migration weakens the rubber-metal bond. In addition to this copper migration, Zn ions (present in the brass) also diffuse and form $ZnO/Zn(OH)_2$ layers. The formation of additional layers like this weakens the rubber-steel bond.

In the humidity aging conditions, the presence of moisture and oxygen affects the adhesion. The water molecules diffuse through the rubber and corrode the steel cords. The moisture and oxygen accelerate the rate of Zn ions diffusion and formation of $ZnO/Zn(OH)_2$ layer. Formation of this layer destroys the integrity of the $Cu_xS$ layer and, therefore, promotes debonding. This process is also known as "dezincification".

The protection of an adhesion layer and avoiding the adhesion loss can be achieved by the proper formulation of rubber stock with the selections of adhesion-promoting components. Not only the rubber composition but also the curing process is important for the initial adhesion of brass plated steel cord to rubber compounds.

### 5.2.3 Rubber Compound Formulation and Adhesion Testing

Rubber compound compositions for bonding brass coated steel wires or cords are formulated with various adhesion-promoting components including methylene donors (capable of releasing formaldehyde) and methylene acceptors (resorcinol and derivatives). An example of a natural skim compound formulation containing resin system that can be used in steel wire bonding is shown in Table 5.13.

#### 5.2.3.1 Preparation of Rubber Masterbatch

The masterbatch used to evaluate the physical, mechanical and dynamic mechanical properties of rubber compounds and also to prepare the steel cord adhesion samples can be prepared employing a three-stage mixing procedure. In the first stage of this procedure, natural rubber, carbon black, zinc oxide,

**Table 5.13.** Rubber compound formulation for steel cord bonding

| Master batch | Parts by weight |
|---|---|
| First stage mixing (Banbury) | |
| 1. Natural rubber | 100 |
| 2. Carbon black (N-326) | 55 |
| 3. Zinc oxide | 8 |
| 4. Stearic acid | 1 |
| 5. Antidegradant (6PPD) | 2 |
| 6. Retarder (PVI) | 0.2 |
| 7. Antioxidant (DHTMQ) | 1 |
| Second stage mixing (two roll mill) | |
| 8. Methylene acceptor | Vary |
| 9. Cobalt salt | 0.1 |
| Third stage mixing (two roll mill) | |
| 10. Insoluble sulfur | 6 |
| 11. Accelerator (DCBS) | 1 |
| 12. Methylene donor (HMT/HMMM) | Vary |

silica, stearic acid, antidegradants, retarder, and oil are mixed in a Banbury mixer at about 150 to 160 °C. Then, in the second stage, a methylene acceptor (namely resorcinol, resorcinol derivative or resin) and cobalt salt are mixed into an appropriate amount of masterbatch on the two roll mill at about 121 °C. The insoluble sulfur, accelerator and an appropriate amount of methylene donor, namely hexamethylene tetramine (known as Hexa) or hexamethoxymethyl melamine (known as HMMM), are mixed in the third stage at about 95 °C.

The test rubber compounds thus prepared are conditioned overnight in a constant temperature room at about 23 °C and 50% relative humidity (RH). Then the compounds are tested for rheometer cure, shaped and optimum cured at 150 °C for the evaluation of wire adhesion and mechanical properties.

### 5.2.3.2
### Determination of Cure, Mechanical and Adhesion Properties

Cure properties of the uncured rubber compounds are generally measured in a Monsanto 100 type rheometer at 150 °C, 1° and 1.67 Hz according to the ASTM D-2084 method. Wire pull out adhesion can be determined for each test compound by the ASTM D-2229 method using $3 \times 0.2 + 6 \times 0.35$ mm brass plated steel cords with 63.5% copper plating embedded 19.0 mm in the rubber pad (Figure 5.27). Tensile data can be obtained using an ASTM D-412 method.

Dynamic mechanical properties can be determined with a Rheometric Scientific Mechanical Spectrometer or similar equipment at different torsional shear strains at 1 Hz and 23 °C, and also at higher temperatures. Dynamic stiffness (G′) has been measured at different strains and the tangent delta,

**Figure 5.27.** Adhesion testing of steel cords in instron machine

an indicator of rubber compound hysteresis or heat buildup, has also been measured at different strains.

In the dynamic mechanical testing, strain and temperature sweep measurements are also carried out to determine the dynamic performance of rubber compounds. To determine the wet traction property, the tangent delta values are measured at lower temperatures (– 5 to 10 °C). Similarly, the rolling resistance of the rubber compound can be determined by measuring the tangent delta at elevated temperatures in the range of 60 to 80 °C.

Unaged adhesion is normally tested on samples at room temperature. Steam and humidity aged adhesions are determined on samples aged for 16 to 24 h at 120 °C (for steam aging) and 14 to 21 days at 85 °C and 95% relative humidity (for humidity aging).

These test data are important to determine the strength of steel cord to rubber bonds. From the adhesion data it can be possible to assess and predict the effectiveness of various adhesion promoters in the rubber compound formulations to improve the bond between steel and rubber.

In general, the rubber compound's physical and mechanical properties can also be affected by mixing procedures and cure conditions. Similarly, the unaged and aged steel wire adhesion can be affected by the nature of methylene donors and acceptors and their levels in the rubber compounds.

### 5.2.3.3
### Mixing Rubber Compounds

Resorcinol and resorcinol-based compounds and resins are predominantly used as the methylene acceptors in the steel skim rubber compound formulations as adhesion promoters. Resorcinol in the un-modified state softens at 105 °C and melts at 110 °C. Similarly, the resorcinolic derivatives and resins, used in the rubber compounds, have softening or melting points in the range of 100 to 115 °C temperature. Therefore, resorcinolic compounds should be mixed at a minimum temperature of 110 to 125 °C to ensure complete melting and uniform dispersion into the rubber compounds.

Methylene donors, such as Hexa or HMMM, react readily with resorcinolic methylene acceptor. Since the methylene acceptor-donor system begins to react at the two roll mixing temperatures, it is necessary to mix the methylene donor separately in the final stage and possibly below 105 °C.

With resorcinol as the methylene acceptor, to reduce the fuming of resorcinol and also allow mixing at lower temperatures, pre-dispersions of resorcinol with carriers such as stearic acid, elastomers, silica and plasticizers have been developed.

In a multi-stage Banbury mixing with three or more passes, resorcinolic methylene acceptors may be added in one of the latter passes before the final one to minimize fuming and also provide finished compounds with optimal performance properties.

### 5.2.4
### Advantages of Resorcinolic Resins in Steel Skim Rubber Compounds

Resorcinol is a phenolic compound having two hydroxyl groups attached to the *meta*-positions with respect to each other in the benzene ring (Figure 5.28).

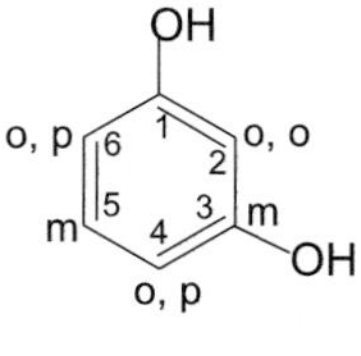


**Figure 5.28.**

The three reactive positions in resorcinol for the methylene donor reactions are 2, 4 and 6 and they are positioned at either *ortho* or *para* to the hydroxyl groups. Therefore, all three positions are activated by the two hydroxyl groups present in the same benzene ring of the resorcinol molecule.

Resorcinol is a white crystalline solid with a melting point of 110 °C and is highly soluble in water. Being a small molecule and possessing an ideal melting point for rubber compound processing, resorcinol can be uniformly dispersed in the rubber compounds without any difficulty at the processing temperatures.

Resorcinol and resorcinolic novolak resins have been widely used by the rubber industry as the reinforcing and bonding agents in rubber compounds. These resorcinolic products are so unique in rubber compounding applications due to the fact they act as thermosetting or vulcanizing plasticizers. With resorcinol in the compound, the uncured rubber mixture exhibits low compound viscosity. This allows easier processing, higher loading and excellent extrusions for the rubber compounds.

The thermosetting properties of resorcinol and resorcinolic compounds on curing allow the vulcanizate to show an increase in hardness, abrasion resistance, aging resistance, solvent and oil resistance and stiffness and also give much improved finishes to the cured rubber stock. This combination of plasticizing and reinforcing action is rare for a single material in rubber compounds. In addition, these resorcinolic products may act as antioxidants in such rubber compounds.

### 5.2.4.1 Methylene Donors in Rubber Compounds

In general, two methylene donors are widely used in rubber compound formulations that are known to react with phenolic or resorcinolic methylene acceptors to enhance cured rubber physical and mechanical properties. The two well-known methylene donors are hexamethylene tetramine (Hexa or HMT) and hexamethoxymethyl melamine (HMMM) and they have the following chemical structures (Figure 5.29).

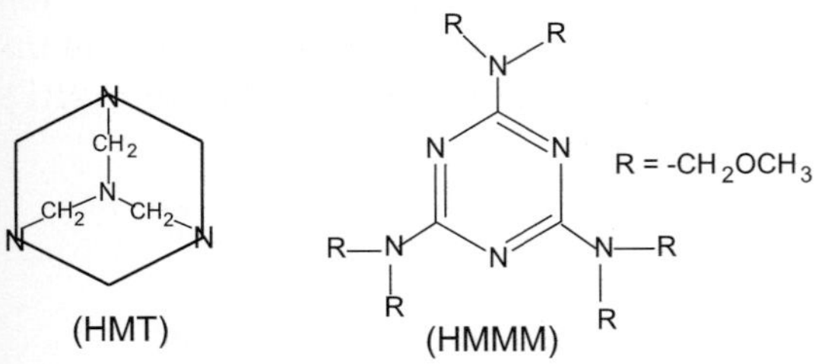


**Figure 5.29.**

Both HMT and HMMM are very effective in their reactions towards resorcinolic compounds under rubber curing conditions to produce highly crosslinked network structures. The presence of resorcinolic networks throughout the rubber matrix provides good physical and mechanical properties for cured rubber. Though HMT has been used as the methylene donor in the natural rubber steel skim compound formulation, the release of ammonia from the HMT was found to affect steel cord adhesion due to corrosion. Therefore, HMT is no longer in use in compound formulations used for steel wire bonding.

On the other hand, HMMM methylene donor reacts with resorcinolic methylene acceptors more effectively under rubber curing conditions and provides higher unaged adhesion and maintains this adhesion even in heat, humidity and steam aged conditions. Now HMMM has been widely used as methylene donor in steel cord skim compounds.

#### 5.2.4.2 Use of Cobalt Salts

It has been observed that the use of an organic cobalt compound such as cobalt naphthenate promotes adhesion in the presence of resorcinol (acceptor) and HMMM (donor) compounds. An organic cobalt compound can effectively promote the initial unaged wire adhesion and resorcinol and the HMMM system provides satisfactory steam and humidity aged adhesion. Based on the extensive studies performed, it has been well documented that a combined wire bonding system (adhesion promoter system) based on resorcinol, HMMM and cobalt should give the best possible balance of initial and aged adhesion properties.

#### 5.2.4.3 Methylene Acceptor/Donor Ratios

To achieve the highest performance in steel cord adhesion and rubber compound properties, the ratio and amount of methylene donor and acceptors should be chosen at the appropriate levels. The methylene donor/acceptor ratio depends on the type of donor used in the formulations. For an optimum performance, a typical resorcinol (R) and resorcinol-formaldehyde resin (RF) bonding systems are given in the following Table 5.14.

**Table 5.14.** Resorcinol and resorcinolic resin bonding system

| Resin system | R/HMT | R/HMMM | RF resin/HMMM | RF resin/HMMM |
|---|---|---|---|---|
| Weight ratio | 65/35 | 50/50 | 75/25 | 60/40 |
| Loading level (phr) | 4 | 4 | 5 | 5 |

Though Table 5.14 data show an optimum ratio and levels that can be used for obtaining a good steel cord adhesion, both should be optimized for particular rubber compound formulations and applications.

In the case of cobalt salt, the cobalt metal level is usually kept below 0.3 phr (parts per hundred rubber) to obtain a satisfactory balance of initial and humidity aged steel cord adhesion.

## 5.2.4.4 Role of Resorcinolic Bonding Systems in Steel Cord Adhesion

Brass plated steel cords are normally embedded in a rubber compound having the composition, for example, as shown in Table 5.13. When the embedded wire compound is vulcanized, various curing and bonding reactions take place. During this process, the resorcinol-HMMM curing reaction, sulfur cross-linking with rubber and $Cu_XS$ layer formation at the brass surface should be synchronized otherwise proper adhesion could not be achieved. It is well known that the formation of a proper thickness $Cu_XS$ layer is responsible for the adhesive strength between the rubber compound and steel cords. In order to determine the role of resorcinolic bonding systems in the adhesion of steel cords studies were conducted and reported [2, 3].

Resorcinol and resorcinolic resins are considered as highly polar molecules in the rubber compounds. Therefore, due to incompatibility, they are basically insoluble in rubber. During vulcanization, the polar molecules migrate out of the rubber and move towards the adhesion interface between the rubber and brass surface. At this interface the resorcinolic resin forms tightly cross-linked network structures from the reaction with a methylene donor as shown in the following reaction schemes (Figure 5.30).

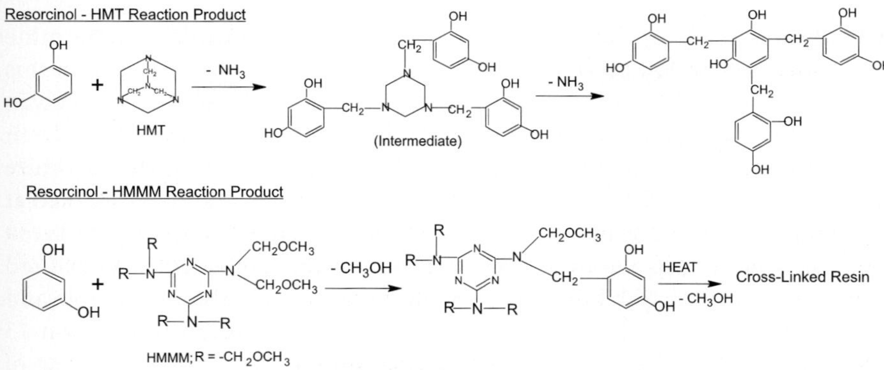


**Figure 5.30.**

Depending upon the type of methylene donor compound, HMT or HMMM, the cross-linked network structure formed at the interface may vary. HMT has been synthesized from the reaction of ammonia and formaldehyde. With HMT methylene donor, ammonia is produced from the reaction with resorcinolic methylene acceptor compounds. The intermediate structure thus formed can further undergo decomposition reaction to produce primarily the resorcinol-formaldehyde type network structure.

This might not be the case with HMMM methylene donor. HMMM is produced from the reaction of melamine and formaldehyde and subsequent etherification with methanol.

The amino group ($NH_2$) present in the melamine is not considered as an amine but an amide type functional group. With HMMM, after the initial reaction of one of the methylol groups with resorcinol, the second methylol group may also react with resorcinolic hydroxyl group in the following way to produce benzoxazine type structure under suitable conditions (Figure 5.31).

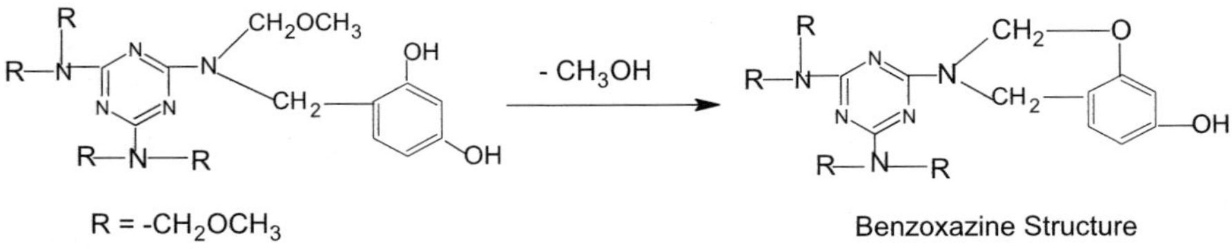


**Figure 5.31.**

Formation of benzoxazine and hydroxyphenylmethylmelamine type reaction products from the HMMM and phenolic resins have already been reported [4, 5].

## 5.2.4.5 Reaction of Resorcinol with HMMM

A study was made of the reaction of resorcinol with HMMM to determine the nature of reaction products that could be formed from these two. In this work, resorcinol (0.30 mol) and HMMM (0.01 mol) were dissolved separately in a methanol solvent and mixed together at room temperature. Then the solvent methanol was evaporated under atmospheric conditions till the temperature reached 80 °C. After this, the reaction was heated to 140 °C and maintained at this temperature for 30 min. Vacuum was applied and the solvent present in the reaction product was removed. The reaction product appeared as a dark red semi-solid material. Solubility of this material in acetone showed soluble and insoluble fractions. Both the fractions were analyzed by proton and Carbon-13 NMR analysis. Analysis of acetone soluble portion showed the presence of unreacted resorcinol and methylene bridged resorcinol (RF) structures. The methylene bridge distribution of 4,4′-, 2,4′- and 2,2′- were present at 63, 33 and 4 area percent (from NMR analysis) for the methylene bridged RF type product.

NMR analysis clearly indicated the absence of $-N-CH_2-O-$ and $-CH_2-N-$ type structure in the reaction mixture. Infrared (IR) determination on the insoluble portion indicated that this material was melamine.

This data strongly suggests that the resorcinol-HMMM reaction mainly leads to the formation of RF novolak type product.

Another study was similarly made with a molar ratio of resorcinol/HMMM/cobalt naphthenate of 0.30:0.02:0.008 to determine the interaction of resorcinol with cobalt in the reaction product. The reaction again produced a material containing acetone soluble and insoluble portions. The acetone extracts (approximately 60 wt. %) have been identified by FT-IR and proton-NMR as unreacted resorcinol, methylene bridged resorcinols and cobalt naphthenate. FT-IR absorption near 2594 wave number is evident of a strongly hydrogen bonded –OH structure, and suggests an interaction of cobalt ion with resorcinolic hydroxyls. The acetone insoluble solids isolated from this sample exhibited IR absorptions characteristics of very highly substituted, high molecular weight resorcinolic compounds consisting of resorcinol-formaldehyde reaction products.

Resorcinol-formaldehyde resins have been used as adhesives in the bonding of wood to provide moisture resistance and waterproof properties in the laminated wood products. This property clearly predicts that the cross-linked RF resins are moisture resistant and also thermally stable due to the presence of methylene bridges between the resorcinol rings. The formation of a resorcinolic network structure at steel adhesion interface makes the protective layer much more water and heat resistant.

It is important to protect the bonding $Cu_xS$ layer from moisture attack under humidity and steam aging conditions. Under aging conditions, the production of more $Cu_xS$ or $Cu_2S$ and $ZnO/Zn(OH)_2$ at the brass-rubber interface increases and thus makes the $Cu_xS$ layer non-bonding. This drastically affects the adhesion performance of the bonding layer. The presence of a highly moisture and heat resistant resin layer on the top of the $Cu_xS$ layer can avoid any further chemical and physical changes at the interface. This, ultimately, enhances the heat, humidity and steam aged adhesion properties of brass plated steel cord to rubber compounds [6].

A possible bonding mechanism proposed for the adhesion of brass plated steel cords with rubber compounds containing resorcinolic bonding system can be seen in the following figure (Figure 5.32).

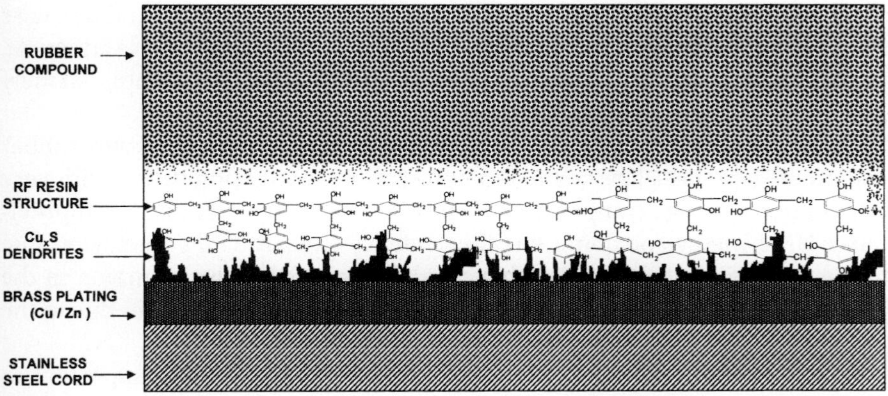


**Figure 5.32.** Bonding system mechanism for steel cords adhesion with resorcinolic bonding

## 5.2.5
## Different Resorcinolic Chemistry for Improving Steel Wire Adhesion

### 5.2.5.1
### Use of Resorcinol

Resorcinol, in the unmodified state, has been successfully used by the rubber industry as a bonding agent to improve the adhesion of brass plated steel wires or cords to rubber for various applications. There are several advantages in using resorcinol in the rubber compound formulation as outlined before. The major advantage is its high reactivity towards the methylene donor present in the compound. Due to this reactivity, the resorcinol-methylene donor reaction can be synchronized with the sulfur cure during vulcanization. If resorcinol is less reactive and the resorcinol-methylene donor reaction is not synchronized with sulfur cross-linking then this may result in the presence of unreacted resorcinol in the cured rubber compounds. The presence of such materials will be expected to affect the performance of cured rubber physical and mechanical properties.

In the steel cord adhesion phenomena, not only the adhesion of brass plated steel cords to skim rubber compounds but also the cured rubber physical, mechanical and dynamic mechanical properties are important. If the additives present in the rubber compounds exhibit high hysteresis, then the performance of cured rubber could be ultimately deteriorated. This in turn will affect adhesion of steel cords to rubber compounds during the service life of tires. The cross-linked structure produced from the reaction of resorcinol with a methylene donor compound improves the hysteresis (low hysteresis and less heat generation) properties of cured rubber. This could be one of the advantages of using resorcinol in the steel skim rubber compound, which is the most critical area of steel belted radial tire production.

The following Table 5.15 illustrates details of the use of resorcinol in a wire skim rubber compound formulation and the data obtained on wire adhesion, physical and dynamic mechanical properties. Two methylene donors, namely HMT and HMMM are used in the formulations [7].

From the adhesion data, it is very clear that HMT (or Hexa) cured rubber compound produces low steam and humidity aged adhesion values with very low rubber coverage compared to HMMM cured system. As discussed previously, HMT methylene donor has the tendency to corrode the steel wires due to the formation of ammonia. This clearly suggests that the reduction in the aged adhesion might be associated with the corrosion of steel wires due to the presence of ammonia and water at the steel-rubber interface.

In determining the cured rubber compound properties, it is important to determine the dynamic mechanical properties such as the storage modulus ($G'$), loss modulus ($G''$) and tangent delta (hysteresis) properties from the dynamic mechanical analysis (DMA). These measurements will provide valu-

**Table 5.15.** Rubber compound properties

| Resorcinol compound/methylene donor | R/HMT | R/HMMM |
|---|---|---|
| Acceptor/donor level | 2.0/2.0 | 2.0/2.0 |
| Rheometer cure at 150 °C (ASTM-D-2084) | | |
| MH, dN-m | 69.5 | 67 |
| ML, dN-m | 8.6 | 9 |
| ts2, min | 2.8 | 2.8 |
| t′90, min | 17 | 13.3 |
| Steel cord adhesion (ASTM D-2229), N (% rubber coverage) | | |
| Unaged | 1526(95) | 1281(90) |
| Steam, 24 h at 120 °C | 520(30) | 1157(60) |
| Humidity, 21 days, 85 °C,95% RH | 342(5) | 1303(80) |
| Tensile property (ASTM D-412) | | |
| 300% Modulus, MPa | 23.03 | 19.76 |
| Tensile strength, MPa | 25.64 | 25.13 |
| Ultimate elongation, % | 361 | 403 |
| Dynamic mechanical[a] | | |
| unaged | | |
| G′ at 0.2% Strain, MPa | 15.9 | 14.48 |
| G″ at 2.0% Strain, MPa | 2.04 | 1.86 |
| Heat aged (three days, 100 °C in air) | | |
| G′ at 0.2% strain, MPa | 15.57 | 18.56 |
| G″ at 2.0% strain, MPa | 19.43 | 22.32 |

[a]RMS 800 Instrument, 0.1 Hz and RT
Rubber compound does not contain silica
Data taken from [7]

able information about the long term use of rubber compounds. The storage modulus (G′) measurement could provide information about the hardness of material. Loss modulus (G″) value will show the ability of a material to dissipate the energy. Lower loss modulus value indicates low energy dissipation, hysteresis and heat buildup of the rubber compounds. Similarly, the tangent delta value can predict the energy absorption characteristics of materials. The measurement of tangent delta from the DMA analysis is a qualitative tool to determine the hysteresis or heat buildup behavior of rubber compounds.

The DMA properties are expected to be affected by the presence of various rubber components including the methylene donor and acceptor compounds. For the good performance of rubber compounds in tires and other applications, the cured rubber compounds should show higher storage modulus (G′) and lower tangent delta values.

Resorcinol, resorcinolic derivatives and resins when used as the methylene acceptor compound in rubber compound formulations can provide improved dynamic mechanical properties for the cured rubber.

### 5.2.5.2
### Resorcinol-Formaldehyde Novolak Resins

Though resorcinol, in the unmodified state, provides good adhesion properties with curing agents, fuming of resorcinol at Banbury mixing often poses some concern for the worker exposures and health problems. The fuming of resorcinol is associated with the sublimation character of resorcinol at its melting temperature. To minimize the fuming of resorcinol, resorcinol-formaldehyde (called RF) resins were developed and successfully used in steel cord adhesion with rubber compounds.

Different manufacturers produce various grades of RF resins, and these resins are sold under different commercial names. Indspec Chemical Corporation, a leading manufacturer of RF resins, has been producing RF resins and selling under the commercial names Penacolite B-1-A, B-18-S and B-19-S resins. These are brittle resins and have been successfully used by various tire, belt and hose manufactures to improve the adhesion of different reinforcing materials with rubber compounds. A generalized structure for the resorcinolic novolak B-18-S and B-19-S resins is shown below (Figure 5.33).

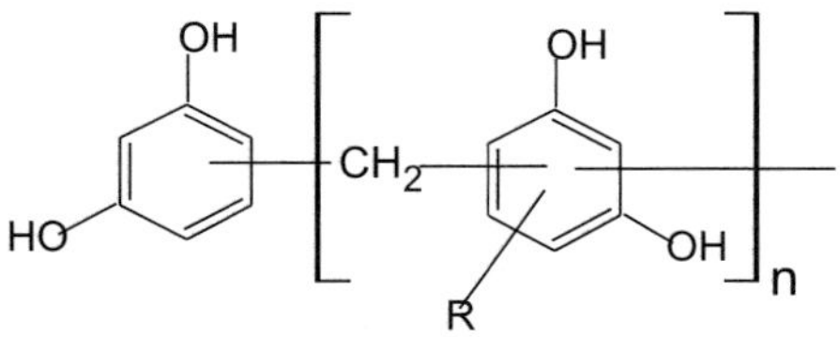


**Figure 5.33.**

Details of the physical and fuming properties (measured from the Thermogravimetric Analysis, TGA) of B-18-S and B-19-S resins are shown in Table 5.16.

**Table 5.16.** Penacolite resin properties

| Resin properties | Penacolite B-18-S | Penacolite B-19-S |
|---|---|---|
| 1. Softening point (°C) | 105 | 107 |
| 2. Moisture content (wt%, max) | 0.5 | 0.4 |
| 3. Free resorcinol (wt%) | 18 | 11 |
| TGA weight loss results (%) (Heating rate = 10 °C/min in nitrogen) | | |
| 125 °C | 0.3 | 0.1 |
| 150 °C | 1 | 0.3 |
| 175 °C | 6 | 3.9 |

From Table 5.16 results, RF resin can exhibit lower fuming under the rubber processing temperature conditions than resorcinol.

The uses of B-18-S and B-19-S resins in steel skim rubber compounds and their effects on improving the cured rubber physical and mechanical

**Table 5.17.** Penacolite resins in steel wire bonding applications

| Rubber compound properties | Penacolite | Penacolite |
|---|---|---|
| Resorcinol compound/methylene donor | B-18-S/HMMM | B-19-S/HMMM |
| Acceptor/donor level | 3.5/2.5 | 3.5/2.5 |
| Rheometer cure at 150 °C (ASTM-D-2084) | | |
| $M_H$, dN-m | 63.2 | 64.1 |
| $M_L$, dN-m | 8 | 8 |
| ts2, min | 2.9 | 3 |
| t′90, min | 14.8 | 15.4 |
| Adhesion (ASTM D-2229), Newtons (% rubber coverage) (Wire=63.5% copper) | | |
| Unaged | 1312(90) | 1312(90) |
| Steam, 16 h at 120 °C | 1290(90) | 1290(90) |
| Tensile property (ASTM D-412) | | |
| 50% Modulus, (MPa) | 2.82 | 2.91 |
| Tensile strength, (MPa) | 25.01 | 23.87 |
| Ultimate elongation, % | 447 | 441 |
| Hardness (Shore A) | 85 | 85 |
| Dynamic mechanical[a] (unaged) | | |
| G′ at 0.2% Strain, MPa | 34.03 | 35.68 |
| Tangent delta at 2.0% strain | 0.205 | 0.204 |

[a]RMS 800, 1.0 Hz at RT

properties, and also the brass plated steel cords adhesion properties with rubber are given in Table 5.17.

Similar to resorcinol, RF resins are also known to provide good brass plated adhesion with rubber compounds. These resins are well known in the rubber industry as effective adhesion promoters for the reinforcement of different fibers and cords with rubbers.

In order to investigate the adhesion properties associated with the use of RF resin (Penacolite B-18S) in the brass plated cords of different cord constructions, a study has been conducted. Two rubber compounds were used, in which one compound contained only cobalt and the second compound contained cobalt and the resin system. The rubber compound formulations employed in this study are presented in Table 5.18 and the results obtained are shown in Table 5.19.

Compared to the cobalt only rubber compound, the presence of resin system in the compound formulation appeared to improve the humidity aged adhesion of brass plated steel cords.

Similar to resorcinol, the RF resins on curing with methylene donors in rubber compounds are known to produce highly cross-linked network structures that can be moisture and heat resistant, on top of the $Cu_xS$ adhesion layer.

**Table 5.18.** Rubber compound formulation for steel cord bonding

| Master batch (parts by weight) | Control | B-18S resin |
|---|---|---|
| First stage mixing (Banbury) | | |
| 1. Natural rubber | 100 | 100 |
| 2. Carbon black (N-326) | 55 | 55 |
| 3. Zinc oxide | 8 | 8 |
| 4. Stearic acid | 1 | 1 |
| 5. Antidegradant (6PPD) | 2 | 2 |
| 6. Retarder (PVI) | 0.2 | 0.2 |
| 7. Antioxidant (DHTMQ) | 1 | 1 |
| Curatives | | |
| 8. Crystex OT-20 | 4.69 | 4.69 |
| 9. Santocure NS | 0.8 | 0.8 |
| Bonding system | | |
| 10. Penacolite resin B-18S | 0 | 3.5 |
| 11. Cyrez 963 (72%) | 0 | 3.47 |
| 12. Cobalt nphthenate (12%) | 0.83 | 0.83 |
| Rheometer cure at 150 °C, 1°, 1.67 Hz | | |
| $M_H$, in-lb | 50.8 | 64 |
| $M_L$, in-lb | 8.8 | 11.8 |
| Ts 2, min | 5 | 3.1 |
| $T_90$, min | 12.8 | 14 |

**Table 5.19.** Goodyear SWAT adhesion test data on the steel cords of different constructions

| Bakaert steel cord wire construction | Embedment | Pull out force, N (% Rubber coverage) Unaged adhesion Control | Unaged adhesion B-18 S | Humidity adhesion Control | Humidity adhesion B-18 S |
|---|---|---|---|---|---|
| 1. 7 × 4 × 0.175 | 3/4″ | 516 (80) | 685 (90) | 222 (5) | 787 (90) |
| 2. 3 + 9 + 15 × 0.175 + 1 | 3/4″ | 582 (90) | 756 (95) | 197 (0) | 782 (80) |
| 3. 12 × 0.22 + 1 | 3/4″ | 436 (50) | 524 (60) | 182 (0) | 698 (90) |
| 4. 3 × 0.20 + 6 × 0.35 | 3/4″ | 538 (80) | 751 (90) | 182 (5) | 934 (90) |
| 5. 2 + 2 × 0.25 | 1/2″ | 289 (80) | 333 (80) | 93 (5) | 333 (90) |
| 6. 2 × 0.30 HT | 1/2″ | 249 (50) | 320 (90) | 71 (5) | 213 (90) |

Brass plated steel, copper content = 63.5%, Humidity Aged: 21 Days at 85 °C/95% RH

RF resins in the uncured state have the tendency to absorb moisture under humidity conditions. Therefore, efforts were made to develop highly processable, non-hygroscopic resorcinolic novolak resins for rubber compounding applications. Various developmental activities have been consistently made throughout the resin industry to modify resorcinol-formaldehyde resin chemistry for an improved performance required by the tire and other industries. In these new developmental efforts, resin manufacturers have been making efforts to develop new high performance resorcinolic products that do not affect the cure, physical, mechanical and adhesion properties of rubber compounds.

Some interesting resorcinolic resin chemistries were developed to improve the cured rubber compound and steel cord adhesion properties.

### 5.2.5.3 Alkyl Substituted Resorcinolic Novolak Resins

The fuming of resorcinol and the hygroscopicity of RF resins are often considered as problems in the handling and use of these materials in rubber compounds. The hygroscopic problem of RF resin can be overcome or reduced by the incorporation of hydrophobic aliphatic hydrocarbon chains in the resorcinol structure. This can be achieved by the alkylation of resorcinol or RF resin with an unsaturated hydrocarbon compounds. By making this modification, the volatility of resorcinolic derivative or resin can also be lowered due to the presence of alkyl substituents.

With the introduction of an alkyl chain in the resorcinol molecule, a significant improvement in the miscibility or compatibility of the substituted resorcinolic compounds with rubber compounds is expected. In addition, improved tack properties with natural and synthetic rubber products can also be expected with the use the of such alkylated resorcinols.

In the development of alkyl substituted resorcinolic novolak type resins, a good synthetic methodology should be followed. It is important to know and understand the reactivity of resorcinol with unsaturated organic compounds and formaldehyde or other aldehydes for developing resins having well defined chemical structures and physical properties. The best mode of developing an alkyl-substituted resorcinolic novolak resin can be done by reacting resorcinol first with an olefinic compound followed by aldehyde reaction. In this way, the reaction can be well controlled and the formation of undesirable reaction products may be minimized.

Alkyl modified resorcinolic resins containing various alkyl substituted products for improving the steel cord adhesion properties were developed based on the following reaction scheme (Figure 5.34) [8].

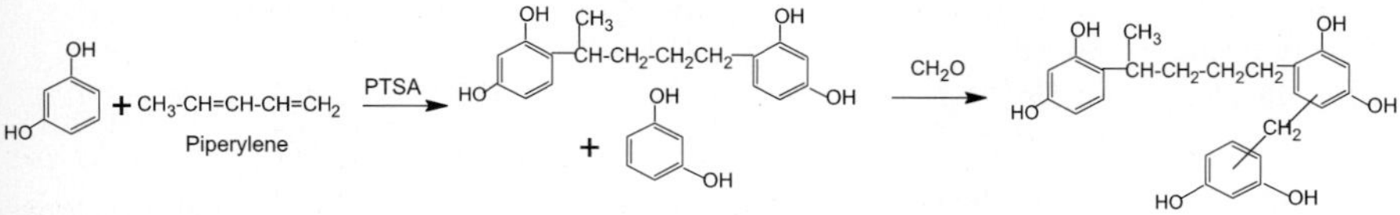


**Figure 5.34.** Piperylene modified resorcinolic novolak resin synthesis

A synthetic procedure used to develop this alkyl-substituted resorcinolic novolak resin from the resorcinol-piperylene-formaldehyde reaction is outlined in the following procedure.

**Preparation of Alkyl-Substituted Resorcinolic Novolak Resin [8]** Into a 1-l reaction kettle equipped with a stirrer, thermometer, reflux condenser and an addition funnel, 220.2 g of resorcinol (2.0 mol) were charged and heated to 120 °C. *p*-Toluene sulfonic acid (2.0 g) was added at 120 °C followed by the slow addition of 40.0 g of piperylene (65 – 70% active) for about 60 min. The temperature of the reaction mixture was maintained between 135 – 145 °C during the piperylene addition. After the piperylene addition has been completed, 100 g of methanol were added first and then 54.8 g of methylformcel (55% W/V, 1.0 mol) were slowly added into the piperylene-resorcinol reaction product for about 30 min at 70 – 80 °C. The reaction mixture was refluxed for 3 h more and the solvent was removed under reduced pressure to produce a novolak resin having a softening point of 99.5 °C.

When developing a resorcinolic novolak resin, the molar ratios and synthesis conditions should be carefully maintained. Otherwise, the reaction will lead to the formation of gel type material which cannot be used in compounding applications. The test data obtained on the use of piperylene modified resorcinolic resins in the rubber compounding formulation is shown in Table 5.20.

**Table 5.20.**

| Rubber compound properties<br>Resorcinol compound/methylene donor | (Pip+R+F)/<br>HMMM | (Control)<br>(R+F)/<br>HMMM | (Control)<br>Cobalt<br>only |
|---|---|---|---|
| Acceptor/donor level | 2.0/2.0 | 2.0/2.0 | No Resin |
| Rheometer cure at 150 °C (ASTM-D-2084) | | | |
| MH, dN-m | 66.2 | 57.9 | 49.9 |
| ML, dN-m | 10 | 10.2 | 8.2 |
| ts2, min | 3.1 | 3.2 | 5.2 |
| t′90, min | 14 | 13.7 | 12.8 |
| Adhesion (ASTM D-2229), Newtons (% Rubber coverage) (Wire = 63.5% Cu) | | | |
| Unaged | 1481(90) | 1285 (90) | 1036 (70) |
| Steam, 16 h at 120 °C | 1503 (95) | 1370 (95) | 387 (5) |
| Humidity, 21 days, 85 °C, 95% RH | 1325(90) | 1250(85) | 556(40) |
| Tensile property (ASTM D-412) | | | |
| 300% Modulus, (MPa) | 19.22 | 19.69 | 18.87 |
| Tensile strength, (MPa) | 26.5 | 25.79 | 26.76 |
| Ultimate elongation, % | 423 | 409 | 451 |
| Hardness (Shore A) | 85 | 82 | 77 |
| Dynamic mechanical[a] (unaged) | | | |
| G′ at 0.2% strain, MPa | 37.7 | 28.14 | 15.57 |
| G″ at 2.0% strain, MPa | 41.27 | 33.57 | 19.36 |

[a]RMS 800, 1.0 Hz at RT

Pip = Piperylene; R = Resorcinol and F = Formaldehyde

Data taken from [8]

By comparing the performance of cured rubber compound and brass plated steel cord adhesion properties of piperylene modified resorcinolic novolak resin against the RF resin and cobalt only compound, improved steam and humidity aged adhesion properties were observed for the modified resin. This data suggests that the adhesion enhancement could be due to the formation of a highly moisture resistant protective layer on the top of $Cu_xS$ bonding layer. Also, the high dynamic storage modulus ($G'$) data indicated that highly cross-linked and higher molecular weight products were produced from the reaction of this alkyl-modified resorcinolic resin with HMMM curing agent.

In the absence of a resorcinolic resin system, the cobalt only rubber compound showed lower humidity and steam aged adhesion values which indicates that the adhesion performance of cobalt system can be improved only with the use of resin/HMMM in the compound formulations.

#### 5.2.5.4 Alkyl Phenol Modified Resorcinolic Novolak Resins

Resorcinolic novolak resins were also modified with alkyl phenols to improve the rubber compatibility and enhance the adhesion performance [9–12]. In the development of resorcinolic resins for the rubber compounding applications to improve steel cord adhesion and other properties, the reactivity of resorcinol in the final resin structure should be maintained as much as possible. In the synthesis of alkylphenol modified resins, the phenolic compound should be reacted first with an aldehyde or mixture of aldehydes before carrying out the resorcinol-aldehyde reaction. In this way, a minimum number of resorcinolic reactive positions are utilized in the resin forming reaction. Resins containing more reactive resorcinolic sites can react faster with HMMM curing agent present in the rubber compound during the vulcanization. Due to the high reactivity of resorcinol towards aldehydes under both the acidic and basic conditions, the use of phenolic compounds after the resorcinol reaction could result in large quantities of unreacted phenolic compounds in the final resin product. The presence of more unreacted material in the final resin can affect the cured rubber compound physical and mechanical properties.

Depending upon the type of alkyl substitution in the phenolic molecule, the modified novolak resins can improve the hygroscopic behavior of RF resins. An example of an alkyl-phenol modified RF resin synthesized to improve the fuming, hygroscopic and adhesion properties of rubber compounds with steel cords is illustrated in Figure 5.35.

A procedure used to synthesize a modified resorcinolic novolak resin is given below.

**Synthesis of Nonylphenol modified Resorcinolic Novolak Resin [9]** The reaction kettle equipped with a stirrer, thermometer and reflux condenser was charged with 310.7 g (1.41 mol) of, nonylphenol, 104.8 g (55% W/W; 1.92 mol) of methyl

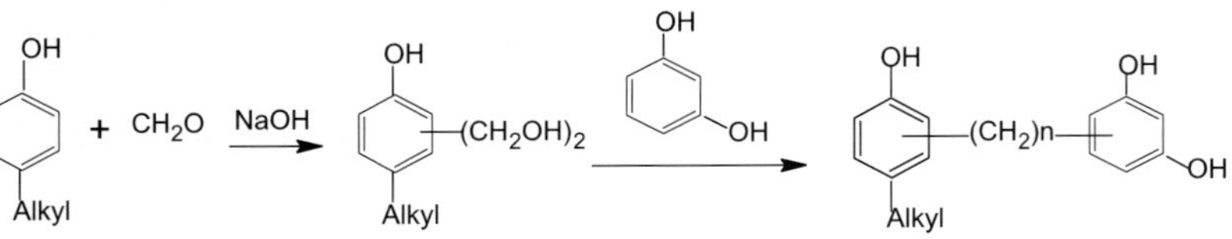


**Figure 5.35.** Alkylphenol modified resorcinolic novolak resin synthesis

FORMCEL 28.8 g (0.3 mol) of furfural and 4.0 g of sodium hydroxide (50% W/W) solution. The reaction was carried out at 80–85 °C for 10 h. After this, the reaction mixture was cooled to room temperature and 105 g of methanol was added followed by the slow addition of 148.7 g of resorcinol (1.35 mol). Again, the reaction mixture was refluxed for 5 h and finally the solvent was removed under reduced pressure to give a novolak type resin having a softening point of 110.5 °C.

Two novolak resins have been developed and were based on the use of furfural (another aldehyde) in addition to formaldehyde in the first synthesis and cashew nut shell liquid (another alkyl phenol) in the second.

In order to determine the fuming and hygroscopic properties of nonylphenol modified resorcinolic resins, a weight loss data was obtained from TGA analysis and moisture pick up determination was done under controlled conditions. These results are shown in Tables 5.21 and 5.22.

**Table 5.21.** Thermogravimetric analysis (TGA) of resins

| Resins | Weight loss (%) at Temperature °C | | |
|---|---|---|---|
| | 125 | 150 | 175 |
| (R+F) resin | 0.2 | 0.8 | 2.2 |
| (NP+F+FF+R) resin | 0 | 0.1 | 0.2 |
| (NP+F+CNSL+R) resin | 0 | 0.2 | 0.8 |

NP=Nonyl phenol; FF= Furfural; CNSL = Cashew nut shell liquid
Data taken from [9]

**Table 5.22.** Moisture absorption of nonylphenol modified R/F resins

| Resins | Moisture absorbed (weight %) in day | | | |
|---|---|---|---|---|
| | 1st | 3rd | 5th | 7th |
| (R+F) Resin | 2.57[a] | 5.14[a] | 6.24[a] | 6.65[a] |
| (NP+F+FF+R) Resin | 0.76 | 1.14 | 1.36 | 1.45 |
| (NP+F+CNSL+R) Resin | 0.75 | 1.21 | 1.47 | 1.57 |

NP=Nonyl phenol; FF= Furfural; CNSL = Cashew nut shell liquid
Conditions: RH at 71% and RT; [a]Resin agglomerated
Data taken from [9]

Based on the results in Tables 5.21 and 5.22, these alkylphenol modified resins showed less fuming and hygroscopic properties than the standard RF resin. Therefore handling of these resins in rubber compound formulations could be easier.

### 5.2.5.5 Aralkyl Modified Resorcinol Novolak Resins

RF resins have been modified in different ways to achieve the desired properties for the rubber compounding applications. Unlike an alkyl substitution, the aralkyl groups present in the resorcinol molecule can improve the cured rubber compound modulus property due to bulkier substitution.

Various aralkyl-substituted resorcinolic novolak resins were developed with the use of different styrene derivatives [13, 14]. This novolak resin chemistry was based on the following reaction scheme (Figure 5.36).

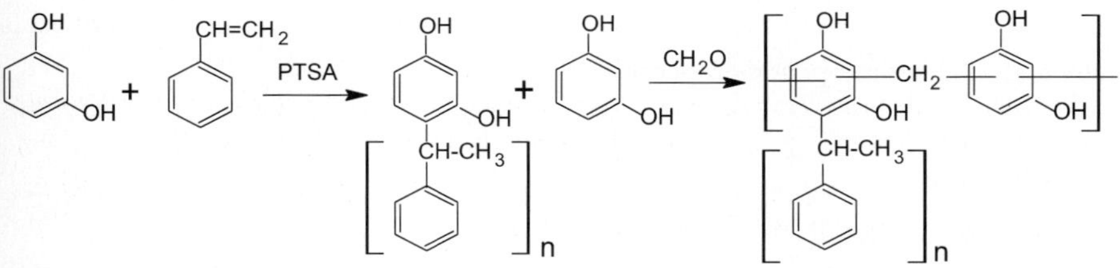


**Figure 5.36.** Reaction scheme for styryl modified resorcinolic novolak resin synthesis

By varying the molar ratios of resorcinol, styrene and formaldehyde, novolak resins having different softening points and free resorcinol content were prepared. A synthesis procedure for the preparation of styryl substituted resorcinol-formaldehyde novolak resin is outlined below.

**Procedure for the Preparation of Styryl Substituted Resorcinolic Novolak Resin [13]**
Into a 2-l reaction kettle equipped with a stirrer, thermometer, reflux condenser and addition funnel, 420.0 g of resorcinol were charged and heated to 120 – 130 °C. *p*-Toluene sulfonic acid (1.45 g) was then added at 120 °C and mixed for 5 min. Then 280.0 g of styrene were charged to the molten resorcinol stream wise over a period of about 90 min. The temperature of the reaction mixture was maintained between 115 – 135 °C during the styrene addition. After all the styrene has been added, the reaction mixture was maintained at this temperature for an additional 60 – 90 min. Then 210.0 g of 37% aqueous formaldehyde solution was added slowly into the kettle over a period of 60 – 90 min at 95 – 120 °C. The reaction mixture was held at reflux for 15 – 30 min more. After this reflux period, 40.0 g of additional resorcinol and 52.5 g of 80% aqueous RM-441 solution were added at once and continued the reaction for additional 60 min. Finally water was removed under reduced pressure to give the styryl substituted resorcinolic resin having a softening point of 102.6 °C.

The presence of styryl group in the novolak resin structure improves the hygroscopic property of resorcinolic resin due to its hydrophobic character. This could allow the handling of this resin much easier in the rubber compounding applications. Also, the use of styryl modified resin in the steel skim rubber compound was thought to improve its compatibility. On curing the rubber compound with HMMM methylene donor, the cross-linked resin structure produced from the resorcinol based novolak resin-HMMM reaction on the top of $Cu_xS$ layer could be more hydrophobic. Therefore, the steam and humidity adhesion properties of brass plated steel cords can be expected to improve with the use of this modified resin.

The rubber compound and adhesion properties associated with the use of styryl substituted resorcinolic novolak resin in the steel skim rubber formulation are shown in Table 5.23.

Good unaged adhesion property and its retention after aging was observed with styryl modified resins. This resin showed a dramatic improvement in the

**Table 5.23.** Rubber compound properties

| Resorcinol compound/ methylene donor | Sty-R-F resin/HMMM | RF/HMMM |
|---|---|---|
| Acceptor/donor level | 3.5/2.5 | 3.5/2.5 |
| Rheometer cure at 150 °C (ASTM-D-2084) | | |
| $M_H$, dN-m | 56.9 | 52.3 |
| $M_L$, dN-m | 8.8 | 9.1 |
| ts2, min | 3.9 | 3.5 |
| $t'_{90}$, min | 16.8 | 14.2 |
| Adhesion (ASTM D-2229), Newtons (% rubber coverage) (Wire = 63.5% copper) | | |
| Unaged | 1174 (89) | 1241 (89) |
| Steam, 24 h at 120 °C | 1263 (95) | 1085 (90) |
| Humidity, 21 days, 85 °C, 95% RH | 1103 (90) | 1018 (90) |
| Tensile property (ASTM D-412) (unaged) | | |
| 200% Modulus, (MPa) | 8.9 | 8.75 |
| Tensile strength, (MPa) | 24.57 | 22.97 |
| Ultimate elongation, % | 476 | 452 |
| Heat aged (three days at 100 °C) | | |
| 200% Modulus, (MPa) | 7.72 | 8.05 |
| Tensile strength, (MPa) | 11.12 | 12.6 |
| Ultimate elongation, % | 144 | 162 |
| Dynamic mechanical[a] (unaged) | | |
| G′ at 0.2% strain, MPa | 40.13 | 31.87 |
| G″ at 2.0% strain, MPa | 45.66 | 38.6 |

[a]RMS 800, 1.0 Hz at RT
Data from [9]

dynamic storage modulus (G′) property compared to RF resin in the cured rubber compounds. This could be associated with the presence of bulkier aralkyl groups in the resorcinolic structure.

Indspec Chemical Corporation is currently manufacturing and marketing styryl-modified resorcinolic novolak resin under the trade name Penacolite B-20-S resin.

The effect of varying HMMM level on the adhesion of brass plated steel cords in the natural rubber steel skim compound formulation is shown in Figure 5.37. From this study, it was observed that, for a known level of Penacolite B-20S resin (3 phr) use in the formulation, the unaged adhesion remained constant for an increased HMMM level. On the other hand, the steam aged adhesion was observed to increase and reaching a maximum before starting to decline slightly [15].

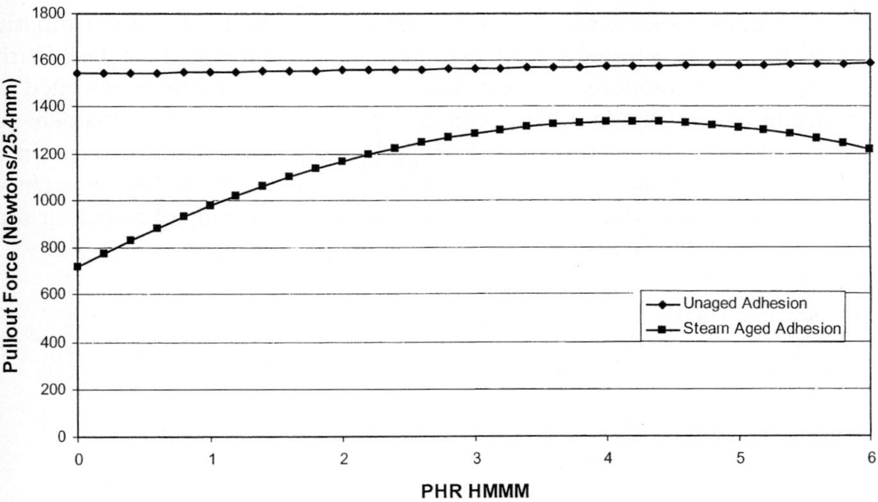


**Figure 5.37.** Effect of HMMM level on steel cord adhesion with Penacolite B-20S resin. (resin at 3 phr and cobalt at 0.1 phr)

Penacolite B-20-S resin has been successfully used by the tire and rubber compounding industry to improve the steel cord adhesion and rubber compound properties.

### 5.2.5.6 Non-volatile Resorcinolic Resins

Due to health and environmental concerns on the fuming of resorcinol (unreacted or free) present in resorcinolic resins, the tire industry demands the need for non-volatile resins that can provide good overall cured rubber compound properties. In the development of resorcinolic resins, in general, the molar ratios of resorcinol to formaldehyde determine the free resorcinol and softening points of the final material. Resins with high softening points (> 115 °C) may not be suitable for rubber compounding applications. Similarly, resins containing higher levels of free resorcinol (> 5 wt. %) are expected to fume during the compounding process. Therefore, in order to reduce the free resorcinol in the resin materials, the formaldehyde level in the resin synthesis should be increased. This could result in high softening point materials due to increase in molecular weights. The use of > 0.75 mol of formaldehyde per mol of resorcinol in the resin synthesis may lead to gel formation, which may cause the resin to be unusable for the desired applications.

Lowering the free resorcinol content in novolak resins without increasing the softening point was a challenge. Resin producers have been developing and implementing various synthetic methodologies to lower the free resorcinol content of the resorcinolic resins.

Resorcinol is a reactive compound towards its reaction with vinyl aromatic compounds in the presence of a small amount of strong acid catalyst. With the use of styrene monomer, various resorcinolic resins have been modified to obtain almost "no free resorcinol"-containing resins [16]. A non-formaldehyde type resorcinolic resin was also developed from the reaction of resorcinol with bis-phenol-A epoxy. This resin contained about 10 wt. % free resorcinol and showed 84.3 °C softening point. This resin was developed based on the following reaction (Figure 5.38).

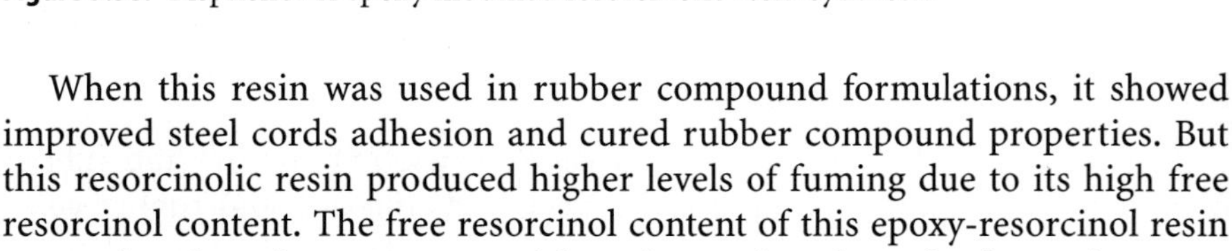


**Figure 5.38.** Bisphenol-A epoxy modified resorcinolic resin synthesis

When this resin was used in rubber compound formulations, it showed improved steel cords adhesion and cured rubber compound properties. But this resorcinolic resin produced higher levels of fuming due to its high free resorcinol content. The free resorcinol content of this epoxy-resorcinol resin was reduced to almost 0 wt. % without increasing the softening point to an appreciable level by reacting with styrene monomers. The details are presented in Table 5.24.

A reaction scheme illustrating the synthesis of a styryl modified resin to reduce the free resorcinol content of bisphenol-A epoxy modified resorcinolic resins is illustrated below (Figure 5.39).

Even after the styrene modification, the epoxy-resorcinol resin showed steel cord adhesion properties similar to the commercially available styryl modified resorcinolic resins, namely Penacolite B-20-S and VPN-1755 resins. This resin

**Table 5.24.** Styrene modified resorcinol-bisphenol-A epoxy resin

| Experiment: | 1 (Control) | 2 | 3 | 4 | 5 |
|---|---|---|---|---|---|
| Raw materials (g) | | | | | |
| 1. Resorcinol | 14.15 kg | 55 | 55 | 55 | 110.1 |
| 2. Triphenylphosphine | 0.13 kg | 0.5 | 0.5 | 0.5 | 1 |
| 3. Bisphenol epoxy | 25 kg | 98.6 | 98.1 | 97.4 | 195.2 |
| 4. PTSA (catalyst) | No | 0.7 | 0.7 | 0.7 | 1.4 |
| 5. Styrene | No | 13 | 26 | 52 | 156 |
| 6. NaOH (50%) | No | 0.4 | 0.4 | 0.4 | 0.8 |
| Distillate removed at 150 – 155 °C/28″ Hg vac | None | None | None | None | None |
| Resin properties | | | | | |
| Softening point (°C) | 84.3 | 89 | 89.4 | 90.7 | 92.2 |
| Free resorcinol (wt.%, GC) | 10 | 4.5 | 2.5 | 0.3 | < 0.01 |

Data taken from [16]

also exhibited higher Shore-A hardness and dynamic storage modulus (G′) values in the cured rubber compounds as shown in Table 5.25.

Demanding applications require various performance improvements in the rubber compound properties. This depends on the raw materials used in the compound formulations. With the use of resorcinol and resorcinol based materials in the rubber compounds, higher performance could be achieved in the rubber compounds reinforced with synthetic fibers. Therefore, there is always a need for improved resorcinolic compounds or resins that could be used in rubber compounds for special applications. This requires the need for the development of advanced resorcinolic resin chemistries and commercial technologies to produce them.

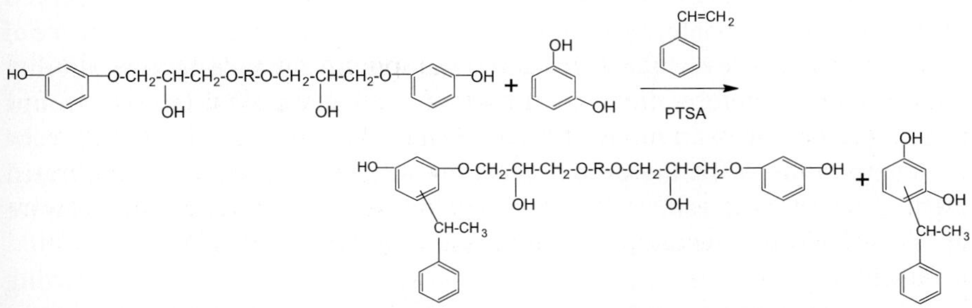


**Figure 5.39.** Styryl modified bisphenol epoxy – resorcinol resin synthesis

**Table 5.25.** Rubber compound properties of styryl modified resorcinol-bisphenol epoxy resin

| Resorcinol compound | (R + Epoxy + Sty) | B-20-S | VPN-1755 |
|---|---|---|---|
| Methylene donor | HMMM | HMMM | HMMM |
| Acceptor/donor level | 2.4/1.6 | 2.4/1.6 | 2.4/1.6 |
| Rheometer cure at 150 °C (ASTM-D-2084) | | | |
| MH, dN-m | 42.6 | 43.2 | 40.7 |
| ts2, min | 5.3 | 4.6 | 5.4 |
| $t'_{90}$, min | 22.8 | 19.2 | 20.5 |
| Fuming at 120 °C | None | Slight | None |
| Adhesion (ASTM D-2229), Newton (% rubber coverage) (Wire = 63.5% Cu) | | | |
| Unaged | 1114 (85) | 1246 (88) | 1146 (83) |
| Steam, 16 h at 120 °C | 1282 (90) | 1271 (90) | 1266 (100) |
| Humidity, 14 days, 85 °C, 95% RH | 1182 (90) | 1243 (95) | 1172 (90) |
| Tensile property (ASTM D-412) | | | |
| 100% Modulus, (MPa) | 4.75 | 4.56 | 4.55 |
| Tensile strength, (MPa) | 28.06 | 28.18 | 28.23 |
| Ultimate elongation, % | 458 | 458 | 455 |
| Hardness (Shore A) | 93 | 81 | 77 |
| Dynamic mechanical[a] | | | |
| G′, MPa (at 0.2% strain) | 26.6 | 23.1 | 19.4 |
| Tan delta (at 2% strain) | 0.206 | 0.198 | 0.209 |

[a] Rheometrics Mechanical Spectrometer, 1.0 Hz and RT
Data taken from [16]

### 5.2.5.7
### Non-fuming and Non-formaldehyde Resorcinolic Resins

The free resorcinol and fuming of RF resins were reduced considerably by the use of styrene in the resin technology. Styryl modified resorcinol was observed to improve its processability with rubber compounds. Similarly, the presence of an alkyl group in the resorcinol molecule appeared to enhance its compatibility with rubber. Therefore, the combination of alkyl and aralkyl (styryl) groups in the resorcinolic resin molecules can enhance their processability with rubber compounds. With the proper selection of alkyl and aralkyl unsaturated organic derivatives, resins containing very low amounts of free resorcinol were developed without the use of formaldehyde in the reaction.

Resins developed in the absence of formaldehyde are based on the following reaction scheme (Figure 5.40) [17].

Resins, in general, are classified as non-fuming if they do not produce any volatiles during rubber processing conditions. This can happen only when the resins contain less than 1.0 wt. % free resorcinol. As a reactive compound, resorcinol can react readily with dicyclopentadiene (DCPD) in the presence of an acid catalyst at elevated temperature conditions. The ratio of resorcinol to

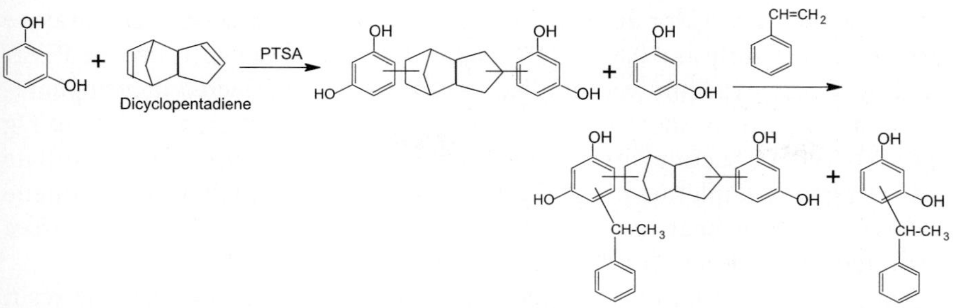


**Figure 5.40.** Non-formaldehyde resorcinolic resin synthesis

DCPD is very critical in developing a resin with the desired softening point to use in rubber compounds. Non-fuming resins were developed employing resorcinol, styrene and DCPD as the reactants and the results are presented in Table 5.26.

**Table 5.26.** Synthesis of non-fuming and non-formaldehyde resorcinolic resins

| Raw materials (mol) | 1 | 2 | 3 | 4 | 5 | 6 |
|---|---|---|---|---|---|---|
| Resorcinol | 1 | 1 | 1 | 1 | 1 | 1 |
| DCPD | 0.5 | 0.4 | 0.4 | 0.35 | 0.3 | 0.25 |
| Styrene | 0.7 | 0.8 | 1 | 1.1 | 1.2 | 1.25 |
| Physical properties | | | | | | |
| Free resorcinol (wt.%, by LC) | 1.8 | 2.3 | 0.95 | 1.2 | 0.75 | 1 |
| Softening point (°C) | 113.6 | 88.9 | 95.2 | 84.9 | 78.2 | 71.3 |

DCPD = Dicyclopentadiene
Data taken from [17]

A procedure used to develop DCPD modified resorcinolic novolak resin is illustrated below.

**Preparation of Resorcinol-Dicyclopentadiene-Styrene Resin [17]** Into a four-necked reaction kettle equipped with a mechanical stirrer, thermometer, reflux condenser and an addition funnel, 1.0 mol (110.1 g) of resorcinol was charged and heated to 125–135 °C. When the resorcinol melted completely, 1.0 g of PTSA catalyst was added and stirred for 5.0 min. After this, 0.5 mol (66.2 g) of dicyclopentadiene was added from the addition funnel over a period of 60–90 min at 135–145 °C. Following the dicyclopentadiene addition, the reaction was continued at 140–145 °C for an additional 60 min. Then, a second charge of 0.5 g of PTSA catalyst was added and 0.3 mol (31.2 g) of styrene was added slowly from the addition funnel at 140–145 °C over a period of 60–90 min. Following the addition of styrene, the reaction was continued at this temperature for 60 min more before heating to 155–160 °C. To complete the reaction, stirring was continued at 155–160 °C for an additional 60 min. Next the reaction temperature was lowered slightly and 0.8 g of 50% NaOH was added to neutralize the PTSA catalyst. Finally, vacuum was applied to remove any unreacted dicyclopentadiene and styrene monomers. The conditions for distillation were 28″ of Hg and 150–160 °C. Vacuum distillation did not produce any appreciable distillate indicating the completion of the reaction of resorcinol with dicyclopentadiene and styrene. The final resin showed a softening point of 127.2 °C and free resorcinol content of 6.9 wt. % by LC/GC analysis.

The TGA weight loss data is given in Table 5.27, and shows that the resin produced from the resorcinol, DCPD and styrene reaction had exhibited very low volatility compared to resorcinol.

**Table 5.27.** Thermogravimetric analysis of resins

| | Weight loss (%) at Temperature (°C) | | | | |
|---|---|---|---|---|---|
| Resorcinolic product | 125 | 150 | 175 | 200 | 225 |
| Resorcinol | 1.7 | 8.5 | 32.6 | 96.4 | 99.1 |
| R/DCPD/Sty (1/0.4/1 mol) | 0.4 | 0.6 | 1 | 1.7 | 3.3 |

Heating rate $m = 10\,°C/min$; in nitrogen
Data taken from [17]

Rubber compound properties associated with the use of DCPD modified resorcinolic resin are summarized in Table 5.28.

The rubber compound dynamic storage modulus (G′) and tear properties were enhanced when this DCPD modified resorcinolic resin was used.

### 5.2.5.8
### Resorcinolic Derivative in Rubber Compounds

Resorcinolic compounds used in rubber compounds are known to form crosslinked resin structures with methylene donors. The network formation during rubber curing is more effective with *meta*-substituted phenolic compounds than the ortho and para substituted phenols due to higher reactivity towards the methylene donors. Examples of such *m*-substituted phenolic compounds include resorcinol, phloroglucinol and *m*-amino phenol. The use of *m*-amino phenol and phloroglucinol in rubber compounds are limited due to higher costs and high melting points.

The two hydroxyl groups present at the *meta*-positions in the resorcinol molecule activate the three reactive sites, namely 2, 4 and 6, for the methylene donor reactions. Modified resorcinolic compounds which can maintain the resorcinol reactivity, and at the same time reduce the fuming, are always desired in rubber compounding applications.

When the resorcinol molecule is derivatized by attaching functional groups either at the benzene ring or hydroxyl groups, the reactivity of resorcinol is

**Table 5.28.** Rubber compound properties

| Resorcinol compound | R/DCPD/STY Resin ( 1/0.4/1 mol) | Resorcinol |
|---|---|---|
| Methylene donor | HMT | HMT |
| Acceptor/donor level | 3/0.5 | 3/0.5 |
| Fuming in rubber compounding | None | High |
| Tensile property (ASTM D-412) | | |
| 100% Modulus, (MPa) | 4.43 | 4.23 |
| 300% Modulus, (MPa) | 17.23 | 17.15 |
| Tensile strength, (MPa) | 26.83 | 26.57 |
| Ultimate elongation, % | 459 | 451 |
| Energy to break, N-m | 24.36 | 23.94 |
| Tear strength (Die-C); kN/M | 129.1 | 118.3 |
| Dynamic mechanical properties[a] | | |
| At 0.2% strain | | |
| G′, MPa | 23.83 | 18.92 |
| Tan delta | 0.079 | 0.089 |
| At 2.0% Strain | | |
| G′, MPa | 13.25 | 9.96 |
| Tan delta | 0.228 | 0.239 |

[a] Rheometric Mechanical Spectrometer, 1.0 Hz and RT
DCPD = Dicyclopentadiene, Sty= Styrene and R=Resorcinol
Data taken from [17]

greatly reduced. For example, substituents present at the 2-, 4- and 6-positions of the benzene ring of resorcinol can reduce the number of reactive sites for the methylene donor reaction. Similarly, the monoether and monoester derivatives of resorcinol have lower reactivity compared to resorcinol. Any modifications made on the resorcinol molecule are expected to lower the fuming but also affect the reactivity to form the cross-linked structures with methylene donors.

*m*-Amino phenol is also a phenolic compound and has exhibited a reactivity about five to ten times more reactive than resorcinol towards the formaldehyde reaction. Therefore, a derivative of resorcinol similar to *m*-aminophenol type product with all three positions available for the reaction are expected to increase the rate of cross-linking reaction with the formaldehyde donor during rubber vulcanization. The use of such a compound could result in the enhancement of rubber compounds physical and mechanical properties.

A resorcinol derivative having all three reactive sites, with enhanced reactivity, available for methylene donor reaction, was developed based on the following scheme (Figure 5.41 [18].

This resorcinol derivative, namely *m*-hydroxy diphenylamine, resembles the structure of *m*-aminophenol and therefore, can be expected to show higher reactivity than resorcinol.

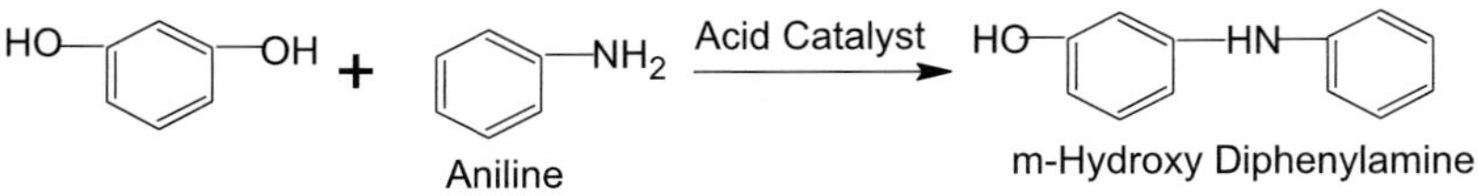


**Figure 5.41.**

The rubber compound properties, given in Table 5.29, showed higher cure rate and dynamic storage modulus (G′) suggesting higher reactivity for this diphenylamine compound compared to resorcinol. The dispersability of *m*-hydroxy diphenylamine in rubber compounds appeared to be good due to its lower melting point (80 – 85 °C).

**Table 5.29.** Rubber compound properties

| Resorcinol compound | *m*-Hydroxy-diphenylamine | Resorcinol |
|---|---|---|
| Methylene donor | HMT | HMT |
| Level (resorcinol/HMT), phr | 3/0.5 | 3/0.5 |
| Rheometer cure at 125 °C (ASTM-D-2084) | | |
| $M_L$, dN-m | 2.86 | 2.8 |
| Rheometer cure at 150 °C (ASTM-D-2084) | | |
| $M_H$, dN-m | 32.77 | 29.45 |
| $M_L$, dN-m | 2.46 | 2.37 |
| ts2, min | 2.41 | 1.63 |
| $t'_{90}$, min | 7.47 | 7.66 |
| Cure rate | 5 | 3.71 |
| Tensile property (ASTM D-412) | | |
| 100% Modulus, (MPa) | 5.32 | 4.84 |
| 300% Modulus, (MPa) | 21.43 | 20.25 |
| Tensile strength, (MPa) | 29.76 | 28.54 |
| Ultimate elongation, % | 423 | 423 |
| Energy to break, N-m | 27.44 | 26.3 |
| "Tear strength (Die-C); kN/M" | 132.3 | 132.3 |
| Hardness (Shore A) | 80 | 78 |
| Dynamic mechanical properties[a] | | |
| At 0.2% Strain | | |
| G′, MPa | 20.08 | 17.16 |
| Tan delta | 0.066 | 0.069 |
| At 2.0% Strain | | |
| G′, MPa | 12.02 | 10.13 |
| Tan delta | 0.187 | 0.188 |

[a]Rheometric Mechanical Spectrometer, 1.0 Hz and RT
Data taken from [18]

## 5.2.5.9
## Lower Molecular Weight Resorcinolic Compound

The use of lower molecular weight and reactive resorcinolic derivative in rubber compound applications has several advantages. Low molecular weight products can be dispersed easily and uniformly in the rubber compound matrix and expected to show low uncured rubber viscosity. This can be related to an improved processability and, possibly, easier handling if they are solids. The derivatives containing bulkier groups are known to provide high dynamic stiffness to cured rubber.

Lower molecular weight resorcinolic derivatives have been prepared using styrene and *N*-methylol caprolactam monomers in the reaction, and the synthesis is based on the reaction scheme shown below (Figure 5.42) [19].

The procedure used to develop this derivative is given below.

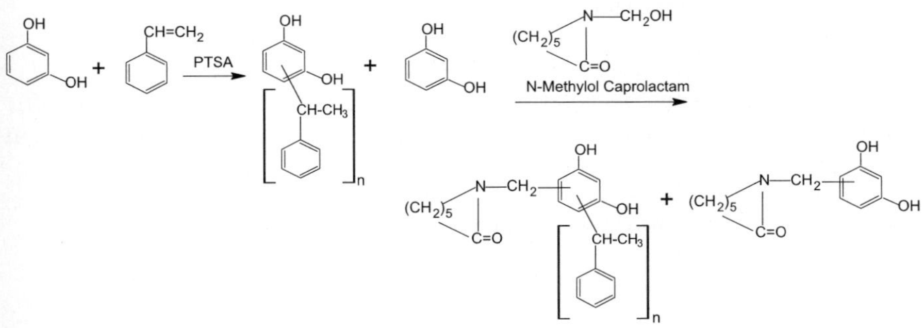


**Figure 5.42.** *N*-Methylol caprolactam modified resorcinol-styrene resin synthesis

**Synthesis of *N*-Methylol Caprolactam Modified Resorcinol-Styrene Resin [19]**
A 500 ml reaction kettle equipped with a stirrer, thermometer, reflux condenser and an addition funnel was charged with 1.0 mol (110.1 g) of resorcinol and heated to 120 to 130 °C. The catalyst, PTSA (1.0 g) was then added and mixed for 5 min. After this, 0.8 mol (83.3 g) of styrene was added slowly over a period of about 90 min. The temperature of the reaction mixture was maintained between 125–135 °C during the styrene addition. After all the styrene had been added, the reaction mixture was maintained at this temperature for an additional 60 to 90 min. Then, 0.75 mol (143.3 g) of 75 wt. % aqueous solution of *N*-methylol caprolactam was added very slowly into the kettle over a period of 60 to 90 min at 95 to 120 °C. The reaction mixture was held at this condition for an additional 60 to 90 min. Finally, 0.5 g of 50% aqueous sodium hydroxide solution was added and then the resin was dehydrated under vacuum (28″ of Hg) at 155 to 160 °C to remove the distillate. The resulting resin had a softening point of 89.4 °C and a free resorcinol content of 0.7 wt. %. Molecular weight determined by the Vapor Pressure Osmometer (VPO) method indicated 329 number average molecular weight.

Table 5.30 shows the details of different caprolactam modified resorcinol derivatives exhibiting lower softening points with very low free resorcinol content. These compounds were prepared from the use of different molar ratios of resorcinol, styrene and methylol caprolactam in the reactions.

**Table 5.30.** Synthesis of resorcinol-styrene-methylol caprolactam resins

| Raw materials (mol) | 1 | 2 | 3 | 4 | 5 |
|---|---|---|---|---|---|
| Resorcinol | 1 | 1 | 1 | 1 | 1 |
| Styrene | 0.6 | 1 | 0.4 | 1 | 0.25 |
| NMC | 0.75 | 0.5 | 0.75 | 0.75 | 1 |
| Physical properties | | | | | |
| Free resorcinol (wt.% by GC) | 1.6 | 0.9 | 3.3 | 0.3 | 1.8 |
| Softening point (°C) | 89 | 77 | 89 | 90 | 100.4 |

NMC = *N*-methylol caprolactam
Data taken from [19]

TGA data showing the volatility of a caprolactam modified resorcinolic compound compared to resorcinol and Penacolite resin B-20-S is presented in Table 5.31.

**Table 5.31.** Thermogravimetric analysis of resins

| | Weight loss (%) at Temperature(C) | | | |
|---|---|---|---|---|
| Resorcinolic product | 125 | 150 | 175 | 200 |
| Resorcinol | 1.7 | 8.5 | 32.6 | 96.4 |
| Penacolite resin B-20-S | 0.4 | 0.7 | 1.2 | 2.4 |
| R/Sty/NMC (1/0.6/0.75 mol) | 0.2 | 0.5 | 1.1 | 2.7 |

Heating rate = 10 °C/min; in nitrogen
NMC = *N*-Methylol caprolactam
Data taken from [19]

The rubber compounding properties obtained are summarized in Table 5.32.

The above data suggest that these lower molecular weight resorcinolic products exhibited performances equal to resorcinol, in providing good mechanical properties to cured rubber compounds and steel cord adhesion properties. This type of compounds is most desired by the rubber compounding industries, and in particular, by the tire industry.

## 5.2.6 Zinc Plated Steel Cord Adhesion

The performance of steel reinforced rubber articles strongly depends on the interfacial adhesion between the reinforced cords or wires and rubber, in

**Table 5.32.** Rubber compound properties

| Resorcinol compound/ methylene donor | Resorcinol/ HMMM | B-20-S/ HMMM | R-Sty-NMC/ HMMM |
|---|---|---|---|
| Acceptor/donor level | 2.0/2.0 | 2.0/2.0 | 2.0/2.0 |
| Rheometer cure at 150 °C (ASTM-D-2084) | | | |
| MH, dN-m | 37.25 | 35.84 | 33.88 |
| ML, dN-m | 2.12 | 2.42 | 2.16 |
| ts2, min | 3.09 | 3.25 | 3.52 |
| $t'_{90}$, min | 21.9 | 25.18 | 27.75 |
| Adhesion (ASTM D-2229), Newtons (% rubber coverage) (Wire = 63.5% Cu) | | | |
| Unaged | 1097 (90) | 1245 (90) | 1144 (90) |
| Steam, 16 h at 120 °C | 1237 (95) | 1262 (90) | 1178 (90) |
| Humidity, 21 days, 85 °C, 95% RH | 757 (70) | 1023 (90) | 1016 (90) |
| Tensile property (ASTM D-412) | | | |
| 100% Modulus, (MPa) | 6.67 | 6.77 | 6.45 |
| Tensile strength, (MPa) | 23.26 | 24.05 | 23.4 |
| Ultimate elongation, % | 333 | 340 | 349 |
| Hardness (Shore A) | 85 | 84 | 84 |
| Die-C Tear, kN/m | 92.7 | 91.2 | 96.7 |
| Dynamic mechanical | | | |
| At 0.2% Strain | | | |
| G′, MPa | 24.74 | 23.66 | 26.12 |
| Tan delta | 0.075 | 0.072 | 0.073 |
| At 2.0% Strain | | | |
| G′, MPa | 14.79 | 14.78 | 15.94 |
| Tan delta | 0.211 | 0.193 | 0.216 |

Rheometrics Mechanical Spectrometer, 1.0 Hz and RT
NMC = *N*-Methylol caprolactam, STY= Styrene and R=Resorcinol
Data taken from [19]

addition to the mechanical properties of the cords and rubber compounds. A great amount of research and developmental efforts have been conducted to improve the adhesion between steel wires and rubber. With the application of brass plating on the steel wires, adhesion has been improved dramatically. Brass plated steel cords are vulnerable to corrosion due to the galvanic coupling of brass and steel, in which brass may act as cathode and iron as anode. Though brass plated steel cords have provided good adhesion, they showed insufficient protection with respect to corrosion. Efforts have been continuously made to develop and use zinc plated steel wires in the place of brass plated cords because of their tendency to be more corrosion resistant.

Zinc plated steel cords have been increasingly used in the manufacture of heavy duty conveyer belts. Unlike brass plated cords, where the adhesion phenomena was due to the formation of $Cu_xS$ layer at the surface, zinc plated steel cords can form $ZnO/Zn(OH)_2$ layer on the surface. This layer has inherently very low tendency for an adhesion with the conventional sulfur cured natural and synthetic rubber compounds. Adhesion of zinc coated wires with rubbers was improved by the addition of very high amounts of cobalt salts and/or lead oxide in the rubber compounds. The cobalt salts appeared to improve the adhesion through the formation of zinc sulfide on the top of steel cord surface.

Resorcinol, in combination with silica and hexamethylene tetramine (HMT or Hexa), a system previously known as the HRH, was used to enhance the zinc plated steel wires to natural rubber compounds [20]. The adhesion performance of zinc plated wires towards the natural rubber compounds in the presence and absence of HRH system was determined, and the results are presented in Table 5.33.

**Table 5.33.** Rubber compound formulation for zinc plated steel cord adhesion

| Master batch (parts by weight) | Formulation | |
|---|---|---|
| | No "HRH" | HRH system |
| 1. Natural rubber | 100 | 100 |
| 2. Carbon black (N-660) | 50 | 50 |
| 3. Zinc oxide | 10 | 10 |
| 4. Stearic acid | 1 | 1 |
| 5. Santoflex 13 | 2 | 2 |
| 6. Aromatic oil | 3 | 3 |
| 7. Monobond C16 | 3 | 3 |
| 8. Sulfur/CBS | 2.5/0.6 | 2.5/0.6 |
| 9. Silica/resorcinol/HMT (HRH) | 0 | 7/1.4/1.4 |
| Zinc wire adhesion, kgs (cure at 150 °C) (rubber coverage) | | |
| 1. Adhesion at optimum cure | 100 (0) | 680 (6) |
| 2. Adhesion after 225 min cure | 93 (0) | 200 (5–6) |

Data taken from [20]

As can be seen from the results in Table 5.33, a dramatic improvement in adhesion was achieved when the resorcinolic bonding system was used in the rubber compound formulation.

Due to the corrosion of steel wires associated with the use of HMT in the rubber compounds, hexamethoxymethylmelamine (HMMM) compounds have been used as the methylene donors along with resorcinolic methylene acceptors in compound formulations. Using resorcinol and HMMM and varying the weight ratios of sulfur to accelerator in the rubber compound formulations, the brass and zinc coated wire adhesion properties were determined [21]. The purpose of this developmental work was to avoid the use of higher levels of cobalt and also to eliminate lead oxide from the compound formulations.

Results from this study can be seen in Table 5.34.

**Table 5.34.** Zinc plated steel wire adhesion using resorcinol

| Experiment: | 1 | 2 | 3 | 4 |
|---|---|---|---|---|
| Bonding system (phr) | | | | |
| 1. Resorcinol | 2 | 2 | 2 | 2 |
| 2. HMMM | 2.8 | 2.8 | 2.8 | 2.8 |
| 3. Sulfur/accelerator (weight) | 1:1 | 0.71:1 | 0.55:1 | 0.36:1 |
| Wire adhesion ( $2 + 2 \times 0.22$) | | | | |
| 1. Zinc coated wire (N) | 287 | 269 | 315 | 463 |
| 2. Brass coated wire (N) | 347 | 285 | 251 | 540 |

Data from [21]

The resorcinol based rubber compounds appeared to provide good adhesion for the zinc plated steel cords, and the application of such encapsulated materials could be used in the manufacture of hoses, belts and tires.

Conveyer belt rubber compounds containing novolak type resin synthesized from the resorcinol and dicylopentadiene (DCPD) raw materials appeared to show an improved adhesion with zinc plated steel wires [22]. In addition to this novolak resin, resorcinol and Penacolite B-20S resin were also used in the evaluation study. The results are shown in Table 5.35. In general, all the resorcinolic products showed good adherence properties when they were employed for the reinforcement of zinc plated cords with vulcanizing rubber materials.

**Table 5.35.** Zinc plated steel cords adhesion using resorcinol and resins

| Rubber compounds containing resorcinol or resin: | Resorcinol | Penacolite B-20 S | R + DCPD Resin |
|---|---|---|---|
| Zinc plated steel cord adhesion (N/50 m) | | | |
| 1. Optimum compound cure at 150 °C + 5 min | 5965 | 6040 | 5595 |
| 2. Over cure for 150 min at 150 °C | 5515 | 5711 | 5111 |
| 3. Loss in adhesion (%) | 7.5 | 5.4 | 8.7 |

R = Resorcinol and DCPD = Dicyclopentadiene
Data from [22]

Studies have shown that the use of resorcinolic materials consistently provided good adhesion for the zinc plated steel cords with rubber compounds. Resorcinolic products showed improvements in the initial and aged adhesion properties for the coated steel materials. The cobalt and cobalt-Penacolite B-20S resin bonding system were evaluated in the natural rubber compound formulations to determine their efficiency to improve the adhesion performance of zinc plated steel wires. Formulations and results are presented in Tables 5.36 and 5.37 [23].

**Table 5.36.** Rubber compound formulation for zinc plated steel cord adhesion

| Master batch (parts by weight) | Formulation A | B |
|---|---|---|
| 1. Natural rubber | 100 | 100 |
| 2. Carbon black (N-326) | 45 | 35 |
| 3. Zinc oxide | 9 | 9 |
| 4. Stearic acid | 1 | 1 |
| 5. Santoflex 13 | 2 | 2 |
| 6. Retarder | 0.2 | 0.2 |
| 7. Oil | 3 | 3 |
| 8. Silica | 20 | 15 |
| Curatives | | |
| 8. Insoluble sulfur | 4 | 4 |
| 9. DCBS | 1 | 1 |

Data from [23]

**Table 5.37.** Zinc plated steel wire adhesion using resorcinolic system

| Master batch used: | A | B | B | B |
|---|---|---|---|---|
| Bonding system (phr) | | | | |
| 1. Cobalt level | 0.4 | 0.4 | 0.4 | 0.2 |
| 2. Resorcinol | 0 | 3.2 | 0 | 0 |
| 3. Penacolite B-20S | 0 | 0 | 4.8 | 4.8 |
| 4. HMMM | 0 | 3.2 | 3.2 | 3.2 |
| Rheometer cure at 150 °C | | | | |
| MH, dN-m | 47.2 | 47.9 | 50.6 | 48.9 |
| ML, dN-m | 8.9 | 6.8 | 7.3 | 7.8 |
| TS, min | 4 | 4.3 | 6.9 | 7.9 |
| $T'_{90}$, min | 12.2 | 13.8 | 18.6 | 25.8 |
| Zinc wire adhesion, 7 × 0.6 mm, N (% rubber coverage) | | | | |
| 1. Initial | 1379 (80) | 1284 (82) | 1460 (87) | 1423 (90) |
| 2. Reheat 2 h at 150 °C | 597 (10) | 967 (80) | 1137 (85) | 1280 (90) |
| 3. Steam age 16 h at 120 °C | 278 (10) | 510 (50) | 625 (45) | 1138 (90) |
| Shore A hardness | 77 | 77 | 81 | 78 |
| Tensile properties | | | | |
| 1. 100% modulus, MPa | 3.64 | 3.97 | 4.3 | 3.96 |
| 2. Tensile strength, MPa | 24.46 | 24.99 | 25.67 | 24.81 |
| 3. Ultimate elongation, % | 466 | 477 | 484 | 479 |

Data from [23]

The initial heat and humidity aged adhesion values were higher for the combined cobalt-B-20S-HMMM system. Similar to the brass plated steel cord adhesion, a reduction in the cobalt content coupled with the addition of resorcinol based bonding agents, such as Penacolite B-20S and HMMM, to the rubber compounds provided significant adhesion improvements for the zinc coated steel cords.

### 5.2.7 Summary and Outlook

Resorcinol and resorcinolic derivatives and resins-based compounds have been proven to be the best bonding agents to improve the adhesion performance of brass and zinc plated steel cords to rubber compounds. High performance tires, hoses and belts reinforced with steel fibers or cords require the use of resorcinolic compounds to strengthen the brass and zinc plated steel cords and rubber compounds interfaces. Due to high reactivity and enhanced tendency to produce highly cross-linked, moisture resistant and thermally stable network structure on the top of bonding $Cu_xS$ layer, resorcinolic products are more valuable for the manufacture of advanced steel belted radial tires. Environmental and worker exposure conditions may require non-fuming and non-volatile resorcinolic resins and compounds in various applications. In future, the need for non-fuming and highly reactive resorcinolic products to enhance the adhesion performance of lower cost steel cords to rubber compounds may grow.

## References

1. W. J. van Ooij, Rubber Chem. Technol., Vol. 52, p 605 (1979).
2. G. R. Hamed and J. Haung, Rubber. Chem. Technol., Vol. 64, p 285 (1991).
3. G. Seo, J. Adhes. Sci. Technol., Vol. 11, p 1433 (1997).
4. W. W. Schloman, US Patent, 4,436,853 (1984).
5. M. Maciejewski et al., Polymer Bull., Vol. 48, p 251 (2002).
6. G. R. Hamed and R. Paul, Rubber. Chem. Technol., Vol. 70, p 541 (1997).
7. B. Durairaj, A. Peterson and G. Salee, US Patent, 4,892,908 (1990).
8. B. Durairaj, A. Peterson and G. Salee, US Patent, 4,889,891 (1989).
9. B. Durairaj, US Patent, 5,030,692 (1991).
10. S. Tanimura et al., US Patent, 4,257,926 (1981).
11. S. Tanimura, Y. Terada and H. Fukuhara, US Patent, 3,963,652 (1976).
12. H. Fukuhra and T. Tsuyuri, US Patent, 4,361,677 (1982).
13. B. Durairaj, A. Peterson, R. M. Lamars and R. T. Hood, US Patent, 5,021, 522 (1991).
14. W. Hesse, E. Leicht and R. Sattelmeyer, US Patent, 5,426,152 (1995).
15. A. Peterson et al., 131st ACS Rubber Div. Meeting, May 1987.
16. R. B. Durairaj and A. Peterson, US Patent, 5,945,500 (1999).
17. R. B. Durairaj and A. Peterson, US Patent, 6,472,457 (2002).
18. R. B. Durairaj and A. Peterson, US Patent, 6,541,551 (2003).
19. R. B. Durairaj and M. A. Lawrence, US Patent, 6,605,670 (2003).
20. J. R. Davies, 7th South African National Conference, Johannesburg, Nov. 1983.
21. R. N. Beers, T. W. Starinshak and D. A. Benko, US Patent, 6,120,911 (2000).
22. C. Morrel-Fourrier, T. Dreyfus, M. Bekhieh and B. L. Stuck, US Patent, 6,265,495 (2001).
23. A. Peterson, Tire Technology International, 1996.

## 5.3 Aerogels

### 5.3.1 Introduction

Aerogels are one class of foams of extremely light and porous solid materials. In lightness they are somewhat less than four times as dense as dry air. Aerogels are open cell polymers with pores less than 50 nanometers in diameter. The open cell foams are characterized by having pore volumes not sealed and they are interconnected. They have the highest internal surface area per gram of any known material. Aerogels are transparent and low density materials and produced by sol-gel polycondensation reactions. In general, an aerogel material results from a supercritical drying process.

There are three types of aerogels known, namely inorganic, organic and carbon aerogels. Inorganic aerogels are usually produced from the highly cross-linked and transparent hydrogels synthesized by polycondensation of metal alkoxides. Organic aerogels are synthesized by the sol-gel polycondensation reaction of resorcinol with formaldehyde in aqueous solutions. Carbon aerogels are synthesized via the preparation of an organic aerogel from the resorcinol and formaldehyde reaction and subsequent pyrolysis in an inert atmosphere.

### 5.3.2 Organic Aerogels

The first organic aerogel was synthesized by the aqueous polycondensation of resorcinol with formaldehyde using sodium carbonate as the base catalyst [1]. Resorcinol was considered as an ideal material to produce an organic aerogel due to its ability to form a gel material at lower temperatures with formaldehyde. The resorcinol molecule has three sites, namely 2, 4 and 6, and is about 10 to 15 times more reactive than phenol. Resorcinol undergoes most of the typical reactions of phenol and reacts with formaldehyde under alkaline and acid conditions to produce a mixture of addition and condensation products. In the presence of an alkaline catalyst, the formation of methylol ($-CH_2OH$ group), methylene bridged ($-CH_2-$) and benzylether bridged ($-CH_2OCH_2-$) resorcinol compounds are observed.

#### 5.3.2.1 Resorcinol-Formaldehyde Aerogel Preparation

Resorcinol can react with formaldehyde to produce a resorcinol-formaldehyde polymer containing methylene and benzylether bridges, in the form of a chemically cross-linked network structure [2–4]. This reaction is generally referred to as the RF reaction. In the synthesis of RF gel, resorcinol is dissolved in an aqueous formaldehyde solution (37 wt. %) at the appropriate ratio to form the cross-linked network. An excess of formaldehyde is used to maximize the cross-link density of the gel material. A suitable catalyst for the resorcinol-formaldehyde reaction is, primarily sodium carbonate, and added at the desired level. For this reaction, the initial pH is in the range of 6.5 to 7.4, and should be maintained to form a transparent gel. The resulting solution is then gently warmed to start the gelation reaction and also to form a chemically cross-linked RF gel.

The chemistry is similar to the sol-gel chemistry in inorganic materials [5]. After the formation of methylol groups from the reaction of resorcinol with formaldehyde, they condense with each other to form nanometer sized clusters, which then crosslink to produce a gel. The formation of clusters is influenced by typical sol-gel parameters such as the temperature, pH and concentration of reactants.

The chemistry outlining the formation of RF aerogels from the resorcinol-formaldehyde is shown in Figure 5.43.

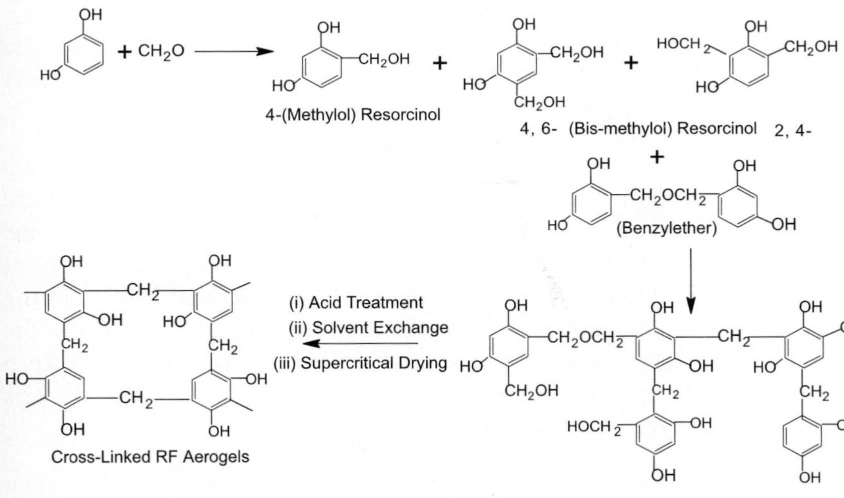


**Figure 5.43.** Chemistry of resorcinol-formaldehyde (RF) aerogols

The gels formed are then treated with a strong acid to promote additional cross-links from the condensation of methylol groups present in the gel molecules. The acid treated gels are solvent exchanged with water miscible solvents like acetone and methanol. Finally, the gels are dried in supercritical $CO_2$ liquid.

### 5.3.2.2
### Process for the Synthesis of Resorcinol Aerogel [6]

**Gel Formation** Resorcinol-formaldehyde (RF) based gels with varying resorcinol/catalyst (R/C) ratios were prepared as follows. The reactants, namely resorcinol, formaldehyde and water, were added to a 250 ml beaker in the order listed in Table 5.38, and then mixed until a clear, homogeneous solution was formed. The sodium carbonate catalyst was then added. The solution was then cast into molds, e.g., glass vials. The gels were cured according to the following cycle: 24 h at room temperature, followed by 24 h at 50 °C, and 72 h at 95 °C. Densities of the resulting aerogels were about 0.3 – 0.4 g/cc.

**Table 5.38.** Preparation of resorcinol-formaldehyde aerogels

| Formulations<br>Reactants (g) | 1 | 2 | 3 |
|---|---|---|---|
| Resorcinol (R) | 12.3 | 12.35 | 12.35 |
| Formaldehyde (F, 37%) | 17.9 | 17.91 | 17.91 |
| Deionized Water | 15.3 | 13.9 | 18.1 |
| Sodium carbonate (0.1 M; C) | 5.58 | 22.32 | 2.79 |
| Resorcinol, catalyst (R/C) | 200 | 50 | 400 |

Data from [6]

**Solvent Exchange** Upon completion of the cure cycle, the gels prepared were removed from their molds and placed in an organic solvent, e.g., acetone. During a first wash, trifluoroacetic acid was usually added to the solvent at a concentration of about 0.1 wt. % to promote additional cross-linking of the gels. With good agitations, the solvent diffused into the gels, replacing the water occupying the gel pores. After several additional exchanges with fresh acetone, the pores were completely filled with the organic solvent and the gels were ready for supercritical extraction.

**Supercritical Extraction** Acetone-filled gels prepared were then placed in open glass containers and covered with additional acetone to ensure the gels were submerged. Submersion in the solvent was necessary throughout the extraction process to ensure that the gels did not crack. The containers were then placed in a jacketed pressure vessel. Trapped air was slowly bled from the vessel as the vessel was filled with liquefied carbon dioxide at about 900 psi and a jacket temperature of 14 °C.

The gels were allowed to stand in the liquefied carbon dioxide for a minimum of 1 h. After this initial induction period, the vessel was flushed with fresh carbon dioxide for approximately 15 min. The carbon dioxide was then drained from the vessel to a level just above the crosslinked gels. The vessel was then refilled with liquefied carbon dioxide. The drain/refill procedure was conducted at least six times daily. In this process, diffusion is solely responsible for the displacement of acetone from the pores of the gel by carbon dioxide. Generally, two days of the drain/refill procedure were required to displace completely the acetone with carbon dioxide.

In the supercritical drying process [$T_C = 31\,°C$; $P_C = 1100\,psi$], the carbon dioxide level was drained to a level just above the gels. All valves to the pressure vessel were then closed and the vessel was heated to 50 °C. At this temperature, a pressure of 1800 psi was usually recorded. The pressure vessel was held at these conditions for a minimum of 4 h; after this the pressure was slowly reduced by bleeding over a period of 8 h while maintaining the jacket temperature at 50 °C. The resulting material was an organic RF aerogel.

Resorcinol-formaldehyde aerogels showed different densities depending upon the molar ratios of the reactants used in the formulation and the details are given in Table 5.39 [3].

**Table 5.39.** Resorcinol-formaldehyde (RF) aerogel properties

| Formulation | % Solids (weight) | [$CH_2O$] (mol) | [Resorcinol] (mol) | $Na_2CO_3$ (mol) | Theoretical density (g/cc) | Actual density (g/cc) |
|---|---|---|---|---|---|---|
| 1 | 5 | 0.29 | 0.58 | 0.0029 | 0.05 | 0.079 |
| 2 | 5 | 0.29 | 0.58 | 0.0019 | 0.05 | 0.073 |
| 3 | 5 | 0.29 | 0.58 | 0.0015 | 0.05 | 0.07 |
| 4 | 4 | 0.24 | 0.47 | 0.0012 | 0.04 | 0.054 |
| 5 | 3.5 | 0.21 | 0.41 | 0.001 | 0.035 | 0.044 |
| 6 | 2 | 0.12 | 0.24 | 0.0006 | 0.02 | 0.03 |

Data from [3]

### 5.3.2.3
### Morphology of Resorcinol Aerogels

The morphology of the aerogels was determined by Scanning Electron Microscopy (SEM) and Transmission Electron Microscopy (TEM) methods.

The aerogels produced according to the process described above have continuous porosity, ultra-fine pore sizes (less than 100 nanometers (nm)), high surface areas (400 to 1000 $m^2/g$) and a solid matrix composed of interconnected colloidal-like particles or polymeric chains with characteristic diameters of 10 nm.

A critical parameter impacting the morphology of the RF aerogel is the molar ratio of resorcinol (R) to catalyst (C), (R/C), used in the aerogel synthesis. At high catalyst concentrations (namely, R/C = 50), very small RF particles (9 nm size) are formed, whereas at low catalyst concentrations (R/C = 900), large RF particles (65 nm size) are formed. Similarly, at high catalyst concentrations (R/C = 50), particles with high surface areas (900 $m^2/g$) are found, whereas at low catalyst concentrations (R/C = 300), lower surface areas ($390\,m^2/g$) are found. The high surface areas found with high catalyst concentrations may be explained by the formation of a large number of very small particles. At higher catalyst concentrations (R/C < 300–400), the RF gels are transparent and homogenous and the dried RF aerogels are dark red in color and transparent to visible light. At lower catalyst concentrations ( R/C > 300–400), the organic gels are typically opaque and the resultant dried organic aerogels are not transparent to visible light (due to the formation of large particles and pore spaces which scatter visible light). These later materials must have particle or pore sizes larger than 100 nm and, therefore, are often no longer considered to be aerogels.

In the formation of RF aerogels, all the reaction parameters are carefully controlled so that the resulting gel can be successfully dried to form an aerogel which retains much of the morphology of the original gel.

The thermal conductivity of RF aerogels tends to decrease with decreasing density. When the density of RF aerogel is lower than $67\,kg\,m^{-3}$, the thermal conductivity tends to increase. The RF aerogels obtained by a resorcinol/catalyst (R/C) ratio of 300 shows a lower thermal conductivity compared to other R/C values [7]. Another study also indicated that RF aerogels showed minimum thermal conductivity values at the density between 0.12 to $0.16\,g/cm^3$ [8].

### 5.3.2.4 Organic Aerogel Microspheres

RF aerogel microspheres having a density in the range of about 0.05 to 1.2 g/cc and diameters ranging from about 1 micron to 3 mm are prepared by stirring the aqueous organic phase of resorcinol and formaldehyde in mineral oil at elevated temperatures until the dispersed phase polymerizes. The size of the microspheres formed depended on the collision rate of the liquid droplets and the reaction rate of the monomers used. These aerogel microspheres can be used in capacitors, batteries, thermal insulation, adsorption/filtration media and chromatographic packing applications [9, 10].

### 5.3.2.5 Applications of Organic Aerogels

Organic aerogels are interesting materials for different applications due to their thermal, optical and mechanical properties. They are stiffer and stronger than silica aerogels and are measurably better insulators. These materials are among the best thermal insulating materials known. Silica aerogels have thermal conductivities of 0.02 W/m. K (about R10/inch) and stable up to 650 °C. On the other hand, organic aerogels have thermal conductivities as low as 0.0045 W/m. K after they have been evacuated and can withstand temperatures up to 3650 °C if pyrolyzed [11]. Organic aerogels have been suggested as an alternative insulator to opacified silica aerogels [2, 12].

RF aerogels have lower intrinsic and radiative conductivities, and are less brittle than the silica materials [13]. Organic aerogels have extremely high thermal resistance and six times more resistance than fiber glass insulation.

Monolithic RF aerogels are the precursor for the carbon nanofoams or aerogel materials.

### 5.3.3 Carbon Aerogels

Carbon aerogels are a special class of air-filled foams prepared by pyrolyzing resorcinol-formaldehyde (RF) organic aerogels in an inert atmosphere. This unique class of material has monolithic structure with high surface area (400 to 1000 $m^2/g$), controllable pore size, high purity and electrical conductivity (25 to 100 S/cm). Carbon aerogels are the first electrically conductive aerogels produced. The properties of carbon aerogels are responsible for the revolutionary developments in super capacitors with power densities to 4000 W/kg.

Carbon aerogel produced in powder form has provided surface areas approaching an astonishing 2500 $m^2/g$.

#### 5.3.3.1 Carbon Aerogel Synthesis and Properties

Carbon aerogel materials are generally produced from the pyrolysis of organic aerogels, which can be synthesized by the supercritical drying of gels obtained by the sol-gel poly-condensation reaction of resorcinol with formaldehyde in aqueous solutions, at an elevated temperature (about 1050 °C) in an inert atmosphere [2, 14]. The structure and properties of carbon aerogels depends on the molar ratios of resorcinol and catalyst used in the organic aerogels synthesis and the temperature used in the pyrolysis of organic aerogels.

The formation of the hydrogel structures during the sol-gel transition is important in RF aerogels. At the initial stage of the synthesis of RF hydrogels, nanometer sized small clusters consisting of branched polymeric species are formed [15]. The size of the clusters is regulated by the concentration of catalyst, namely sodium carbonate, in the resorcinol-formaldehyde solution. More specifically, the molar ratio of resorcinol (R) to catalyst (C), R/C, controls the size, number of RF clusters and electrochemical properties of the resulting gel. When the R/C ratio is low, the RF clusters formed are small. This leads to a high degree of cross-linking which results in high surface areas and better inter-connectivity. On the other-hand, larger R/C ratios produce structures similar to strings of colloidal particles with less connectivity.

Organic aerogel densities can be varied by adjusting the resorcinol concentration in the solution. The density of gels can be adjusted by altering the polymerization and drying conditions to affect solid volume fraction and pore size. Electrical conductivity of the gels increases significantly with increasing density. Higher density materials have more inter-connected particles per unit volume.

Carbon aerogel surface area can be increased through the proper selection of purge gases employed during the pyrolysis cycle.

The carbon aerogel materials available commercially have the properties shown in Table 5.40.

**Table 5.40.** Properties of carbon aerogels

| Density ($g/cm^3$) | 0.1 – 0.6 |
|---|---|
| Surface area ($m^2/g$) | 400 – 1000 |
| Average pore size (nm) | 4.0 – 30 |
| Electrical conductivity (S/cm) | 1.0 – 10 |

Data from Aerogel Composite, LLC 2003 Web Site

Gas permeability measurements made on the carbon aerogels under the nitrogen gas flow, for potential catalyst supports and gas filters applications, showed the permeabilities on the order of $10^{-12}$ to $10^{-10}$ $cm^2$ over the density range from 0.05 to 0.44 $g/cm^3$ [16].

The microporosity features of carbon aerogels prepared via the RF gels (R/C = 200) as function of pyrolysis temperature (1050 to 2100 °C) have been studied by $CO_2$ adsorption at 273 K and small angle X-ray scattering (SAXS). This study was aimed to analyze the contribution of the total microporosity, detectable by SAXS that can be accessible to $CO_2$. From this study, it was determined that with increasing pyrolysis temperature, the portion of the total microporosity accessible to $CO_2$ molecules was significantly decreasing [17].

### 5.3.3.2 Carbon Aerogels as Electrodes in Aerocapacitors

A capacitor is an energy storage device. Charge is stored electrostatically in polarized liquid layers between an ionically conducting electrolytes and a conducting electrode. Even though it is an electrochemical device, there are no chemical reactions taking place in the energy storage mechanism. The rate of charge and discharge of energy depends on the physical properties of capacitors.

An aerocapacitor is also known as an ultracapacitor which can release energy at a much faster rate. This is a high power-density version of a conventional electrolytic capacitor holding hundreds of times more energy per unit volume or mass than the later. The micro-pores in the carbon aerogel based electrode material of an aerocapacitor provide enormous surface area and very high capacitance unattainable by a conventional capacitor. Carbon aerogels are very ideal electrode materials for this application due to their low electrical resistivity (< 25 mohm – cm), interconnected and open pore structure with controlled pore size distribution (5 to 500 A) and high surface areas (about 1000 $m^2/g$).

Carbon aerogels have the ability to store and discharge a large amount of energy very quickly, which is different than a battery.

#### 5.3.3.3
#### Mesoporous Carbon Aerogels

Carbon aerogel materials produced from the pyrolysis of RF aerogels, which are produced from the supercritical drying process, are generally expensive. The carbon aerogels made by this method have microporous structures (pore diameter of < 2 nm). Pore sizes are important if the materials are used as adsorbents to remove the toxic organic compounds from liquid or vapor streams. Mesoporous materials, on the other hand, have pores having diameter in the range of 2 to 50 nm. With the use of emulsion polymerization process, mesoporous carbon aerogels were produced without the use of supercritical $CO_2$ drying [18].

In the modified process, mesoporous RF polymer has been prepared first in the presence of a cationic surfactant. The size of the micelles present in the solution controlled the pore size of RF gel, which ranged in diameter from 4 nm to micron size. This surfactant templating synthetic procedure used in the RF gel preparation predominantly produced mesoporous polymers. The pore size of the gel can be varied by adjusting the formulations. Mesoporous RF gels were dried under normal drying conditions and carbon aerogels could be produced by pyrolyzing the dried RF gels.

#### 5.3.3.4
#### Gas Diffusion Electrodes

Gas diffusion electrodes are used in car batteries and polymer electrolyte membrane (PEM) fuel cells. With the help of fuel cell technology, the energy stored chemically in the hydrogen and oxygen is released conveniently from the reaction of these two materials. The energy released is then converted in to an electrical energy by means of an electrochemical process. The PEM fuel cells have a central membrane/electrode unit comprising a thin proton conducting solid state electrolyte on both sides of which a smooth, hydrophobic and porous gas diffusion electrodes having a catalyst coating. These electrode materials should have good electrical conductivity, gas permeability, mechanical stability and smooth surface. Typical electrodes for this application are made of a gas diffusion layer, normally porous carbon paper, supporting a layer of finely dispersed platinum catalyst. It is important for the electrodes to have hydrophobic character, since this prevents the water formed in the electrochemical reactions between oxygen and hydrogen from accumulating in the pores. A carbon aerogel prepared for this type of application was hydrophobicized by the treatment with a silane or perflourinated polymer [19].

#### 5.3.3.5 Applications of Carbon Aerogels

Carbon aerogels have been used as electrodes in energy storage devices known as double-layer capacitors. These devices can deliver power faster than conventional batteries and therefore have potential application in electric vehicles, load leveling devices, telecommunications and microelectronics. In addition, other future applications abound for the carbon aerogels are in thermal insulation, chromatographic packings, adsorbents, sensors, filters, gas storage media, catalyst supports and molding material for the precision casting and dead-mold casting of metals or metal alloys [20].

### 5.3.4 Carbon Foam/Carbon Paper Composites

Thin films of porous carbon electrodes exhibiting lower resistance, high efficiency and power capability are required in batteries and capacitors. Thin films can provide short electrolyte diffusion distances and better access to the carbon. These electrodes are utilized in electrochemical storage applications such as super capacitors. In the method of making this, a thin, flat, highly porous and low density ($< 0.2$ g/cc) carbon paper was infiltrated with an RF gel and cured to form a crosslinked polymer [21–24]. The crosslinked RF polymer was then dried under atmospheric conditions without the need for a complex, intensive and expensive supercritical drying. Finally, the carbon paper impregnated RF foam composite was pyrolyzed to produce a carbon aerogel/carbon paper composite material.

A synthetic procedure outlining the method of preparing a carbon aerogel/carbon paper composite is described below.

**Preparation of an Air-Dried Resorcinol-Formaldehyde Low Density Open Cell Carbon Foam/Carbon Substrate Composite (R + F = 50% w/v and R/C 1000) [23]** Resorcinol (24.7 g, 0.224 mol), 35.820 g of 37% aqueous formaldehyde (13.25 g, 0.441 mol), and 21.52 g of deionized water were stirred together until the resorcinol was dissolved. Then 2.24 g of a 0.10 mol/l sodium carbonate solution (about 0.224 mmol $Na_2CO_3$ assuming a specific gravity of 1 g/cm$^3$) was added with stirring. The final volume was measured to be 75.9 ml. The solution was warmed at 50 °C for 30 min. A lightweight carbon fiber paper (Technimat.RTM. 6100-050, Lydall Technical Papers, Rochester, N.Y.) was then infused with this solution. The infused paper was placed between two glass plates, sealed with a plastic over-wrap, and allowed to stand at room temperature for 15 h to gel. The RF infused paper was then placed in an 85 °C oven for 24 h to cure. After cooling, the plastic over-wrap and the glass plates were then removed. The RF composite was placed in a humid horizontal flow oven and dried for 12 h at about 75 °C. The RF composite was a thin opaque sheet (about 10 mil. to 0.25 mm thick) with a mottled light brown appearance and a smooth, shiny surface. It had a density of about 0.6 mg/cm$^3$ and a surface area of 494 m$^2$/g.

Pyrolysis of the RF composite at 1050 °C under an inert atmosphere ($N_2$) (room temperature to 250 °C in about 4 h, 250 °C for 2 h, 250 to 1050 °C in about 9 h, 1050 °C for 4 h, cool to room temperature in about 12 h) resulted in a black carbonized sheet with dimensions comparable to the RF composite. This material had a density of 0.419 mg/cm$^3$ and a surface area of 397 m$^2$/g. This material had a pore size distribution for mesopores of 20 nm. A specific capacitance of about 20.6 F/g was determined when this material was used as an electrode in an aqueous-based (5 mol/l KOH) super capacitor.

Unlike the standard procedure used in the RF gel synthesis, this method used lower catalyst concentration, R/C = > 1000, and lower pH (less than 6.0). With this modification in the formulation, the new cured organic gel materials were found to be stronger. Surface analysis measurements showed that these carbon foams exhibited low density open cells with large cell size, high porosity, high surface area and high electrical capacitance (Table 5.41) [23].

**Table 5.41.** Carbon foam/carbon paper composite made using different R + F concentarions Low density open cell carbon foam

| Example | R + F (% W/V) | R/C | Density (mg/cm$^3$) | Capacitance (F/g) | Surface area (m$^2$/g) | Pore size (nm) |
|---|---|---|---|---|---|---|
| 1 | 50 | 1000 | 419 | 20.6 | 397 | 20 |
| 2 | 50 | 2000 | 413 | 17.8 | 497 | 17 |
| 3 | 60 | 1000 | 453 | 15.7 | 414 | 5 |
| 4 | 60 | 2000 | 471 | 19.1 | 408 | 5 |

Data from [22]

The carbon aerogel/carbon paper electrode material produced by this method has several advantages including its flexibility.

## 5.3.5 Metal Doped Carbon Aerogels

### 5.3.5.1 β-Resorcylic Acid in Metal Doped Carbon Aerogels Synthesis

Incorporation of metal ions into a carbon structure can modify the conductivity and catalytic activity of the aerogel materials. The distribution of desired metal ions in the carbon matrix was effectively done by the use of β-resorcylic acid in the gel formation process. A new method was developed for the preparation of metal doped carbon aerogels with the use of potassium salt of $\beta$-resorcylic acid in the sol-gel process, producing potassium doped hydrogels. The potassium in the gel were then replaced with the desired metal ions, such as copper, cobalt and nickel, through an ion exchange process. Finally the gels were dried and pyrolyzed to obtain these novel metal doped carbon aerogels [25–28].

The chemistry is outlined in the following reaction scheme (Figure 5.44).

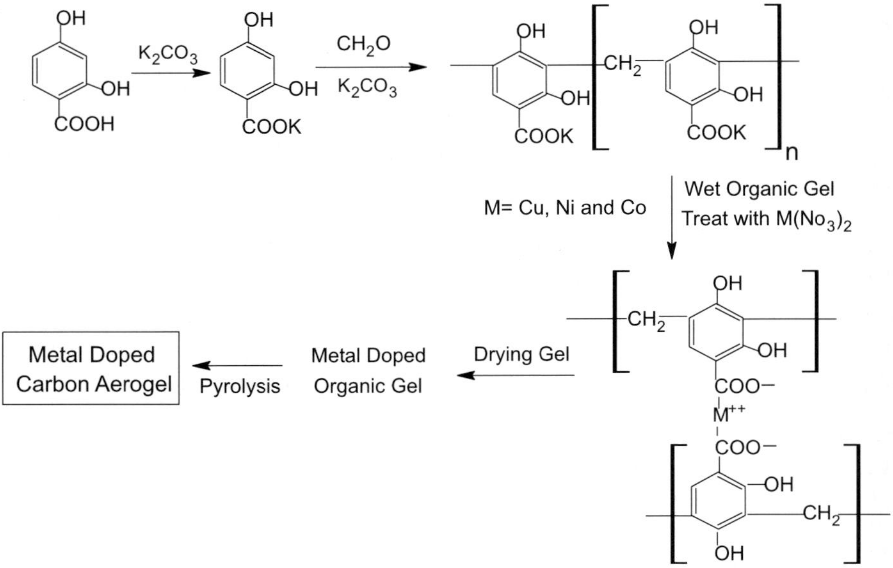


**Figure 5.44.** Synthesis of metal-doped carbon aerogol from resorcylic acid

The metal doped carbon aerogels contained carbon nanoparticles having a diameter of 15 to 30 nm with a surface area of 712 $m^2/g$. Due to the presence of different metal ions, these new carbon aerogel materials can be expected to show novel catalytic and transport properties and, therefore, could be used as new catalysts, electrode materials and substrates for the growth of carbon nanotubes.

### 5.3.5.2
### Metal Doped Aerogels from Resorcinol

Metal-containing carbon aerogels were prepared from the RF aerogel doped with metal ions [29]. In the preparation of RF aerogel, nitrates of chromium, iron, cobalt and nickel salts were added into the resorcinol-formaldehyde gel solution. Then the metal doped and dried RF gel was pyrolyzed to obtain the carbon aerogel. The final aerogel material contained metals in the range of 1.4 to 5.4 weight percent.

### 5.3.6
### Summary and Outlook

Potential applications for the resorcinol based aerogel materials are believed to be limitless. The possibilities are in the composite structures, as reinforcing agents for organic rubbers, in pigments for ink-jet printers, as insulators for electrochemical storage devices, in media for gas separation, high temperature insulation, fuel cell electrodes and metal catalyst supports.

## References


1. R. W. Pekala , Polym. Prepr., Vol. 39, pp 221–223 (1989).
2. R. W. Pekala, J. Mater. Sci., Vol. 24, pp 3221–3227 (1989).
3. R. W. Pekala, US Patent 4,873,218 (1989).
4. R. W. Pekala, US Patent 4,997,804 (1991).
5. H. Tamon, H. Ishizaka, M. Mikami and M. Okazaki, Carbon, Vol. 35, p 791 (1997).
6. J. L. Kaschmitter, S. T. Mayer and R. W. Pekala, US Patent 5,529,971 (1996).
7. K. P. Yoo, S. Y. Kim, D. H. Yeo and J. W. Lim, High Temp.-High Pressures, Vol. 33, p 441 (2001).
8. X. Lu, M.C. Arduini-Schuster, J. Kuhn, O. Nilsson, J. Frickie and R. W. Pekala, Science, Vol. 255, p 971 (1992).
9. S. T. Mayer et al., US Patent, 5,908,896 (1996).
10. S. T. Mayer et al., US Patent, 5,508,341 (1996).
11. L. Sibille et al., Space Resources Roundtable II (2000).
12. X. Lu et al., Science, Vol. 255, p 971 (1992).
13. L. W. Hrubesh and R. W. Pekala, J. Mater. Res., Vol. 9, p 731 (1994).
14. R. W. Pekala et al., US Patent, 5,260,855 (1993).
15. H. Tamon and H. Ishizaka, J. Colloid Interface Sci., Vol. 206, p 577 (1998).
16. F. M. Kong, J. D. LeMay, S. S. Hulsey, C. T. Alviso and R. W.Pekala, J. Mater. Res., Vol. 8, p 3100 (1996).
17. G. Reichenauer, A. Emmerling, J. Fricke and R. W. Pekala ('Microporosity in carbon aerogel' from the internet).
18. W. Bell and S. Dietz, US Patent 6,297,293 (2001).
19. R. Petricevic, J. Frickie, R. Leuschner and M. Lipinski, US Patent, 6,503,655 (2003).
20. L. Ratke and J. Fricke, US Patent, 6,599,953 (2003).
21. S. T. Mayer, R. W. Pekala and J. L. Kaschmitter, US Patent, 5,898,564 (1999).
22. R. W. Pekala, S. T. Mayer, L. Kaschmitter and R. L. Morrison, USPatent, 5,932,185 (1999).
23. M. W. Drege, US Patent, 5,945,084 (1999).
24. L. Hrubesh, US Patent Appln, 2003/0,134,916 (2003).
25. J. H. Satcher and T. Baumann, US Patent Appln, 2003/0,100,619 (2003).
26. D. Suh, T. Park, Y. Yoon, K. Lee and I. Kim, US Patent Appln, 2003/0,176,277 (2003).
27. J. Satcher and T. Baumann, US Patent, 6,613,809 (2003).
28. J. Satcher and T. Baumann, US Patent Appln, 2003/0,216,482 (2003).
29. Maldonado-Hodar et al., Langmuir, Vol. 16, p 4367 (2000).

# 6 Resorcinol Formaldehyde Latex (RFL) Adhesives and Applications

## 6.1 Adhesive Formulations, Testing and Mechanism

### 6.1.1 Introduction

Rubber articles such as tires, hoses and belts are composite materials. They are made using various natural and synthetic rubbers reinforced with different reinforcing materials such as carbon black, fibers and other reinforcing fillers. With an increasing demand for the longer lasting high performance products, constant improvement is needed in the reinforcing and elastomeric materials. In the case of tire applications, higher strength reinforcements could allow the construction of thinner carcass plies which in turn could reduce the heat build up and enhance the fatigue resistance in the final products. These requirements have resulted in the development of high tenacity fibers such as the nylon, polyester, steel and aramid for the reinforcement of rubber goods for tires and other applications. In general, synthetic fibers have the difficulty of adhering to rubbers because of their smooth polymer surfaces and low reactivity. The low fiber surface activity is due to lower polarity and reactivity of the polymer molecules. Therefore an establishment of physical and or chemical bonding between the fiber and rubber is considerably reduced. Since a direct bond cannot be established between the rubber and fiber, an adhesive which can promote the adhesion between these two was necessary.

The first adhesive system developed for rayon, which was a blend of casein and natural rubber latex, gave acceptable levels of adhesion for rayon with rubber compounds. But the introduction of synthetic fibers as reinforcements in rubber composites identified the need for an improved adhesives and adhesion promoters for the enhancement of strength between rubbers and fibers. An improved adhesion requirement for the tire, conveyer and transmission belt applications had led to the development of resorcinol (R)-formaldehyde (F)-latex (L) adhesives [1].

## 6.1.2
## Resorcinol Formaldehyde Latex (RFL) Adhesive Formulations

Synthetic fibers used in applications such as tires, hoses and belts, have high strength and low elongation. On the other hand, rubbers exhibit high elongation with relatively low strength. Therefore, the adhesive must bridge the gap between these two material properties for the overall best performance.

RFL is a composite material, consisting of several components including resorcinol (R), formaldehyde (F) and latex (L), that provides the bonding of synthetic fibers and rubber compounds. The major function of the adhesive is to transfer the load stress from the rubber matrix to the reinforcing fiber materials. To accomplish this, the adhesive system should develop a high level of adhesion with a modulus intermediate between the fiber and rubber compounds. In the case of an RFL adhesive, the RF resin component of this adhesive preferably bonds to the fiber whereas the rubber-like latex component bonds to the rubber through co-vulcanization. The low viscosity of RFL solution provided good fiber wetting and thereby, increasing the interaction between the fiber and adhesive.

### 6.1.2.1
### Preparation of RFL Solutions

The current RFL dip systems can be complex, and many different formulations have been developed to fulfil the needs for various applications. But the process of developing a RFL solution essentially consists of an in-situ RF resin or a pre-formed or pre-condensed RF resin dispersed in synthetic rubber latex, namely the styrene-butadiene-2-vinyl pyridine latex (called VP latex).

A pre-formed resin dissolved in water is available in various concentrations from Indspec Chemical Corporation as Penacolite resins and the details are given in Table 6.1.

**Table 6.1.** Resorcinol-formaldehyde (RF) pre-formed resins Penacolite resin solution properties

| Property | R-50 | R-2200 | R-2170 |
|---|---|---|---|
| Color | Orange-red | Orange-red | Orange-red |
| Resin solids (%) | 50 | 70 | 75 |
| pH | 1.6 | 1.1 | 1 |
| Viscosity (Poise at 23 °C) | 1.07 | 13 | 50 |
| Free resorcinol (wt.%) | 10 | 14 | 15 |

Source: Indspec Chemical Corporation Technical Bulletin

Similarly, the pre-formed RF resin solutions available from other commercial sources can also be used in the RFL preparations.

A standard RFL is prepared in two major steps. The first step consists of a reaction between resorcinol and formaldehyde to produce a "resole" type molecules. The second step is primarily the introduction of latex into the first step RF resin.

### 6.1.2.2 First Step Resorcinol-Formaldehyde Reaction Chemistry

Resorcinol has been chosen as the phenolic molecule of choice among other derivatives in the adhesive formulation of fiber-rubber composite applications, due to its water solubility, enhanced reactivity with formaldehyde at room temperature and basic conditions. Resorcinol reacts with formaldehyde to produce partially condensed reaction products.

A typical RF resin formulation (Table 6.2) and the procedures to prepare this resin solution are shown below.

**Table 6.2.** Solutions prepared with Penacolite resins

| Formulations | R-2170 (75%) | | R-2200 (70%) | R-50 (50%) |
|---|---|---|---|---|
| Resin solution | Dry | Wet | Wet | Wet |
| Penacolite resin | 12.6 | 16.8 | 18 | 25.2 |
| Water (soft) | | 176.2 | 175 | 167.8 |
| Sodium hydroxide (50%, aqu) | 0.45 | 0.9 | 0.9 | 0.9 |
| Formaldehyde (37%, aqu) | 4.5 | 12.2 | 12.2 | 12.2 |
| Total | 17.55 | 206.1 | 206.1 | 206.1 |

Source: Indspec Chemical Corporation Technical Bulletin

### RF Solution Preparation Procedure

1. Add sodium hydroxide solution into water at room temperature.
2. Blend the required amount of resorcinol or resorcinol-formaldehyde resin (RF) solution into the sodium hydroxide water solution. Agitate the solution for 5–10 min to dissolve the resin.
3. Now add the formaldehyde to the resin solution slowly. The reaction between resorcinol and formaldehyde is "exothermic". Therefore, care must be taken to remove the heat from the solution so that the temperature remains essentially at room temperature. With RF resin, the temperature rise accompanying formaldehyde addition is minimal.
4. With RF resin, no aging of resin solution is required and can be used in immediately in the next step. On the other hand, with resorcinol the solution should be aged for at least 4–6 h before further use.

Resorcinol reacts with formaldehyde in the presence of an acid or alkaline catalyst conditions to produce novolak type resins. When the formaldehyde concentration is less than one mol per mol of resorcinol, the resin contains predominantly novolak type structures (Figure 6.1).

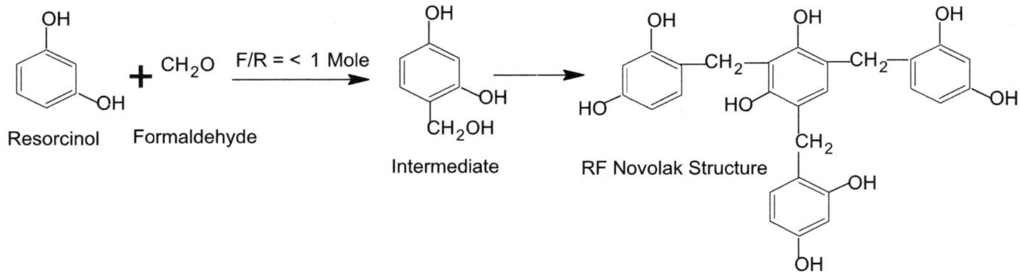


**Figure 6.1.**

But, with an increase in formaldehyde concentration, low temperature (preferably room temperature or lower) and alkaline conditions, the formation of methylol groups in the resin structure from the resorcinol-formaldehyde reaction can be seen (Figure 6.2).

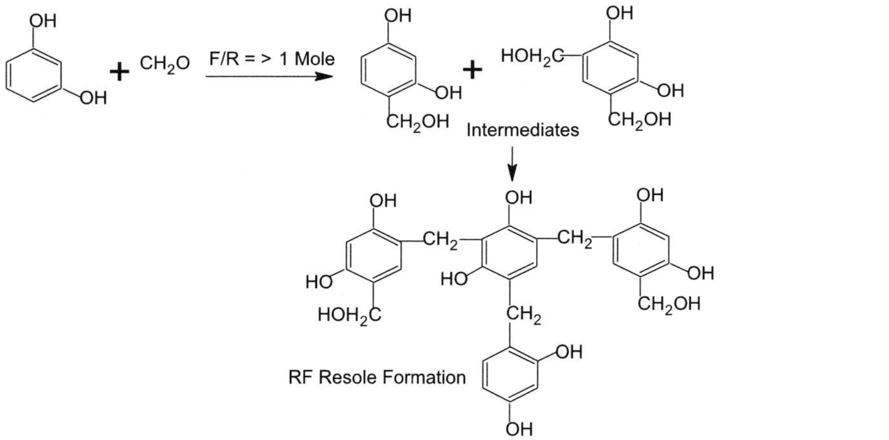


**Figure 6.2.**

In an in-situ resorcinol-formaldehyde solution preparation in the presence of sodium hydroxide, the solution predominantly contains resorcinol "trimers" having methylol groups (called resoles) with some mono and dimers of resorcinol. The term trimer means that three resorcinol molecules are connected through a methylene ($-CH_2-$) bridge.

### 6.1.2.3 RFL System or Final Formulation

When the preparation of RF solution is completed, then it can be added to a natural rubber latex or synthetic rubber latex (L) solution for the final RFL formulation ready to be used to treat the synthetic fiber materials. The final resin latex mix solution is called the "dip solution" or simply the "RFL dip". This solution can be used immediately but the adhesion is improved if the dip is aged for at least 12 h at 18–24 °C temperature conditions.

The following Table 6.3 contains a typical "final mix" RFL formulation and the properties.

**Table 6.3.** Typical RFL dip formulations

| Resin solutions | Precondensed resin | | In situ resorcinol | |
|---|---|---|---|---|
| | Dry | Wet | Dry | Wet |
| Penacolite R-2200 (70%) | 12.6 | 18 | | |
| Resorcinol (R) | | | 11 | 11 |
| Water (Soft) | | 175 | | 238.2 |
| Sodium hydroxide (50%) | 0.45 | 0.9 | 0.3 | 0.6 |
| Formaldehyde (37%) (F) | 4.5 | 12.2 | 6 | 16.2 |
| **Total** | 17.55 | 206.1 | 17.3 | 266 |
| | Aging not required | | Age 4–6 h | |
| Final RFL dip | | | | |
| Vinyl pyridine latex (41%) | 100 | 244 | 100 | 244 |
| Resin solution | 17.55 | 206.1 | 17.3 | 266 |
| Water (soft) | | 72.2 | | 60 |
| Ammonium hydroxide (28%) | | 12 | | 11.3 |
| **Total** | 117.55 | 534.3 | 117.3 | 581.3 |
| Final dip properties | | | | |
| Dip solids (%) | 18–26 | | | |
| pH | 9 to 10 | | | |
| F:R molar ratio | 1.2–2:1 | | | |
| Resin/latex ratio | 15–20/100 | | | |

### 6.1.3 Factors Affecting the RFL Performance

Numerous parameters of the RFL dip have a marked influence on the performance and quality of the adhesion developed directly or indirectly between the synthetic fibers and rubber compounds.

#### 6.1.3.1 Method of RFL Preparation

In the RFL formulation, the RF resin containing the methylol groups (resole) can be prepared directly from the use of resorcinol (in situ) or a pre-condensed resorcinol-formaldehyde (RF) novolak resin. There are several advantages in using the pre-condensed RF resin in the RFL dip preparation. In the case of an in situ method, resorcinol is allowed to react with formaldehyde directly at room temperature in the presence of an alkaline catalyst, primarily sodium hydroxide. The resorcinol-formaldehyde reaction is exothermic. Therefore, controlling the reaction temperature, crucial for the formation of methylol groups, should require monitoring of the addition rate of formaldehyde to maintain the temperature. With the pre-condensed resin, the amount of formaldehyde usage is low compared to the in-situ method. Therefore, less heat is generated from the RF-formaldehyde reaction.

The resorcinol-formaldehyde reaction in the in situ method at room temperature and alkaline conditions takes a long time to obtain the resin molecules of desired molecular weights. The formation of such desired structures are important for the good adhesive performance of RFL. Resin solutions prepared by the in situ method must be aged before the latex solution is added to the system. This aging time allows the resorcinol and formaldehyde to form some resin structures and usually takes about 6 h at 23 °C or at room temperature. The progress of the resorcinol and formaldehyde could be followed by several techniques that include calorimetry, bromination of active sites on the resorcinol, solution viscosity changes and liquid chromatography.

Liquid chromatography was used to monitor the disappearance or reaction of resorcinol at a regular rate at 23 °C. The obtained results are presented in Table 6.4 [2].

**Table 6.4.** Free resorcinol content in an "in situ" RF resin

| Aging time (h) | Free resorcinol wt.% | mol/l, C | $\text{Log}(C_0/C)$ |
|---|---|---|---|
| 0 | 5.67 | $C_0 = 0.515$ | None |
| 1 | 2.1 | 0.191 | 0.431 |
| 2 | 1.8 | 0.163 | 0.5 |
| 4 | 1.5 | 0.136 | 0.578 |
| 6 | 1.2 | 0.109 | 0.674 |
| 8 | 0.95 | 0.089 | 0.777 |
| 10 | 0.8 | 0.073 | 0.848 |

$C_0$ = Molar concentration of resorcinol at time, $t_0$
Data taken from [2]

At the end of 6-h aging time, the amount of resorcinol had fallen from the initial level of 5.67 down to 1.2 wt. %. A plot of log $(C_0/C)$ vs time is shown in Figure 6.3 [2]. From this plot, a straight line curve indicated that the resorcinol consumption obeyed a first-order reaction rate (at least it was pseudo first-order). This showed that the resorcinol was consumed independently of its concentration under the dilute solution conditions. The controlling factor was time. The rate constant obtained from this plot was $K_r = 0.106\,h^{-1}$ or $2.94 \times 10^{-5}\,s^{-1}$.

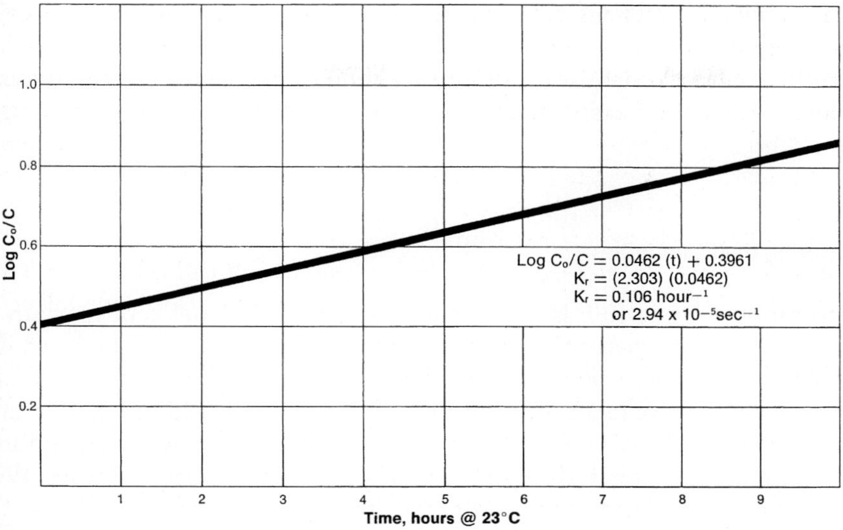


**Figure 6.3.** Reaction of resorcinol and formaldehyde: first order reaction [2]

Aging the resin solution at higher temperature accelerates the reaction rate and therefore significantly reduces the time necessary to reach the optimum resin age.

This problem can be eliminated by the use of a pre-condensed RF resin. In addition, the pre-condensed RF solution has consistent quality of uniform molecular weight structures before mixing with the latex solution. Because of this, all the RFL raw materials can be accurately metered by volume when preparing the dip.

Resorcinol-formaldehyde (RF) pre-condensed resin is normally prepared under acidic conditions. Therefore, it should be first neutralized with sufficient base, such as ammonium hydroxide or sodium hydroxide, to prevent the coagulation of the latex when the resin solution is mixed with the latex.

### 6.1.3.2
### Resorcinol-Formaldehyde (R/F) Ratio

The formaldehyde to resorcinol (F/R) ratio is an important factor in determining the maturating time. The rate of methylol formation is accelerated with an increase in formaldehyde concentration [3]. The optimum molar ratio of resorcinol to formaldehyde is about 1 to 2 to obtain a complete methylol reaction. Studies also indicated that, with RFL system, maximum adhesion could be achieved when the F/R ratio is about 2/1 [4, 5]. But this ratio depends on the nature of fiber and adhesion level required in these fibers. The R/F ratio for achieving an optimum adhesion was determined to be about

1/2 for nylon, 1/1.4 for adhesive activated polyester and 1/1.6 for Kevlar fibers [6].

When the formaldehyde concentration is increased, the viscosity of the resin solution is also increased due to the formation of higher molecular weight products.

#### 6.1.3.3
#### Maturing of RFL on Adhesion

Maturation time of RFL is important to ensure the partial condensation of the RF resin. The maturing conditions necessary to prepare a stabilized dip depends on the R/F solution concentration, amount and type of catalyst and type of latex. Longer maturation for RF reaction or after blending with the latex is detrimental because condensation of resorcinol and formaldehyde continues and decreases progressively with the number of reactive sites or groups. With the pre-condensed resin based dip, too long maturation may decrease the adhesion.

#### 6.1.3.4
#### Effect of pH on Adhesion

It was determined that the adhesion of rubber to fibers depends on the pH of RFL [5]. If the pH is too low or high, the adhesion decreases. The type of base, either sodium hydroxide or ammonium hydroxide, used in RFL also affects the adhesion [4, 7].

With sodium hydroxide catalyst, the reaction between resorcinol and sodium hydroxide results in the formation of a resorcinate ion. This product may not directly interfere with the reaction of formaldehyde (Figure 6.4).

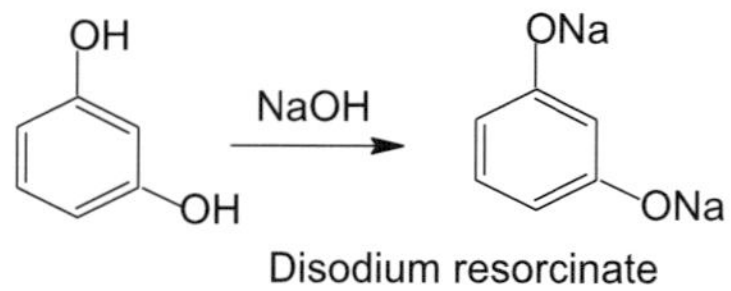


**Figure 6.4.**

However, with an ammonium hydroxide solution the situation can be quite different. When the resin solution is prepared with ammonium hydroxide, formaldehyde reacts with ammonia to produce methylolated amines first, which in turn reacts with resorcinol or resorcinolic (RF) resin to produce an insoluble material [8] (Figure 6.5).

Ammonia is more reactive towards formaldehyde, and therefore the formation of insoluble material cannot be avoided. If ammonium hydroxide has to be employed in the RFL, then it must be added after the resorcinol and

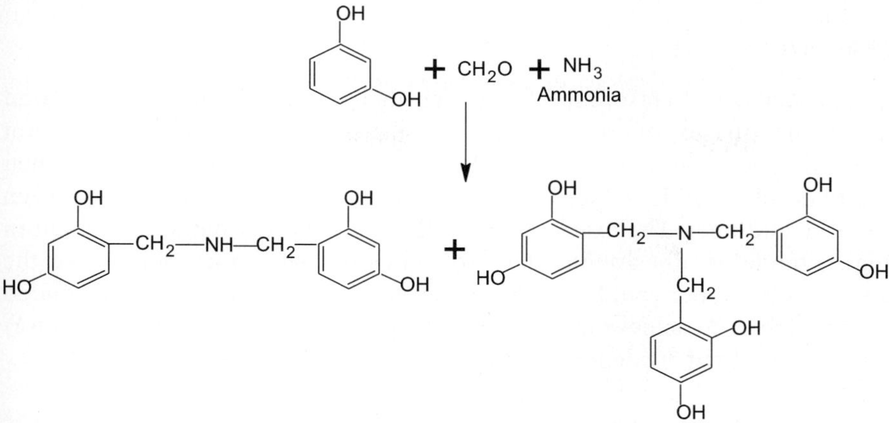


**Figure 6.5.**

formaldehyde react sufficiently. Resin solution precipitation can be avoided by using a small amount of sodium hydroxide in combination with ammonium hydroxide.

Adhesion performance of different synthetic fibers depends on the pH of dip. For example, an optimum adhesion of polyester fiber can be achieved with 10 to 11 pH of the dip solution, and similarly for nylon fiber the pH of dip may be about 8 to 9.

#### 6.1.3.5 Choice of Latex

The kind of latex used in the RFL adhesive formulation depends on the type of rubber to be bonded [3]. Latex must be compatible with rubber to ensure vulcanization. The superior adhesion of RFL adhesive is due to the interaction between the latex and textile fibers. Though great efforts were made to develop and use various rubber lattices, the latex based on the ter-polymer 2-vinyl pyridine-styrene-butadiene has been used widely in the RFL dip. The presence of pyridine nuclei in the rubber latex molecule facilitates the interaction between the RF resin and fiber through hydrogen bonding mechanism. It was also observed that a maximum adhesion could be achieved with a 2-vinyl pyridine content of about 15%. A typical vinyl pyridine ter-polymer composition contains 70% butadiene, 15% vinyl pyridine and 15% styrene.

### 6.1.3.6 RF Resin to Latex Ratio

The ratio of latex to resorcinol-formaldehyde is also important to obtain good adhesion. Adhesion of synthetic fiber can be increased with increasing amount of resin up to about 25 wt. %. If the RFL contains small amount of resin, then the fiber adhesion is reduced due to smaller number of cross links between the RFL and fiber. The excess resin in RFL, on the other hand, may result in higher modulus adhesive film that produces stiff coated fiber. In this case, the adhesion to rubber could be reduced because the latex preferentially bonded to the rubber. To achieve the maximum adhesion, the latex to resin ratio was found to be about 5:1 depending upon the requirement [7].

## 6.1.4 Fiber Dipping and Heat Treating Conditions

In a standard single-step dip process, the fiber, cord or fabric has been passed through the stabilized or aged RFL dip solution and then heat treated in the drying and curing ovens at optimal time, temperature and tension conditions to obtain the maximum balance of strength and adhesion properties.

Adhesion performance of the dipped fibers depended on the amount of RFL dip pickups, treating temperature and time conditions and also on the rubber compound composition of the rubber to be bonded.

### 6.1.4.1 RFL Dip Pickup

When the fibers or cords are passed through an adhesive dip bath, depending on the nature of fiber surface and concentrations of the dip solution, fibers pick up a certain amount of adhesive material. The dip pick up is the relative amount of adhesive deposited on the fiber or cord. Adhesive dip pick up and its ability to penetrate the fiber surface determine the adhesion performance.

Adhesion is normally increased with dip pick up and reaches a maximum. Excess of an adhesive on the fiber or cord affects the cord stiffness and therefore affects the adhesive strength. For achieving maximum adhesion performance, the RFL dip-pick up should be between 6 to 8% of the cord or fiber weight. Normally the dip pick up depends on whether the fiber surface is hydrophilic or hydrophobic in nature. For example, polyester fiber surface is less hydrophilic than nylon and therefore can expect to pick up less dip compared to nylon.

### 6.1.4.2 Heat Treatment of RFL Dipped Fibers or Cords

After the fibers or cords are dipped into an RFL solution, they are dried first to remove or evaporate the solvent in the drying oven or zone and then heat treated in the curing oven. The temperature of the drying and heat treating ovens depends on the nature of fibers to be treated and also the adhesive type and performance required. Drying oven temperatures are normally between 130 to 170 °C and curing temperatures can range from 190 to 240 °C. During the heat treatment a highly cross-linked network structure is formed in the RFL. Strong interaction between the fibers and adhesive is achieved in the curing process. Adhesion properties of the RFL adhesive are reaching highest at the 200 to 240 °C treating conditions and are reduced at both under cure and over cure conditions [9].

## 6.1.5 Adhesion Testing of RFL Treated Fibers or Cords

The RFL treated fibers are embedded in a formulated and uncured rubber compound and cured for adhesion testing. An example of a rubber stock or compound used to prepare the adhesion test samples is given below. But this composition can vary depending on the applications and properties required (Table 6.5).

**Table 6.5.** Rubber compound used for adhesion testing

| Composition | Parts by weight |
|---|---|
| 1. Natural rubber | 70 |
| 2. SBR | 30 |
| 3. Carbon black (N-660) | 50 |
| 4. Zinc oxide | 4 |
| 5. Stearic acid | 2 |
| 6. Naphthenic oil | 5 |
| 7. Antioxidant (DHTMQ) | 1.8 |
| 8. Sulfur | 2.5 |
| 9. 2,2′-Dibenzothiazyl disulfide | 0.8 |
| Cure properties (MDR rheometer, at 160 °C) | |
| $M_H$, dN-m | 11.45 |
| $M_L$, dN-m | 0.85 |
| $t_s$ 2, min | 4.13 |
| t′90, min | 10.15 |

Source: Indspec Product Application Guide No. 2

The components present in rubber compositions and vulcanization conditions also play a major role in determining the adhesive strength of the fiber to rubber compounds. For example, in the case of sulfur, adhesion levels are found to be higher with rubber compounds containing 2–3 parts sulfur. High sulfur level enhances the opportunity for co-vulcanization with the RFL adhesive layer on the cord.

Similarly, longer scorch time associated with the use of scorch safety accelerators such as dibenzothiazyl disulfide (MBTS) and $N,N'$-(dicyclohexyl)-2-benzothiazyl sulfenamide (DCBS) increases fiber to rubber interaction [10]. For achieving higher levels of adhesion, the rubber and adhesive on the fiber surface should contact intimately one another. This can be achieved by using a rubber compound with longer scorch time [11].

Not only the scorch time but also the polarity of the rubber compound contribute to the adhesive strength of the fiber to rubber. The polarity of a rubber compound can be enhanced by adding suitable fillers such as the silica. For obtaining a good adhesion between synthetic fibers and rubber compounds, the two most important requirements are the long scorch time and high polarity of rubber compounds [11].

#### 6.1.5.1
#### Adhesion Test Methods

After vulcanizing the RFL treated fiber-rubber composite, the resultant cured samples are evaluated for adhesion performance. In general, static and dynamic adhesion tests are normally performed before and after heat, steam, humidity and salt water aging conditions. These tests usually provide valuable information regarding the durability and service life of the fabricated articles with these fiber reinforcements.

Static adhesion tests are normally conducted by ASTM standards D 2229, D 2138 and D 2630 for the T-Pull, H-Pull and peel adhesion methods.

#### 6.1.5.2
#### H-Test Adhesion Method

Testing of cord adhesion by this method is also known as the H, H-block, H-pull or H-test. This method has been used to determine the static adhesion (or force) required to pull a cord (fabric or steel) in the direction of its axis, from a strip of rubber (shaped like an "H") in which the ends of the cords are embedded. An overall adhesion of cord to rubber is measured by this method. From this method, high rubber coverage of cords indicates that the failure is within the rubber. Failure at the rubber to adhesive interface can be seen by the low rubber coverage of cords. Cord breakage can be considered as a failure of cord.

#### 6.1.5.3
#### Peel Adhesion Test Method

Peel adhesion measures properties similar to the "H" test method except that no cord breakage occurs in this method. This test has been used to determine the force required to separate two laminated layers of cords (fabric or steel) bonded together by an intermediate layer of rubber. Good rubber to cord adhesion can be determined by a high separation force and high rubber coverage. The property measured from this test is the overall cord to rubber adhesion.

#### 6.1.5.4
#### Cord Stability Test

The effect of various rubber compound formulations and adhesive dip systems on synthetic cord stability (such as polyester) can be measured by cord strength retention after curing rubber stocks or compounds (with embedded cords) at elevated temperatures, namely 100 °C or greater, for 2 h or more. This test is designed to stimulate the thermal degradation of synthetic cords when the tires or rubber compounds are over cured in the mold. Greige or treated synthetic cords can be used to test their stability with this method.

A dynamic adhesion test based on ASTM D 430 for the Scott flex tester gives an indication of the fatigue life of the fiber-rubber composite after exposure to cyclic strain. These tests are helpful in identifying the failure modes in fiber-reinforced composites.

### 6.1.6
### Adhesion Mechanisms for RFL Systems

The performance of fiber reinforced rubber composite depends on how well the RFL treated fibers are bonded to the rubber compounds. In this respect, it is the RFL film responsible for the bonding of fibers to the rubber compounds. In order to investigate how effectively the RFL film can bond these two, one should know the interfacial reactions or interactions occurring between the RFL, fibers and rubbers compounds.

In the case of an interaction between the RFL and rubber compound, one should consider the diffusion of sulfur from the rubber compound into the RFL dip film. Because of this, the sulfur can interact with the rubber portion of the RFL and rubber matrix and can form a sulfur cross-link. In this way, a direct chemical bond can be produced between the rubber latex present in the RFL and the rubber matrix.

The RF resin present in the RFL can also react with the rubber. During the RFL formulation and subsequent stabilization of resorcinol-formaldehyde resins, the formation of "methylol" groups can be expected. This resorcinolic methylol groups can react with the rubber present in the matrix and latex. The methylol groups may react with rubber molecules in two different ways. According to Van der Meer, a methylene bridge type reaction can take place between a phenolic methylol group and the rubber [12]. Based on the work of Van der Meer, the resorcinolic resole can form a methylene bridge between the RF resin and rubber according to the following reaction scheme (Figure 6.6).

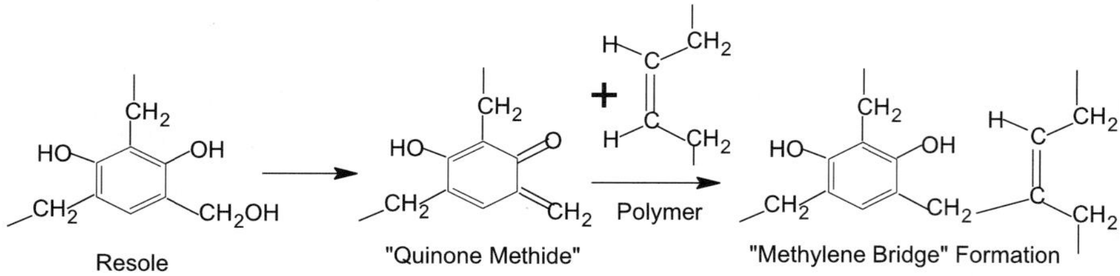


**Figure 6.6.**

Separately, Greth proposed the formation of a "chroman" type structure when the rubber was reacted with a phenolic, such as the resorcinolic, methylol group [13] (Figure 6.7).

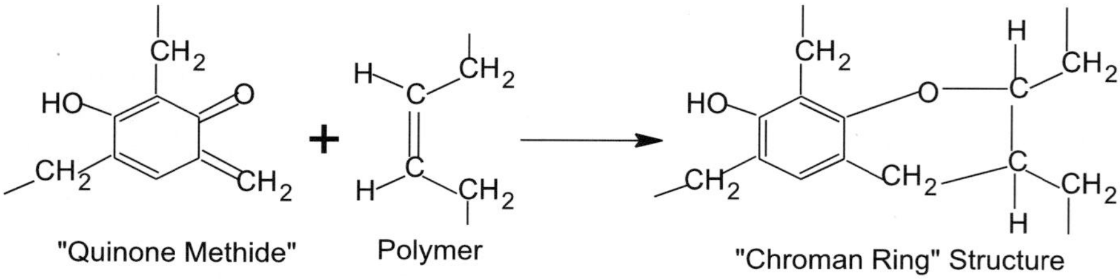


**Figure 6.7.**

Chemical reactions between the methylolated RF and rubber matrix, according to Van der Meer and Greth, can produce covalent type bonds. The formation of such chemical bonds can strengthen the interface between the RFL and rubber matrix. However, this chemical bond may not be formed if the RFL treated cords are cured at elevated temperature conditions before their embedment into unvulcanized rubber compounds. This may be due to the possible non-existence of resorcinolic methylol groups after exposing them to high temperatures. Resorcinolic methylol groups are highly reactive and very unstable at elevated temperature and high pH conditions. Therefore, these groups can react and predominantly produce cross-linked resorcinolic networks, which may result in the consumption of all the available methylol groups. The availability of methylol groups for the rubber reaction may not exist after the high temperature treatment of the RFL treated fibers.

Model studies have been conducted to determine the nature of reaction and the mechanism of vulcanization between a phenolic resole (methylol) and unsaturated elastomers [14]. From these studies, it was shown that "chroman" type structures were formed from the reaction of methylol phenols with aliphatic unsaturated hydrocarbons (Figure 6.8).

From the methylol phenols, chroman ring structures were formed with the aliphatic unsaturated compounds in the presence of an acid or base [15,16]. The formation of a chroman structure might have taken place through the "quinone

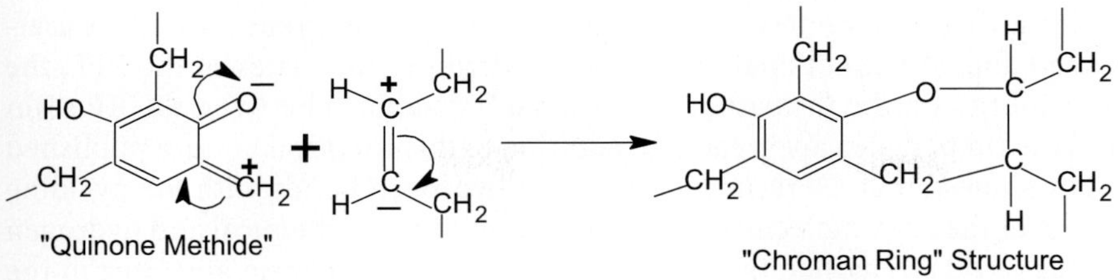


**Figure 6.8.**

methide" intermediate produced from the methylol phenols. In the presence of an acid catalyst, the quinone methide structure may be produced by the formation of a carbonium ion according to the following scheme (Figure 6.9).

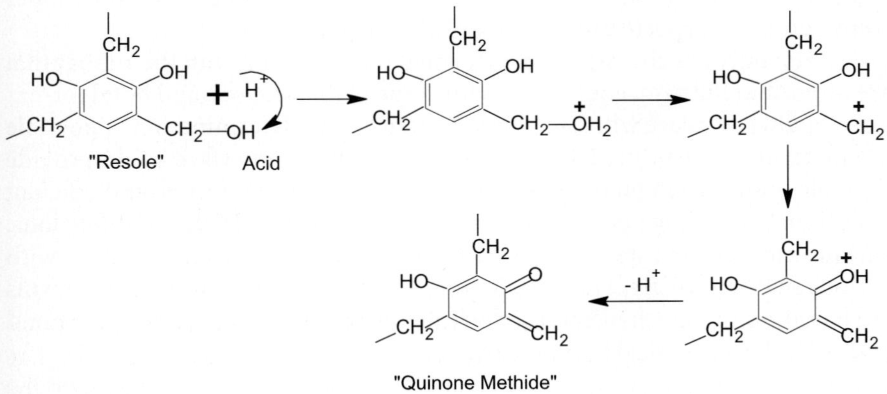


**Figure 6.9.**

Similarly, the quinone methide structure can also be formed by the base catalyzed reaction of phenolic methylol group, and may be based on the following reaction mechanism (Figure 6.10).

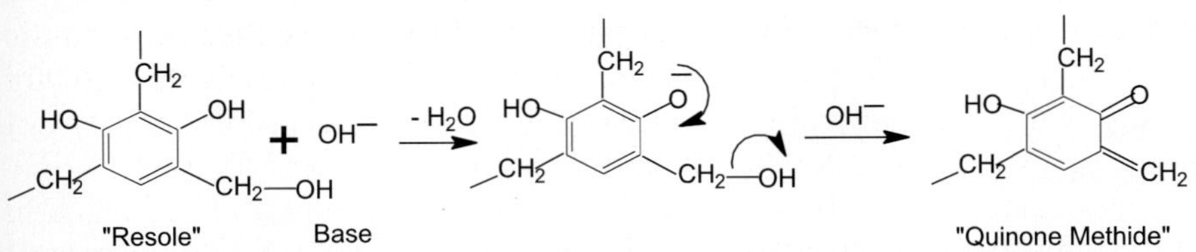


**Figure 6.10.**

Experimental determinations have proved, under RFL preparative conditions and also in the presence of sodium hydroxide catalyst, the formation and existence of methylol resorcinols from the resorcinol and formaldehyde reaction [17].

Since these resorcinolic methylol groups are in co-existence with the unsaturated aliphatic functional groups of the vinyl pyridine latex in the RFL, the possibilities on the formation of "chroman" structures between the RF resin and the rubber latex are greater. In addition to this chemical bond, a published study showed that RF resin can also form hydrogen bonds with the pyridine nuclei of the latex molecules [18]. The combination of chemical and hydrogen bonds could strengthen the interaction between the RF resin and latex in the RFL system.

Even after the formation of "chroman" type chemical bonds, the butadiene portion of the latex has more "unsaturated" carbon–carbon double bonds available for the sulfur cross-linking with the rubber molecules of the rubber compounds. Because of sulfur reactions, chemical bonds can be formed between the latex and the rubber matrix. Due to these interactions and, physical and chemical bond formations, the interface between RFL and rubber compounds are expected to be stable and stronger.

It is not only the chemical bonds formed from the RFL with the rubber that are important but also, equally, the same type of interactions and bond formation should take place with the fibers. Therefore, the determination of possible interactions between the RFL films and fibers at the interface could provide valuable information on the strength of the bond for an effective and efficient adhesion. Depending upon the nature of fiber surfaces and their modifications, various mechanisms were proposed for the adhesion of RFL adhesives with various synthetic fibers. The most important adhesion promoting mechanisms are based on the mechanical interlocking, diffusion, primary chemical bond and secondary chemical bond formations.

#### 6.1.6.1 Mechanical Interlocking

Based on this mechanism, the penetration of RFL and rubber into the yarn or fabric structure should take place. In the case of rubber, very little penetration can take place into the fiber due to its incompatibility. The penetrability of the RFL components depends on the solvent and the nature of additives.

#### 6.1.6.2 Diffusion

Diffusion is favored when the RFL adhesive and the fiber are thermodynamically compatible. This occurs when the solubility parameter of the components present in the RFL match with the fiber. Not only the solubility parameter but also the molecular size determine the diffusion across the interface into the fiber. RF resin containing low molecular weight oligomers and the additives present in the RFL formulations can diffuse and make either a chemical or physical bond with the functional groups present at the fiber surface.

### 6.1.6.3
### Primary Chemical Bonding

A chemical reaction between the RFL and fiber can lead to the formation of covalent type bonds between these two. If these reactions take place, then this will contribute to the main mechanism of adhesion. The formation of covalent bonds depends on the functional groups present on the polyester or other fiber surface and their reactivity towards the components present in the RFL formulations.

### 6.1.6.4
### Secondary Chemical Bonding

The formation of secondary chemical bonds, such as the hydrogen bond, provides a major contribution in the overall bonding of the RFL to the fiber. With the RFL adhesive, fiber surface containing more functional groups, by chemical activation, can form more hydrogen bonds from the interactions with pyridine nuclei and resorcinolic hydroxyl groups.

In general, with synthetic fibers, more than one single mechanism may be operative in the adhesion development of fibers to rubber compounds with RFL adhesives.

## References


1. W. H. Charch and D. B. Marney, US Patent, 2,218,228 (1938).
2. J. P. Noe, B. J. Brown and D. A. Lederer, Koppers Company Inc, Technical information on resorcinol based adhesives for the tire industry (1977).
3. T. Takeyama and J. Matsui, Rubber Chem. Technol., Vol. 42, p 159 (1969).
4. N. K. Porter, J. Coated Fabrics, Vol. 23, p 34 (1993).
5. A. L. Miller and S. B. Robinson, Rubber World, Vol. 137, p 397 (1957).
6. R. Iyengar, Presentation at ACS Rubber Div. Meeting (1987).
7. M. I. Dietrick, Rubber World, Vol. 136, p 847 (1957).
8. G. E. van Gils, J. Appl. Polym. Sci., Vol. 13, p 835 (1969).
9. Y. Iyengar, J. Appl. Polym. Sci., Vol. 13, p 353 (1969).
10. K. D. Albrecht, Rubber Chem. Technol., Vol. 46, p 981 (1973).
11. Y. Iyengar, Rubber World, Vol. 148, p 39 (1963).
12. S. Van der Meer, Rubber Chem. Technol., Vol. 18, p 853 (1945).
13. A. Greth, Angew. Chem., Vol. 51, p 719 (1938).
14. R. P. Lattimer et al., Presentation at ACS Rubber Div. Meeting, Dallas, TX, April 19 (1988).
15. R. P. Quirk, Presentation at ACS Rubber Div. Meeting, Cleveland, OH, October 6 (1987).
16. R. A. Knowroski, J. P. Shocker, E.G. Gregg and J. L. Savoca, Rubber Chem. Technol., Vol. 59, p 328 (1986).
17. G. E. van Gils, Ind. Eng. Chem., Prod. Res. Div, Vol. 7, p 151 (1968).
18. G. Xue, J. Macromol. Sci. Chem., A-24, p 1107 (1987).

## 6.2 Bonding Nylons to Rubber Compounds

### 6.2.1 Nylon Chemistry

Aliphatic polyamides, also known as nylons, have a wide range of industrial applications and are manufactured in large quantities. Even though a wide variety of polyamides are available for various uses, the two polyamides, namely nylon-6 and nylon-66, have found applications in the tire and hose industries.

Nylon-6 has been synthesized from the ring opening polymerization of caprolactam according to the following reaction scheme (Figure 6.11).

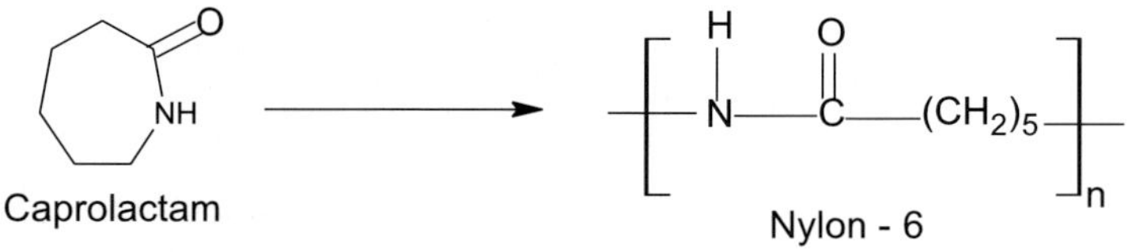


**Figure 6.11.**

This polymer is semi-crystalline in nature and has a melting point of 220 °C and glass transition temperature of 47 °C. Nylon-6 is most commonly used in tire cords.

Nylon-66 is also a semicrystalline polyamide and commonly used in fiber applications such as tire cords. Unlike nylon-6, this polyamide is synthesized from the condensation polymerization reaction between hexamethylene diamine and adipic acid. The following reaction scheme illustrates the synthesis (Figure 6.12).

**Figure 6.12.**

Nylon-66 has a melting point of 255 °C and glass transition temperature of 50 °C.

### 6.2.2 Application of Nylon-6 and Nylon-66 in Tire Cords

Fiber reinforced rubber composites have been widely used in industrial products such as tires, belts and hoses. These rubber products undergo repeated deformations and particularly require resistance to fatigue from flexing. In pneumatic tires, the flex-life of the carcass is one of the most critical factors affecting the life of the tire. Long flex-life is dependent not only upon the inherent flexibility of the carcass structure but also the strength of the bond between the tire fabric or fiber and adjacent rubber. Materials typically used to prepare the carcass cords for radial tires are polyester, rayon or nylon multi-filament yarns. With nylon, the two kind of monofilaments suitable as the reinforcement of carcass ply or plies are poly (hexamethylene adipamide) (nylon-66) and poly (caproamide) (nylon-6). The strength and toughness properties of nylons make them ideal for their applications in the tire industry.

### 6.2.3 RFL Chemistry and Nylon Bonding

Nylons contain amide (-CO-NH-) groups in the polymer main chain or backbone. These amide groups are responsible for the adhesion of nylon to carcass ply rubber compounds through an RFL. In the preparation of an RFL, either by the in situ or pre-condensed RF resin method, methylol groups are produced and stabilized in the presence of latex and alkaline conditions. The reaction of a methylol group with an amide group is well known. Therefore, a primary chemical bond from the reaction of resorcinolic methylol and nylon amide group can be expected (Figure 6.13).

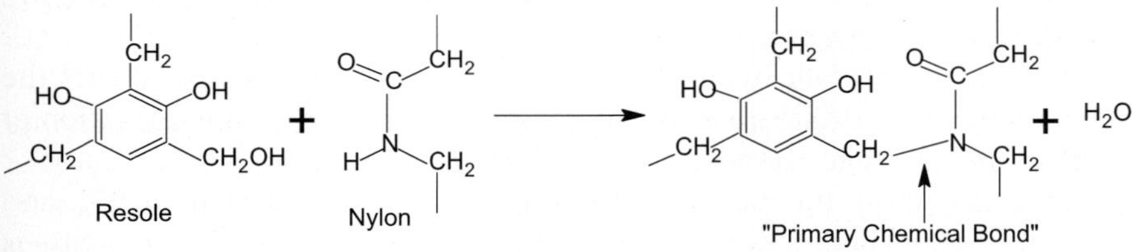


**Figure 6.13.**

This reaction can take place only when a resorcinolic methylol can penetrate the nylon surface and reacts. According to a published report, the solubility parameter ($\delta$) of resorcinol is 15.9 and that of nylon-6 or nylon-66 is 16 [1]. Since these two solubility parameter closely match, both nylon and resorcinol or methylolated RF resins are much more compatible for the penetration and chemical interaction. It was also reported that resorcinol itself can swell polyamide fibers. In addition, the RFL treating conditions are well above the glass transition temperature (Tg) of polyamides. This increases the polyamide molecular chain mobility to interact with the components, especially the RF resins, in the RFL. Due to all these factors, a chemical reaction for the formation of a primary chemical bond between RF and nylon is mostly favored. Also, the resorcinolic hydroxyl groups can form hydrogen bonds with the nylon amide groups in two different path ways (Figure 6.14).

The formation of strong hydrogen bonds strengthens the nylon fiber and RFL interface more effectively. Adhesive systems containing high resorci-

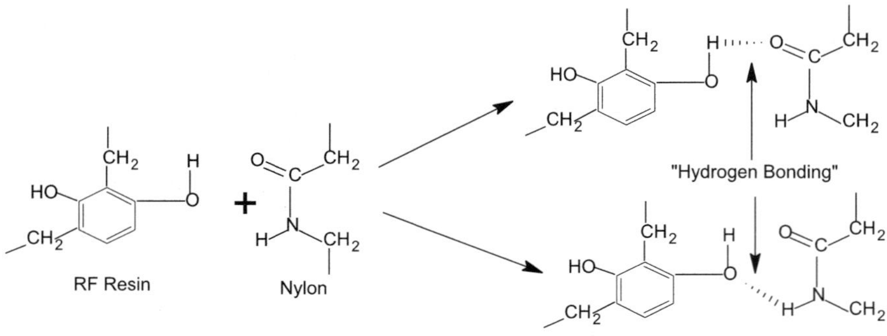


**Figure 6.14.**

nol content were observed to develop higher adhesion values due to an increased number of hydroxyl groups for more hydrogen bonding with the nylon amide groups. Primary chemical and secondary chemical bonds, therefore, strengthen the fiber and RFL interface by using standard RFL formulations and treating conditions.

When considering an interaction between RF and latex, it may be quite possible that a chemical bond could be formed from the methylol resorcinol and the latex's carbon–carbon double bond. As a result, a chroman type structure can be expected from the reaction.

Also, the resorcinolic hydroxyl groups can form a hydrogen bond with the nitrogen of the pyridine nucleus. Therefore, one can expect the formation of covalent and hydrogen bonds between RF resin and latex.

For determining the nature of bonding mechanism between an RFL and the rubber, a detailed understanding of the latex composition and structure is important. The vinyl pyridine latex, normally, is made from 70% butadiene, 15% styrene and 15% 2-vinyl pyridine monomers. Therefore, one can expect higher levels of un-saturation (C=C bond) in the copolymer structure. After the RFL dipped fibers are heat treated, even after the chroman ring formation of some unsaturated groups with RF, the latex still may contain a fair amount of unsaturation. This is because, in the presence of basic conditions, the alkylation type reaction of an unsaturated compound with resorcinol (an alkyl resorcinol) is less likely. With the presence of carbon–carbon double bonds in the rubber latex, a chemical bond associated with the sulfur cross-linking can also be expected between the latex and rubber during the vulcanization process.

The formation of covalent and hydrogen bonds between the RFL, nylon fiber and rubber matrix with the RFL adhesive system clearly indicates that this is an excellent adhesive for nylon to rubber adhesion. Previous studies had shown that the adhesive strength of nylon-rubber was due to the specific contribution of different bonding mechanisms (Table 6.6) [2, 3].

To determine the strength of the bond between the RFL treated nylon and natural black rubber compound, a strap peel adhesion test was performed

**Table 6.6.** Contribution of different mechanisms for nylon bonding with RFL adhesives

| Mechanism | Contribution (%) |
|---|---|
| Direct mechanical bonds | 15 |
| Diffusion | 5 |
| Primary chemical bonds | 60 |
| Hydrogen bonds | 20 |

Data from [2, 3]

after immersing the samples in water for 72 h [2]. From the test results, a 95% adhesion was retained for nylon. This indicates that the contribution from the hydrogen bonding mechanism is smaller when compared to covalent bonding. An adhesion evaluation of heat, steam or water aged samples of cured fiber reinforced rubber compounds could provide valuable information on the nature and strength of adhesive bonds which, in turn, determine the durability of rubber products for prolonged applications.

Based on experimental observations and results, a possible mechanism for the bonding of RFL dipped nylon fibers to natural rubber compounds can be proposed as follows (Figure 6.15).

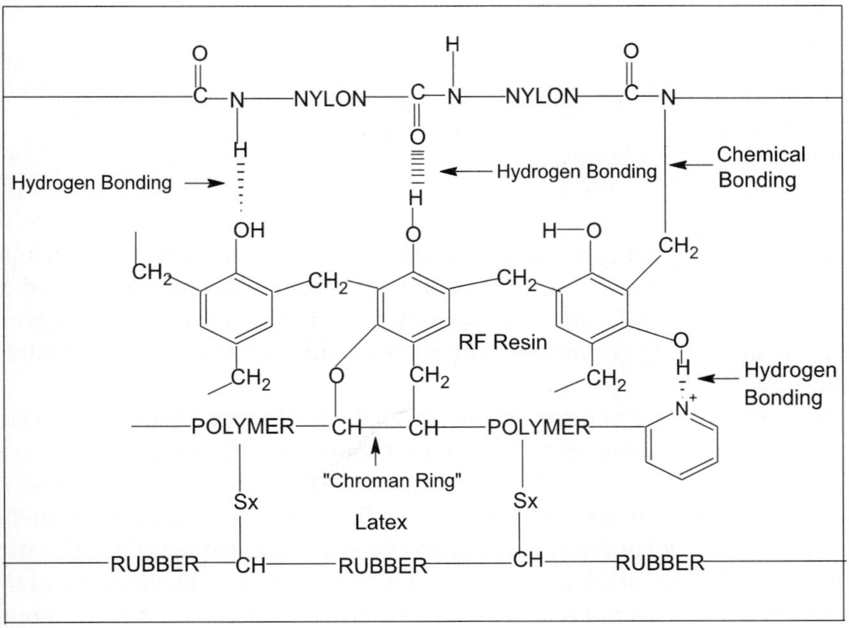


**Figure 6.15.** Possible bonding mechanism for nylons to rubber compounds

### 6.2.4 Rubber Compounding Test Results

A conventional RFL formulation to treat the nylon fibers or tire cords is shown in Table 6.7.

**Table 6.7.** Typical Nylon-6 and Nylon-66 RFL dip formulation and properties

| Resin solutions | Preformed resin Wet (parts) | In situ resorcinol Wet (parts) |
|---|---|---|
| Penacolite R-2170 (75%) | 16.8 | |
| Resorcinol (R) | | 11 |
| Water (soft) | 176.2 | 178 |
| Sodium hydroxide (50%) | 0.9 | 0.9 |
| Formaldehyde (37%) (F) | 12.2 | 16.2 |
| Total | 206.1 | 206.1 |
| | Aging not required | Age 1 to 4 h |
| Final RFL Dip | | |
| Vinyl pyridine latex (41%) | 244 | 244 |
| Resin solution | 206.1 | 206.1 |
| Water (soft) | 72.2 | 72.2 |
| Ammonium hydroxide (28%) | 12 | 12 |
| Total | 534.3 | 534.3 |
| Nylon treating conditions | Temperature (°C) | Dwell time (s) |
| Drying oven | 135 | 120 |
| Curing oven | 200 | 90 |
| Tempering oven | RT | 12 |
| Dip pickup (%) | 5 to 6 | |

The treated nylon cords were embedded in a rubber compound (composition given in Table 6.9), cured and adhesion properties were measured for the untreated and RFL treated cord embedded samples at room temperature and also after aging conditions. The results are summarized in the following Table 6.8.

As can be seen from the results in Table 6.8, the RFL treatment improved the nylon and rubber compound adhesion. Though the adhesion was improved considerably between the nylon and rubber, the 80 to 90% rubber coverage on the nylon cords after the heat and aged conditions clearly shows that further improvement could be made to obtain the highest strength possible. This can be achieved by the use of resorcinol and resorcinolic resins as the bonding agents in the rubber compound formulations. An example of one rubber compound formulation is shown in Table 6.9 [4, 5].

The adhesion test data are shown in Table 6.10 [4, 5].

Nylon cords rubber coverage (100%) and high aged adhesion values obviously show that the bond between the nylon fibers and rubber compounds were highly strengthened. Therefore, with the use of single dip RFL formulations, nylon fibers or cords can be effectively bonded to rubber compounds.

**Table 6.8.** Effect of RFL adhesives on nylon adhesion

| Adhesion property (ASTM 4776) | Newtons (% rubber coverage) |
|---|---|
| Untreated Nylon-6 (1260/2 cord) | |
| Unaged | 33.9 (5) |
| Steam aged 8 h at 120 °C | 29.5 (5) |
| Humidity aged, 7 days at 85 °C/95% RH | 21.2 (5) |
| RFL treated Nylon-6 (1260/2 cord) | |
| Unaged | 168.3 (90) |
| Steam aged 8 h at 120 °C | 160.5 (100) |
| Humidity aged, 7 days at 85 °C/ 95% RH | 109 (95) |
| Untreated Nylon-66 (1260/2 cord) | |
| Unaged | 32.6 (5) |
| Steam aged 8 h at 120 °C | 30.5 (5) |
| Humidity aged, 7 days at 85 °C/95% RH | 24.3 (5) |
| RFL treated Nylon-66 (1260/2 cord) | |
| Unaged | 143.7 (80) |
| Steam aged 8 h at 120 °C | 130.2 (90) |
| Humidity aged, seven days at 85 °C/95% RH | 93 (90) |

**Table 6.9.** Rubber compound containing bonding agent

| Composition | Parts by weight |
|---|---|
| 1. Natural rubber | 70 |
| 2. SBR | 30 |
| 3. Carbon black (N-660) | 50 |
| 4. Zinc oxide | 4 |
| 5. Stearic acid | 2 |
| 6. Naphthenic oil | 5 |
| 7. Antioxidant (DHTMQ) | 1.8 |
| 8. Sulfur | 2.5 |
| 9. 2,2′-Dibenzothiazyl disulfide | 0.8 |
| 10. Penacolite B-20-S | 3 |
| 11. Hexamethoxymethyl melamine | 2 |
| Cure properties (MDR rheometer, at 160 °C) | |
| $M_H$, dN-m | 12.1 |
| $M_L$, dN-m | 0.62 |
| $t_s$ 2, min | 3.3 |
| t′90, min | 23.8 |

**Table 6.10.** Adhesion improvements of RFL treated nylons using rubber compounds containing resorcinolic bonding agent system

| Adhesion property, Newtons (% rubber coverage) | Control compound | Penacolite B-20-S compound |
|---|---|---|
| Untreated Nylon-6 (1260/2 cord) | | |
| Unaged | 33.9 (5) | 164.4 (50) |
| Steam aged 8 h at 120 °C | 29.5 (5) | 177.4 (60) |
| Humidity aged, 7 days at 85 °C/95% RH | 21.2 (5) | 144.1 (95) |
| RFL treated Nylon-6 (1260/2 cord) | | |
| Unaged | 168.3 (90) | 199.2 (100) |
| Steam aged 8 h at 120 °C | 160.5 (100) | 181.6 (100) |
| Humidity aged, 7 days at 85 °C/95% RH | 109 (95) | 131.5 (100) |
| Untreated Nylon-66 (1260/2 cord) | | |
| Unaged | 32.6 (5) | 154.1 (60) |
| Steam aged 8 h at 120 °C | 30.5 (5) | 157.3 (70) |
| Humidity aged, 7 days at 85 °C/95% RH | 24.3 (5) | 142.3 (95) |
| RFL treated Nylon-66 (1260/2 cord) | | |
| Unaged | 143.7 (80) | 186.5 (90) |
| Steam aged 8 h at 120 °C | 130.2 (90) | 167.6 (100) |
| Humidity aged, 7 days at 85 °C/95% RH | 93 (90) | 121.1 (100) |

Data from Indspec Chemical Product Application Guide No. 2

## References


1. Y. Iyengar, J. Appl. Polym. Sci., Vol. 11, p 2311 (1967).
2. D. B. Wootton, Presentation at ACS Rubber. Div. Meeting, Houston, TX, October 24 (1983).
3. Th. G. F. Schoon and L. Zierler, Kaut. Gummi. Kunstst., Vol. 23, p 615 (1970).
4. Product Application Guide No. 1, Indspec Chemical Corporation (1999).
5. Products Application Guide No. 2, Dipping Systems, Indspec Chemical Corporation (1999).

## 6.3 Bonding Polyester to Rubber Compounds

### 6.3.1 Polyester Chemistry

Poly(ethylene terephthalate), also known as PET, is synthesized from the condensation reaction of ethylene glycol with dimethyl terephthalate (Figure 6.16).

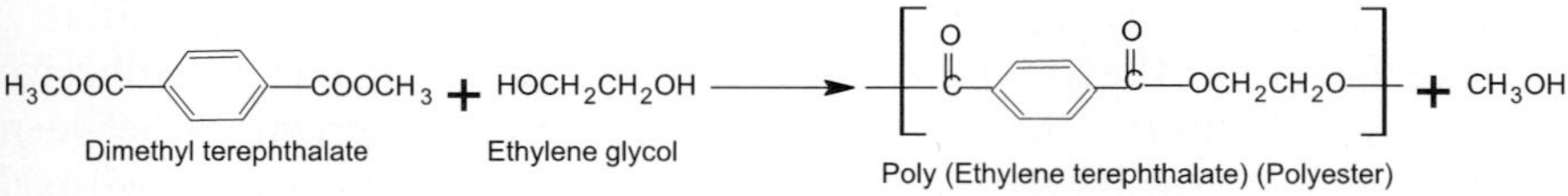


**Figure 6.16.**

This is a semi-crystalline material and has a melting point of 250 °C and glass transition temperature of 76 °C. This polymer (polyester) is primarily used in the packaging and fiber applications.

Polyester fibers have superior physical and chemical properties and are produced in large quantities. This material's high modulus and low elongation, particularly in the tire applications, reduce tire deformation and thus lead to better high speed performance and tread wear, reduced stress cracking and better steering characteristics.

Polyester is a very ideal material for use as the reinforcing material in the manufacture of various rubber articles including the tire. In the tire manufacturing, polyester has been widely used as body ply fabric in bias, belted bias and radial passenger tires.

### 6.3.2 Application of RFL in Polyester to Rubber Compounds Adhesion

Though polyester is an excellent reinforcement material to be used in carcass ply rubber compounds, this fiber has the difficulty of achieving a good bond with rubber. The poor bondability of polyester fibers with rubber is attributed to the hydrophobic nature of fiber due to the limited number of reactive or polar hydrogen bonding groups available at the surface. Unlike nylons, where the amide groups are located on the backbone of the polymer main chain, polyester fibers have the reactive hydroxyl and carboxylic acid functional groups at the end of polymer molecules. These functional groups are also available in limited numbers due to high molecular weight of this polymer. Therefore, this polyester behaves as an inert chemical compound. Consequently, when a simple or single adhesive is applied to polyester surface to improve the bond to rubber, the results are not good due to poor compatibility.

Molecular inter-diffusion between the adhesive and substrate, namely the fiber or rubber, is an important factor in the adhesion development of polyester material. Diffusion and subsequent interactions are much more favored when the substrate and adhesive are thermodynamically compatible. This can be possible only when the solubility parameter of these two matches closely. In the case of RFL, the resorcinol component has the solubility parameter of 16 ($\delta = 16$). The presence of two strongly hydrogen bonding polar hydroxyl groups in the resorcinol molecule is responsible for the high solubility parameter ($\delta$) value [1].

For polyester, the solubility parameter is 10.3. Due to large difference in the solubility characteristics, both the polyester and RF resins are incompatible.

Therefore, when the RFL adhesive is applied to untreated fiber, the adhesive may not wet the fiber. This could result in no chemical interactions between the hydroxyl and carboxyl groups of polyester with the methylol or hydroxyl groups of resorcinol or RF resin. Unless a chemical reaction takes place between the fiber and the adhesive, no covalent bond formation can be expected. The covalent bond formation between the adhesive and fiber could be mainly responsible for maintaining high adhesion levels under various heating and humidity aged conditions.

The covalent bond of polyester fibers with an adhesive can be developed if the polymer has higher concentrations of carboxylic acid groups. Higher carboxylic acids content could provide maximum adhesion under milder treating conditions than with lower carboxyl containing polyester. The adhesion of PET can be improved further if they are treated well above their glass transition temperature ($T_g$). Under this condition, due to an increased configurationally degree of freedom of molecular chains, the fiber and RFL adhesive interactions are further enhanced. This can lead to higher adhesion, possibly through hydrogen bonding and other mechanisms.

### 6.3.3 Single Step RFL Formulation and PET Adhesion

An example of a single step RFL formulation used to treat the polyester fibers is given in the following Table 6.11 [2].

The treated polyester fibers or cords are embedded in an uncured rubber compound, and cured to $t'90$ at 160 °C. The adhesion test results are shown in Table 6.12 [2,3].

From the results of Table 6.12, it is obvious that RFL treated polyester has good unaged adhesion performance compared to the untreated material. But the adhesion and rubber coverage dropped dramatically under steam and humidity aged conditions. These results strongly suggest that the bond formed between the fiber and adhesive is not strong enough and may not be a chemical bond.

In order to explain the initial adhesion and the drop in adhesion values associated with steam and humidity aged conditions, a good understanding of the chemical structure of the adhesive and fibers is important. With RFL,

**Table 6.11.** Typical single step RFL dip formulation for polyester

| Resin solution | Preformed resin wet (parts) |
|---|---|
| Penacolite resin R-2170 (70% solids) | 25.3 |
| Water (soft) | 269.1 |
| Sodium hydroxide (50%) | 2.8 |
| Formaldehyde (37%) | 7.8 |
| Total | 305 |
| Final RFL dip | Aging not required |
| Vinyl pyridine latex (41%) | 244 |
| Resin solution | 305 |
| Water (Soft) | 36.9 |
| Caprolactam blocked MDI (50%) | 20 |
| Total | 605.9 |
| Resin solution solids (%) | 7.6 |
| Total solids (%) | 22 |
| Formaldehyde/resorcinol (mol) | 1.2 |
| (F + R)/Latex | 22/100 |
| Polyester treating conditions | Temperature (°C) |
| Drying oven | 110 |
| Curing oven | 225 |

Product Application Guide No. 1 (June 1999) from Indspec Chemical Corporation

**Table 6.12.** Single step RFL treated polyester adhesion

| Polyester adhesion (ASTM D4776) | Newtons, (% rubber coverage) |
|---|---|
| Untreated polyester (1000/2 Cord) | |
| Unaged | 28.5 (5) |
| Steam aged 8 h at 120 °C | 19.6 (5) |
| Humidity aged 7 days at 85 °C/95% RH | 16.6 (5) |
| RFL treated polyester (1000/2 Cord) | |
| Unaged | 121.1 (70) |
| Steam aged 8 h at 120 °C | 77.9 (20) |
| Humidity aged 7 days at 85 °C/95% RH | 73.5 (20) |

Product Application Guide No. 2 (September 1999) from Indspec Chemical Corporation

there may be enough evidence to show that the RF resin has methylol groups. In the RFL formulations a caprolactam blocked isocyanate is very often used, and has the following chemical structure (Figure 6.17):

The unblocking temperature for the caprolactam blocked isocyanate to liberate diphenylmethane diisocyanate (MDI) is in the range of 158 – 162 °C [4].

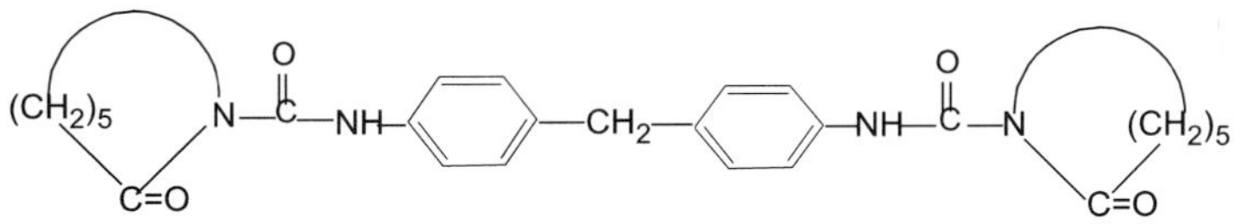


Caprolactam Blocked MDI

**Figure 6.17.**

Therefore, under the normal RFL preparation and drying conditions, there may not be any chemical interactions or reaction between the blocked isocyanate and RF methylol groups. Since the methylol groups are very reactive even at the drying temperature conditions (about 110 °C), there is a high probability that these groups could react with the available reactive sites in the resorcinol molecule to form methylene bridged structures. In addition, they can also react with the C=C double bond present in the rubber portion of the latex molecule to form the "chroman ring" type structure through the formation of "quinone methide" intermediate. Due to these reactions, one might expect a cross-linked RF novolak resin structure formed on the polyester fiber surface. This might result in a weak hydrogen bond formation between the polyester and resorcinolic hydroxyl groups (Figure 6.18).

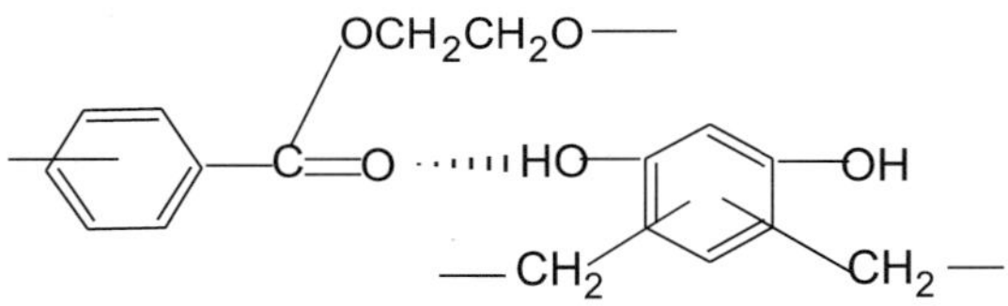


Hydrogen Boding of PET with RF Resin

**Figure 6.18.**

The formation of a chemical bond between the carboxylic acid groups present at the fiber surface and resorcinolic hydroxyl groups may be very difficult in the presence of basic conditions.

After the drying of RFL treated fibers, the heat treating temperature reaches as high as 225 °C. At this temperature conditions, the blocked isocyanate can unblock and, therefore, the regeneration of diisocyanate, namely the MDI, and caprolactam molecules from the blocked isocyanate can occur. MDI is a highly reactive molecule and can undergo the following type of reactions at higher temperatures.

### 6.3.3.1 Dimerization of Isocyanates

Aromatic isocyanates are known to dimerize and produce "uretdione" type structures [5, 6] (Figure 6.19).

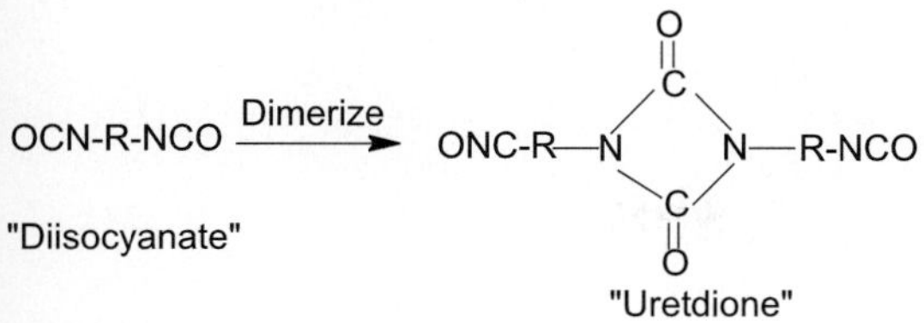


**Figure 6.19.**

The known catalysts used in the dimerization of isocyanates are amines, such as pyridine. Therefore, the presence of pyridine nuclei in the rubber latex may act as a catalyst for this type of reaction. Uretdione is not a stable structure. This can decompose at high temperatures to liberate isocyanate group again (Figure 6.20).

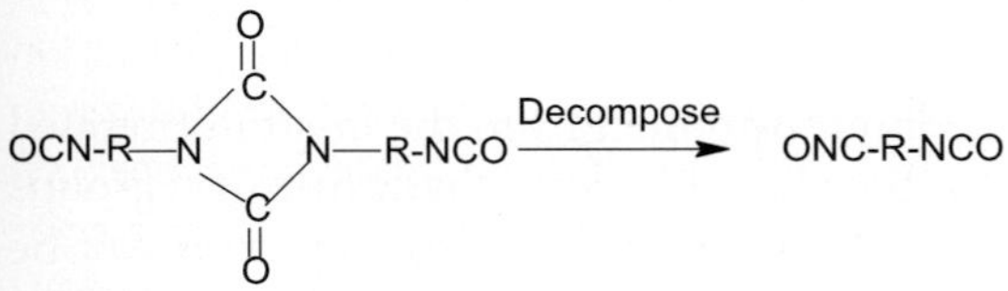


**Figure 6.20.**

### 6.3.3.2 Trimerization of Isocyanates

Isocyanate groups can also undergo trimerization reactions to produce "isocyanurate" type structures [7,8] (Figure 6.21).

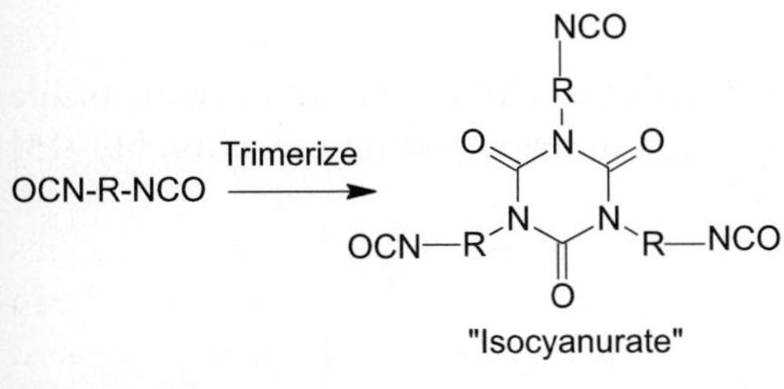


**Figure 6.21.**

### 6.3.3.3 Formation of Carbodiimides from Isocyanates

A very interesting type of reaction that the isocyanate group can undergo is the formation of a carbodiimide type structure based on the following reaction scheme [9–11] (Figure 6.22).

R-NCO $\overset{\text{Heat}}{\rightleftharpoons}$ R-N=C=N-R + $CO_2$

Carbodiimide

**Figure 6.22.**

Similar to the mono-isocyanate reaction with MDI, polymeric carbodiimide products can be expected to be formed. Depending upon the nature of catalyst, reaction time and temperature, variety of products with different molecular weights having a very low volatility could be formed from the MDI monomer (Figure 6.23).

OCN-R-NCO ⟶ OCN-[R-N=C=N-]n-R-NCO

MDI Monomer

**Figure 6.23.**

The formation of such a carbodiimide product from the in-situ generated isocyanate group can react with carboxylic acid and aliphatic hydroxyl groups present in the polyester molecules. With carboxylic acids, acylureas can be formed, which in turn can undergo decomposition to form an amide and isocyanate groups as shown in the following reaction scheme [8,12] (Figure 6.24).

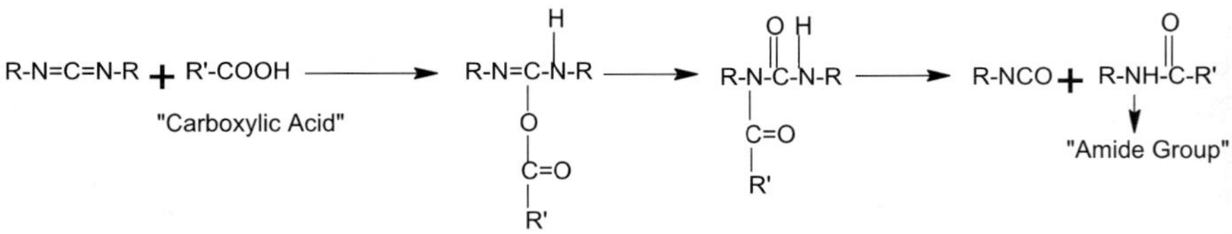


**Figure 6.24.**

Aromatic isocyanates can also react with carboxylic acids to form an amide compounds with the elimination of carbon dioxide from the reaction [13–15] (Figure 6.25).

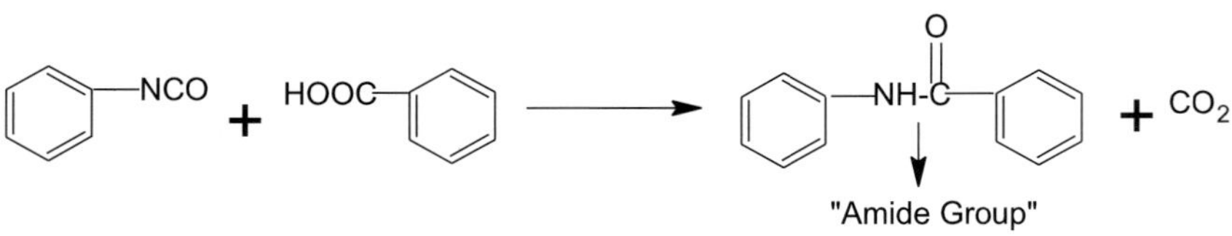


**Figure 6.25.**

The aliphatic alcohol can also react with an aromatic isocyanate to form a urethane type structure according to the following scheme (Figure 6.26).

For the reaction of aromatic isocyanates with an acid or alcohol, amines are the well known catalysts.

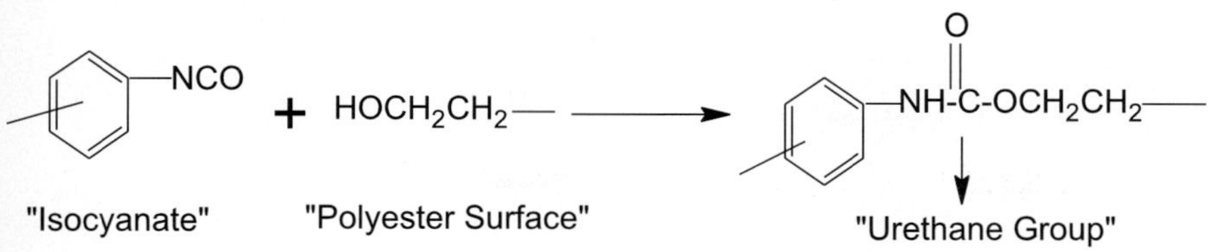


**Figure 6.26.**

Based on the above information, there may be a possibility that stable and very reactive carbodiimide type reaction products could be formed from the in-situ generated MDI from the caprolactam blocked isocyanate. If the MDI produces lower molecular weight carbodiimide and isocyanate products, as shown above, then they can diffuse into the polyester fiber surface and react with the terminal functional groups, such as the carboxylic acid and hydroxyl groups. This may result in the formation of an amide and urethane bonds and strengthen the interface between the fiber and the adhesive.

The caprolacatam liberated from the unblocking reaction can undergo a homopolymerization reaction to produce nylon-6 type products. Previous studies have shown that isocyanates and caprolactam blocked isocyanate can be employed as catalysts for the polymerization of caprolacatam [16].

In order to make the polyester fiber and adhesive interface stronger, the chemical reactions of terminal functional groups present in the polyester molecules with the adhesive should take place. With the use of caprolactam blocked MDI in the single dip RFL formulations, the formation of such chemical bonds between the polyester and adhesive is less likely. This might be the reason why the adhesion force and rubber coverage values dropped dramatically on steam and humidity aging of single step RFL dipped polyester material. These results clearly suggest that a physical, hydrogen bonding, and not chemical bonding mechanism is responsible for the adhesion of polyester with the single dip process.

### 6.3.4
### Chemistry and Mechanisms of Vulcabond-E Adhesion

Vulcabond-E, also known as Pexul or H-7, has been successfully used in RFL formulations to treat the polyester fibers, cords or fabrics for use in the manufacture of tires, belts and hoses [17, 18].

Vulcabond-E chemistry was developed from the reaction of *p*-chlorophenol, formaldehyde and resorcinol. The first step in the synthesis of Vulcabond is the reaction of *p*-chlororesorcinol with formaldehyde to produce the intermediate, namely 2,6-bis(hydroxymethyl)-4-chlorophenol (Figure 6.27).

This compound is a crystalline material with a melting point of 161 – 163 °C. Then, this was allowed to react with resorcinol at various molar ratios. The final

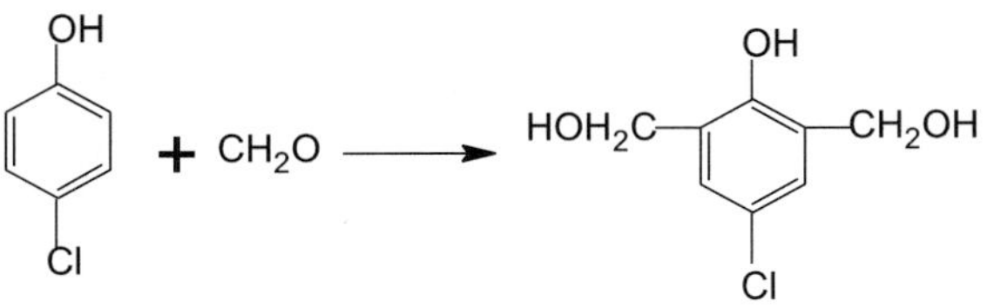

**Figure 6.27.**

reaction product, predominantly containing 2,6-bis-(2,4′-dihydoxyphenyl)-4-chlororesorcinol, was obtained based on the following chemical reaction (Figure 6.28).

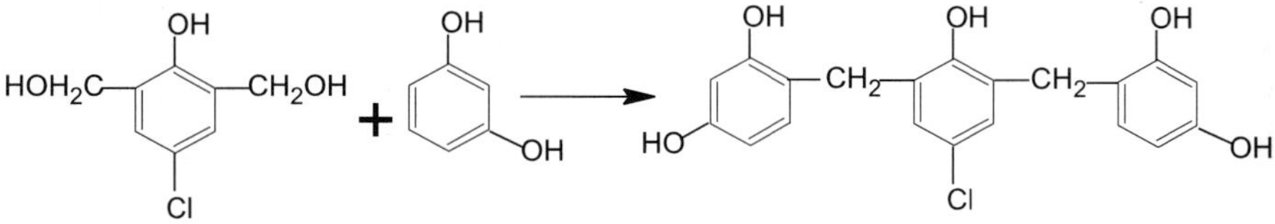

**Figure 6.28.**

Depending upon the molar ratios, fusible oligomeric products having different melting points in the range of 150 – 200 °C were produced. The resultant products are soluble in aqueous ammonia. Therefore, a dilute ammoniacal solution of Vulcabond-E was prepared and used as an additive in the RFL formulations.

A single step RFL dip formulation prepared using Vulcabond-E is presented in Table 6.13 [19].

**Table 6.13.** Single step RFL formulation using Vulcabond-E

| Component | Parts by weight |
|---|---|
| Penacolite R-2170 (70%) | 3.4 |
| Sodium hydroxide (50%) | 0.2 |
| Water | 18.8 |
| Vulcabond-E (20%) | 30 |
| Vinylpyridine latex (38%) | 31.8 |
| Water | 12.6 |
| Formaldehyde | 1 |
| Water | 1 |
| Total | 98.8 |

Data from [19]

This adhesive produced an acceptable level of adhesion determined from the pull-out force measured at 120 °C. Also, the adhesion values were found to be dependent on the molar ratios of resorcinol (R) and 2,6-bis(hydroxymethyl)-

**Table 6.14.** Adhesion performance of Vulcabond-E materials

| R/BHMCP | M.Pt. (or soft point) of condensate (°C) | Adhesion values, in lbs |
|---|---|---|
| 1.6/1.0 | 220–235 | 32.7 |
| 1.4/1.0 | 240–250 | 27.3 |
| 1.2/1.0 | 240–250 | 25 |

Data from [18]

4-chlororesorcinol (BHMCP) used in the synthesis of Vulcabond material (Table 6.14) [18].

It was believed that the adhesion mechanism of Vulcabond with polyester material was due to the sorption or penetration of the resorcinolic derivative into the polyester fiber surface. In the diffusion process, it has been well known that small molecules or lower molecular weight products can penetrate or diffuse faster, thereby enhancing the probability of contacting the functional groups. As can be seen from the results in Table 6.14, the higher softening point Vulcabond material, which has higher molecular weight in nature, showed lower adhesion values than the lower softening point material.

Infrared (IR) spectral studies made on the fiber surface indicated the carbonyl stretching frequency at 1722 $cm^{-1}$ (for the polyester) shifted to 1695 $cm^{-1}$ and peak broadening [20]. This study suggested a hydrogen bonding mechanism for the Vulcabond-containing RFL adhesive. In addition, based on this study, the resorcinolic, and not the phenolic, hydroxyl groups are responsible for the hydrogen bond formed between the polyester and the adhesive as shown in the following figure (Figure 6.29).

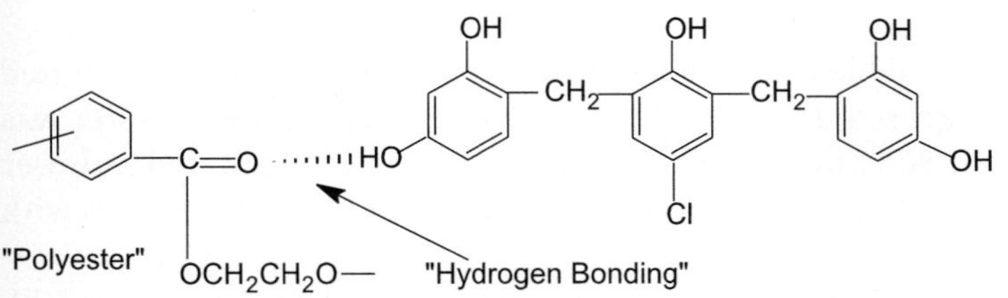


**Figure 6.29.**

Another study also indicated a hydrogen bonding mechanism between the Vulcabond-E based adhesive and the polyester [41]. In this study, the Vulcabond treated polyester cords were immersed in water for 72 h. The adhesion test made on the water immersed polyester samples showed greater loss of adhesion. Since water molecules would replace, reversibly, the adhesion forming hydrogen bonds, the loss in adhesion suggest a larger contribution from the secondary chemical bond, such as hydrogen bonding, for this adhesion system.

### 6.3.5 Resorcinolic Derivatives in Single Dip RFL Formulations

The major advantage of using resorcinol in RFL is its reactivity towards the formaldehyde to produce methylol groups at room temperature and the ability to form a permanent and stable bond with latex and rubber compounds. Since the solubility parameter of resorcinol ($\delta = 16.0$) is much higher than of the polyester ($\delta = 10.3$), both polyester and resorcinol are not compatible with each other. The high solubility parameter of resorcinol is due to the two hydroxyl groups present in the molecule, which are hydrophilic in nature. If the hydrophilic character of resorcinol is reduced then this will lower the solubility parameter of the resorcinol molecule. One such modification is to alkylate the resorcinol. The alkyl substituted resorcinols have been shown to exhibit a lower solubility parameter compared to resorcinol (Table 6.15) [1].

**Table 6.15.** Solubility parameters of mono-substituted resorcinol derivatives

| Substitution | Water solubility | Solubility parameter, $\delta$, $(cal/cc)^{1/2}$ |
|---|---|---|
| Resorcinol | Soluble | 15.9 |
| 2-Methyl | Soluble | 15.9 |
| 4-Hexyl | Insoluble | 12.5 |
| Octyl | Insoluble | 11.5 |
| 4-$COCH_3$ | Insoluble | 11.5 |
| 4-$COCH_2CH_3$ | Insoluble | 11.5 |

Data from [1]

As can be seen in Table 6.15, when the alkyl chain length of the substituted resorcinol was increased, the solubility of resorcinol derivative in water was decreased, and the solubility parameter was observed to be lowered. A lower solubility parameter can enhance the interaction of resorcinol derivatives with polyester material.

In order to observe the effect of replacing resorcinol in the single step RFL formulation with a resorcinol derivative having lower solubility parameter, *n*-hexyl resorcinol was substituted and tested for polyester adhesion performance. Hexyl resorcinol-formaldehyde-latex adhesive formulation containing a high resin/latex rubber ratio showed 60% higher adhesion than the conventional RFL formulation. This result strongly suggests that the wetting characteristics of the RFL adhesive system could be enhanced by the modification of resorcinol molecule. It will be important and also advantageous if the modified resorcinolic derivative or resin is either water soluble or dispersible to use in the RFL formulations to enhance the adhesive-fiber interactions.

### 6.3.6
### Adhesion Difficulties of Single Step RFL Dip for Polyesters

The major problem in achieving good adhesion of polyester with single step RFL formulations is the ability to establish a stable and permanent covalent bond between the RFL and polyester. The RF resin and the additives present in the RFL adhesive, either alone or in combination, do not chemically interact with the carboxyl and/or hydroxyl groups present in the polyester fiber. Unless chemical bonding or reaction takes place, the interface between the polyester and adhesive may not be stronger. This will ultimately result in adhesive-fiber bond failure. Since the single step RFL adhesive dip did not adequately provide sufficient fiber-adhesive bond strength, a double step was developed. In the double dip process, the second step or dip (also called top coat) always contained an RFL formulation. The first step (called the sub-coat) of the double dip process is to activate the polyester surface either physically or chemically so that it can interact effectively with the RFL top coat formulations.

### 6.3.7
### Different Approaches in Double Dip Adhesives for Polyesters

#### 6.3.7.1
#### Epoxy Resin in the Sub-Coat Formulation

Epoxy groups present in the epoxy resins are known to react with aromatic carboxylic acids to form the following structure (Figure 6.30):

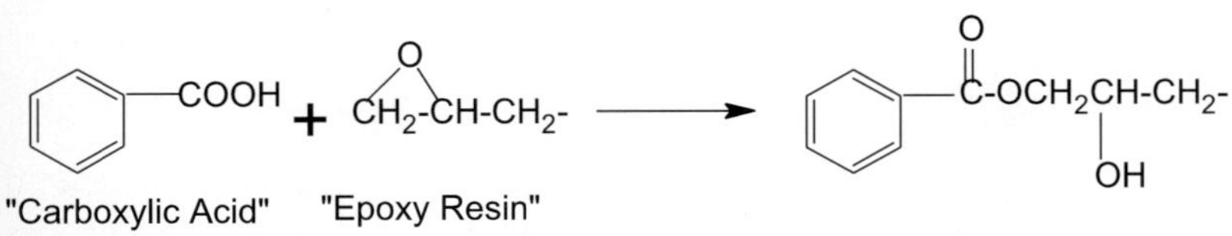


**Figure 6.30.**

Therefore, water soluble or dispersible aliphatic and aromatic poly-epoxy compounds are used in the sub-coat to react and activate the polyester fibers for further interaction with the RFL [21]. In this type of processes different oils or emulsifiers are used not only to disperse the epoxy but also to wet the fiber surface for the epoxy resin to penetrate and interact with carboxylic acid groups. The terminal hydroxyl groups of the polyester molecules most likely may not take part in any chemical reaction with the epoxy groups.

After treatment, the treated polyester surface may contain un-reacted epoxy groups present in the sub-coat for further reaction. When the RFL is applied as the top coat, these epoxy groups now react with the resorcinolic hydroxyl groups to obtain the following type of reaction product (Figure 6.31).

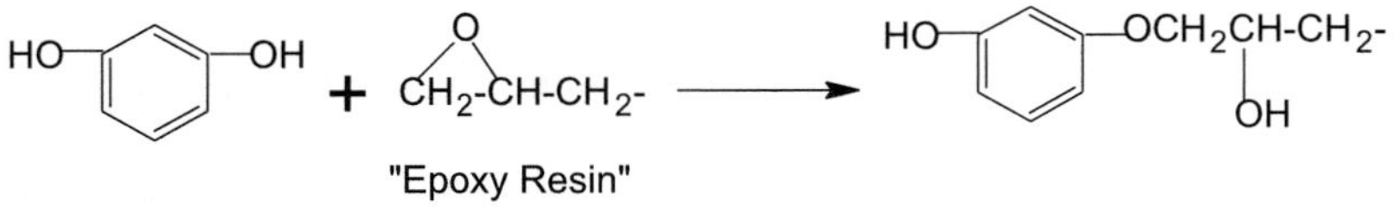


**Figure 6.31.**

In addition to the above reaction, the resorcinolic methylol groups can also react with the primary or secondary aliphatic hydroxyl groups present in the sub-coat (Figure 6.32).

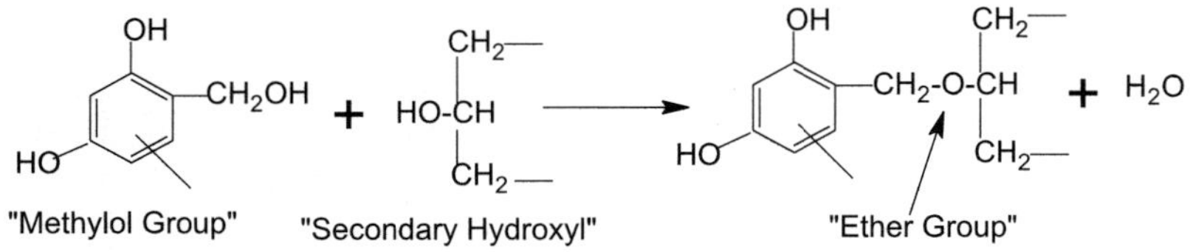


**Figure 6.32.**

The formation of these reaction products can strengthen the interface between the fiber and adhesive.

Aliphatic epoxy compounds such as the diglycidylether of glycerol, polyglycidylether of sorbitol and the like are known to be used in the polyester fiber treatment due to their very low volatility and possessing multifunctional epoxy groups to enhance the chemical reactions. Structures of glycidyl ethers of glycerol and sorbitol are shown below (Figure 6.33).

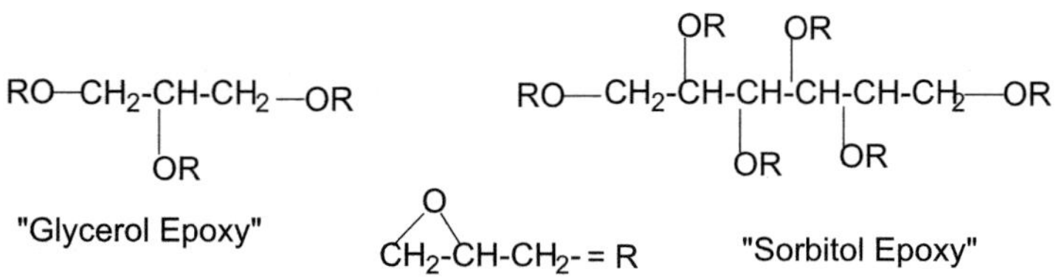


**Figure 6.33.**

## 6.3.7.2 Silane and Aliphatic Epoxy Resins in the Sub-Coat

Instead of using an aliphatic epoxy resins in the sub-coat formulations, extensive work has been carried out to apply these resins as a spin finish on the fiber surface after the fibers are spun and before drawing. In this process, typically, aliphatic epoxies, such as the glycerol or sorbitol, are sometimes mixed with a silane epoxy to enhance the fiber surface and RFL interactions [22–24].

Silane compounds are well known to wet surfaces of various kinds to protect and make reactions with various functional groups, including aliphatic hydroxyl groups. For this reason, the use of epoxy silanes in the sub-coat is expected to wet the polyester fiber surface, penetrate and react with the hydroxyl groups of polyester molecules according to the following schemes (Figure 6.34).

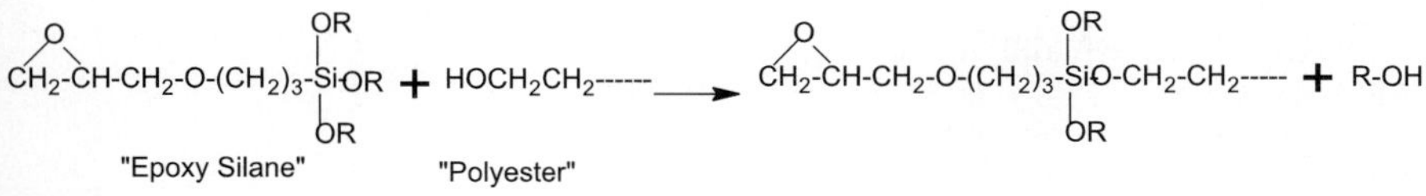


**Figure 6.34.**

Epoxy silanes used in the sub-coat can be expected to react simultaneously with the carboxyl and hydroxyl groups of the polyester to develop a chemical bond between the polyester surface and the sub coat. Thus, the sub-coat treated polyester fibers have more surface activity due to the presence of additional functional groups and capable of strengthening the polyester and RFL bonds through chemical and physical interactions. Polyester fibers or cords previously coated with an adhesive like the epoxy are called "adhesive activated" cords and are ready to treat with RFL adhesives.

### 6.3.7.3 Mixed Silanes in the Polyester Sub-Coat

An interesting technology was developed to treat the polyester surface with a mixture of silanes as a sub-coat before applying the RFL containing caprolactam blocked MDI. In this method, two silanes, namely aminoethylaminopropyl trimethoxysilane and methacryloxypropyl trimethoxysilane, having the following chemical structures, are mixed in the presence of water and then applied to the polyester fiber or fabric and dried at 177 °C [25, 26] (Figure 6.35).

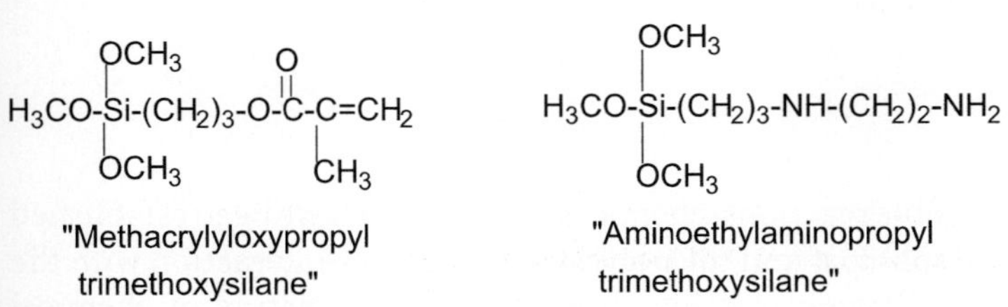


**Figure 6.35.**

The silanes coated polyester materials were then treated with RFL formulations containing a blocked isocyanate. The polyester fabric was embedded in a SBR rubber, cured and the peel adhesion data was obtained (Table 6.16).

**Table 6.16.** Adhesion performance of mixed silanes in the sub-coat

| Adhesive applications | Peel strength, lbs/in |
|---|---|
| First mixed silanes coat + second RFL coat | 36 |
| First mixed silanes coat alone and no RFL coat | 5 |
| No mixed silanes coat and RFL coat only | 6 |

Data from [25]

Adhesive failure was reported to be cohesive and occurred within the rubber. These results suggested that the bond strength developed between the polyester and RFL adhesive was stronger and could be explained based on the following possible mechanisms of interactions.

When the two silanes are mixed together, a possible chemical reaction takes place between the amino and vinyl groups to form the "Michael" addition type reaction products (Figure 6.36).

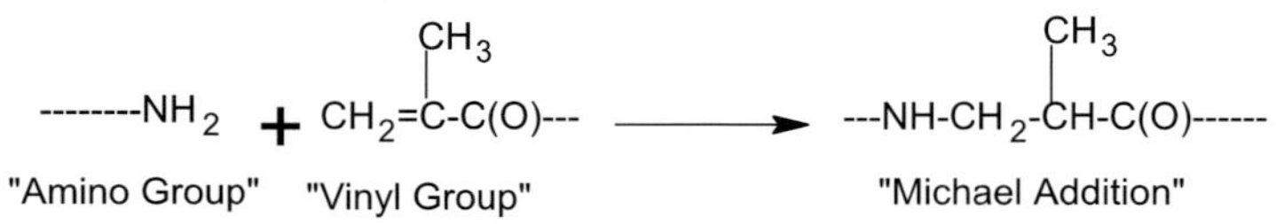


**Figure 6.36.**

On the application of mixed silanes to polyester, due to excellent wetting characteristics, the silane functional groups could have reacted with the hydroxyl groups of the polyester. Simultaneously, the polyester carboxyl group might be reacting with the amino group of the silane molecules (Figure 6.37).

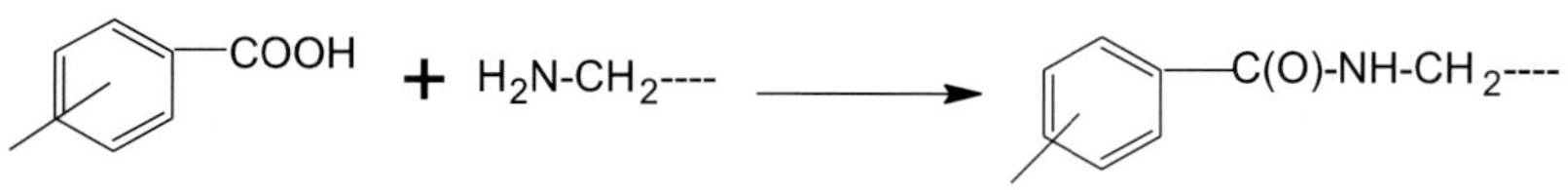


**Figure 6.37.**

Due to these possible reactions, chemical bonds might have been established between the silane sub-coat and the polyester. On further interaction with the RFL top coat, the following reactions leading to the formation of chemical bonds could be produced (Figures 6.38 and 6.39).

The formation of secondary amine, tertiary amine and urea groups can provide extensive hydrogen bonding with polyester fibers and RF resins, which could make the interface much stronger.

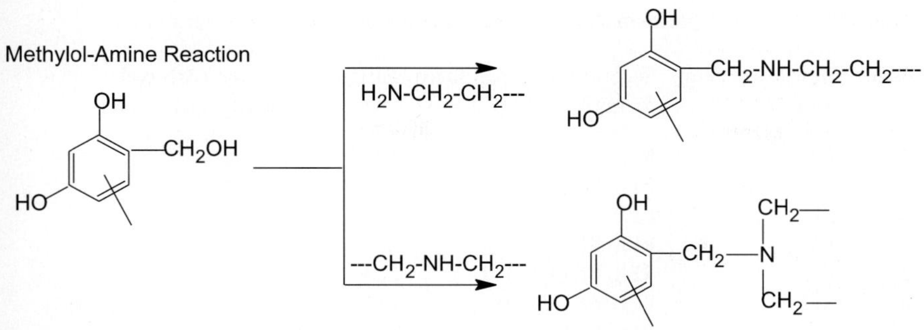


**Figure 6.38.**

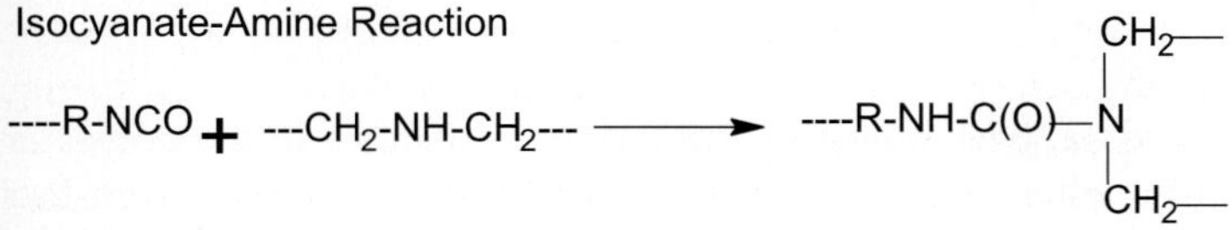


**Figure 6.39.**

### 6.3.7.4
### Isocyanate Chemistry in the Sub-Coat

Aromatic isocyanates alone can be used in the sub-coat to treat the polyester before the RFL for enhancing the adhesion. In a two-step process, polyester cords were first treated with polymethylene polyphenylene isocyanate (PAPI) dissolved in tri-chloromethane solvent and then dried at about 177 °C temperature [27,28].

The polyphenylene isocyanate (PAPI) has the following structure (Figure 6.40):

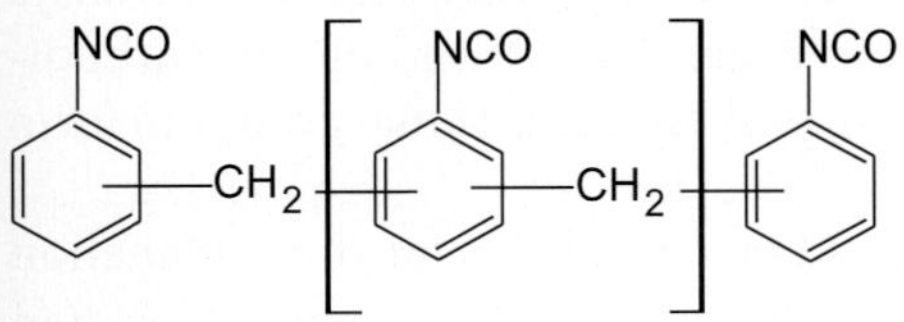


**Figure 6.40.**

The dried cords were then treated with an RFL solution containing RF blocked PAPI (BNCO) and dried at about 218 °C for $2\frac{1}{2}$ minutes. Then the cords were embedded in a natural rubber compound stock, cured and the peel adhesion and fatigue resistance properties were measured. The details are presented in Table 6.17.

**Table 6.17.** Adhesion performance of polyester cords on isocyanate treatment

| Expt. | First dip pickup, solids (%) | Second dip formulation | Peel force | Fatigue | Polyester carboxyl content, equiv./106 g |
|---|---|---|---|---|---|
| 1 | 4.8 | RFL Control | 100 | 100 | |
| 2 | 4.8 | RFL/BNCO | 102 | 112 | |
| 3 | 0.4 | RFL | 131 | 125 | |
| 4 | 0.4 | RFL/BNCO | 111 | 201 | |
| 5 | 0.25 | RFL/BNCO | 101 | 228 | |
| 6 | 0.12 | RFL/BNCO | 101 | 268 | |
| 7 | 0.4 | RFL/BNCO | 110 | 110 | 35 |
| 8 | 0.4 | RFL/BNCO | 100 | 100 | 18 |
| 9 | 0.4 | RFL/BNCO | 99 | 149 | 5 |

Data from [27]

From Table 6.17 data, peel adhesion and fatigue resistance of the polyester have been improved when the RFL adhesive contained isocyanate. From this study report, it was not known how the blocked isocyanate functioned to provide the unexpected high resistance to fatigue property measured under the flex conditions.

The effect of carboxyl group concentration on the peel adhesion and fatigue resistance is also shown in Table 6.17. This work theorized that the decrease in peel adhesion was due to a decrease in the carboxyl content associated with fewer sites available for the adhesive bonding.

In order to explain the unexpected results, careful understanding of isocyanate chemistry is necessary. Aromatic isocyanates are known to produce, at elevated temperatures, dimers (uretdiones), trimers (isocyanurates) and carbodiimide structures. Isocyanate dimers, though produced at high temperatures, can also decompose to liberate isocyanate groups again. The formation of trimers (isocyanurates) from the isocyanate monomers can produce highly cross-linked network type structures, which can be brittle in character. Isocyanurate production may be associated with the concentrations of isocyanate groups present in the system. On the other-hand, the formation of carbodiimide at higher temperature can react with polyester functional groups to form stable covalent type structures.

All of the reactions of isocyanate should be considered for the explanation of results obtained from this work.

The higher dip pick up during the first step (sub-coat) might have produced highly cross-linked isocyanurate network type structure on the fiber surface, and might have resulted in the lower peel adhesion. When the isocyanate dip pickup (first step) was low, the formation of carbodiimide type reaction product might have occurred. This, on further reaction with carboxylic acid, might have produced a stable amide type bond. The high fatigue resistance values may be associated with the phenomena.

The data from this work strongly suggest that chemical bonds, which were more stable and flexible, might have been established between the isocyanate and polyester end functional groups for the improved adhesion performance of polyester.

### 6.3.7.5
### Non-Epoxy Activation of Polyester Surface

Polyester fiber, after being synthesized, often contained higher levels of carboxyl end groups ranging from 30 to 40 meq./g. To improve the hydrolytic stability of polyester, various reactive chemical compounds such as the ethylene oxide, ethylene carbonate and polyglycidylether, are added to react with carboxyl group and reduce its content (Figure 6.41).

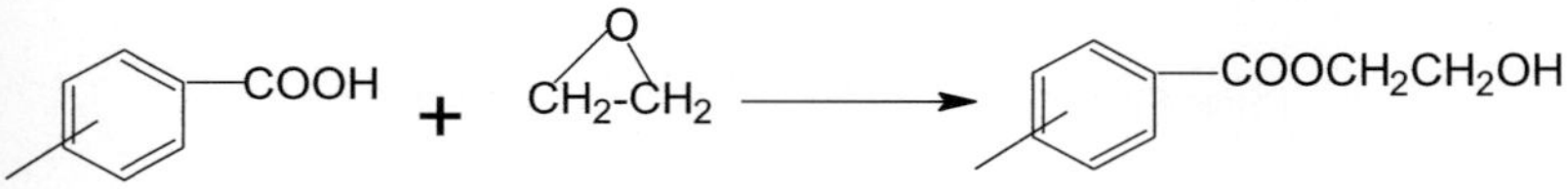


**Figure 6.41.**

By means of this treatment, the polyester is stabilized and the carboxyl content is typically reduced to about 10 to 18 meq./g [29,30]. The chemically stabilized polyester is now spun; drawn and two ply cords are made. Due to the introduction of more hydroxyl groups at the end of polyester molecules, the RFL treated cords appeared to improve the adhesion.

In order to improve the adhesion further, these chemically modified polyester fibers are again treated with an epoxy or silane epoxy to activate the surface for the RFL application. In the case of a spin finish of polyester fiber with an aliphatic epoxy resin, hazards (toxicity) associated with the handling of epoxy resins are realized. With an epoxy treatment, the chemistry is associated with the formation of an ester group from the carboxyl and epoxy reaction (Figure 6.42).

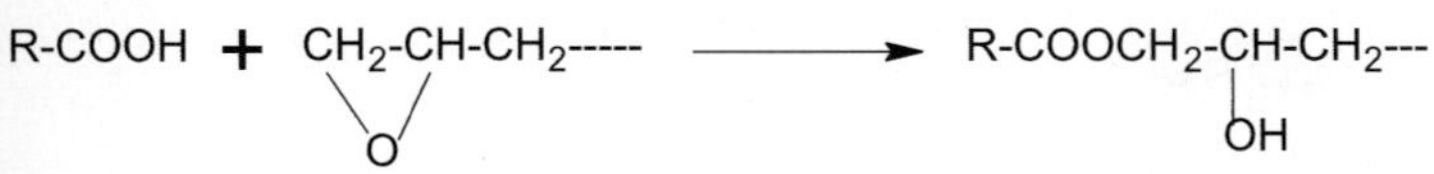


**Figure 6.42.**

To overcome the toxicity problem of epoxy resins, a non-toxic chemical that brings similar kind of chemistry has been developed. The non-toxic chemical developed and used in the activation of polyester fibers before the RFL treatment is based on the following reaction scheme [31–35] (Figure 6.43).

This halogenated polyhydroxy ether compound is considered non-toxic, if not less toxic compared to the corresponding epoxy compound, and contained predominantly hydroxyl groups with a very small amount of halogens present.

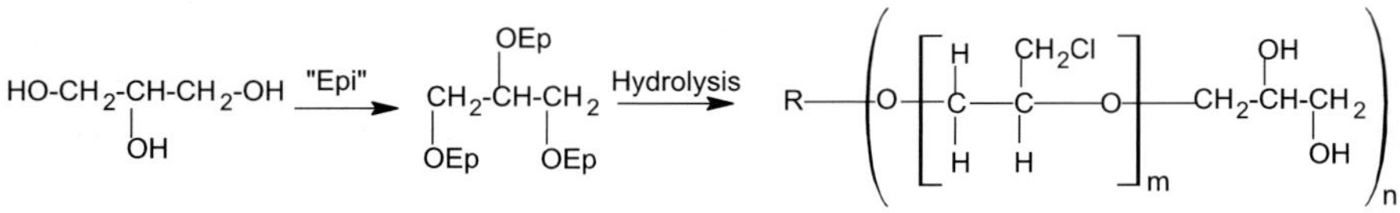


**Figure 6.43.**

A solution was prepared using this polyhydroxy compound and applied to the polyester fiber as a finish immediately after the spinning and before drawing. During the heat treatment in the drawing process, a possible chemical reaction can occur between the polyester carboxyl and the finish hydroxyl groups (Figure 6.44).

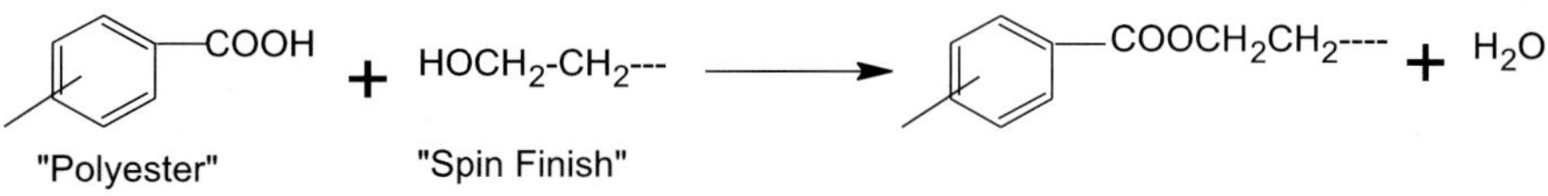


**Figure 6.44.**

The polyester fibers or cords thus prepared are called adhesive activated cords and are ready for the RFL treatment. The adhesive activated cords have more hydroxyl groups on the fiber surface. Therefore, a chemical reaction might be expected between the reactive functional groups present in the RFL formulations and the hydroxyl groups as outlined in the following schemes (Figures 6.45, 6.46 and 6.47).

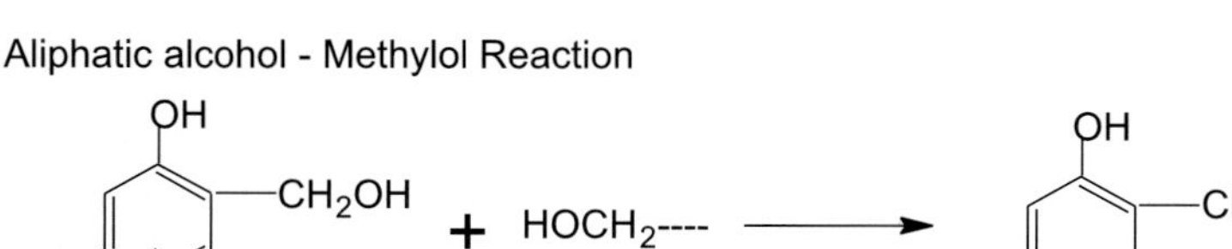


**Figure 6.45.**

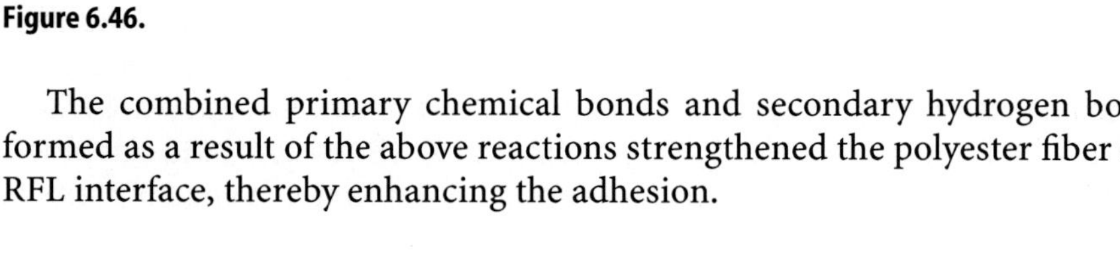


**Figure 6.46.**

The combined primary chemical bonds and secondary hydrogen bonds formed as a result of the above reactions strengthened the polyester fiber and RFL interface, thereby enhancing the adhesion.

Hydrogen Bonding

**Figure 6.47.**

### 6.3.7.6 Epoxy Finish with Epoxy and Blocked Isocyanate in RFL

Another study showed the use of an epoxy resin and blocked isocyanate together in the RFL formulation to enhance adhesion of an epoxy activated polyester cords [36].

In this process, epoxy activated polyester cords are treated with two different RFL solutions containing an epoxy and (epoxy and caprolactam blocked MDI) respectively. The two epoxy resins used in the RFL formulations are sorbitol epoxy (X-614 B from Nagase Chemical) and cresol-novolak epoxy (ECN-1400 from CIBA). After the treatment, the cords were embedded separately in two different rubber stocks of different modulus and tested for peel adhesion and rubber coverage. The results are summarized in Tables 6.18 and 6.19.

**Table 6.18.** Adhesion performance of RFL treated polyester cords in low modulus rubber stock

| Epoxy used | Epoxy concn. in RFL, (wt.%) | Isocyanate concn. in RFL, (wt.%) | Peel force (N) RT, (% rubber cov.) | Peel force (N) 106°C, (% rubber cov.) |
|---|---|---|---|---|
| Sorbitol | 9.1 | 2.3 | 249.8 (100) | 182.1 (100) |
| Sorbitol | 4.4 | NO | 246 (100) | 194.3 (100) |
| Cresol-novolak | 1.9 | 4.8 | 238.5 (100) | 198.7 (100) |
| Cresol-novolak | 3.9 | NO | 252.1 (100) | 196 (100) |

Data from [36]

The excellent adhesion and 100% rubber coverage data (Tables 6.18 and 6.19) clearly indicate that chemical bonds could have formed between the fiber and RFL adhesive. To propose a mechanism for the above adhesion, an understanding on the interactions between the substrate (fiber), RFL and the rubber is necessary.

Polyester yarn was activated with an epoxy adhesive. Therefore, a reaction of the polyester carboxyl with an epoxy group of the epoxy resin can be expected. During the RFL preparation, RF resin was formed first, and then the latex solution was added. Finally epoxy and blocked isocyanate additives were added to obtain the final RFL adhesive. Under these preparation conditions, it

**Table 6.19.** Adhesion performance of RFL treated polyester cords in high modulus rubber stock

| Epoxy used | Epoxy concn. in RFL, (wt.%) | Isocyanate concn. in RFL, (wt.%) | Stock cure 320 F/20 min. Peel force (N) (% rubber cov.) | Stock cure 320 F/60 min. Peel force (N) (% rubber cov.) |
|---|---|---|---|---|
| Sorbitol | 4.6 | 4.6 | 383.7 (100) | 325.8 (100) |
| Sorbitol | 4.9 | NO | 369.1 (90) | 311.1 (95) |
| Cresol-novolak | 1.9 | 4.8 | 436.4 (100) | 327 (100) |
| Cresol-novolak | 2 | NO | 409.3 (100) | 314 (100) |

Data from [36]

can be expected that the epoxy and blocked isocyanate functional groups are stabilized. After the application of RFL on the polyester cord and during the drying (at 359°F) and curing (at 475°F) conditions, an interaction between the RFL and fiber surface can take place. The epoxy functional group in the RFL might react primarily with the resorcinolic hydroxyl groups to form the hydroxyether type structures (Figure 6.48).

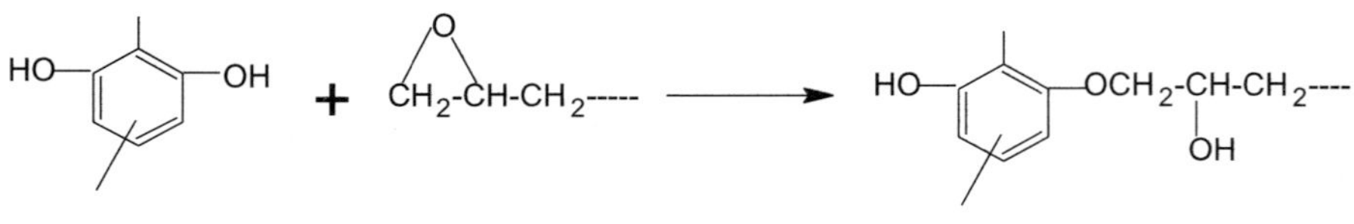


**Figure 6.48.**

The partially reacted epoxy (sub-coat) present on the polyester surface may also react with the resorcinolic hydroxyl groups to form the hydroxyether structure. On unblocking, the isocyanate group from the MDI either directly or indirectly, through the carbodiimide formation and its reaction, reacts with the secondary aliphatic hydroxyl groups to produce urethane type products. The methylol groups of the resorcinolic molecule interact, through the quinone methide intermediate, with the unsaturated portion of the latex molecules to produce chroman ring structures. Another possible reaction for the methylol groups is to react with epoxy groups present in the formulations.

Based on the above possible chemical reactions that might take place during the RFL treatment of polyester fibers and their subsequent cure with the rubber compounds, the following probable mechanism can be proposed for the polyester adhesion (Figure 6.49).

From this work, in order to obtain good adhesion between the polyester and adhesive, chemical bonds should be formed between all the components. With an epoxy activated polyester cords, good adhesion could be achieved with the use of multifunctional epoxy alone in the RFL without the blocked isocyanates.

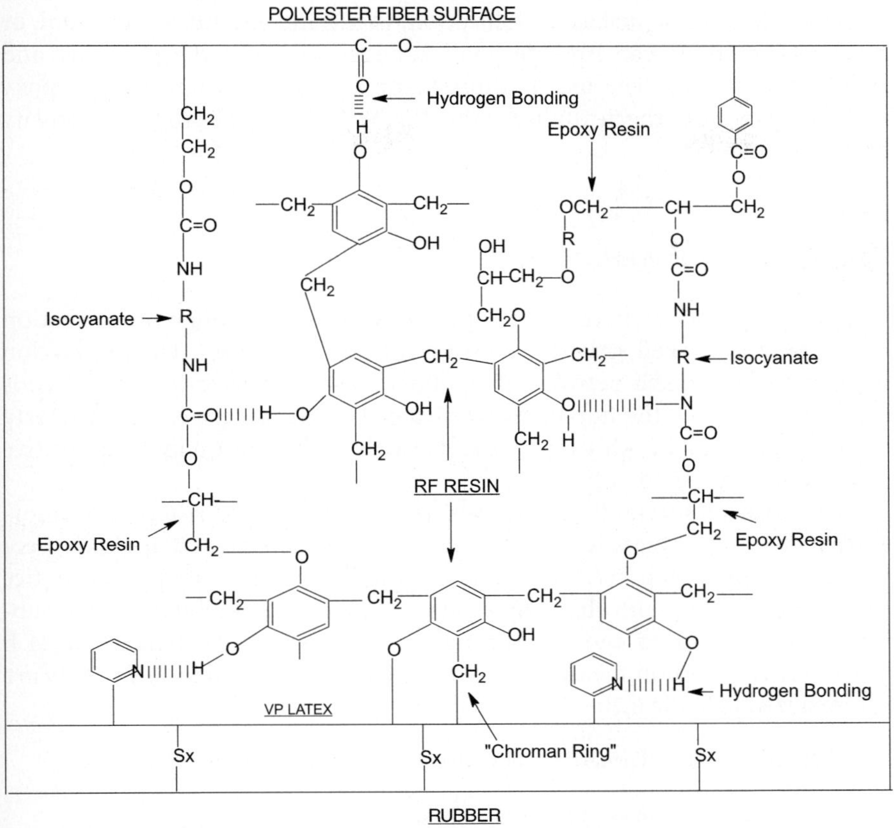


**Figure 6.49.** Possible mechanism for the bonding of polyester fibers with rubber compounds

If an epoxy silane is used in the activation of polyester yarn, the following reaction can be expected at the fiber surface [37] (Figures 6.50 and 6.51).

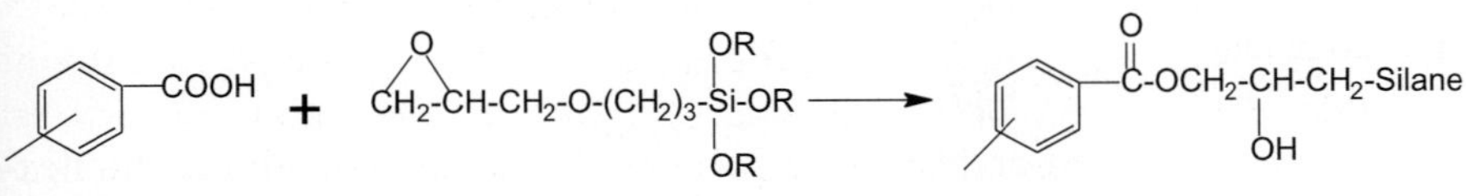


**Figure 6.50.**

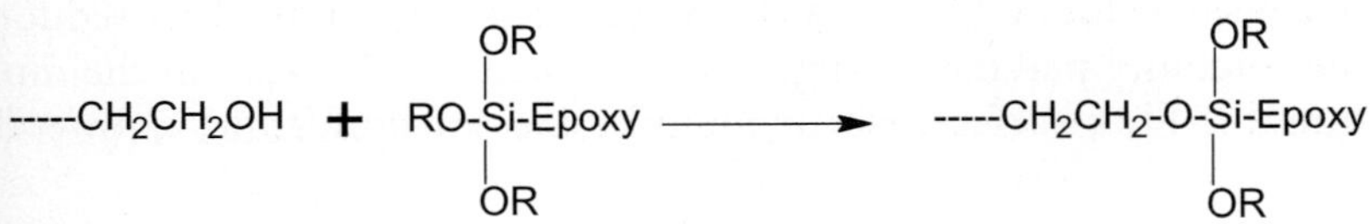


**Figure 6.51.**

With this kind of surface modification, the RFL-containing epoxy and/or blocked isocyanate can provide good adhesion between the polyester and rubber compounds. The use of aliphatic or higher molecular weight epoxy resins may improve the flexibility of the RFL by reducing the cross-link density of RF resins.

### 6.3.7.7 Flexibilizing RF Resin in RFL

The presence of RF resin in the RFL provides modulus through the formation of highly cross-linked structures during the curing process. The production of such a cross-linked network at the fiber surface or fiber-RFL interface is expected to reduce the flex-fatigue resistance of the adhesive. This property can be improved if a highly reactive and also flexiblized resorcinolic derivative can be employed in the RFL.

Flexibilized resorcinolic derivatives developed from the reaction of resorcinol with an allylchloride and other unsaturated aliphatic compounds were used in the single step RFL formulations [38]. The non-activated polyester tire cords were treated with this single step RFL, and then embedded in a rubber compound, cured and tested for the unaged and aged (heated for 24 h at 100 °C) H-adhesion measured at 100 °C temperature. Adhesion results are summarized in Table 6.20.

**Table 6.20.** Adhesion performance of flexibilized RFL treated polyester cords

| RFL formulation | Unaged adhesion, force (lbs) | Aged (100 °C/24 h) force (lbs) |
|---|---|---|
| RFL (control) | 14 | 13.6 |
| RFL dip prepared using resorcinol-allylchloride reaction product | 20.5 | 23.3 |

Data from [38]

In-spite of the non-activation of polyester cord, RFL containing the flexiblized resorcinolic derivative provided good adhesive strength, and maintained or improved its strength after heat treatment. The flexiblized resorcinol derivative has an alkyl substitution and, therefore, the solubility parameter might have been reduced considerably compared to resorcinol. This reduction in the solubility parameter might have allowed the RF resin in the dip to penetrate the fiber surface, which could be the reason for the improved adhesion.

In response to improve the RF resin flexibility, several alkyl substituted resorcinol-formaldehyde resin solutions are made from resorcinol, unsaturated alcohol and formaldehyde reactions and used in the place of RF resin solution in the RFL formulations [39]. With allyl alcohol, resorcinol and formaldehyde as the reactants, the following reaction products are produced (Figure 6.52).

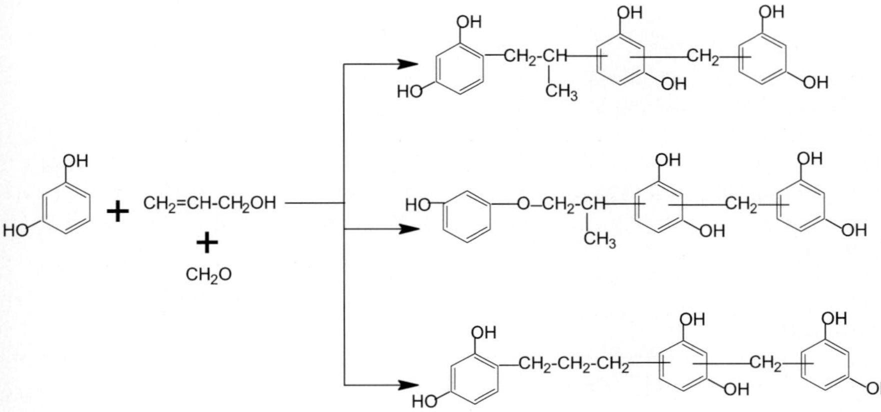


**Figure 6.52.**

Similarly, flexibilized resorcinolic resins are prepared using 1,4-butene-diol and formaldehyde. The structures of flexibilized RF resin obtained from the reaction are outlined in the following reaction scheme (Figure 6.53):

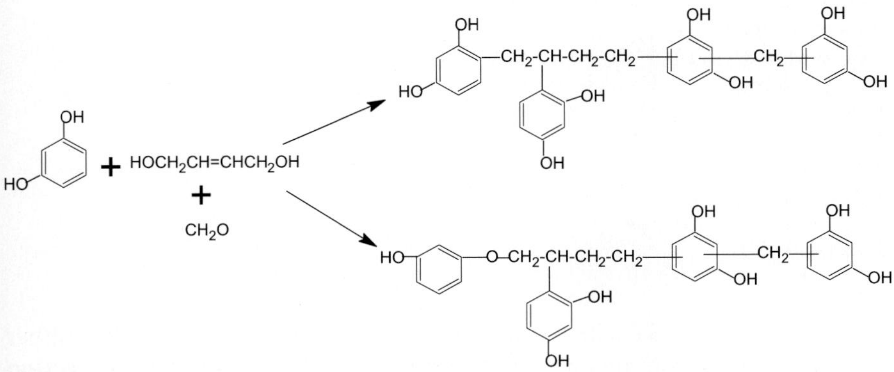


**Figure 6.53.**

Flexibilized RFL formulations prepared using the resorcinol, butenediol and formaldehyde reaction products are shown in Table 6.21.

Adhesive activated (epoxy) polyester tire cords are treated with the flexibilized RFL formulation and tested for H-adhesion performance. The adhesion results are presented in Table 6.22.

Compared to the RF resin (control), the flexibilized RF novolak based RFL showed improved adhesion values. This might be due to improved compatibil-

**Table 6.21.** RFL formulation prepared using resorcinol-butenediol-formaldehyde reaction product

| Components<br>resin solution (weight in grams) | R-2170 (control) | Butenediol-RF resin |
|---|---|---|
| Water | 269.1 | 251.2 |
| Sodium hydroxide (50%) | 2.8 | 2.8 |
| Penacolite resin R-2170 (75%) | 25.3 | NO |
| Butenediol-RF solution (50%) | NO | 40.1 |
| Formaldehyde (37%) | 7.8 | 10.8 |
| Latex mix | | |
| Vinylpyridine SBR latex (41%) | 244 | 244 |
| Water | 36.9 | 36.9 |
| Total parts | 585.9 | 585.9 |
| Final RFL solution properties | | |
| Resin solution solids | 7.6 | 8.5 |
| Total solids | 21 | 21.5 |
| pH | 9.3 | 10 |

Data from [39]

**Table 6.22.** Adhesion performance of flexibilized RFL formulation

| Adhesion test | R-2170 (control) | Butenediol-RF resin |
|---|---|---|
| Unaged (3/8″ mold, 160 °C cure) | | |
| Adhesive force, N | 67.4 | 102.8 |
| Rubber coverage (%) | 7 | 40 |
| Heat aged (3 days at 100 °C) | | |
| Adhesive force, N | 50.1 | 80.6 |
| Rubber coverage (%) | 5 | 10 |
| Humidity aged | | |
| Adhesive force, N | 54.7 | 78.7 |
| Rubber coverage (%) | 5 | 20 |

Data from [39]

ity of the flexibilized RF with the adhesive activated polyester cords. The rubber coverage observed on the unaged polyester also indicated a better interaction of the flexibilized dip system with the fiber compared to the standard RFL.

The adhesion of flexibilized RF resins with polyester and other fibers can be improved further by the use of other additives such as epoxy and blocked isocyanates in the RFL formulations.

### 6.3.8 Three Step Dip Process

Though the two step RFL dip process adequately provides good adhesion of polyester materials with rubber compounds, demanding applications such as power transmission belts require higher adhesion. In order to enhance the adhesion still further, an adhesive activated polyester fibers are treated in a three step dip process [40].

In the first step, polyester cords are dipped in an adhesive liquid containing sorbitol epoxy and caprolactam blocked isocyanate and heat treated. Then the cords are dipped into an RFL and heated. In the final stage, the cords are once again dipped into an RFL containing the rubber latex.

All these studies clearly indicate the importance of improving the chemical and physical interactions between the polyester, RFL and rubber compounds for obtaining the maximum adhesion. Achieving the highest polyester adhesion with the rubber compounds is very important for the long term durability of polyester reinforced rubber articles and products.

## References

1. Y. Iyengar, J. Appl. Polym. Sci., Vol. 11, p 2311 (1967).
2. Product Application Guide No. 1, Indspec Chemical Corporation (1999).
3. Products Application Guide No. 2, Dipping Systems, Indspec Chemical Corporation (1999).
4. T. Anangnostou and E. Jaul, J. Coat. Technol., Vol. 53, p 673 (1981).
5. R. Richter and H. Ulrich, Synthesis, p 463 (1975).
6. H. Ulrich, Cycloaddition reactions of heterocumulenes, Academic Press, New York (1967).
7. Z. W. Wicks, Prog. Org. Coat., Vol. 9, p 3 (1981).
8. D. A. Wicks and Z. W. Wicks, Prog. Org. Coat., Vol. 36, p 148 (1999).
9. R. Stolle, Chem. Ber, Vol. 41, p 1125 (1905).
10. S. Patai, The chemistry of cyanates and thio derivatives, Part 2, Wiley (1977).
11. W. Newmann and P. Fischer, Angew. Chem. Int. Edn., Vol. 1, p 621 (1962).
12. H. G. Khorana, Chem. Reviews, Vol. 53, p 145 (1953).
13. R. L. Zapp, G. E. Sernivk and K. S. Minckler, Rubber Chem. Technol., Vol. 43, p 1154 (1970).
14. B. A. Bolton, US Patent, 3,592,789 (1971).
15. M. Rudolf and M. Gerhard, British Patent, 1,058,236 (1962).
16. D. Petit, R. Jerome and Ph. Teyssie, J. Polym. Sci. Polym. Chem. Edn, Vol. 17, p 2903 (1979).
17. E. R. Alexander and A. D. Harry, British Patent, 1,140,528 (1969).
18. R. A. Edington and D. H. Aldred, US Patent, 3,660,202 (1972).
19. T. S. Solomon, Rubber Chem. Technol., Vol. 58, p 561 (1985).
20. J. Mather, Br. Polym. J., Vol. 3, p 58 (1971).
21. J. A. Cardina, US Patent, 3,247,043 (1966).
22. K. Dardoufas and R. Marshall, US Patent, 3,730,892 (1973).
23. R. Arrowsmith, US Patent, 3,793,425 (1974).

24. K. Chakravarti, US Patent, 4,348,517 (1982).
25. M. D. F. Maria and S. Li, US Patent, 6,046,262 (2000).
26. M. D. F. Maria and S. Li, US Patent, 6,333,281 (2001).
27. R. S. Bhakuni and G. W. Rye, US Patent, 4,031,288 (1977).
28. W. D. Havens and G. W. Rye, British Patent, 975,377 (1962).
29. N. S. Anderson and J. D. Gibbon, US Patent, 4,751,143 (1988).
30. N. S. Anderson, G. E. Gillberg-La Force, E. J. Powers and A. W. Yankowsky, US Patent, 4,933,236 (1990).
31. N. S. Anderson and A. L. Promislow, US Patent, 5,525,424 (1996).
32. N. S. Anderson and A. L. Promislow, US Patent, 5,328,765 (1994).
33. T. Hirakawa and R. C. Marrs, US Patent, 5,483,100 (1996).
34. T. E. Strangman, US Patent, 5,562,998 (1996).
35. K. G. Berndt, US Patent, 6,074,753 (2000).
36. C. D. Fisher, US Patent Appln., 2002/0,122,938 (2002).
37. R. B. Moore and K. C. Dardoufas, US Patent, 4,054,634 (1977).
38. R. Wright, US Patent, 3,835,082 (1974).
39. R. B. Durairaj and M. A. Lawrence, US Patent Appln., 2004/0,147,712 (2004).
40. H. Watanabe, S. Makino and T. Kuroda, US Patent, 6,528,113 (2003).
41. D.B. Wooten, Paper Presented at ACS Rubber Div. Meeting, Houston, TX, October 24 (1983).

# 6.4
# Bonding Aramid to Rubber Compounds

## 6.4.1
## Aramid Chemistry and Properties

The Dupont Company invented aramid fibers in the 1960s and commercialized them in 1972 under the trade name Kevlar [1]. The chemical composition of aramid or Kevlar is poly (*p*-phenylene terephthalamide) and was synthesized from the reaction of terephthaloyl chloride and *p*-phenylene diamine (Figure 6.54).

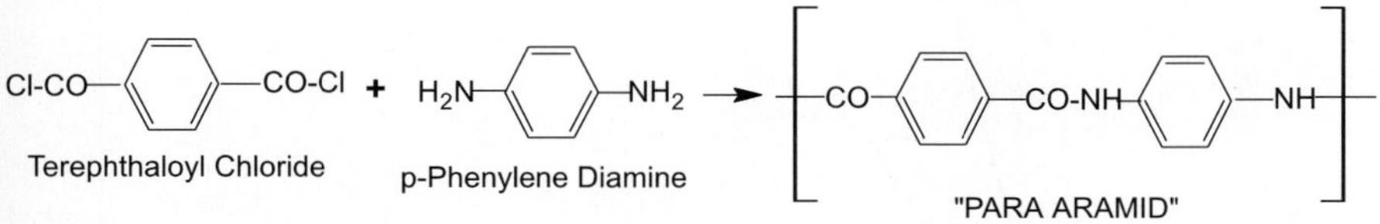


**Figure 6.54.**

The polymer chains are highly oriented in a highly crystalline state and are fixed by hydrogen bonds.

Some of the general features of the aramid fibers are as follows:

- High tensile strength at low weight
- Low elongation to break and high modulus
- High chemical resistance
- Excellent dimensional stability
- High toughness and cut resistance
- Flame resistance and self-extinguishing

Aramid fibers are about five times stronger weight for weight than steel but light and also half the density of glass. Stress-strain curves of industrial yarns are illustrated in the following Figure 6.55 [2].

## 6.4.2
## Aramid as Reinforcement in Rubber Compounds

Aramid fibers are strong, flexible and have good dimensional stability in high temperature environments as compared to other organic fibers. Aramid fibers offer exceptionally high strength and modulus and also textile processability. These fibers in the form of cords or fabrics have been used as reinforcements with various rubber compounds in the manufacture of tires, hoses, conveyer belts and transmission belts.

The initial market for aramid fibers was as reinforcement in tires. Steel wires were commonly employed in radial tire belts and radial truck tire carcasses.

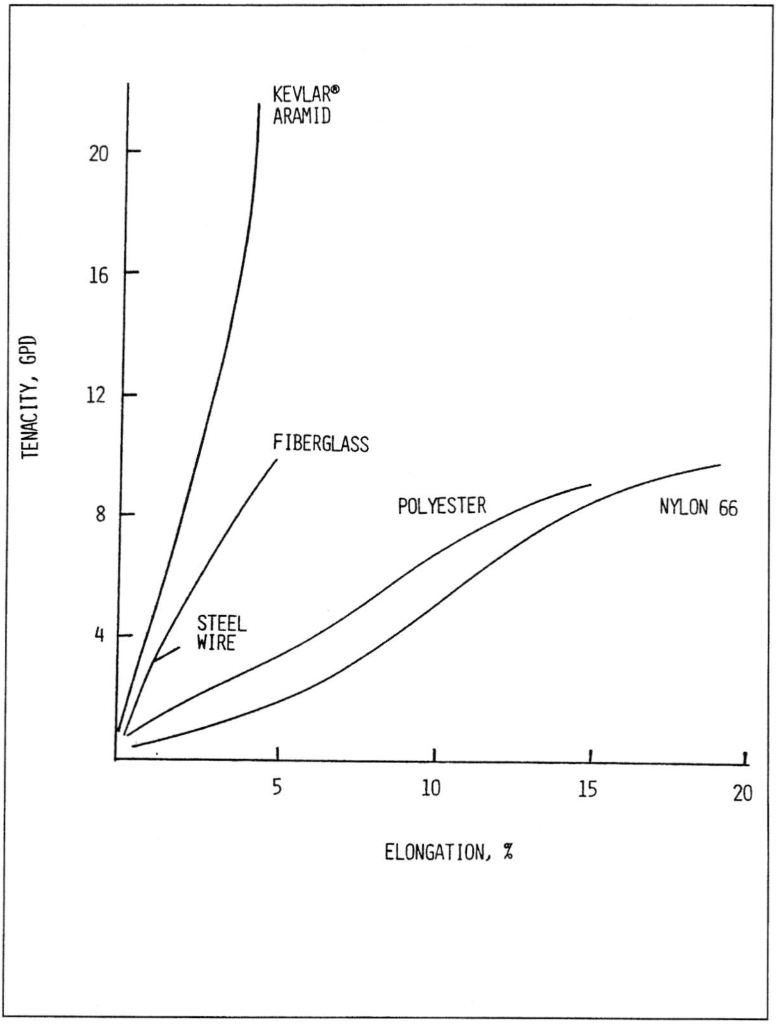


**Figure 6.55.** Stress–strain curves of industrial filament yarns

Aramid fibers were originally developed for the replacement of steel cords in radial tires because of their light weight and exceptional strength. Although the tires made with aramid reinforcements performed quite well, they were not able to displace steel reinforcements due to cost.

Aramid fibers have been successfully used as reinforcements in traditional rubber compounds and composite materials. In order to reinforce rubber compounds more effectively with aramid fibers, a great deal of effort has been devoted to develop adhesive systems that provide a good bond between the fibers and rubbers.

## 6.4.3
## Bonding Difficulties of Aramid Fibers to Rubber

The bonding of synthetic fibers or cords to rubbers can take place through the interactions of functional groups present in the molecular backbone of the fiber structures. Good adhesion development can be achieved by physical and chemical interactions of these groups with the adhesive components. In the case of aramid, -NH and -C=O functional groups are present in the molecular chain. These functional groups are hidden between the bulkier aromatic groups and are therefore not accessible for any interactions with the adhesive systems [3]. The reactivity of an amide group (-C(=O)-NH-) of the aramid fiber to adhesive is low. Therefore, the aramid cord is considered to be fundamentally a poor adhesion cord.

## 6.4.4
## Enhancing Aramid Adhesion

Aramid fibers have difficulty in exhibiting their superior physical characteristics such as strength and modulus as rubber reinforcements because they are not sufficiently strongly bonded to rubber when compared to other synthetic fibers. To enhance the bond between aramid and the adhesive that bonds the fiber and rubber, the solubility parameter of the adhesive should match that of aramid for hydrogen bonding and chemical interactions [4]. The formation of primary chemical bonds between the aramid and adhesive, in addition to hydrogen bonding, makes the aramid-adhesive interface stronger. With aramid, a primary chemical bond can be formed only from the reaction of the -CONH- group of aramid with other reactive functional groups of the dip system.

Many attempts were made to modify the aramid fiber surface and promote adhesion [5]. In one of the studies, it was proposed that an aramid fiber surface might react with an isocyanate group present in the organc molecule according to the following reaction [6] (Figure 6.56).

**Figure 6.56.**

Another published report indicated that an aromatic amide group can react with an aromatic isocyanate at elevated temperature conditions [7] (Figure 6.57).

The formation of such a product should have resulted from the following initial reaction of an amide group with the isocyanate group (Figure 6.58).

Based on the above information, an adhesive containing an isocyanate component having the isocyanate functional groups can form a chemical bond

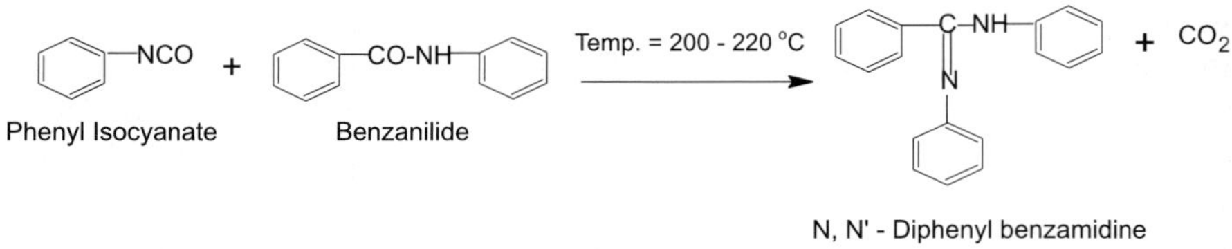


**Figure 6.57.**

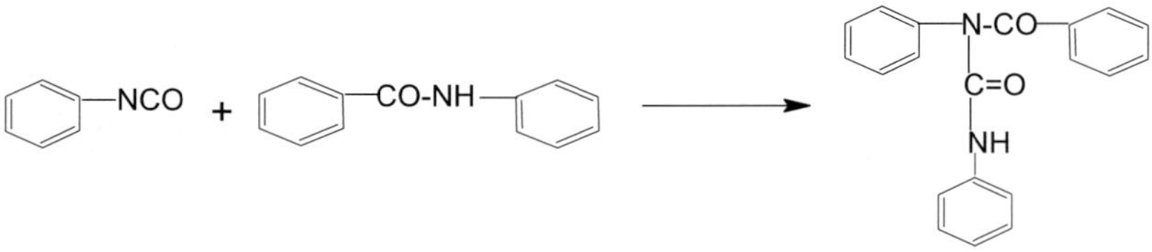


**Figure 6.58.**

through chemical reaction with aramid fibers. Similarly, an epoxy compound such as the diglycidyl ether of glycerol may interact with the aramid fibers at elevated temperatures. When epoxies are used in the subcoat, due to lower molecular weight and highly reactive compound they may react to form hydrogen bonds or chemical bonds with aramid.

The solubility parameter of aramid was found to be higher ($\delta = 29$) than most of the epoxy and other sub-coat compounds ($\delta = 22\text{–}34$) [4]. The surface activity at the aramid fiber surface is very low for any interactions with the adhesives. Therefore, a significant contribution from the direct chemical bonding may not happen with aramid fibers. A contribution from secondary bonding, such as hydrogen bonds, may also be low. Because of this, the bonding mechanism with aramid fibers is the true diffusion of low molecular weight sub-coat adhesives at elevated treating conditions [8].

In order to strengthen the fiber-adhesive interface, the adhesive system should contain functional groups such as isocyanate and epoxy, to react or interact with the aramid fiber. In the same way, the adhesive should interact with the rubber through some chemical as well as physical means to make the adhesive-rubber interface strong enough to avoid failure at this interface.

## 6.4.5 RFL Adhesive Systems for Aramid

Many of the adhesive systems proposed for aramid or Kevlar and rubber bonding were two-step dip or double dip systems with the first step consisting of a primer or subcoat followed by the application of a resorcinol-formaldehyde-latex (RFL) top coat. In the subcoat/top coat systems, the sub-coat formulation would consist of diglycidyl ether of glycerol, triglycidyl isocyanurate, trimethylol phenol and polyethyleneimine composition and the RFL top coat with or without the HAF carbon black. In addition, a caprolactam blocked diphenylmethane diisocyanate (MDI) in combination with diglycidyl ether of glycerol was also employed and reported to enhance the aramid adhesion with rubber compounds [4,8,9].

The Dupont Company recommended that a double dip system based on diglycidyl ether of glycerol (subcoat) and Penacolite R-2170 (RFL topcoat) should be used to improve the Kevlar adhesion with rubber compounds. Dupont recommended formulations are presented in Table 6.23.

**Table 6.23.** Two-step RFL dip for Kevlar aramid cord

| IPD-31 subcoat dip formulation | Parts by weight | |
|---|---|---|
| | Dry | Wet |
| Diglycidyl ether of glycerol | 2.22 | 2.22 |
| Water (soft) | | 96.94 |
| Sodium hydroxide (10%) | 0.03 | 0.28 |
| 5% aerosol OT (75%) | 0.02 | 0.56 |
| Sub-coat dip pH | 10.5–11.0 | |
| IPD-39 top coat dip formulation | | |
| Water | | 141 |
| Ammonium hydroxide (28%) | | 6.1 |
| Penacolite resin R-2170 (75%) | 16.5 | 22 |
| Mix and add without aging to | | |
| Vinyl pyridine latex (41%) | 100 | 244 |
| Mix and add formaldehyde (37%) | 4.1 | 11 |
| Water | | 51.4 |
| Age 12 h at 25 °C and add when ready to use | | |
| HAF black dispersion (25%) | 15.1 | 60.3 |
| Haveamul M-1118 wax dispersion (45%) | 5.4 | 11.9 |
| Total | 141.1 | 547.7 |
| Topcoat dip properties | | |
| Solids (%) | 25.8 | |
| pH | 9.5–9.9 | |
| Subcoat curing | | |
| Temperature (°C)/time (min) | 238–243/1.0 | |
| Topcoat curing | | |
| Temperature (°C)/time (min) | 215–230/1.0 | |

A typical natural rubber compound formulation used as the radial belt compound having the composition shown in Table 6.24 was used to test the aramid cords adhesion [4].

The two-ply adhesion test results obtained and reported are presented in Table 6.25 [4].

Table 6.25 results suggested that low sulfur-containing rubber compounds provided good levels of two-ply pull-out adhesion with aramid cords. Heat

**Table 6.24.** Natural rubber compound for aramid cords adhesion

| Components | Parts by weight |
|---|---|
| 1. Natural rubber | 100 |
| 2. HAF carbon black | 60 |
| 3. Stearic acid | 2 |
| 4. Zinc oxide | 8 |
| 5. Pine tar | 5 |
| 6. Tackifier | 3 |
| 7. 2,2′-Dihydro-2,2,4′-trimethyl quinoline | 1 |
| 8. *N*-Cyclohexyl-2-benzothiazole sulfenamide | 0.2 |
| 9. Oiled sulfur (20%) | 3.13 |
| 10. *N*-(Cyclohexylthio) phthalimide | 0.3 |

**Table 6.25.** Adhesion of aramid cords with natural rubber compound

| Two-ply adhesion | Pull-out adhesion (1.9 cm) at 25 °C, N |
|---|---|
| Unaged | 489 |
| Heat aged, 100 °C/72 h | |
| Air atmosphere[a] | 422 |
| Nitrogen atmosphere[a] | 489 |

[a]Complete rubber coverage was observed
Data from [4]

aged aramid cords were completely covered with rubber compounds which suggested that a slight reduction in the heat aged adhesion (aged under the air) might be associated with the rubber degradation [4].

The aramid fiber adhesion was shown to improve when the modulus of the topcoat adhesive formulation (RFL) was increased. High adhesive stiffness was found to be beneficial in improving the aramid adhesion by creating a more favorable modulus gradient from the rubber to fiber. Also, the application of preformed (RF) resin in the place of an in-situ resin appeared to improve the aramid adhesion by the higher levels of dip pickup [10].

### 6.4.6
### Aramid Adhesion with Nitrile and Neoprene Compounds

In the manufacture of automobile tires, hoses, power transmission belts and conveyer belts, safety and reliability are the most important requirements. Due to high strength, modulus and dimensional stability, aramid cord or textile reinforcements have been used with the highly loaded, high modulus nitrile and neoprene rubber compounds. Aromatic polyamide fibers, in general, are very poorly wettable to RFL solutions because of their inactive fiber surface compared to the aliphatic polyamides. Various developmental efforts have been made to enhance the adhesion between aramid fibers and different synthetic rubbers. RFL formulations containing different lattices were evaluated to enhance the aramid adhesion. A double dip RFL formulation containing two different rubber latex compounds appeared to show good adhesion with neoprene and nitrile rubber compounds [11–13].

RFL formulations, treating conditions and the adhesion properties are presented in Table 6.26. Rubber compounds and their compositions used in the adhesion evaluations are given in Table 6.27.

**Table 6.26.** RFL formulations and aramid cords adhesion with synthetic rubbers

| Formulations (parts by weight) | Control | RFL-1 | RFL-2 |
|---|---|---|---|
| A. First step dip | | | |
| Resorcinol | 5 | 5 | 5 |
| Formalin (37%) | 3.1 | 3.1 | 3.1 |
| PVC latex (48%) | 61.6 | 61.6 | 61.6 |
| Water | 84.3 | 84.3 | 84.3 |
| First dip treating, temp./time: | | 200 °C/2 min | |
| B. Second step dip | | | |
| Resorcinol | 5 | 5 | 5 |
| Formalin (37%) | 3.1 | 3.1 | 3.1 |
| Natural rubber latex (60%) | 61.6 | None | None |
| Chloroprene rubber latex | None | 61.6 | None |
| Nitrile rubber latex | None | None | 61.6 |
| Water | 84.3 | 84.3 | 84.3 |
| First dip treating, temp./time: | | 200 °C/2 min | |
| Neoprene rubber compound adhesion (kg/cord) | | | |
| Peeling adhesion strength | 10.7 | 13.8 | 11.5 |
| Mode of failure | A-R | R | R |
| Nitrile rubber compound adhesion (kg/cord) | | | |
| Peeling adhesion strength | 1.9 | 13.2 | 14.8 |
| Mode of failure | A-R | R | R |

A-R = Adhesive and rubber interface ; R = Rubber cohesive failure
Data from [11]

The use of rubber latex compatible with the rubber compound in the top coat dip of RFL was observed to improve the aramid cord adhesion. The presence of rubber compound compatible rubber latex in the RFL lets the system attain fixed, strong and durable adhesion between the aramid fibers and rubbers without deterioration of resistance to flexing fatigue. Adhesion of aramid cords, from Table 6.26, with natural rubber, nitrile rubber and neoprene rubber compounds were improved by the use of natural rubber, nitrile rubber and chlorinated rubber lattices respectively in the RFL [11].

**Table 6.27.** Synthetic rubber compound formulations for aramid cords adhesion

| Components (parts by weight) | Neoprene compound | Nitrile compound |
|---|---|---|
| 1. Neoprene rubber | 100 | |
| 2. Nitrile rubber | | 100 |
| 3. Magnesium oxide | 4 | |
| 4. Zinc oxide | 5 | 5 |
| 5. SFR carbon black | 40 | 40 |
| 6. Stearic acid | 0.5 | 1 |
| 7. Process oil | 10 | |
| 8. 2-Mercaptoimidazoline | 0.5 | |
| 9. *N*, *N*′-Di-*β*-naphthyl-*p*-phenylene diamine | 0.5 | |
| 10. 2,2,4-Trimethyl-1,2-dihydroquinoline | 0.5 | |
| 11. Sulfur | | 0.5 |
| 12. Tetramethylthiuram disulfide | | 2 |
| 13. Mercapto benzothiazole | | 0.5 |

Data from [11]

For the improvement of adhesion between aramid fibers and hydrogenated nitrile rubber compounds, fibers were dipped in a three-step dip process and the details are shown in Table 6.28 [14].

In the first step, aramid cords were treated with an epoxy compound and heat treated. RFL containing nitrile rubber latex was employed in the second

**Table 6.28.** RFL formulation for aramid-nitrile rubber adhesion

| Formulations | Parts by weight |
|---|---|
| A. First step dip | |
| Diglycidyl ether of glycerol (5%) | |
| Treating condition, temp./time: | 200 °C/5 min |
| B. Second step dip | |
| Resorcinol | 5 |
| Formalin (37%) | 3.1 |
| NBR latex | 61.6 |
| Water | 84.3 |
| Treating condition, temp./time: | 200 °C/2 min |
| C. Third step dip | |
| Chlorinated rubber (58% chlorine) | 10 |
| Poly(phenylene isocyanate) | 10 |
| Carbon black | 5 |
| Toluene | 7.5 |
| Treating condition, temp./time: | 110 °C/2 min |

Data from [14]

treatment. After heat treatment, RFL dipped cords were subjected to polyisocyanate compound treatment in the third step of the process. Finally, they were embedded in a nitrile rubber compound, cured and tested for peeling adhesion. RFL formulation having nitrile rubber latex gave an improved adhesion value compared to the control formulation without the nitrile latex.

All these studies have clearly shown that adhesion performances of aramid fibers, cords or textile materials with different rubber compounds can be greatly improved by employing the respective rubber lattices in the RFL formulations.

## References


1. S. L. Kwolek, US Patent, 3,600,350 (1971).
2. E. R. Baron, Paper Presented at Clemson Tire Technology Conference, October 30, 1985.
3. R. Iyengar, Paper Presented at Tire Cord Technology Conference, Greenville, SC, October 28, 1987.
4. Y. Iyengar, J. Appl. Polym. Sci., Vol. 22, p 801 (1978).
5. A. G. Andreopoulos, J. Appl. Polym. Sci., Vol. 38, p 1053 (1989).
6. W. C. Wake, J. Adhes., Vol. 3, p 315 (1972).
7. P. F. Wiley, J. Am. Chem. Soc., Vol. 71, p 3746 (1949).
8. T. S. Solomon, Rubber Chem. Technol., Vol. 58, p 561 (1985).
9. M. Ikemoto et al., Japanese Patent 51/23586 (1976).
10. Y. Iyengar, ACS Rubber Division Meeting, Akron, OH, p 100, April 1978.
11. A. Fujiwara and K. Hata, US Patent, 5,306,369 (1994).
12. A. Fujiwara and K. Hata, US Patent, 5,484,497 (1996).
13. A. Fujiwara and K. Hata, US Patent, 5,728,245(1998).
14. A. Fujiwara, H. Matsuoka, S. Onoe and K. Kusaka, US Patent, 4,978, 409 (1990).

## 6.5 Bonding Poly (Benzoxazole), PBO, Fibers with Rubber Compounds

### 6.5.1 PBO Chemistry and Properties

Poly(benzoxazole), also called PBO, fiber is a new high performance fiber developed by Toyobo company in Japan and marketed under the trade name "Zylon".

This polymer was synthesized from the reaction of 4,6-diamino resorcinol with terephthaloyl chloride (Figure 6.59).

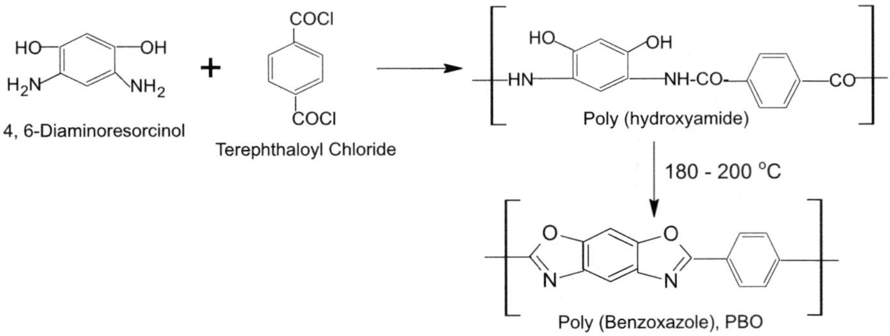


**Figure 6.59.** Synthesis of Poly (benzoxazole) polymer (PBO)

PBO is a rigid-rod isotropic crystalline polymer. The PBO fiber has superior tensile strength and modulus compared to Kevlar aramid fibers. A comparison of mechanical properties of Zylon, Kevlar and high strength steel fibers are presented in Table 6.29.

**Table 6.29.** Comparision of fiber properties

| Properties | Poly(benzoxazole) "Zylon" | Aramid (Kevlar) | High tensile steel |
|---|---|---|---|
| Tensile strength (GPa) | 5.8 | 3.6 | 1.3 |
| Tensile modulus (GPa) | 180 | 60 | 210 |
| Density (g/cm$^3$) | 1.54 | 1.44 | 7.8 |
| Specific strength (N/Tex) | 2.54 | 2.08 | 0.1 |

PBO has excellent flame resistance and thermal stability, withstanding temperatures up to 650 °C, properties among organic fibers. These fibers are much more impact-resistant than steel or carbon fibers. The moisture absorption is very low and the fibers are dimensionally stable against humidity.

This fiber is quite flexible in-spite of its extremely high mechanical properties. The excellent mechanical properties of this fiber are found to be suitable as a reinforcement material in the development of high strength and light weight reinforced rubber composites.

### 6.5.2 RFL Bonding of PBO Fibers with Rubbers

Timing belts used in the automobile applications have to withstand high temperatures and high load environments. High heat resistant rubbers employed to develop these belts are based on hydrogenated nitrile and chloroprene rubber materials. Though high strength fibers, such as the aramid, have been used in these applications, much higher performance requirements are prompting to use PBO fibers due to their superior heat resistant and dimensional stability than the aramid.

PBO fiber has high crystallinity and close-knit structure and therefore, a direct bondability with the resorcinol-formaldehyde-latex (RFL) can be poor. With the application of an epoxy compound as the primer, similar to the aramid

**Table 6.30.** RFL dip formulations for PBO adhesion

| | Example 1 | Example 2<br>Parts by weight | Example 3 |
|---|---|---|---|
| A. First step dip | | | |
| Diglycidyl ether of glycerol | 5 | 5 | 2.22 |
| NBR latex | 11.1 | 11.1 | |
| Sodium hydroxide (10%) | 5 | | 0.28 |
| 2-Methyl imidazole (10%) | | 5 | |
| Aerosol OT (5%) | | | 2.24 |
| Water | 78.9 | 78.9 | 95.26 |
| Total | 100 | 100 | 100 |
| Treating condition, temp./time: | | 250 °C/100 s | |
| B. Second step dip | | | |
| Resorcinol | | 1.88 | |
| Formalin (37%) | | 2.76 | |
| Vinyl pyridine-SBR latex (40.5%) | | 42.1 | |
| Sodium hydroxide (10%) | | 0.51 | |
| Ammonium hydroxide (28%) | | 3.32 | |
| Water | | 49.43 | |
| Total | | 100 | |
| Treating condition, temp./time: | | 250 °C/100 s | |
| Adhesion with chloroprene compounds | | | |
| Bond strength (N/15 mm) | 110 | 140 | 50 |
| Peeling portion or mode of failure | Rubber-cord interface | Rubber | Rubber-cord interface |

Data from [1]

adhesion, RFL formulations were found to enhance the PBO fiber adhesion with chloroprene rubber compounds [1].

The RFL formulations and the adhesion performance are presented in Table 6.30.

From peel adhesion results, it was determined that the presence of 2-methyl imidazole catalyst in the primer and the application of vinyl pyridine containing RFL made the adhesion stronger between the PBO fiber and chloroprene rubber compounds.

With the proper RFL formulations, the adhesion performances of high strength PBO fibers towards various synthetic rubber compounds can be improved [2].

## References

1. H. Sogabe, S. Isshiki and H. Yamamoto, US Patent Appln 2002/0,142,147 A1 (2002).
2. J. Chu, US Patent Appln, 2004/0,096,658 (2004)

## 6.6
## Bonding Carbon Fibers

### 6.6.1
### Carbon Fiber Chemistry

Carbon fibers have been widely used as reinforcement materials in many high performance polymeric and elastomeric composites. Carbon fibers are also known as graphite fibers. Graphite technically refers to fibers that are greater than 99% carbon composition whereas carbon fibers have the carbon content of about 93–95%. Carbon fibers are made from organic precursors, namely poly(acrylonitrile) (PAN), rayon and coal tar pitches. The rayon and pitches are generally used to produce low modulus fibers. Carbon fibers are known for their excellent strength and stiffness properties. High temperature performance is outstanding for carbon fibers.

### 6.6.2
### Synthesis of Carbon Fibers from PAN

Starting from poly(acrylonitrile) fiber, carbon fibers are made by oxidation and carbonization processes and this can be schematically illustrated in the following reaction (Figure 6.60).

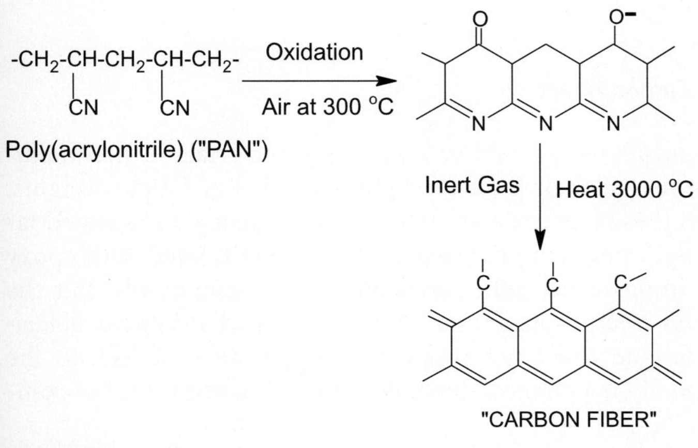


**Figure 6.60.**

During the oxidation process, the aliphatic polymer is converted into an aromatic structure with the elimination of hydrogen from the PAN fibers. Cross-links are formed with the addition of oxygen to the polymer. The color of the fiber changes from white to black. In the carbonization step, the concentrations of oxygen, nitrogen and hydrogen decrease with increasing temperature. The oxidized polymer is completely converted into carbon structures.

## 6.6.3 Comparison of Fiber Properties

Carbon fibers have superior properties compared to steel and other synthetic fibers. The comparison of mechanical properties of carbon fiber with Kevlar and high tensile steel are presented in Table 6.31.

**Table 6.31.** Performances of various fibers

| Fiber properties | Standard grade carbon fiber | Kevlar fiber | High tensile steel fiber |
|---|---|---|---|
| Tensile strength (GPa) | 3.5 | 3.6 | 1.3 |
| Tensile modulus (GPa) | 230 | 60 | 210 |
| Density (g/cc) | 1.75 | 1.44 | 7.87 |
| Specific strength (GPa) | 2 | 2.5 | 0.17 |

Carbon fibers have lower density, higher modulus of elasticity and higher strength compared to steel and are therefore considered an excellent reinforcement for rubber compounds. But carbon fibers have poor adhesion properties to rubbers and rubber compounds. The degree of adhesion between the unmodified or untreated carbon yarn and the rubber is not satisfactory for most applications. RFL treatments of carbon fibers enhance the adhesion of these fibers with natural and synthetic rubber compounds.

## 6.6.4 RFL Formulations for Carbon Fibers

Carbon fibers are considered as rubber reinforcing materials in the manufacture of tires because of their high strength, modulus and light weights. As reinforcing cords these materials are inferior in resistance to repeated fatigue due to their high modulus property. Carbon fibers treated with epoxy resins appeared to improve the adhesion with rubber compounds. But the epoxy treated carbon fibers were found to be stiffer and exhibited deterioration in flexing fatigue resistance properties. Application of RFL to the epoxy treated carbon fibers enhanced the adhesion with natural rubber compounds [1].

RFL formulation and the adhesion performance of carbon fibers are given in Table 6.32.

High fatigue strength retention associated with the flexing fatigue resistance clearly demonstrated that the adhesion between the carbon fibers and natural rubber compounds was enhanced.

Reinforcing yarns coated with an elastomer for use in dynamic rubber articles, such as tires, should have high tenacity. Multi-filament yarns of carbon have high modulus, high tenacity, good dimensional stability and chemical

**Table 6.32.** RFL formualtion for carbon fiber adhesion

| First step dip | Parts |
|---|---|
| Carboxy terminated butadiene-acrylonitrile rubber (CTBN) | 100 |
| Glycidylamine epoxy resin (Araldite MY720) | 18 |
| React CTBN and MY720 at 100 °C for 2 h | |
| Dissolve in MEK solvent. | |
| Add 2-ethyl-4-methylimidazole catalyst | |
| Conditions | |
| Dry (temperature (°C)/min) | 120/3 |
| Cure (temperature (°C)/min) | 200/2 |
| Second step dip | |
| Resorcinol | 23.1 |
| Formaldehyde (37%) | 25.6 |
| NaOH (10%) | 6.3 |
| Water | 387.6 |
| Vinylpyridine latex (40%) | 543.5 |
| Ammonium hydroxide (28%) | 13.9 |
| Conditions | |
| Dry (temperature (°C)/min) | 85/2 |
| Cure (temperature (°C)/min) | 200/2 |
| Adhesion performance of carbon fibers | |
| Drawing strength (kg/8 mm) | 19.5 |
| Two-ply peel strength (kg/25 mm) | 25.9 |
| Flexing fatigue strength retention (%) | 85 |

Data from [1]

resistance for the tire and high performance applications. RFL coated carbon yarns showed good tenacity and exhibited fatigue resistance. The treatment of original carbon yarn with the water soluble epoxy resin before the impregnation with RFL has been effective for increasing the adhesion between carbon yarn and rubber [2].

RFL adhesives prepared to treat the epoxy coated carbon fibers are presented in Table 6.33.

Compared to the RFL prepared by the in situ method, the pre-condensed resin based RFL adhesive was effective in increasing the adhesion of coated carbon yarns to rubber compounds [3–7].

These studies and published work clearly suggest that impregnating carbon yarns or cords first with an epoxy adhesive before the RFL composition can provide good adhesion of carbon fibers to various rubbers and rubber compounds.

**Table 6.33.** RFL formulations for treating carbon fibers

| Components in parts | In situ RFL | Precondensed resin-RFL |
|---|---|---|
| Resorcinol | 50.4 | |
| Penacolite R-50 | | 200 |
| Formaldehyde (37%) | 44.7 | 16 |
| NaOH (10%) | 50.7 | 13 |
| Water | 425.6 | 316 + 1145 |
| NBR latex (41%) | 245 | 862 |
| Vinylpyridine latex (42%) | 245 | 862 |
| Surfynol SE (acetylenic diol) | | 8 |
| Hamposyl (lauroyl sarcosine salt) | | 7 |
| Potassium stearate | | 36 |
| Urea | | 36 |
| Aging of adhesive at RT (h) | 24 | 24 |
| Cure: temperature (°C/min) | 177/3 | 177/3 |
| Carbon cords adhesion (J) | 2 | 2.9 |

Data from [3]

## References

1. H. Ogawa, K. Kogo, S. Takahashi and Y. Suzuki, US Patent, 4,883,712 (1989).
2. M. Toki, N. Kusatsu, Y. Kohno and K. Matsugasako, US Patent, 4,044,540 (1977).
3. J. G. Gillik and C. Bennett, European Patent Appln, EP 0,902,046 A2 (1999).
4. J. G. Gillik and C. Bennett, US Patent, 6,077,606 (2000).
5. J. G. Gillik and C. Bennett, US Patent, 6,350,492 (2002).
6. P. S. Knutson, US Patent Appln, 2003/0,130,077 (2003).
7. C. Tonon, US Patent Appln., 2004/0,043,213 (2004).

## 6.7 Bonding EPDM and Nitrile Rubber Compounds with Fibers

### 6.7.1 Ethylene-Propylene-Diene Monomer (EPDM) Rubbers

EPDM is a terpolymer of ethylene, propylene and a diene monomer. These rubber materials are the most widely used and fastest growing synthetic rubbers in specialty and general purpose applications. They have an outstanding resistance to oxygen, ozone and sunlight due to their stable saturated polymer backbone structures. As non-polar polymers, they are resistant to polar materials such as phosphate esters, ketones, acids, alkalies and alcohols. EPDM polymers have good electrical resistivity and heat resistance up to 130 °C. Compounds containing EPDM rubbers show good high temperature compression set resistance when cured with sulfur or peroxide curing systems.

EPDM polymers can be loaded with higher filler and plasticizer compounds, thereby providing economical compounds for various applications. These compounds provide high tensile and tear properties as well as excellent abrasion and oil swell resistance.

Since EPDM materials have excellent properties and compounding characteristics, they are used in the manufacture of hoses, tubing, belts, electrical insulation, mechanical goods and thermoplastic vulcanizates.

#### 6.7.1.1 EPDM Rubbers in Hoses

In the manufacture of automobile hoses, they are used as fuel hoses, fuel cell hoses (methanol fuel cell hose or hydrogen fuel cell hose in fuel cell cars), engine cooling system hoses (radiator hoses, heater hoses) and refrigerator coolant transport hoses. The current stainless steel hoses are heavier in weight, poor in fuel efficiency and costly. Fuel cells with superior gas barrier properties, lighter weight and lower cost are made using EPDM polymers. In the production of hoses, nylon fibers, sheets and fabrics are often used as the reinforcing materials with EPDM compounds. With nylon reinforcements, resorcinol and resorcinolic resins often provide good adhesion with EPDM rubber compounds as the methylene acceptors along with methylene donors such as the hexamethoxymethyl melamine (HMMM) or hexamethylene tetramine (HMT).

By employing resorcinolic products in the EPDM compounds, there is no need to treat the nylons with RFL formulations before embedment. The presence of resorcinol and resorcinolic resin materials with methylene donor materials can enhance the adhesion of nylons with EPDM compounds based on the following chemistry (Figure 6.61).

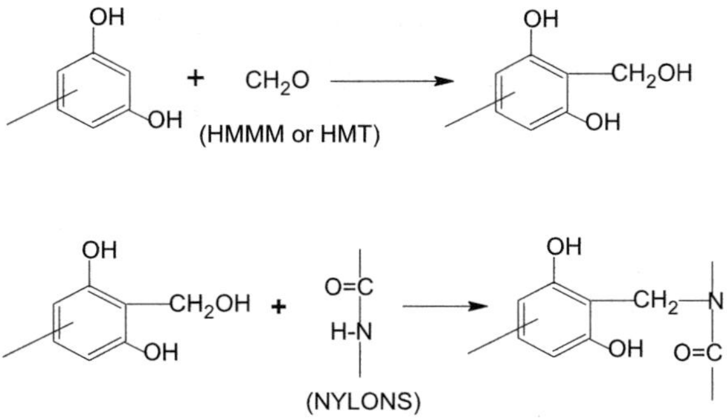


**Figure 6.61.**

The direct covalent bond that could be formed between the resorcinolic bonding systems and nylon reinforcements can enhance the EPDM compounds adhesion [1–4].

Good tensile and adhesion properties were achieved when the resorcinolic resin and HMMM materials were used in EPDM rubber hose compound formulations (Table 6.34).

**Table 6.34.** Bonding of nylons with EPDM rubber compounds

| | Experiment number | | | | | |
|---|---|---|---|---|---|---|
| Components in parts | 1 | 2 | 3 | 4 | 5 | 6 |
| EPDM rubber | 100 | | | | | |
| Carbon black | 100 | | | | | |
| Processing oil | 60 | | | | | |
| Peroxide | 4.2 | | | | | |
| Resorcinol resin (Sumikanol620) | 1 | 5 | 10 | 0.1 | 0 | 0 |
| HMMM (Sumikanol507 A) | 0.77 | 3.85 | 3.8 | 0.05 | 0 | 1 |
| Tensile properties | | | | | | |
| Tensile break (MPa) | 13 | 10.5 | 8.2 | 13.8 | 14 | 14 |
| Elongation at break (%) | 250 | 250 | 230 | 260 | 260 | 270 |
| Adhesion properties | | | | | | |
| Adhesive force (kg/25 mm) | 8.8 | 8.6 | 9.2 | 6.4 | 0.2 | 1 |

Data from [1]

Rubber compounds without the resorcinolic bonding system exhibited poor nylon cord to EPDM compounds adhesion. With the bonding system present, the adhesion failure was in the rubber compound itself and the adhesion was reported to be stronger at the nylon and rubber interface. The gas barrier properties of nylon reinforced EPDM compounds have been dramatically improved by the excellent adhesion of nylon to the laminate materials.

EPDM polymer reinforced with nylon-66 has also been used as an automobile vibration insulating rubber composite material at the engine mount. The use of resorcinolic resin in the EPDM rubber compound formulation of this application greatly enhanced the adhesion as can be seen in Table 6.35 [2].

**Table 6.35.** Nylon-66 and EPDM adhesion using resorcinolic resins

| | Experiments | | | | |
|---|---|---|---|---|---|
| Components in parts | 1 | 2 | 3 | 4 | 5 |
| EPDM rubber | | | 100 | | |
| Carbon black | | 100 | | 45 | |
| Processing oil | | 60 | | 20 | |
| Dicumyl peroxide | 0 | 4.2 | 2 | 2 | 2 |
| Resorcinol resin (Sumikanol620) | 3 | 0 | 0.1 | 3 | 1 |
| HMMM (Sumikanol507A) | 3 | 0 | 0.1 | 3 | 0.5 |
| Tetramethyl thiuram disulfide | 1.5 | | | | |
| Zinc dimethyldithio carbamate | 1.5 | | | | |
| Sulfur | 0.5 | | | | |
| Cured rubber properties | | | | | |
| Hardness (HS) | 56 | 57 | 57 | 60 | 59 |
| Tensile break (MPa) | 18 | 16 | 16.5 | 16.8 | 16.7 |
| Tensile elongation (%) | 663 | 635 | 628 | 612 | 620 |
| Adhesive strength (N/mm) | 0 | 0 | 12 | 13.3 | 13 |
| Mode of failure | I | I | R | R | R |

Note: I = Nylon - rubber interface; R = Rubber breakage
Data from [2]

In the absence of a peroxide (vulcanizing agent) and resorcinolic resin system, the cured rubber compound showed poor adhesion with nylon-66 reinforcement.

#### 6.7.1.2 EPDM Rubbers in Belts

In the automobile industry, an increase in demand for energy savings has led to the design of compact automotive engines. This opted for the engine compartment to become smaller which resulted in the engine operating at relatively higher temperatures. The power transmission belts employed in the automobiles are, therefore, expected to operate and perform reliably at these high temperatures.

EPDM polymers exhibited good heat and cold resistance properties and were found to be suitable for the manufacture of power transmission belts. These belts are often reinforced with high strength and high modulus fibers such as polyester and aramid. In order to enhance the adhesion of EPDM rubber compounds with these fibers, RFL adhesives are often used. RFL formulations containing different lattices, suitable for bonding EPDM or other rubbers, were found to be effective for improving the adhesion of reinforcement fibers with rubber compounds.

### 6.7.1.3 Polyester Adhesion to EPDM Compounds

Polyester (PET) fibers or cords have been used to reinforce EPDM rubber compounds for the manufacture of power transmission belts. In a standard procedure, normally, these fiber materials have been immersed in a dipping solution of resorcinol-formaldehyde-vinyl pyridine latex (RFL) and then embedded in EPDM rubber compounds and vulcanized. With this method, the adhesion between polyester and EPDM elastomer was observed to be poor. When the transmission belts are subjected to repetitive flexing, the load- carrying reinforced cords may prematurely separate from the embedded rubber. In order to enhance the adhesion of PET cords with EPDM rubber, improved adhesive formulations have been developed [5–7].

For increasing the PET-EPDM adhesion, the PET cords were treated first with an isocyanate compound as primer, then RFL, and finally a solution containing two different rubbers. The latex used in RFL preparation was NBR latex. Chlorinated and EPDM rubbers dissolved in toluene solvent were used to treat the PET cords, after the RFL, before being embedded into the rubber compounds. The composition of EPDM compound used is presented in Table 6.36.

**Table 6.36.** EPDM rubber compound for PET bonding

| Component | Parts |
|---|---|
| EPDM rubber | 100 |
| Zinc oxide | 5 |
| Stearic acid | 1 |
| Carbon black | 35 |
| Hydrated silica | 20 |
| RF resin | 2 |
| HMMM | 2 |
| Antioxidant | 2 |
| Accelerator | 1 |
| Sulfur | 1 |

Data from [5]

The adhesive formulation and adhesion performance data obtained from this work are given in Table 6.37.

In addition to RFL treatments, the application of isocyanate primer on the PET cords and the use of resorcinolic resin bonding system in the rubber compound formulation appeared to enhance the adhesion of polyester with EPDM rubber.

**Table 6.37.** RFL formulations for the adhesion of PET with EPDM rubber

| Components in parts | Experiment | | |
|---|---|---|---|
| 1. First step dip | 1 | 2 | 3 |
| Polymeric isocyanate | | 10 | |
| Toluene | | 90 | |
| Treating conditions, temp./time | | 180 °C/4 min | |
| 2. Second step dip (RFL) | | | |
| Resorcinol | | 14.6 | |
| Formaldehyde (37%) | | 9.2 | |
| NaOH (10%) | | 1.5 | |
| Water | | 262.5 | |
| NBR latex (40%) | | 100 | |
| Treating conditions, temp./time | | 230 °C/2 min | |
| 3. Third step dip | | | |
| Chlorinated rubber | 3 | 5 | 8 |
| EPDM Rubber | 3 | 5 | 8 |
| Toluene | 94 | 90 | 84 |
| Treating conditions, temp./time | | 100 °C/3 min | |
| Peel adhesion properties with EPDM rubber compounds (N/25 mm) | | | |
| Room temperature | >500 | >500 | >500 |
| 120 °C | 85 | 100 | 120 |

Data from [5]

### 6.7.1.4 Aramid Adhesion with EPDM Compounds

RFL adhesives have been used in different ways to improve the adhesion of high strength fibers, such as aramid, with EPDM rubber compounds. In one of the methods of use, aramid fiber adhesion has been improved when the RFL contained polybutadiene latex [8]. By comparing it to vinyl pyridine and chlorosulfonated polyethylene lattices based RFL adhesives, polybutadiene based RFL had shown enhanced adhesion, and can be seen in Table 6.38. This might be due to the compatibility of polybutadiene with EPDM rubber.

In another method, aramid cords were first treated with polymeric isocyanate material before treating with RFL adhesive [9]. RFL contained rubber compound compatible EPDM latex and the formulations are given in Table 6.39.

The application of non-volatile polymeric isocyanate in the first step of fiber treating processes might have strengthened the fiber-adhesive interface by means of a chemical reaction between the isocyanate and amide group of aramid fibers. RFL adhesives applied after the isocyanate treatment had shown superior bond strength. For the application of transmission belts, EPDM rubber compounds were reinforced with RFL treated aramid cords. The aramid treatment is illustrated in the scheme shown in Figure 6.62 [10].

**Table 6.38.** Adhesion of aramid cords with EPDM rubber compounds

| Components (parts by weight) | Experiment | | |
|---|---|---|---|
| 1. First step dip-primer | 1 | 2 | 3 |
| Diglycidyl ether of glycerol | | 16 | |
| NaOH (10%) | | 2 | |
| Aerosol OT (5%) | | 4 | |
| Water (soft) | | 698 | |
| Treating conditions, temp./time | | 246 °C/1 min | |
| 2. Second step dip | | | |
| RF preparation | | | |
| Resorcinol | | 11 | |
| Formaldehyde (37%) | | 16.2 | |
| NaOH (1 N) | | 7.5 | |
| Water | | 23.1 | |
| Aging of RF solution | | 6 h at 23 °C | |
| RFL preparation | | | |
| RF solution (prepared above) | | 55 | |
| Polybutadiene latex | 100 | | |
| Vinylpyridine latex | | 100 | |
| Chlorosulfonated polyethylene latex | | | 100 |
| Ammonium hydroxide (28%) | | 2 | |
| Aging of RFL before dipping | | 16 h at 23 °C | |
| Cord adhesion-H-pull (Newtons) | 221 | 142 | 135 |

Data from [8]

**Table 6.39.** Adhesion of aramid cords with EPDM compounds

| 1. First step dip | Parts |
|---|---|
| Polymeric isocyanate | 10 |
| Toluene | 90 |
| Total | 100 |
| Treating conditions, temp./time | 180 °C/4 min |
| 2. Second step dip (RFL) | |
| Resorcinol | 11 |
| Formaldehyde (37%) | 16.2 |
| NaOH (10%) | 0.3 |
| Water | 299.5 |
| EPDM latex (40%) | 244 |
| Treating conditions, temp./time | 230 °C/2 min |
| Kevlar cord adhesion (1670 dtex/1×4) | |
| Flat peel strength (N/25 mm) | 380 |

Data from [9]

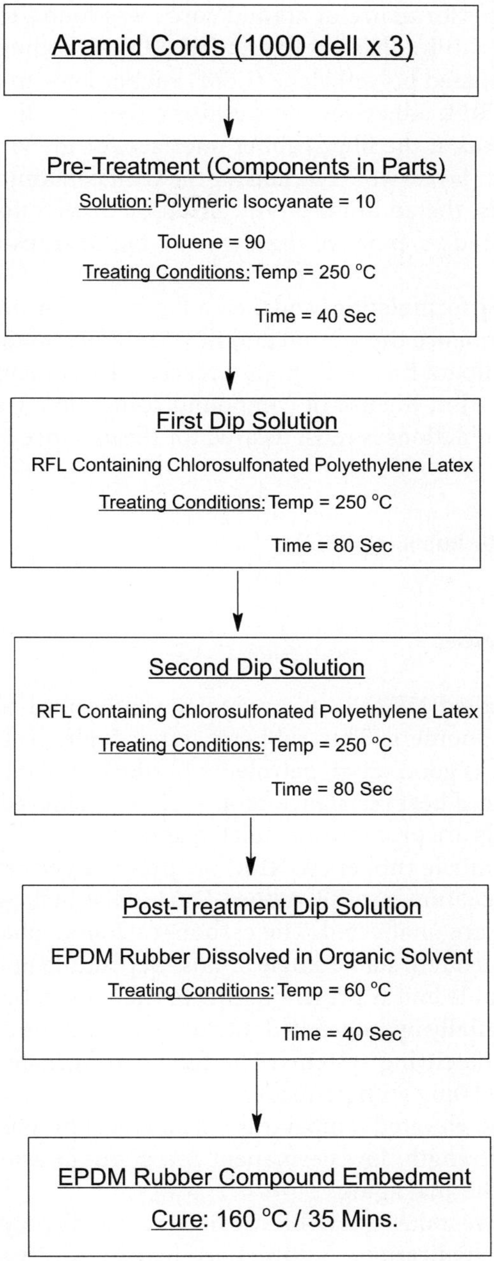
Aramid Cords (1000 dell x 3)
Pre-Treatment (Components in Parts)
Solution: Polymeric Isocyanate = 10
Toluene = 90
Treating Conditions: Temp = 250 °C
Time = 40 Sec
First Dip Solution
RFL Containing Chlorosulfonated Polyethylene Latex
Treating Conditions: Temp = 250 °C
Time = 80 Sec
Second Dip Solution
RFL Containing Chlorosulfonated Polyethylene Latex
Treating Conditions: Temp = 250 °C
Time = 80 Sec
Post-Treatment Dip Solution
EPDM Rubber Dissolved in Organic Solvent
Treating Conditions: Temp = 60 °C
Time = 40 Sec
EPDM Rubber Compound Embedment
Cure: 160 °C / 35 Mins.

**Figure 6.62.**

The adhesion performance of aramid cords was found to be improved by isocyanate and two RFL adhesives treatments [10]. According to this study, the use of chlorosulfonated polyethylene (CSM) rubber latex in the place of vinyl pyridine latex in RFL adhesives increased the dynamic life of the belt. The adhesion failure was at the fiber-rubber interface for the vinyl pyridine latex whereas the rubber layers were cracked for the CSM containing RFL adhesives. In addition to this, the addition of vulcanization accelerator and zinc oxide in the RFL appeared to enhance the dynamic life of transmission belt even further.

Examples of RFL formulations and their adhesive performances clearly show that, in order to enhance the aramid and EPDM rubber compounds adhesion, the treatment of fibers first with highly reactive isocyanate compounds are highly necessary. Also, the use of compound compatible rubber latex in the RFL adhesive formulations is most desired for the enhanced adhesion.

### 6.7.2
### Hydrogenated Nitrile Rubbers (HNBR)

#### 6.7.2.1
#### HNBR Rubber Properties

Nitrile rubber (called NBR) is a copolymer synthesized from acrylonitrile and butadiene monomers. This rubber has good physical and mechanical properties as well as good water, petroleum products and oil resistance. Good low temperature and heat resistance properties are achieved when the nitrile rubber compounds are properly formulated and used.

Hydrogenated nitrile rubbers (HNBR) are produced either by the complete or partial hydrogenation of NBR rubber. During the hydrogenation process, the nitrile groups are unaffected. The carbon–carbon double bonds of nitrile rubber are converted into stable single bonds. Depending upon the conditions employed, the double bonds present in nitrile rubber can be partially or fully hydrogenated. Partially hydrogenated nitrile rubber can be cross-linked using sulfur and peroxide curing systems. The fully hydrogenated nitrile material can be cross-linked only with peroxides.

HNBR has good elevated temperature mechanical properties. This rubber has high tensile strength, low permanent set, good abrasion resistance and stability towards thermal aging.

In the automotive industry, HNBR has been replacing polychloroprene rubber in timing belt applications. Sulfur cross-linked HNBR rubber compounds can withstand higher temperatures longer than chloroprene based compounds and are therefore found to be suitable materials for the timing belts. HNBR belts are known to bring greater safety and give a longer service life.

## 6.7.3
## RFL in Bonding Polyester to HNBR Compounds

Polyester fiber materials, after the conventional RFL treatment, when incorporated into HNBR belt compounds showed good initial adhesion and then the adhesive strength was lowered after use at high temperatures. In order to attain high initial adhesion and high heat-resistant adhesive strength, RFL adhesives were formulated with Vulcabond-E materials and applied twice onto the surface of polyester materials before reinforcement in the HNBR compounds [11]. RFL formulations and the adhesive properties obtained from this work are shown in Table 6.40.

**Table 6.40.** RFL formulations for bonding HNBR with polyester

| Dip formulations (parts) | Vinylpyridine latex (Dip 1) | HNBR latex (Dip-2) |
|---|---|---|
| Resorinol | 16.6 | 11 |
| Formaldehyde (37%) | 14.6 | 16.2 |
| NaOH | 1.3 | 0.3 |
| Water | 333.5 | 238.5 |
| Time for RF reaction (hours at 23 °C) | 2 | 6 |
| Latex added to prepare RFL | 100 | 100 |
| RFL reaction time (h) | 20 | 20 |
| Vulcabond E added in RFL | 80 | 0 |
| Ammonium hydroxide (28%) | 0 | 22.6 |
| Final water | 0 | 47.9 |
| RFL treating conditions | | |
| Temperature (°C) | 245 | 245 |
| Time (min) | 1 | 1 |
| Polyester-EPDM compound adhesion | | |
| Adhesion values (kg/25 mm) | Initial | Heat treatment in air (168 h at 120 °C) |
| 1. Dip-1 alone | 2.4 | 2 |
| 2. Dip-2 alone | 2.5 | 2.1 |
| 3. Dip-1 and Dip-2 used together | 25.5 | 15.2 |

Data from [11]

The initial and heat-resistant adhesion results were observed to be excellent when the two RFL dips were employed in treating polyester cords.

Another published work indicated that polyester fibers were dipped three times before being used in the reinforcement [12, 13]. The RFL formulations and adhesion results are presented in Tables 6.41 and 6.42.

In this work, polyester fibers were dipped first in a toluene solution containing an isocyanate compound and then, in the second step, treated with RFL adhesive containing NBR latex. In the third step, the dip or processing liquid

**Table 6.41.** RFL for bonding polyester with HNBR compounds

| Components in parts | Formulation |
|---|---|
| 1. First step dip (dip-1) | |
| Polymeric isocyanate (PAPI-135) | 10 |
| Toluene | 90 |
| Treating conditions: | |
| Temperature (°C) | 180 |
| Time (min) | 5 |
| 2. Second step dip (RFL) (dip-2) | |
| Resorcinol | 11 |
| Formaldehyde (37%) | 16.2 |
| NaOH (10%) | 3 |
| Water | 235.8 |
| RF aging time (hours at 25 °C) | 6 |
| NBR latex (40%) | 100 |
| Ammonium hydroxide (28%) | 11.3 |
| Water | 59.2 |
| Total | 586.5 |
| RFL aging time (hours at 25 °C) | 20 |
| Treating conditions: | |
| Temperature (°C) | 230 |
| Time (min) | 2 |
| 3. Third step dip (Dip-3) | |
| EPDM rubber | 100 |
| Zinc oxide | 5 |
| Stearic acid | 1 |
| Carbon black | 15 |
| Hydrated silica | 30 |
| Resorcinol resin | 2 |
| HMMM | 4 |
| Promoter (MBTS) | 1.5 |
| Sulfur | 2 |
| Prepare the mix in a Banbury and blend with polymeric isocyanate in toluene | |
| Treating conditions: | |
| Temperature (°C) | 160 |
| Time (min) | 5 |

Data from [12]

contained a rubber compound (called dip-3 compound prepared separately in a Banbury) and polymeric isocyanate. The dip-3 compound and polymeric isocyanate were mixed in different weight ratios, dissolved in a solvent and applied to polyester fibers after the RFL adhesive application. With this approach, excellent adhesion values of polyester with EPDM compounds were seen.

**Table 6.42.** Adhesion performance of RFL treated polyester with HNBR compounds

| Polyester cord adhesion | Experiment | | | | |
|---|---|---|---|---|---|
| | 1 | 2 | 3 | 4 | 5 |
| Dip-3 solids content (wt.%) | 3 | 5 | 7 | 1 | 9 |
| Polysocyanate/dip-3 compound (wt) | 1.0:2.0 | 1.0:2.0 | 1.0:2.0 | 1.0:2.0 | 1.0:2.0 |
| Adhesion, peeling force (N/25 mm) | | | | | |
| Room temperature | 530 | 550 | 550 | 300 | 200 |
| At 100 °C | 240 | 230 | 200 | 100 | 70 |
| After heat age at 120 °C/8 days | 329 | 360 | 280 | 150 | 140 |

Data from [12]

## References


1. H. Ikeda, K. Katayama, M. Hibino and Y. Okado, US Patent Appln, 2002/0,074,050 (2002).
2. J. Suzuki, K. Katayama, H. Ikeda and A. Ikemoto, US. Patent Appln, 2002/0,142,166 (2002).
3. A. Ikemoto, M. Hibino, H. Ikeda and K. Katayama, US Patent Appln, 2003/0,124,284 (2003).
4. A. Ikemoto and K. Senda, US Patent, 6,552,128 (2003).
5. H. Hasaka, US Patent, 6,607,828 (2003).
6. H. Hasaka, K. Takano and T. Kinoshita, US Patent, 6,524,417 (2003).
7. T. Takada, K. Ishida, Y. Hineno and I. Yoshida, US Patent, 6,361,462 (2002).
8. O. C. Elmer, US Patent, 4,285,756 (1981).
9. H. Hasaka, K. Takano and T. Kinoshita, US Patent Appln, 2001/0,034,282 (2001).
10. H. Fujimoto, US Patent Appln, 2003/0,088,7715 (2003).
11. H. Hisaki, O. Mori and M. Oyama, US Patent, 5,112,397 (1992).
12. T. Kinoshita and H. Hasaka, US Patent, 6,358,609 (2002).
13. Y. Fujii and A. Ikeda, US Patent, 6,489,385 (2002).

# 7 Resorcinol Based Polymers

## 7.1 Polyarylates

### 7.1.1 Introduction

Polyarylates or aromatic polyesters essentially consist of aromatic ester groups in the polymer main chain or backbone. The introduction of aromatic groups in the polymer chain results in high bond energies, low degree of reactivity, and rigidity of the polymer chain structure. Therefore, polyarylates are considered as high temperature, high performance thermoplastic polymers with a good combination of thermal and mechanical properties [1–5]. Polyarylates, in general, have inherent flammability and combustion resistance, good unnotched toughness as well as good weatherability properties. These polymers have exhibited good melt stability at high temperatures and good color retention. They show good processability which makes them suitable for various molded article applications.

Aromatic polyester is made from an aromatic dihydroxy compound and an aromatic dicarboxylic acid. Though a large number of aromatic dihydroxy compounds have been utilized in the synthesis and manufacture of polyarylates, the use of resorcinol and resorcinolic compounds in the development of polyesters has been the main focus in this chapter.

Resorcinol and resorcinol derived compounds have been employed to synthesize aromatic polyester materials, which are used in various applications. With resorcinolic derivatives, homo and copolyesters were developed and studied. In this chapter, the different synthesis procedures used in the preparation of resorcinol based aromatic homo and copolyesters, their compositions, properties, liquid crystalline polyesters and the potential applications are discussed.

### 7.1.2 Synthesis of Aromatic Polyesters or Polyarylates

Polyarylates of resorcinol and its derivatives can be synthesized by the reaction of aromatic dicarboxylic acids with resorcinol (Direct Esterification), diacetates of resorcinol with aromatic dicarboxylic acids (Diacetate Method), resorcinol and aromatic dicarboxylic acid diaryl esters (Diphenate Process), and also resorcinol with chlorides of aromatic dicarboxylic acids (Solution and Interfacial Polycondensation Processes).

### 7.1.2.1 Direct Esterification Method

Resorcinol polyarylates can be obtained by the direct condensation of resorcinol with aromatic dicarboxylic acids, such as terephthalic acid using the following reaction (Figure 7.1).

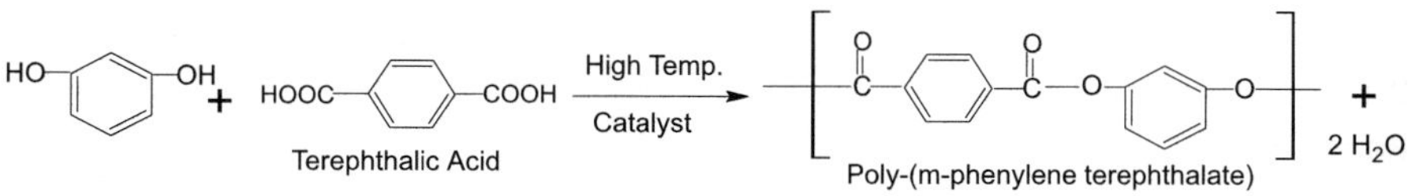


**Figure 7.1.**

In order to obtain high molecular weight from this method, longer reaction times and elevated temperature conditions are often required. This method, even when catalyzed, tends to produce lower molecular weights, and therefore has limited commercial applications.

### 7.1.2.2 Diacetate Method

Polyarylates can be made by the reaction of an aromatic dicarboxylic acid with resorcinol diacetate compound, and this process is also known as the melt polycondensation method [6] (Figure 7.2).

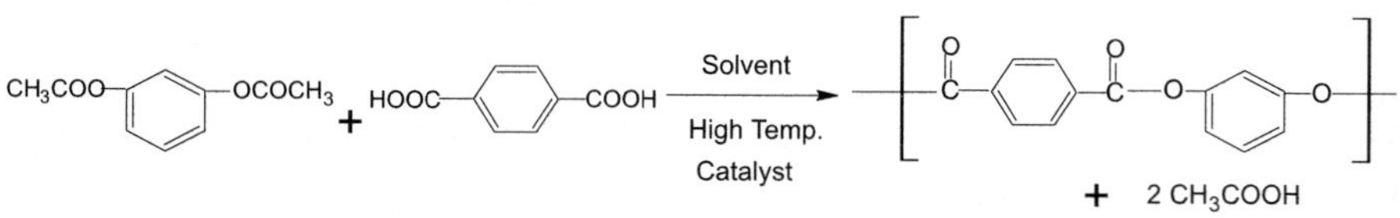


**Figure 7.2.**

In this method, the production of high molecular weight polyesters is a problem due to the potential loss of reagents out of the reaction vessel associated with their sublimation characters, which can upset the reaction stoichiometry. To overcome this difficulty, a variety of solvents, namely, sulfolane, *N*-methyl pyrrolidone, hexamethyl phosphoramide, diphenyl ether, biphenyls and sulfones, have been employed in the polymerization reactions.

A preparative procedure used in the synthesis of homo polyarylates from resorcinol is given below [6].

**Preparation of Poly-(*m*-phenylene terephthalate) by the Diacetate Process [6]** In a three-necked flask equipped with a stirrer, thermometer, and Dean–Stark take off column, were placed 19.61 g of *m*-phenylene diacetate (0.101 mol), 16.61 g of terephthalic acid (0.100 mol) and 218 g of tetrachloro biphenyls. The reaction mixture was stirred and heated with a heating mantle. After 16 min the reaction temperature reached 281 °C and acetic acid started to distill. The reaction mixture was refluxed at 327 to 332 °C for a total of 6.72 h. During this time the acetic acid was gradually distilled and collected. The resulting and slightly viscous brown solution was allowed to cool to precipitate the polymer. The acetic acid obtained was a total of 98.5% yield of acetic acid. The polymer was isolated and washed four times with 1-l portions of boiling acetone. It was filtered and dried to give 24.01 g (99.0% yields) of poly-(*m*-phenylene terephthalate) melting at 288 – 291 °C. The polymers produced by this method showed an intrinsic viscosities in the range of 0.5 to 0.6 determined in 2,4,6-trichloro phenol at 75 °C.

The rate of reaction of diacetate of resorcinol and aromatic dihydroxy compounds with aromatic dicarboxylic acids can be enhanced by employing low levels (about 50 – 150 ppm) of esterification catalysts based on antimony, titanium, magnesium and zinc salts. Instead of preparing the resorcinol diacetate separately and using it in the polymerization reaction, very often this was prepared in situ in the polymerization reactor in the presence of dicarboxylic acids and then continued with the polymerization [7]. This method has been widely used in the preparation of liquid crystalline and amorphous polyesters.

### 7.1.2.3
### Diphenyl Ester Process

Another melt polycondensation process has been used in the synthesis of resorcinol polyarylates, and this method is also known as the diphenyl ester or diphenate process.

Based on this process, the reaction between resorcinol and diaryl ester of terephthalic acid resulted in a high molecular weight polyester product. Though the ester interchange reaction of the diphenate process can proceed in the absence of a catalyst, the reaction rate can be increased by using catalysts [8] (Figure 7.3).

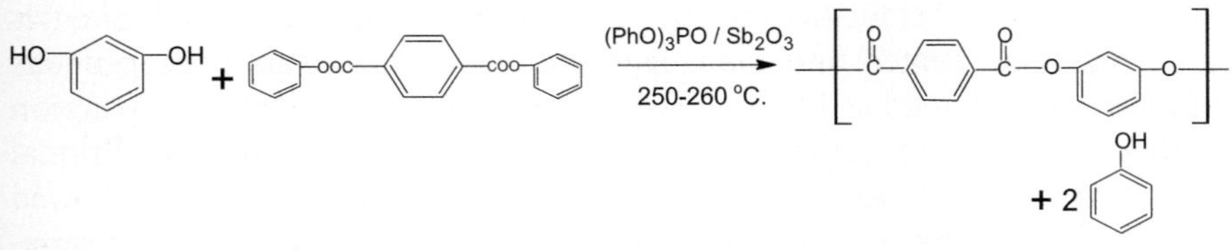


**Figure 7.3.**

The following procedure was used in the preparation of resorcinol polyarylates.

**Preparation of Poly-(*m*-phenylene terephthalate) by Diphenyl Ester Process [8]** A reactor was charged with 1000 parts of diphenyl terephthalate, 346 parts of resorcinol, 1.03 parts of triphenyl phosphate and 0.18 part of antimony trioxide, and they were reacted at 250 °C for 22 h while distilling off the resulting phenol out of the reaction system. Then, 15 parts of resorcinol was added to the reaction mixture, and reacted at 260 °C for 4 h. The pressure of the reaction system was gradually reduced over the course of 4 h, and the pressure was finally adjusted to 0.7 mm Hg. At the same time, the temperature was raised to 345 °C. There was obtained poly-1,3-phenylene-terephthalate having an intrinsic viscosity of 0.67 and a melting point of 290 °C.

Polyarylates prepared by the diphenate process, in general, compare favorably with the diacetate process with respect to product quality. But, the limiting factor for this process is the preparation of diphenyl ester of terephthalic acid, which could be very expensive.

### 7.1.2.4 Dicarboxylic Acid Chloride Methods

Polycondensation of aromatic dicarboxylic acid chlorides with resorcinol can be accomplished in three different ways, namely, high temperature solution, low temperature solution and interfacial methods.

**High Temperature Solution Process** The high temperature reaction of resorcinol and terephthaloyl chloride was performed in the presence of high boiling solvents to obtain poly-(*m*-phenylene terephthalate) [6] (Figure 7.4).

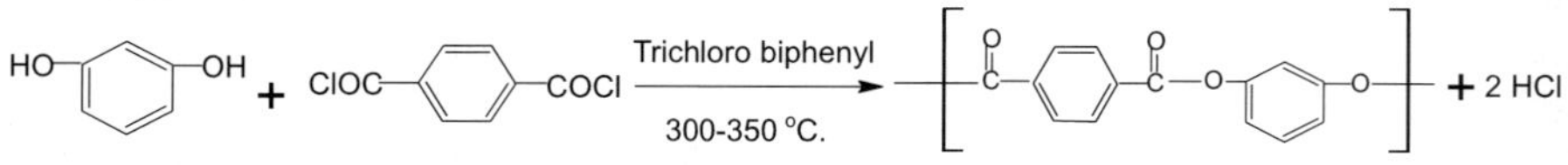


**Figure 7.4.**

The following procedure was employed in the polyester preparation.

**Preparation of Poly-(*m*-phenylene terephthalate) by High Temperature Solution Process [6]** A mixture of 11.2 g (0.102 mol) of resorcinol, 20.3 g (0.100 mol) of redistilled terephthaloyl chloride, and 209 g of redistilled mixed trichloro biphenyls was stirred and heated under nitrogen atmosphere. A clean, homogeneous solution was obtained at 140 °C. After 9 min, the temperature of the reaction reached 335 °C. The polymerization was allowed to continue for an additional 3 min at 330–335 °C to form a viscous solution. The mixture was allowed to cool whereby the polymer precipitated at 185 °C. The polymer was separated by adding acetone and filtering. The solid was washed three times with

1-l portions of acetone, filtered and dried to yield 20.8 g (86%) of white poly-(*m*-phenylene terephthalate) melting at 281 – 295 °C. A sample of this polymer had an intrinsic viscosity of 0.66 in 2,4,6-trichloro phenol at 75 °C.

Strong, flexible, transparent films of poly-(*m*-phenylene terephthalate) were obtained by pressing this polymer between aluminum foil at 350 – 380 °C and pressures of 7 – 15 MPa. An amorphous film prepared from this polyester exhibited tensile strength of 75.9 MPa, and an elongation of 31% and a yield point of 69.4 MPa.

**Low Temperature Solution Polycondensation Process** The preparation of resorcinol polyarylates from the low temperature solution process could be generally carried out with a stoichiometric balance of resorcinol and thaloyl chlorides (terephthaloyl or isophthaloyl and the like) in the presence of an inert solvent (Figure 7.5).

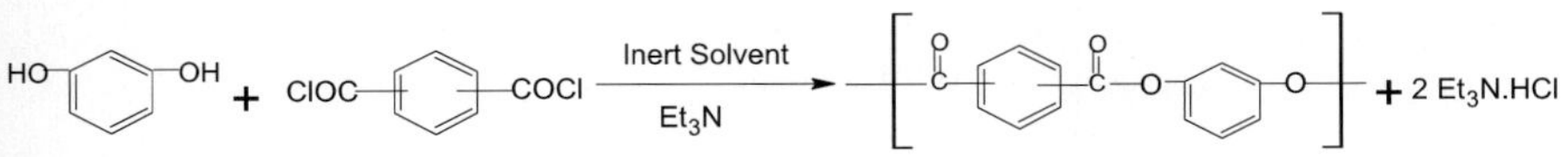


**Figure 7.5.**

This polymerization reaction could be done at room temperature or low temperatures.

**Interfacial Polycondensation Process** In the interfacial polycondensation process, the disodium salt of resorcinol was contacted with a solution of terephthaloyl chloride in a water-immiscible solvent to obtain resorcinol polyarylates [8] (Figure 7.6).

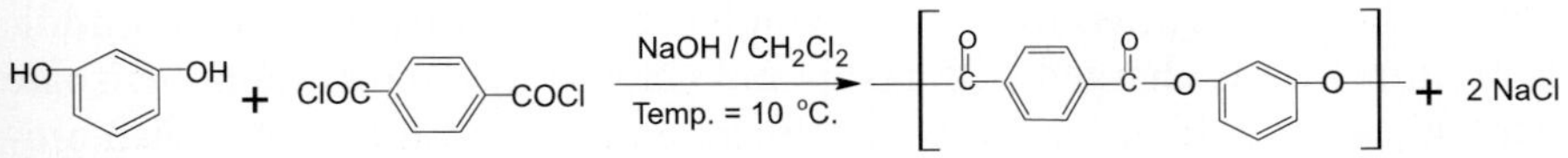


**Figure 7.6.**

An emulsion or surfactant is normally added, and then the solution is rapidly mixed in this process. This method has been the versatile one to produce high molecular weight polyarylates. By employing phase transfer catalysts, such as benzyltrimethylammonium chloride, high molecular weight polyesters are produced, and may be due to increase in the rate of polymerization.

The following interfacial polycondensation procedure was used to synthesize resorcinol arylate polymers.

**Preparation of Poly-(*m*-phenylene terephthalate) by Interfacial Polycondensation Process [8]** An alkaline aqueous solution of resorcinol was prepared from 110.1 parts of resorcinol, 82.7 parts of sodium hydroxide, 60 parts of sodium laurylsulfate, 2 parts of sodium hydrosulfite and 6600 parts of water, and cooled to 10 °C. The aqueous solution was vigorously stirred, and a solution of 211.2 parts of terephthaloyl dichloride in 3030 parts of methylene chloride was added to the aqueous solution over the course of 30s. With stirring, the mixture was reacted at 10 °C for 1 h. After the reaction, the reaction mixture was poured into a large amount of acetone to precipitate the resulting poly-1,3-phenylene terephthalate. The precipitate was recovered by filtration, and repeatedly washed with water. The purified poly-1,3-phenylene terephthalate had an intrinsic viscosity of 1.30 and a melting point of 290 °C.

The preparation of resorcinol arylate polymers containing both *m*-phenylene isophthalate and *m*-phenylene terephthalate groups in the polymer main chain were carried out based on the following reaction scheme and the procedure given below [9] (Figure 7.7).

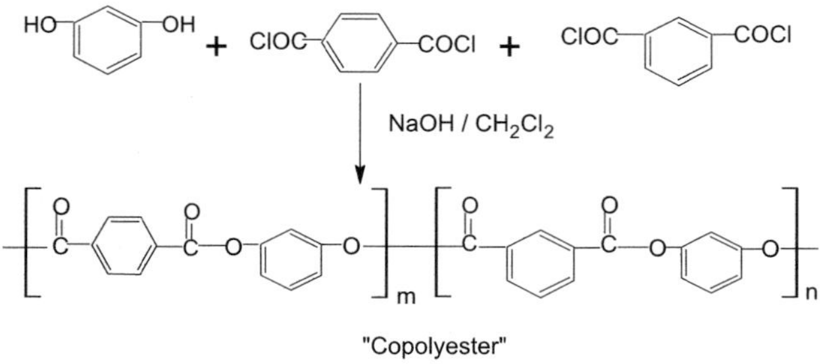


**Figure 7.7.**

**Synthesis of Resorcinol Arylate Polyester Containing Iso- and Terephthalates [9]** To a 1-l, five neck, Morton flask blanketed with nitrogen and equipped with a mechanical stirrer, pH electrode, reflux condenser, two pressure equalizing addition funnels, was charged with resorcinol (21.8 g), resorcinol monobenzoate capping agent (1.07 g; 2.5 mol%), triethylamine (0.274 ml; 1 mol%), dichloromethane (150 ml), and water (100 ml). One addition funnel was charged with sodium hydroxide pellets (16.84 g; 0.42 mol) and water (32 ml, while a solution of isophthaloyl dichloride (20.3 g; 0.1 mol), terephthaloyl dichloride (20.3 g; 0.1 mol), and dichloromethane (150 ml) was added to the second. The pH of the reaction mixture was adjusted to 7.5 with sodium hydroxide prior to the addition of acid chloride solutions, which were added over 6 min. The pH of the reaction was maintained between 7.25 and 7.75 for the first 10 min of reaction. At 11 min the reaction the pH was raised to about 10 with the addition of sodium hydroxide and held for an additional 10 min. The stirring was stopped and the aqueous layer was removed. The resulting gray organic layer was washed with 1 N hydrochloric acid, 0.1 N hydrochloric acid, water (three times), and the polymer isolated by precipitation into boiling water yielding a white, fibrous material which was dried in vacuum at 110 °C overnight. The isolated polymer was the desired resorcinol arylate polyester.

## 7.1.3
## Structure and Properties of Polyarylates

### 7.1.3.1
### Homopolyesters of Resorcinol

Homopolyarylates synthesized from resorcinol and isophthalic acid, and terephthalic acid have the following structures (Figure 7.8).

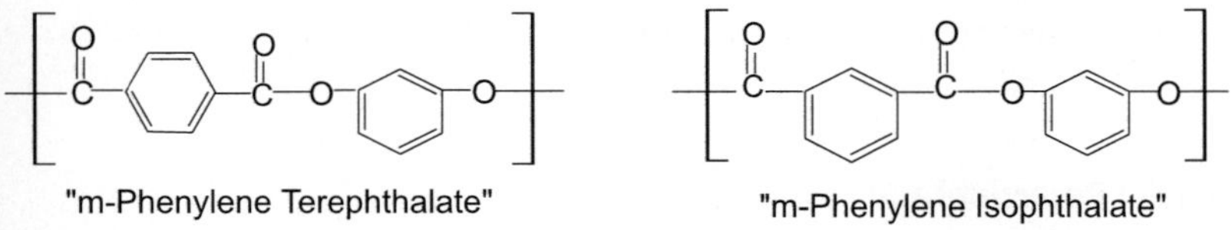


**Figure 7.8.**

The use of resorcinol in the polyester preparations introduces *m*-phenylene linkages into the polymer main structure, which can increase the polymer chain flexibility. As a result, the polyesters could exhibit lower melt temperatures, low glass transition temperatures, improved processability and enhanced solubility in various solvents.

Poly-(*m*-phenylene isophthalate) obtained from the interfacial polycondensation process showed polymer inherent viscosity of 1.38 dL/g (measured in tetrachloro ethane/phenol solvents at 30 °C) and exhibited very tough film properties with a polymer melt temperature of 245 °C [10].

Mono-axially stretched film of poly-(*m*-phenylene terephthalate) had a tensile strength value of 141.1 MPa and an elongation of 36% at break [8]. When this stretched film was subjected to heat treatment for 24 h at 150 °C and tensile tested, it showed a strength value of 140.1 MPa. This data strongly suggests that the films produced from resorcinol arylate polymers have good heat resistance property. Though the presence of *m*-phenylene structure is expected to decrease the crystalline character of the polyester, the film produced from the poly-(*m*-phenylene terephthalate) showed about 25% crystallinity with a melting point in the range of 281 – 295 °C [6,8].

The hot films of poly-(*m*-phenylene terephthalate) were observed to be amorphous and completely transparent, and also flexible and quite tough. It was suggested that fibers could be drawn either from the hot melt or by extruding the polyester through a die to form monofilaments [4]. Strength and toughness of these fibers could be enhanced by stretching them to orientation and crystallization process at 200 – 250 °C temperature conditions.

The combustion characteristics of poly-(*m*-phenylene terephthalate) was determined to be V-0 from a flame resistance test performed according to UL-94746 method. This polyester sheet was carbonized and determined to be self-extinguishing.

Aromatic polyesters are soluble only in limited number of solvents. The major factor influencing the polyester solubility is believed to be its crystallinity. Solubility improvements in polyarylates have been achieved by synthesizing polymers with increased structural irregularity in the main chain, either by introducing long or bulkier substituents or by copolymerizing with different monomers. With resorcinolic monomers, polymer chains can be flexible, and therefore the resultant polyesters could be more soluble than hydroquinone based polymers. Copolyesters synthesized from 2-methyl- and 4-benzyl resorcinol have shown that the presence of methyl substitution on the resorcinol ring did not change the solubility characteristics of the polyesters [11].

### 7.1.3.2
### Copolyesters of Resorcinol and Properties

Copolyarylates of resorcinol and its derivatives were prepared by employing two or more aromatic dicarboxylic acids or using another aromatic dihydroxy compound as a comonomer with resorcinol in the synthesis. The properties and applications of copolyesters are strongly depended on the nature of comonomers used.

Copolyesters were prepared using resorcinol, terephthalic and isophthalic acid esters by the melt polycondensation process [8]. They were then compression molded, monoaxially stretched to obtain films of various thicknesses and tested for tensile properties. Table 7.1 presents the copolyester compositions and properties.

**Table 7.1.** Copolyester composition and monoaxially stretched film properties

| Polymer composition (mol%) | Polyesters | | | |
|---|---|---|---|---|
| | 1 | 2 | 3 | 4 |
| Resorcinol | 50 | 50 | 50 | 50 |
| Terephtahlic acid | 50 | 49 | 47.5 | 42.5 |
| Isophthalic acid | 0 | 1 | 2.5 | 7.5 |
| Tensile properties before heat treating | | | | |
| Strength (MPa) | 185.2 | 205.8 | 186.2 | 176.4 |
| Elongation (%) | 12 | 17 | 23 | 25 |
| After heat treating at 150 °C/24H | | | | |
| Strength (MPa) | 186.2 | 200.9 | 185.2 | 177.4 |

Data from [8]

As the resorcinol isophthalate content of the polyester was increased, tensile elongation was increased with minimal effect on the strength properties. Poly-(*m*-phenylene phthalates) showed excellent heat resistance, fire retardancy, electrical, high tensile strength and elongation at break, and high degree of surface smoothness properties.

Copolyesters containing *m*-phenylene terephthalate groups interspersed with *p*-phenylene terephthalate groups were synthesized using resorcinol and hydroquinone monomers [6] (Figure 7.9).

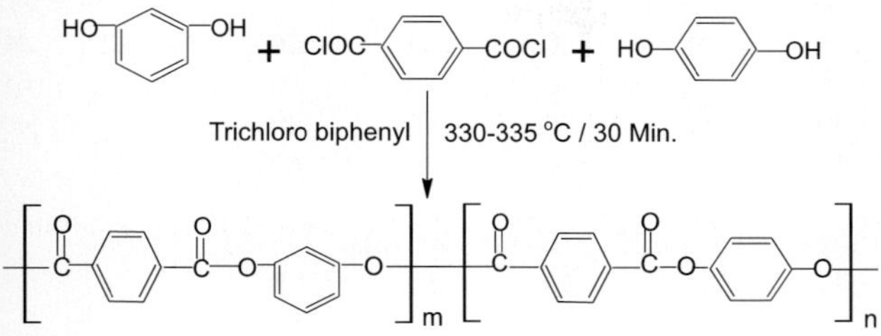


**Figure 7.9.**

These polymers were considered as super polyesters, since the impact strength, flexibility and tensile properties obtained from these materials were higher than the regular polymers. Polyester compositions and properties are shown in Table 7.2.

**Table 7.2.** Homo and copolyesters of resorcinol prepared by the high temperature solution method using terephthaloyl chloride

| Diphenol monomer (mol%) | Polyesters | | | | |
|---|---|---|---|---|---|
| | 1 | 2 | 3 | 4 | 5 |
| Resorcinol | 100 | 90 | 80.4 | 75 | 67.6 |
| Hydroquinone | 0 | 10 | 19.6 | 25 | 32.4 |
| Polyester properties | | | | | |
| Yield (%) | 86 | 89 | 95 | 89 | 79 |
| Intrinsic viscosity (dL/g)[a] | 0.66 | 0.6 | 0.53 | 0.95 | 1.23 |
| Melting point (°C) | 281–295 | 273–362 | 261–284 | 295–336 | 331–386 |

[a]Determined in 2,4,6-trichlorophenol at 75 °C
Data from [6]

In order to obtain super polyesters from resorcinol and hydroquinone monomers, it was suggested that the terephthaloyl chloride reaction be carried out first with resorcinol and then hydroquinone or the terephthaloyl chloride reaction be carried out simultaneously with a mixture of resorcinol and hydroquinone. As the hydroquinone content of the polyester was increased, the polymer melting point was also increased, which might be due to the presence of higher levels of rigid and symmetrical *p*-phenylene terephthalate structure in the polymer chain. These polyesters could be structurally oriented by cold drawing techniques to provide films and fibers, and expected to be more flexible and to show higher strengths.

Isophthalic acid, in addition to terephthalic acid, was employed in the synthesis of copolyesters containing resorcinol and hydroquinone by the high temperature solution polycondensation process (Table 7.3) [12, 13].

**Table 7.3.** Copolyesters of resorcinol prepared from the solution polycondensation method

| Polyarylate composition | Polyester | | | | |
|---|---|---|---|---|---|
| Monomer used (mol%) | 1 | 2 | 3 | 4 | 5 |
| Resorcinol | 75 | 75 | 50 | 50 | 60 |
| Hydroquinone | 25 | 25 | 50 | 50 | 40 |
| Isophthaloyl chloride | 80 | 60 | 50 | 40 | 40 |
| Terephthaloyl chloride | 20 | 40 | 50 | 60 | 60 |
| Reaction conditions | | | | | |
| Temperature (°C) | 300–315 | 310–325 | 310–325 | 305–320 | 315–332 |
| Time (min) | 21 | 15 | 20 | 10 | 5 |
| Polymer properties | | | | | |
| Yield (%) | 98.3 | 98.3 | 95 | 93.8 | 86.8 |
| Intrinsic viscosity (dL/g)[a] | 0.76 | 0.75 | 0.71 | 0.88 | 0.93 |
| Melting point (°C) | 295–323 | 287–319 | 229–285 | 320–341 | 315–332 |

[a]Measured in 2,4,6-trichlorophenol at 75 °C
Data from [12], [13]

Polyesters containing higher resorcinol content did not crystallize and formed amorphous and transparent films when the hot films were quenched or allowed to cool slowly in the mold. When the films were subjected to elevated temperature conditions, they exhibited excellent tensile strength properties as shown in Table 7.4.

**Table 7.4.** Elevated temperature tensile properties of poly-(*m*-phenylene-*p*-phenylene-isophthalate terephthalate)

| | Tensile properties | |
|---|---|---|
| Temperature (°C) | Strength (MPa) | Elongation (%) |
| 25 | 62.2 | 11 |
| 100 | 43.4 | 28 |
| 150 | 24.7 | 68 |
| 200 | 5.1 | 390 |

Molar ratios: resorcinol/hydroquinone = 50:50
Isophthaloyl/terephthaloyl chloride = 40:60
Data from [12], [13]

Resorcinol derivatives were also used to prepare polyesters by the interfacial polycondensation process based on the reaction shown in Figure 7.10 and the details are presented in Table 7.5 [10, 11].

The copolyesters prepared from the resorcinol derivatives were observed to be more soluble than the resorcinol homopolyarylates.

Copolyesters of resorcinol terephthalate containing terephthalate esters of various aromatic dihydroxy compounds were synthesized and their film properties were studied (Table 7.6) [8].

**Table 7.5.** Polyesters synthesized from resorcinol derivatives and properties (isophthaloyl chloride/terephthaloyl chloride (mol ratio) = 50:50)

| Resorcinol derivatives (mol%) | Polyester[a] 1 | 2 | 3 | 4 | 5 | 6 | 7 | 8 |
|---|---|---|---|---|---|---|---|---|
| Resorcinol | 100 | 75 | 50 | | | | | |
| 2-Methyl resorcinol | | 25 | 50 | 100 | | | | |
| 4-Chloro resorcinol | | | | | 100 | | | |
| 4-Acetyl resorcinol | | | | | | 100 | | |
| 2,4-Dihydroxy benzophenone | | | | | | | 100 | |
| 2,4-Dihydroxy benzaldehyde | | | | | | | | 100 |
| Polyester properties | | | | | | | | |
| Yield (%) | 97 | 88 | 96 | 97 | NA | 61 | NA | NA |
| Inherent viscosity[b] | 0.91 | 0.7 | 0.47 | 0.57 | 1.47 | 0.18 | 0.38 | 0.23 |
| Appearances[c] | | | | | | | | |
| Film | F | F | F | F | T | B | B | B |
| Resin | C | C | C | Sl.Y | NA | C | NA | NA |

[a] Polyesters prepared by the interfacial polycondensation method
[b] Measured at 30 °C in 70/30 (v/v) phenol/tetrachloro ethane solution
[c] C = Colorless, F = Flexible, B = Brittle, T = Tough and Sl.Y = Slightly yellow
Data from [10], [11]

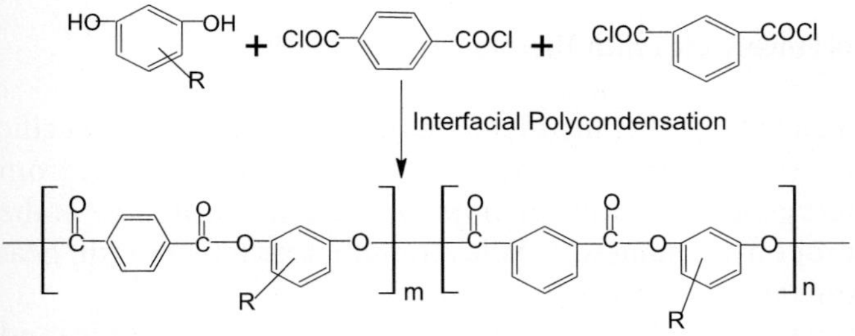


**Figure 7.10.**

Stretched copolyester films have shown excellent tensile properties before and after heat treatments. Possible applications for polymers possessing these properties are in electrical insulations, magnetic tapes, photographic films and metal vapor-deposited films. Specifically, these polyesters could be useful for insulation in electrical machines such as motors and transformers, electric cables and wires, condensers, insulative adhesive tapes and printed circuit boards' applications.

**Table 7.6.** Copolyarylate composition and monoaxially stretched film properties (resorcinol and different diphenol compounds)

| Copolyester composition (mol%) | Polymer 1 | 2 | 3 | 4 | 5 | 6 | 7 | 8 |
|---|---|---|---|---|---|---|---|---|
| Terephthalic acid | 50 | 50 | 50 | 50 | 50 | 50.2 | 49.6 | 47.6 |
| Resorcinol | 50 | 47.5 | 42.5 | 47.4 | 49 | 46.3 | 49.1 | 47.6 |
| 4,4′-Dihydroxydiphenyl thioether | | 2.5 | 7.5 | | | | | |
| 3-Hydroxyphenyl-4-hydroxy-benzoate | | | | 2.6 | | | | |
| 1,3,5-Trihydroxybenzene | | | | | 1 | 0.9 | | |
| 4-Hydroxybenzoic acid | | | | | | 2.6 | | |
| Bisphenol-A | | | | | | | 0.9 | |
| Hydroquinone | | | | | | | | 2.4 |
| Isophthalic acid | | | | | | | 0.4 | 2.4 |
| Tensile properties | | | | | | | | |
| Before heat treating | | | | | | | | |
| Tensile Strength (MPa) | 185.2 | 155.8 | 151.9 | 176.4 | 178.4 | 180.3 | 154.8 | 168.6 |
| Elongation (%) | 12 | 29 | 35 | 23 | 25 | 25 | 27 | 32 |
| After heating at 150 °C/24 h | | | | | | | | |
| Tensile strength (MPa) | 186.2 | 156.8 | 149 | 173.5 | NA | NA | 152.3 | NA |

Data from [8]

### 7.1.4 Blending of Resorcinol Polyarylates with Thermoplastics

Polymers are often blended together in order to optimize properties for specific applications at the lowest cost. More often, the compatible blends made from two different polymers generally exhibit an improvement in one or more of the following physical properties, namely, tensile strength, impact strength, heat distortion temperature and higher melting transitions.

Polyarylates, in general, are known for their high temperature stability and excellent mechanical property retention after long term UV exposure. Poly(ethylene terephthalate), PET, is a semicrystalline aliphatic polyester known for its outstanding mechanical, thermal and solvent resistance properties. Semicrystalline PET was blended with amorphous copolyesters obtained from resorcinol, bisphenol-A and isophthalic acid monomers. Thermal transition properties obtained from the DSC analysis on these blend materials are shown in Table 7.7 [14].

Resorcinol arylate polyester and PET blends have showed an improved Tg over PET, while retaining the crystallinity of PET. Tg values determined after annealing the blend samples indicated that the presence of resorcinol arylates in the blends did not interfere with the crystallinity of PET.

Polycarbonates are known for their excellent physical properties including high impact strength, but can lose their gloss, transparency and color on expo-

**Table 7.7.** Resorcinol based polyarylate and PET blends and their properties

| Polyarylate composition (mol%) | Polymer 1 | 2 | 3 | PET |
|---|---|---|---|---|
| Isophthalic acid | 50 | 50 | 50 | |
| Resorcinol | 25 | 30 | 50 | |
| Bisphenol-A | 25 | 20 | | |
| Glass transition temperature (°C) | 168 | 155 | 150 | |
| Polyarylate/PET blends | | | | |
| PET content in blend (wt.%) | 50 | 50 | 50 | 100 |
| DSC analysis results of blends | | | | |
| Glass transition temperature (°C)[a] | 125 | 116 | 106 | 87 |
| Crystallization temperature ($T_c$), °C | 189 | NA | NA | 178 |
| Melting temperature (°C) | 241 | 221 | 232 | 255 |

[a]Determined after annealing the sample at 150 °C for 2 h
Data from [14]

sure to UV radiation. The weatherability of polycarbonates could be improved by blending them with resorcinol arylate polymers [15].

## 7.1.5 Weatherable Polyesters from Resorcinol

Most polymeric materials are known to degrade on continuous exposure to sunlight or light sources associated with UV radiation. Upon exposure to UV sources, polymers could attain changes in their colors, appearances, loss of gloss and shape. The yellowing of polymers caused by UV light or radiation of sunlight is also known as photoyellowing. In polymers, photoyellowing may result in the degradation of polymers' physical properties, such as impact strength and heat distortion temperatures. In order to protect these polymers against the effects of UV radiation, they are often coated with good weatherable polymers or coatings containing the UV absorbers.

Polyarylates were known to possess UV stability and reported as good weatherable polymers for the protection of plastic materials [11]. Resorcinol based arylate polymers, prepared using the isophthalic and terephthalic acid monomers, were observed to show excellent weatherability properties and provided protection against photoyellowing of thermoplastics, and have been studied extensively [16–22]. The aromatic ester group of resorcinol polyarylates may undergo a photochemical Fries rearrangement to form an *o*-hydroxybenzophenone structure (Figure 7.11) [23].

Resorcinol based benzophenone compounds are well known UV absorbers and have been used in most of the thermoplastic materials to protect them from the harmful effects of UV radiation. From the resorcinol arylate group, the formation of mono as well as the dibenzoyl type structures by the Fries

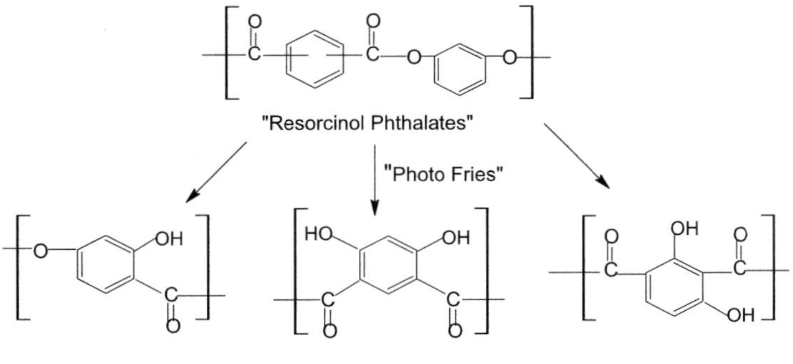


**Figure 7.11.**

reaction could enhance the UV stabilizing properties of resorcinol based polyarylates. Coating formulations made from the resorcinol polyarylates were known to protect various polymers against photoyellowing. When exposed to UV light, the *o*-hydroxybenzophenone structure formed from the resorcinol arylate groups at the polymer surface may provide a thin uniform skin, which may be opaque in the UV light but transparent to visible light. This skin, thus formed in situ during the irradiation, may act as a protective coating by preventing the interior from UV degradation. In this way, the UV sensitive polymers or components are protected from the damaging actions of UV radiations with highly effective resorcinol arylate containing compositions.

Weatherable resorcinol arylate polymers may provide a clear, soluble, and linear and non-crosslinked polyesters, which could easily rearrange in the presence of UV light to form a clear, stable barrier coating for ultraviolet light. Block copolyesters containing resorcinol arylates and soft blocks of resorcinol dicarboxylates were prepared from resorcinol, isophthaloyl chloride, sebacoyl chloride, 1,4-cyclohexane dicarboxylic acid chloride and 2,6-naphthalene dicarboxylic acid chloride monomers [24]. These block copolyesters exhibited little change in color upon exposure to UV light. Table 7.8 presents the copolyester compositions and their glass transition temperatures.

Table 7.8 results show that the Tg value of the copolymer can be controlled by the nature and concentrations of the soft block groups. All the resorcinol based polyester compositions gave strong, essentially transparent, and colorless films. After exposure to accelerated weathering conditions, these films exhibited a superior color retention compared to bisphenol-A-terephthalate polymer.

Coating compositions developed from weatherable resorcinol arylate polymers could be useful in the applications for the protection of multilayer articles such as thermoplastic polymers, thermoset polymers, cellulosic materials, glass and metals against discoloration, loss of gloss and physical and mechanical properties due to UV radiation.

**Table 7.8.** Weatherable resorcinol based soft blocks containing copolyesters

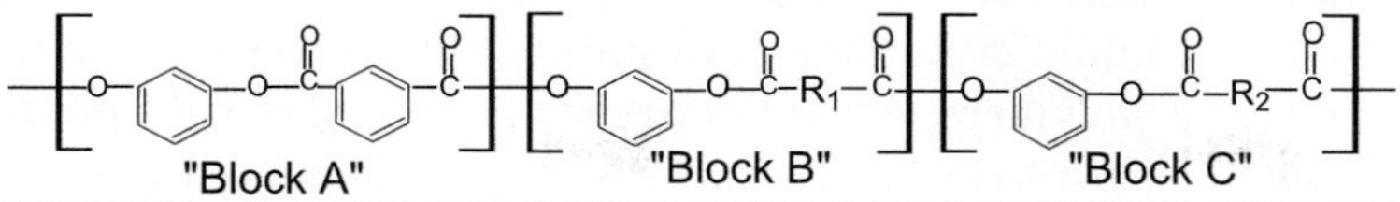


| Isophthalate (block A) (mol%) | Copolyester 1 (block A and B) R1 = Sebacate | | Copolyester 2 (block A and B) R1 = 1,4-CHD | | Copolyester 3 (blocks A, B and C) | | | Copolyester 4 (blocks A, B and C) | | |
|---|---|---|---|---|---|---|---|---|---|---|
| | | | | | R1 = TA | R2 = S | | R1 = NDA | R2 = S | |
| | mol% | Tg (°C) | mol% | Tg (°C) | mol% | mol% | Tg (°C) | mol% | mol% | Tg (°C) |
| 90 | 10 | 116 | 10 | 143 | | | | | | |
| 80 | 20 | 99 | 20 | 130 | 10 | 10 | 120 | 10 | 10 | 122 |
| 70 | 30 | 78 | 30 | 129 | 20 | 10 | 120 | 20 | 10 | 123 |
| 60 | 40 | 59 | 40 | 126 | 30 | 10 | 121 | 30 | 10 | 133 |
| 50 | 50 | 44 | 50 | 122 | 40 | 10 | 120 | 40 | 10 | 139 |
| 50 | | | | | 48 | 2 | 142 | | | |
| 50 | | | | | 45 | 5 | 131 | | | |
| 40 | | | | | 50 | 10 | 123 | 50 | 10 | 145 |
| 30 | | | | | 60 | 10 | 126 | | | |

Note: TA = Resorcinol terephthalate, S = Resorcinol sebacate, IA = Resorcinol isophthalate
1,4-CHD = Resorcinol-1,4-cyclohexane dicarboxylate, NDA = Resorcinol-2,6-naphthalate
Data from [24]

## 7.1.6 Liquid Crystalline Polyesters from Resorcinol

A wide variety of polymeric materials has been constantly under development to fulfil the need for various applications. As the demand for high performance polymers was required, aromatic character of the polymer was increased, which provided higher thermal stability and greater mechanical properties. The search for polymers with exceptional mechanical properties resulted in the break through technology of developing liquid crystalline polymers. These polymers possess a parallel ordering of the molecular chains and are prepared from monomers, which are generally long, flat and fairly rigid along the axis of the molecule. Liquid crystalline polyesters have inflexible (rigid) molecular structure and anisotropic melt phase.

Liquid crystalline (LC) polyesters have several advantages over the general polyesters which can form isotropic melts. Anisotropic melts of LC polymers can show greatly reduced viscosities compared to the isotropic melts of similar polymers. As a result, LC polymers could be easily fabricated from their melts. In addition, melts of LC polymers orient much more readily during flow than from melts of isotropic polymers. Upon molding, the inflexible polymer molecules highly orient to give molded parts with excellent mechanical properties as well as high heat resistance. Due to outstanding properties, liquid crystalline polymers are known as super engineering plastics.

Liquid crystalline polymers can be classified as either lyotropic or thermotropic. Lyotropic liquid crystalline polymers (LLCPs) exhibit liquid crystallinity in solutions. Thermotropic liquid crystalline polymers (TLPs), on the other hand, show liquid crystallinity in the melt phase.

Thermotropic liquid crystalline polyesters are well known [25–29]. These are thermoplastic polymers, which can be melt processed to yield products of exceptionally high performance characteristics [30]. Some of the advantages of LCP materials are:

- Outstanding melt processability, even in thin sections
- Excellent chemical resistance, even at high temperatures
- High impact strength
- Extremely high strength and modulus
- Excellent dimensional stability
- Low moisture absorption

In general, liquid crystalline polyarylates could be derived from one or more of the following monomers: *p*-hydroxybenzoic acid, terephthalic acid, isophthalic acid, hydroquinone and its derivatives, resorcinol, 4,4′-biphenol, 2,6-naphthalene diol, 2,6-naphthalene dicarboxylic acid and 6-hydroxy-2-naphthoic acid.

In the case of liquid crystalline polyesters, the increase in aromatic character and linearity of the polymer chain not only enhances the overall balance of mechanical properties, but also increases the melting points and processing temperatures. Polyesters with lower melting points are required to melt process them into conventional articles, such as fibers and molded articles, using conventional equipments. Higher melting points of liquid crystalline polyesters could be lowered by employing resorcinol as the flexible monomer in the preparation. Resorcinol use in the LC polymer introduces non-linear structure in the polymer main chain, which could lower polymer melt temperature as well as liquid crystallinity. Therefore, the amount of resorcinol used was depended upon the nature and amounts other liquid crystalline monomers used in the polyester synthesis.

Resorcinol was employed as a comonomer for the preparation of highly crystalline aromatic polyesters. The *meta*-orienting property of resorcinol reduced the melting points of the resulting copolyesters and greatly improved their processability. Aromatic polyesters having resorcinolic structures in the polymer main chain were prepared (Figure 7.12) and the details are presented in Table 7.9 [31,32].

The homopolymer of resorcinol, namely poly-(*m*-phenylene terephthalate), did not show liquid crystalline property, whereas the polyesters made with *p*-hydroxybenzoic acid were liquid crystalline. Heat distortion temperatures, polymer flow temperatures and flexural modulus properties were increasing when the hydroxybenzoic content was increased, whereas the flexural strength values decreased.

Liquid crystalline copolyesters were also synthesized from *p*-hydroxybenzoic acid, resorcinol, terephthalic acid and 4,4′-dihydroxy biphenyl monomers and the details are shown in Table 7.10.

**Table 7.9.** Liquid crystalline polyesters containing resorcinol groups

| Polymer composition<br>Monomer used (mol ratio) | Control | Polyester<br>1 | 2 | 3 |
|---|---|---|---|---|
| *p*-Hydroxybenzoic acid | None | 350 | 450 | 800 |
| Terephthalic acid | 100 | 100 | 100 | 100 |
| Resorcinol | 100 | 100 | 100 | 100 |
| Polymer properties | | | | |
| Flow temperature (°C)[a] | 262 | 259 | 261 | 309 |
| Molten state property | NLC | LC | LC | LC |
| Heat distortion temperature, °C | 151 | 166 | 167 | 207 |
| Flexural properties | | | | |
| Strength (MPa) | 64.7 | 103.9 | 77.4 | 68.9 |
| Modulus (GPa) | 2.93 | 3.16 | 3.72 | 7.06 |

[a]Temperature at which polymer shows a melt viscosity of 48 000 poises
LC = liquid crystalline and NLC = not liquid crystalline
Data from [31], [32]

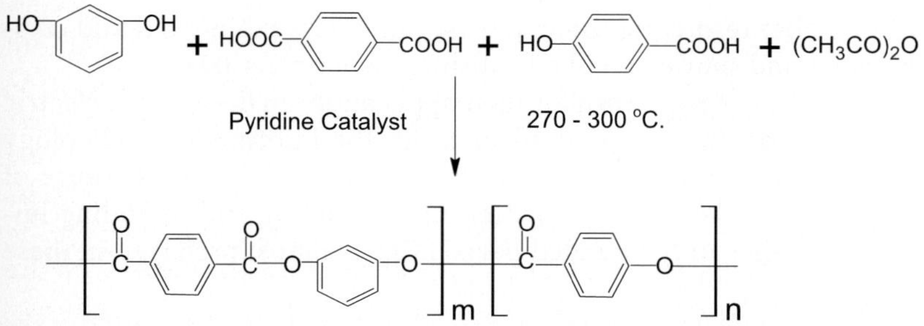


**Figure 7.12.**

The combined use of resorcinol and another aromatic diol appeared to strengthen the liquid crystalline property of the aromatic polyesters. Since resorcinol use in polyesters lowers the polymer flow temperatures, these polyesters can be molded at lower temperatures.

Copolyesters were prepared from the reactions of resorcinol diacetate, hydroquinone diacetate and terephthalic acid monomers by the acidolysis method [33]. These copolyesters have shown exceptional mechanical properties compared to general polymers. In these polyesters, as the resorcinol content was increased, the flexural modulus property was decreased (Table 7.11) [33, 34].

Mechanical properties obtained from these polyesters were sufficiently high, and therefore, could be suitable for making high strength articles without any reinforcing materials.

Polyester resins exhibiting optical anisotropic properties in the molten state were prepared using resorcinol as a comonomer (Table 7.12) [35–37].

**Table 7.10.** Liquid crystalline copolyesters containing resorinol groups

| Polymer composition | Copolyester | | | | | |
|---|---|---|---|---|---|---|
| Monomer used (mol ratio) | 1 | 2 | 3 | 4 | 5 | 6 |
| *p*-Hydroxybenzoic acid | 250 | 300 | 300 | 350 | 375 | 400 |
| Terephthalic acid | 100 | 100 | 100 | 100 | 100 | 100 |
| Resorcinol | 50 | 25 | 50 | 50 | 50 | 50 |
| 4,4′-Dihydroxy biphenyl | 50 | 75 | 50 | 50 | 50 | 50 |
| Polymer properties | | | | | | |
| Flow temperature (°C)[a] | 280 | 306 | 276 | 287 | 285 | 293 |
| Heat distortion temperature, °C | 128 | 254 | 155 | 187 | 213 | 231 |
| Izod impact strength (J/m) | 255 | 785 | 275 | 336 | 343 | 294 |
| Flexural properties | | | | | | |
| Strength (MPa) | 152.9 | 147 | 109.8 | 144.1 | 155.8 | 163.7 |
| Modulus (GPa) | 11.3 | 13 | 6.2 | 9.2 | 12.9 | 12 |

[a]Temperature at which polymer shows a melt viscosity of 48 000 poises
Data from [31], [32]

In spite of higher resorcinol content, these polyesters exhibited liquid crystalline character and showed good heat distortion temperatures.

LC polymers are widely known for their applications in the areas of electric and electronic fields. Recent rapid advances in the information technology required smaller and thinner devices in the electronic industry. Because of good flow properties, less flash formation upon molding and providing improved heat resistance to the molded parts, LCPs are most preferred in these applications.

Aromatic polyesters forming optical anisotropic melts can be used to make melt spin oriented filaments. Polyesters derived from *p*-oriented dihydric phenols and *p*-oriented dicarboxylic acids often exhibit anisotropic melts and therefore used for spinning the fibers. The use of resorcinol in these polyesters lowers the melting points suitable for melt spinning and thereby making the melts of polyesters more amenable to processing.

**Table 7.11.** Liquid crystalline polymers from resorcinol and hydroquinone monomers [poly-(*m*-phenylene terephthalate-*p*-phenylene terephthalate)]

| Polymer composition | Polyester | | | | |
|---|---|---|---|---|---|
| Dihydroxy monomer (mol ratio) | 1 | 2 | 3 | 4 | 5 |
| Resorcinol | 30 | 50 | 60 | 25 | 65 |
| 2-Chloro hydroquinone | 70 | 50 | 40 | | |
| 2-Methyl hydroquinone | | | | 75 | 35 |
| Mechanical property | | | | | |
| Flexural modulus (GPa) | 11.4 | 8.4 | 7 | 8.6 | 4.8 |

Data from [33], [34]

**Table 7.12.** Liquid crystalline polyesters made using resorcinol diacetate monomer

| Polymer composition Monomer (mol%) | Polyester 1 | 2 | 3 | 4 | 5 |
|---|---|---|---|---|---|
| *p*-Acetoxybenzoic acid | 50 | 40 | 70 | 86 | 88 |
| 2,6-Acetoxynaphthoic acid | 3 | 4 | 4 | 4 | 2 |
| Terephthalic acid | 23.5 | 28 | 13 | 5 | 5 |
| Resorcinol diacetate | 23.5 | 28 | 13 | 5 | 5 |
| Polymer properties | | | | | |
| Melting point (°C) | 310 | 304 | 311 | 320 | 350 |
| Heat distortion temperature (°C) | 210 | 205 | 180 | 230 | 240 |
| Torsional rigidity (MPa)[a] | 0.17 | 0.17 | 0.18 | 0.2 | 0.23 |

[a]The rigidity at 250 °C was measured in a Rheometer instrument
Data from [35–37]

Normally the concentration of resorcinol in these polyesters is in the range of about 2 to 10 mol%. If the resorcinol amount is less than 2 mol%, then this will have little effect on the polymer properties. On the other hand, when the amount of resorcinol exceeds 10 mol%, then this could produce deleterious effect on filament tenacity and modulus. Filaments melt spun from the resorcinol containing LC polyester materials can be heat treated to obtain high tenacity and high modulus fibers.

In various thermotropic liquid crystalline polyester formulations, resorcinol has been utilized as a comonomer to improve their processability [38–44]. Table 7.13 shows the fiber properties of resorcinol arylate containing liquid crystalline polyesters [45].

The highly oriented thermotropic liquid crystalline polyesters often exhibit good chemical resistance, excellent barrier properties, and also flame retardancy.

**Table 7.13.** Copolyesters of resorcinol and fiber properties

| Polymer composition (mol%) Monomer used | Polymer 1 | | Polymer 2 | |
|---|---|---|---|---|
| Chlorohydroquinone diacetate | 40 | | 37.3 | |
| Resorcinol diacetate | 5 | | 5.4 | |
| Terephthalic acid | 45 | | 50 | |
| 4-(4′-Hydoxyphenyl)-benzoicacid acetate | 10 | | | |
| 4,4′-Dihydroxybiphenyl diacetate | | | 7.4 | |
| Fiber properties | As Spun | Heat Treat | As Spun | Heat Treat |
| Linear density (Tex) | 0.9 | 0.89 | 0.83 | 0.78 |
| Tenacity (dN/Tex) | 4.5 | 20.9 | 3.5 | 14.1 |
| Elongation (%) | 1.6 | 4.3 | 1 | 3.6 |
| Tensile modulus (dN/Tex) | 279 | 507 | 416 | 404 |

Data from [45]

### 7.1.7 Summary and Outlook

Polyarylates are extremely important and promising class of polymers, which can be used successfully in various applications where a high heat resistance, photo stability, effective electrical insulation and mechanical properties are needed. The use of resorcinol in polyarylates offers the greatest advantage of improving their processability which could lead to enormous prospects for numerous applications.

## References

1. V. Korshak and S. Vinogradova, "Polyester", Pergamon, New York, 1965.
2. L. Robeson and J. Tibbitt, "History of polyarylates" in High Performance Polymers: Their Origin and D evelopment, R. Seymour and G. Krishenbaum, Eds., Elsevier, New York, 1986.
3. S. Vinogradova, V. Vasnev and P. Valetskii, Russ. Chem. Rev., Vol. 63, p 833 (1994).
4. G. Bier, Polymer, Vol. 15, p 527 (1974).
5. S. Cottis, Modern Plastics, p 62 (1975).
6. F. Holub and S. Kantor, US Patent, 3,160,604 (1964).
7. I. Nitta, K. Asai and H. Harada, Euro. Pat., 0,504,426 B1 (1999).
8. T. Yatsu, A. Todo and T. Sotoyama, US Patent, 4,390,683 (1983).
9. D. Brunelle, J. Suriano, T. Siclovan and J. Pickett, US Patent, 6,306,507 (2001).
10. W. Earecson, J. Polym. Sci., Vol. 11, p 399 (1959).
11. C. Cohen, R. Young and A. Markhart, J. Polym. Sci., Part-A1, Vol. 9, p 3263 (1971).
12. E. Holub and S. Kantor, US Patent, 3,062,992 (1962).
13. E. Holub and S. Kantor, US Patent, 3,160,602 (1964).
14. M. Akkabeddi and S. Chung, US Patent, 5,552,463 (1996).
15. J. Webb, R. Carter, T. Siclovan and J. Pickett, US Patent, 6,143,839 (2000).
16. R. Young, S. Cohen and A. Markhart, US Patent, 3,460,961 (1969).
17. A. Shakhnovich, US Patent, 6,228,910 (2001).
18. D. Brunell et al., US Patent, 6,291,589 (2001).
19. D. Brunell et al., US Patent, 6,306,507 (2001).
20. J. Pickett et al., US Patent, 6,607,814 (2003).
21. J. Pickett et al., US Patent, 6,610,409 (2003).
22. J. Pickett et al., US Pat. Appln., 2003/0,072,945 (2003).
23. S. Maerov, J. Polym. Sci., Part-A, Vol. 3, p 487 (1965).
24. J. Webb et al., US Patent, 5,916,997 (1999).
25. W. Jackson and H. Kuhfuss, J. Polym. Sci. Polym. Chem. Edn., Vol. 14, p 2042 (1976).
26. F. McFarlane et al., Liquid crystal polymers II, Contemporary Topics in Polymer Science, Vol. 2, Plenum Press, 1977.
27. W. Wooten et al., Ultra high modulus fibers, A. Cifferi, Ed., Applied Science, p 362, London (1999).
28. A. Blumstein et al., Liquid crystalline order in polymers, Academic Press, New York (1978).
29. A. Cifferi, W. Krigbaum and R. Meyer, Eds., Polymer liquid crystals, Academic Press, New York (1982).
30. J. Dole, Chemtech, Vol. 17, p 242 (1987).

31. I. Nitta, K. Asai and H. Harada, US Patent, 5,399,656 (1995).
32. I. Nitta and K. Asai, US Patent, 5,488,138 (1996).
33. W. Jackson and J. Morris, US Patent, 4,156,070 (1979).
34. W. Jackson and J. Morris, US Patent, 4,201,856 (1980).
35. N. Hayashi and K. Hijikata, US Patent, 4,920,197 (1990).
36. N. Hayashi and K. Hijikata, US Patent, 4,918,154 (1990).
37. N. Hayashi and K. Hijikata, US Patent, 4,937,310 (1990).
38. G. Calundann, H. Davis, F. Gorman and R. Mininni, US Patent, 4,083,829 (1978).
39. R. Irwin, US Patent, 4,188,476 (1980).
40. R. Siemionko, US Patent, 4,347,349 (1982).
41. A. East and G. Calundann, US Patent, 4,318,841 (1982).
42. H. Kock, M. Portugall, B. Hisgen and J. Mertes, US Patent, 4,749,769 (1988).
43. S. Shen and L. Charbonneau, US Patent, 5,656,714 (1997).
44. H. Linstid et al., US Patent, 6,132,884 (2000).
45. R. Siemionko, US Patent, 4,412,058 (1983).

## 7.2 Polycarbonates

### 7.2.1 Introduction

Polycarbonates are versatile thermoplastic materials known for their excellent impact strength, high transparency, toughness, good electrical, rigidity and dimensional stability properties. Because of these outstanding properties, they have been used in a wide range of industrial and household applications, which include automotives, packaging, electrical, electronic, medical and health care, and optical media. Several books and review articles have been already published on the chemistry and applications of polycarbonates developed from the use of various bisphenol monomers, and also using different polymerization techniques [1–9]. Though numerous polycarbonate materials were reported in the literature, the vast majority of polycarbonates are produced from bisphenol-A (BPA) monomer. In this chapter, the advantages of using resorcinol in the polycarbonates synthesis, and their potential applications are discussed.

### 7.2.2 Resorcinol Monomer in Polycarbonates

From the history of polycarbonates development, resorcinol was one of the first monomers employed in the synthesis of polycarbonate using phosgene in the presence of pyridine solvent [10]. The first polycarbonate material obtained from the resorcinol was observed to be brittle and glassy solid, which melted at about 200 °C (Figure 7.13).

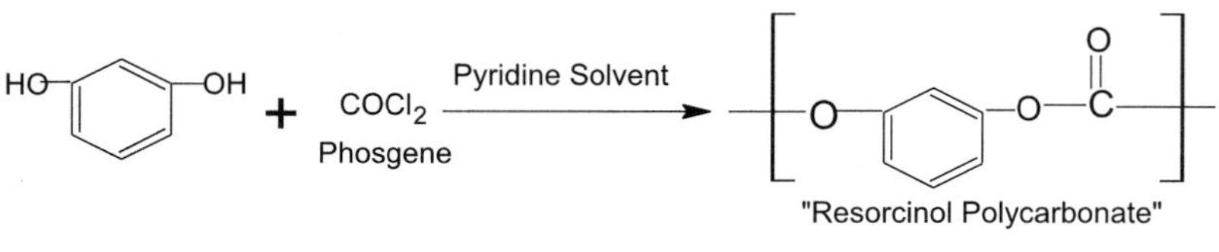


**Figure 7.13.**

Resorcinol is a *meta*-substituted dihydroxy benzene compound, and therefore the polymers that could be synthesized from this monomer can be expected to possess non-symmetrical structures and be more flexible. The resorcinol based polymers can be amorphous in character, and therefore the molded articles from these polymers may be transparent and tough. Polymers' thermal stability may have been enhanced because of the introduction of aromatic rings into the polymer backbone structures. In addition, due to chain flexibility, the glass transition and melting temperatures of polymers obtained from resorcinol could be lower.

Aromatic polycarbonates obtained from the BPA monomer are the most commercially important polycarbonates because of their low cost, good transparency, excellent impact resistance, low moisture absorption and good mechanical properties. Employing resorcinol in the place of BPA, homo polycarbonate of resorcinol could be produced from the polymerization reaction with phosgene. But resorcinol polycarbonate has not been produced commercially, possibly due to higher monomer cost, and also lower physical and mechanical properties compared to BPA.

In general, the mechanical properties of polycarbonates can be improved by increasing their molecular weights. As the polymer molecular weight increases, the processability of polycarbonates has been decreased because of higher melt viscosities. Polycarbonates with exceptionally high melt viscosities are difficult to fabricate from their melts. Various attempts have been made to overcome the high melt viscosity of polycarbonates. In this respect, the incorporation of resorcinol derived carbonate groups into BPA polycarbonate structure was observed to improve the melt flow and molding properties of polymers. Therefore, resorcinol has been employed as a comonomer to enhance BPA polycarbonates processability. Resorcinol derived polycarbonates enhanced the solvent resistance and heat resistance properties while maintaining the excellent mechanical properties and inherent transparency of BPA polycarbonates.

## 7.2.3 General Methods of Polycarbonates Preparations

Homo and copolycarbonates of resorcinol can be synthesized by any one of the following polycondensation processes.

- Melt process
- Solution process
- Interfacial process

### 7.2.3.1 Melt Polycondensation Process

Aromatic polycarbonate can be produced by polymerizing an aromatic dihydroxy compound, for example resorcinol, with diphenyl carbonate in the molten state (Figure 7.14).

In this transesterification process a high degree of polymerization can be achieved by distilling out the by-product phenol at elevated temperatures (250 – 300 °C) and high vacuum (< 10 mm Hg) conditions. In order to enhance or increase the rate of transesterification reaction, small amounts of polymerization catalysts such as the hydroxides, oxides, carbonates and acetates of alkali

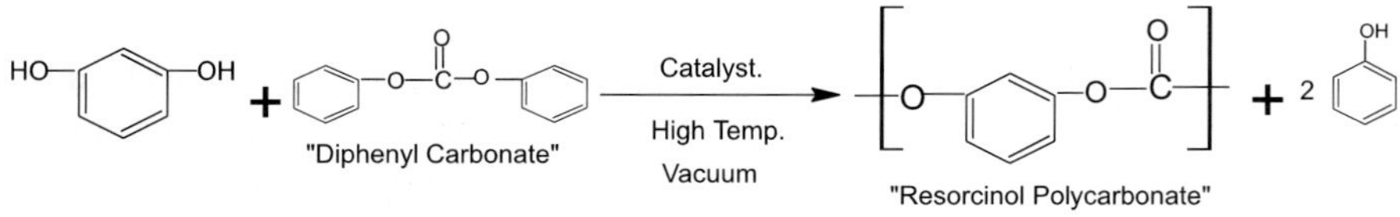


**Figure 7.14.**

and alkaline earths, and tetraalkylammonium or tetraalkylphosphonium hydroxides have been used. An oligomeric polycarbonate is first produced from the aromatic dihydroxy compound, and then it is converted into a high molecular weight polymer by increasing the temperature and reducing the pressure. From this process, polymer molecular weights as high as 100 000 could be achieved.

Various studies made on the transesterification process have shown that this is an environmentally friendlier method since there is no volatile organic solvent used during the polymerization. A high quality aromatic polycarbonate can be obtained from this process.

## 7.2.3.2 Solution Polycondensation Process

This process involves the reaction of phosgene with a difunctional phenol, like resorcinol, in an organic solvent such as dichloromethane in the presence of an organic base (pyridine or triethylamine) to accept the by-product hydrogen chloride (Figure 7.15).

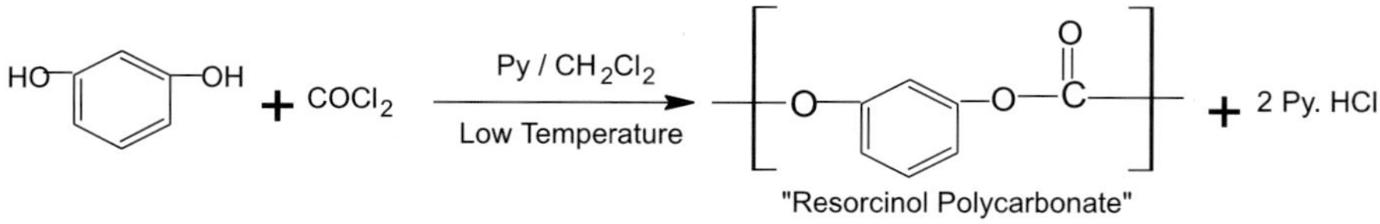


**Figure 7.15.**

## 7.2.3.3 Interfacial Polycondensation Process

According to the interfacial polycondensation process, polycarbonate resins are obtained by the reaction of sodium salt of aromatic dihydroxy compounds with phosgene (Figure 7.16).

In this process, the dihydroxy aromatic compound is introduced as a sodium salt in an aqueous solution and reacted with phosgene. An organic solvent such as dichloromethane, which can serve as a solvent for the polymer but not for the aromatic dihydroxy salt has been used in the reaction. A phase transfer catalyst

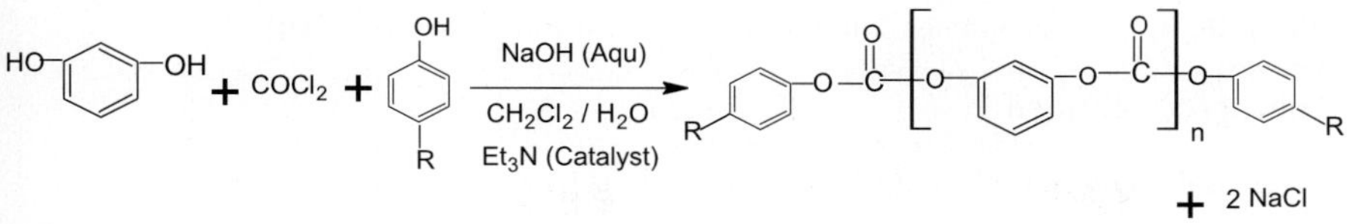


**Figure 7.16.**

such as triethylamine may be used to react with the by-product hydrogen chloride from the condensation reaction in the organic phase, and then to transfer the HCl to the basic aqueous phase where it is neutralized.

The polycarbonate molecular weights can be controlled by the addition of mono-functional phenolic derivatives such as *p-tert*-butyl phenol and *p*-cumyl phenol as chain stoppers. Polymer melt viscosity and processability can be adjusted by controlling its molecular weight. Polycarbonates with a weight average molecular weight (Mw) range of 25 000 to 65 000 can be obtained by controlling the reaction conditions and chain stopper concentrations in the interfacial process.

## 7.2.4 Homopolycarbonate from Resorcinol

Resorcinol based homopolycarbonate was obtained by the transesterification process of heating the bis-phenyl carbonate of resorcinol in the presence of sodium salt of bisphenol-A [11] (Figure 7.17).

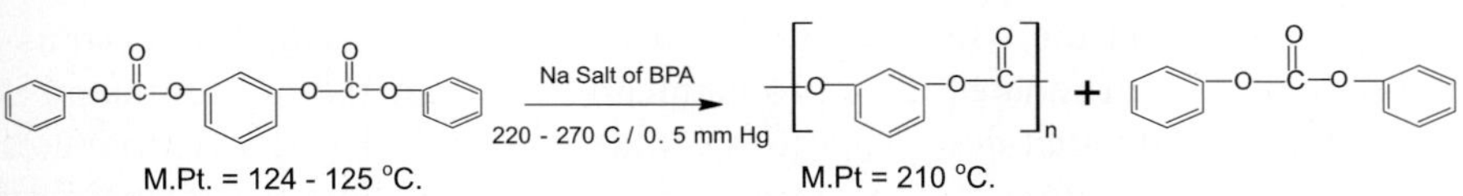


**Figure 7.17.**

## 7.2.5 Copolycarbonates Containing Resorcinol

Considerable efforts were made to improve the melt flow properties of bisphenol-A polycarbonates. Though the flow characteristics of polycarbonates were achieved by the introduction of aliphatic ester groups in the polymer chain, the thermal properties were reduced. A substantial increase in polymer flow properties with high thermal stability was achieved by the use of resorcinol in the polycarbonate synthesis (Table 7.14) [12, 13].

The incorporation of resorcinol slightly reduced the glass transition temperature (Tg) of polycarbonate. But the polymer processability, as measured by

**Table 7.14.** Copolycarbonates containing resorcinol and their properties[a]

| Polymer composition | 1 | 2 | 3 |
|---|---|---|---|
| Bisphenol-A/resorcinol (mol%) | 100/0 | 90/10 | 80/20 |
| Polymer properties | | | |
| $M_w \times 103$ | 27.6 | 26 | 24.6 |
| $M_n \times 103$ | 10.5 | 9.3 | 8 |
| $M_w/M_n$ | 2.61 | 2.8 | 3.06 |
| Glass transition temperature (°C) | 150 | 144 | 137 |
| TGA results | | | |
| 5 wt. % Loss temperature (°C) | 440 | 427 | 427 |
| Weight loss at 250 °C | 2.2 | 2.9 | 4.2 |
| Melt viscosity index (cc/10 min) | | | |
| 280 °C | 6.5 | 9.9 | 17.8 |
| 300 °C | 15 | 21.2 | 39.9 |

[a]Prepared by the interfacial polycondensation process
Data from [12], [13]

melt viscosity index, has been dramatically increased even though resorcinol is an aromatic compound. This might be associated with the chain flexibility of *m*-phenylene linkages present in the polymer chain. TGA analysis results showed good thermal stability of resorcinol based polycarbonate.

Polycarbonates, due to excellent impact strength, heat resistance and transparency properties, are widely used in the fabrication of machine parts, optical disks and automobile parts. During the molding operations, polycarbonates are heated at high temperatures and kept under the molten conditions for a longer period of time, which resulted in discoloration and reduction in transparency of polycarbonates. This problem can be minimized or eliminated if the polycarbonate melt flow properties could be improved. Copolycarbonates having excellent heat resistance, transparency, improved flow and formability properties were obtained by the use of resorcinol as a comonomer in the melt polycondensation process [14, 15] (Figure 7.18).

The results are presented in Table 7.15.

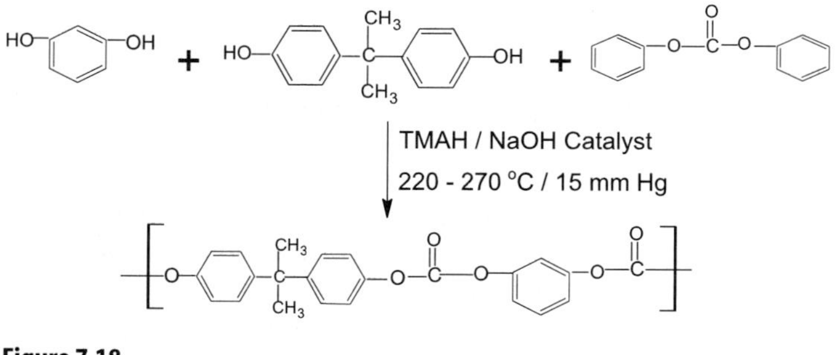


**Figure 7.18.**

**Table 7.15.** Melt flow improvements in copolycarbonates

| Polymer composition | 1 | 2 | 3 | 4 | 5 |
|---|---|---|---|---|---|
| Bisphenol-A/resorcinol (mol%) | 90/10 | 85/15 | 75/25 | 50/50 | 100/0 |
| Polymer properties | | | | | |
| Intrinsic viscosity (dL/g) | 0.49 | 0.49 | 0.49 | 0.49 | 0.49 |
| Melt flow index (g/10 min.) | 12 | 13 | 15 | 20 | 10 |
| Color tone (YI) | 1.8 | 1.9 | 2 | 2.1 | 1.8 |
| Light transmission (%) | 90.8 | 90.8 | 90.8 | 90.8 | 90.8 |
| Haze (%) | 0.3 | 0.4 | 0.3 | 0.4 | 0.3 |
| Heat resistance, HDT (°C) | 131 | 130 | 123 | 111 | 135 |

Data from [14], [15]

The incorporation of resorcinol into BPA polycarbonate has increased the melt flow without affecting the transparency at similar polymer viscosities.

## 7.2.6 Blends of Resorcinol Copolymer with ABS and SAN Resins

In the manufacture of interiors and exteriors of automobiles and housings for office automation devices, PC/ABS blends have been widely used because of their superior impact resistance and dimensional stability properties. The blends and alloys of PC normally provide performance enhancements such as melt flows, solvent resistance and low temperature impact resistance. With an increase in demands for PC/ABS materials to produce parts with reduced thicknesses, polycarbonates possessing enhanced hot melt flow and good mechanical strength properties are needed. Copolycarbonates having the following structural compositions (present at a molar ratio of 50:50) were produced from the polycondensation reactions of resorcinol and bisphenol-A with diphenyl carbonate monomer [16–19] (Figure 7.19).

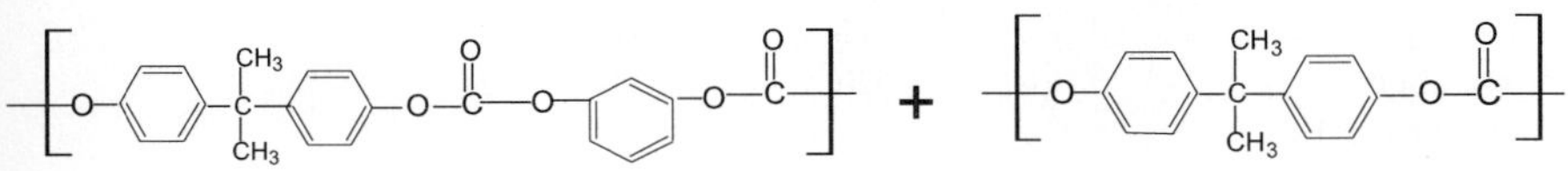

**Figure 7.19.**

This resorcinol copolycarbonates (RS-PC) was blended with ABS and SAN resins and tested for flow, impact strength and thermal stability properties (Table 7.16).

Compared to the commercial BPA polycarbonate, RS-PC showed much improved hot melt properties with the retention of high impact strength values.

Glass fiber reinforced molding compounds were produced from the RS-PC containing ABS and SAN resins, and tested for their surface appearances and flow properties (Table 7.17) [16–18].

**Table 7.16.** Blends of resorcinol copolycarbonate with ABS and SAN polymers

| Blend composition | 1 | 2 |
|---|---|---|
| Resorcinol copolycarbonate (RS-PC) | 50 | |
| Lexan PC ($\mu = 0.48\,dL/g$) | | 50 |
| ABS resin | 25 | 25 |
| SAN resin | 25 | 25 |
| Blend properties | | |
| Melt flow index (g/10 min.) | 18 | 7.2 |
| Notched Izod impact strength (J/m) | 588 | 588 |
| Heat distortion temperature (°C) | 100 | 104 |

Lexan PC – produced by General Electric Co., Japan
ABS – product of Ube Cycon Co. Ltd, Product Name UX 050
SAN resin – product of Ube Cycon Co. Ltd, Product Name SR 30 B
Data from [16–18]

The presence of resorcinol in the copolycarbonates has greatly enhanced hot melt flow as well as surface appearance properties of glass fiber filled molding compounds. Studies have also showed that the use of RS-PC in place of BPA-PC in the PC/ABS/SAN blends dramatically increased the low temperature (– 40 °C) impact strength property [19].

**Table 7.17.** Glass reinforced molded products and properties resorcinol based copolycarbonate blends

| Blend composition | 1 | 2 | 3 | 4 |
|---|---|---|---|---|
| Resorcinol copolycarbonate (RS-PC) | 40 | | 40 | |
| Lexan PC ($\eta = 0.48\,dL/g$) | | 40 | | 40 |
| ABS resin | 20 | 20 | | |
| SAN resin | 20 | 20 | 40 | 40 |
| Glass fiber | 20 | 20 | 20 | 20 |
| Blend properties | | | | |
| Melt flow index (g/10 min) | 7.5 | 4.5 | 40 | 21 |
| Notched izod impact strength (J/m) | 177 | 147 | 98 | 88 |
| Heat distortion temperature (°C) | 120 | 125 | 120 | 125 |
| Surface appearance | | | | |
| Glossiness | 74 | 48 | 78 | 54 |
| Surface roughness (R-Max, ηm) | 3 | 9.8 | 2.8 | 7.8 |

Lexan PC – Produced by General Electric Co., Japan
ABS – Product of Ube Cycon Co. Ltd, Product Name UX 050
SAN Resin – Product of Ube Cycon Co. Ltd, Product Name SR 30 B
Glass Fiber – Asahi Fiber Glass, Product Name FT-116
Data from [16–18]

### 7.2.7 Summary and Outlook

Resorcinol offers the greatest advantage of enhancing the melt flow properties of BPA based thermoplastic polycarbonate materials for the production of high performance products with reduction in costs. The incorporation of resorcinol into polycarbonates may improve their ductility and high flow properties because of chain flexibility. In general, polymers obtained from the resorcinol monomer could exhibit amorphous character, and therefore the optical clarity of polycarbonates may not be affected.

## References

1. D. Fox and W. Christopher, Polycarbonates, Reinhold Publishing, New York, 1962.
2. H. Schnell, Chemistry and physics of polycarbonates, Wiley-Interscience, New York, 1964.
3. D. Fox, In: Encylopedia of chemical technology, Vol. 18, 3rd Edn., Wiley, p 479 (1982).
4. L. Bottenbruch, Ed., Polycarbonates, Hanser Publishers, New York, 1996.
5. D. Legrand and J. Bender, Eds., Handbook of polycarbonate science and technology, Marcel Dekker Inc., New York, 2000.
6. D. Brunelle and G. Kailasam in Polycarbonates, Technical Information Series from General Electric Company, 2002.
7. J. Schmidhauser and P. Sybert, J. Macromol. Sci. Polym. Rev., Vol. C41, p 325 (2001).
8. R. Sandler and W. Karo, Eds., Polymer syntheses, Academic Press, New York, 1974.
9. K. Sehanobish, H. Pham and C. Bosnyak, Polycarbonates, In: Polymeric materials encyclopedia, CRC Press, p 5697 (1996).
10. A. Einhorn, Liebigs Ann. Chem., Vol. 300, p 315 (1898).
11. Bayer Corp., British Patent, 808,488 (1959).
12. P. Raymond, US Patent, 5,336,751 (1994).
13. P. Raymond, US Patent, 5,418,317 (1995).
14. T. Sakashita, T. Shimoda and T. Nagai, Euro. Patent, 0,508,774 B1 (1998).
15. T. Sakashita, T. Shimoda and T. Nagai, US Patent, 5,384,388 (1995).
16. K. Ishiwa and H. Ito, Euro. Patent, 0,647,679 A1 (1995).
17. K. Ishiwa and H. Ito, US Patent, 5,420,192 (1995).
18. K. Ishiwa and H. Ito, US Patent, 5,565,515 (1996).
19. T. Fujiguchi, A. Saito and H. Ito, Euro. Pat. Appln., 0,663,425 A2 (1995).

## 7.3 Polyestercarbonates

### 7.3.1 Polycarbonates

Aromatic polycarbonates (PC) are the engineering thermoplastics with outstanding impact strength, optical transparency, high gloss, dimensional stability and excellent paintability properties. Polycarbonates have been used in a wide range of industrial, household and medical applications because of their toughness and glass-like transparency. In the outdoor applications, polycarbonates' attractiveness and long term color stabilities are known to deteriorate in the presence of sun light or UV radiation. Although polycarbonates are highly transparent initially, after the sun light exposure their appearance, color and gloss have been slowly degraded. Therefore, polycarbonates' use in the outdoor applications has been limited.

### 7.3.2 Weathering Exposure and Polycarbonate Stability

Upon exposure to the sun's UV radiation, polycarbonates can change their color from colorless to yellow. The effect of UV light, particularly in combination with oxygen and moisture (weathering conditions), can lead polycarbonates to undergo photo-oxidative degradation, which results in yellowing and reduction in molecular weights. The yellowing of polymers associated with the action of UV radiation is often called "photoyellowing" [1]. As a result of photoyellowing, loss of gloss, transparency and reduction in mechanical properties were seen in polycarbonates.

In order to overcome the photoyellowing, UV absorber compounds were used in the polycarbonate compositions. Resorcinol based benzophenone and triazine derivatives have been successfully used to protect polycarbonates against the effects of UV radiation. The use of UV absorbers in the polycarbonate compositions, as additives, should be limited to the minimum level (up to about one weight percent), otherwise their impact strength and thermal stability properties could be affected.

In one another method, polycarbonates were coated with formulations containing weatherable polymers or UV absorber compounds. Coating compositions containing the dibenzoyl resorcinol derivatives have been successfully employed to protect the polycarbonates against their loss of gloss and yellowing. Similarly, resorcinol based polyarylates were found to be very effective and good weatherability properties against the yellowing of polymers [2].

### 7.3.3 Advantages of Resorcinol Polyarylates as UV Stabilizers

Poly(resorcinol arylates) could be conveniently synthesized from the reaction of resorcinol with either isophthaloyl chloride or terephthaloyl chloride or mixtures containing these two phthaloyl chlorides. These polyarylates were found to be amorphous and colorless, and the films obtained were determined to be tougher. After exposure to UV light, resorcinol based polyarylates films remained almost colorless, whereas the arylate polymers obtained from the bisphenol-A monomer changed into a colored film. The resorcinol arylate groups present in the polyester were known to undergo the Fries rearrangement in the presence of UV light, and being converted into *o*-hydroxybenzophenone groups, which could protect the polycarbonates against UV degradation (Figure 7.20).

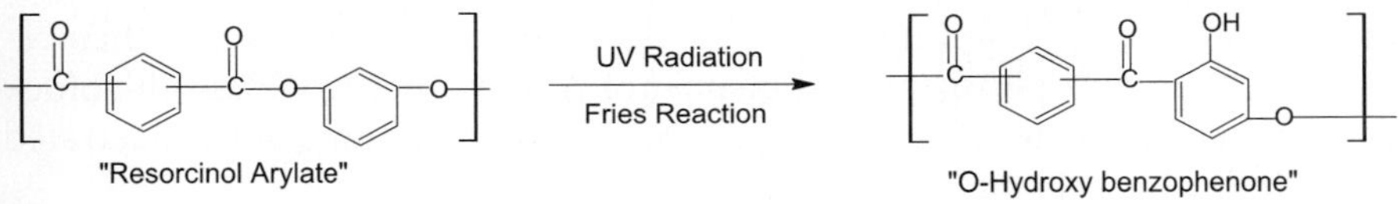


**Figure 7.20.**

It is well known that resorcinol based benzophenone compounds are the best and effective UV stabilizers for polycarbonates and other polymers. Therefore, the use of resorcinol arylate ester containing polymers in the form of an additive or coating compositions could be very effective to protect polycarbonates against the damaging effects of UV radiation.

### 7.3.4 Poly(resorcinol arylates) and Polycarbonates

Since poly(resorcinol arylates) showed good UV stabilization with good mechanical, clear and colorless film properties, the incorporation of these polyesters into polycarbonates was expected to exhibit polymers having a combination of improved properties. For example, the heat resistance, solvent resistance and UV absorption properties were expected to improve with a copolymer containing resorcinol arylate and carbonate groups. Aromatic polymers containing arylate ester and carbonate groups are generally called polyestercarbonates or copolyestercarbonates. With the incorporation of resorcinol arylate esters in the polyestercarbonates, the optical properties, such as transparency, haze and gloss, were expected to be unaffected, because of amorphous polymers realized from the resorcinol monomer.

The presence of resorcinol arylate groups in the polymer chain could impart UV stability to the resulting polyestercarbonates. In the polymer chain, resorcinol arylates are part of the molecule, and their presence could function as latent stabilizer. Therefore, copolymers containing resorcinol arylates do not need additional UV stabilizers, which could affect the impact strength and gloss properties, to protect them against UV radiation. For these reasons, copolyestercarbonates with resorcinol arylate groups can find widespread utility, particularly for glazing and transparent applications demanding high UV resistance, impact strength, scratch resistance and solvent resistance properties.

### 7.3.5
### Generalized Copolyestercarbonate Synthesis and Properties

Copolyestercarbonate resins were conveniently prepared by a two-step polymerization process, in which resorcinol was reacted with a mixture of isophthaloyl chloride and terephthaloyl chloride in the presence of an acid acceptor, such as triethylamine. The dihydroxy terminated polyester intermediate produced from the resorcinol and phthaloyl chlorides reaction was then reacted further with phosgene and bisphenol-A monomer to obtain the block copolyestercarbonates [3–5]. This polymerization reaction can be illustrated in the following scheme (Figure 7.21).

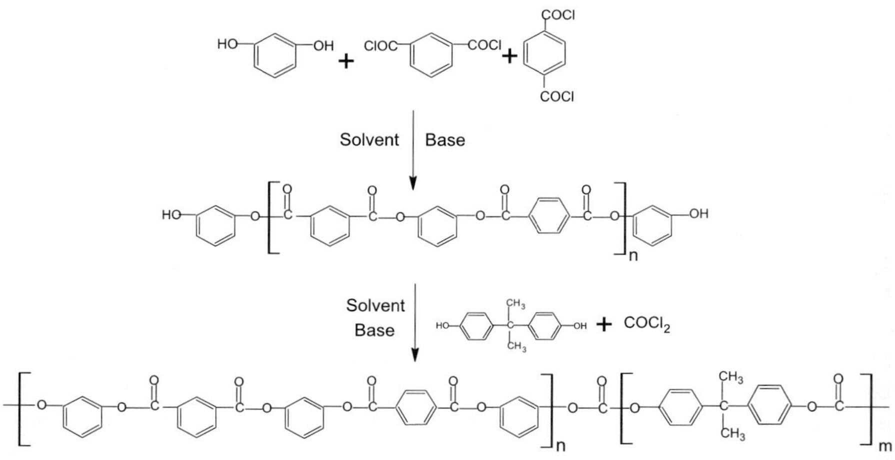


**Figure 7.21.** Synthesis of resorcinol based block copolyestercarbonates

The properties of copolyestercarbonates largely depend on the ratio of carbonate groups and carboxylate groups as well as the isophthalate to terephthalate groups in the polymer chain. Moldability, glass transition temperature (Tg) and heat distortion temperature of copolymers also depend on the composition of isophthalate and terephthalate structures in the polymer. Polymers containing high proportions of resorcinol terephthalate structures could show lower melting points, melt viscosities and exhibit substantially improved moldability. The mechanical properties such as tensile strength, flexural strength, impact strength, creep resistance and physical properties such as solvent resistance and boiling water resistance of copolyestercarbonates could be controlled by the mix ratios of two phthaloyl chlorides in the polyester synthesis.

As the ester content of the polyester is increased, the heat distortion temperature can be increased, which may affect the impact resistance of copolymers. Therefore, in order to improve or maintain the excellent impact strength properties of copolyestercarbonates, the phthalate/carbonate ratios should be controlled. This, in turn, may depend upon the number average molecular weight (Mn) of the polyester block used in the copolymer synthesis.

Poly(resorcinol arylates) can show amorphous character due to the *m*-phenylene structure of resorcinol. Therefore, the copolyestercarbonates synthesized from the copolymerization reaction of resorcinol arylates may be amorphous. The optical properties obtained from these amorphous copolymers can be similar to BPA-polycarbonate.

### 7.3.6
### Copolyestercarbonates from Resorcinol – Methods of Preparation

Aromatic copolyestercarbonates have been synthesized by an interfacial polymerization or trans-esterification process.

#### 7.3.6.1
#### Interfacial Polymerization Process

In this process, copolyestercarbonates were prepared by first mixing an aqueous alkaline solution of resorcinol with a solution of mixed isophthaloyl chloride and terephthaloyl chloride monomers in an organic solvent to produce low molecular weight polyester oligomers having terminal hydroxyl groups. Thereafter, the oligomeric resorcinol arylate polyester was reacted with phosgene in the presence of bisphenol-A monomer to obtain the final copolymer [3] (Figure 7.22).

A synthesis procedure used in the preparation of iso/terephthalate-block-copolycarbonate is outlined below [3].

To a 30-l glass reactor equipped with a glass impeller, centrifuge recirculation pump, reflux condenser and pressure-equilibration addition funnel were charged resorcinol (605.6 g, 5.5 mol), methyltributylammonium chloride (82.5 g of a 75 wt% aqueous solution; 0.275 mol), dichloromethane (6.5 l), and water (2.5 l). The recirculation pump was turned on and the mixture was degassed with nitrogen while stirring. The pH of the aqueous phase was adjusted to 7 with 50% aqueous sodium hydroxide solution. Reaction was carried out by adding a solution of acid dichlorides (507.5 g each of iso- and terephthaloyl dichlorides; 5.00 mol total in 2.0 l of dichloromethane solution) while stirring and simultaneously adding 50% sodium hydroxide solution at such a rate that

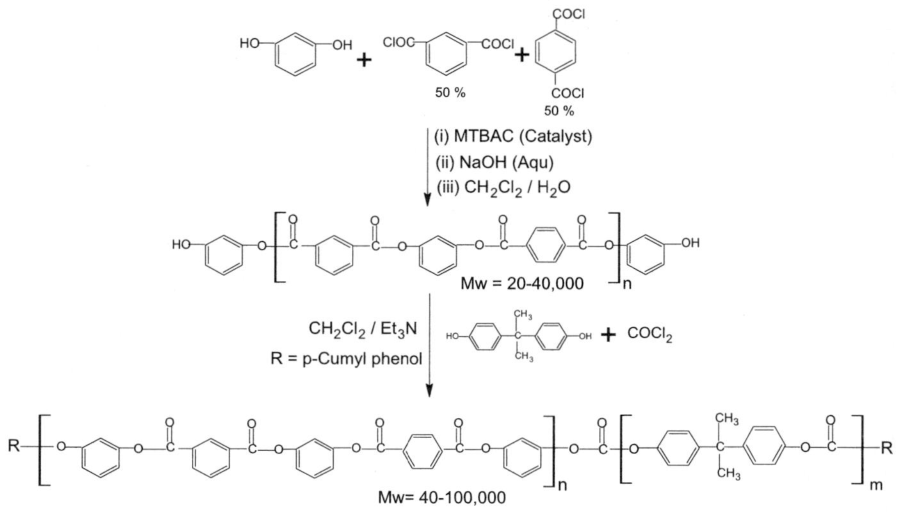


**Figure 7.22.** Interfacial polymerization process for copolyestercarbonate Preparation

the pH was maintained between 6 and 8. The acid dichlorides were added using a three-step program with the rate of addition declining with each step. A timer was started at the beginning of acid dichloride addition. The pH was maintained at 8 while adding acid dichlorides over 8 min using the following protocol: 40% of total acid dichlorides were added over the first 2 min; 30% of total was added over the next 2 min; the remaining 30% of total was added over the next 4 min. The recirculation loop was running during the entire time.

Following complete addition of acid dichlorides, the pH was slowly raised to 11–12 over two to three minutes, and the reaction mixture was stirred for 10 min. The polymer formed was a hydroxy-terminated oligomer with weight average molecular weight (Mw) of approximately 20 000.

Bisphenol A (1102 g 4.83 mol), dichloromethane (4.0 l), triethylamine (17 ml, 0.12 mol), *p*-cumyl phenol capping agent (60 g), and water (6.0 l) were then added and the recirculation loop was turned on. Phosgene was introduced at pH 7.5–8.5, then increasing slowly to pH 10–10.5 using a 15% excess over the theoretical amount of phosgene.

The reaction mixture was separated; the organic phase washed with 1 N hydrochloric acid, 0.1 Nhydrochloric acid, and twice with water, and the polymer was isolated by precipitation in methanol. The polymer obtained was the desired resorcinol iso/terephthalate-*co*-BPA polycarbonate copolymer.

In general, the reaction of resorcinol with phthaloyl chlorides often produced high molecular weight polyesters under the normal polymerization conditions. This might be associated with the high solubility of resorcinol in the aqueous phase of the process solvent. In order to control the polyester molecular weights, steps were taken to adjust the mode of caustic additions, catalyst levels and temperature conditions [6,7]. By using polyesters with controlled molecular weights, copolyestercarbonates possessing improved processability, thermal stability and color were produced (Table 7.18).

**Table 7.18.** Copolyestercarbonates from the interfacial polymerization process

| Polyester Intermediate | | | Copolyestercarbonate | |
|---|---|---|---|---|
| Catalyst | Mn | Mw | Polyester block (%) | Mw |
| MTBAC | 12 000 | 24 300 | 20 | 75 100 |
| TBAB | 8300 | 16 900 | 35 | 95 000 |
| TBAB | 6200 | 12 500 | 50 | 75 200 |
| MTBAC | 8600 | 17 000 | 50 | 99 000 |
| MTBAC | 7200 | 14 500 | 80 | 78 100 |

TBAB = Tetra-*n*-butylammonium bromide
MTBAC = Methyl tri-*n*-butylammonium chloride
Data from [1]

### 7.3.6.2
### Melt Polycondensation Process

The formation of hydroxy terminated, low molecular weight resorcinol polyarylate oligomers from the conventional solution process was found to be difficult due to high water solubility of resorcinol. Even when a large excess of resorcinol was employed, higher molecular weight polyesters were produced. Melt polycondensation method was successfully used to obtain low molecular weight poly(resorcinol arylates), which were subsequently used for the production of copolyestercarbonates by a solution process [8,9] (Figure 7.23).

The procedure used for the preparation of copolyestercarbonate according to reaction scheme shown in Figure 7.23 is described below [8,9].

A glass reactor equipped with a stir shaft, Dean Stark trap with Vigreux column and vacuum attachment was charged with resorcinol (3.03 g, 0.0275 mol), diphenylisophthalate (3.98 g, 0.0125 mol), diphenylterephthalate (3.98 g, 0.0125 mol), lithium hydroxide (3.5 mg), and tetramethylammonium hydroxide (10 microliters of 25 wt. % solution). A nitrogen atmosphere was established (three vacuum purges) and slow stirring began. The reactor was lowered into a salt bath controlled at about 200 °C. After 15 min, the pressure was reduced to 100 Torr. This pressure and temperature were maintained for 45 min during which time about 2.2 ml of phenol was collected. The pressure was then lowered to 50 Torr and maintained for 30 min, collecting an additional 0.8 ml of phenol. The temperature was then raised to 220 °C and after 30 min the cumulative total of phenol was about 3.3 ml. For the next 30 min the reactor was held at 220 °C and 25 Torr then 240 °C and 25 Torr for 30 min. Finally, the temperature held at 240 °C and full vacuum applied for 60 min, the final

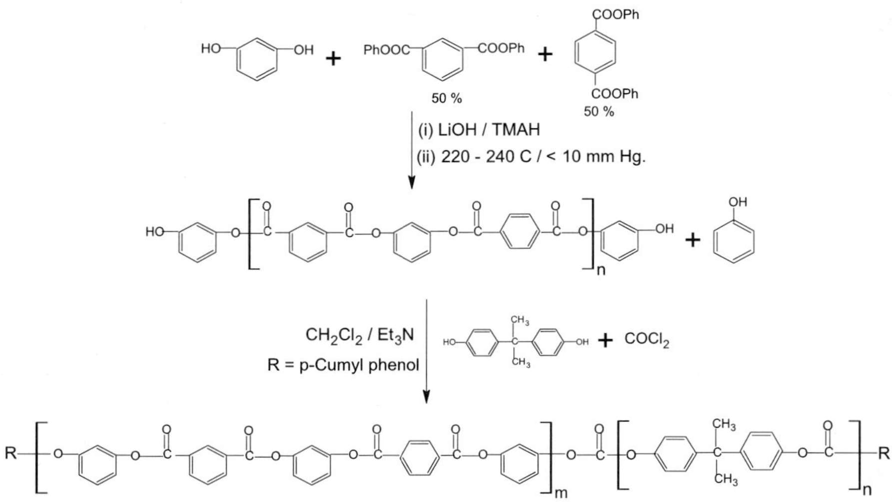


**Figure 7.23.** Melt polymerization process for copolyestercarbonate preparation

amount of phenol collected being about 4.1 ml (93% of theory). The viscous amber colored material was poured onto an aluminum pan, yielding 4.9 g. GPC analysis indicated Mw of 14 400.

Into a five-neck Morton round bottomed flask equipped with a pH electrode, phosgene addition tube, condenser, caustic addition port, and a stirring shaft was placed polyester oligomers (4.75 g), bisphenol A (3.01 g, 0.0132 mol), methylene chloride (100 ml), deionized water (100 ml), and triethylamine (200 microliters. The pH of the reaction mixture was taken to 10 and phosgene added (3.0 g, 0.75 g/min). The pH was maintained at 10 for the entire phosgenation. The reaction mixture was separated, the organic phase washed with hydrochloric acid (1 N), hydrochloric acid (0.1 N) and twice with water, and the polymer was isolated by precipitation in hot water. The Mw of the isolated material was 35 861.

Table 7.19 presents the details on the polyester and copolyestercarbonate molecular weights and the ratios of ester to carbonate groups in the copolymers.

To achieve the controlled resorcinol arylate polyester molecular weights, phase transfer catalysts were employed in the melt condensation reaction. In this way the copolymer molecular weights were optimized for improving their melt flow and moldability properties.

## 7.3.7
## UV Stabilization and Mechanism of Protection

The weatherability property of resorcinol arylates containing copolyestercarbonates was attributable to the formation of *o*-hydroxybenzophenone groups

**Table 7.19.** Resorcinol based copolyestercarbonates

| Melt condensation Polyester intermediate | | | Solution process polyestercarbonate | | |
|---|---|---|---|---|---|
| Excess resorcinol (mol%) | Phenol removed (%) | Mw of oligomer | End cap phenol (%)[a] | Mw | Composition (%) ester/carbonate |
| 10 | 80 | 8175 | 20 | 13 202 | 100/0[b] |
| 10 | 86 | 9891 | 8.3 | 24 513 | 60/40 |
| 10 | 93 | 14 410 | 4.2 | 35 861 | 60/40 |
| 32 | 98 | 5511 | 4.7 | 32 158 | 80/20 |
| 10 | 95 | 11 080 | 5.7 | 19 219 | 80/20 |
| 100 | 93 | 1306 | 7 | 52 325 | 83/17 |
| 32 | 95 | 5209 | 3.6 | 50 637 | 80/20 |
| 10 | 93 | 7541 | 5.3 | 21 328 | 80/20 |

[a] Includes the additional end cap phenol plus the phenol left behind from melt condensation
[b] Contains no bisphenol-A polycarbonate
Data from [8], [9]

from the arylate blocks by the thermally or photochemically induced Fries rearrangement reaction [1] (Figure 7.24).

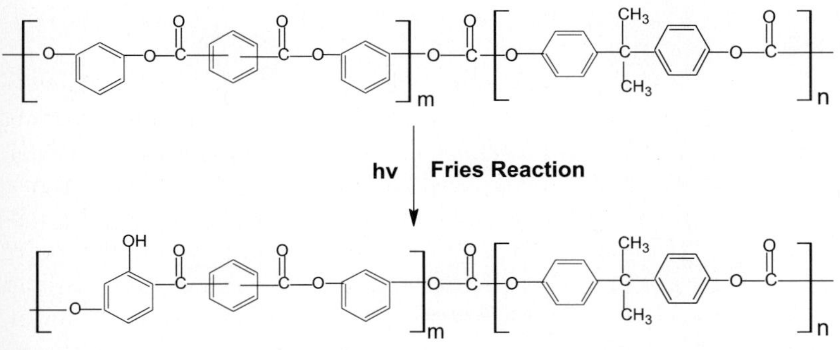


**Figure 7.24.**

The new polymeric structure formed on the top surface of copolyestercarbonates, as a result of this rearrangement, was inherently a resorcinol based benzophenone type UV absorber structure, which makes this copolymer self-protecting against UV radiation. A transmission spectra of resorcinol arylate polyester based copolyestercarbonate film before and after the Fries reaction is shown (Figure 7.25).

Before the UV exposure, poly(resorcinol arylate) was acting as a latent stabilizer for the copolymers with no Fries rearrangement. After the sun light exposure, the polymer became an active UV stabilizer as a result of the formation of *o*-hydroxybenzophenone groups. The newly formed *o*-hydroxybenzophenone groups absorb the harmful UV energy, undergo structural changes and then dissipate this in the form of harmless heat energy as illustrated in Figure 7.26.

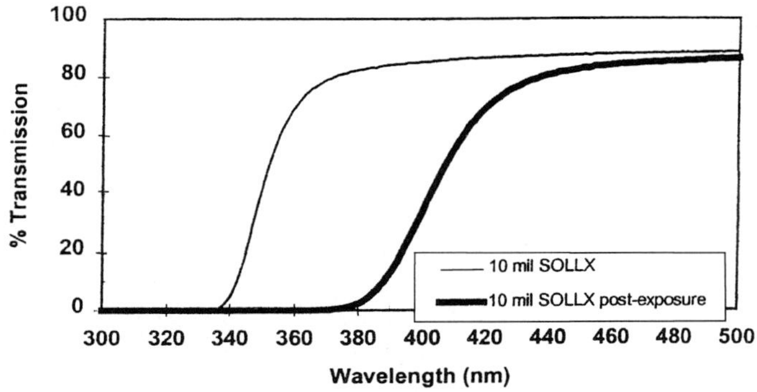


**Figure 7.25.** Spectra of copolyestercarbonate film before and after UV exposure [10]

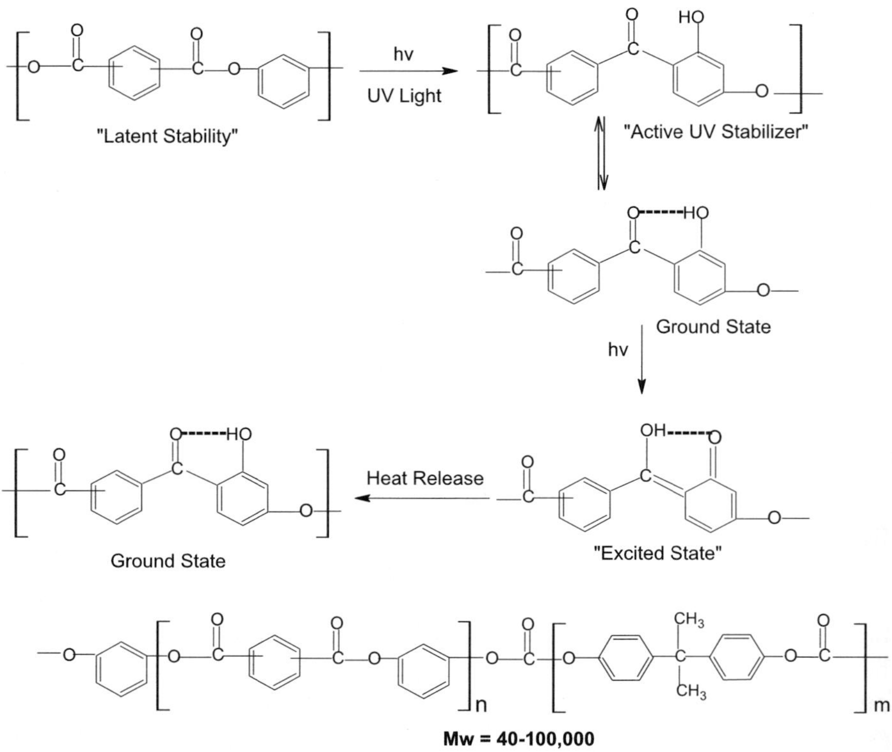


**Figure 7.26.** Mechanism of UV stabilization of resorcinolic polyestercarbonates

In this way, the copolyestercarbonate was functioning as a self-protecting UV stabilizer polymer capable of protecting various other substrates known to be affected by the sun's harmful UV radiation.

## 7.3.8
## Blends of Copolyestercarbonate with Poly(butylene terephthalate)

When two polymers are blended together, the resulting compatible blend may show properties somewhere between the properties of the individual blend components. In the case of polyestercarbonate resins, blends obtained with certain aromatic polycarbonates have shown higher impact strength properties than either of the components. Blending of polymers may also tend to change the failure mode from the brittle to be more ductile.

Resorcinol based copolyestercarbonates were blended with various thermoplastic polymers such as polyesters, styrene-acrylonitrile copolymers, poly (methyl methacrylate) and poly(carbonate)-*co*-poly(dimthyl siloxane), and their physical and mechanical properties were studied [11, 12].

Table 7.20 presents the data obtained from the blends of copolyestercarbonate with poly(butylene terephthalate), PBT, material.

**Table 7.20.** Blends of coployestercarbonate with poly(butylene terephthalate)

| PBT (wt.%) | Tg (°C) | Notched izod impact strength (J/m) | Haze (%) |
|---|---|---|---|
| 0 | 143 | 800 | 5.12 |
| 10 | 129 | 1008 | 5.51 |
| 20 | 115 | 1062 | 4.45 |
| 35 | 98 | 1399 | 5.32 |
| 50 | 81 | 1089 | 7.05 |
| 65 | 64 | 1350 | 87.4 |
| 80 | 50 | NA | NA |
| 90 | 48 | 53 | NA |
| 100 | 43 | 48 | NA |

PBT had an intrinsic viscosity of 1.17 dl/g at 30 °C and was measured in 1:1 phenol/tetrachloro ethane solvent system
Copolyestercarbonate contained 50 wt. % carbonate block and poly(resorcinol arylate) containing 60:40 tere- to isophthalate groups
Data from [11], [12]

From the observation of a single glass transition temperature for each of the blends shown in Table 7.20, these two polymers were shown to produce compatible blends. Impact strength values were increased for blends containing upto 65 wt. % PBT. Since the melt flow properties of polycarbonates is directly related to their glass transition temperatures (Tg), the observed decrease in the Tg value for the blends indicates an enhanced melt flow properties for these blends.

Haze is also an important property when designing for a transparent sight application. Lower haze values of polymers imply greater transparency. From the haze values (Table 7.20), it can be concluded that these blends could exhibit good transparency. By adjusting the length of resorcinol arylate blocks in the polyestercarbonates, transparent blends could be developed.

### 7.3.9
### Fiber Reinforced Laminates Using Polyestercarbonate Resins

Fiber reinforced thermoplastic resins have been used in a variety of applications. When exposed to outdoor weathering, these fiber reinforced articles suffer from the loss of gloss, appearance, discoloration, and strength reduction. The use of resorcinol arylate based polyestercarbonates in the fiber reinforced articles appeared to retain their gloss and color properties even after exposure to severe weathering conditions as shown in Table 7.21 [13].

**Table 7.21.** Fiber Reinforced Resorcinol Based Copolyestercarbonate Resin Composites Properties

| Composite composition | 1 | 2 | 3 | 4 | 5 | 6 | 7 |
|---|---|---|---|---|---|---|---|
| R-PEC | 95 | 90 | 85 | 80 | 70 | | |
| BPA-PC | | | | | | 80 | 70 |
| GF-1 (epoxy) | 5 | 10 | 15 | 20 | 30 | 20 | 30 |
| Properties | | | | | | | |
| Flexural modulus (GPa) | 3.2 | 4 | 5 | 6.1 | 8.7 | 5.5 | 7.6 |
| Flexural strength (GPa) | 0.14 | 0.16 | 0.19 | 0.21 | 0.23 | 0.13 | 0.16 |
| HDT (°C, 1.82 MPa) | 125 | 129 | 132 | 133 | 134 | 149 | 152 |
| Notched izod strength (J/m) | 64 | 64 | 85 | 96 | 123 | 106 | 106 |
| Gloss (ASTM-D 523) | | | | | | | |
| As molded | 96 | 93 | 89 | 76 | 81 | 26 | 22 |
| Exposure at 5486 $kJ/m^2$ | 104 | 98 | 86 | 65 | 52 | 6 | 2 |
| Retention (%) | 100 | 100 | 97 | 86 | 64 | 23 | 9 |
| Yellowness index (ASTM -D1926) | | | | | | | |
| As molded | 26 | 29 | 30 | 28 | 20 | 19 | 14 |
| Exposure at 5486 $kJ/m^2$ | 25 | 28 | 30 | 28 | 22 | 39 | 36 |
| Change (%) | 4 | 3 | 0 | 0 | 0 | 105 | 157 |

R-PEC: Resorcinol based polyestercarbonate. Composition : 90 mol% resorcinol arylate and 10 mol% BPA carbonate
BPA-PC: Polycarbonate manufactured by General Electric Company as Lexan resin
GF-1 (Epoxy): Glass fiber manufactured by Owens Corning Co
Data from [13]

Compared to polycarbonate compositions, resorcinol based copolymers have shown excellent weatherability in terms of gloss retention and resistance to discoloration. Therefore, copolyestercarbonate resins could be useful in the outdoor applications such as benches, seating, doors, window frames, outdoor lighting, transport devices, green houses, siding materials, architectural hardware, mail boxes, post, fencing, enclosures and communication equipment.

### 7.3.10
### Commercial Applications of Resorcinol Based Polyestercarbonates

Resorcinol arylates containing polyestercarbonates were commercially introduced as weatherable polymers by General Electric Company (GE) under the trade names Sollx and Lexan SLX. According to GE, the development of Sollx material was a "breakthrough" technology, since no other polymer provided the combination of weatherability, high gloss, toughness, transparency, scratch resistance, durability and chemical resistance properties. This breakthrough in the Sollx technology could be possible only when resorcinol was used as the monomer in the polyarylate part of the final polyestercarbonate material.

#### 7.3.10.1
#### Sollx Film Properties

Sollx is a weatherable thermoplastic polymer film produced from the resorcinol arylate polyester containing polyestercarbonate resin. Sollx film's physical properties are presented in Table 7.22 [10, 14].

**Table 7.22.** Resorcinol arylates containing polyestercarbonate (Sollx) resin physical properties

| Property | Value |
|---|---|
| Glass transition temperature (°C) | 145 |
| Melt flow index (load: 1.2 kg at 300 °C) | 3.2 |
| Notched Izod impact strength (J/m) | 779 |
| Dynatup (J/m) | 3043 |
| Tensile strength (MPa) | 75.9 |
| Tensile elongation (%) | 100 |
| Density (g/cm$^3$) | 1261 |
| Coefficient of thermal expansion (40 to 90 °C, m/m °C) | $6.8 \times 10^{-5}$ |
| Refractive index | 1603 |

Data from [10, 14]

In general, Sollx film offers good heat resistance, tensile strength, rigidity, and excellent impact resistance properties. Sollx film's high refractive index could provide a high level of gloss and exceptional depth of image.

#### 7.3.10.2
#### Weatherability and Gloss Properties of Sollx

Sollx film was observed to retain 95% of its gloss when exposed to more than 25 000 kJ/m$^2$ of Xenon-arc accelerated exposure, which was equivalent to ten years of Florida (USA) weather conditions [10]. This result showed that a molded-in-color (MIC) Sollx glossy panel requires no paint and could retain a paint like gloss appearance even after ten years of sunlight exposure like Florida state (USA) weather. Sollx film can be used on colored thermoplastic and thermoset substrates to provide weather resistant class-A glossy finish as an alternate to painting and clear coating [15,16]. This film can be applied over a variety of plastic substrates through an in-mold decoration (IMD) process, which could be used in the production of large parts such as an exterior automobile body panels. Plastics covered with Sollx film appeared like freshly painted metal.

### 7.3.10.3 Scratch Resistance Property

Figure 7.27 compares the scratch resistance of Sollx film to some other well known systems [10].

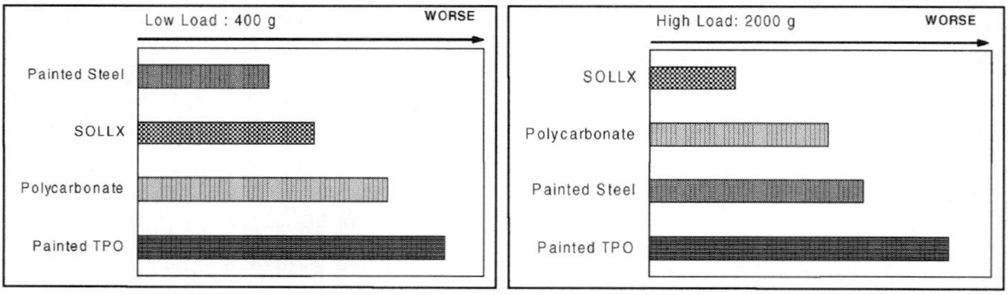


**Figure 7.27.** SollX film scratch resistance property [10]

At higher applied loads, Sollx film was observed to show more resistance property than the painted metal. The ductile failure from the scratch of Sollx was less severe compared to painted steel, where the failure was observed to be a brittle fracture.

In addition to scratch resistance, Sollx film also showed improved performance in the resistance of aggressive agents such as the brake fluids, acid rain, gasoline and other chemicals.

### 7.3.11 Sollx Applications

Copolyestercarbonate film (Sollx) has shown the property of protecting various substrates against UV degradation, and possessed an excellent scratch resistance, good depth of image, excellent impact strength properties, and maintained good dimensional stability over a wide temperature range. The primary target for Sollx film was in the automotive paint replacement in in-mold decoration (IMD) applications. In the outdoor weatherable applications,

Sollx film could be useful in front fascias, door panels, mirror housings, lawn and garden equipment, construction and boats. Clear sheet of Sollx laminated with polycarbonates could find applications in lighting, signs, displays, and industrial screens.

Lexan Slx resins possess chemical resistance, weatherability, thermal stability, impact resistance, processability, and knit line/flow properties. These injection moldable opaque and transparent resins, introduced by GE, were useful in the commercial production of industrial backhoe, fenders, the housing for a wireless infra structure, sporting goods, and telecommunication enclosures.

### 7.3.12
### Summary and Outlook

The "breakthrough" polymer technology developed using the resorcinol based chemistry, according to GE Plastics, has a bright future with the automotive industry to replace painted automobile exterior trim and body panels. In the non-automotive markets, greater potential may exist for these copolyestercarbonates where the plastic materials are exposed to substantial outdoor weathering conditions. The outstanding weatherable polymers developed from the resorcinol chemistry strongly suggests that this chemical has the potential for the development of new and more breakthrough technologies in future.

## References

1. T. Siclovan, B. Brown, D. Buckley and J. Pickett, PCT Appln., WO/026,275 (2000).
2. C. Cohen, R. Young and A. Markhart, J. Polym. Sci., Part A-1, Vol. 9, p 3263 (1971).
3. T. Siclovan et al., US Patent, 6,559,270 (2003).
4. D. Brunelle et al., US Patent, 6,291,589 (2001).
5. D. Bruneele et al., US Patent, 6,294,647 (2001).
6. J. Pickett et al., US Patent, 6,689,474 (2004).
7. J. Silva et al., US Patent, 6,664,366 (2003).
8. J. Suriano et al., US Pat. Appln., 2003/0,050,400 (2003).
9. J. Suriano et al., US Patent, 6,538,065 (2003).
10. E. Brister et al., Paper Presented at the International Waterborne, High Solids and Powder Coatings Symposium, Feb. 6–8, New Orleans, LA (2002).
11. P. Vollenberg et al., US Patent, 6,583,256 (2003).
12. T. Siclovan et al., US Pat. Appln., 2003/0,216,539 (2003).
13. R. Galluci, X. Li and P. Sybert, US Patent, 6,627,303 (2003).
14. J. Suriano et al., Polym. Prepr., Vol. 44, p 748 (2003).
15. J. Pickett et al., US Pat. Appln., 2003/0,207,096 (2003).
16. J. Pickett et al., US Patent, 6,689,474 (2004).

## 7.4 Polymers from Cyanate Esters

### 7.4.1 Introduction

Cyanate ester compounds and resins can be classified as high performance thermoset polymers, which became the predominant resin materials for use in advanced electronic and aerospace applications [1,2]. Compared to epoxy and bis-maleimides (BMI), cyanate ester resins have been reported to be tougher, show significantly good electrical properties and also, have shown low moisture absorption properties. In general, cyanate esters have two or more cyanate ester (-OCN) functional groups attached to bridged or non-bridged aryl groups. Upon heating or curing, cyanate ester groups undergo polymerization reaction via a thermally activated addition reaction to form three-dimensional network structures by the cyclotrimerization process. The thermoset polymers formed from the cross-linking reaction of three cyanate ester groups containing oxygen-linked triazine rings are known as polycyanurates. Due to the presence of triazine ring in the cured cyanate ester resins, they exhibit extremely low polarity and more hydrophobic character.

Polycyanurates contain polymeric network structures with high aromatic content, which could provide high glass transition temperatures (Tg). The absence of strong hydrogen bonding groups in the cured resin could show less water absorption and low dielectric properties. During the curing process, cyanate esters do not produce any volatile by products and therefore void-free castings and fiber reinforced composites with excellent physical and mechanical properties could be obtained. The excellent hot/wet and dielectric properties of cyanate esters make them very suitable for antennas and radomes applications. The glass and aramid fiber reinforced cyanate ester based composites have been employed for high speed multilayer circuit board applications due to their high Tg, low dielectric constant and good peel strength properties.

Though bisphenol-A based cyanate esters have been widely employed as matrix resins for the advanced composites and prepreg applications, resorcinol based cyanate esters have also shown improved properties in the cured resin compositions. In this chapter, the chemistry and applications of resorcinol based cyanate ester compounds have been briefly reviewed.

### 7.4.2 Synthesis of Cyanate Ester Compounds and Resins

Aryl cyanates, in general, can be synthesized from the reaction of phenols with cyanogen halides in the presence of a base [3] (Figure 7.28).

Similarly, a highly pure resorcinol dicyanate ester compound was obtained in high yields by reacting resorcinol with cyanogens chloride in the presence of triethylamine as the base and isopropyl alcohol (IPA) solvent [4] (Figure 7.29).

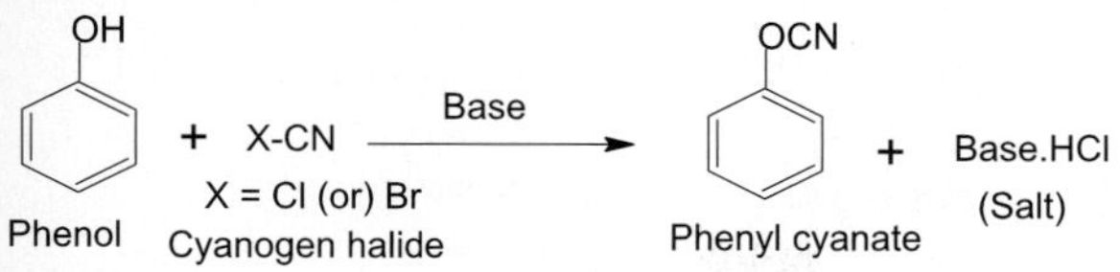


**Figure 7.28.**

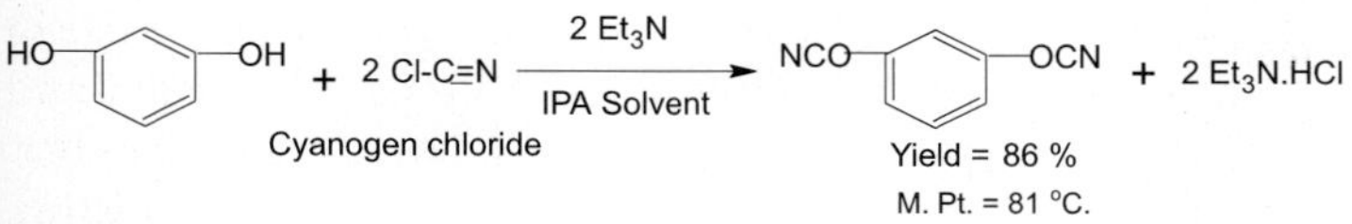


**Figure 7.29.**

The procedure used in the preparation of resorcinol dicyanate ester can be illustrated in the following procedure [4].

**Procedure for the Synthesis of Resorcinol Dicyanate Ester [4]** 271 g (4.4 mol) of cyanogen chloride in 1.1 l of isopropanol are introduced at – 5 °C into a 3-l glass vessel equipped with a stirrer, a thermometer and an inlet pipe.

By means of a metering pump, a solution of 220 g (2 mol) of resorcinol and 430 g (4.04 mol) of triethylamine in 750 ml of isopropanol is pumped in below the surface with stirring. The reaction temperature is best kept at – 5 to + 3 °C. On completion of the reaction, the crystalline sludge formed is filtered off under suction from the isopropanol and the filter residue is washed with 100 ml of isopropanol. The residue is thoroughly stirred with 1.1 l of water and filtered off under suction again. The filter residue is then taken up in 800 ml of methylene chloride and washed with water until free from chloride. Following the addition of 600 ml of perchloroethylene, the methylene chloride distils off initially at 45 to 50 °C and subsequently at 60 °C and at about 160 Torr. The residue is left to cool with stirring to 5 – 10 °C and the resorcinol dicyanate which has crystallized out is filtered off under suction, washed twice with 50 ml of perchloroethylene and dried in a stream of air. 270 g (84% of the theoretical) of resorcinol cyanate are obtained. Mp: 81 °C.

In a modified procedure, the use of stoichiometric quantities of sodium hydroxide in the presence of catalytic amount of triethylamine also produced resorcinol dicyanate ester in higher yields (92%) and high purity (M.Pt. = 82 °C) [5].

Resorcinol cyanate ester having a free hydroxyl group, namely 3-hydroxyphenyl cyanate, was prepared by reacting resorcinol with cyanogen chloride in the presence of a water immiscible organic solvent and an inorganic base [6] (Figure 7.30).

By employing cyanogen halide, 1,3-dicyanato-4-acetyl benzene (Fig. 7.31) [7], 4,4′-bis (*m*-cyanato phenoxy) diphenyl ether (Figure 7.32) [8], and 4,4′-bis(*m*-cyanato phenoxy) diphenyl (Figure 7.33) [8] compounds have been synthesized.

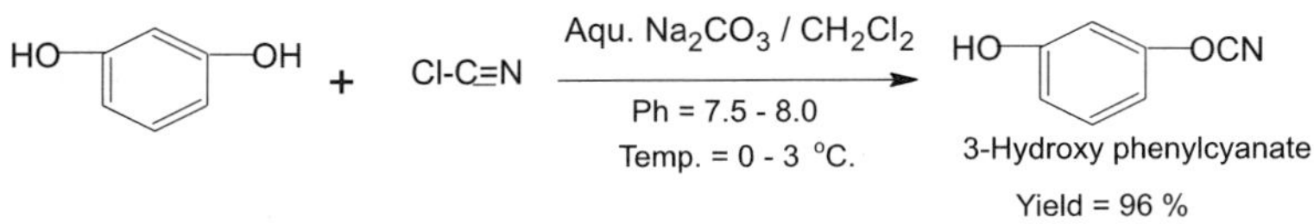


**Figure 7.30.**

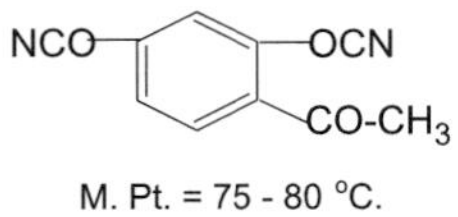


**Figure 7.31.**

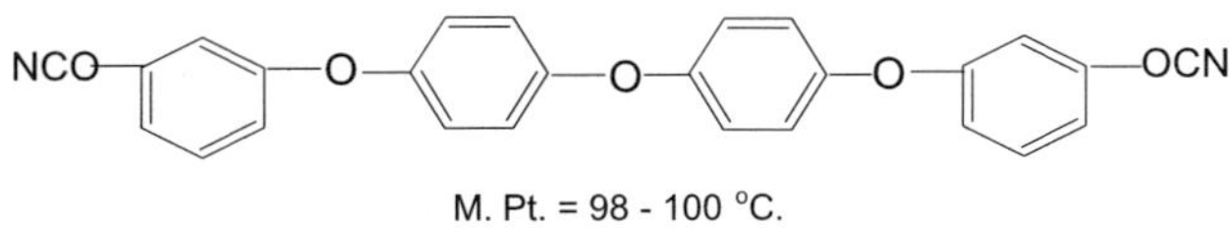


**Figure 7.32.**

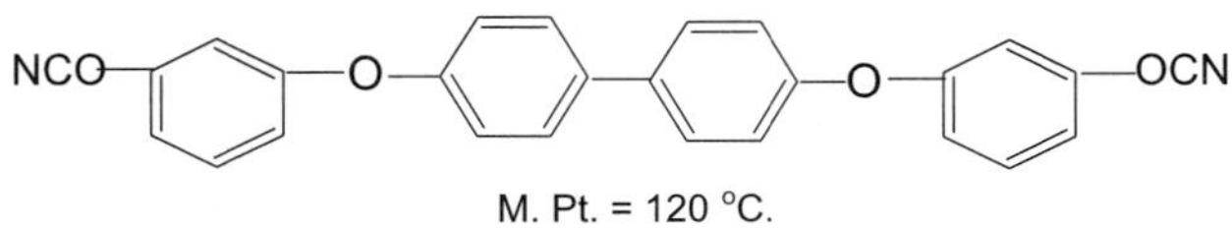


**Figure 7.33.**

Polyarylene ether compounds terminated with cyanato phenyl groups can be thermally polymerized to obtain polyarylene ether cyanurates having outstanding strength, toughness and flexibility, and can be useful in adhesives, coatings and binders [8].

Resorcinolic novolak resin, prepared from the resorcinol and formaldehyde reaction, was utilized for the production of cyanate ester resins having more cyanate groups per molecule [9] (Figure 7.34).

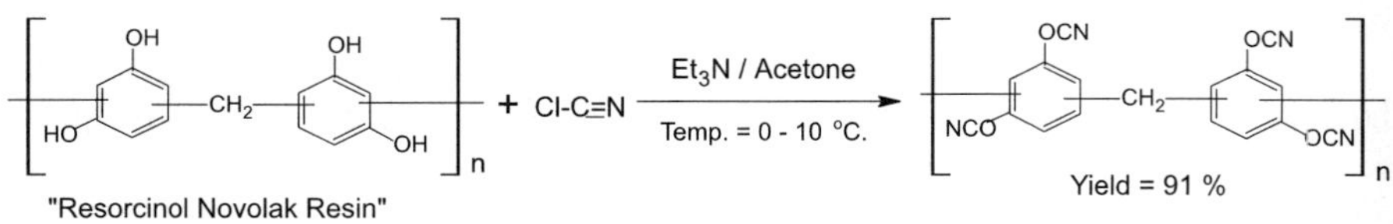


**Figure 7.34.**

## 7.4.3 Properties of Resorcinol Cyanate Esters

In general, most aryl dicyanates are stable and less reactive compared to aromatic diisocyanates. Due to the presence of two electro-negative atoms

(O and N) attached to the carbon atom of the cyanate group (-O-C≡N), cyanate ester group acts strongly as electrophilic. Therefore, nucleophilic reagents can react with aromatic cyanate ester compounds under mild conditions and also, can be catalyzed by acids or bases [3].

When the phenols are reacted with arylene cyanates, bisaryl imidocarbonates are produced in the presence of base catalysts. Polymers containing imino carbonic ester groups in the back bone were obtained from the reaction of resorcinol dicyanate with resorcinol [10] (Figure 7.35).

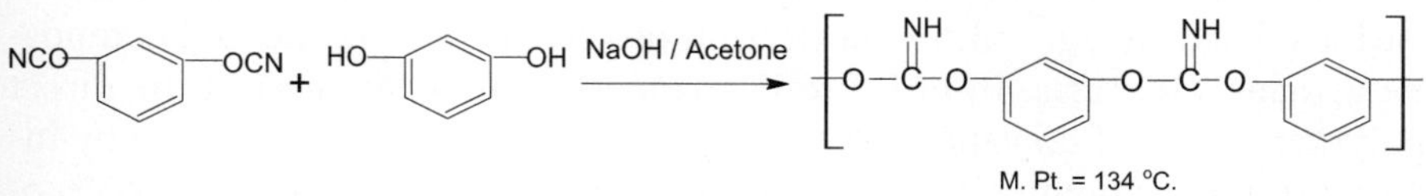


**Figure 7.35.**

The formation of the imino ether group from the above reaction was confirmed by observing an absorption in the IR spectrum at between 5.9 and 6.0 micron.

Polymeric *N*-cyano-isourea ether was obtained in quantitative yields when resorcinol dicyanate was reacted with di-*N*-cyanamide compound, which was then converted into an extremely hard, light brown polytriazine by heating at 150 °C temperature [11] (Figure 7.36).

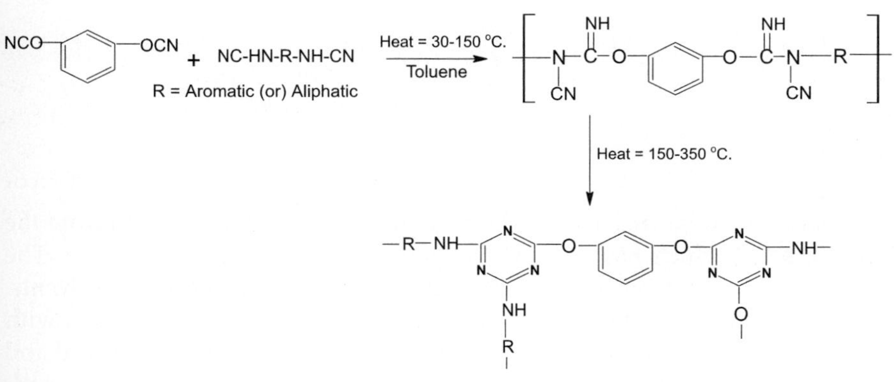


**Figure 7.36.**

## 7.4.4 Prepolymers of Resorcinol Dicyanate Ester

Di- and poly-functional cyanate esters have been used in various industrial applications, including the printed circuit boards, antenna coatings, adhesives, structural composites, encapsulating resins and matrix resins for abrasives.

The di-functional dicyanate esters can be converted into poly-functional compounds by polymerizing them in the presence of heat and also catalysts.

It has been well known that di-functional or poly-functional cyanate esters can be polymerized to form high molecular weight cylotrimerized polycyanurates or polytriazines. The polymerization reaction of cyanate esters are highly exothermic, and may go through a 'prepolymer' stage before producing high molecular weight triazines. By the time the pre-polymerization reaction has taken place from the dicyanate esters, the heat of reaction and heavy shrinkage associated with the trimerization have been already occurred. This might give various disadvantages in the fabrication of various fiber reinforced composite articles and parts. In order to overcome the disadvantages of the direct polymerization of dicyanate esters, prepolymers have been developed by interrupting the polymerization reaction by cooling after about 30 to 65% of cyanate groups have reacted.

The prepolymers formed from the dicyanate esters are still soluble in organic solvents, fusible at temperatures greater than 100 °C and can be stored indefinitely at room temperature conditions.

By heating resorcinol dicyanate ester, prepolymers having a molecular weight greater than 400 could be achieved (Figure 7.37).

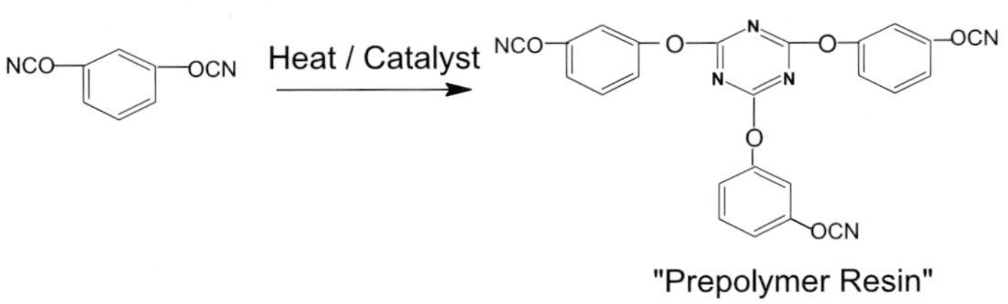


**Figure 7.37.**

Prepolymer mixture contains the cylotrimerized reaction product from the cyanate ester, which has the triazine ring in the molecular structure. The prepolymer can exhibit reduced volatility and be soluble in organic solvents. In addition, the partially trimerized prepolymer can be easily blended with other resins, catalysts and additives to achieve good cured resin physical and mechanical properties.

The prepolymer of cyanate esters contains highly reactive terminal cyanate esters and therefore, by thermal polymerization these cyanate groups could cyclotrimerize and produce a three dimensional network structures with polyarylene linkages between cyanurate rings (Figure 7.38).

Resorcinol cyanate ester prepolymer was conveniently prepared by heating a mixture of resorcinol dicyanate and *p*-tricresyl phosphate for a period of 5 h at 160 °C [12]. On heating this prepolymer at 180 °C for 4 h in the presence of zinc octoate catalyst, a honey colored, transparent, elastic and non-flammable cured polycyanurates or polytriazine material was obtained.

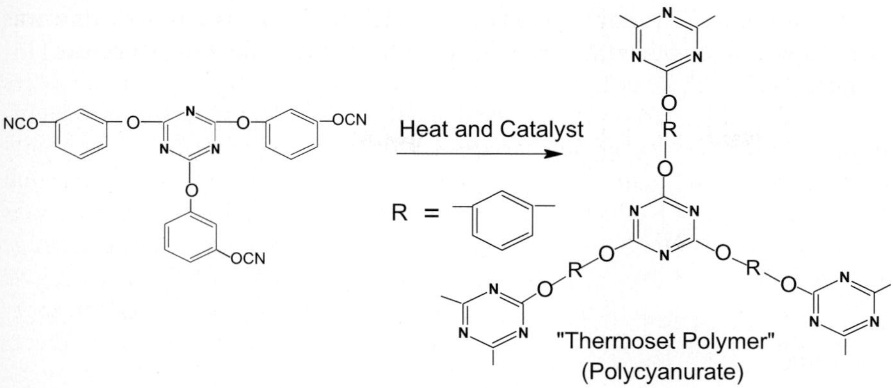


**Figure 7.38.**

Zinc octoate catalyst was employed for the preparation of prepolymer and also, to produce polytriazine compound from 2,4,6-trimethyl resorcinol dicyanate [13].

In the manufacture of molding materials from resorcinol dicyanate, this must be heated to 150 °C for more than 20 h to achieve the B-stage prepolymer resin. By adding 1.0 wt. % 3-hydroxyphenyl cyanate (Figure 7.30) to the pure resorcinol dicyanate, the corresponding prepolymer was obtained in 3 to 5 h at 150 °C [6].

The triazine prepolymer was also advantageously and conveniently prepared by condensing the resorcinol with cyanuric chloride, and subsequently reacting the free hydroxyl groups of the resultant product with cyanogen halide in the presence of a base [14] (Figure 7.39).

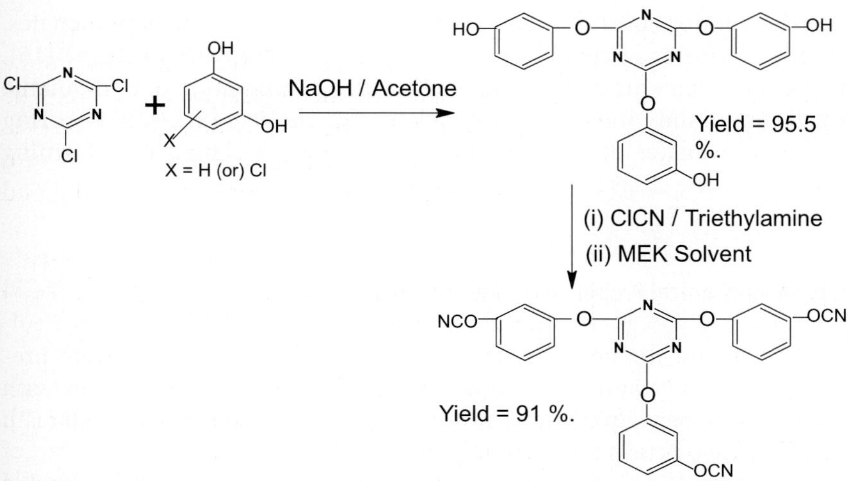


**Figure 7.39.**

In addition to cyanuric chloride, 4,6-dichloro-2-phenoxy-*S*-triazine was also employed to obtain the dicyanate ester having the following structure [15] (Figure 7.40).

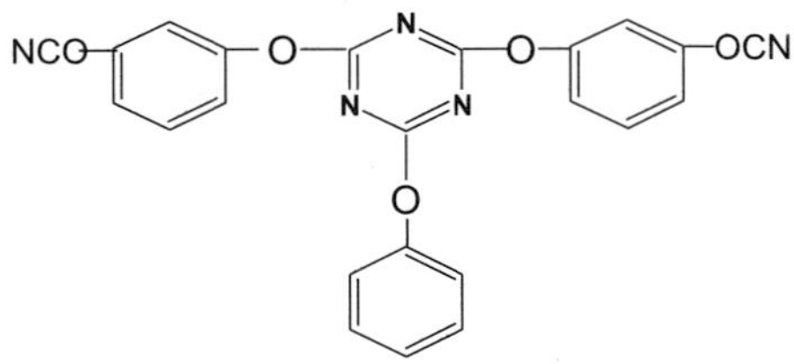


**Figure 7.40.**

The triazine prepolymer obtained from the cyanuric chloride process do not contain any monomeric dicyanate esters. This is particularly useful with resorcinol dicyanate because, it has been reported that this monomer was irritant. Since the polymerization of triazine prepolymer results in little or no shrinkage, dimensionally stable composite moldinging parts could be produced. Also, these high molecular weight prepolymers may not produce any volatile monomers during their processing.

### 7.4.5 Curing of Cyanate Ester Resins

The curing or trimerization reaction of cyanate ester monomers or prepolymers can be done in the presence of heat and catalysts. The rate of polymerization or trimerization of cyanate esters is dependent upon the temperature and catalyst concentration. Suitable curing catalysts for the cyanate ester trimerization are the solutions of zinc, cobalt and copper carboxylates and naphthenates. The acetlyacetonates of cobalt was also employed as the curing catalyst [16]. Aromatic dicyanates are also polymerized with a combination of non-volatile hydroxyl compounds and metal carboxylates and chelates [2,17]. Post-curing of the cured cyanates often found to improve the performance by forming additional triazine rings from the terminal cyanate groups.

### 7.4.6 Neat Resin Mechanical Properties of Resorcinol Dicyanate

In order to obtain the neat resin mechanical properties, castings were prepared by catalyzing the molten resorcinol and bisphenol-A dicyanates with nonylphenol-copper naphthenate solutions, cured and then post-cured [17]. The performance details are presented in Table 7.23.

Compared to bisphenol-A, resorcinol dicyanate homopolymer showed much improved tensile strength and modulus properties. The increased cross-

**Table 7.23.** Properties comparison: resorcinol vs bisphenol-A cyanate ester

| Dicyanate ester | Resorcinol | Bisphenol-A |
|---|---|---|
| Melting point (°C) | 78 | 79 |
| Vapor pressure, °C (at 0.1 Torr) | 114 | 182 |
| Toxicity | Irritant | Low |
| Homopolymer properties | | |
| HDT, °C (dry) | 251 | 255 |
| HDT, °C (moisture conditioned)[a] | 170 | 172 |
| Moisture absorption (%) | 2.3 | 1.9 |
| Tensile properties | | |
| Strength, MPa | 119.3 | 82.8 |
| Modulus, GPa | 4.8 | 3.3 |
| Strain (%) | 3 | 3.6 |
| Methylene chloride absorption (%) (3 h at RT) | 0.03 | 5.8 |
| Flammability, UL-94 | Intumeses | Burns |

[a] Conditioned for 64 h at 92 °C and > 95% RH
Catalyst: nonylphenol and copper naphthenate
Cure: 177 °C/1 h, 200 °C/1 h and 250 °C/2 h
Data from [17]

link density of the network structures formed from the resorcinol dicyanate exhibited solvent resistant properties and relatively non-flammable material.

Cured neat resin castings were obtained from the resorcinol dicyanate ester prepolymer by heating them at elevated temperature conditions and their bulk resin properties were determined (Table 7.24) [18].

As can be seen from these results, resorcinol based cyanate resins showed a combination of excellent mechanical properties at low and high temperature conditions.

### 7.4.7 Thermal and Electrical Properties of Resorcinol Cyanate Resins

Polycyanurates obtained from the cure of pure cyanate monomers, in general, have excellent thermal and dielectric properties. For the development of advanced composites, cyanate resins can offer the advantages as matrix resins because of their non-volatile, low moisture absorption, radiation resistance and high thermal stability properties. In the manufacture of printed circuit boards (PCB) for the computer industry, PCBs with lower dielectric constant (Dk) and dissipation factor (Df) are needed to permit faster signal speeds and signal integrity. The thermal and dielectric properties of resorcinol dicyanate are presented in Table 7.25 [16].

**Table 7.24.** Mechanical properties of cured resorcinol dicyanate ester

| Property | Strength (MPa) | Modulus (GPa) | Elongation (%) |
|---|---|---|---|
| Tensile | | | |
| RT | 124.1 | 4.8 | 3.2 |
| 350 °F | 88.3 | 3.9 | 4.2 |
| 450 °F | 43.4 | 3 | 9.2 |
| Flexural | | | |
| – 65 °F | 188.3 | 5.6 | NA |
| RT | 201.4–224.1 | 4.5–5 | NA |
| 350 °F | 153.8–157.9 | 3.7 | NA |
| 450 °F | 69 | 2.7 | NA |
| Compressive yield | | | |
| – 65 °F | 294.5 | 3.1 | NA |
| RT | 229.7 | 2.9 | NA |
| 350 °F | 111.7 | 1.9 | NA |
| 450 °F | 75.9 | 1.5 | NA |

NA = Not available
Data from [18]

The combination of high thermal stability, low dielectric constant and non-burning properties of resorcinol dicyanate could provide good matrix resin for the printed PCB applications.

**Table 7.25.** Thermal and electrical properties of cured resorcinol dicyanate ester resin

| Thermal property | Values |
|---|---|
| Heat distortion temperature, °C | 260 |
| Flammability | Non-burning, Non-dripping |
| Coefficient of linear expansion ($\times 10^{-6}$ /°C) | 49 |
| Mold shrinkage | Approx. 0 |
| Electrical properties | |
| Dielectric constant (RT, Dry) | |
| At 50 Hz | 3.06 |
| At 20 kHz | 3.33 |
| Power factor (RT, Dry) | |
| At 50 Hz | <0.001 |
| At 20 kHz | <0.0001 |
| Volume resistivity ($\rho$) (Ohm-cm) | |
| At RT | $1.7–2.7 \times 10^{16}$ |
| At 450 °F | $> 4 \times 10^{16}$ |

Data from [18]

### 7.4.8 Carbon Fiber-Resorcinol Dicyanate Composite Properties

Polytriazines obtained from the di or polyfunctional cyanate esters and their prepolymers offer favorable mechanical, dielectric and thermal stability properties. Cyanate resins have been increasingly used in the printed wiring board and structural adhesives/composites for aerospace applications due to their low dielectric and improved thermal performance properties. Carbon fiber reinforced cyanate ester composites can be useful in primary structures, high impact resistance and space applications because of their radiation resistance and thermal stability. Resorcinol dicyanate prepolymer based carbon fiber reinforced composites were developed and their flexural and interlaminar shear properties were evaluated (Table 7.26) [18].

**Table 7.26.** Resorcinol cyanate ester based carbon fiber composites

| Composite system | System A | System B |
|---|---|---|
| Resorcinol cyanate resin (mol%) | 100 | 100 |
| Carbon fiber | Modmor II | Thornel 300 |
| Carbon fiber volume (%) | 61 | 75 |
| Flexural properties | | |
| At RT, strength (GPa) | 1.6 | 1.9 |
| Modulus (GPa) | 141.4 | 165.5 |
| At 350 °F, strength (GPa) | 1.4 | 1.5 |
| Modulus (GPa) | 132.4 | 158.6 |
| Interlaminar shear strength (MPa) | | |
| Initial | | |
| RT | 128.3 | 84.1 |
| 350 °F | 97.2 | 64.1 |
| After humidity aging[a] | | |
| RT | 93.1 | 73.1 |
| 350 °F | NA | NA |
| Moisture pickup (%) | 1.26 | 1.3 |

[a] After 1000 h at 120 °F and 95% RH
Data from [18]

Good flexural and shear strength properties were retained after the humidity exposure of 1000 h at 12 °F and 95% RH conditions for the carbon fiber-resorcinol cyanate ester composites.

In order to improve the toughness and compression after-impact (CAI) properties, cyanate esters have been often toughened with various engineering thermoplastic polymers such as polyethersulfone (PES), polyetherimide (PEI), polyimide (PI), polyarylates (PA), elastomers and MEK-soluble copolyesters.

## 7.4.9 Summary and Outlook

Aryldicyanate esters, in general, are a fascinating and versatile class of thermosetting polymers. Resorcinol based cyanate prepolymers could be valuable in applications such as structural composites, semiconductor encapsulants, adhesives, printed circuit boards and powder coatings. Though a limited amount of progress has been reported on the synthesis and applications of resorcinolic cyanate ester resins, greater potential still exists on exploring this chemistry further because of the unique structures and properties that could be expected from the resorcinol based compounds.

## References

1. I. Hamerton, Ed, Chemistry and technology of cyanate ester resins, Blackie Academic Publishers, London, 1994.
2. V. McConnell, Tough promises from cyanate esters, Adv. Compos., May/June, 28 (1992).
3. E. Grigat and R. Putter, Angew. Int. Edn, Vol. 6, p 206 (1967).
4. G. Rottoff, R. Sundermann, E. Grigat and R. Putter, US Patent, 4,028,393 (1977).
5. G. Rottoff, R. Sundermann, E. Grigat and R. Putter, US Patent, 4,046,796 (1977).
6. R. Sundermann, US Patent, 4,060,541 (1977).
7. E. Grigat and R. Putter, US Patent, 3,553,244 (1971).
8. B. Loudas and H. Vogel, US Patent, 3,595,900 (1971).
9. E. Grigat, H. Schminke and R. Putter, US Patent, 3,448,079 (1969).
10. H. Schminke et al., British Patent, 1,060,933 (1965).
11. R. Sundermann, G. Rottloff and E. Grigat, US Patent, 4,066,577 (1978).
12. R. Sundermann, G. Rottloff, E. Grigat and R. Putter, US Patent, 4,094,852 (1978).
13. R. Sundermann, G. Rottloff and E. Grigat, British Patent, 1,526,035 (1978).
14. R. Sundermann, G. Rottloff and E. Grigat, US Patent, 3,978,028 (1976).
15. R. Sundermann, US Patent, 4,042,567 (1977).
16. D. Shimp, 32nd International SAMPE Symp, April 6–9, p 1063 (1987).
17. D. Shimp, Polym. Mater. Sci. Eng., Vol. 54, p 107 (1986).
18. C. Delano and E. Harrison, 20th National SAMPE Symp and Exhibition, San Diego, CA, April (1975).

## 7.5
## Poly(Benzoxazoles)

### 7.5.1
### Introduction

Polybenzoxazoles, PBO, are a class of polymers within the larger class of poly (benzazoles), (PBZ), which includes poly(benzoxazoles) or poly(benzobisoxazoles) (PBO), poly(benzimidazoles) (PBI) and poly (benzothiazoles) (PBT). In the case of PBO polymers, oxazole ring bonded to either aromatic or aliphatic groups, which are present in the polymer backbone or main chain. Generally, poly(benzoxazoles) refers to poly(*p*-phenylene benzobisoxazoles), in which the oxazole rings, namely benzo-[1,2-*d*:5,4-*d*′]-bisoxazole-2,6-diyl group, are connected to *p*-phenylene groups. PBO polymers, for example poly{[benzo-(1,2-*d*:5, 4-*d*′)-bisoxazole-2,6-diyl]-1,4-phenylene}, possess high thermo oxidative stability due to their all-para and rod-like character. They are capable of forming liquid crystalline phases in concentrated solutions of methanesulfonic acid. Ultra-high strength and modulus fibers having high degree of orientation and crystalline character can be obtained from the solutions of PBO polymer.

Poly(*p*-phenylene-2,6-benzobisoxazole), PBO, is a rigid rod isotropic crystalline polymer. Considerable research and developmental efforts were made on the development and commercialization of PBO polymer, due to the fact that it was expected that extremely high strength and high modulus fibers could be obtained from the liquid crystalline solutions of PBO [1–4]. PBO fiber has shown outstanding high flame resistance and thermal stability among organic fibers. Also, this polymer showed excellent performance in properties such as creep, mechanical resistance and cut/abrasion resistance, and is quite flexible in spite of its extremely high mechanical properties.

PBO is a high performance fiber and commercially manufactured by Toyobo Company in Japan, and sold under the trade name Zylon. The derivative of resorcinol, namely 4,6-diaminoresorcinol (DAR), is the chief raw material for the production of Zylon (PBO) polymer. In this chapter, the synthesis and applications of 4,6-diaminoresorcinol in the manufacture of PBO and other polymers are discussed. The properties and applications of PBO fiber are briefly outlined.

### 7.5.2
### Synthesis of 4,6-Diaminoresorcinol (DAR)

Various synthetic routes have been developed to synthesize 4,6-diaminoresorcinol. Employing resorcinol as the starting raw material, DAR was obtained from the reduction reactions of dinitro and diazoresorcinol compounds.

### 7.5.2.1
### Synthesis of DAR from Dinitro Resorcinols

Resorcinol is a very reactive compound for the nitration reactions. DAR can be conveniently prepared by reducing the 4,6-dinitroresorcinol with molecular hydrogen and in the presence of noble metal hydrogenation catalysts. But the synthesis of high purity 4,6-dinitroresorcinol is often very difficult to obtain due to the formation of 2,4-dinitro and 2,4,6-trinitroresorcinol (styphnic acid) compounds from the nitration reaction of resorcinol.

In order to avoid the formation of 2-nitro derivative, several approaches have been made.

The preparation of 4,6-dinitroresorcinol derivative was obtained by nitrating the 2-chlororesorcinol with the nitrating mixture [5, 6] (Figure 7.41).

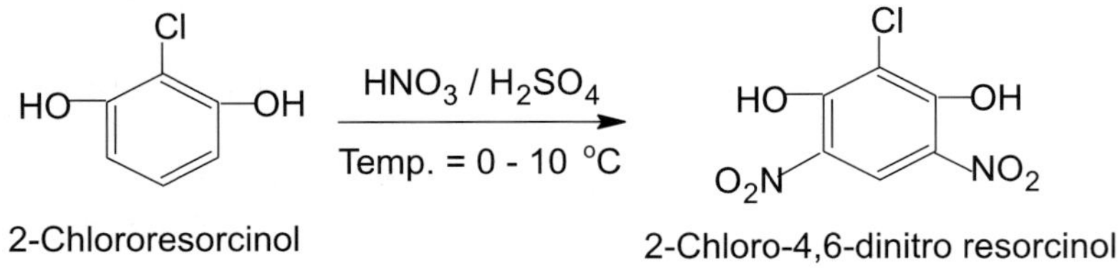


**Figure 7.41.**

On hydrogenation, the 2-chloro-4,6-dinitroresorcinol was converted into 4,6-diaminoresorcinol and isolated as the hydrochloride salt [6,7] (Figure 7.42).

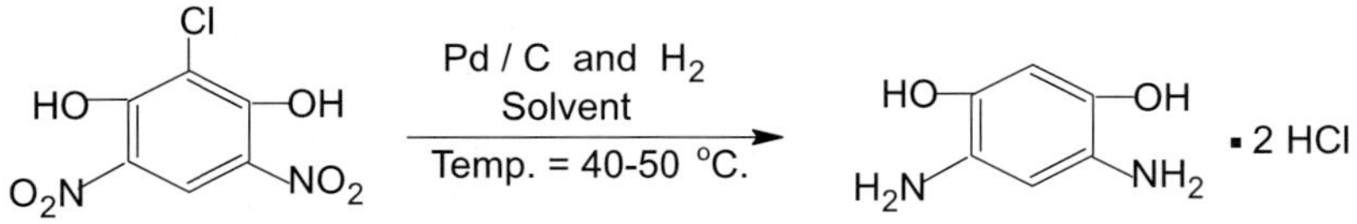


**Figure 7.42.**

4,6-Dinitroresorcinol was also synthesized from resorcinol through 1,3-bis-(methylcarbonato) benzene [8] and 1,3-dibenzyloxy benzene intermediates [9] (Figures 7.43 and 7.44).

Since the nitration of resorcinol tends to produce 2-substituted undesirable nitration intermediates, there is always a danger of potentially producing 2,4,6-trinitroresorcinol (styphnic acid), which is a powerful explosive. Therefore, several other safer processes were developed for obtaining 4,6-diaminoreorcinol in higher yields and purity.

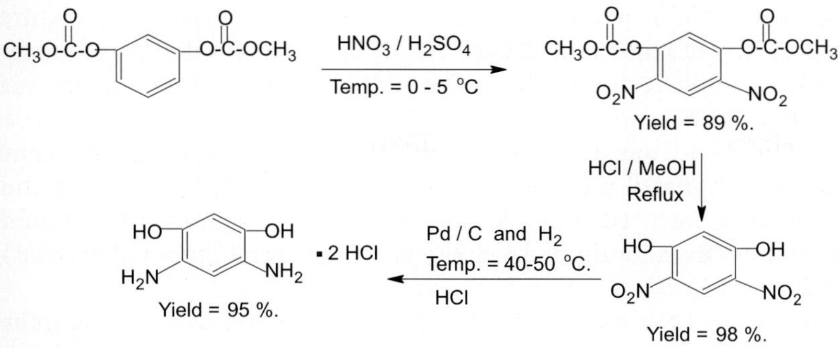


**Figure 7.43.**

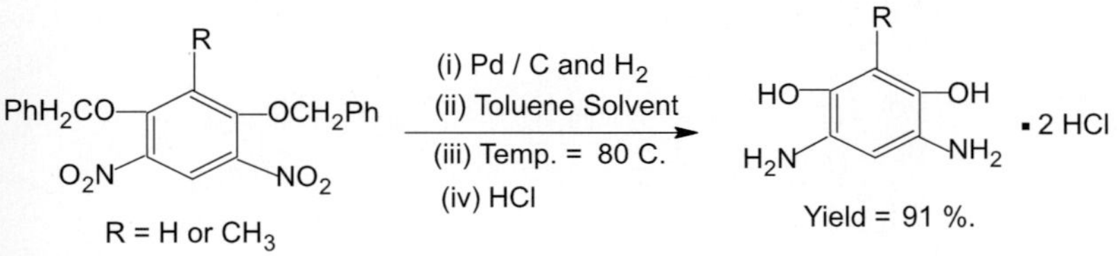


**Figure 7.44.**

### 7.5.2.2
### Synthesis of DAR Using Diazo Resorcinols

DAR was produced by reducing the 4,6-diphenylazo resorcinol, which was prepared from the reaction of resorcinol or 2-methylresorcinol with benzenediazonium chloride under basic conditions [10,11] (Figure 7.45).

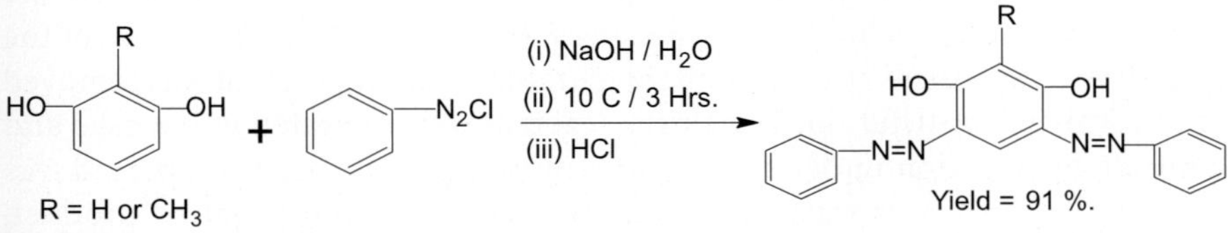


**Figure 7.45.**

A synthesis procedure used in producing 4,6-bisphenylazo resorcinol is described below [11].

**Synthesis of 4,6-Bisphenylazo Resorcinol** A solution of 3.04 g of sodium nitrite dissolved in 5.5 g of water at 0 to 5 °C was added drop-wise to a mixture of 4.1 g of aniline, 9.8 g of 35% hydrochloric acid and 10 g of water to synthesize a benzenediazonium chloride aqueous solution. This benzenediazonium chloride aqueous solution and a solution composed of 4.9 g of NaOH and 12.3 g of water were sent by pump in the same molar ratio, and were mixed in line while cooling at 10 °C.The resulting mixture was added drop-wise to a mixture composed of 2.2 g of resorcinol, 2.4 g of NaOH and 11 g of water at 10 °C over 40 min. After completion of the drop-wise addition, the resulting mixture was stirred at 10 °C for 3 h.

Hydrochloric acid was added to the reaction mixture to make the same acidic. Solid precipitated was collected by filtration, washed with water and dried to obtain a dark red solid. As a result of determination with a liquid chromatography, the amount of 4,6-diphenylazo resorcinol product was 5.8 g and the yield was 91%.

By reducing the 4,6-bisphenylazo resorcinol in a selected solvent and in the presence of noble metal catalyst using hydrogen, 4,6-diaminoresorcinol was obtained in high yields [12] (Figure 7.46).

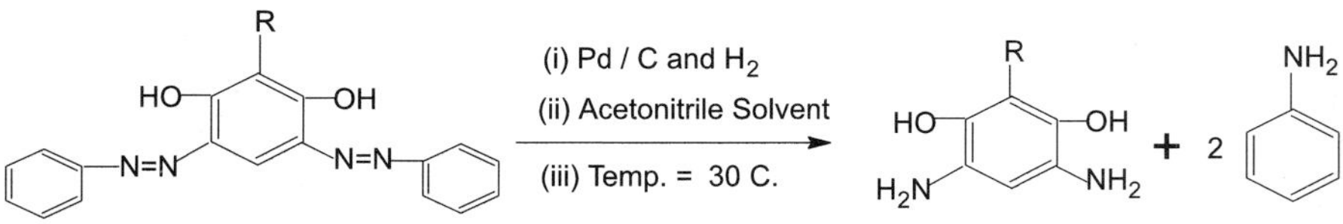


**Figure 7.46.**

The preparation of DAR from the phenylazo resorcinol was based on the following procedure [12].

**Synthesis of 4,6-Diaminoresorcinol** A weight of 10.0 g (0.025 mol) of 4,6-bisphenylazoresorcinol (purity; 79.5%), 50 g of acetonitrile and 2.0 g of 5% Pd/C and 1.0 g of active carbon were introduced in a 100 ml autoclave made by Hastelloy, to react at a stirring rate of 600 rpm and at 30 °C with hydrogen under a constant pressure of 300 kPa from a hydrogen pressurized vessel. A pressure decrease in the hydrogen pressurized vessel was stopped after the reaction period of 2.5 h. After the stirring was stopped, the temperature of the autoclave was returned to the room temperature and the reactant was removed after nitrogen substitution. The slurry reactant was separated into a cake and a filtrate by filtration under a nitrogen atmosphere. The filtration period was 10 min. The cake was washed with acetonitrile to obtain a wetted product. A yield of DAR in the wetted product was 93.1% and a yield in the filtration wash liquid was 4.9%, so that a total yield was 98.0%.

### 7.5.2.3 Synthesis of DAR from Diacetyl Resorcinol

In addition to the above methods, a novel method was also developed for DAR synthesis, and was based on the following reaction scheme [13] (Figure 7.47).

In the above reaction scheme, the conversion of dioxime into diamino compound was taking place through the Beckmann rearrangement in the presence of polyphosphoric acid (Figure 7.48).

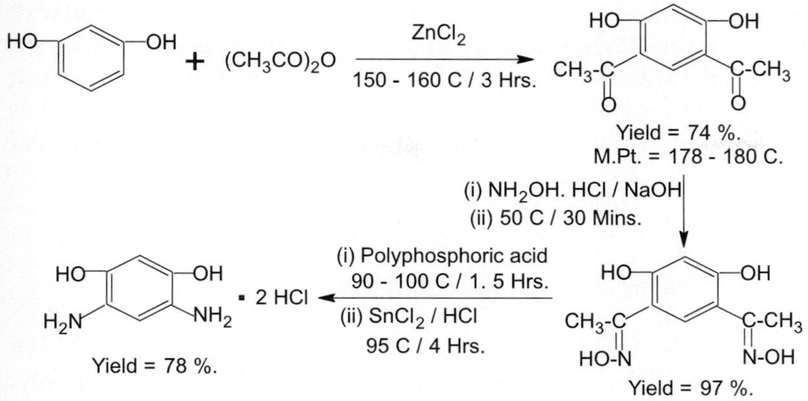


**Figure 7.47.**

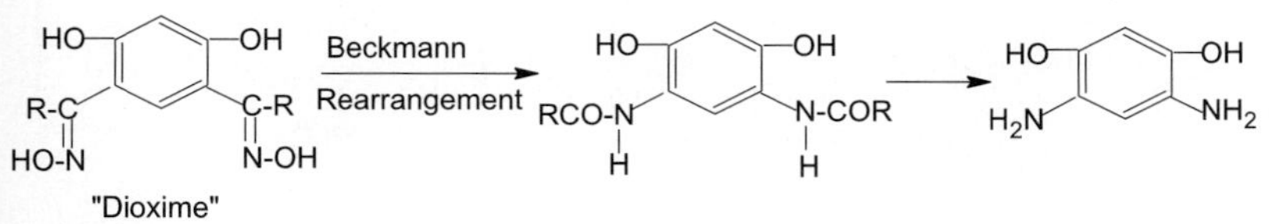


**Figure 7.48.**

## 7.5.3 Stabilization of 4,6-Diaminoresorcinol

DAR is the necessary raw material for the synthesis of PBO polymer, which can be processed into a very strong fiber having high elastic modulus. To obtain the fibers with good mechanical properties, the PBO polymer should possess high molecular weight, which could be possible only by using monomers of high purity in the synthesis.

In the case of DAR monomer, it is susceptible to oxidation. DAR can be easily discolored when this compound is left in the atmospheric conditions for several days and sometimes, in several hours when the oxidation proceeds faster. With the use of impure DAR, the PBO polymer appeared green to purple in color and also, lower in molecular weights. It was theorized that the first oxidation product of DAR has the following structure (Figure 7.49).

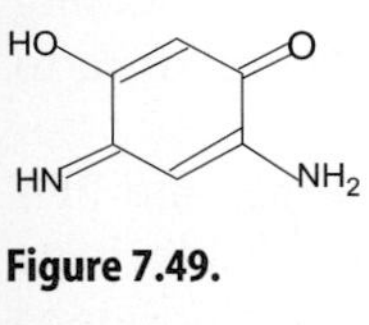


**Figure 7.49.**

This oxidation product can be more easily reduced back to DAR before it oxidizes further by employing reducing agents such as stannous chloride or tin (II) chloride as monomer stabilizers [14,15]. DAR monomer stabilized with 2000 ppm level of tin (II) chloride and, stored at 30 °C in the presence of air for three months did not show any degradation of DAR. When this stabilized DAR was used in the PBO synthesis, the resulting polymer dope appeared yellow in color and exhibited high intrinsic viscosity associated with high molecular weight.

## 7.5.4 Synthesis of Poly(*p*-Phenylene Benzobisoxazole) (PBO)

PBO has been synthesized by the solution polycondensation reaction of DAR with *para*-aromatic dicarboxylic acid or its chloride in the presence of polyphosphoric acid (PPA) solvent [16–24] (Figure 7.50).

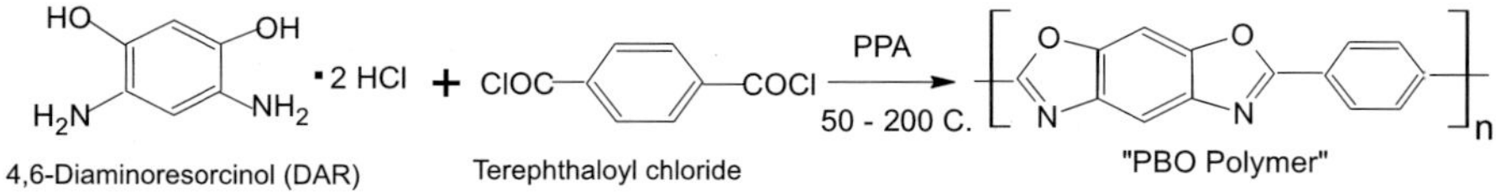


**Figure 7.50.**

Generally, a monomer having an electron deficient carbon groups attached to the benzene ring can be used with DAR to obtain PBO polymer. For example, monomers containing carboxylic acid, carboxylic acid halide, carboxylic acid ester and carboxylic acid salt were employed in the PBO synthesis. In addition to these monomers, compounds having trihalomethyl groups attached to the aromatic ring, which are not moisture sensitive compared to the carboxylic acid halides, were also employed for producing PBO [25,26] (Figure 7.51).

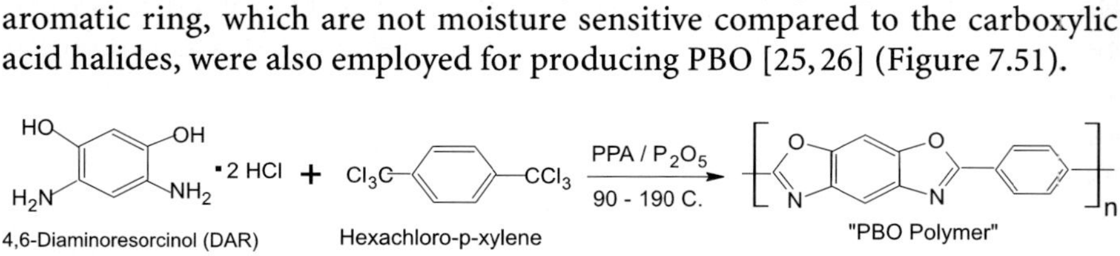


**Figure 7.51.**

The polymerization reaction of 4,6-diaminoresorcinol and terephthalic acid chloride is a condensation reaction. This reaction has been conducted in the presence of polyphosphoric acid (PPA), which could be obtained by adding phosphorus pentoxide to orthophosphoric acid and heating the mixture to about 300 °C (Figure 7.52).

The reaction mixture containing DAR, terephthalic acid chloride and PPA was normally heated prior to the polymerization, in order to dehydrochlorinate the mixture. During this process, PPA reacts with terephthalic acid chloride to form poly(phosphoric terephthalic anhydride), which in turn reacts with DAR to yield PBO [25,27] (Figure 7.53).

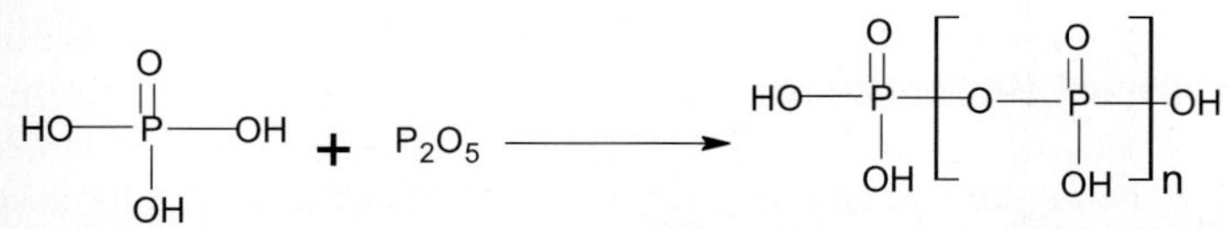


**Figure 7.52.**

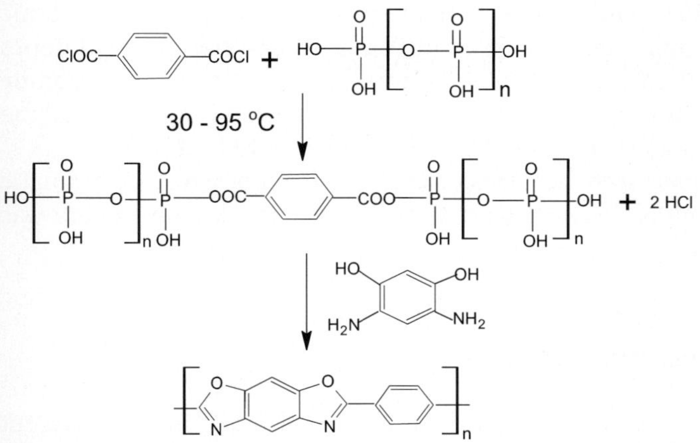


**Figure 7.53.**

The formation of higher molecular weight polymers from the monomers was observed to be dependent upon the shear rate of the PPA solutions in the reactor. As the shear rate was increased, the speed of polymerization reaction was found to be increased. A preparative procedure for the synthesis of poly(*p*-phenylene benzobisoxazole) from 4,6-diaminoresorcinol is illustrated below [22].

**Synthesis of Poly(benzoxazole) Using 4,6-Diaminoresorcinol [22]** 4,6-Diaminoresorcinol dihydrochloride (9.1 g, 42.8 mmol), terephthalic acid (7.1 g, 42.8 mmol), 116% polyphosphoric acid (43.3 g), phosphorus pentoxide (15.0 g) and iron(II) phosphate octahydrate ($Fe_3(PO_4)_2 .8 H_2O$, 0.245 g, iron(II) ion 3000 ppm relative to 4,6-diaminoresorcinol dihydrochloride) were weighed and placed in a polymerization apparatus equipped with a stirrer, and the mixture was heated at 70 °C for 1 h and 15 min then for 3.5 h to 120 °C while stirring under dry nitrogen. The mixture was heated to 135 °C over 15 min and reacted for 12 h at the same temperature. Finally, the mixture was heated to 200 °C over 1 h and reacted for 2 h at the same temperature. The obtained polybenzazole had an intrinsic viscosity of 64.4 dL/g and its color was yellow.

To achieve the highest molecular weight, the molar ratios of DAR to dicarboxylic acid or acid chloride should be kept about 1:1 and, also a stepwise heating of the polymerization reaction mixture was necessary.

### 7.5.5 Spinning of Polymer and Fiber Formation

The poly(*p*-phenylene benzobisoxazole) obtained from the DAR monomer was used to produce fibers and ribbons having very high tensile strength and modulus properties. PBO polymer was observed to be soluble in strong acids such as 100% sulfuric acid ($H_2SO_4$), chlorosulfonic acid (CSA), methane sulfonic acid (MSA) or a 97.5/2.5 mixture of methane sulfonic acid and chlorosulfonic acids [17]. Microscopic examination of PBO in methane sulfonic acid dopes showed that this polymer existed in an anisotropic nematic liquid crystalline state at room temperature. Fibers were spun using a lyotropic liquid crystalline PBO and PPA dope and their properties were characterized [28, 29].

When the PBO fibers were heat treated at elevated temperatures, a dramatic increase in the modulus of heat treated fibers over the as-spun fibers was observed [17].

### 7.5.6 Poly(benzoxazole) Fiber Properties

PBO fibers are commercially produced by Toyobo of Japan in two different grades, namely the regular (as-spun, AS) and high modulus (HM). The production of these two fiber grades is thought to involve a dry-jet wet spinning process. The regular fiber grade may be commercialized literally "as-spun", whereas the "HM" fiber might have given a subsequent heat treatment process to improve the fiber's Young's modulus. The PBO fiber properties are summarized in Table 7.27.

High performance and high temperature resistant fibers are in constant demand for unique and challenging applications. Some of the important prop-

**Table 7.27.** Properties of poly(*p*-phenylene benzobisoxazole) fibers (PBO, Zylon)

| Fiber properties | Zylon AS | Zylon HM |
|---|---|---|
| Density (g/cm$^3$) | 1.54 | 1.56 |
| Tensile strength, cN/dtex | 37 | 37 |
| GPa | 5.8 | 5.8 |
| Tensile modulus, cN/dtex | 1150 | 1720 |
| GPa | 180 | 270 |
| Elongation at break, % | 3.5 | 2.5 |
| Decomposition temperature in air, °C | 650 | 650 |
| Limiting oxygen index | 68 | 68 |
| Dielectric constant at 100 kHz | NA | 3 |
| Dissipation factor | NA | 0.001 |
| Moiture regain at 65% RH, % | 2 | 0.6 |

AS = As-spun, HM = High modulus

erties of high performance fibers are high tensile strength, operating temperatures, limiting oxygen index (LOI) and chemical resistance. Tensile strength is often the determining factor in choosing a fiber for a specific application. A major advantage of high strength fiber, such as PBO, over steel is its superior strength to weight ratio.

A performance comparison of PBO to other synthetic high performance fibers is presented in Table 7.28.

**Table 7.28.** Comparative properties of various synthetic fibers

| Synthetic fibers | Tensile properties Tenacity (g/de) | Modulus (g/de) | Elongation (%) | Continuous use temperature, °F |
|---|---|---|---|---|
| Polybenzoxazole | 42 | 1300 | 3.5 | 550–600 |
| Polyamide (*meta*-aramid) | 3.8–7.2 | NA | 25–40 | 400 |
| Polyamide (*para*-aramid) | 22–26 | 460–1100 | 2.4–4.4 | 375 |
| Poly(phenylene sulfide) | 3.5–4.5 | NA | 32–49 | 500 |
| Poly(benzimadazole) | 2.7 | 32 | 29 | 482 |
| Polyimide | 4.2 | NA | 30 | 500 |
| Carbon fiber (pitch) | 14–30 | 1000–3850 | 0.2–1.3 | 570–1000 |
| Glass fiber (E-glass) | 35 | 524 | 4.8 | NA |

Data source: manufactures brochures and internet sources

PBO fiber has superior tensile strength and modulus properties compared to aramid (Kevlar) and carbon fibers. It also has outstanding high thermal stability among organic fibers. It has been reported that Zylon (PBO) is known as the strongest fiber commercially available. The unique combination of high tensile strength, modulus and thermo oxidative stability properties makes PBO fibers suitable for many aerospace applications. The high modulus of PBO makes it an excellent candidate for composite reinforcements.

### 7.5.7 Flammability Properties of PBO and Other Fibers

Flame retardant character of polymeric materials, including the fibers, is generally determined by the Limiting Oxygen Index (LOI) method. LOI value shows the amount of oxygen needed in the atmosphere to support the combustion of materials. Fibers with a limiting oxygen index value (LOI) greater than 25 are considered to be flame retardant. The LOI values of PBO and other high performance fibers are presented in Table 7.29.

Compared to Kevlar and Nomex fibers, the PBO fiber had much higher oxygen index suggesting that PBO has over twice the flame retardant properties of aromatic polyamide fibers. The high LOI value of PBO indicates that this polymer has potentially good flame retardancy properties.

**Table 7.29.** Flammability properties of high performance fibers

| Synthetic fiber | Manufacturer | Trade name | Limiting Oxygen Index, % |
|---|---|---|---|
| Polybenzoxazole | Toyobo | Zylon | 68 |
| Polybenzimadazole | Celanese | PBI | 41 |
| Polyimide | Inspec | P-84 | 38 |
| Polyamide | Dupont | Nomex | 30 |
| Polyamide | Dupont | Kevlar | 25–28 |
| Polytetrafluoroethylene | Dupont | Teflon | 95 |
| Polyphenylenesulfide | Amoco | Ryton | 34 |

### 7.5.8 Thermal Properties of PBO Polymer

The thermal properties of PBO were evaluated by TGA and isothermal aging methods [2, 30]. Based on these studies, it was reported that PBO started decomposing at temperature above 650 °C. From the TGA analysis, it was observed that, in the presence of air, 50 wt. % decomposition occurred at 725 °C and, around 810 – 825 °C temperature, a complete decomposition took place for PBO material. A heat resistance study was made on PBO fiber and compared against Kevlar-49 fibers, and the results showed 160 °C higher heat resistance temperature for PBO fiber than Kevlar [31]. PBO fibers could withstand temperature up to 600 °C, which is the temperature normally found in the hotter areas of typical flames.

### 7.5.9 Poly(benzoxazole) Applications

The high performance PBO fiber has several uses in both the military and commercial applications. In the case of military applications, this material has been used as ballistic jackets on the outside of pressure vessels such as helium tanks on satellites, and used for flak vests in body armor. The Federal Aviation Administration (FAA) is currently considering this material for use as a flame resistant protective cloth between the cabin and outer fuselage of aircraft to guard against engine shrapnel. Insulating belts for high temperature industrial grinders and in protective clothing for fighters are some of the non-military applications for PBO fiber.

PBO has been reported to be used in a variety of commercial products including tires, belts, hoses, plastics, concrete, gaskets, abrasive materials, heat and flame resistant clothing, ballistic flak vests, and cut and abrasion resistant safety gloves.

Power transmission belts with good heat and abrasion resistant properties were achieved with higher strength PBO fibers, previously treated with RFL adhesives, used as reinforcements in chloroprene rubber compositions [32].

In the manufacture of radial tires, steel cords are used in the carcass area of the tire. The rolling resistance of the steel cords reinforced tires is increased due to extra weight, which could lead to increase in fuel consumption. Since the *p*-aramid fibers lacked sufficient tensile strength properties to achieve equal performance similar to steel cords, the high tenacity and light weight PBO fibers were successfully used to improve the the tire rolling resistance [33].

For improving the fatigue resistance and energy saving properties of lighter weight rubber reinforced composites, PBO fibers having tenacity of 35 g/d or higher and elastic modulus of 800 g/d were used in the composite materials [34].

Ballistic resistant articles made using PBO fibers were observed to be lighter in weight, more comfortable to wear and, also exhibited better ballistic resistant properties than UHMW polyethylene fiber [35].

## 7.5.10 Use of DAR in other PBO Polymers and Applications

Advanced aircraft and aerospace systems demand light weight structures, which can be developed using polymers exhibiting superior physical and mechanical properties, high degree of thermo oxidative stability and hydrolytic stability properties. PBO polymer obtained from DAR monomer exhibited liquid crystalline behavior in concentrated inorganic acids. This might be due to the rigid rod structure of the polymer molecules associated with close packing. The use of *p*-terphenylene dicarboxylic acid in the PBO synthesis was expected to improve the solubility of the resulting polymer, and, in fact, was determined to be soluble in less corrosive organic solvents such as *m*-cresol +20–30% dichloroacetic acid or formic acid [36] (Figure 7.54).

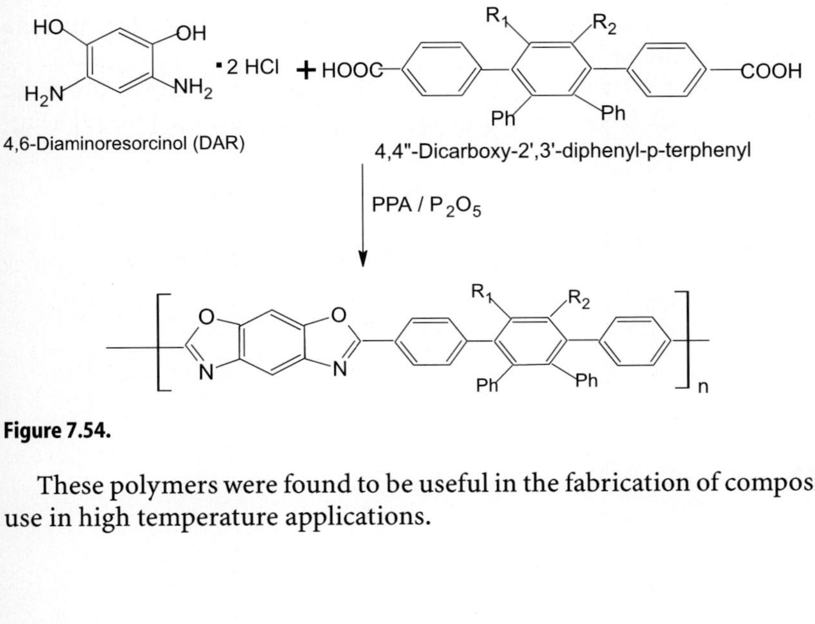


**Figure 7.54.**

These polymers were found to be useful in the fabrication of composites for use in high temperature applications.

Gas separation membranes separate mixtures of gases by permitting one gas in the mixture to diffuse through the membrane from a region of high pressure to a region of low pressure at a faster rate than another gas in the mixture. Separation properties of gas separation membranes are characterized by their permeability, flux and selectivity. In order to function effectively, these membranes should possess sufficient strength to form film, increased high temperature stability and resistance to common organic solvents and vapors. PBO type polymer was obtained from DAR and 1,1,3-trimethyl-3-phenylindan-4′,5-dicarboxylic acid monomers, and was based on the following reaction [37] (Figure 7.55).

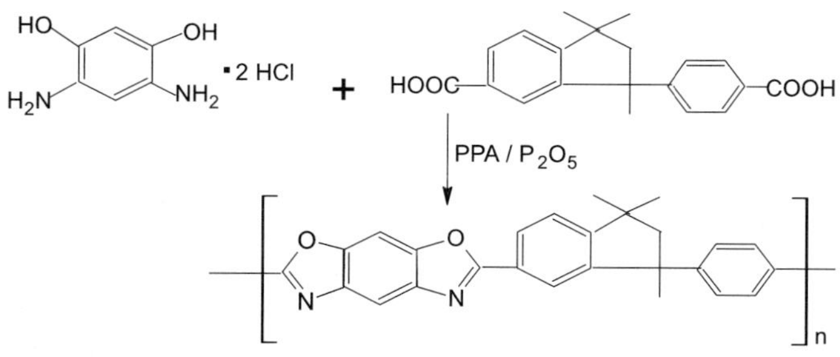


**Figure 7.55.**

The polymer film cast from the *m*-cresol solution was tested for gas permeability, and showed a permeability of 10.45 barrers (Ba) for helium and 0.022 Ba for methane, which corresponded to a selectivity of about 500.

A membrane was also fabricated from the above polymer and tested for its permeability with respect to oxygen, nitrogen, methane, ethylene and helium gases at 0.20 and 0.34 MPa pressures. The membrane had a permeability of 8.3 Ba for nitrogen, 37 Ba for oxygen, 10.9 Ba for methane, 17.2 Ba for ethylene and 121 Ba for helium [37].

Copolymers of PBO containing flexible ether linkages between aromatic rings were obtained from the following scheme of reaction [38, 39] (Figure 7.56).

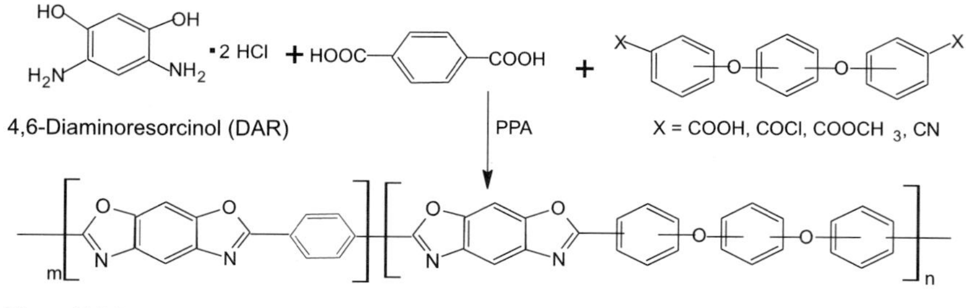


**Figure 7.56.**

The average length of the all-*para* segments between the relatively flexible diphenoxy benzene units was controlled by the stoichiometry of the copolycondensation reaction.

By employing 1,3,5-benzene tricarboxylic acid in the polymer synthesis, a branched PBO polymer was produced [40] (Figure 7.57).

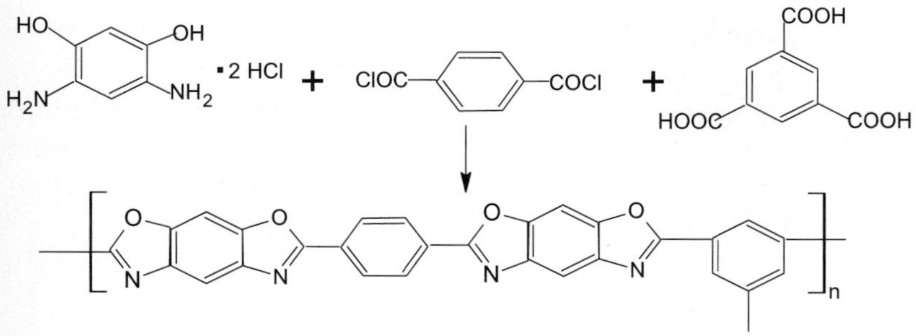


**Figure 7.57.**

The branched polymer can be used to form strong and light articles such as fibers and films.

## 7.5.11 Biscyclobutarene Monomer and Polymers

Diaminoresorcinol was used to synthesize Biscyclobutarene monomer based on the reaction given below [41] (Figure 7.58).

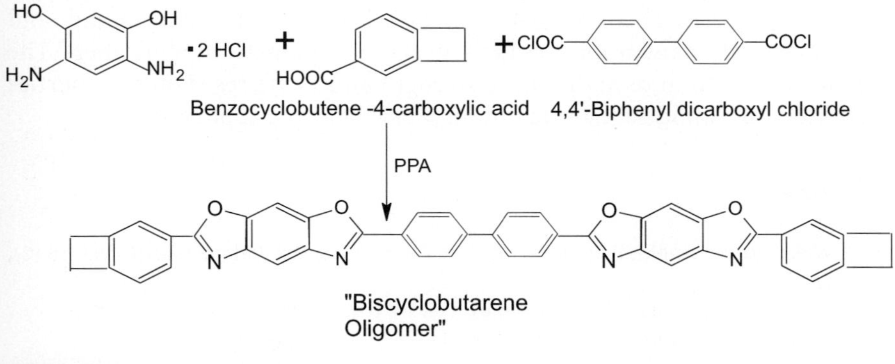


**Figure 7.58.**

This monomer can be copolymerized with maleimides, olefins, acetylenes and cyanate ester monomers. Polymers obtained from the biscyclobutarene monomer exhibited good thermal stability at higher temperatures, chemical resistance to most industrial solvents, physical and mechanical properties, and also low sensitivity to water. Isothermal weight loss data of the polymer showed

90 wt% of the material remained after 900 h at 316 °C and also, after 200 h at 343 °C temperature conditions in an air atmosphere. This monomer could be useful in the fabrication of advanced composites, coatings and films, and also in adhesives.

### 7.5.12 Dihydroxy-Pendant PBO Polymer

Novel rod-like heterocyclic aromatic polymers having semi-conducting properties and could be useful for replacing copper wires in larger aircrafts were obtained from the diaminoresorcinol and 2,5-dihydroxy terephthalic acid chloride [42,43] (Figure 7.59).

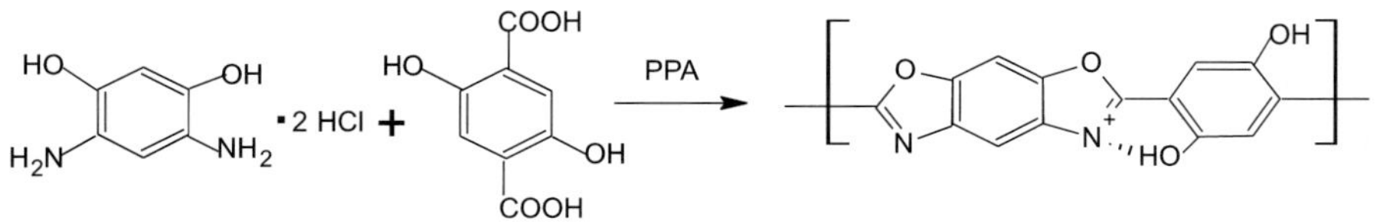


**Figure 7.59.**

### 7.5.13 Summary and Outlook

PBO has been known as the strongest synthetic organic fiber and the development of this fiber could be possible with the use of diaminoresorcinol in the polymer synthesis. This polymer exhibits an excellent thermal stability and better flame retardant properties than most other synthetic organic fibers. The unique combination of high tensile strength and flame resistance properties makes the PBO an ideal material for high performance applications.

## References

1. J. Wolfe and F. Arnold, Polym. Prepr., Am. Chem. Soc., Div. Polym. Chem., Vol. 18 (2), p 414 (1977).
2. J. Wolfe and F. Arnold, Macromoecules, Vol. 14, p 909 (1981).
3. J. Wolfe and F. Arnold, AFML-TR-78-197 (1978).
4. J. Wolfe "Polybenzothiazoles and polybenzoxazoles", Encyclopedia of Polymer Science and Engineering, Wiley, Vol. 11, p 601 (1988).
5. B. Nader, US Patent, 5,371,291 (1994).
6. P. McGrath, US Patent, 5,360,932 (1994).
7. B. Nader, US Patent, 5,410,083 (1995).
8. Z. Lysenko and C. Rand, US Patent, 4,982,001 (1991).
9. H. Behre, H. Fiege, H. Blank and W. Eymann, US Patent, 6,093,852 (2000).
10. T. Morgan, B. Nader, P. Vosejpka, W. Wu and A. Kende, US Patent, 5,453,542 (1995).

11. K. Akimoto et al., US Patent, 5,962,740 (1999).
12. K. Tokunaga et al., US Patent, 6,222,074 (2001).
13. J. Kawachi, H. Matsubara, Y. Nakahara and Y. Watanabe, US Patent, 5,892,118 (1999).
14. Z. Lysenko, S. Rosenberg and W. Harris, US Patent, 5,142,021 (1992).
15. G. Matsuoka and F. Kubota, US Patent, 5,911,908 (1999).
16. T. Helminiak and F. Arnold, US Patent, 4,051,108 (1977).
17. E. Choe, US Patent, 4,423,202 (1983).
18. Z. Lysenko, US Patent, 5,021,580 (1991).
19. W. Harris and Z. Lysenko, US Patent, 5,151,490 (1992).
20. K. Hotta, F. Kubota, Y. Araki and M. Ohguchi, US Patent, 5,919,890 (1999).
21. W. Sweeny and S. Kwolek, US Patent, 4,608,427 (1986).
22. F. Kubota, M. Fukushima and M. Koyama, US Patent, 6,169,165 (2001).
23. J. Wolfe, P. Sybert and J. Sybert, US Patent, 4,533,693 (1985).
24. J. Wolfe et al., US Patent, 4,703,103 (1987).
25. Y. So, US Patent, 5,235,023 (1993).
26. Y. So, US Patent, 5,422,416 (1995).
27. E. Choe and S. Kim, Macromolecules, Vol. 14, p 920 (1981).
28. R. Bubeck et al., US Patent, 5,286,833 (1994).
29. R. Bubeck et al., US Patent, 5,356,584 (1994).
30. E. Orndoff, NASA Technical Memorandum 104814, September 1995.
31. W. Bin, J. Zhihao, Q. Zheming and L. Mu, J. Xian Jiaotong Univ., No. 11, p 1189 (2001).
32. T. Takehara et al., US Pat. Appln, 2004/0, 005,447 A1 (2004).
33. K. Yabuki, US Patent, 5,948,186 (1999).
34. Y. Okuyama, H. Hirahata and K. Yabuki, US Patent, 5,922,259 (1999).
35. D. Dalman, C. Weber and G. LaCasse, US Patent, 6,268,301 (2001).
36. F. Arnold and J. Wolfe, US Patent, 4,108,835 (1978).
37. J. Summers and R. Wessling, US Patent, 5,034,206 (1991).
38. R. Evers, F. Arnold and T. Helminiak, US Patent, 4,229,566 (1980).
39. R. Evers, F. Arnold and T. Helminiak, Macromolecules, Vol. 14, p 925 (1981).
40. N. Madison and W. Harris, US Patent, 5,124,432 (1992).
41. A. Shrock, W. Harris and N. Madison, US Patent, 4,864,010 (1989).
42. T. Dang, H. Chuah, L. Tan and F. Arnold, US Patent, 5,041,522 (1991).
43. T. Dang, H. Chuah, L. Tan and F. Arnold, US Patent, 5,106,940 (1992).

## 7.6 Poly(arylene ethers)

### 7.6.1 Introduction

Poly(arylene ethers) are well known class of engineering thermoplastic materials. These organic polymers contain etheric oxygen atoms covalently linked to aromatic nuclei, which may be a mono or polynuclear aromatic compound. The aromatic portion of this class of polymers is thought to contribute to thermal stability and good mechanical properties, and the ether linkages are considered to facilitate polymer processing while maintaining or improving the oxidative stability. In addition to ether linkages, arylene ether polymers can also contain other functional groups such as the sulfone ($SO_2$) and ketone (CO) in their molecular back bone. Polymers containing these functional groups are called poly(arylether sulfones) or poly(arylether ketones) depending upon the presence of either sulfonyl or keto groups. Poly(arylene ethers) offer exceptional balance of properties; namely high melting point, exceptional thermal stability, excellent hydrolytic stability, high stiffness and strength, good toughness and excellent solvent and environmental stress rupture resistance. Because of these exceptional properties, poly(arylene ethers), in general, are found to be useful for the electric and electronic apparatuses and instruments, machinery and chemical industry applications.

### 7.6.2 General Methods of Poly(arylene ethers) Synthesis

A suitable synthetic method for the preparation of poly(arylene ethers) is the nucleophilic polycondensation of dihydric, mononuclear or polynuclear phenols with activated dihalo aromatic compounds in aprotic solvents and at high temperature conditions. The halogen groups are activated by the electronegative substituents, such as nitro($NO_2$), sulfonyl($SO_2$), carbonyl(CO) and nitrile (CN) groups, present in the *ortho*- or *para*-positions of the benzene ring. In the nucleophilic polycondensation reaction, the actual reactant is the phenate ions formed from the dihydric phenols by the action of alkalies. The formation of phenate ions can be achieved by the reaction of phenolic compounds with alkali metal hydroxides or with alkali metal carbonates. The alkali metal carbonates chiefly used are potassium carbonate. With potassium carbonate, the phenolate formation proceeds very rapidly, and consequently high molecular weight polyethers can be obtained readily.

Polymers containing sulfonyl and ether groups are known as poly(sulfone ethers) [1–4]. Linear poly(arylenesulfone ethers) are prepared by reacting an alkali metal-dihydric phenol double salt with an equimolecular amount of dihalobenzene compound having sulfonyl group in the presence of a sulfoxide

(DMSO) or dimethyl acetamide (DMAC) solvent. Using resorcinol as the dihydric phenol compound, the polymerization reaction can be represented as follows (Figure 7.60).

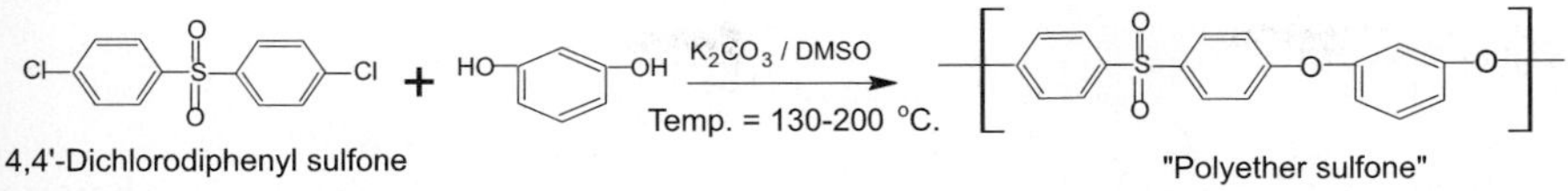


**Figure 7.60.**

Poly(ether sulfones) offer a combination of properties, including high glass transition temperatures (Tg), solvent resistant, good flexural and tensile strengths.

Poly(aryleneether ketones) having ketone and ether functional groups in the polymer main chain are prepared, in a similar way like poly(aryleneether sulfones), by the substantially equimolar one-step reaction of a double alkali metal salt of a dihydric phenol with a dihalogeno compound having a ketone group (Figure 7.61).

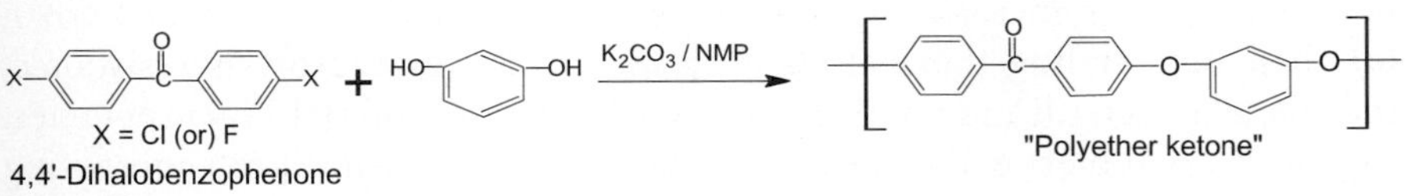


**Figure 7.61.**

Poly(aryleneether ketones) can be either crystalline or semi-crystalline compound, which could depend on the nature of the dihydric phenol employed in the polymer synthesis. Since poly(ether ketones) show excellent heat resistance, chemical resistance and mechanical strength properties, they can be used as molding materials in the manufacture of electrical and automobile parts.

Resorcinol has been employed as the dihydric phenol in the development of poly(arylene ethers) in order to improve their processability. Due to *meta*-phenylene structure, resorcinol based thermoplastic polyethers can be expected to be amorphous rather than crystalline in character. But crystalline thermoplastic polyether materials have been prepared and reported from the resorcinol monomer [5]. In this chapter, poly(arylene ethers) synthesized from resorcinol and its derivatives have been discussed.

### 7.6.3 Poly(cyanoaryl ethers) from Resorcinol

Poly(arylene ethers) having an aryl group substituted with cyano group are known as poly(cyanoaryl ethers) or poly(arylether nitriles) (PEN), and can be represented by the following general formula (Figure 7.62).

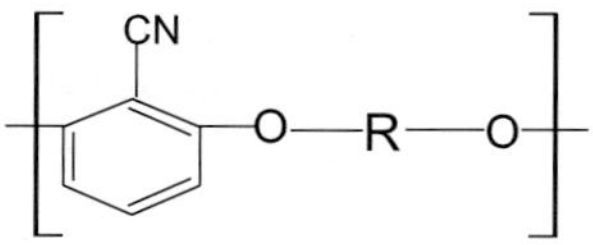


**Figure 7.62.**

In the above formula, "R" represents a variety of arylene compounds having 1 to 2 aromatic rings including resorcinol. Polymers containing cyano group on an aromatic ring can promote adhesion to many substrates, possibly through polar interaction with other functional groups; and also serves as a potential site for polymer cross-linking [6–9].

Poly(cyanoaryl ethers) prepared using 2,6-dihalobenzonitrile can provide polymers with *meta*-phenylene ether structures in the back bone. In addition to 2,6-dihalobenzonitrile, when resorcinol (*meta*-dihydroxy benzene) was used as the dihydric phenol monomer in the polyether synthesis, the repetition of *meta*-phenylene ether structure was seen in the polymer. Consequently, due to enhanced structural order in the polymer chain, the crystallinity of the polymer has been increased [10]. Polymers with high crystallinity are known for their outstanding thermal stability, chemical and hydrolysis resistance, mechanical strength and toughness as well as dimensional stability properties. In general, polymers exhibiting these properties are called super engineering plastics since, they provide excellent performance than the regular engineering plastics.

### 7.6.3.1 Synthesis of Poly(cyanoaryl ether)

The poly(cyanoarylene ether) was prepared by the nucleophilic substitution reaction of 2,6-dichlorobenzonitrile with resorcinol using potassium carbonate in the presence on *N*-methyl pyrrolidone (NMP) solvent [11–21] (Figure 7.63).

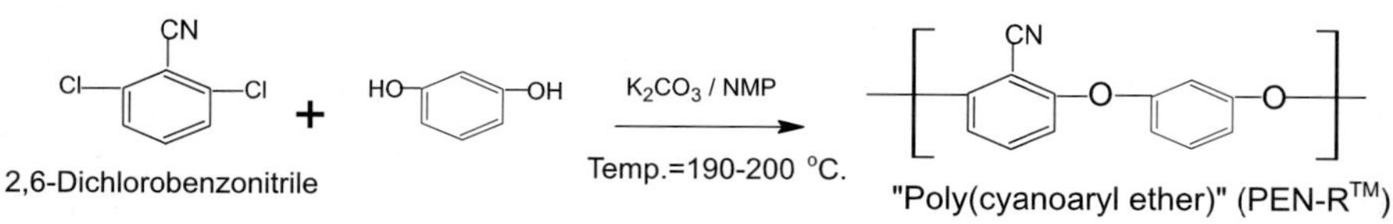


**Figure 7.63.**

A synthesis procedure used in the preparation of resorcinol based cyanoaryl ether polymer is given below [14].

**Preparation of Polycyanoaryl Ether [14]** In a separable flask of 5 l inner volume equipped with a stirrer, a fractionating device and an argon gas introducing pipe, 237.74 g of 2,6-dichlorobenzonitrile, 154.14 g of resorcinol, 203.16 g of potassium carbonate, and, as a solvent, 2 l of *N*-methylpyrrolidone (NMP) and 1 l of toluene were charged. While blowing argon gas, the temperature of the mixture was elevated to 195 °C over 1 h and at this temperature, the mixture was allowed to react for 3 h. After completion of the polymerization reaction, the product was cooled and pulverized by use of a blender produced by Warning Co., followed by washing once with 5 l of water containing 5 g of oxalic acid, three times with 5 l of water and further once with 5 l of methanol. As a result, 274 g of a polymer was obtained.

This polymer was found to have a reduced viscosity ($\eta$ sp/c) of 1.05 dl/g. As for the thermal properties of this polymer, it had a glass transition temperature of 145 °C, a melting point of 340 °C and a thermal decomposition initiating temperature of 495 °C (air).

### 7.6.3.2
### Poly(cyanoarylene ether) from Cyclic Oligomers

Thermoplastic poly(arylene ether) materials were also synthesized from low melt viscosity cyclic oligomers through the ring opening polymerization process. Cyclic cyanoaryl ether oligomers were obtained from the reaction of resorcinol with 2,6-difluorobenzonitrile [22, 23] (Figure 7.64).

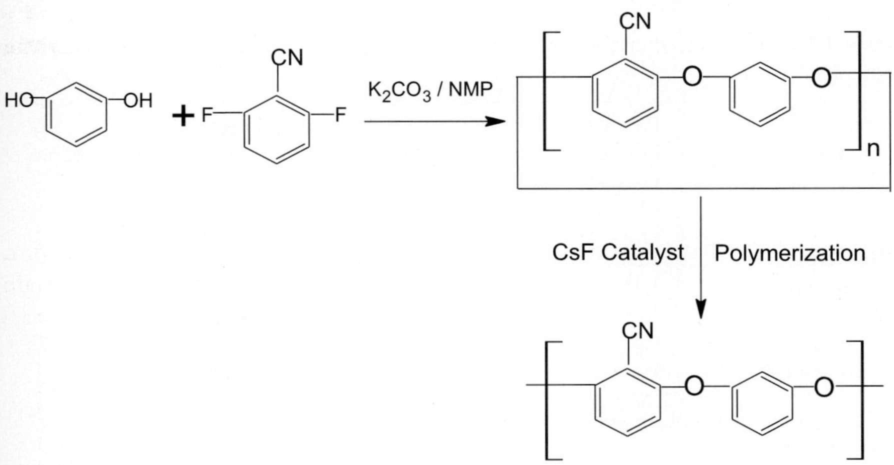


**Figure 7.64.**

The following procedure can be used to prepare the cyclic cyanoaryl ether oligomers from resorcinol [22].

**Cyclic Poly(aryl ether nitrile) from Resorcinol and 2,6-Difluorobenzenenitrile [22]**
A 1 l three-necked flask equipped with a Dean-Stark trap and condenser, a thermometer, a nitrogen inlet, and magnetic stirring was charged with 400 ml *N*-methylpyrrolidinone, 150 ml of toluene, and 3.45 g (0.025 mol) potassium

carbonate. After the solvents were heated to reflux (166 °C pot temperature), two separate 0.40 mol/l solutions of the monomers (resorcinol and 2,6-difluorobenzenenitrile) in *N*-methylpyrrolidinone were added simultaneously using a syringe pump. The addition continued (rate 15 ml/h) until 50 ml of both solutions were added. During an additional hour of reflux the pot temperature was raised to 200 °C by draining toluene from the Dean-Stark trap. The reaction mixture was poured into 1 l water which was then neutralized with aqueous 1 mol/l HCl. The precipitate was filtered, washed with methanol, and treated with boiling chloroform. A pure cyclic oligomer with a DP of 4 (0.592 g, m.p. 397 °C) was precipitated from the chloroform solution on cooling. Slow evaporation of the mother liquor afforded a second crop of crystals (0.21 g) which was a pure cyclic oligomer with a DP of 3 (m.p. 451 °C).

### 7.6.3.3
### Synthesis and Properties of Poly(cyanoarylene ethers)

Poly(cyanoaryl ethers) were synthesized from the reaction of 2,6-dichlorobenzonitrile with resorcinol and other dihydric phenolic compounds. Polymer viscosity, thermal and thermal properties obtained for these polymers are given in Table 7.30 [10].

**Table 7.30.** Synthesis, thermal and mechanical properties of homo poly(cyanoarylene ethers)

| Dihydroxy monomer | Polymer inherent viscosity (dL/g) | Thermal properties (DSC) Tg (°C) | Tm (°C) | Td (°C) | Tensile properties Strength (MPa) | Modulus (GPa) | Elongation (%) |
|---|---|---|---|---|---|---|---|
| Resorcinol | 1.38 | 148 | 340 | 480 | 137 | 4.4 | 94 |
| Hydroquinone | 1.35 | 182 | 348 | 535 | 150 | 3.9 | 35 |
| 4,4′-Biphenol | 1.68 | 216 | 347[a] | 556 | 155 | 3.5 | 35 |
| 2,7-Naphthalenediol | 0.82 | 214 | 348[a] | 527 | 122 | 4.3 | 35 |
| Bisphenol-A | 1 | 180 | NA | 506 | 89 | 3.5 | 37 |

[a] Melting endotherm observed in the first scan only
Data from [10]

From the viscosity data, it can be seen that the poly(cyanoarylene ethers) obtained from the nucleophilic substitution reaction are of sufficiently high molecular weights. Depending upon the polymer structural rigidity, the glass transition temperature of these polymers was varied. High polymer melting temperature was observed for resorcinol based polyether. It has been reported that the polymer derived from the resorcinol monomer was observed to crystallize more readily than the other polymers [10]. From the DSC analysis, it was determined that the energy associated with the endothermic melting of resorcinolic cyanoaryl ether polymer (energy = 44 mJ/mg) was considerably greater than the hydroquinone based polymer (energy = 13 mJ/mg). This data was strongly suggesting that the polymer containing an extended *meta*-linkage structure was more crystallizable than the corresponding *para*-linked polymer [10].

The neat resin tensile properties data (from Table 7.30) clearly show that resorcinol based polymer showed an excellent mechanical strength properties.

The poly(cyanoarylene ether) or poly(ether nitrile) (PEN) obtained from the resorcinol and 2,6-dichlorobenzonitrile monomers has been commercially produced by Idemitsu Kosan Co., Japan and sold under the trade name PEN-R.

## 7.6.4 Copoly(cyanoaryl ethers)

In order to improve the solubility, thermal and mechanical properties of resorcinol based poly(cyanoaryl ethers), copolymers have been synthesized using hydroquinone as the co-monomer and characterized [5,24] (Figure 7.65).

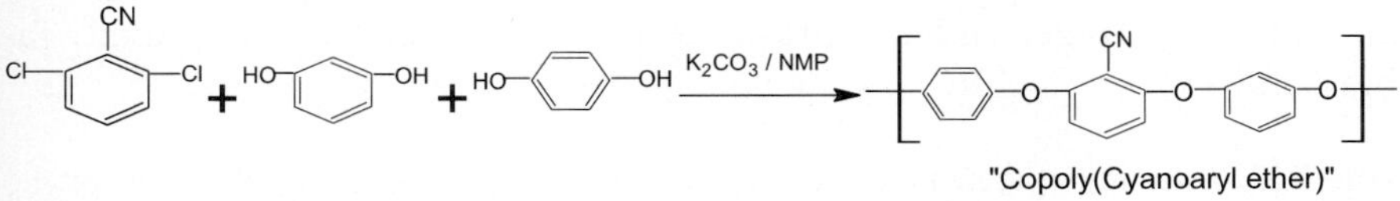


**Figure 7.65.**

By employing different molar ratios of resorcinol and hydroquinone monomers, copolymers were prepared. The copolymer composition, viscosity data and thermal properties obtained from the DSC and TGA analyses are presented in Table 7.31 [5].

**Table 7.31.** Copolymers of resorcinol based poly(cyanoaryl ethers)

| Dihydroxy monomers | | Viscosity[a] | DSC analysis | | | TGA analysis | |
|---|---|---|---|---|---|---|---|
| Resorcinol | Hydroquinone | | Tg | Tm | Degree of crystallinity | IDT[b] | Char yield[c] |
| (mol%) | (mol%) | (dL/g) | (°C) | (°C) | (%) | | |
| 100 | 0 | 0.49 | 145 | 336 | 30 | 480 | 50 |
| 75 | 25 | 0.76 | 153 | 266 | 16 | 415 | 53 |
| 50 | 50 | 0.49 | 158 | 200 | 3 | 430 | 52 |
| 25 | 75 | 1.11 | 169 | 307 | 11 | 470 | 42 |
| 0 | 100 | 0.79 | 175 | 366 | 28 | 490 | 48 |

[a] Inherent viscosity measured at 60 °C in *p*-chlorophenol
[b] IDT – Initial decomposition temperature
[c] Measured at 900 °C in nitrogen atmosphere
Data taken from [5]

From these results, the Tg values of the copolymers were increasing when the concentration of hydroquinone monomer in the polymer was increased. This phenomenon might be associated with the greater asymmetry present in the copolymer due to the incorporation of hydroquinone group in the polymer chain. The copolymers of resorcinol exhibited less crystallinity compared to the homo-polymers due to the random distribution of *m*-arylene and *p*-aryleneether groups in the copolymers, which might have affected the crystallization behavior because of the distorted version of molecular symmetry. Poly(ether nitrile) (PEN-R) derived from the resorcinol retained the crystalline property similar to poly(ether ketones) [5]. These copolymers showed excellent thermal stability properties based on the TGA results.

### 7.6.5 Blending of PEN-R with Thermoplastic Materials and Properties

The thermoplastic and crystalline PEN-R was blended with other well-known thermoplastic polymers such as polycarbonate (PC), polyetheretherketone (PEEK) and polysulfone (PS), and the heat distortion temperatures (HDT) and Limiting Oxygen Index (LOI) properties of the blends were studied (Table 7.32) [13].

**Table 7.32.** Blending of thermoplastic polymers with resorcinol based poly(ether nitrile) (PEN-RTM) resin

| Thermoplastic polymer (TP) | TP/PEN (wt) | Heat distortion temperature (°C) | Limiting Oxygen Index (%) |
|---|---|---|---|
| Polycarbonate (PC) | 100/0 | 131 | 30 |
| Polycarbonate | 50/50 | 146 | 39 |
| Polyetheretherketone (PEEK) | 100/0 | 145 | 35 |
| Polyetheretherketone | 50/50 | 150 | 43 |
| Polyethersulfone (PES) | 100/0 | 205 | 38 |
| Polyethersulfone (PES) | 50/50 | 160 | 43 |
| Polyethernitrile (PEN-R) | 0/100 | 165 | 42 |

Data from [13, 14]

The incorporation of PEN-R in the PC and PEEK materials appeared to enhance the HDT and LOI values of the blended polymers.

### 7.6.6 PEN-R Based Molding Compounds and Properties

The presence of polar cyano groups (–CN) in the poly(ether nitriles) is expected to improve their adhesion properties with glass fibers, which in turn can result in excellent mechanical properties. PEN-R has been used as a matrix resin for fiber reinforced composite materials. Molding compounds were prepared using PEN-R as the matrix resin, and glass and carbon fibers as the reinforcing materials. The HDT and tensile properties measured on the molding compounds are presented in Table 7.33 [13, 14].

**Table 7.33.** Molding compounds from resorcinol based poly(ether nitrile) resin and properties

| Reinforcing material | | Tensile properties | | | HDT |
|---|---|---|---|---|---|
| Fiber type | Loading (wt.%) | Strength (MPa) | Modulus (GPa) | Elongation (%) | (°C) |
| Carbon[a] | 10 | 176.6 | 5.4 | 6 | 220 |
| Carbon[a] | 20 | 186.4 | 8.8 | 4 | 290 |
| Carbon[a] | 30 | 206 | 14.7 | 4 | > 300 |
| Glass[b] | 10 | 166.8 | 4.9 | 7 | 200 |
| Glass[b] | 20 | 181.8 | 9.8 | 4 | 280 |
| Glass[b] | 30 | 206 | 11.8 | 3 | > 300 |
| None | 0 | 137.3 | 3.2 | 40 | 140 |

[a]Carbon fiber produced by Toray K.K, Trade name: Toreca
[b]Glass fiber produced by Asahi Fiber Glass Co
Data from [13, 14]

Glass and carbon fibers reinforced molded materials showed excellent heat resistance and mechanical strength and stiffness properties. These molding compounds could be useful for developing various kinds of articles including electric or electronic parts, car parts, machine parts, and aerospace parts.

### 7.6.7 Performance Comparison of PEN-R versus PEEK Resin

Thermal and mechanical properties of resorcinol based poly(aryleneether nitrile) (PEN-R) and polyetheretherketone (PEEK) materials were compared for their performance with respect to their neat resin and glass fiber reinforced composite properties (Table 7.34) [10].

Although the neat resin HDT for PEN-R was lower than PEEK, the HDT for 30% glass reinforced material was 30 °C higher for PEN-R, suggesting that this was due to the higher melting and crystallinity of PEN-R polymer [10]. Tensile and flexural strength properties of PEN-R were also determined to be better than those of PEEK.

**Table 7.34.** Comparison of polyethernitrile (PEN-R) vs polyetheretherketone PEEK resins

| Property | Polyethernitrile | | Polyetheretherketone | |
|---|---|---|---|---|
| | Neat resin | 30% glass | Neat resin | 30% glass |
| Melting (by DSC, Tm, °C) | 340 | NA | 335 | NA |
| Glass transition (Tg, °C) | 148 | NA | 145 | NA |
| Heat distortion at 1.81 MPa (°C) | 165 | 330 | 160 | 300 |
| Continuous service temp. (°C) | NA | 225 | NA | 230 |
| Tensile | | | | |
| Strength (MPa) | 132 | 173 | 95 | 153 |
| Modulus (GPa) | 3.3 | 7.7 | 3.2 | 8.6 |
| Elongation (%) | 10 | 3 | 80 | 3 |
| Flexural | | | | |
| Strength (MPa) | 194 | 214 | 153 | 224 |
| Modulus (GPa) | 3.8 | 8.4 | 3.6 | 9.1 |
| Limiting oxygen (3.2 mm sample) | 42 | 114 | 35 | 98 |

PEN-R – Available from Idemitsu Kosan, Japan
PEEK – Victrex PEEK from ICI
Data from [10]

## 7.6.8 Laminated Products and Properties

PEN-R was used as the matrix resin in the fabrication of carbon and glass fibers based laminated materials. In these laminated products, the concentration of polycyanoaryl ether was set at 80% and the fiber mat was at 20% by weights. The tensile, flexural and HDT properties determined for the laminated materials are presented in Table 7.35 [13, 14].

**Table 7.35.** Laminates made using resorcinol based poly(cyanoaryl ether) resin and properties

| Matrix resin | Reinforcing Fiber | | Tensile properties | | | Flexural properties | | HDT |
|---|---|---|---|---|---|---|---|---|
| | Fiber type | Loading (%) | Strength (MPa) | Modulus (MPa) | Elongation (%) | Strength (MPa) | Modulus (MPa) | (°C) |
| PEN-R | Carbon | 20 | 313.8 | 6.3 | 3 | 343.2 | 10.8 | 300 |
| PEN-R | Glass | 20 | 294.2 | 5.9 | 3 | 294.2 | 9.8 | 305 |
| PEEK | Carbon | 20 | 205.9 | 5.2 | 3 | 245.2 | 9.8 | 300 |

PEN-R – Resorcinol based poly(cyanoaryl ether) produced by Idemitsu Kosan
PEEK – Polyetheretherketone produced by ICI
Carbon fiber mat: Toreca Mat B0030 produced by Toray K.K
Glass fiber mat: CSM-M9600 produced by Asahi Fiber Glass
Data from [13, 14]

PEN-R based laminates exhibited an excellent tensile and flexural properties compared to PEEK resin. Also, the laminated products obtained from PEN-R showed good flame retardancy and non-dripping behavior in the UL-94 type testing.

### 7.6.9 Applications of PEN-R Resin

The homo and copolymers of resorcinol based poly(cyanoaryl ether) resins show crystalline, sufficient heat resistance, solvent resistance and good mechanical properties. In order to improve and maintain the heat resistance and stiffness properties at high temperatures, PEN-R was mixed with inorganic fillers such as silicon nitride, silicon carbide, silica, alumina and zirconia, and then used for the production of various molded articles [25].

Resin compositions containing PEN-R resin and metal or alloy powder of iron, brass, nickel, stainless steel or aluminum were developed for powder molding compounds. The powder molded parts obtained from these compositions have shown excellent dimensional stability against heat, chemical resistance and mechanical strength properties [26, 27].

PEN-R resin compositions containing carbonized phenolic resin showed excellent abrasion resistance properties. These compositions were useful in the production of various molded articles including the sliding components of electric and electronic instruments and machines [28].

For improving the impact strength properties of thermoplastic materials, fluorinated elastomers have been incorporated into the compositions. These elastomers, due to their poor wetting characteristics associated with the low surface energy, produce thermoplastics compositions with poor mechanical properties. In addition, at high temperatures, the fluorinated elastomers partially decompose and produce hydrogen fluoride vapors, a strong acid. With the use of PEN-R as the thermoplastic material, the HF gas production was appeared to be less, compared to other thermoplastic polymers [29].

### 7.6.10 Poly(arylether sulfones) and Poly(arylether ketones)

Poly(arylethers) containing sulfone or sulfonyl and ketone groups have been synthesized by reacting di-alkali metal salts of resorcinol with 4,4′-dichlorodiphenyl sulfone and 4,4′-difluorobenzophenone [5, 30–32] (Figure 7.66).

4,4′-Bis-(hydroxy aryloxy)-phenyl sulfone was synthesized by reacting resorcinol with 4,4′-dichlorodiphenyl sulfone in the presence of potassium hydroxide and strongly polar inert organic solvents [33] (Figure 7.67).

The procedure used in the preparation of bis-(hydroxy phenoxy) compound from the resorcinol is described below.

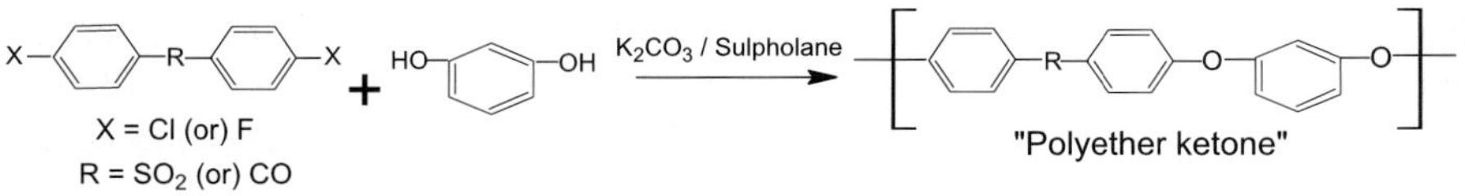


**Figure 7.66.**

**Figure 7.67.**

**Preparation of 4,4′-Bis-(3-hydroxy phenoxy)-phenyl sulfone [33]** Resorcinol (11.0 g, 0.1 mol) was placed in a glass flask of 250 ml capacity and dissolved in 50 ml dimethyl sulfoxide (DMSO). The glass flask was fitted with a gas feed tube, stirrer, thermometer, reflux condenser and a water receiver filled with benzene, and a slow nitrogen current was passed through the apparatus in order to create an inert atmosphere. 11.22 g (0.2 mol) potassium hydroxide and 20 ml benzene were added. The reaction mixture was then heated at a temperature of 140 °C for 4–5 h, while the water resulting from the salt formation was constantly distilled off azeotropically with benzene. To a solution of the di-potassium resorcinate in DMSO, there were added 14.36 g (0.05 mol) of 4,4′-dichlorodiphenyl sulfone. The reaction mixture was then heated at 140 °C for 4 h, whereby the potassium chloride formed in the reaction rapidly separated. After cooling, the mixture was slowly poured into 300 ml of water which had been acidified with sulfuric acid. The reaction product was precipitated in solid form. It was filtered off with suction and then redissolved in a 5% aqueous sodium hydroxide solution to remove any unreacted dichlorodiphenyl sulfone. The solution was filtered and subsequently poured with rapid stirring into 5% sulfuric acid, whereby the dihydroxyaryl ether was precipitated in the form of a white powder. The powder was washed with hot water to remove any unreacted resorcinol. The yield was 86% of theory and the product showed a melting point of 86–87 °C.

The resorcinol based dihydroxyaryl ether compound could be used as an intermediate for the production of synthetic polymers such as polyethers, polyesters, epoxies and polycarbonates.

The glass transition temperature (Tg) and tensile properties of resorcinol based poly-(aryleneether nitrile) was compared with the corresponding sulfone and ketone containing polymers, and the results are shown in Table 7.36 [10].

PEN-R exhibited an excellent tensile strength and modulus properties compared to sulfone and ketone ether based polymers, suggesting that this might be attributable to the dipole-dipole interactions of the nitrile groups [10].

**Table 7.36.** Resorcinol based poly(arylethers) and properties

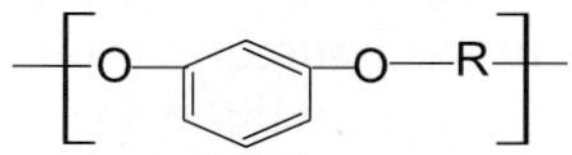

| R group | Polymer inherent viscosity (dL/g) | Tg (°C) | Tensile properties Strength (MPa) | Modulus (GPa) | Elongation (%) |
|---|---|---|---|---|---|
| 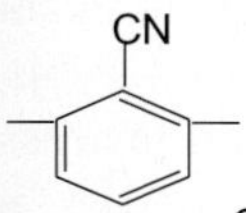 | 1.38 | 148 | 137 | 4.4 | 94 |
| 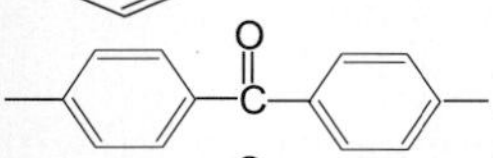 | 0.96 | 130 | 82 | 3.1 | 100 |
| 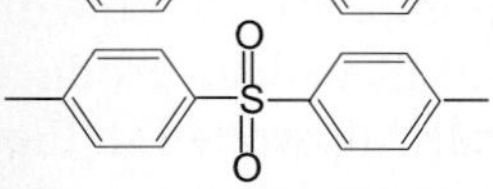 | 0.54 | 174 | 79 | 2.9 | 46 |

Polymer viscosity measured in *p*-chlorophenol
Data from [10]

## 7.6.11 Arylether Phthalonitrile Monomers and Polymers

Bisphthalonitriles or phthalonitrile monomers have been utilized in the polymerization process to form strong, high temperature thermosetting resins. Resorcinol was used to synthesize aryloxy bisorthodinitrile compound based on the following reaction scheme [34] (Figure 7.68).

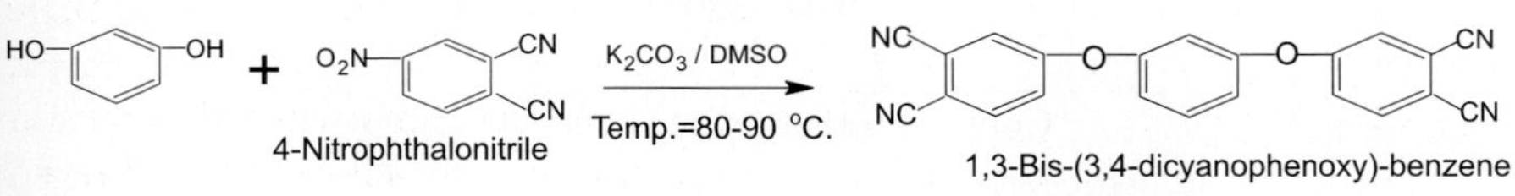


**Figure 7.68.**

A synthesis procedure used in the preparation of 1,3-bis-(3,4-dicyanophenoxy)-benzene from resorcinol has been shown below [34].

**Synthesis of 1,3 Bis-(3,4-Dicyanophenoxy) Benzene [34]** A mixture of 5.5 g (0.05 mol) of resorcinol, 17.3 g (0.10 mol) of 4-nitrophthalonitrile, 20.8 g (0.15 mol) of anhydrous potassium carbonate and 60 ml of dry dimethyl sulfoxide was stirred at 80–90 °C for 5 h under a nitrogen atmosphere. After cooling, the resulting dark mixture was poured into 300 ml of cold dilute hydrochloric acid solution. The solid precipitate which separated was washed with water until the washings were neutral. The product was then collected by suction filtration. Recrystallization from acetone-water yielded 10.7 g (60%) of the desired product, m.p. 183–186 °C.

This resorcinol based bisphthalonitrile monomer was utilized in various applications. When this monomer was polymerized, polyether bridged poly (phthalocyanine) suitable for use as structural resin was produced. The polymerized material showed good thermal stability at 310 °C in an air atmosphere, exhibited self extinguishing property, and also produced high char yields [34].

The prepolymers prepared from 1,3-bis-(3,4-dicyanophenoxy)-benzene and ether based diamines such as 1,3-bis-(3-aminophenoxy)-benzene and bis-[4-(4-aminophenoxy)-phenyl]-sulfone were used as the curing agents in the preparation of composite laminate samples from carbon fiber and 4,4′-bis-(3,4-dicyanophenoxy)-biphenyl. After the curing and carbonization process, the density of the resultant carbon-carbon composite was determined to be 1.65 g/cc [35, 36].

Resorcinol based diether linked bisorthodinitrile monomer was mixed with 4,4′-methylene dianiline (MDA) curing agent, and then heated to produce the conducting polymer. The process used is schematically illustrated in Figure 7.69 [37].

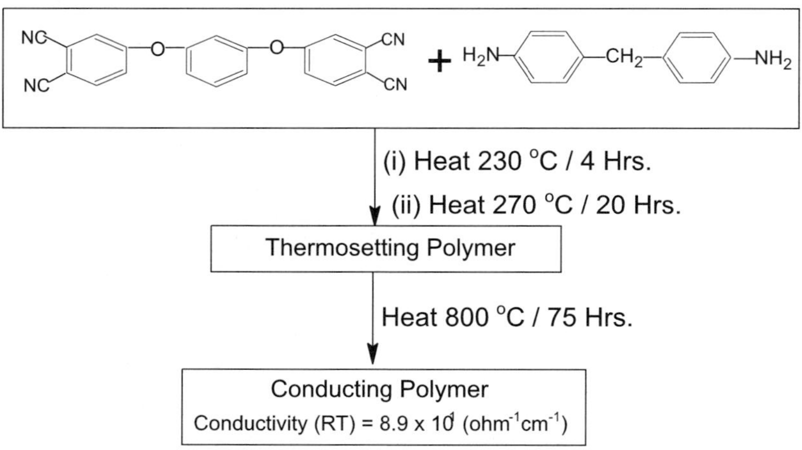


**Figure 7.69.**

The non-doped, conductive organic polymer produced from the diether linked bisorthodinitrile had sufficient structural strength and thermal stability properties, and therefore could replace conductive materials, particularly metals. Potential use for this type of conductive polymers include light weight batteries, electrical devices, low cost solar cells, electromagnetic interference shields for computers and sensitive circuitry, and other electronic applications [37].

Phthalonitrile monomers containing aromatic ether oligomer spacers were synthesized by reacting resorcinol with dihalobenzene compounds in the presence of CuI and $Cs_2CO_3$ catalysts, and then reacting the resulting product with 4-nitrophthalonitrile [38]. The reactions are illustrated in the following figure

(Figure 7.70).

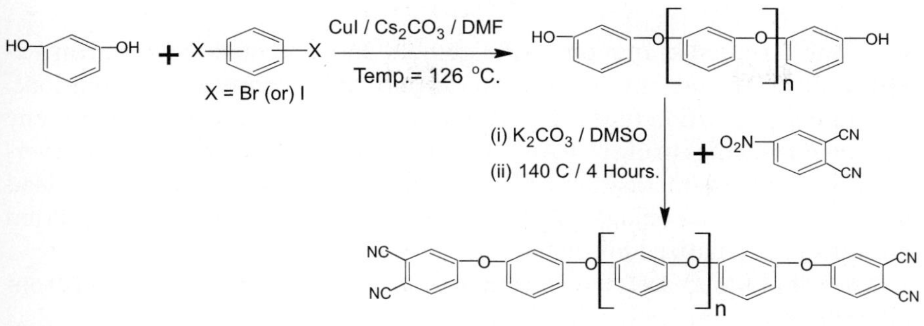


**Figure 7.70.**

Resorcinol based aromatic ether bridged phthalonitrile monomers were expected to show lower melting points, and therefore useful as thermoset resins in the fabrication of composite materials. The cross-linked produced from the curing of ether bridged phthalonitrile monomers could provide sufficient flexibility, high thermal and oxidative stability properties.

### 7.6.12 Bismaleimide Resins Containing Ether Bridges

Arylether bridged diamine compound was synthesized from the dinitro compound produced from resorcinol by the Ullmann's reaction, and then converted into bismaleimide resin [39] (Figure 7.71).

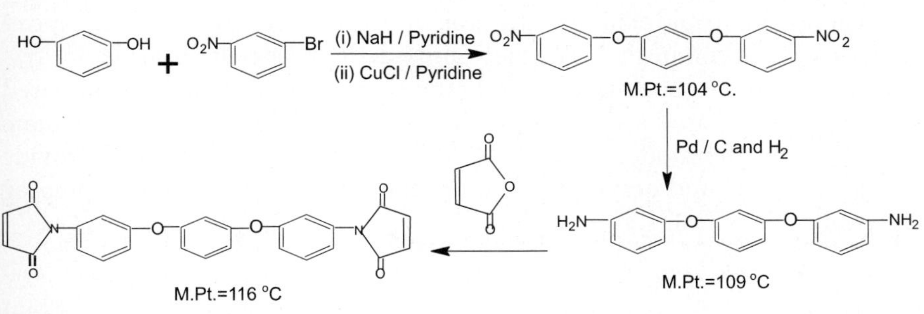


**Figure 7.71.**

The ether bridged diamine compound obtained from the resorcinol can be used in the synthesis of polyamides, polyimides and also as curing agent to cure epoxy resins. This bismaleimide could be utilized as a matrix resin in the development of advanced composites with improved thermal and mechanical properties.

## 7.6.13 Acetylene Terminated Poly(phenylene ethers)

Acetylene terminated monomers, oligomers and polymers have shown excellent shelf life and can be cured under moderate temperature conditions. They undergo addition type polymerization reactions and do not produce any volatiles during the curing process. These materials are known to undergo thermally induced polymerization by the free radical mechanism, which can lead to the formation of conjugated phenylene structures first and then converted into various aromatized products.

Resorcinol based acetylene terminated aryl ethers were synthesized from the reaction scheme presented in Figure 7.72 [40].

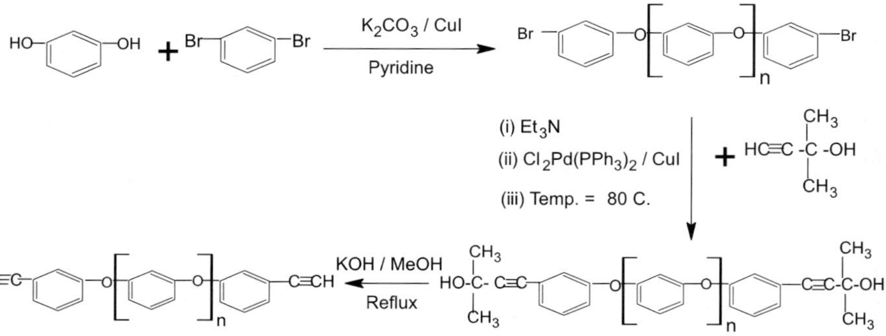


**Figure 7.72.**

The acetylene terminated ether bridged compounds can be useful as structural adhesives and matrix resins for advanced carbon fiber reinforced composites. These resins could exhibit better moisture resistance properties than epoxy resins in certain composites. The thermo-oxidative properties of arylether systems were found to be excellent and better than epoxy resins. In the uncured state, resorcinol based acetylene terminated resin exhibited amorphous state with Tg below the room temperature, which suggests that this could provide an excellent tack and drape properties in the fabrication of graphite prepreg materials.

## 7.6.14 Halogens Containing Poly(arylether) Compounds

Oligomeric brominated poly(phenylene ether) compound was prepared from the resorcinol and decabromo diphenyl ether according to the following reaction [41] (Figure 7.73).

When used as flame retardant additives in the thermoplastic polymer compositions such as polyamides and polyesters, this flame retardant did not

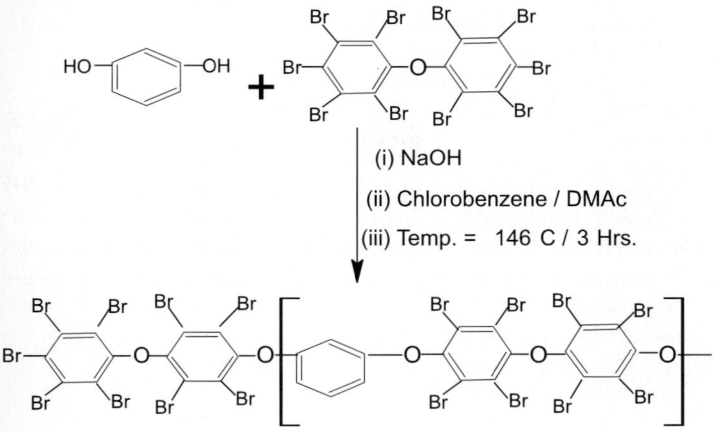


**Figure 7.73.**

exhibit any tendency to migrate towards the surface of the molded articles, and provided V-0 rating in UL-94 testing.

Similarly, aromatic polyethers containing repeating perchloro aromatic units were prepared from resorcinol and used [42].

### 7.6.15 Summary and Outlook

Resorcinol based arylether monomers and polymers offer the advantages of providing materials with improve processability and outstanding properties. The super engineering thermoplastic poly (cyanoaryl ether) resin developed from the resorcinol has shown excellent heat resistance and mechanical properties as compared to other high performance engineering thermoplastics.

## References

1. R. Johnson, A. Farnham, R. Clendinning, W. Hale and C. Merriam, J. Polym. Sci., A15, p 2375 (1967).
2. R. Johnson and A. Farnham, US Patent, 4,108,837 (1978).
3. R. Viswanathan, B. Johnson and J. McGrath, Polymer, Vol. 25, p 1827 (1984).
4. R. Clendinning, A. Farnham, and R. Johnson. High performance polymers: their origin and development, Elsevier, New York (1987).
5. M. Rao, V. Rao, T. Radhakrishnan and S. Ramachandran, Polymer, Vol. 33, p 2834 (1992).
6. J. Verbort and C. Marvel, J. Polym. Sci. Polym. Chem. Edn., Vol. 11, p 261 (1973).
7. K. Sivaramakrishnan and C. Marvel, J. Polym. Sci. Polym. Chem. Edn., Vol. 12, p 651 (1974).
8. T. Keller, J. Polym. Sci. Polym. Chem. Edn., Vol. 26, p 3199 (1988).
9. T. Keller, Polymer, Vol. 34, p 952 (1993).

10. S. Matsuo, T. Marukami and R. Takasawa, J. Polym. Sci. Polym. Chem. Edn., Vol. 31, p 3439 (1993).
11. D. Heath and J. Wirth, US Patent, 3,730,946 (1973).
12. S. Matsuo and T. Murakami, Euro. Patent, 0,192,177 (1986).
13. S. Matsuo, T. Murakami, K. Nagotoshi and T. Bando, Euro. Patent, 0,243,000 (1987).
14. S. Matsuo, T. Murakami, K. Nagotoshi and T. Bando, US Patent, 4,977,234 (1990).
15. S. Matsuo, T. Murakami, T. Bando and K. Nagatoshi, US Patent, 5,089,340 (1992).
16. S. Matsuo, T. Murakami, T. Bando and K. Nagatoshi, US Patent, 4,812,507 (1989).
17. S. Matsuo, T. Murakami, T. Bando and K. Nagatoshi, US Patent, 4,853,443 (1989).
18. R. Takazawa, Jap. Patent, 3,281,530 (1991).
19. S. Matsuo and T. Murakami, Jap. Pat. Publ., 62-223,222 (1987).
20. T. Ishii and H. Kawahigashi, Jap. Pat. Publ., 03-275,758 (1991).
21. J. Shinoda and R. Takazawa, Jap. Pat. Publ., 02-302,435 (1990).
22. M. Mullins and E. Woo, US Patent, 5,264,520 (1993).
23. M. Mullins, E. Woo, K. Balon, D. Murray and C. Chen, US Patent, 5,264,538 (1993).
24. S. Matsuo, T. Murakami and R. Takasawa, US Patent, 4,703,104 (1987).
25. H. Kawato and T. Ishii, Euro. Pat. Appln., 0,420,245 (1991).
26. H. Kawato and T. Tomioka, Euro. Pat. Appln., 0,437,851 (1991).
27. H. Kawato and T. Tomioka, US Patent, 5,350,558 (1994).
28. T. Murakami, US Patent, 5,147,922 (1992).
29. T. Bando, US Patent, 5,389,725 (1995).
30. S. Matsuo, N. Yamukai and C. Kayano, Jap. Pat. Appln., 63-045,432 (1988).
31. G. Blinne and C. Cordes, US Patent, 4,200,727 (1988).
32. J. Simmons, US Patent, 5,393,324 (1995).
33. H. Schnell and G. Darson, US Patent, 3,579,591 (1971).
34. T. Keller and J. Griffith, US Patent, 4,259,471 (1981).
35. S. Sastri, J. Armistead and T. Keller, US Patent, 5,965,268 (1999).
36. S. Sastri and T. Keller, Polym. Compos., Vol. 17, p 816 (1996).
37. T. Keller, US Patent, 4,587,325 (1986).
38. T. Keller and D. Domingguez, US Pat. Appln., 2003 / 0,229,198 (2003).
39. T. Pascal, R. Mercier and B. Sillion, Polymer, Vol. 30, p 740 (1989).
40. G. Loughran, B. Reinhardt and F. Arnold, Div. Org. Coat. Plast. Chem., Vol. 43, p 777 (1980).
41. R. Jones, US Patent, 4,272,429 (1981).
42. E. Novak, US Patent, 3,736,293 (1973).

## 7.7 Epoxy Resins

### 7.7.1 Introduction

Epoxy resins are an important class of thermosetting polymers that exhibit properties such as high tensile strength and modulus, chemical and corrosion resistance and dimensional stability. Due to excellent mechanical properties, epoxy resins have been used in a wide range of industrial applications, including structural adhesives, coatings and matrix resins in fiber reinforced composites.

Epoxy resins, based on bisphenol-A, are well known compounds and used extensively in various industrial applications [1,2]. This epoxy resin has been made by the reaction of bisphenol-A with an excess of epichlorohydrin in the presence of sodium hydroxide (Figure 7.74).

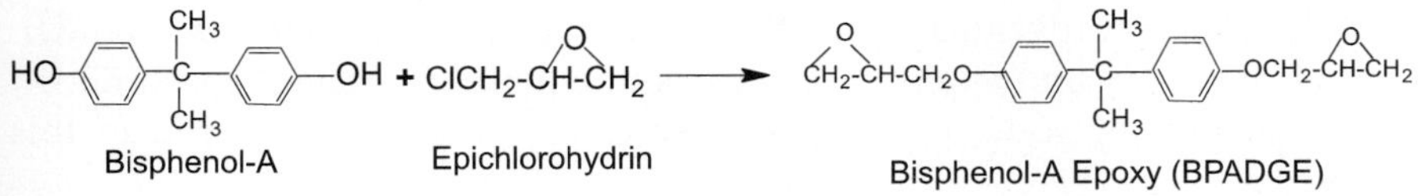


**Figure 7.74.**

Bisphenol-A is an isopropylidene substituted diphenolic compound. The bisphenol-A epoxy resin contains two epoxy groups located one on each of the benzene ring structure of the molecule.

Resorcinol, on the other hand, is mono nuclear dihydroxy benzene. The preparation of an epoxy resin using resorcinol, in place of bisphenol-A, can be expected to produce a chemical compound having the structure shown below (Figure 7.75).

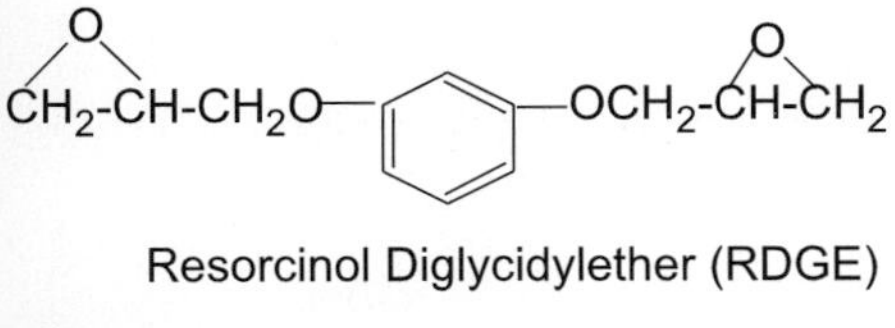


**Figure 7.75.**

In this epoxy resin structure, the two glycidyl ether groups are attached to the same benzene ring at the *meta*-positions to each other. The resonance and inductive effects resulting from the *meta*-substitution can make the epoxy groups more reactive than bisphenol-A epoxy. The higher the epoxy content per unit weight may result in faster reactivity with amine and other curing agents. The highest epoxy content in an aromatic diepoxide compound is available in diglycidyl ether of resorcinol (RDGE). Due to low molecular weight, the RDGE is exhibiting very low viscosity and high reactivity.

The inherent low viscosity of RDGE and high order of reactivity with curing agents permitted a wide selection of curing conditions and offered great advantages in many applications.

Epoxy resins, in general, are extensively used in adhesive formulations and matrix resins for advanced composite applications. Fabrication methods for these demanding applications ideally require materials with low viscosity for easier processing and relatively fast cures to minimize processing costs during the manufacturing operations. Once fabricated, it is necessary for the manufactured products to possess good physical properties such as glass transition temperatures (Tg) and high flexural, tensile and toughness properties to maximize the products life and minimize in-service failures.

The epoxy compounds or resins derived from resorcinol, resorcinolic derivatives and polymers have their unique molecular structures and cure characteristics towards a variety of curing agents. The *meta*-substituted diepoxide compounds that could be obtained from the resorcinol based chemical derivatives have several advantages than the other aromatic diepoxide compounds.

The chemistry and applications of resorcinol and resorcinol based di, tri, tetra and poly-epoxide or glycidyl ether compounds are discussed in this chapter.

## 7.7.2 Diglycidyl Ether Compounds (Di-epoxides)

### 7.7.2.1 Resorcinol Diglycidyl Ether (RDGE)

Resorcinol diglycidyl ether (RDGE) has been prepared by the reaction of resorcinol with epichlorohydrin and followed by the dehydrohalogenation reaction of the resultant product [3–5] (Figure 7.76).

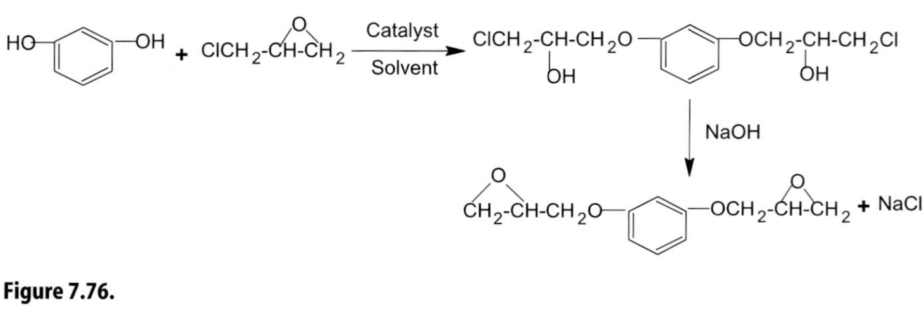


**Figure 7.76.**

In the synthesis of resorcinolic epoxy resin, caustic (NaOH) can also be used as a catalyst for the nucleophilic ring-opening of the epoxide group on the primary carbon atom of the epichlorohydrin by the resorcinolic hydroxyl groups (Figure 7.77).

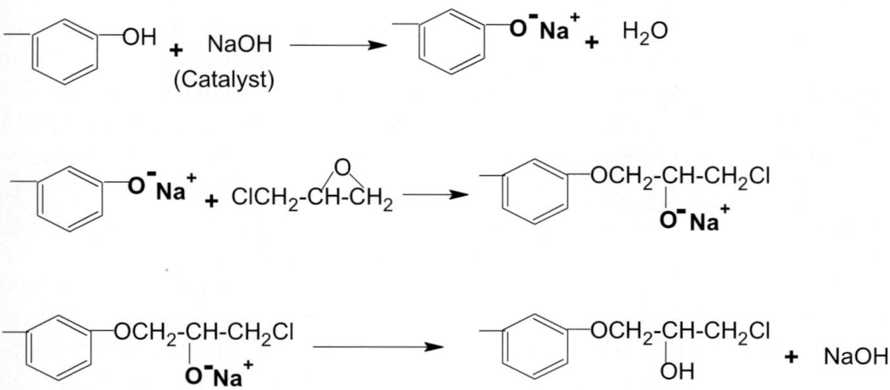


**Figure 7.77.**

In the dehydrohalogenation step, the addition of stoichiometric amount of NaOH converts the halohydrin group into an epoxy group (Figure 7.78).

A synthesis procedure for the preparation of RDGE is illustrated below [4].

**Figure 7.78.**

**Procedure for the Synthesis of Diglycidyl Ether of Resorcinol (RDGE) [4]** Resorcinol – 220 parts – were dissolved in 1480 parts of epichlorohydrin and this solution heated to 98 °C. Addition of an alcoholic solution of potassium hydroxide composed of 160 parts of potassium hydroxide in 1200 parts of 95% ethanol was started dropwise. At the end of 5 min the reaction temperature dropped to less than 90 °C. Addition of the alcoholic potassium hydroxide was accompanied by removal of alcohol at the same rate in order to maintain a constant volume. The entire solution of potassium hydroxide was added in 1 h and the reaction mixture refluxed for 2 h. The reaction mixture was filtered while hot to remove sodium chloride and the filtrate vacuum distilled. From the residue was recovered a glycidyl ether of resorcinol in a yield of 393 parts corresponding to 88.5% of theory based on resorcinol. The product analyzed 14.2% oxirane oxygen compared to the theoretical oxirane oxygen content of 14.4%.

In the preparation of RDGE, an excess of epichlorohydrin has been always used to minimize the polymerization of reactants to obtain high molecular weight compounds. RDGE has been manufactured by several chemical companies and marketed for numerous applications. Some of the physical properties of RDGE are presented in Table 7.37.

**Table 7.37.** Physical properties of resorcinol diglycidyl ether (RDGE)

| Properties | Values/information |
|---|---|
| Physical state | Clear to yellow viscous liquid |
| Viscosity at 25 °C, Cps | 250 – 500 |
| Melting point (°C) | 40 – 45 |
| Boiling point at 0.8 mm Hg, (°C) | 172 |
| Epoxy equivalent weight, EEW, g/eq | 120 – 135 |
| Solubility | Insoluble in water<br>Miscible with benzene, toluene, methanol, acetone, MEK, MIBK, ethyl acetate |
| Specific gravity at 25 °C | 1.21 |

RDGE has shown the viscosity in the range of 250 to 500 centipoises at room temperature. This viscosity is much lower when compared to a commercial grade bisphenol-A epoxy (BPADGE) resin, which has an average viscosity of 11,000 to 15,000 centipoises at room temperature. The low viscosity and excellent wetting characteristics of RDGE can provide thorough penetration of coil and reinforcing fiber materials for the efficient adhesion and bonding in its applications.

**Viscosity Modifier** RDGE is one of the most fluid aromatic diepoxide resins available to industry. In selected applications, it may be desirable to use it in conjunction with other resins to produce formulations with low inherent viscosities. This viscosity reduction effect is illustrated in Table 7.38, by several mixtures of RDGE with two epoxy resins, one liquid and one solid.

RDGE has been found to provide improved properties in castings, potting, encapsulation and tooling formulations, in adhesives, in reinforced laminates and in surface coatings.

In the case of adhesive formulations, this epoxy was observed to bond with almost any materials presenting either polar or porous surface. These improved properties were characterized by good surface hardness, toughness, dimensional stability, resistance to moisture and chemicals, and high dielectric strength and electrical resistivity.

The high reactivity of RDGE suggests the possibilities of reducing cure times and/or cure temperatures without reduction in other properties. The broad range of compatibility of RDGE with resins and solvents as well as the range of curing conditions obtained with various curing agents makes this epoxy particularly suited for the modification and improving the existing epoxy formulations.

**Table 7.38.** Effect of RDGE on viscosity reduction of commercial epoxy resins (weights in parts)

| RDGE used (liquid, EEW = 128) | BPADGE (liquid, EEW = 190–210) | Solid epoxy resin (m.p. = 64–67 °C) (EEW = 450–525) | Viscosity (Poise) |
|---|---|---|---|
| 0 | 100 | | 150 |
| 10 | 90 | | 70 |
| 20 | 80 | | 36 |
| 30 | 70 | | 23 |
| 60 | | 40 | 150 |
| 80 | | 20 | 36 |

Data from [7]

**Curing Reactions** RDGE can be cured with all the curing systems which are commonly used for curing other epoxy resins. With mono, di- and poly-functional curing agents, reaction products containing linear, branched and three-dimensional cross-linked structures are produced. RDGE is capable of reacting with active hydrogen compounds and forms reaction products having the generalized structure (Figure 7.79).

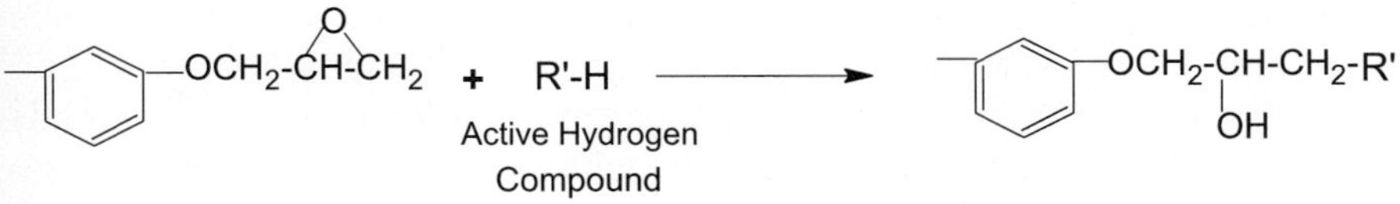


**Figure 7.79.**

Some of the well known curing reactions of aromatic epoxy group with amines, amides, organic acids and anhydrides, phenols and thiols are illustrated in the reaction scheme shown below (Figure 7.80).

A wide range of compounds could be used to cure RDGE. Because of enhanced reactivity, RDGE may be expected to cure more rapidly or at lower temperatures than many other epoxies. The choice of curing agents and conditions depends on the applications and properties needed in the final cured products.

**Advanced Composites** Epoxy resins are employed in the fabrication of composites by the wet and dry lay-up techniques for electrical, aerospace, filament winding and hardware industries. RDGE offers the unique advantage of having very low viscosity and high reactivity towards the curing agents used in these applications. Compared to bisphenol-A epoxy, RDGE has been known to produce cured resins with strengths and modulus at least 15 to 20% higher. The aromatic diamines such as 4,4′-methylene dianiline (MDA) and 4,4′-diaminodiphenyl sulfone (DDS) curing agent compounds have been

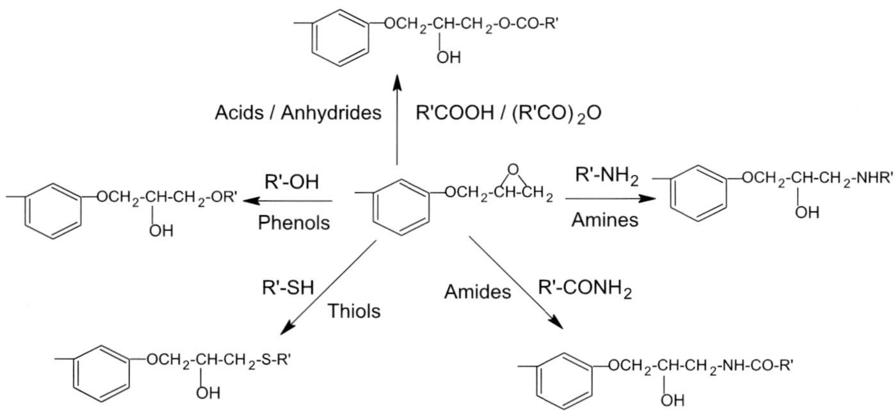


**Figure 7.80.** Some curing reactions of resorcinolic epoxy compounds

widely used in the castings with epoxy resins, since they are providing good thermal stability and mechanical properties for the final cured resins.

It has been known that industries prefer to use DDS as the curing agent for epoxies in the fabrication of composites and coating applications. The major advantages of this curing agent are (1) it is considered non-toxic and (2) it showed improved mechanical and physical properties in cured resin compounds, such as tensile, flexural and glass transition temperatures (Tg).

Casting of neat resins, without any fillers and reinforcements, is generally carried out by mixing or blending epoxies with curing agents at elevated temperatures and then, void-free castings are made from the molten resin blend. Resin castings were made by blending the stoichiometric amounts of RGDE with DDS and 4,4′-[1,4-phenylene-(1-methylidene)]-bis-(2,6-dimethyl benzenamine)

(Curing Agent 1062-M from Shell Chemical) at about 130 to 140 °C, then poured into the mold and cured. After the cure, neat resin properties were determined and the results are presented in Table 7.39 [6].

The DDS cured resin tensile, flexural, Tg and toughness properties were found to be higher than the 1062-M cured system. This data suggests that DDS might be a better curing agent for RDGE to provide improved physical and mechanical properties that can be used in the formulations for advanced composites development. Castings obtained using RDGE and aromatic diamine compounds showed good resistance to organic solvents and enhanced thermal stability.

**Adhesive Formulations** RDGE exhibits a high degree of versatility in adhesives which can be formulated to provide high strengths when bonding a wide range of substrates at curing temperatures ranging from room temperature to about 175 °C. Since, RDGE is compatible with a wide variety of resins

**Table 7.39.** Neat resin properties of RDGE cured with DDS and 1062-M

| Curing agent used | DDS | | 1062-M | |
|---|---|---|---|---|
| Structure | 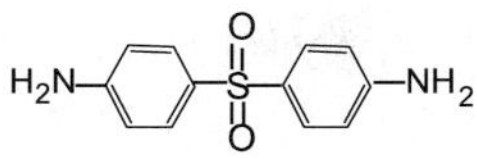 | | 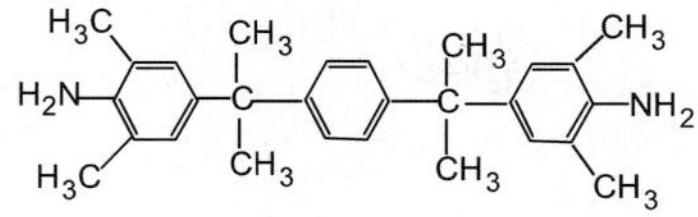 | |
| Cure cycle | 2h/150 °C 4h/200 °C | 2h/180 °C 2h/210 °C | 2h/150 °C 4h/200 °C | 2h/180 °C 2h/210 °C |
| Flexural properties | | | | |
| Strength (MPa) | 171.7 | 176.6 | 139.3 | 130.3 |
| Modulus (GPa) | 3.97 | 3.93 | 3.06 | 3.43 |
| Strain (%) | 7.4 | 6.59 | 5.6 | 4.43 |
| Tensile properties | | | | |
| Strength (MPa) | 99.3 | 93.8 | 79.3 | 72.4 |
| Modulus (GPa) | 3.93 | 4.26 | 3.63 | 3.33 |
| Strain (%) | 5.08 | 4.9 | 4.3 | 3.6 |
| Fracture toughness ($K_{1C}$, MPa $m^{1/2}$) | 1.13 | 1.01 | 0.67 | 0.64 |
| Fracture energy ($G_{1C}$), ($J/m^2$) | 325.3 | 247.5 | 123.8 | 119.3 |
| Moisture pickup (%) | 3.6 | 3.4 | 1.6 | 1.6 |
| Glass transition temp (Tg, °C) | 182.3 | 178.3 | 154.7 | 154.3 |

Curing Agent 1062-M supplied by Shell Chemical Corporation
Data taken from [6]

and modifiers, adhesive formulations and properties can be adjusted to meet many special requirements. The low viscosity of RDGE may permit the use of high concentrations of inert fillers while maintaining workable adhesive characteristics.

The high reactivity of RDGE makes it particularly suitable for two-part adhesive formulations which can cure at room temperature. The curing behavior of RDGE (100 parts) with various aromatic amines curing agents at room temperature conditions is illustrated in Table 7.40 [7].

RDGE can also be employed and readily formulated into two-part adhesive systems which give high strength bonds when heat cured. Adhesives exhibiting high bond strengths at elevated temperatures were formulated with the use of pyromellitic anhydride (PMDA) and RDGE. Table 7.41 contains formulations illustrating the results that were obtained when PMDA and maleic anhydride (MA) curing agents were employed. The curing systems were cured for 1.0 h at 175 °C. The bond or adhesion strength values shown, in Table 7.41, were obtained at elevated temperatures indicated and were for aluminum bonds $1' \times 0.5'$ overlap broken in tensile shear [7].

**Table 7.40.** Room temperature cure of RDGE with aromatic amine curing agents

| Aroamtic amine | Amount in parts | Pot life in hours |
|---|---|---|
| Triaminotriphenyl methane | 48.2 | 2 to 3 |
| Diaminophenol | 24.5 | 6 to 7 |
| *m*-Phenylene diamine | 21.5 | 7 to 8 |
| Diaminopyridine | 21.6 | 8 to 9 |
| Diaminoazo benzene | 42.1 | 8 to 9 |
| Diaminodiphenyl sulfone (DDS) | 49.5 | 11 to 12 |

Data from [7]

**Table 7.41.** High temperature adhesives and properties (RDGE = 100 parts)

| Curing agents used in parts | | | Adhesive strength, MPa at | | |
|---|---|---|---|---|---|
| PMDA | MA | DAT | 300 °F | 400 °F | 500 °F |
| 162 | 27 | | 11.72 | 13.1 | 12.4 |
| 144 | 24 | | 11.72 | 12.41 | 12.4 |
| 72 | 12 | | 8.97 | 7.59 | 7.59 |
| 72 | 12 | 6 | 20.69 | 19.31 | 5.51 |
| 72 | 12 | 12 | 20.69 | 13.1 | 2.06 |
| 72 | 12 | 18 | 23.45 | 11.72 | 2.06 |

PMDA = Pyromellitic dianhydride
MA = Maleic anhydride
DAT = 2,4-Diamino toluene
Data from [7]

**Coating Formulations and Applications** In applications requiring chemical and solvent resistance combined with good adhesion and flexibility, RDGE can be combined with other resin formulations and used. Since RDGE was known to be compatible with wide selections of solvents, resins and curing agents, this can be useful as a reactive modifier in various surface coating formulations. RDGE imparts greater heat resistance and higher adhesion to metals and wood. In some formulations, small amount of RDGE could replace major amounts of epoxy resins to obtain faster cure and better color in surface coating applications.

**Toxicity of RDGE** The hazards associated with the handling of RDGE may be similar to other liquid epoxy materials. RDGE has slight acute oral toxicity in animals with an $LD_{50}$ for rats value of 2 to 3 g/kg. It has moderate acute toxicity by intraperitoneal injection with an $LD_{50}$ varying from 0.13 to 0.24 g/kg [8]. Studies had shown that RDGE was moderately irritating to the skin and severely irritating to the eyes. Repeated or prolonged contact with the skin can cause severe injury. Certain biological reactions indicated that some of the un-cured bis-epoxy compounds of low molecular weights might have carcinogenic properties, but the potency appeared to be low [8, 9]. The cured epoxy resins were considered to be non-toxic.

In studying the carcinogenity of 20 epoxy compounds, RDGE was found to cause skin tumors in rats and mice [10]. Though RDGE was reasonably anticipated to be a human carcinogen, no adequate data were available to evaluate the carcinogenity of this epoxy in humans [11, 12].

It is thought that the toxicity or carcinogenic character of RDGE may be associated with its low molecular weight and low viscosity. Compounds having low volatility and enhanced penetration into the skin can have potential toxicity problems. The toxicity of low molecular weight epoxy compounds is greater than that of high molecular weight epoxy materials. Once absorbed into the skin, these low molecular weight epoxies can cause irritation and other toxic effects. If the molecular weight of epoxy resin is increased, then the resulting resin may be expected to show higher viscosity and lower toxicity due to lower skin penetration [13].

In order to minimize the RDGE's toxicity while maintaining its good reactivity and excellent cured resin physical and mechanical properties, several substituted resorcinol based epoxy compounds were synthesized and reported [14]. In this work, it was anticipated that the presence of various substitutions, such as alkyl, aralkyl and aroyl groups, in the aromatic nucleus of the RDGE molecule could increase the viscosity sufficient enough to reduce the toxicity of the resulting epoxies.

If the substitutions are present in the benzene ring of resorcinol molecule and predominantly at the 4- and/or 6-positions, then the diepoxide compounds from these compounds may perform similar to RDGE with minimal toxic potential.

The chemistry on the synthesis and properties of various di-, tri-, tetra- and poly-epoxy or glycidyl ether compounds obtained from the reactions of resorcinol with different reactive functional monomers are discussed below.

#### 7.7.2.2
#### Diglycidyl Ether of 4-Benzoyl Resorcinol (BRDGE)

4-Benzoyl resorcinol (also known as 2,4-dihydroxy benzophenone) has been extensively used as an UV stabilizer for the plastics, coatings and cosmetics. The presence of bulky substituent, such as benzoyl group, in the RDGE molecule was expected to increase the molecular weight of this epoxy compound, and therefore expected to increase the resin viscosity and lower the toxicity. 4-Benzoyl resorcinol has been manufactured from the reaction of resorcinol and benzotrichloride and is commercially available from various sources. The synthesis of 4-benzoyl resorcinol diglycidyl ether (BRDGE) was made from the reaction of 4-benzoyl resorcinol, epichlorohydrin and NaOH, which can be schematically shown below [14] (Figure 7.81).

A synthesis procedure used in the preparation of BRDGE is given below.

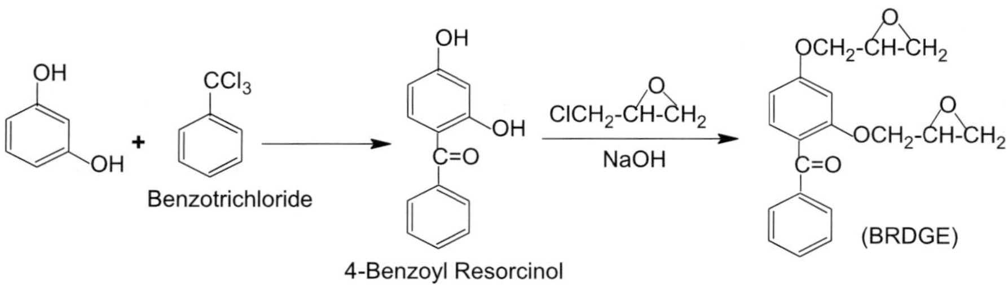


**Figure 7.81.**

**Synthesis of Epoxy Resin from Benzoylresorcinol (BRDGE) [14]** Benzoylresorcinol (240.8 g; 1.125 mol) and epichlorohydrin (10.1 g; 11.25 mol) were placed in a 3-l round-bottomed flask fitted with a stirrer, thermometer, Dean Stark condenser and an addition funnel. The contents of the flask were heated to about 100–125 °C and an aqueous sodium hydroxide solution (50% W/W; 1.3 g; 2.29 mol) was slowly added into the refluxing solution of benzoylresorcinol and the epichlorohydrin. While continuously removing the azeotropic water, the separated epichlorohydrin was continuously returned to the reaction flask. The addition time of sodium hydroxide was between about 1 to 1.5 h and the time required for the complete removal of water (1.2 g) was about 2 to 2.5 h. After this, the excess epichlorohydrin was distilled out first under atmospheric pressure and then under vacuum distillation conditions [maximum pot temperature about 130–135 °C and vacuum about 27–28 inches of mercury (Hg)]. Then about 1200 ml of acetone was added to the reaction flask and the contents were refluxed for about 15 min to dissolve the epoxide. Finally, the salt, sodium chloride was filtered and the solvent acetone was distilled out using both atmospheric and vacuum distillation conditions (temperature about 95–97 °C and vacuum about 27–28 inches of Hg) to obtain 3.2 g (yield = 96%) of epoxide having an epoxide equivalent weight (EEW) of 180 and weight percent oxirane-oxygen content of 8.9 (Theory = 9.8).

The viscosity of BRDGE resin was found to be higher than the unmodified resorcinol epoxy RDGE as shown in Table 7.42 [6].

**Table 7.42.** Physical properties of RDGE vs BRDGE

| Epoxy resin: Properties | RDGE | BRDGE[a] |
|---|---|---|
| Epoxy Equivalent Weight, EEW | 130 | 180–195 |
| Viscosity (Pa.S, at 25 °C) | 0.5 | 25–30 |
| Epoxy groups per molecule | > 1.9 | 1.9–1.95 |

[a]Crystallizes upon standing at RT for long period of time
Data from [6]

From the IR/NMR spectroscopic analysis results, it was determined that BRDGE contained greater than 1.9 epoxy groups per molecule, and therefore can be expected to show reactivity similar to RDGE towards different curing agents.

To determine the potential toxicity of BRDGE, sensitivity and irritation studies have been conducted on mice and rats and the performance was compared against RDGE (Table 7.43) [15].

**Table 7.43.** Toxicity study of RDGE and BRDGE (mouse and rat test results)

| Epoxy resin used: toxicity study | RDGE | BRDGE |
|---|---|---|
| Dermal irritant | (i) Well defined redness and swelling in 1 h<br>(ii) No appreciable recovery over 72-h observation<br>(iii) Score: 3.7 | (i) Slight degree of redness and barely perceptible swelling in 1 h<br>(ii) Complete recovery at 24 h<br>(iii) Score: 1 |
| Eye irritant | (i) Severe degree of irritation involving the cornea and deeper tissue<br>(ii) No recovery at 72 h | (i) Light irritation to conjuctiva but not the cornea or iris<br>(ii) Full recovery at 48 h |
| Time | Scores | |
| 24 h | 20 | 5.3 |
| 48 h | 22.3 | 2 |
| 72 h | 19.8 | 0 |
| 7 days | 9.3 | |
| 14 days | 6.8 | |

Data from [15]

Table 7.43 data strongly suggests that the benzoyl substitution at the 4-position of resorcinol molecule undoubtedly reduced the toxicity of RDGE type epoxy resin.

In the development of carbon fiber reinforced composites, epoxy resins are often cured with aromatic diamines to achieve high glass transition temperatures (Tg) and enhance cured resin mechanical properties. For the proper selection and use of epoxy resins for the advanced composite applications, the cured resin properties are often determined.

**BRDGE Cured with Curing Agents DDS and 1062-M** BRDGE was evaluated and determined the cured mechanical and physical properties with curing agents DDS and 1062-M and the results are presented in Table 7.44 [6].

**Table 7.44.** Neat resin properties of BRDGE cured with DDS and 1062-M

| Curing agent used | DDS | | 1062-M | |
|---|---|---|---|---|
| Structure | 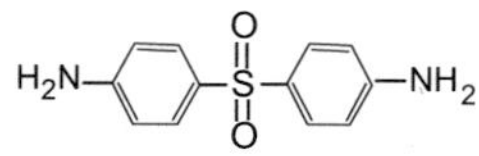 | | 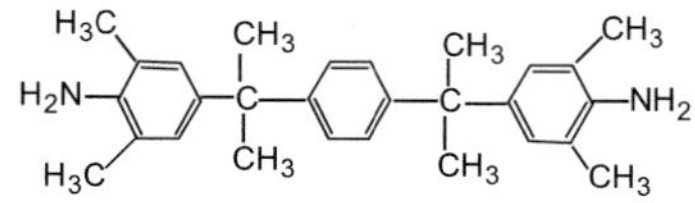 | |
| Cure cycle | 2 h/150 °C<br>4 h/200 °C | 2 h/180 °C<br>2 h/210 °C | 2 h/150 °C<br>4 h/200 °C | 2 h/180 °C<br>2 h/210 °C |
| Flexural properties | | | | |
| Strength (MPa) | 175.1 | 174.4 | 121.4 | 117.2 |
| Modulus (GPa) | 4.1 | 4.2 | 3.9 | 4 |
| Strain (%) | 5.9 | 5.3 | 3.3 | 3.1 |
| Tensile properties | | | | |
| Strength (MPa) | 87.6 | 92.3 | 73.8 | 69 |
| Modulus (GPa) | 4.2 | 3.9 | 4.1 | 3.7 |
| Strain (%) | 2.9 | 3.5 | 2.9 | 2.6 |
| Fracture toughness ($K_{1C}$, MPa m$^{1/2}$) | 0.54 | 0.59 | 0.6 | 0.67 |
| Fracture energy ($G_{1C}$), (J/m$^2$) | 69.1 | 88.6 | 88.5 | 122 |
| Moisture pickup (%) | 2.8 | 2.9 | 1.4 | 1.5 |
| Glass transition temp (Tg, °C) | 173.2 | 171 | 167.6 | 163.1 |

Curing agent 1062-M supplied by Shell Chemical Corporation
Data taken from [6, 15]

Mechanical properties and the glass transition temperatures were observed to be uniformly higher with the DDS curing agent compared to 1062-M. Fracture toughness is slightly better for the 1062-M curing systems. Higher mechanical properties, moisture pickup and Tg for the DDS cured systems are associated with the presence of sulfone group in the molecule. In addition to the improved fracture toughness, the 1062-M cured samples experienced about half the moisture absorption compared to DDS system. This might be due to the presence of hydrophobic groups in the 1062-M curing agent [16].

**Properties of BRDGE at Different Cure Conditions** The mechanical properties of BRDGE cured with DDS at different cure conditions are presented in Table 7.45.

High flexural modulus properties were maintained irrespective of the cure conditions. The Tg of the cured compounds did not change, which indicated that the resin system was forgiving of varying cure conditions.

**Elevated Temperature Mechanical Properties** The flexural properties measured at elevated temperatures on the BRDGE-DDS cured resin samples are given in Table 7.46.

**Table 7.45.** Neat resin properties of BRDGE cured with DDS at different cure conditions

| Epoxy/curing agent: | BRDGE: 1.0 g; DDS: 35–36 g | | | |
|---|---|---|---|---|
| Cure cycle | 2 h/180 °C 2 h/210 °C | 4 h/150 °C 7 h/200 °C | 2 h/150 °C 4 h/200 °C | 1.0 h/180 °C 3 h/225 °C |
| Flexural properties | | | | |
| Strength (MPa) | 184.1 | 195.1 | 197.9 | 169.6 |
| Modulus (GPa) | 4.2 | 4.1 | 4.1 | 4.3 |
| Strain (%) | 6.5 | 6.4 | 6.9 | 4.7 |
| Moisture pickup (%) (After 48 h water boil) | 3 | ND | 3.1 | 3 |
| Glass transition temp (Tg, °C) | 171 | 166 | 175 | 175.5 |
| Impact strength (notched Izod, ft. – lb/in) | 0.5 | ND | 0.38 | 0.4 |

Data taken from [6, 15]

**Table 7.46.** Elevated temperature mechanical properties of BRDGE cured system

| Epoxy/curing agent: Cure cycle | BRDGE: 100 g; DDS: 35–36 g 2 h/150 °C and 4 h/200 °C | | |
|---|---|---|---|
| Flexural properties | 25 °C | 65.5 °C | 121.1 °C |
| Strength (MPa) | 186.4 | 158.6 | 106.9 |
| Modulus (GPa) | 4.5 | 3.3 | 2.3 |
| Strain (%) | 4.6 | 6.9 | 5.9 |

Data taken from [6, 15]

As can be expected, both flexural strength and modulus properties were decreased as the temperature of measurements increased. At 121 °C, 57% of the room temperature flexural strength and 50% of the flexural modulus were retained, which was similar to the performance of other high performance multifunctional epoxy resins.

**Wet Properties of BRDGE-DDS Cured System** Flexural properties obtained after 48 h water boil on the BRDGE-DDS cured samples are presented in Table 7.47.

From the results of these experiments, it was observed that BRDGE retained about 80% of its strength and 86–95% modulus values even after the water boil.

**Comparison of BRDGE against High Performance Epoxies** Neat resin flexural and fracture toughness properties of BRDGE-DDS cured resins were compared to the well known commercial epoxy resins used in the high performance

**Table 7.47.** Mechanical properties of BRDGE cured resin after water boil[a]

| Epoxy/curing agent: | BRDGE: 100 g; DDS: 35 – 36 g | | | |
|---|---|---|---|---|
| Cure cycle | 2 h/180 °C<br>2 h/210 °C | | 2 h/150 °C<br>4 h/200 °C | |
| Flexural Properties | Dry | Wet | Dry | Wet |
| Strength (MPa) | 173.7 | 141.3 | 186.2 | 144.8 |
| Modulus (GPa) | 4.4 | 3.8 | 4.3 | 4.1 |
| Strain (%) | 5.4 | 4.1 | 6.2 | 4.3 |

[a]48 h water boil
Data taken from [6, 15]

composites (Table 7.48).

**Table 7.48.** Cured epoxy neat resin tensile properties comparison[a]

| Epoxy resin used | BRDGE<br>BRDGE | RDGE<br>(Heloxy 69) | DGEBA<br>(n = 0) | TGMDA<br>(MY-720) | HPT-1071<br>HPT-1071 |
|---|---|---|---|---|---|
| Epoxy resin source | Indspec | Shell Chemical | Dow | Ciba-Geigy | Shell Chemical |
| Epoxy functionality | 2 | 2 | 2 | 4 | 4 |
| Curing agent used | DDS | DDS | DDS | DDS | DDS |
| Tensile properties | | | | | |
| Strength (MPa) | 87.6 | 99.3 | 89.6 | 62 | 117.2 |
| Modulus (GPa) | 4.2 | 3.93 | 3.2 | 3.7 | 3.9 |
| Strain (%) | 2.6 | 5.08 | 3 | 1.8 | 3.7 |
| Fracture toughness (MPa $m^{1/2}$) | 0.54 | 1.13 | 0.87 | 0.44 | 0.52 |
| Moisture pickup (%) | 3 | 3.6 | 4 | 4.1 | 3.6 |

DGEBA = Diglycidyl ether of Bisphenol-A
TGMDA = Tetraglycidyl Methylene Dianiline
[a]Haung et al., Mater. Sci. Eng., Vol. 65, p 225 (1991)
Data taken from [6, 15]

The performance of BRDGE compared well with the other multifunctional epoxy resins. When comparing to RDGE, the benzoyl resorcinol based epoxy resin showed similar performance indicating that BRDGE can be substituted for RDGE in applications where high performance epoxy with reduced toxicity is needed most.

### 7.7.2.3
### Diglycidyl Ether of Styryl Substituted Resorcinol

Styrenation of resorcinol can produce aralkyl substituted resorcinol, from which, when used in the resin preparation, an epoxy resin with improved moisture absorption characteristics can be expected. This property might be due to the presence of hydrophobic groups, such as the styryl, in the resorcinol molecule. In the cured epoxy system, the styryl substituted resorcinol epoxy resin can exhibit hydrophobic or low moisture absorption properties compared to RDGE.

Diglycidyl ether of styryl substituted resorcinol was synthesized based on the chemistry illustrated below [14] (Figure 7.82).

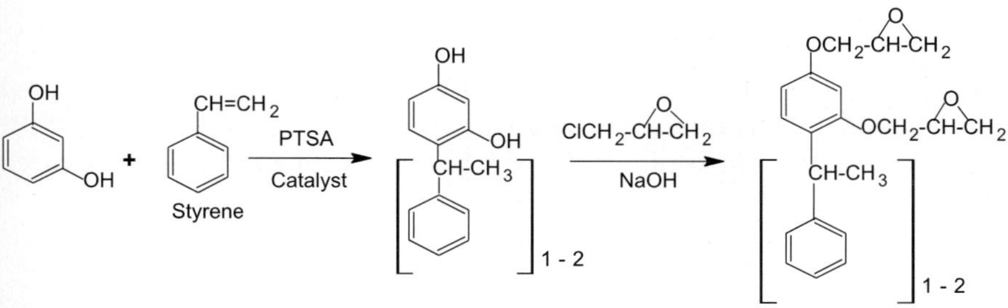


**Figure 7.82.**

The epoxy resin obtained based on the above reaction chemistry showed an EEW value of 215 and contained 1.97 epoxy groups per molecule as determined by IR/NMR analyses.

Similar to styryl substituted resorcinol epoxy, *alpha*-methyl styrene was employed in the synthesis of bis-(cumyl)-substituted resorcinol epoxy resin [14] (Figure 7.83).

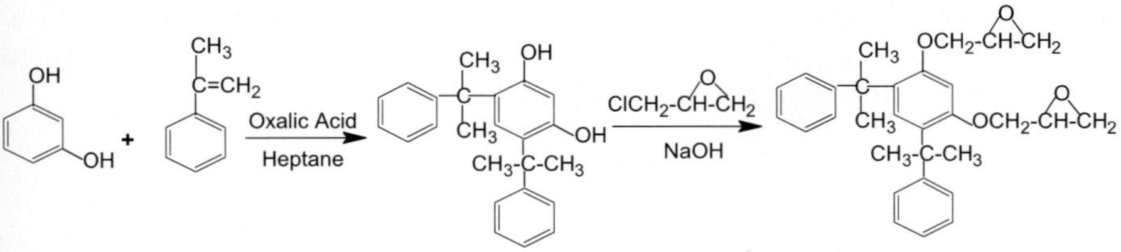


**Figure 7.83.**

## 7.7.2.4
## Diglycidyl Ether of Isopropyl-substituted Resorcinol

Isopropyl substituted resorcinol epoxy was prepared from the corresponding isopropyl derivative obtained by the reaction of resorcinol with isopropyl alcohol in the presence of sulfuric acid catalyst (Figure 7.84).

The diepoxide compound obtained based on the above reaction scheme exhibited EEW value of 161 and viscosity of 600 centipoises at room temperature. Structural characterization by IR/NMR analysis of this epoxide showed 1.82

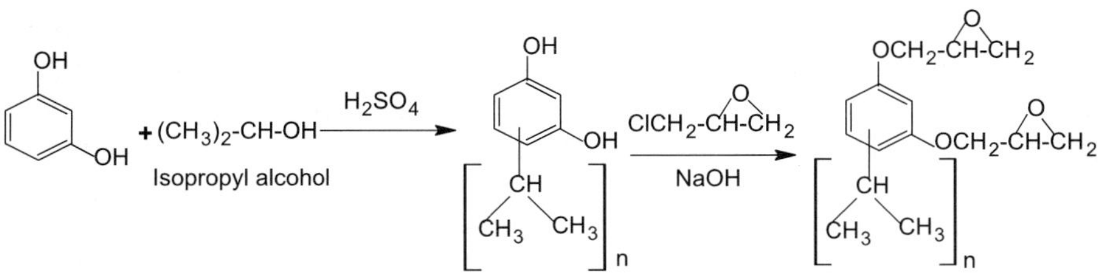


**Figure 7.84.**

epoxy groups per molecule. When cured with DDS curing agent, the cured product showed flexural and tensile strength values of 133.1 MPa and 81.4 MPa respectively and Tg value of 179.1 °C which was very close to RDGE-DDS cured system.

### 7.7.2.5 Diglycidyl Ether of Tribromo Resorcinol

The incorporation of bromine substitution in the resorcinol molecule will result in mono, di- and tribromo resorcinol derivatives. By using these derivatives in the epoxy resin synthesis, the diglycidyl ether derivatives of resorcinol could be obtained. The use of bromo-resorcinol epoxy in the conventional epoxy system could yield a superior fire-resistant or flame retardant resin, perhaps with better properties or handling characteristics than now obtainable with the diepoxide compound of tetrabromo bisphenol-A. The diglycidyl ether of tribromo resorcinol, which can function in a similar way and equivalent to the commercially available tetrabromo bisphenol-A derivative can be obtained from the following chemistry [17] (Figure 7.85).

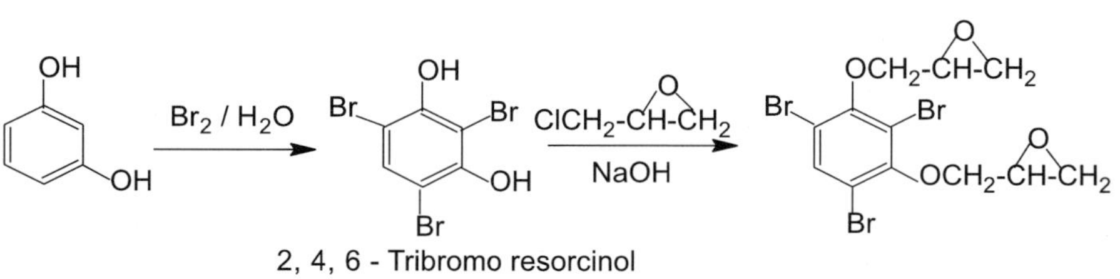


**Figure 7.85.**

A procedure for the synthesis of diglycidyl ether resin from tribromo resorcinol is outlined below.

**Synthesis of Diglycidyl Ether of 2,4,6-Tribromo Resorcinol** A 2-l reactor equipped with a stirrer, thermometer, condenser and a nitrogen bubbler submerged in the liquid, was charged with 347 g (1.0 mol) of tribromo resorcinol, 1850 g

(20 mol) of epichlorohydrin and 3.5 g of dimethyl formamide (DM). The solution was heated to reflux (114 °C) within 2 h and maintained at reflux for an additional 1.5 h under a nitrogen atmosphere. The excess epichlorohydrin and DMF were removed by distillation to a pot temperature of 135 °C/4 mm; 1270 g (13.6 mol) of fresh epichlorohydrin was charged to the reactor and the solution was heated to 90 °C. A 50% aqueous caustic solution contained 82 g (2.05 mol) of NaOH was added in 2 h at reflux. The water from the reaction medium along with epichlorohydrin was removed by distillation. After the caustic solution was added the mixture was dehydrated during a period of 1 h. Excess epichlorohydrin was stripped to a pot temperature of 150 °C/5 mm and the reactor charged with 1500 ml of toluene. After distillation of 250 ml of solvent, the hot solution was filtered through a filter-aid and devolatilized to 150 °C/5 mm. A quantitative yield (463 g) of diglycidyl ether of tribromo resorcinol compound having the melting point of 123 – 125 °C was obtained. Recrystallization of the crude product in toluene gave 429 g (94% yield) having melting point = 125–126 °C, Br = 51.7% (Theory = 52.3%) and ox–ox = 6.8% (Theory = 7.0%).

### 7.7.2.6
### Diglycidyl Ether of 4-Vinyl Resorcinol

4-Vinyl substituted resorcinol was prepared by reacting resorcinol with paraldehyde (acetaldehyde trimer) in the presence of hydrogen chloride dissolved in methanol. In the synthesis, paraldehyde was added slowly over a period of 3.5 h at 65 °C into a methanol solution containing resorcinol and hydrogen chloride (HCl gas). After the addition of paraldehyde, the reaction was continued for an additional 30 min and then the vinyl resorcinol was obtained by the evaporation of solvent methanol as a light tan solid material. This product did not show any sharp melting but charred somewhat at 280 °C. IR analysis showed a definite peak at 7.00 micron, indicating the presence of vinyl group. In addition, the peaks appeared at 10.10 and 11.18 micron regions of the IR spectrum further confirmed the vinyl substitution [18].

Diglycidyl ether resin was prepared and used in the advanced composite formulations (Figure 7.86).

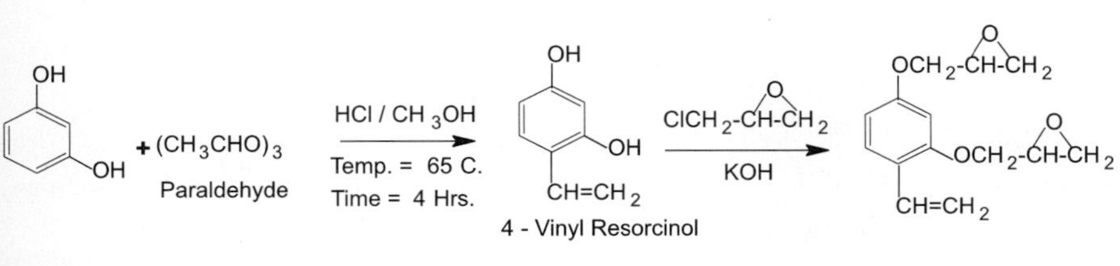


**Figure 7.86.**

### 7.7.2.7 Diglycidyl Ether of 4-Methylol Resorcinol

Epoxy resins have been employed as potting resins, encapsulants and protective coatings due to their high mechanical strengths and good electrical insulating ability. These properties make them ideal for protecting delicate electronic components. With heat sensitive electronic components, the cure of epoxy resins should be done at lower temperatures. Highly reactive epoxies, such as RDGE, could be modified to enhance its reactivity towards relatively slow reacting curing agents. The rapid-curing or highly reactive epoxy resin could be employed as adhesives for the bonding of small heat-sensitive electrical components. The presence of methylol groups increases the reactivity of epoxy group multi-fold. These resins can exhibit faster cure at lower temperatures.

The *beta*-resorcylaldehyde was synthesized from resorcinol and subsequently converted into 4-methylol resorcinol diglycidyl ether compound [19].

Formylation of resorcinol using a modified Gattermann reaction produced *beta*-resorcylaldehyde. The diglycidyl ether of *beta*-resorcylaldehyde was then synthesized by reacting with epichlorohydrin in the presence of a base (Figure 7.87).

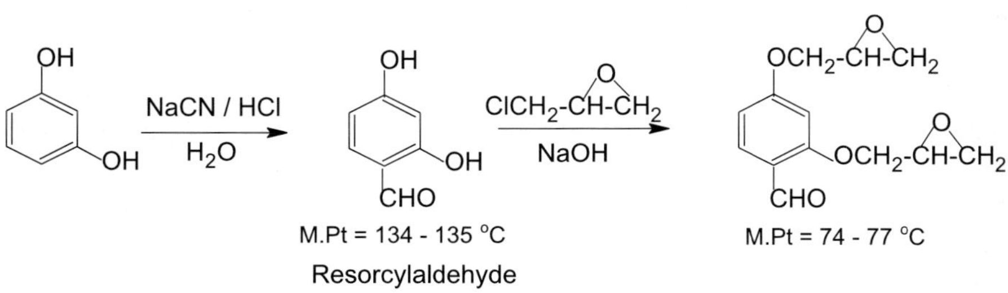


**Figure 7.87.**

Subsequently, the aldehyde was converted into methylol group by the catalytic hydrogenation process. When Pd/C catalyst was employed, 10–15% of methylol group was converted into methyl group, thus producing 4-methyl resorcinol diglycidyl ether. The use of sodium borohydride in methanol, on the other hand, converted the aldehyde exclusively into the methylol group (Figure 7.88).

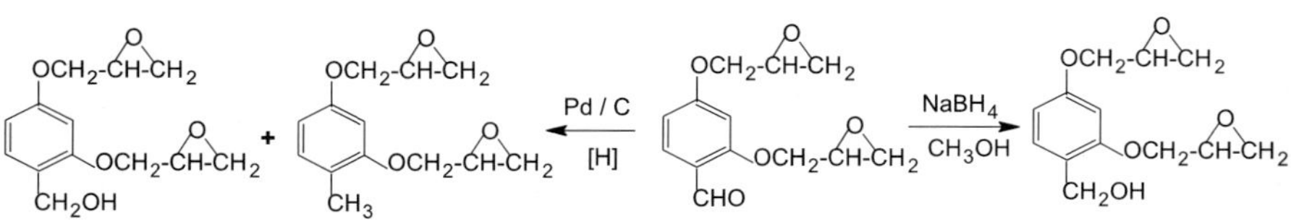


**Figure 7.88.**

This epoxy compound, when cured with diethylene triamine curing agent, cured faster than the diglycidyl ethers of resorcinol and bisphenol-A at room temperature conditions.

### 7.7.2.8
### Diglycidyl Ether of 4-Acetyl Resorcinol

4-Acetyl resorcinol, also known as resacetophenone or 2,4-dihydroxy acetophenone, was used in the synthesis of its epoxy resin [20] (Figure 7.89).

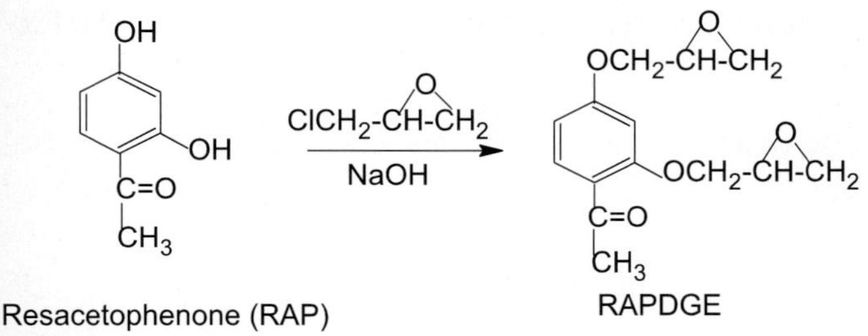


**Figure 7.89.**

The EEW determined for this epoxy was 140 g/eq. This was mixed with diamine curing agents and used to prepare prepreg materials. Glass prepregs were made by immersing the glass mats in the epoxy-diamine reaction mixture and dried. Then, ten prepregs were stacked together, compression molded at 260 °C and tested for their physical, mechanical and dielectric properties (Table 7.49).

**Table 7.49.** Glass reinforced composites mechanical and electrical properties

| Diamine curing agent | Epoxy/ amine ratio | Flexural strength (MPa) | Compressive strength (MPa) | Impact strength (MPa) | Dielectric strength in air ($kV\,mm^{-1}$) |
|---|---|---|---|---|---|
| 4,4′-Diaminodiphenyl ether | 1.0 : 1.0 | 240 | 220 | 276 | 12 |
| 4,4′-Diaminodiphenyl sulfone | 1.0 : 1.0 | 152 | 140 | 145 | 14 |

Data from [20]

From this study, it was observed that the glass reinforced composite properties were not affected by immersion in polar organic solvents which indicated the high cross-link density of epoxy-diamine curing system. The dielectric strength values of these composites were low.

## 7.7.3
## Triglycidyl Ether Compounds

The trifunctional epoxy resins can permit the curing reaction performed at low temperatures and also expect to develop excellent elevated temperature properties. Various trihydroxy aromatic compounds which could be developed from the resorcinol and resorcinol derivatives can be effectively utilized to synthesize multifunctional epoxy resins.

The distillation residue, higher boiling point material, left after the resorcinol distillation from the crude reaction product obtained in the resorcinol manufacture by the benzene sulfonation-fusion process, mainly contained *o*-,*p*-,*m′*-trihydroxy diphenyl having the following chemical structure [21] (Figure 7.90).

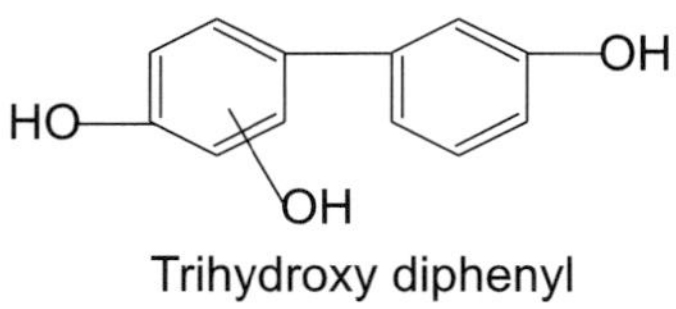


**Figure 7.90.**

This trihydroxy compound when reacted with epichlorohydrin yielded *o*,*p*,*m′*-triglycidyl ether of biphenyl compound (Figure 7.91).

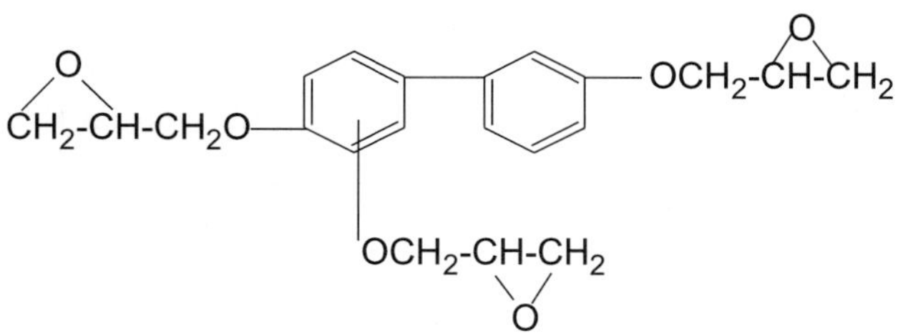


**Figure 7.91.**

The triglycidyl ether resin obtained was observed to cure rapidly with aliphatic amine compounds at room temperature. Potential uses for this epoxy were in joint sealers and protective coatings.

Resorcinol can produce organic compounds, having multi-hydroxy groups, by reacting with various functional groups containing chemical compounds. In this aspect, the reaction of resorcinol with cyclohexanone in the presence of an acid catalyst produced a trihydroxy derivative (shown below) which on further reaction with epichlorohydrin formed the triglycidyl ether compound [22] (Figure 7.92).

This triepoxide compound was mixed with a novolak resin, $Ph_3P$ (catalyst) and other additives, transfer molded and post cured for 5 h at 180 °C. The cured

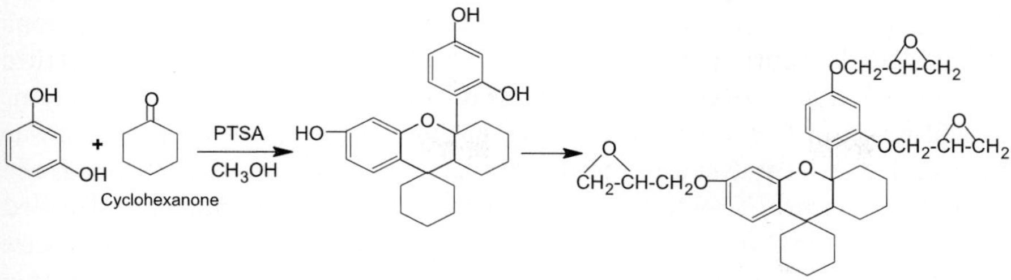


**Figure 7.92.**

material showed a glass transition temperature of 208 °C, bending strength of 2 kg/mm$^2$ and flexural modulus of 90 kg/mm$^2$. Humidity aging for 72 h at 85% RH showed moisture absorption of 0.40%.

### 7.7.4 Tetraglycidyl Ether Resins (Tetraepoxides)

Poly-glycidyl ethers (epoxies) can react with curing agents at an exceptional speed and therefore, employed for certain important applications where short curing schedules are most desirable. Curing speeds can be adjusted by proper selection of the amount and kind of curing agent to meet any reasonable demand. The epoxidation of the substantially tetra functional novolak of resorcinol and formaldehyde having a low viscosity at casting temperatures is based on the chemistry and preparative procedures given below [23] (Figure 7.93).

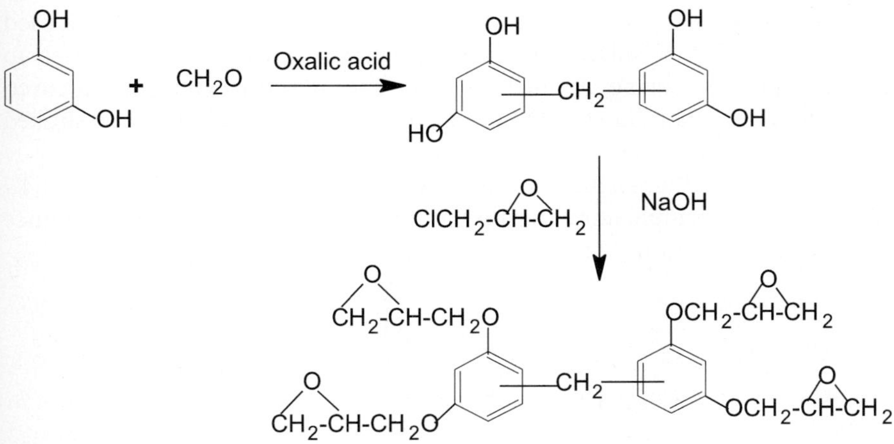


**Figure 7.93.**

**Synthesis of Tetraglycidyl Ether of Resorcinol-Formaldehyde Novolak Resin [23]**
A novolak was prepared by dissolving 7700 g of (70 mol) of resorcinol in

1500 ml of water at 60–65 °C. To this was added 1136 g (14 mol) of 37% aqueous formaldehyde solution with stirring at 60 °C. The temperature of the mixture rose to 100–105 °C over the next 10–15 min and this temperature was maintained for 30 min. A solution of 44.4 g of oxalic acid in 300 ml of water was then added as a catalyst and the reaction continued for another 30 min at the same temperature. Thereafter the water and excess resorcinol were distilled from the novolak at temperature below 170 °C and at an absolute pressure below 3 mm Hg. The yield was substantially 100% of theory of tetra functional novolak.

The novolak was then dissolved in epichlorohydrin, using a 5 : 1 mol ratio of epichlorohydrin to hydroxyl group. A 50% aqueous sodium hydroxide solution at 100–105 °C was added over a period of about 2.5 h. The amount of sodium hydroxide added was equal to about 1.1 times the equivalents of phenolic hydroxyl groups. As the reaction proceeded, an azeotropic water and epichlorohydrin were removed, condensed, separated and the epichlorohydrin recycled back to the reactor. Upon completion of the epoxidization, the byproduct sodium chloride was removed by filtration then the product was water-washed and excess epichlorohydrin was removed by distillation at 145 °C at 2 mm Hg absolute pressure.

The epoxy resin thus prepared had the following properties.

1. Epoxy equivalent weight (EEW) = 135
2. Durran's softening point (°C) = 42
3. Viscosity at 75 °C (Centipoise) = 4.720
4. Molecular weight = 560

Due to low melt viscosity of this tetra epoxide compound, they were used for potting and encapsulation applications where the resin could flow freely to fill voids prior to curing. A stoichiometric amount of this epoxy when cured with methylene dianiline (MDA) curing agent produced a highly crosslinked and dense product having a heat distortion temperature of 300 °C.

Tetraglycidyl ether resins were also made from bisresorcinol-F and 2,2′,4,4′-tetrahydroxy benzophenone derivatives and used for the high performance composite applications [24] (Figures 7.94 and 7.95).

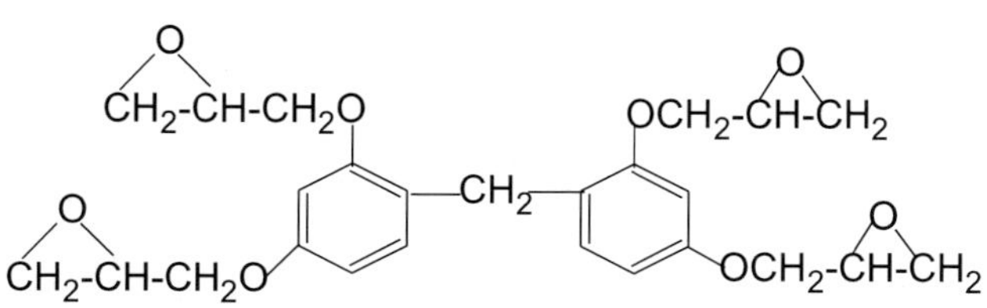


Tetraglycidyl ether of Bis-resorcinol-F

**Figure 7.94.**

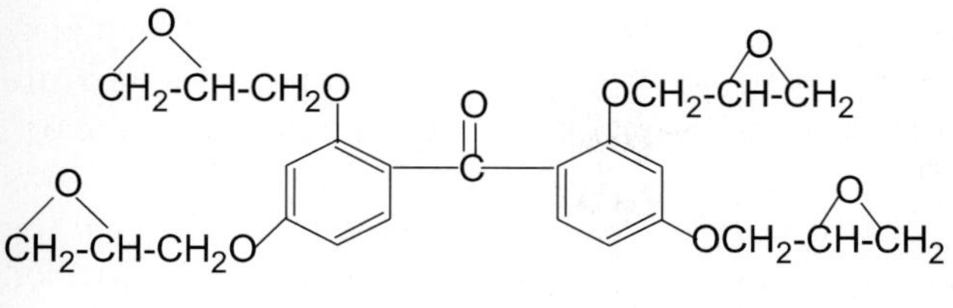


**Figure 7.95.**

Tetra epoxy compounds having aryl or aralkyl substitutions, when cured by heating with curing agents, such as amines, amides or anhydrides of dicarboxylic acids, may provide cured resin compounds with high heat distortion temperatures. These tetra epoxide materials are in many respects highly valuable and have enjoyed extensive commercial uses in applications such as coatings, embedment of electrical apparatus and impregnation of fabrics and fibrous mats.

Aralkyl substituted resorcinol based tetraglycidyl ether resins were synthesized from the reaction products with 4,6-bis-(chloromethyl)-*m*-xylene and 1,4-bis-(chloromethyl)-benzene compounds [25] (Figures 7.96 and 7.97).

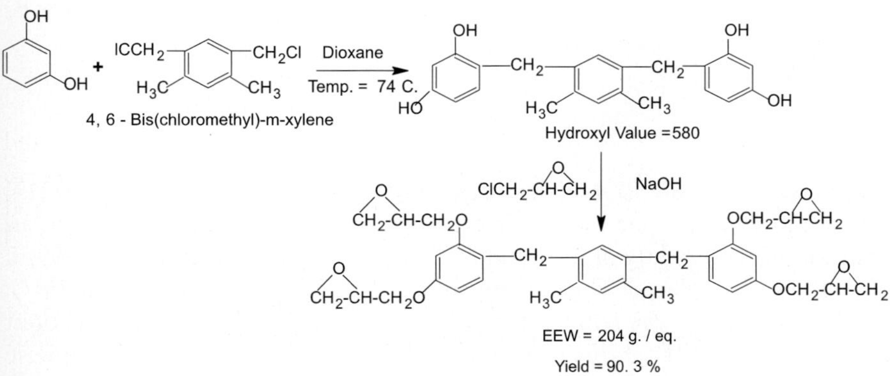


**Figure 7.96.**

Resorcinol based tetra epoxy compounds was developed from the corresponding derivative of 1,3-bis-[(2,4-dihydroxyphenyl)-*alpha*-methyl ethyl]-benzene which was prepared from the resorcinol and 1,3-diisopropenyl benzene reaction. This epoxy could show cured resin properties such as low moisture absorption, high reactivity, low volatility, high glass transition temperatures and mechanical properties [14, 15] (Figure 7.98).

The synthetic procedures used for the preparation of the above tetraglycidyl ether compound starting from the resorcinol are outlined below.

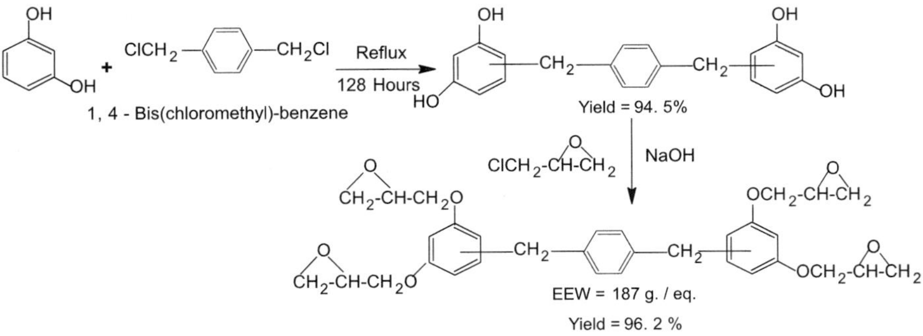


**Figure 7.97.**

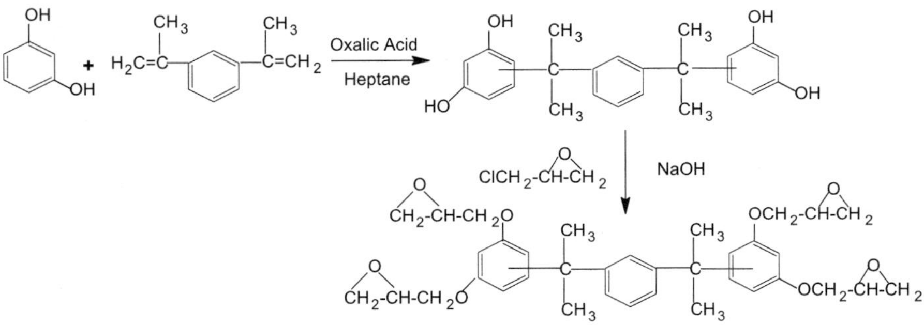


**Figure 7.98.**

**Preparation of 1,3-Bis-[(2,4-di-hydroxyphenyl)-*alpha*-methyl ethyl] Benzene [14]**
Into a 2-l kettle equipped with stirrer, thermometer, reflux condenser and an addition funnel, 264.0 g of resorcinol (2.4 mol) 12.0 g of oxalic acid and 900 ml of heptane were placed. The contents of this flask were heated to about 75 °C. While maintaining the temperature of the resorcinolic slurry between about 75–85 °C, 126.4 g of 1,3-diisopropenyl benzene (0.8 mol) was added dropwise for about 2 h. After the completion of the above addition, the temperature of the reaction mixture was raised to and maintained at about 95 °C for about an additional 2.0 h, while the reaction mixture was continuously stirred. The reaction mixture was then cooled and the separated heptane layer was removed by decantation. Finally the solidified mass was treated with excess dichloromethane to separate the reaction product. The white precipitate that separated from the solution was filtered, washed with dichloromethane and then washed with distilled water. After air drying at about room temperature the precipitate weighed 215 g. Structural characterization by IR/NMR analysis of the reaction product confirmed the structure of 1,3-bis-[(2,4-di-hydroxyphenyl)-*alpha*-methyl ethyl] benzene.

**Preparation of an Epoxy Resin from 1,3-Bis-[(2,4-di-hydroxyphenyl)-*alpha*-methyl ethyl] benzene [15]** 1,3-Bis-[(2,4-di-hydroxyphenyl)-*alpha*-methyl ethyl] benzene (15.1 g; 0.04 mol) and epichlorohydrin (74.0 g; 0.8 mol) were placed in a 500 ml round bottomed flask fitted with a stirrer, thermometer, Dean Stark condenser and an addition funnel. The contents of the flask were heated to about 100 – 125 °C and an aqueous sodium hydroxide solution (50 wt. %; 13.4 g; 0.168 mol) was added slowly into the refluxing solution. While continuously removing the azeotropic water, the separated epichlorohydrin was continuously returned to the reaction flask. The addition time of sodium hydroxide was about 1.0 h and the time required for the complete removal of water (9.6 g) was about 1 1/2 to 2 h. After this, the excess epichlorohydrin was distilled out first under atmospheric pressure and then vacuum distillation conditions (maximum pot temperature about 140 – 150 °C and vacuum about 27–28″ of Hg). Then, about 130 ml of acetone was added to the reaction flask and the contents were refluxed for about 15 min to dissolve the epoxide. Lastly, sodium chloride was filtered out and the solvent acetone was distilled out using both atmospheric and vacuum distillation conditions (temperature about 95 – 97 °C and vacuum about 27–28″ of Hg) to obtain 17.4 g of a light yellow 1,3-bis-[(2,4-di-(2,3-epoxypropoxy) phenyl)-*alpha*-methylethyl] benzene having an epoxy equivalent weight of 210. Structural characterization by IR/NMR of this epoxide showed 3.76 epoxide groups (Theory $= 4.0$) per molecule. Differential scanning calorimetry (DSC) analysis showed that this resin had a softening point of 59.8 °C.

The above tetra epoxy compound was blended with DDS curing agent at the stoichiometric ratio and cured for 2 h at 150 °C and then post-cured for 4 h at 200 °C. The thermal mechanical analysis (TMA) measurement made on the cured epoxy resin showed a Tg of 244.7 °C.

### 7.7.5 Polyglycidyl Ether Compounds (Polyepoxides)

Epoxy resins prepared from the resorcinolic novolak resins may combine the high reactivity of epoxy group and thermal stability of the resorcinolic resin back bone. They are generally synthesized by reacting epichlorohydrin with RF novolak resins. RF resins, in turn, are obtained by condensing resorcinol with formaldehyde under acidic or basic conditions and at formaldehyde to resorcinol molar ratio of $< 0.8:1$. RF based epoxy compound can be represented by the following structural formula (Figure 7.99).

A preparative procedure to synthesize RF novolak based epoxy is outlined below [26].

**Procedure for the Preparation of Epoxy Resin from Resorcinol-Formaldehyde (RF) Novolak Resin [26]** To a 50-gallon pilot plant resin kettle, there was charged 60 parts by weight resorcinol-formaldehyde (RF) resin (1 mol of resorcinol

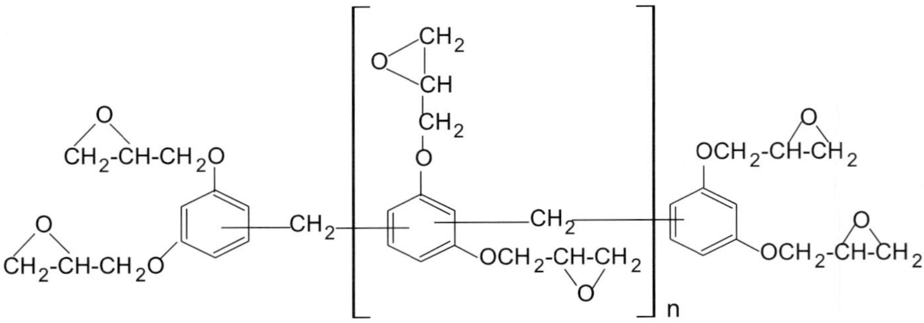


**Figure 7.99.**

condensed with 0.62 mol of formaldehyde) and 378 parts by weight epichlorohydrin. The charge was heated to reflux (116 °C) in 42 min. Trace water was removed by azeotropic distillation over a 10 min period. There was then added 0.6 part dimethyl formamide and the kettle contents were heated to refluxing for 38 min. The reaction mixture was maintained at reflux for 120 min, and one equivalent of 70 percent by weight aqueous caustic was added in a thin stream. After approximately half of the molecular equivalent of the caustic was added, azeotropic distillation was begun to remove the water. The distillation was continued for 28 min after all of the caustic was added. The kettle contents were cooled to room temperature over a period of 50 min. The reaction mixture was then filtered to remove the precipitated salt and reheated to the reflux temperature in 47 min. The catalyst and excess epichlorohydrin were removed by devolatilization to 148 °C at 8 mm of Hg in 122 min. There was obtained 103.5 parts by weight of a completely soluble and fusible epoxidized resorcinol-formaldehyde resin. The product resin had an oxirane oxygen content of 9.5 wt. % and a total chlorine content of 3.1 wt. %.

The number of epoxide groups present in the RF novolak resin based epoxy compounds varies from 2 to 4. Higher epoxide functionality can provide higher crosslink density and improved heat resistance properties.

Novolak resins are prepared from phenolic compounds, including resorcinol, and aldehyde reactions. In the case of *p*-substituted phenols, the bismethylol compounds, (2,6-disubsituted), can be synthesized by their reactions with formaldehyde in the presence of basic catalysts. The use of bismethylolated phenolic derivatives in the resorcinolic novolak resins preparation can lead to linear rather than three-dimensional structures having uniform molecular weights. Molecular structures and degree of condensation or polymerization may be controlled by the proper selection of molar ratios of resorcinol and *p*-substituted-2,6-bismethylol phenol compounds. For example, 2,6-bis-(2,4-dihydroxybenzene)-*p*-cresol was obtained from resorcinol and 2,6-dimethylol-*p*-cresol reaction in which higher molar excess of resorcinol was used to avoid the high molecular weight formation [27] (Figure 7.100).

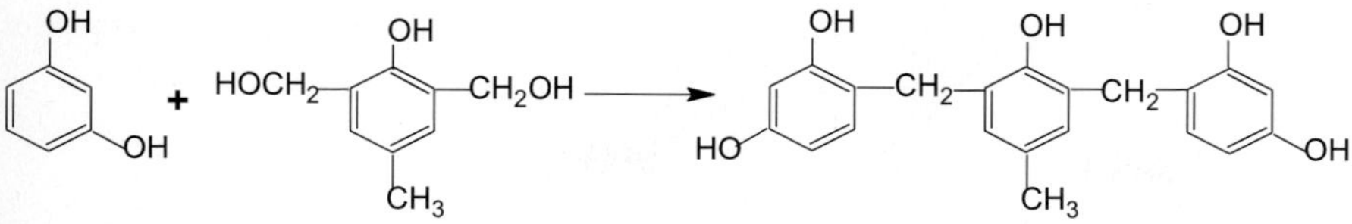


**Figure 7.100.**

The above bis-resorcinol was synthesized using the procedure given below.

**Synthesis of 2,6-Bis-(2,4-dihydroxy benzene)-*p*-cresol [27]** A reaction vessel was charged with 1980 g (18 mol) of resorcinol, 504 g (3 mol) of 2,6-dimethylol-*p*-cresol, 9 g (0.05 mol) of *p*-toluene-sulfonic acid and 4500 g of ethanol, and reaction was carried out for 4 h under reflux. Under a reduced pressure of 300 mm Hg, ethanol was removed and the reaction mixture was concentrated. The content in the reaction vessel was poured into 26 l of water to precipitate a reaction product. The product was recovered by filtration, washed with water and dried to obtain 740 g of a resin.

The melting point of the resin was 128–135 °C as determined according to the microscope method, and the number average molecular weight Mn of the resin was 360 as determined by gel permeation chromatography. In the IR spectrum of this compound, characteristic absorptions of the phenolic – OH group, the aromatic nucleus and the methyl group were observed at about 3500 $cm^{-1}$, about 1600 and 1500 $cm^{-1}$ and about 1450 $cm^{-1}$, respectively.

When the molar ratio of resorcinol to 2,6-dimethylol-*p*-cresol was 2 : 1, then higher molecular weight products were obtained (Figure 7.101).

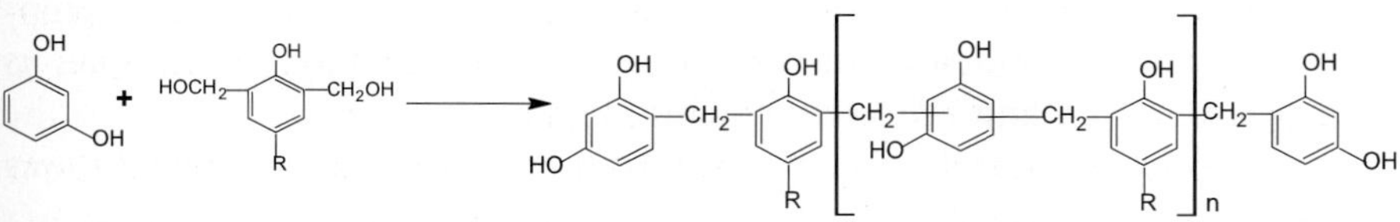


**Figure 7.101.**

By employing different *p*-substituted phenolic bis-methylol compounds, novolak resins having different chemical structures were synthesized and documented (Table 7.50) [27].

Polyepoxide compounds having the following idealized structures were prepared by reacting with epichlorohydrin in the presence of NaOH as the dehydrohalogenating agent, as outlined in the synthesis procedure given below (Figure 7.102).

**Procedure for the Preparation of Epoxy Resin of 2,6-Bis-(2,4-dihydroxy benzene)-*p*-cresol [27]** A reaction vessel was charged with 615 g (1.7 mol) of 2,6-bis-(2,4-dihydroxy benzene)-*p*-cresol, 4023 g (43.5 mol) of epichlorohydrin, 57.2 g

**Table 7.50.** Novolak resins prepared from *p*-substituted-bis-methylol phenols and resorcinol

| *p*-Substitution in phenol | Group | Molecular weight, Mn (GPC) | Degree of polymerization (n) | Melting point (°C) |
|---|---|---|---|---|
| Methyl | $CH_3$ | 900 | 2 | 157–172 |
| *tert*-Butyl | $C(CH_3)_3$ | 960 | 2 | 150–165 |
| Chloro | Cl | 810 | 2 | 175–190 |
| Hydrogen | H | 600 | 1 | 140–155 |
| Phenyl | $C_6H_5$ | 1000 | 2 | 200–215 |

Data from [27]

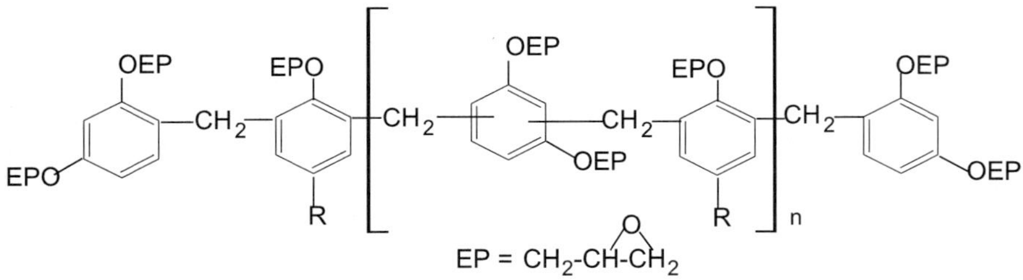


**Figure 7.102.**

(0.26 mol) of a 50% aqueous solution of tetramethyl ammonium chloride and 64 g of water, and reaction was carried out with stirring at 90 °C for 4 h.

Then 798 g (9.6 mol) of a 48% aqueous solution of sodium hydroxide was added dropwise to the reaction mixture over a period of 1.5 h. During this dropwise addition, the condensed distillate was separated, and the upper aqueous layer was removed and the lower epichlorohydrin layer was returned to the reaction vessel, so that the water concentration in the reaction mixture was maintained at about 2%.

After completion of the reaction, the reaction mixture was thrown into 2250 l of water and the mixture was sufficiently stirred to wash the organic layer. Then, the organic layer was separated and 2250 l of water was added thereto, and the mixture was sufficiently stirred to wash the organic layer. Then, the organic layer was separated and 240 ml of a 17% solution of sodium phosphate was added to effect neutralization. Water was removed from the organic layer by heating and the organic layer was then filtered. The filtrate was concentrated to obtain 960 g of an epoxy resin.

From the results of gel permeation chromatography, it was found that the number average molecular weight Mn was 1120 and the molecular weight distribution Mw/Mn was 1.42. The resin had a melting point of 53 to 58 °C as determined by the microscope method, and the epoxy equivalent of the resin was 150 g/eq as determined by the hydrochloric acid-dioxane method.

The epoxide materials prepared from resorcinol and *p*-cresol bis-methylol compounds, when cured with methylene dianiline (MDA) curing agent, produced excellent heat resistant, high Tg and high heat distortion temperature compounds (Table 7.51) [27].

**Table 7.51.** Cured neat resin mechanical properties curing agent: methylene dianiline (MDA)

Cure cycle: (i) 2 h/100 °C and (ii) 4 h/150 °C

| R Group in epoxy resin | Flexural strength ($kg/mm^2$) | Flexural modulus ($10^3\,kg/mm^2$) | Rockwell hardness (M) | Glass transition temperature (Tg) | Heat distortion temperature (°C) |
|---|---|---|---|---|---|
| Methyl | 6.3 | 0.32 | 121 | 327 | 286 |
| *tert*-Butyl | 3.7 | 0.28 | 120 | 327 | 288 |
| Chloro | 3.9 | 0.28 | 120 | 320 | 275 |
| Hydrogen | 6.5 | 0.33 | 120 | 300 | 265 |
| Phenyl | 7 | 0.29 | 118 | 310 | 273 |
| Bisphenol-A Epoxy (control) | 11.2 | 0.26 | 102 | 163 | 158 |

Data from [27]

Due to excellent mechanical and heat resistant properties, they could be employed in the molding materials, varnishes, laminates for printed circuits and advanced composites.

Epoxy resins, in general, provide the best overall properties in fiber reinforced composites. The disadvantages of many epoxy resins are in their ability to absorb high levels of moisture under hot/wet conditions, which can lead to deterioration of mechanical properties. Therefore, epoxies that are capable of enhancing physical and mechanical properties with low moisture absorption characteristics are needed by the composite industries. Epoxy resins having high reactivity, low moisture absorption, chemical resistance and internal plasticizability properties were obtained from the following chemistry [28–31] (Figure 7.103).

A glass cloth or mat impregnated with the epoxy resin obtained from the resorcinol-dicyclopentadiene (structure shown above) showed excellent moisture resistant and mechanical properties which could be utilized in insulation materials development. Compositions based on these materials showed excellent heat resistant properties and found to be suitable for copper clad laminates.

## 7.7.6 Toughening of Resorcinolic Epoxy Resin Systems

Epoxy resins have various desirable properties such as high tensile strength and modulus, good thermal and chemical resistance, dimensional stability

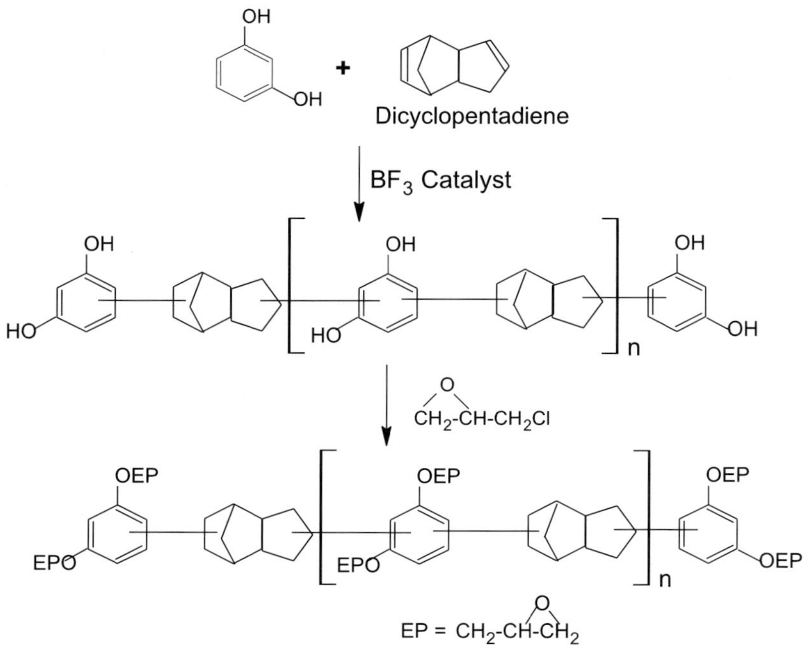


**Figure 7.103.**

and therefore, most widely used as matrix resin materials for the advanced composites. But, they are very brittle and as a result, have poor thermal cycling behavior and fairly low impact and fracture strengths. Various methods have been employed to toughen the epoxy systems to enhance their toughness characteristics. Reactive butadiene-acrylonitrile oligomers having both carboxyl and amine end groups and functionally terminated siloxane oligomers have been utilized as toughening agents for epoxy resins [32–35].

Similarly, tough, ductile, chemically and thermally stable engineering thermoplastics, such as poly(ether sulfones) and poly(ether-imides), have been utilized as toughening agents [36, 37].

In addition, amine terminated poly (ether sulfone) oligomers have also been used to chemically bond the modifier into epoxy networks [38].

Impact modification of epoxy resin systems was carried out and used in graphite fiber reinforced composites. With toughener modifications, the epoxy networks showed considerable improvement in fracture toughness while maintaining high modulus and chemical resistance. Most commercial epoxy resins based prepreg systems, for advanced carbon fiber reinforced composite applications, are based on tetraglycidyl ether derivatives of 4,4′-diaminodiphenyl methane (MDA) and diaminodiphenyl sulfone (DDS) (hardeners) systems. RDGE and diaminodiphenyl ether system also showed high flexural and tensile modulus properties and therefore, were utilized as matrix resins for advanced composites.

Amine terminated thermoplastic tougheners based on the reaction scheme shown below were synthesized and used to enhance the toughness characteristics of BRDGE and RDGE resin based systems [15] (Figure 7.104).

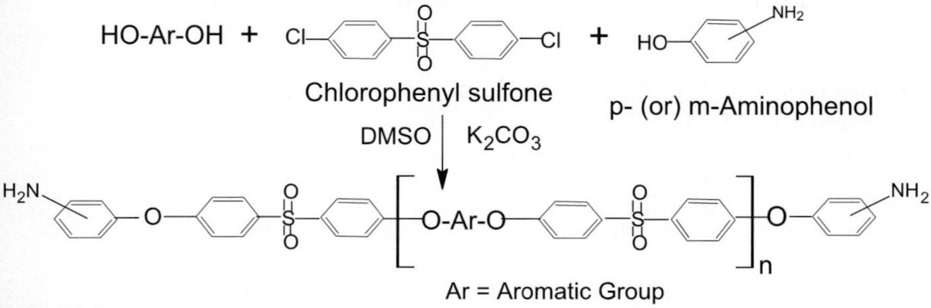


**Figure 7.104.**

The toughening agents synthesized using resorcinol, benzoyl resorcinol and bisphenol-A compounds and their Tg values are presented in Table 7.52.

**Table 7.52.** Synthesis of toughening agents for epoxy resins[a]

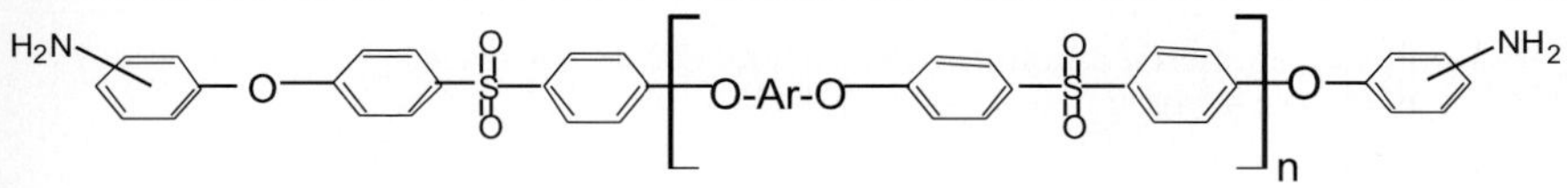


| Diphenol used | Designation | Ar group | Glass transition temp (Tg, °C) n = 8 | n = 40 |
|---|---|---|---|---|
| Bisphenol-A | BPA-PES | 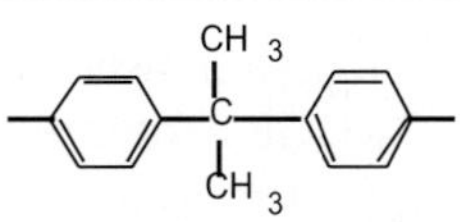 | 125 | 162 |
| Resorcinol | RES-PES | 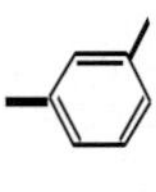 | 138 | 165 |
| Benzoyl resorcinol | BR-PES | 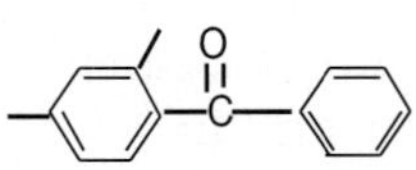 | 109 | 112 |

PES = Polyether sulfone
[a]After the method of M.J. Jurek and J.E. McGrath, Polymer, Vol. 30, p 1552 (1989)

By adjusting the molar ratios of reactants, reactive tougheners with the degree of polymerization (n) 8 and 40 were synthesized.

Neat resin castings of cured epoxy resin materials containing DDS and amine terminated tougheners were prepared using low and high molecular weight tougheners. The toughener concentrations employed in the resin castings were 5, 10 and 15 wt. % of the final cured materials. The resin properties of BRDGE and RDGE toughened epoxy resins are presented in Tables 7.53 and 7.54.

**Table 7.53.** Toughened RDGE-DDS cured resin mechanical properties

| Toughener used | BPA-PES, M.Wt. = 3550 | | | | BPA-PES, M.Wt. = 13270 | | |
|---|---|---|---|---|---|---|---|
| Toughener level Used (wt.%) | 0 | 5 | 10 | 15 | 5 | 10 | 15 |
| Flexural properties | | | | | | | |
| Strength (MPa) | 171.7 | 173.8 | 174.4 | 170.3 | 162.7 | 159.3 | 140 |
| Modulus (GPa) | 4 | 3.9 | 3.9 | 3.9 | 4 | 3.8 | 3.9 |
| Strain (%) | 7.4 | 7.4 | 7.2 | 6.2 | 5.7 | 6.1 | 4.5 |
| Tensile properties | | | | | | | |
| Strength (MPa) | 99.3 | 99.3 | 104.8 | 102.7 | 106.2 | 99.3 | 91 |
| Modulus (GPa) | 3.9 | 4.5 | 4.3 | 4.3 | 4.3 | 4.1 | 4.1 |
| Strain (%) | 5.1 | 5.2 | 6.1 | 4 | 6.1 | 5.8 | 5 |
| Fracture toughness ($K_{1C}$, MPa $m^{1/2}$) | 1.13 | 1.05 | 1.24 | 1.26 | 1.29 | 1.32 | 1.33 |
| Fracture energy ($G_{1C}$, $J/m^2$) | 325.3 | 247.5 | 362.6 | 372.7 | 388.9 | 428 | 429.4 |
| Moisture pickup (%) | 3.6 | 3.3 | 3.2 | 3.1 | 3.2 | 2.3 | 2.9 |
| Glass transition Temp (Tg, °C) | 182.3 | 182.6 | 172.5 | 164.5 | 175.2 | 160.1 | 163.7 |

Cured with stoichiometric amounts of RDGE and DDS. Cure cycle:
(i) 2 h/150 °C (ii) 4 h/200 °C
Data taken from [15]

From the results of Tables 7.53 and 7.54, it was observed that the high flexural and tensile modulus values of RDGE and BRDGE cured systems remained essentially unchanged. The toughener used in this study showed the most positive overall effect on the BRDGE compared to RDGE resin system. In the case of BRDGE, fracture energy increased from 69.1 $J/m^2$ in the unmodified resin to 269 $J/m^2$ with 15% loading of high molecular weight toughener. The fracture toughness ($K_{1C}$) values also increased almost four times but the Tg was maintained.

Another published work showed that amine-functionalized poly (arylether sulfone) oligomeric modifiers with Mn =< 5000 were developed using bisphenol-A, chlorophenyl sulfone and *m*- and *p*-aminophenol compounds. The tougheners were used up to 40% level to toughen RDGE (Heloxy 69) and DDS cured systems [39, 40]. The results are presented in Table 7.55.

Based on Table 7.55 results, RDGE toughened epoxy resin systems showed significant improvements in the toughening properties for them to be utilized in the advanced composites development. From the Transmission Electron Microscopic (TEM) studies, these materials were observed to possess a thermoplastic-rich continuous phase accounting for the RDGE-DDS system's improved toughness properties.

**Table 7.54.** Toughened BRDGE-DDS cured resin mechanical properties

| Toughener used | BPA-PES, M.Wt. = 3550 | | | | BPA-PES, M.Wt. = 13 270 | | |
|---|---|---|---|---|---|---|---|
| Toughener level Used (wt.%) | 0 | 5 | 10 | 15 | 5 | 10 | 15 |
| Flexural properties | | | | | | | |
| Strength (MPa) | 175.1 | 165.5 | 170.3 | 151 | 124.8 | 132.4 | 121.4 |
| Modulus (GPa) | 4.1 | 4.1 | 4.1 | 3.9 | 4 | 4 | 4.2 |
| Strain (%) | 5.9 | 5.6 | 6.3 | 5.7 | 3.4 | 3.7 | 3 |
| Tensile properties | | | | | | | |
| Strength (MPa) | 87.6 | 91.7 | 88.3 | 88.9 | 89.6 | 80 | 69.6 |
| Modulus (GPa) | 4.2 | 4.1 | 4.3 | 3.9 | 4 | 4.1 | 4 |
| Strain (%) | 2.9 | 3.9 | 3.75 | 4 | 3.5 | 3 | 2.3 |
| Fracture toughness ($K_{1C}$, MPa $m^{1/2}$) | 0.54 | 0.72 | 0.94 | 0.97 | 0.78 | 0.88 | 1.02 |
| Fracture energy ($G_{1C}$, $J/m^2$) | 69.1 | 126.8 | 173.3 | 242 | 149.4 | 190.4 | 269 |
| Moisture pickup (%) | 2.8 | 2.3 | 2.3 | 2.2 | 2.4 | 1.9 | 1.9 |
| Glass transition Temp (Tg, °C) | 173.2 | 167 | 171.7 | 178 | 182 | 182 | 174 |

Cured with stoichiometric amounts of BRDGE and DDS. Cure cycle:
(i) 2 h/150 °C (ii) 4 h/200 °C
Data taken from [15]

With this toughened resin systems, tough RDGE based matrix materials could be easily achieved. These resins were highly processable in the uncured stage due to the low viscosity of RDGE. Considerable improvements in the toughness properties, such as $K_{1C}$ and $G_{1C}$, were achieved by the higher loading of toughening agents.

### 7.7.7 Diglycidyl Ether of Alkoxylated Resorcinol

In the development of carbon fiber reinforced composites, multi-functional epoxies in combination with aromatic diamine curing agents have been used to achieve high Tg and modulus properties which are associated with the high crosslink density of materials. Highly crosslinked epoxies often produced low impact strength due to brittleness. If the crosslink density of the system is decreased, then the brittle character may be minimized. This can be achieved by increasing the distance between the two epoxy groups in the molecule by means of introducing linear alkyl or aralkyl ether groups. The introduction of an alkylene ether groups in the epoxy molecule can be done by alkoxylating the aromatic hydroxyl compounds before the glycidyl ether reaction with epichlorohydrin. The use of epoxy resins and epoxy hardeners or curing agents

**Table 7.55.** Toughened RDGE-DDS cured resin mechanical properties (Cure cycle: (i) 2 h/177 °C and (ii) 2 h/200 °C)

| Toughening agent (PES-APT) | *p*-Aminophenol based | | | | | *m*-Aminophenol based | |
|---|---|---|---|---|---|---|---|
| Experiment (parts) | 1 | 2 | 3 | 4 | 5 | 6 | 7[a] |
| Epoxy/curing/ toughening system | | | | | | | |
| RDGE (Heloxy 69) | 40.8 | 21.5 | 21.6 | 20.8 | 41.58 | 38.5 | 44.1 |
| Diaminodiphenyl sulfone (DDS) | 19.2 | 17 | 16.8 | 18.4 | 23.42 | 21.5 | 20.97 |
| Toughening agent (PES-APT) | 40 | 40 | 40 | 40 | 35 | 40 | 35 |
| BPADGE (DER 332) | | 21.5 | | | | | |
| BPFDGE (Epiclon 830) | | | 21.6 | | | | |
| TGMDA (Araldite MY 720) | | | | 20.8 | | | |
| Properties | | | | | | | |
| Fracture toughness ($K_{1C}$, MPa $m^{1/2}$) | 1.71 | 1.55 | 1.8 | 1.54 | 2.01 | 2.3 | 2.19 |
| Fracture energy ($G_{1C}$, $J/m^2$) | 833 | 674 | 938 | 692 | 1134 | 1420 | 1269 |
| Shear storage modulus, (G′ w) | 1.3 | 1.32 | 1.28 | 1.27 | 1.32 | 1.38 | 1.4 |
| Glass transition temp. (Tg, °C) | 163 | 185 | 170 | 195 | 176 | 175 | 180 |

PES-APT = Aminophenol terminated poly(ether sulfone)
BPADGE = Diglycidylether of bisphenol-A
BPFDGE = Diglycidylether of bisphenol-F
TGMDA = Tetraglycidylether of methylene dianiline
[a] 3,3′-Diaminodiphenyl sulfone used as the curing agent
Data taken from [39,40]

having flexible alkyl ether groups improves the toughness of the cured epoxy system by reducing the brittleness.

A resorcinol derivative made from the reaction of resorcinol and ethylene carbonate produced alkyl ether groups in the molecule. This was used to develop diglycidyl ether of alkoxylated resorcinol, the use of which is expected to improve the impact resistance of epoxy polymers. The synthesis of this epoxy is based on the following chemistry [41,42] (Figure 7.105).

The molecular flexibility of DGER may be expected to improve the toughness properties of the cured epoxy resin systems. Unlike many other epoxy materials, DGER was considered non-toxic. Acute dermal irritation results indicated that DGER was non-irritating to the skin, which may compliment the material's expected enhanced performance in its applications. DGER can be utilized in various high performance applications such as coatings, adhesives and composites.

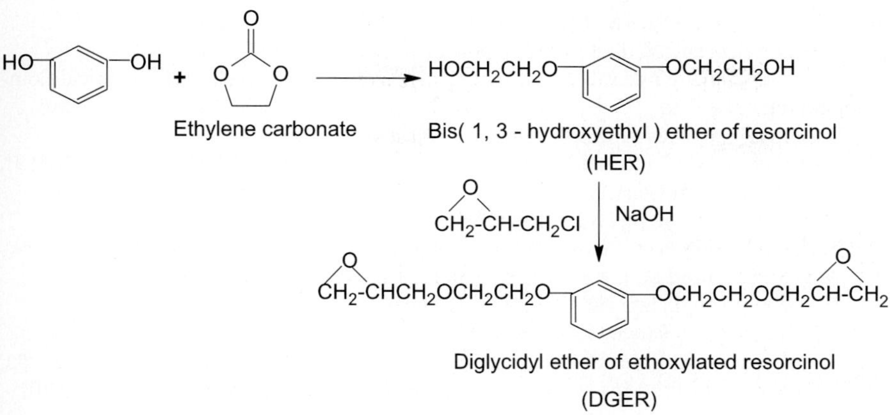


**Figure 7.105.**

## 7.7.8 Summary and Outlook

The epoxy resins prepared from resorcinol and its derivatives have shown excellent cured resin properties for use in various high performance industrial applications. The toxicity associated with RDGE can be overcome by employing substituted resorcinolic derivatives in the epoxy synthesis. Published work clearly show enormous potential still exists for resorcinol derived epoxy resins in numerous potential applications.

## References


1. C. A. May and Y. Tanaka, Eds., Epoxy resins – chemistry and technology, Marcel Dekker, New York (1973).
2. H. Lee and K. Neville, Handbook of epoxy resins, McGraw-Hill, New York (1982).
3. G. Werner and E. Farenhorst, US Patent, 2,467,171 (1949).
4. W. E. St. Clair, US Patent, 2,892,849 (1959).
5. B. T. Larkin and W. E. St. Clair, US Patent, 2,995,583 (1961).
6. R. Durairaj and M. Tackie, 39th SAMPE Symosium, Vol. 39, p 262 (1994).
7. Resorcinol diglycidyl ether. In: Koppers Company Technical Bulletin, October (1959).
8. C. H. Hine et al., The toxicity of epoxy resins, A.M.A. Arch. of Ind. Health, Vol. 17, p 129 (1958).
9. C. H. Hine et al., An investigation of the oncogenic activity of two representative epoxy resins, Cancer Res., Vol. 18, p 20 (1958).
10. P. Kotin and H. L. Falk, Cancer, Vol. 9, p 910 (1956).
11. IARC Monographs on the Evaluation of the Carcinogenic Risk of Chemicals to Humans, Vol. 36, p 369 (1985).
12. IARC Monographs on the Evaluation of the Carcinogenic Risk of Chemicals to Humans, Vol. 71, p 1589 (1999).
13. H.H. Borgstedt, In: C. A. May and Y. Tanaka, Eds., Epoxy resins – chemistry and technology, Marcel Dekker, New York (1973).

14. R.B. Durairaj, US Patent, 5,300,618 (1994).
15. R. B. Durairaj and M. Tackie, PCT Application, WO 94/15986 (1994).
16. Epon HPT Curing Agent 1062-M, Technical Bulletin, SC : 525-88, Shell Chemical Company, Houston, Tx.
17. K. Leonid, et al., Soviet Union Patent, 1,824,403 (1993).
18. Armed Services Technical Information Agency, Arlington, VA, Report, AD:268134, WADD TR 60-791 (Part 1).
19. J. J. Schmid, US Patent, 3,925,315 (1975).
20. K. D. Patel, High Performance Polym., Vol. 8, p 301 (1996).
21. W. E. St. Clair and M. J. Waldron, US Patent, 3,133,033 (1964).
22. T. Morimoto et al., Jap. Pat, 06,157,717 (1994).
23. T. S. Boozalis, US Patent, 3,406,150 (1968).
24. A. L. Cupples et al., In: H. Lee, Ed., Epoxy Resins Advances in Chemistry Series, 92, Synthesis of epoxy resins for room temperature cures, ACS, Washington, DC (1970).
25. R. M. Christensen, M. Wismer and W. R. Hydro, US Patent, 3,001,972 (1961).
26. W. E. St. Clair and B. T. Larkin, US Patent, 3,148,171 (1964).
27. S. Katayama, T. Nakano and T. Takeda, US Patent, 4,614,826 (1986).
28. H. Minoru, Jap. Pat. Publ., 07-252373 (1995).
29. I. Kazuo, T. Kenji and H. Hitoshi, Jap. Pat. Publ., 05-339340 (1993).
30. N. Hiroaki, S. Kazukuni and H. Minoru, Jap. Pat. Publ., 62-007719 (1987).
31. N. Hiroaki and H. Minoru, Jap. Pat. Publ., 62-004720 (1987).
32. E. H. Rowe, R. Siebert and R.S. Drake, Mod. Plast., Vol. 47, p 110 (1970).
33. J. M. Sultan and F. McGarry, Polym. Eng. Sci., Vol. 13, p 29 (1973).
34. C. Tran et al., ACS Org. Coat. Appl. Polym. Sci. Proc., Vol. 49, p 498 (1983).
35. J. S. Riffle et al., Epoxy resin chemistry II, ACS Symp. Ser., No. 221, Ch. 2 (1983).
36. C. B. Bucknall and I. K. Patridge, Polymer, Vol. 24, p 639 (1983).
37. J. Diamant and R. J. Moulton, 29th Natl. SAMPE Symp. Vol. 29, p 422 (1984).
38. J. E. McGrath and J. M. Jurek, Polymer, Vol. 30, p 1552 (1989).
39. H. Jabloner, B. J. Swetlin and S. G. Chu, US Patent, 4,656,207 (1987).
40. H. Jabloner, B. J. Swetlin and S. G. Chu, US Patent, 4,656,208 (1987).
41. A. N. Sorokin et al., Soviet Union Patent, 702,017 (1979).
42. R. B. Durairaj and G. A. Jesionowski, US Patent, 6,586,607 (2003).

# 7.8 Barrier Polymers

## 7.8.1 Introduction

Plastic materials have been extensively used as replacement for glass and metal containers for packaging food and beverages. In these applications, in general, plastics can offer lighter weight, less breakage, high degree of design flexibility and potentially lower costs than the glass. The growth opportunities for plastics with the packaging industry are enormous but success will depend on the development of cost-effective and high performance packaging materials for specific applications. Plastic packaging materials cannot provide comparative shelf life for the packaged food and beverages better than glass or metal. The shelf life of products depends on their protection against gases, such as oxygen ($O_2$) and carbon dioxide ($CO_2$), and moisture. Protection against oxygen is one of the most important requirements in the food packaging since the oxidation of food, in the presence of oxygen, produces off-flavors, colors and nutrient loss.

The sensitivity of certain food and beverage materials is presented in Table 7.56.

**Table 7.56.** Estimated maximum oxygen tolerance levels for selected food and beverages

| Food or beverage | Oxygen tolerance (ppm) |
|---|---|
| Beer | 1 to 2 |
| Canned milk, soups, meats, vegetables | 1 to 3 |
| Wine | 2 to 5 |
| Fresh ground coffee | 2 to 5 |
| Tomato products | 3 to 5 |
| Fruit juices | 8 to 20 |
| Carbonated soft drinks (CSD) | 10 to 40 |
| Peanut butter | 30 to 100 |

Data from [1]

The maximum tolerance levels of some important food and beverage materials fall in the ranges between 1 and 5 parts per million (ppm) levels [1]. Therefore, in order to package and preserve food and beverages for a long period of time, plastics or polymeric materials having high resistance to gas permeabilities, or barrier properties, are required. Products' shelf life is influenced by the barrier properties of the containers made from these polymers. Due to low density and the presence of free volume in the molecular structures of polymeric materials, they are relatively permeable for gases and liquids. Polymers exhibiting high resistance to molecular flow of permeating agents, such as gases or liquids, through the polymeric matrix are generally classified as high barrier polymers [2–4].

The barrier performance of polymers depends on several factors, including their crystallinity, glass transition temperature (Tg), moisture sensitivity, molecular chain packing and cohesive energy density [1]. Polymer polarity is very important in determining the barrier property of a particular polymer. For an example, highly polar polymers containing hydroxyl groups were observed to exhibit high barrier properties against gases but showed poor moisture barrier. Similarly, non-polar hydrocarbon based polymers showed excellent water barrier and poor gas barrier properties. In order to express the gas permeability property of polymers, a wide range of units have been employed [3]. The commercial permeability unit used in the measurement of $O_2$ and $CO_2$ permeation rates is cc.mil $(100\,in^2.\,atm.\,d)^{-1}$, which is also known as Barrier Unit (BU), has been used in this chapter. The conversion to the metric unit, which is cc.mm $(m^2.\,atm.\,d)^{-1}$, is accomplished by multiplying the standard value (BU) by 0.394 [5].

Currently, a wide range of polymers have been employed for packaging a variety of food and beverages. These polymeric materials have both advantages and limitations in the packaging applications, as can be seen in Table 7.57 [5].

**Table 7.57.** Selected polymers used in food and beverage packaging applications

| Polymer | Properties | Limitations |
|---|---|---|
| Poly(ethylene terephthalate) (PET) | Excellent clarity and gloss<br>Easy to fabricate<br>Good toughness<br>Good impact resistance | $P_{O_2} = 5-10\,BU$ |
| Ethylene vinylalcohol (EVOH) | Excellent $P_{O_2}$ $(0.01-2\,BU)$<br>Extensive hydrogen bonding crystallinity | $P_{O_2}$ increases at high humidity. |
| Poly(vinylidene chloride) (PVDC) | Excellent<br>$P_{O_2}$ $(0.02-0.03\,BU)$<br>Crystallinity<br>Good chain packing | Low Tg<br>Residual<br>Monomer |
| Polyamides (Nylon 6 and 66) | $P_{O_2} = 3-5\,BU$ | Humidity<br>Sensitive |

Data from [3,5]

Poly(ethylene terephthalate), PET, is a well known packaging material which was FDA approved for direct food contact applications and considered as a low cost material. The oxygen permeability rate for PET is about 5 to 10 BU, which is considered adequate for holding $CO_2$ in a 2-l bottle for carbonated soft drinks (CSD). However, PET cannot provide long shelf life for highly oxygen sensitive materials, such as beer, due to its oxygen barrier limitations, namely an oxygen permeation rate ($P_{O_2}$) of 5 to 10 BU. Ethylene vinyl alcohol polymer (EVOH), on the other hand, has an excellent oxygen permeability rate of 0.07 BU under dry conditions, but loses this property under the conditions of high humidity. Also, it is important to note that, with containers of 0.5 l or less in volume, the surface to volume ratio is relatively high. Therefore, in these small containers, many food products may have unacceptable shelf life. In addition to these problems, most of the packaging materials or polymers have some restrictions or limitations in specific applications.

The packaging industry has been constantly looking to replace more glass with polymeric materials in beer containers and other important food and beverage applications. Polymers or plastics have the obvious advantage of being lightweight and safe to use. However, when plastics are used, preserving the taste of beverage materials, beer in particular, is a problem. The taste and freshness of food and beverage products can be retained only if the $CO_2$ loss and $O_2$ ingress of the container can be avoided or minimized to the lowest level. This could be achieved with the use of packaging materials and containers, which could be produced, from the polymers having the following properties.

- Very low $P_{O_2}$ to reduce or block oxygen ingress in to the container
- Low $P_{CO_2}$ to retain $CO_2$ gas in the container
- Provide good adhesion to polymers used in multi-layer containers
- Good flexibility and high impact resistance
- High gloss and good clarity

## 7.8.2 Polymers Exhibiting High Barrier Properties

Polymers having high crystallinity can exhibit good gas and liquid barrier properties due to the difficulty of penetrant molecules to penetrate the tightly bound rigid chains. High glass transition (Tg) temperatures associated with chain stiffness can also enhance the polymer barrier performance. Similarly, highly polar polymers having large number of functional groups, which are capable of providing hydrogen bonding type interactions, are expected to possess strong chain to chain attractions, namely high cohesive energy density. Polymers undergoing this type of interactions can provide good barrier properties.

If the polymer chains are packed closely or tightly then the polymer free volume is expected to be low for the penetrant molecule to pass through, which may result in high barrier property.

Resorcinol and resorcinolic derivatives have been used in the development of high barrier polymers. These materials, in turn, have been either utilized or have the potential for applications in the manufacturing of high performance food and beverage packaging materials [6–8].

### 7.8.3 Current Packaging Technologies for Oxygen Sensitive Materials

There are currently three technologies available to develop plastic containers to package and preserve the oxygen sensitive food and beverages. They are, namely the monolayer, multi-layer and barrier coated container technologies. In the case of monolayer containers development, the polymeric materials should possess very good $O_2$ and $CO_2$ permeability properties in order to provide adequate shelf life for the food materials. In addition to gas barrier properties, the polymer must be FDA approved for the direct food contact and, also low cost. Currently, no single polymer exists to package beer that meets these criteria without additional modifications.

Multi-layer container technology, on the other hand, can use low cost PET as the main component along with a high barrier layer as a minor component. This technology has been a standard method to enhance oxygen barrier for more than 25 years. Multi-layer remains an effective means to achieve barrier for different gases. In the case of PET multi-layer technology, the barrier layer may act as physical barrier to gas permeation or chemically to scavenge oxygen from the PET and also can intercept oxygen diffusing in from the outside. With advances in this technology, the multi-layer is offering manufacturing flexibility and out-standing oxygen barrier performance. A three- to eight-layer PET bottle containing a thin layer of high barrier polymer of 2 wt. % of the container is possible with current co-injection technology.

With the proper selection of high barrier coating materials, the gas permeability of monolayer containers can be improved many fold. Since the coating has been applied at the outer surface, there is no direct contact of this coating material with the food or beverage present inside the container. In addition, by the proper treatments, the packaging materials can be recycled after the removal of the coated barrier layer. Barrier coating technologies could provide the longest carbonation shelf life (keep $CO_2$ inside) and high oxygen barrier properties.

### 7.8.4 Plastic Beer Bottles and Production Challenges

Six major breweries, namely Miller, Heineken, Bass, Feldschlossen, Carlton United and Karlsberg, have successfully bottled beer in plastic containers. The plastic beer bottle for Miller Lite was developed by Continental PET Technologies. This bottle contained five layers with a polyamide (Nylon MXD6) as the barrier layer less than 5 wt. % of the container. In Australia, Carlton United introduced the monolayer plastic beer bottle coated with PPG Industries' Bairocade coating. Since the main component of these plastic beer bottles was the low barrier PET, the expected shelf life was expected to be about three months. With the further improvement in $P_{O_2}$ and $P_{CO_2}$ properties of barrier materials, the PET bottles could deliver six to nine months or more shelf life for beer. Therefore, the packaging industry has been racing for the new and improved barrier polymers to maximize the shelf of life beer as well as other food and beverages. In order to achieve the target, new or existing polymeric materials should exhibit very high oxygen ($PO_2$ =< 0.2 BU) and carbon-dioxide ($PCO_2$ =< 1–2 BU) properties [6–8].

### 7.8.5
### Resorcinol Chemistry in High Barrier Polymers

Resorcinol chemistry has been explored greatly in developing polymers having high barrier properties to achieve the desired packaging materials for the highly oxygen sensitive products. Based on the published technical and patent information, three resorcinolic derivatives have been utilized in the barrier materials development. These derivatives are resorcinol dioxyacetic acid (RDOA), bis (hydroxyethyl) ether of resorcinol (HER) and resorcinol diglycidyl ether (RDGE) (Figures 7.106, 7.107 and 7.108).

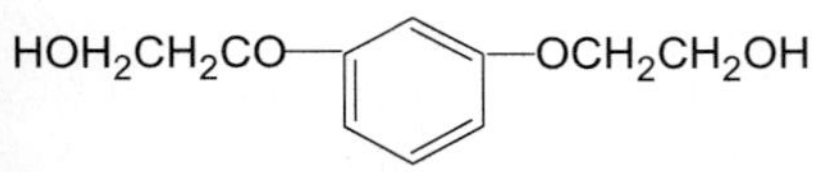


Bis(Hydroxyethyl) Ether of Resorcinol (HER)

**Figure 7.106.**

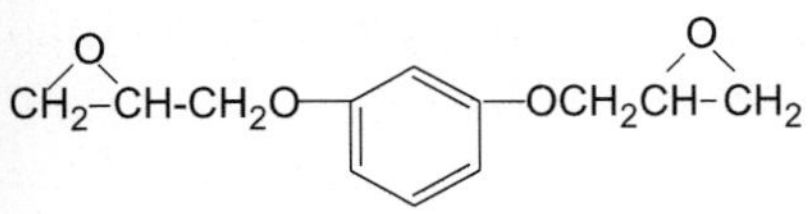


Resorcinol Diglycidyl Ether (RDGE)

**Figure 7.107.**

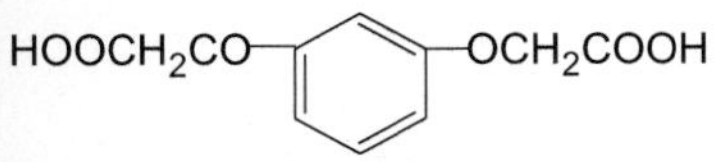


Resorcinol Dioxyacetic Acid (RDOA)

**Figure 7.108.**

In general, the use of resorcinol and its derivatives in polymer synthesis is expected to bring enhanced properties, such as the processing, to the resulting polymers. Based on the chemical structures of RDOA, HER and RDGE, it can be seen that these derivatives are 1,3-di-substituted benzene compounds. Therefore, these derivatives could offer enhanced flexibility to the polymers made from these molecules. Unlike the 1,4-di-substituted benzene derivatives, the polymers made with the 1,3-di-substituted compounds can be expected to be amorphous. The packaging materials, such as the containers or bottles, which could be made with these amorphous polymers, might be highly transparent. These resorcinolic derivatives can provide polymers having compact back bone structures and chain to chain close packing.

The presence of oxy-methylene groups in all three resorcinolic derivatives may provide increased flexibility to the polymer backbone structures. Various polymers synthesized from these compounds are expected to possess improved processability. The functional groups present in these molecules may also provide oxidative and hydrolytic stabilities to the resulting polymers.

With the presence of functional groups such as the carboxylic acid, hydroxyl and oxirane (epoxy), these derivatives can offer a wide range of possibilities to employ them in the synthesis of various polymeric materials with unique structures and properties for the packaging applications. Barrier polymers have been developed with HER, RDOA and RDGE monomers which showed much improved oxygen, carbon-dioxide and moisture barrier properties.

## 7.8.6 HER Based Barrier Polymers

Hydroxyethyl ether of resorcinol (HER), also named as 2,2′-(*m*-phenylenedioxy) diethanol or 1,3-bis(2-hydroxyethoxy) benzene, is a white crystalline solid material with a melting point of 92 °C and can be synthesized from the reaction of resorcinol with ethylene oxide or ethylene carbonate or ethylene chlorohydrin. A synthetic scheme showing the preparation of HER from resorcinol and ethylene carbonate is illustrated below [9] (Figure 7.109).

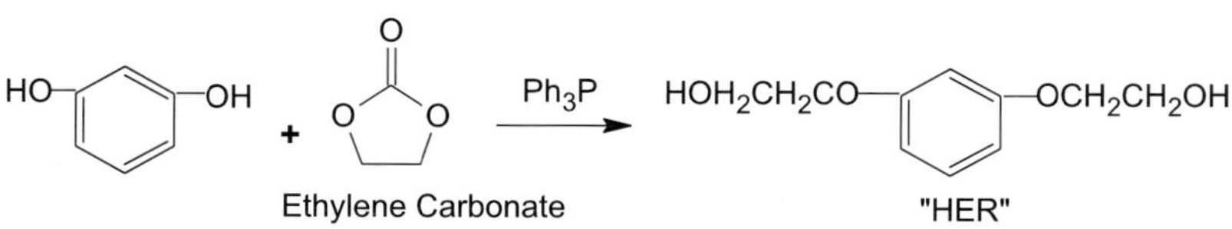


**Figure 7.109.**

The molecular structure of HER shows the presence of two aliphatic type hydroxyl groups. Therefore, HER can be used as a monomer in the synthesis of polyesters. When this material is used as the dihydroxy compound, homopolymers could be produced from the dicarboxylic acid reactions (Figure 7.110).

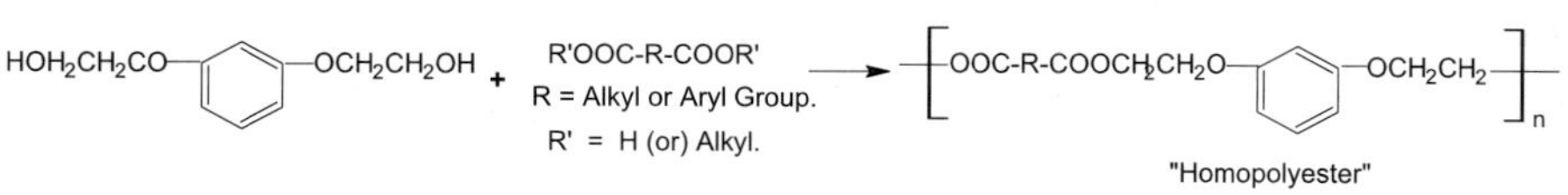


**Figure 7.110.**

The dicarboxylic acids which may be employed in the polyesters synthesis can be either aliphatic or aromatic compounds. In general, the polyesters used in the packaging materials production are predominantly phthalate and naphthalate ester types and are obtained primarily based on the use of the following dicarboxylic acids (Figures 7.111, 7.112 and 7.113).

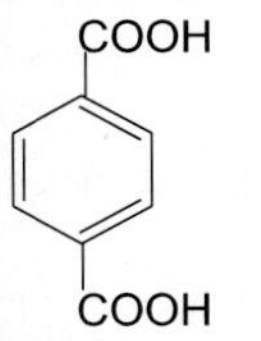


Terephthalic Acid (TA)

**Figure 7.111.**

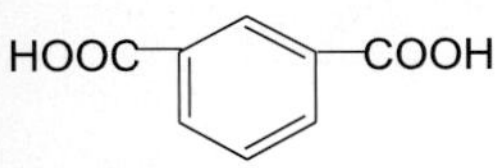


Isophthalic Acid (IA)

**Figure 7.112.** isophthalic acid (IA)

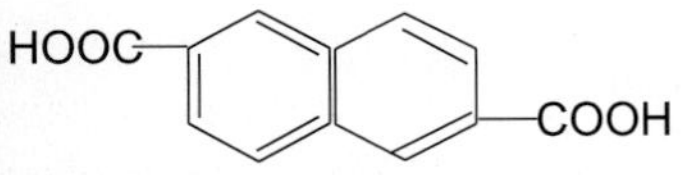


2, 6 - Naphthalene Dicarboxylic Acid (NDC)

**Figure 7.113.**

In the polyester synthesis, these dicarboxylic acids can be used in the form of either carboxylic acids or their corresponding esters.

Poly(ethylene terephthalate), commercially known as PET, has been synthesized from the reaction of ethylene glycol with dimethyl terephthalate (DMT) (Figure 7.114).

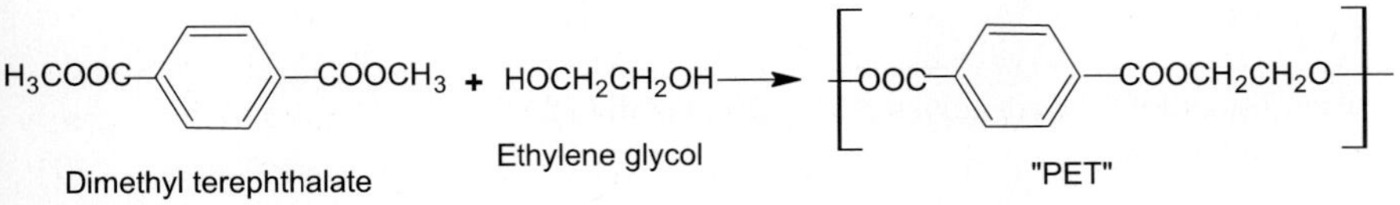


**Figure 7.114.**

By employing isophthalic acid (IA) in the place of terephthalic acid, poly(isophthalate) (PEI) material can be produced. PET is a highly crystallizable material, which is highly desirable for many applications. PEI, on the other hand, is crystallizable with difficulty and has a lower Tg than PET. Both the PET and PEI are satisfactory for many packaging applications, but PEI showed better gas permeability than PET.

In order to improve the barrier performance of PET, HER was used in combination with ethylene glycol (EG) to develop various co-polyester materials. Owens-Illinois Company first developed the co-polyester technology using HER as the co-dihydroxy compound [10–12] (Figure 7.115).

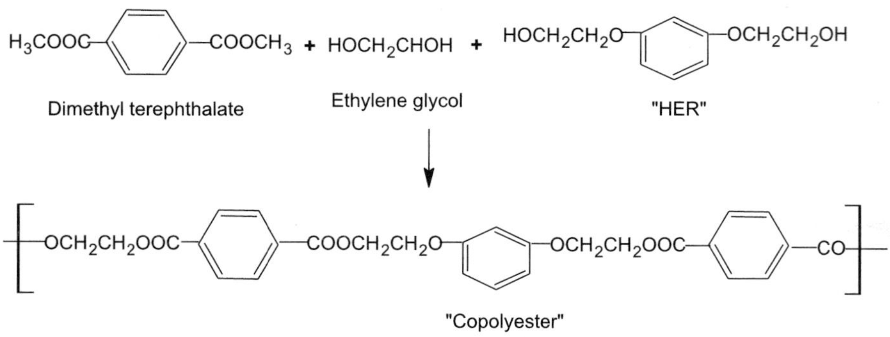


**Figure 7.115.**

Replacing part of EG with HER in the co-polyester synthesis was observed to lower the $PO_2$ and $PCO_2$ values (Table 7.58).

**Table 7.58.** Copolyesters from HER, TA and EG Monomers

| Polyester: | 1 | 2 | 3 |
|---|---|---|---|
| Composition (mol) | | | |
| TA or DMT | 1 | 1 | 1 |
| EG | 2.38 | 2.25 | 1.8 |
| HER | None | 0.25 | 0.7 |
| Properties | | | |
| Glass transition Temp (Tg, °C) | 72 | 72 | 64 |
| Gas permeability (BU) | | | |
| $PO_2$ | 8.5 | 5.9 | 3.8 |
| $PCO_2$ | 55.3 | 35.8 | 22.4 |

Mols of EG and HER were adjusted for 1.0 mol of diacid (TA)
Data from [10–12]

With the use of small amounts of HER in the polyester synthesis, considerable improvements in $PO_2$ and $PCO_2$ were achieved without much change in the Tg of final polyester (compare polyester 1 and 2 in Table 7.58).

Similarly, when part of EG was replaced with HER (0.05 mol) in the copolyester synthesis employing TA and IA monomers, a dramatic decrease in

**Table 7.59.** Copolyesters from HER, TA, IA and EG monomers

| Polyester: Composition (mol) | 1 (PEI) | 2 | 3 | 4 | 5 | 6 (PET) |
|---|---|---|---|---|---|---|
| IA | 1 | 1 | 1 | 0.7 | 0.9 | |
| TA or DMT | | | | 0.3 | 0.1 | 1 |
| EG | 1 | 1.55 | 1.45 | 1.45 | 1.6 | 1 |
| HER | | 0.05 | 0.15 | 0.15 | 0.15 | |
| Properties | | | | | | |
| Glass transition temp (Tg, °C) | 61 | 61 | 60 | 63 | 64 | 72 |
| Gas permeability (BU) | | | | | | |
| $P_{O_2}$ | 3.7 | 1.6 | 1.6 | 2 | 1.6 | 8.5 |
| $P_{CO_2}$ | | 7.1 | 6.6 | 9.7 | 6 | 55.3 |

Barrier property enhancement of PET with copolyester of composition

| Blend ratio (weight) | Improvement (%) | |
|---|---|---|
| | $P_{O_2}$ (BU) | $P_{CO_2}$ (BU) |
| PET/copolyester-5 = 90 : 10 | 17 | 22 |
| PET/copolyester-5 = 80 : 20 | 28 | 41 |

Data from [10–14]

$P_{O_2}$ (from 3.7 to 1.6BU) value of PEI polyester was observed (Table 7.59) [10–14].

By combining TA, IA, EG and HER monomers, a co-polyester exhibiting Tg = 64 °C, $P_{O_2}$ = 1.6BU and $P_{CO_2}$ = 6.0BU was developed (Table 7.59) (Figure 7.116).

This co-polyester material was then blended with PET. The blends containing different levels of co-polyester appeared to enhance the barrier properties of PET significantly.

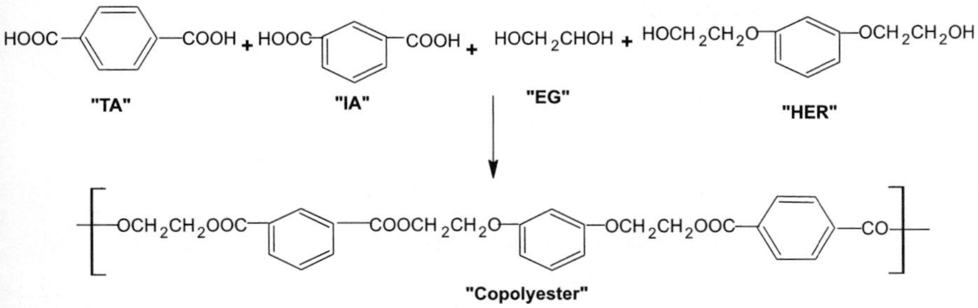


**Figure 7.116.**

The co-polyester technology developed at Owens-Illinois was later licensed to Mitsui Chemical Corporation. A commercial gas barrier polyester resin sold under the trade name Mitsui BO10 may be associated with the 10/90 blend of HER based co-polyester and PET as shown in Table 7.59. The blend containing up to 30 weight percent of this co-polyester is FDA approved for direct food contact, suggesting HER based polyester may be non-toxic [6].

The presence of *meta*-phenylene ring structures in the polyester materials not only improves the barrier performance but also affect the Tg of the resulting polymers. Polyesters having higher Tg, in general, will have very stiff chains and low gas permeability properties [1]. Copolyesters produced from IA and HER monomers have lower Tg than from the TA and EG monomers, due to chain flexibility. In order to increase the Tg of co-polyesters, naphthalene dicarboxylic acid (NDC) monomer is often used. Co-polyesters having higher Tg, excellent gas barrier and surface properties were synthesized using NDC, IA, EG and HER monomers [15] (Figure 7.117).

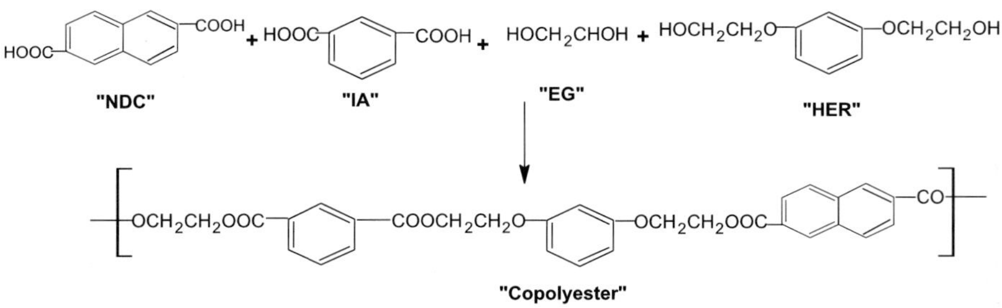


**Figure 7.117.**

Higher Tg and improved $P_{CO_2}$ properties were achieved with different molar ratios of the co-monomers used in the synthesis (Table 7.60).

These co-polyesters exhibited excellent moldability, stretchability, improved $P_{O_2}$ and $P_{CO_2}$ properties, thermal resistance, surface properties, impact strength, good transparency and chemical resistance.

**Table 7.60.** Copolyesters from HER and barrier performance

| Copolyester composition | Monomer ratio mol(%) | Polyester properties Tg (°C) | $P_{CO_2}$ (BU) |
|---|---|---|---|
| IA/NDC/HER, EG | 90/10/15/85 | 73 | 3 |
| IA/NDC/HER, EG | 80/10/15/85 | 78 | 3.1 |
| IA/NDC/HER, EG | 70/30/15/85 | 84 | 3.3 |
| IA/EG (Polyester – PEI) | 100/100 | 67 | 3 |
| NDC/EG (Polyester – PEN) | 100/100 | 124 | 7.2 |

Data from [15]

Due to high processing temperatures involved in the manufacture of PET bottles, a significant quantity of acetaldehyde by-product is formed during the extrusion and molding operations. The presence of acetaldehyde, even at relatively very low concentrations, imparts an undesirable or bad after taste to many food stuffs packaged in PET. Therefore, the presence or use of HER in the co-polyesters may lower the acetaldehyde generation during processing. These copolyesters showed excellent transparency, moldability, enhanced gas barrier property and finally good economical efficiency due to enhanced processability.

### 7.8.7 High Barrier Polymers from RDGE

Resorcinol diglycidyl ether (RDGE) is a colorless liquid with very low viscosity (200–500 centipoise) at room temperature and highly reactive towards various functional groups. This compound has been extensively used in the development of high performance composites, but its use in the high barrier materials development for the packaging industry is a novel finding.

#### 7.8.7.1 Thermoplastic Barrier Polymers

Epoxy resins often contain more than two epoxy groups per molecule, and therefore, expected to produce highly cross-linked network type structures when reacted with difunctional curing agents or reactants. Thermoplastic polymers, namely poly(hydroxyl ethers), prepared from the reaction of diepoxides with dihydroxy compounds were realized as novel class of polymers [16, 17]. The films produced from the poly(hydroxyl ether) materials were found to be completely amorphous. Subsequent to these developments, further work on the poly (hydroxyl ether) chemistry produced materials with a combination of high barrier, good mechanical properties, and clarity and fabrication flexibility [18].

Poly(hydroxyl ether) compounds were synthesized from the stoichiometric reaction of diglycidyl ethers (epoxies), including RDGE, with various aromatic dihydroxy compounds (Figure 7.118).

The results are presented in Table 7.61.

The polymers produced from RDGE showed excellent barrier performance and were highly processable thermoplastic materials. When resorcinol was employed as the dihydroxy compound, the polymer exhibited very low $PO_2$ value. This suggests that high barrier properties could be achieved in polymers containing more *meta*-phenylene structures in the polymer back bone. The presence of *m*-phenylene, compared to *p*-phenylene structure, group can allow the polymer back bone to adopt more favorable chain conformations for the enhanced chain to chain packing [5, 17, 20]

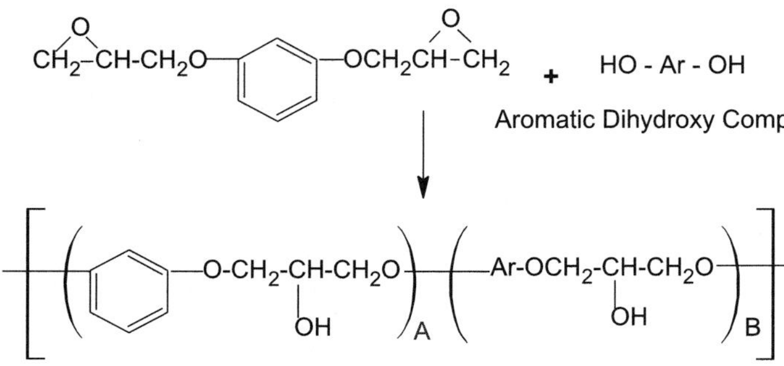


**Figure 7.118.**

**Table 7.61.** Synthesis and barrier performance of poly(hydroxy ethers) from RDGE

| Aromatic dihydroxy compound used | Composition of polymer (A/B) | Polymer properties Tg (°C) | $P_{O_2}$ (BU)[a] |
|---|---|---|---|
| Resorcinol | 0.5 : 0.5 | 65 | 0.22 |
| Hydroquinone | 0.5 : 0.5 | 64 | 0.26 |
| 4,4′-Biphenol | 0.5 : 0.5 | 95 | 0.15 |
| Bisphenol-A | 0.5 : 0.5 | 83 | 1.5 |
| | 0.25 : 0.75 | 89 | 3.8 |
| | 0.1 : 0.9 | 92 | 5.4 |
| | 00 : 01.0 | 102 | 9.4 |

[a]Permeability values measured at 70 – 90% RH of Oxygen Atmosphere
Data taken from [18]

The Tg values of poly(hydroxyl ether) obtained from the RDGE monomer could be increased by using 4,4′-diphenol in the polymer synthesis. Bisphenol-A compound can also be used to enhance the Tg of the polymers, but the introduction of bisphenol-A structure in to the polymer backbone apparently appeared to affect the barrier performance. This might be due to the increase in free volume of the polymer associated with the irregular structure of bisphenol-A in the polymer main chain. Polymers with poor symmetry are expected to pack not so closely and therefore may have high free volume and high gas permeability.

Mono-ethanolamine, when reacted with RDGE, produced a high molecular weight amorphous thermoplastic poly(amino ether) material with good mechanical properties, clarity, fabrication flexibility and very low $P_{O_2}$ (0.04 BU) [21–23] (Figure 7.119).

The gas barrier performance (shown in Table 7.62) observed for the RDGE based poly(amino ether) material was the lowest known ($P_{O_2}$) for any amorphous polymer to date.

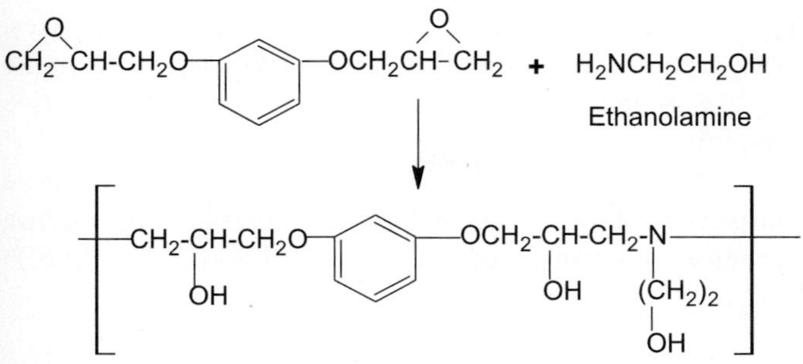

**Figure 7.119.**

**Table 7.62.** Thermoplastic poly(amino ethers) from epoxy and ethanolamine monomers and their barrier performance

| Diglycidyl compound used | Poly(amino ether) Properties Tg (°C) | $P_{O_2}$ (BU)[a] |
|---|---|---|
| Resorcinol | 56 | 0.04 |
| Hydroquinone | 55 | 0.3 |
| Bisphenol-F | 69 | 0.5 |
| Bisphenol-A | 81 | 0.7 |

Tg Determined by the DSC Method
[a]Oxygen Permeability rate measured at 65 – 70% RH conditions
Data from [21–23]

The reaction of RDGE with ethanolamine produced a polymer containing an abundance of functional groups capable of forming strong hydrogen bonds. Groups such as the polar hydroxyl and tertiary amines can participate in hydrogen bonding type interactions, which can lead to strong inter chain cohesion. When the hydrogen bonding interactions are increased between the polymer molecules, the gas permeability, particularly oxygen, is decreased. With ethanolamine as the co-reactant, the polymer chain contained fewer number of methylene groups in the polymer repeat unit. This might have resulted in increase in the density of hydrogen bonding, which lead to stronger inter chain cohesion. In addition, the presence of *m*-phenylene groups in the polymer main chain might have contributed to the close packing of polymer molecules. Therefore, the very low $P_{O_2}$ value observed for the resorcinol based thermoplastic poly(amino ether) was associated with the combination of polymer chain close packing and extensive hydrogen bonds formed between polymer molecules.

Dow Plastics has developed high barrier thermoplastic resins based on the poly(amino ether) chemistry and introduced high barrier resins for the packaging applications under the trade name BLOX [24,25]. According to Dow Plastics products information, BLOX was considered as the high adhesion barrier resins offering outstanding adhesion, excellent gas barrier properties, clarity and mechanical strength and toughness. These properties can make BLOX resins ideal for use in the rigid packaging of $O_2$ and $CO_2$ sensitive beverages, such as beer, juices and carbonated soft drinks.

By employing L-tartaric acid as the reactant, another thermoplastic type polymer containing large number of functional groups capable of producing extensive hydrogen bonds were synthesized and reported [26] (Figure 7.120).

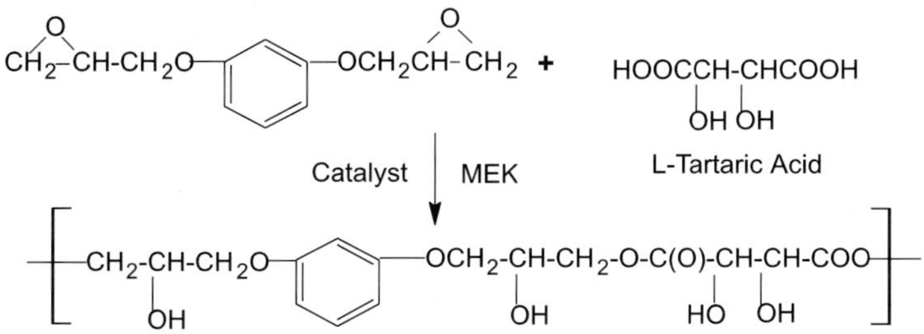


**Figure 7.120.**

## 7.8.7.2 Thermoset Barrier Polymers

Coating of PET and poly(propylene) (PP) bottles to obtain high barrier performance is another way of improving the packaging efficiency. Polymers or packaging materials having very low permeabilities could be easily achieved with RDGE based coating formulations. When RDGE was reacted with polyamine compounds, such as Gaskamine 328 from Mitsubishi Chemical, the cured film showed much improved $P_{O_2}$ properties even at high relative humidity (RH) conditions (Table 7.63) [27] (Figure 7.121).

**Table 7.63.** Thermost polymers synthesized from RDGE and Gaskamine 328 monomers

| Epoxy resin used | NH/epoxy (mol) | $P_{O_2}$ on cured samples at RH (%) conditions 0 | 20–25 | 50–55 | 70–75 |
|---|---|---|---|---|---|
| Resorcinol | 1 | 0.23 | 0.13 | 0.17 | 0.18 |
| Bisphenol-A | 1 | 3.33 | 1.86 | 2.52 | 2.54 |

Coating compositions were prepared and coated on a 2-mil. PET or PP film to produce a cured coating of about 0.5 to 0.6 mil thickness
Gaskamine 328 is a trademark of Mitsubishi Chemical
Data from [27]

Similarly, *m*-xylylene diamine (MXDA) curing agent provided RDGE based coating formulations, which when coated on a 2-mil PET or PP film produced a cured coating of about 0.5 to 0.6 mil thickness (Figure 7.122).

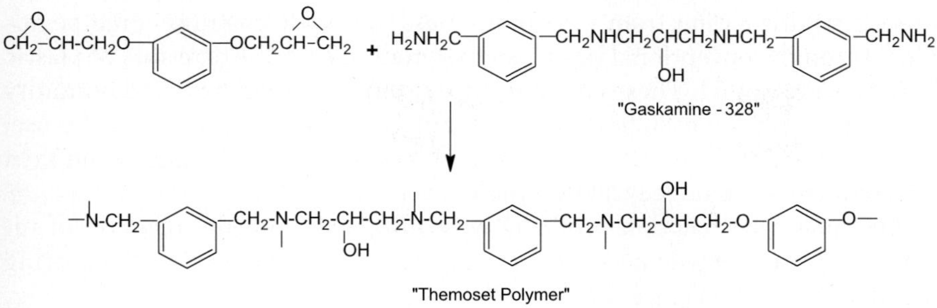


**Figure 7.121.**

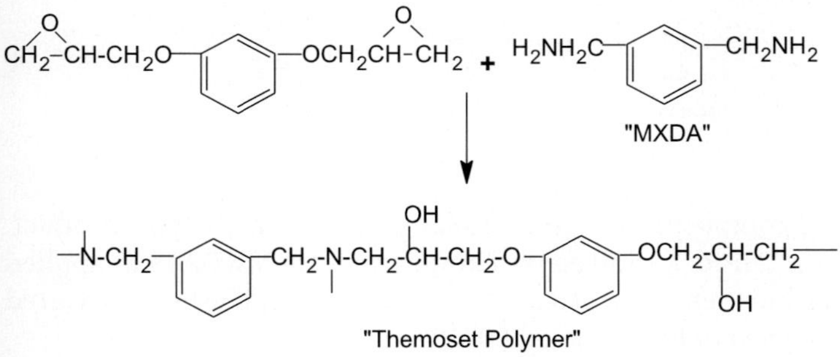


**Figure 7.122.**

The oxygen permeability values of RDGE cured samples were compared against butanediol epoxy cured materials (Table 7.64).

**Table 7.64.** Thermost polymers synthesized from RDGE and MXDA monomers

| Epoxy resin used | NH/epoxy (mol) | $P_{O_2}$ on cured samples at RH (%) Conditions 0 | 20–25 | 50–55 | 70–75 |
|---|---|---|---|---|---|
| Resorcinol | 1.3 | 0.17 | 0.09 | 0.1 | 0.11 |
| 1,4-Butanediol | 1.3 | 10.64 | 17.49 | 16.75 | 24.33 |

Coating compositions were prepared and coated on a 2-mil. PET or PP film to produce a cured coating of about 0.5 to 0.6 mil thickness
Data from [27]

RDGE based coatings maintained their low $PO_2$ values even under high RH conditions compared to bisphenol-A and butanediol epoxies coatings. At least 10- to 15-fold improvement in performance was achieved with RDGE based coatings.

PPG Industries has commercialized Bairocade coatings based on the poly(amino ether) type thermoset polymers obtained from the formulations containing poly(epoxide), such as RDGE, and polyamine compounds. Based on the products information from PPG Industries, Bairocade coatings could be applied through conventional liquid coating methods for the exteriors of plastic bottles, such as PET. These coatings impart physical toughness and humidity resistance with a pleasing clear or colored glossy finish. Also, after the use, these coatings can be easily removed with standard stripping agents and then the containers can be recycled very effectively.

Mannich type reaction products were obtained from the reaction of resorcinol, formaldehyde and MXDA compounds and also used in the coating formulations [29] (Figure 7.123).

**Figure 7.123.**

Barrier coating composition prepared using the Mannich type product, Gaskamine 328 and Tetrad-X (polyepoxy compound from MXDA) was applied to a 2-mil PET film and then cured at 82.5 °C to obtain a film. The $P_{O_2}$ measured for this film showed a very low value of 0.07 BU.

Based on these published information, RDGE based coatings showed outstanding oxygen barrier performances. Therefore, with high barrier coating technology, good opportunities exist for the monolayer PET containers in the packaging applications to enhance the performance and shelf life of beer, wine and oxygen sensitive materials.

## 7.8.8 High Barrier Polymers Based on RDOA

Resorcinol dioxyacetic acid (RDOA) or 1,3-phenylene dioxydiacetic acid is a white crystalline solid material with a melting point of 190 – 192 °C. This compound was synthesized from the reaction of resorcinol with chloro acetic acid (Figure 7.124).

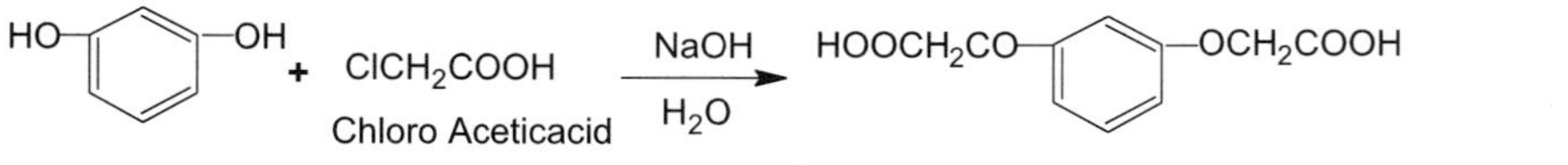


**Figure 7.124.**

This is a dicarboxylic acid compound and therefore can be used alone or in combination with other aliphatic or aromatic dicarboxylic acids to develop polymers of varying structures, physical and mechanical properties for the packaging applications. A large number of patents have been issued on the use of RDOA in the development of various high barrier polymers.

### 7.8.8.1
### Polyesters from RDOA and Aromatic Dicarboxylic Acids

Homopolymers made from the reaction of EG with TA, IA and NDC monomers showed excellent properties such as mechanical strength, chemical strength, transparency and hygienic qualities. But their permeability properties were poor to use them for the production of monolayer container for some important applications. In order to enhance the barrier properties of PET, PEI and PEN polyesters, RDOA was used as a co-monomer in the synthesis of co-polyester materials [20, 30–34] (Figure 7.125).

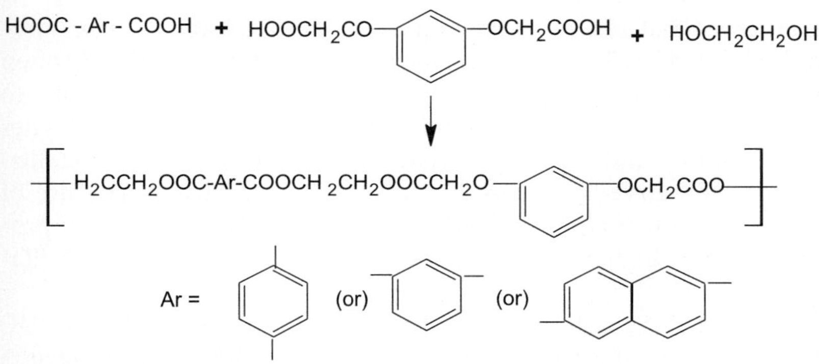


**Figure 7.125.**

The co-polyester permeability properties are presented in Table 7.65.

**Table 7.65.** Barrier performance of copolyesters made from RDOA mononer dicarboxylic acid used as co-monomers in the copolyester

| Terephthalic acid[a] | | | Isophthalic acid[b] | | 2,6-Naphthalene dicarboxylic acid[c] | | |
|---|---|---|---|---|---|---|---|
| mol% | $P_{O_2}$ | $P_{CO_2}$ | mol% | $P_{O_2}$ | mol% | $P_{O_2}$ | $P_{CO_2}$ |
| 100 | 10 | 59 | 100 | 2.5 | 100 | 3.1 | NA |
| 95 | 7.5 | 49 | 90 | 1.55 | 95 | 2.5 | 13 |
| 90 | 5 | 26 | 80 | 0.89 | 90 | 2.2 | 11 |
| 80 | 3.1 | 18 | 70 | 0.56 | 80 | 1.6 | 8 |
| 50 | 0.6 | NA | 60 | 0.41 | 50 | 0.5 | NA |
| 0 | 0.1 | 0.6 | 40 | 0.3 | | | |

[a] Data from [30, 31]
[b] Data from [33, 34]
[c] Data from [20, 32]
Oxygen ($P_{O_2}$) and carbon dioxide ($P_{CO_2}$) permeability values are in barrier units (BU)
Combined mol% of RDOA and dicarboxylic acids in the copolyester is 100

In general, unoriented PET exhibits $PO_2$ value of 8–10 BU. The addition of RDOA as a co-monomer in the polyester synthesis greatly improved the barrier performance of PET. Compared to RDOA homo-polymer $PO_2$ (0.1 BU), the PET showed higher permeability ($PO_2$ = 10 BU) value. This might be due to the presence of *m*-phenylene groups associated with the resorcinolic structure in the polymer chains. Significant improvement in the $PO_2$ value of PET could be achieved by the partial substitution of RDOA monomer for TA in the polyester synthesis. With $PO_2$ approaching 0.6 BU for the co-polyester containing RDOA/TA at 50 : 50 mol% level, this polyester material could be a candidate for the fabrication of mono-layer containers or used as a high barrier layer in a multi-layer container construction.

It was observed that a homo-polymer made from RDOA and EG was amorphous and exhibited low Tg. Containers having low levels of crystallinity often resulted in poor mechanical properties. Therefore, it was advantageous to use RDOA as a co-monomer in the synthesis of co-polyesters having higher levels of crystallinity and superior mechanical properties. Highly crystallizable PET and PEN have Tg values 75 and 124 °C respectively. The use of RDOA as a co-monomer along with NDC in the polyester synthesis can provide materials with higher Tg values and excellent barrier properties (Table 7.65).

Based on Table 7.65 results, co-polyesters made from RDOA, IA and NDC monomers showed good gas barrier properties. This result strongly suggests that considerable improvements in the polyester's barrier performances can be achieved with the use of RDOA. From the published literature, it was clear that the barrier performance of polyester materials were directly related to the amount of RDOA content in the polymer. Also, the co-polyesters made from RDOA showed excellent transparency, gas permeability and good mechanical properties. These polyesters can be used as packaging materials in the form of containers and films.

Mitsubishi Chemical has also developed co-polyester resins using 96 mol% TA and 4 mol% RDOA as the dicarboxylic acid monomers [35,36]. The barrier performance of the bi-axially stretched film showed $PO_2$ = 1.78 BU which, after retort at 120 °C/100% RH conditions exhibited $PO_2$ = 2.77 BU. Comparing this data to a bi-axially stretched PET ($PO_2$ = 4.01 BU), it was determined that even with very low levels of RDOA, the barrier performance of PET could be considerably enhanced. The polyester produced, using a known of amount of RDOA in a specified range, namely 3–4 mol%, was unexpectedly superior in all physical and mechanical properties. This material could be used to produce a molded article for the packaging applications.

### 7.8.8.2 Container Performance with RDOA Based Polyesters

Mitsubishi Chemical has developed and showed the performance of multi-layer hollow containers based on the RDOA containing co-polyester materials [34, 37] (Figure 7.126).

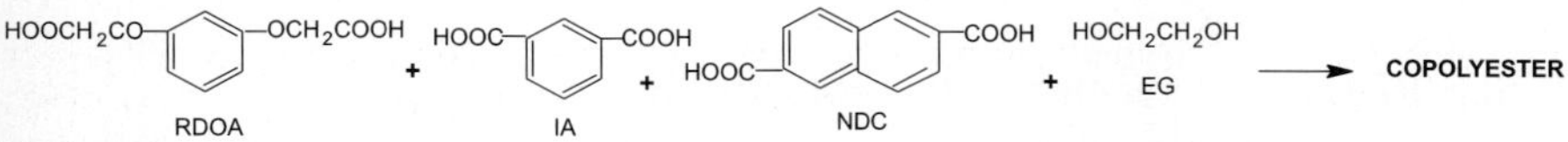


**Figure 7.126.**

Co-polyesters containing RDOA, IA and NDC monomers with different molar ratios were developed first. Then three layered hollow PET containers were made employing the RDOA based co-polyesters as the middle barrier layer at 20 wt. % level. The results are presented in Table 7.66.

**Table 7.66.** Containers made from RDOA copolyesters and their barrier properties

| Copolyester | Copolyester composition Dicarboxylic acid components (mol%) | | | | Copolyester properties | |
|---|---|---|---|---|---|---|
| | IA | RDOA | NDC | TA | Tg (°C) | $P_{O_2}$ (BU) |
| 1 | 20 | 35 | 45 | 0 | 60 | 0.76 |
| 2 | 10 | 50 | 40 | 0 | 53 | 0.36 |
| 3 | 20 | 50 | 30 | 0 | 51 | 0.3 |
| 4 (PEI) | 100 | 0 | 0 | 0 | 60 | 1.98 |
| 5 (PET) | 0 | 0 | 0 | 100 | 72 | NA |

Three layered hollow containers developed
Layer configuration: PET (inner)/copolyester (middle)/PET (outer)
Volume: 500 cc; Wall thickness: 300 μm

| Copolyester used | Copolyester in container (wt.%) | $P_{O_2}$ (cc/bottle.d.atm) |
|---|---|---|
| Copolyester 1 | 20 | 0.08 |
| Copolyester 2 | 20 | 0.05 |
| Copolyester 3 | 20 | 0.04 |
| Polyester 4 (PEI) | 20 | 0.18 |
| Polyester 4 (PET) | 0 | 0.25 |

Data from [37]

Analysis of Table 7.66 results clearly indicates that the $P_{O_2}$ values of the co-polyester materials were greatly improved by the RDOA content. PET multi-layered hollow containers containing the RDOA based co-polyesters as the barrier layer exhibited a better gas barrier property than a single layer PET container or multi-layered container of PET and PEI. The RDOA co-polyester based containers showed approximately six times better performance. The incorporation of RDOA co-polyester in to PET did not affect the PET's mechanical and heat resistance properties.

### 7.8.8.3
### Polyesters from RDOA and Aromatic Diols

When RDOA was reacted with aromatic diol monomers, such as 1,3-bis-(hydroxyl ethoxy) benzene (HER) and 1,4-bis(hydroxyl ethoxy) benzene (HQEE), polyester materials with good $P_{O_2}$ and $P_{CO_2}$ properties were produced (Table 7.67 [38] (Figure 7.127).

**Table 7.67.** Homopolyesters synthesized from RDOA and aromatic diols

| Diol component | Polyester properties | | | | |
|---|---|---|---|---|---|
| | Viscosity | Tensile property | | Gas permeability (BU) | |
| | | Strength | Strain | | |
| | (η inh) | (kg/cm$^2$) | (%) | $P_{O_2}$ | $P_{CO_2}$ |
| HQEE | 0.5–0.58 | 570 | 6 | 0.45 | 4.7 |
| HER | 0.87 | 422 | 4 | 1 | 3.55 |

Data from [38]

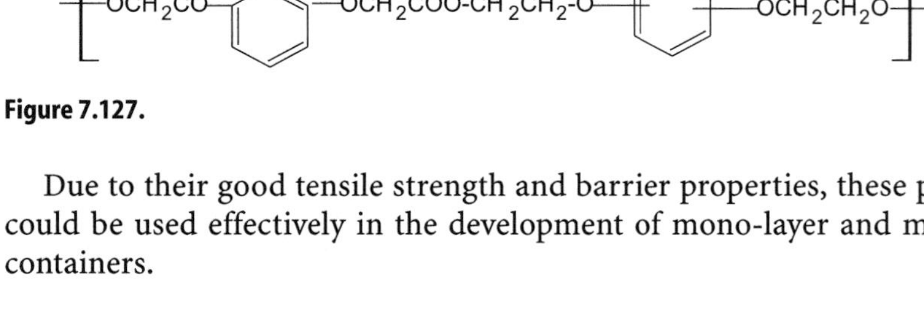


**Figure 7.127.**

Due to their good tensile strength and barrier properties, these polyesters could be used effectively in the development of mono-layer and multi-layer containers.

### 7.8.8.4
### Polyester from Dual Functional Monomer

Homo-polymer was synthesized from *m*-(2-hydroxy ethoxy) phenoxyacetic acid monomer [39] (Figure 7.128).

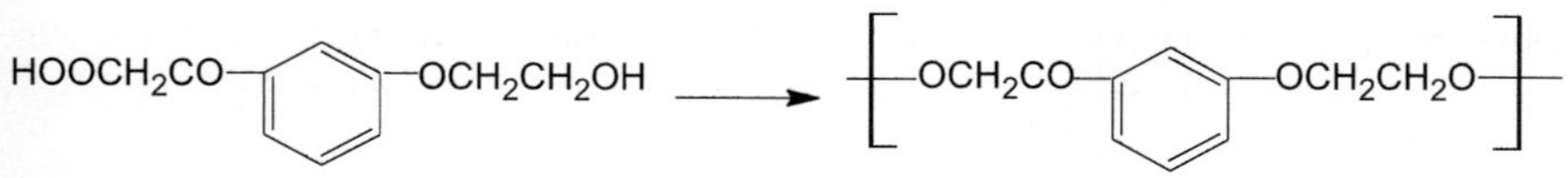


**Figure 7.128.**

This monomer did not require any additional diacid or dihydroxy compound to produce the polyester. The polyester prepared from the above synthesis, exhibited a Tg = 48 °C, Tm = 133 °C, $PO_2 = 0.5$ BU and $PCO_2 = 1.05$ BU properties.

### 7.8.8.5 High Barrier Polyamides from RDOA

Polyamides, in general, have been synthesized from the reactions of aliphatic or aromatic dicarboxylic acids with diamine compounds. Nylon-6, also a polyamide, has been synthesized from the ring opening polymerization of caprolactam monomer. Polyamides, such as nylon 6, nylon 66 and MXD6 (made from MXDA and adipic acid), can provide good to moderate barrier performance against oxygen and carbon-dioxide. However, these polymers are slightly moisture sensitive. In the case of MXD6, the oxygen permeability increases as the humidity is increased. To develop polymers with better properties, RDOA has been utilized in the development of various polyamide synthesis and their barrier properties were investigated.

Homo-polyamides were synthesized using RDOA and different diamine compounds [40] (Figure 7.129).

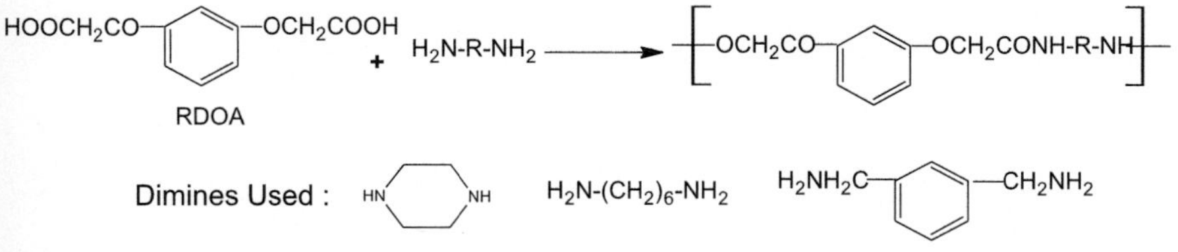


**Figure 7.129.**

The polyamide properties are outlined in Table 7.68.

Similarly, co-polyamides were prepared employing NDC as the co-monomer [41] (Figure 7.130).

The composition of co-polyamides and their properties are shown in Table 7.69.

From the results of Tables 7.68 and 7.69, it has been seen that $PO_2$ values are low enough to consider these polyamides as high barrier polymers. Moreover, the tensile strength properties of polyamides strongly suggest that they could provide opportunities to use them in high barrier packaging materials. The

**Table 7.68.** Synthesis and barrier properties of polyamides from RDOA

| Diamine component | Polyamide properties<br>Tensile properties<br>Strength (MPa) | <br><br>Strain (%) | <br>Barrier properties<br>$P_{O_2}$ (BU) | <br><br>$P_{CO_2}$ (BU) |
|---|---|---|---|---|
| Piperazine | 76 | 4 | 0.2 | NA |
| *m*-Xylylene diamine | 69 | 5 | 0.25 | 1.7 |
| 1,6-Hexamethylene diamine | 80 | 11 | 0.67 | NA |

Data from [40]

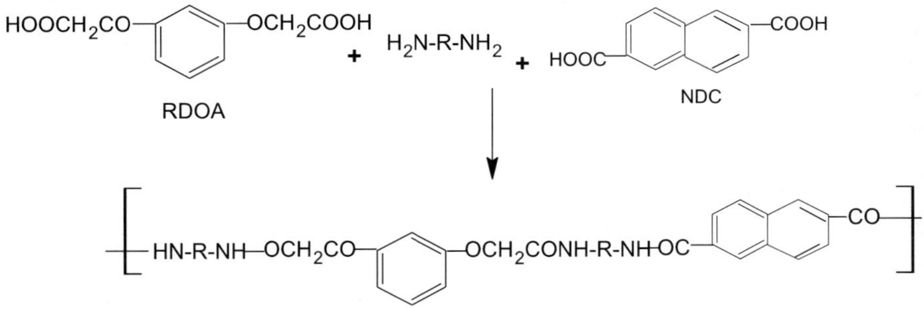


**Figure 7.130.**

**Table 7.69.** Copolyamides from RDOA and NDC monomers

| Dicarboxylic acid (mol%)<br>RDOA | <br>NDC | Diamine | Gas permeability<br>$P_{O_2}$ (BU) | <br>$P_{CO_2}$ (BU) |
|---|---|---|---|---|
| 75 | 25 | 1,6-Hexane diamine | 0.8 | 6.1 |
| 50 | 50 | 1,7-Heptane diamine | 1.7 | 12.8 |
| 50 | 50 | 1,9-Nonane diamine | 5 | 26 |
| 90 | 10 | *m*-Xylylene diamine | 0.2 | 1.2 |
| 90 | 10 | *p*-Xylylene diamine | 0.3 | 1.8 |

Data from [41]

very low $P_{O_2}$ values indicate that mono-layer containers could be made and used for preserving the taste and flavors of oxygen sensitive materials for an extended period of time without further modifications.

Compared to MXD6 ($P_{O_2} = 0.6\,BU$), the polyamides obtained from RDOA and MXDA showed a $P_{O_2}$ value of 0.25 BU, which suggests that multi-layer containers containing RDOA based polyamide could extend the shelf life of beer longer than the MXD6. Due to excellent barrier performance, RDOA based polyamides could play a major role in the development of high barrier packaging materials.

### 7.8.8.6
### High Barrier RDOA Based Polyamide – Clay Nanocomposites

Polyamide based on MXD6 has been utilized as the middle layer in the construction of a tri-layer stretch blow molded container for the packaging industry. Nylon MXD6 is a semi-crystalline polymer. Therefore, during the formation of containers, very high levels of haze or whitening from the crystallization of semi-crystalline polymer might result. In order to reduce or eliminate this problem, amorphous polyamide materials have been used. The application of nanocomposite materials, based on clay materials dispersed in a high barrier amorphous polymer matrix, has been observed to improve the barrier properties 2 to 3 times better and also reduce the haziness. The treated clays dispersed within the polymer matrix create a tortuous path for the passage of permeating gas, which can result in high barrier properties for the composite materials. With the use of clay dispersed polymer matrix as the inner layer of multi-layer containers, the haze of these containers may be reduced. In addition, by reducing or eliminating the haziness, the clay based composite materials may improve the appearance of oriented films and molded articles, including the bottles.

Amorphous polyamides capable of enhancing the haziness of oriented film materials were produced from the reactions of RDOA, MXDA, TA and NDC monomers [42–44] (Figure 7.131).

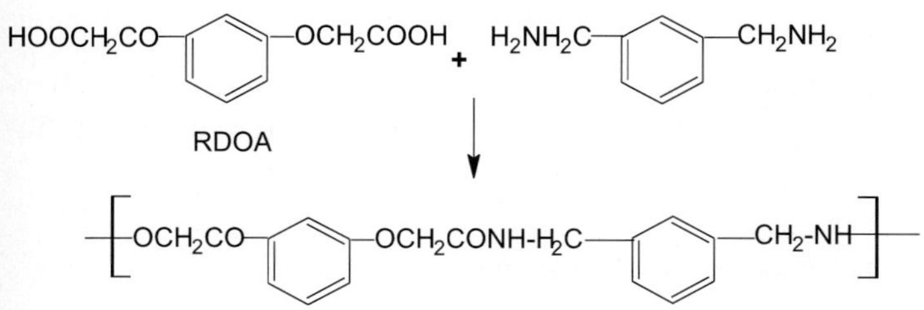


**Figure 7.131.**

The amorphous polyamides thus produced showed high glass transition temperatures (Tg) (Table 7.70).

The nanocomposite polymers produced from the RDOA based amorphous polyamides could be used to manufacture high barrier packaging materials with good clarity.

### 7.8.8.7
### Polyesteramides from RDOA

As a dicarboxylic acid, RDOA can be employed to synthesize various copolymers including the ester-amide type polymers. Polyesteramides were syn-

**Table 7.70.** Amorphous polyamides for nanocomposites

| Copolyamide composition Monomers employed (mol%) | | | Tg of copolyamides (°C) |
|---|---|---|---|
| RDOA | TA | NDC | |
| 100 | 0 | 0 | 96.7 |
| 90 | 10 | 0 | 101.1 |
| 90 | 0 | 10 | 101.1 |

Data from [42–44]

thesized by mixing stoichiometric quantities of RDOA and MXDA monomers with an excess of EG monomer and a catalyst [45] (Figure 7.132).

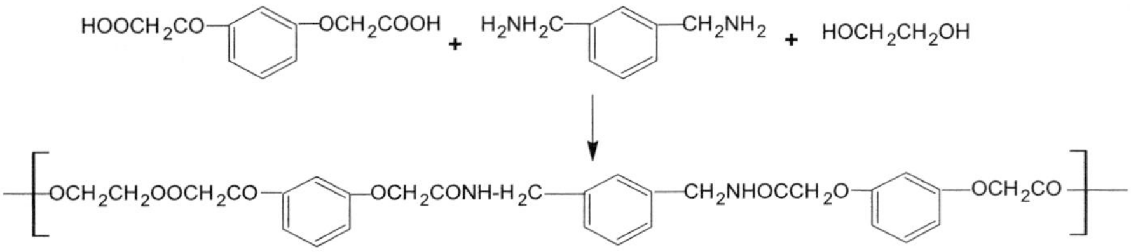

**Figure 7.132.**

By using different molar ratios of MXDA and EG, polyesteramides having different tensile strength and high barrier properties were obtained (Table 7.71).

**Table 7.71.** Synthesis and properties of poly(esteramides) from RDOA

| Polymer composition (mol%) | | Polymer properties $P_{O_2}$ (BU) | Tensile strength (MPa) |
|---|---|---|---|
| MXDA | EG | | |
| 25 | 75 | 0.26 | 51.7 |
| 50 | 50 | 0.12 | 59.3 |

Data from [45]

The presence of both amide and ester groups in the polymer chain provided increased Tg as compared to the previously known high barrier polyester materials. These polymers could be used for dip coating of the containers, such as PET, to improve the barrier performance.

### 7.8.9 Summary and Outlook

Overall, resorcinol and resorcinolic derivatives are unique in their chemistry due to the presence of the *m*-phenylene structure in these compounds. The asymmetrical nature of these molecules offers products and process improvements in the development of various polymeric materials. In the case of barrier polymer applications, resorcinol derived materials are expected to be more amorphous than crystalline in character. The packaging materials made from these amorphous polymers can, therefore, be highly transparent. Due to the presence of *meta*-linkages in these polymers, exceptionally high gas barrier properties were observed. High flexibility associated with the *m*-phenylene groups in the polymer main chain increases the chain to chain close packing, which provides the compact back bone structure.

Therefore, resorcinol based polymers are ideal gas barrier materials in the development and production of packaging materials for the highly oxygen sensitive food and beverages. The polyesters, polyamides and polyesteramides, when used as barrier layer in the multi-layer containers have greater potential to extend the shelf life of beer, beyond the current barrier polymers can provide. In the case of RDGE based thermoplastic polymers, due to an outstanding $PO_2$ performance even under very high humidity conditions, an excellent opportunity exists to compete with EVOH in numerous applications. Similarly, thermoset coatings based on RDGE can provide good potential to modify and improve the low barrier mono-layer PET and PP containers into high performance packaging containers. In this way, efficient and cost-effective packaging materials could be made to package beer and beverage materials.

## References

1. M. Salame, J. Plast. Film Sheeting, Vol. 2, p 321 (1986).
2. W. Koros, In: W. Koros, Ed., Barrier polymers and structures, ACS Symposium Series, No. 423, p 1 (1990).
3. P. DeLassus, Barrier polymers, Encyl. Of Chem. Tech., 4th Edn., Vol. 3, p 931, Wiley, New York (1992).
4. S. Steingiser et al., Barrier polymers, Encyl. Of Chem. Tech., 3rd Edn, Vol. 3, p 480, Wiley, New York (1978).
5. C. H. Silvis, Trends Polym. Sci., Vol. 5, p 75 (1997).
6. R. B. Durairaj, Plast. Additives Compounding, p 30, October (1999).
7. R. B. Durairaj, Int. Bottler Packer, p 34, September (1999).
8. R. B. Durairaj, Eur. Food Drink Rev., p 73, Winter (1999).
9. H. Dressler, US Patent, 5,059,723 (1991).
10. S. W. Go, US Patent, 4,938,017 (1983).
11. S. A. Jabarin, US Patent, 4,578,295 (1986).
12. S. A. Jabarin and G. M. Fehn, US Patent, 4,601,926 (1986).
13. K. Yamamoto et al., US Patent, 5,556,675 (1996).
14. R. R. Smith, US Patent, 4,403,090 (1983).

15. M. Hashimoto and N. Kaneshige, US Patent, 5,039,780 (1991).
16. N. H. Reinking, A. E. Barnabeo and W. F. Hale, J. Appl. Polym. Sci., Vol. 7, p 2135 (1963).
17. N. H. Reinking, A. E. Barnabeo and W. F. Hale, J. Appl. Polym. Sci., Vol. 7, p 2145 (1963).
18. J. E. White and J. W. Ringer, US Patent, 5,164,472 (1992).
19. D. Brennan et al., Polym. Prepr., Vol. 34, p 306 (1993).
20. R. Light and R. Seymour, Polym. Eng., Vol. 22, p 857 (1982).
21. D. J. Brennan et al., Macromolecules, Vol. 28, p 6694 (1995).
22. C. Silvis, C. Brown, S. Kram and J. White, Polym. Prepr., Vol. 36, p 178 (1998).
23. C. Silvis and J. White, US Patent, 5,275,853 (1994).
24. J. E. White et al., Adv. Mater., Vol. 12, p 1791 (2000).
25. BLOX, Thermoplastic resins, products information, Dow Plastics, 2001.
26. T. R. Mallen and T. A. Stevensen, US Patent, 6,346,596 (2002).
27. L. H. Carlblom, US Patent, 5,637,365 (1997).
28. A. L. Brody, Food Technol., Vol. 57, p 75 (2003).
29. L. H. Carlblom, C. Kutchko and K. W. Niederst, US Patent, 6,309,759 (2001).
30. R. B. Barbee and T. H. Wicker, US Patent, 4,440,922 (1984).
31. R. B. Barbee and T. H. Wicker, US Patent, 4,552,948 (1985).
32. R. B. Barbee and T. H. Wicker, US Patent, 4,663,426 (1987).
33. R. R. Smith, US Patent, 4,403,090 (1983).
34. T. Hirahara, M. Nukui and K. Tanaka, US Patent, 5,030,705 (1991).
35. T. Hirahara et al., US Patent, 5,239,045 (1993).
36. F. Keitoku et al., US Patent, 6,239,249 (2001).
37. T. Hirahara, M. Nukui and K. Tanaka, US Patent, 4,959,421 (1990).
38. R. B. Barbee and B. Davis, US Patent, 4,518,763 (1985).
39. J. R. Overton, S. L. Stafford and C. E. Summer, US Patent, 4,956,443 (1990).
40. R. B. Barbee, R. K. Bass and B. Davis, US Patent, 4,482,695 (1984).
41. R. Davis and R. B. Barbee, US Patent, 4,640,973 (1987).
42. S. R. Turner et al., US Patent, 6,417,262 (2002).
43. S. R. Turner et al., US Patent, 6,552,114 (2002).
44. T. Lan et al., US Patent, 6,376,591 (2002).
45. R. B. Barbee et al., US Patent, 4,501,879 (1985).

## 7.9 Polyurethanes

### 7.9.1 Introduction

Polyurethanes are a versatile group of polymers obtained from the reaction of diisocyanates and dihydroxy compounds. These materials have a wide range of industrial applications according to their physical form, such as flexible foam, rigid foam, elastomers, adhesives/sealants/coatings and fibers.

In the case of polyurethane elastomers, they can be classified as RIM, liquid castable, thermoplastic and millable gum. Due to their excellent physical properties, RIM elastomers are used in applications of auto fascia, bumper covers, external body panels and exterior/interior body trim. Cast elastomers can be either heat cured or room temperature (RT) cured. The heat cured cast elastomers have been used in the manufacture of cast wheels, tires, rolls, rollers and mechanical goods. With RT cured elastomers, they can be employed in windows insulation and gymnasium surface applications. Thermoplastic polyurethanes (TPU) are not only used as elastomers but also used in coatings, adhesives and films. Millable gums, on the other-hand, are vulcanizable and therefore, used in specialty automotive applications, roll covers and parts for military vehicles.

Polyurethane elastomers made from prepolymers are versatile engineering materials, and are capable of providing properties which are not available in conventional rubbers, metals and plastics. Urethane elastomers are, in general, more abrasion and corrosion resistant than metals. Compared to most general purpose rubbers and plastics, urethane elastomers possess higher oil resistance, solvent resistance and better aging properties. In addition, these elastomers have tear and abrasion resistance properties better than natural rubber and also, greater load-bearing properties.

The combination of high tensile strength, modulus and elongation properties of polyurethane elastomers provide toughness and durability in fabricated parts. Urethane elastomeric materials can be formulated to a wide range of hardness ranging from Shore-A value of about 10 to Shore-D value of 80, which makes them useful in numerous applications.

Polyurethane elastomers properties are attributable to their polymer structure morphologies in which both the hard and soft segments are located in the same molecular backbone [1,2]. These elastomers are composed of phase separated flexible soft segments and crystalline hard segments. The hard and soft segment contents of urethane elastomers can influence a number of properties such as low temperature flexibility, tensile strength, modulus and high temperature properties such as heat sag and distortion. The soft segment of urethane elastomers is the result of employing either polyester or polyether diol compounds in the polyurethane synthesis. When the aromatic diisocyanate compounds react with low molecular weight diol compounds, such as butanediol, hard segments are produced from the isocyanate and hydroxyl group reaction [3]. The soft segments have low polarity and are longer when compared to the hard segments, which provide a soft and flexible matrix. Hard segments, on the other-hand, are shorter, highly polar and have stronger tendency to aggregate. Due to the thermodynamic incompatibility of the hard and soft segments, they phase separate, which in-turn ultimately influences the physical and mechanical properties of polyurethane elastomers.

## 7.9.2 Cast Polyurethane Elastomers

Cast polyurethane elastomers, or simply cast urethane elastomers, are generally synthesized from the reactions of isocyanate terminated prepolymers with dihydroxy chain extenders (Figure 7.133).

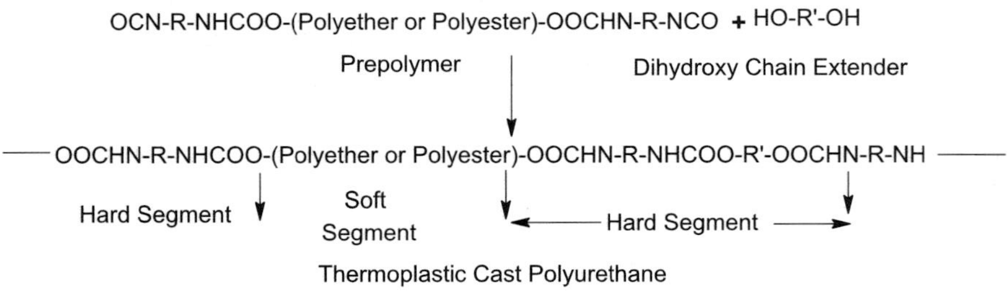


**Figure 7.133.** Cast polyurethane synthesis from prepolymer and diol compound

The prepolymers used in cast polyurethanes are obtained from the diisocyanate and polyol compounds (Figure 7.134).

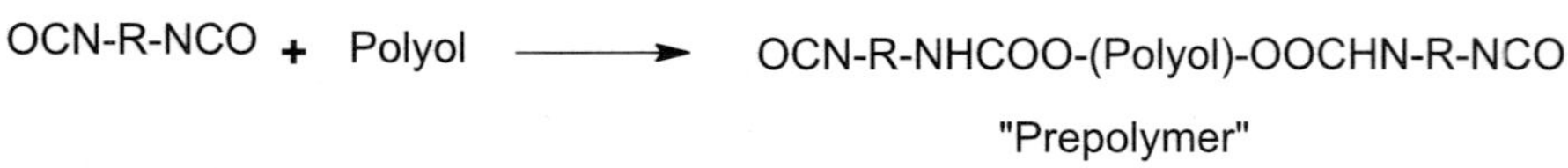


**Figure 7.134.**

Most of the prepolymers are isocyanate terminated which, therefore, are obtained by the reaction of stoichiometric excess diisocyanate compounds with polyols. Depending upon the type of applications and properties needed in the cast elastomers, diisocyanate compounds have been selected and used in the prepolymer synthesis. It has been widely known in the polyurethane industry that toluene diisocyanate (TDI) and 4,4′-diphenylmethane diisocyanate (MDI) are extensively employed in the manufacture of prepolymers. The generalized reaction schemes for the synthesis of TDI and MDI prepolymers are illustrated in Figures 7.135 and 7.136.

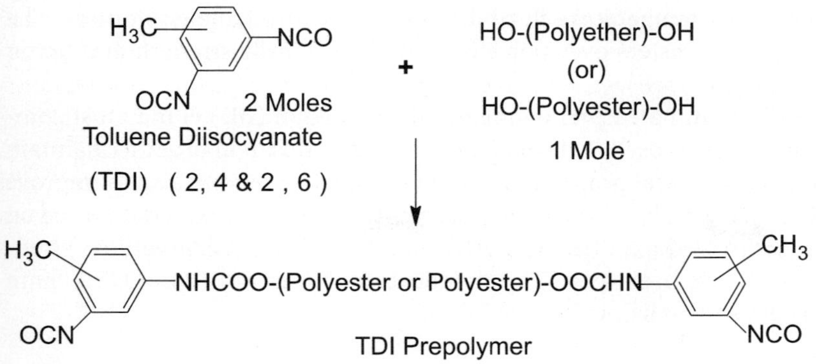


**Figure 7.135.**

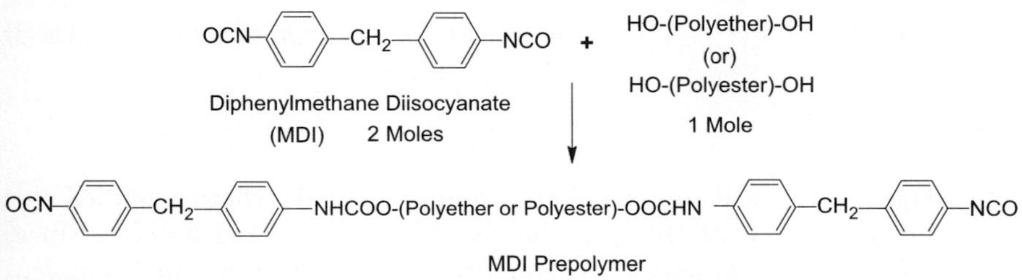


**Figure 7.136.**

In addition to TDI and MDI prepolymers, *p*-phenylene diisocyanate (PPDI) has also been employed in the development of prepolymers for the manufacture of cast elastomers with properties which could not be obtained from the conventional prepolymers (Figure 7.137).

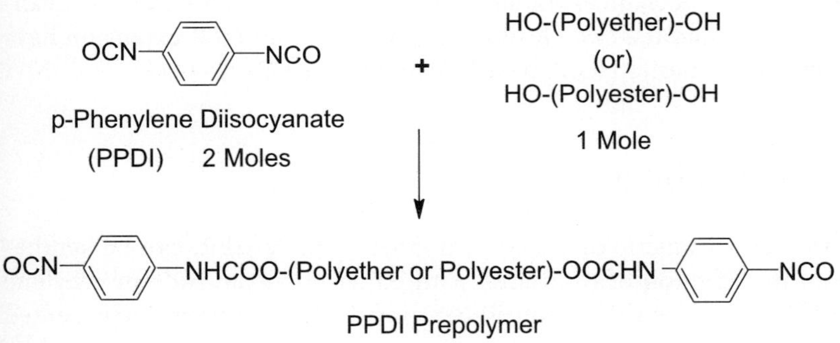


**Figure 7.137.**

The polyols used in prepolymer synthesis can be either polyether or polyester polyols. They are, normally, dihydroxy compounds having molecular weights in the range of 500 to 6000. Polyethers often provide hydrolysis resistance and low temperature flexibility to the polyurethane elastomers. The advantages of polyesters over polyethers are high tensile strength and better heat resistance.

With MDI terminated prepolymers, 1,4-butanediol (BD) is the most common chain extender used to develop cast elastomers. But, in order to maintain the desired mechanical properties at higher temperatures and also to improve hardness and tear resistance properties, aromatic diol chain extenders based on 1,3-bis(2-hydroxyethoxy) benzene (HER) and 1,4-bis(2-hydroxyethoxy) benzene (HQEE) have been used. The chemical structures of BD and HER chain extenders are shown below (Figure 7.138).

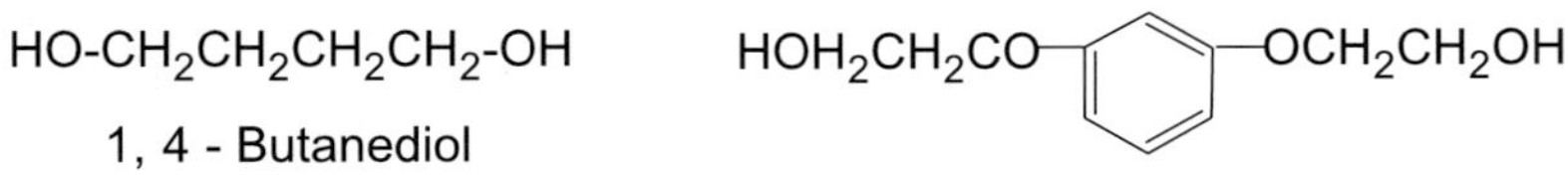


1, 4 - Butanediol (Aliphatic Diol)

1 , 3 -Bis(hydroxyethyl) ether of Resorcinol (HER) (Aromatic Diol)

**Figure 7.138.**

Though the physical and mechanical properties of BD chain extended cast urethane elastomers are excellent, they have limited thermal stability. Since, HER and HQEE are aromatic diol chain extenders and possess similar molecular structures; they are expected to provide similar physical and mechanical properties in cast urethane elastomers. One of the major disadvantages of HQEE in the cast elastomer applications is associated with its high melting point of 102 °C. In order to use HQEE, this chain extender must be processed at higher temperatures than 110 °C and possibly as high as 120 – 130 °C. HER, on the other-hand, has a lower melting point (89 °C) and super cooling behavior, thus facilitating a more forgiving chemistry and processing ease. Studies have shown that HER chain extender could offer significant processing advantages over HQEE and the cast elastomers obtained from HER extension have comparable properties to HQEE based elastomers [3–6].

## 7.9.3 Resorcinol Based Aromatic Diols

The resorcinolic aromatic diol chain extender, namely HER, can be synthesized from the reaction of resorcinol with either ethylene oxide or ethylene carbonate [7–9]. A reaction scheme illustrating the HER synthesis is presented in Figure 7.139.

A preparative procedure for synthesis of HER using resorcinol and ethylene carbonate compounds is given below [8].

**Synthesis of 1,3-Bis(2-hydroxyethyl) Ether of Resorcinol (HER) [8]** In a suitable reactor a mixture of 110.1 g (1.0 m) resorcinol, 179.7 g (2.0 m) ethylene carbonate

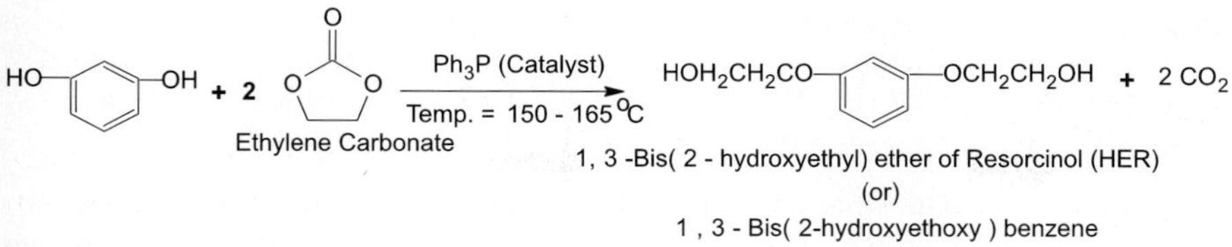


**Figure 7.139.**

(98%) and 1.5 g triphenylphosphine was stirred and heated to 150 °C, where $CO_2$ evolution started. The stirred melt was then held at 150 – 170 °C for 4.5 h. ($CO_2$ evolution ceased after about 4 h.) The melt was poured into a crystallizing dish. After cooling, there were obtained 198.8 g (essentially quantitative yield) of an off-white solid, m.p. 88 °C, which was identified as resorcinol di-(2-hydroxyethyl) ether by IR/NMR analysis; DSC analysis indicated the product was about 96% pure.

Highly pure (purity > 99%) HER can be obtained from the stoichiometric ratios of resorcinol (1 mol) and ethylene carbonate (2 mol) as the reactants. If the mol ratio of ethylene carbonate has been increased beyond 2.0 mol per mol of resorcinol then higher molecular weight products could be formed in the reaction in addition to HER (Figure 7.140).

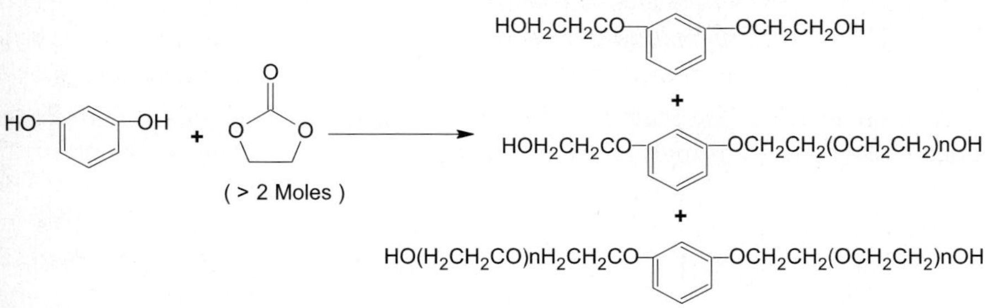


**Figure 7.140.**

By increasing the ethylene carbonates amount, technical grade (TG) HER materials containing different levels of varying molecular structure have been synthesized [10, 11]. Various technical grade materials developed from the ethylene carbonate reactions are available commercially and the details are presented in Table 7.72.

The major difference between HER HP (or simply HER) and HER TG (technical grade) materials is the presence of higher molecular weight (HMW) diol monomers present in the technical grade materials at varying concentrations. Because of the HMW materials, the melting points of TG materials were lowered, thereby their processability in polyurethane applications has been enhanced [12].

**Table 7.72.** Properties of resorcinol based aromatic diols[a]

| Aramatic diols: Properties | HER HP | HER TG-210 | HER TG-250 | HER TG-300 | HER TG-400 | HPR |
|---|---|---|---|---|---|---|
| Appearance | Off-white powder | Off-white solid | Off-white solid | Light yellow liquid | Light yellow liquid | Yellow liquid |
| Purity (HER content, %) | 99 | 93 | 68 | NA | 13 | NA |
| Hydroxyl number (mg KOH/g of sample) | 560 | 552 | 485–520 | 440–470 | 374–394 | 466–496 |
| Melting point (°C) | 89 | 78–90 | 65–70 | NA | NA | NA |
| Melt viscosity at 100 °C (cps) | 13.5 | 16 | 25–50 | NA | NA | NA |
| Viscosity at 23 °C (cps) | | | | | | 19,700 |

TG = Technical grade materials
TG Materials contain: mixtures of 1,3-bis(2-hydroxyethoxy)benzene,
1,3-bis[2-(2-hydroxyethoxy)ethoxy]benzene
and higher oligomers of 1,3-bis[2-hyroxy(ethoxy)(*n*)benzene]
[a]Data from Technical Data Sheets
from Indspec Chemical Corporation, Pittsburgh, Pa, USA
[a]"HER Materials for Polyurethane Applications",
PMA Meeting, Hilton Head, SC, Spring 1999

Similar to HER, the reaction of resorcinol with propylene carbonate produced bis-(hydroxypropyl) ether of resorcinol (HPR) [6] (Figure 7.141).

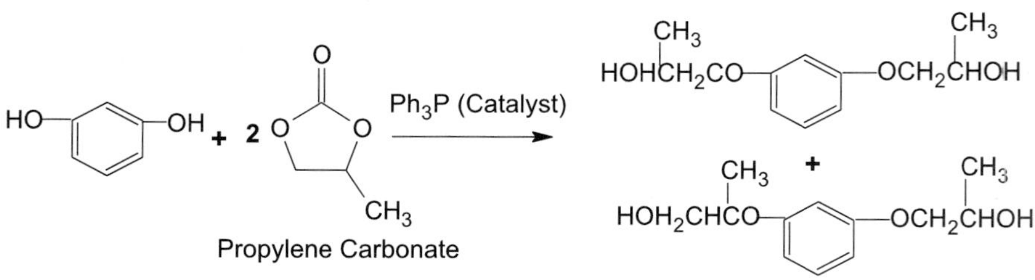


**Figure 7.141.**

HPR can also be used as an aromatic diol chain extender in polyurethane applications, which exists as a liquid at room temperature. A procedure for the synthesis of HPR is described below [13].

**Synthesis of 1,3-Bis(hydroxypropyl) Ether of Resorcinol (HPR) [13]** Liquid propylene carbonate (1472.8 g, 14.4 mol) was added to a 3-l resin kettle, equipped with a stirrer, thermocouple well, nitrogen inlet, heating mantle and reflux condenser attached to a bubbler to monitor nitrogen and/or carbon dioxide evolution rate. The vessel was purged with $N_2$ for 60 min with stirring. At this point, the controller set point was adjusted to 100 °C and 770.7 g (7.0 mol) of resorcinol and 10.5 g (0.040 mol) of triphenylphosphine were added to the reaction flask as solid materials. The total mol ratio of phenolic compound to carbonate was 1 : 2.06. After the temperature stabilized at 110 °C, the temperature was gradually raised to 190 °C. At around 180 to 185 °C, the mixture began to evolve $CO_2$. The pot temperature equilibrated at 190 °C, and a steady flow of $CO_2$ was evolved. The reaction mixture was heated for a period of 4.0 h. The reaction mixture was then cooled to less than 100 °C giving a clear yellow liquid with a green hue. The yield was 1596.7 g (100%) which was confirmed by both compositional analyses by proton-magnetic resonance techniques and hydroxyl number of 475 (mg KOH/g sample). The viscosity was about 200 poise at room temperature (about 25 °C).

Structural characterization of HPR showed that the hydroxyl groups present in the molecule consists of both secondary and primary with secondary/primary ratio of 85 : 15. Since this chain extender material is a liquid at ambient conditions, great potential exists as polyurethane curative for different applications.

## 7.9.4 Cast Polyurethanes Using MDI Prepolymers

In order to study the performance of HER extended cast urethane elastomers, various MDI based prepolymers were obtained from the commercial sources with similar % NCO content and the details are presented in Table 7.73.

**Table 7.73.** Commercial MDI prepolymers for cast elastomers applications

| MDI-polyether prepolymer | Supplier | % NCO |
|---|---|---|
| Baytec ME-050 | Bayer | 5.8 |
| Versathane SME-90A | Air Products | 7.2 |
| Vibrathane B-625 | Uniroyal Chemical | 6.4 |
| MDI-polyester prepolymer | | |
| Baytec MS-242 | Bayer | 6.6 |
| Versathane SMS-85A | Air Products | 6.8 |
| Vibrathane 8520 | Uniroyal Chemical | 6.9 |

The procedures used in the preparation of cast elastomers and the test methods employed in the determination of physical and mechanical properties of elastomers are reported in the literature [6, 10, 12]. Tables 7.74 and 7.75 summarize the cast urethane elastomers obtained from the HER chain extender and their properties.

**Table 7.74.** HER extended cast polyurethanes using polyester prepolymers

| Prepolymer type | MS-242 | B-8520 | SMS-85A |
|---|---|---|---|
| Prepolymer manufacturer | Bayer | Uniroyal | Air Products |
| Prepolymer temperature (°C) | 110 | 90 | 95 |
| Chain extender temperature (°C) | 110 | 100 | 95 |
| Mix ratio (prepolymer/extender) | 100/14 | 100/15 | 100/14.4 |
| Stoichiometry, % theory | 90 | 90 | 90 |
| Cure, h/°C | 16/110 | 16/110 | 16/110 |
| Tensile properties (MPa) | | | |
| 100% Modulus | 12.4 | 10.3 | 11 |
| 200% Modulus | 16.8 | 13.7 | 14.4 |
| 300% Modulus | 21.2 | 17.8 | 18.5 |
| Tensile strength | 31.9 | 29.6 | 23 |
| Elongation (%) | 645 | 483 | 415 |
| Tear strength, Die-C (kN/m) | 134.4 | 111.8 | 121.3 |
| Compression set, % | 30.4 | 39.5 | 30.6 |
| Bashore rebound, % | NA | 41 | 25 |
| Hardness, Shore-D | 45 | 43 | 52 |
| Dynamic mechanical analysis (DMA) | | | |
| Tan delta ($\delta$) at 10 Hz/23 °C (frequency) | 0.073 | 0.07 | 0.114 |
| Tan delta ($\delta$) at 25 °C (temp. sweep) | 0.054 | 0.051 | 0.083 |
| Tg (tan delta peak), °C | – 21.9 | – 25.4 | – 14 |
| Thermal stability from DMA (Temp. max for constant G″) | | | |
| Temperature (°C) | 150 | 150 | 145 |
| Storage modulus (G′), (dynes/cm$^2$) $\times 10^8$ | 1.48 | 1.62 | 0.83 |

Data from [6, 10, 12]

In general, urethane elastomer properties can be correlated with the soft and hard segments present in the elastomers. The soft segment of urethane elastomers primarily affect the resiliency, wear, tear, compression set and low temperature properties, whereas the hard segment content determines the hardness, modulus and tensile properties [14].

In the case of polyester based MDI elastomers, MS-242 and B-8520 prepolymers exhibited higher tensile strength and lower hysteresis (tan $\delta$) properties compared to SMS-85A prepolymer. With polyether cast elastomers, ME-050, B-625 and SME-90A prepolymers showed comparable performance with respect to their physical and mechanical properties when extended with HER.

One of the method by which the thermal stability of cast polyurethane materials are evaluated or determined is Dynamic Mechanical Analysis (DMA) method. In the DMA, the temperature at which the storage modulus (G′) decreases significantly is considered the limit of elastomer's thermal stabil-

**Table 7.75.** HER extended cast polyurethanes using polyether prepolymers

| Prepolymer type | ME-050 | B-625 | SME-90A |
|---|---|---|---|
| Prepolymer manufacturer | Bayer | Uniroyal | Air Products |
| Prepolymer temperature (°C) | 110 | 90 | 95 |
| Chain extender temperature (°C) | 110 | 105 | 100 |
| Mix ratio (prepolymer/extender) | 100/12.5 | 100/14.3 | 100/15.8 |
| Stoichiometry, % theory | 90 | 90 | 90 |
| Cure, h/°C | 16/110 | 16/110 | 16/110 |
| Tensile properties (MPa) | | | |
| 100% Modulus | 9.3 | 10.7 | 11.5 |
| 200% Modulus | 12.1 | 14 | 14.5 |
| 300% Modulus | 15.3 | 17.8 | 17.9 |
| Tensile strength | 23.5 | 24.4 | 24.1 |
| Elongation (%) | 605 | 481 | 569 |
| Tear strength, Die-C (kN/m) | 109.6 | 88 | 106.2 |
| Compression Set, % | 14.6 | 14.2 | 16 |
| Bashore rebound, % | NA | 61 | NA |
| Hardness, Shore-D | 47 | 43 | 54 |
| Dynamic mechanical analysis (DMA) | | | |
| Tan delta ($\delta$) at 10 Hz/23 °C (frequency) | 0.038 | 0.048 | 0.043 |
| Tan delta ($\delta$) at 25 °C (temp. sweep) | 0.032 | 0.04 | 0.036 |
| Tg (tan delta peak), °C | – 62.9 | – 61.3 | – 63.4 |
| Thermal stability from DMA (Temp. max. for constant G″) | | | |
| Temperature (°C) | 160 | 160 | 160 |
| Storage modulus (G′), (dynes/cm²) $\times 10^8$ | 1.55 | 2.11 | 2.24 |

Data from [6, 10, 12]

ity. From the DMA method, it was determined (Tables 7.74 and 7.75) that polyester based elastomers extended with HER were thermally stable upto 150 °C whereas the polyether based materials showed temperature stability of about 160 °C.

Tear strength is also a property commonly used to compare the cut growth propagation in a rubber like material and very important in many applications. Polyurethanes containing the polyester soft segments normally show higher tear strength properties when compared to polyether soft segments. By comparing the tear strength values in Tables 7.74 and 7.75, cast polyester MDI elastomers extended with HER showed high tear strength properties.

Resiliency or rebound is the most important property in skate wheel applications [15]. HER extended polyether MDI urethanes showed higher rebound resilience properties than the polyester system.

### 7.9.4.1
### Comparative Performance of HER versus HQEE

The performance of two widely known aromatic diol chain extenders used in the cast polyurethanes were studied and reported [12]. In order to improve the properties of cast urethanes associated with aromatic diol chain extenders, trimethylol propane (known as TMP) was used at 10% level in such a way that the aromatic diol/TMP ratio was kept 90/10 (weight) in the extender mix. The composition, processing temperatures, cure conditions and cast elastomer properties obtained from the MDI prepolymer are presented in Table 7.76.

In general HER extended elastomers showed tensile strength, modulus and elongation properties comparable to HQEE chain extender in the MDI based system (Figure 7.142).

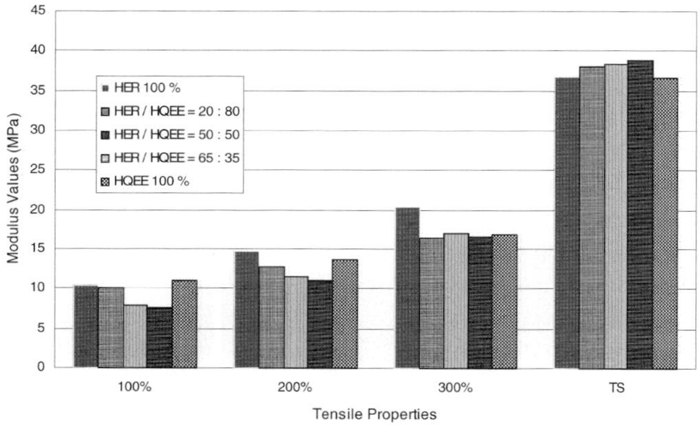


**Figure 7.142.** Cast elastomers made from Baytec MS-242 and HER + HQEE + Triol Extender systems

The hardness of HQEE extended elastomers, in general, exhibited 2–3 units higher than the HER based materials. Varying the HER/HQEE ratio in the elastomer composition affected the physical and mechanical properties of the resulting elastomer. Taber abrasion was improved with HER/HQEE ratio at 20/80 (weight) in the cast elastomer composition. From the DMA results, the polyester MDI elastomers extended with HER and HQEE were found to be thermally stable up to 170 °C (Table 7.76).

Differential Scanning Calorimetry (DSC) analysis can also be used to determine the thermal stability of elastomers. In the case of polyester based cast elastomers, HER, HQEE and HER/HQEE extended elastomers exhibited endothermic peaks in the DSC thermograms at about 189 °C. These DSC curve peaks are associated with the melting behavior of hard segment domains of cast elastomers. Based on the hard segment melting temperatures and endothermic peak energies observed from the DSC analysis, it can be reasonably assumed that the hard segments formed from HER and HQEE chain extenders have similar character and properties.

**Table 7.76.** Cast elastomers from Baytec MS-242 and HER + HQEE/triol extenders

| Chain extenders used | HER-HP | HER HQEE | HER HQEE | HER HQEE | HQEE |
|---|---|---|---|---|---|
| Extender/triol Ratio = 90/10 Wt | 100% | 20/80 | 50/50 | 65/35 | 100% |
| Triol extender used (Multranol) | M4012 | M4012 | M4012 | M4012 | M4012 |
| Prepolymer temp., °C | 100 | 110 | 100 | 100 | 110 |
| Chain extender temp., °C | 100 | 110 | 100 | 100 | 110 |
| Mix ratio (prepolymer/extenders) | 100/15.7 | 100/15.7 | 100/15.7 | 100/15.7 | 100/15.7 |
| Cure, h/°C | 16/110 | 16/110 | 16/110 | 16/110 | 16/110 |
| Tensile property (MPa) | | | | | |
| 100% Modulus | 10.4 | 10 | 7.9 | 7.6 | 11 |
| 200% Modulus | 14.7 | 12.7 | 11.5 | 11 | 13.6 |
| 300% Modulus | 20.4 | 16.4 | 17 | 16.5 | 16.9 |
| Tensile strength | 36.8 | 38 | 38.4 | 38.8 | 36.6 |
| Elongation (%) | 515 | 536 | 475 | 497 | 574 |
| Tear strength, Die-C (kN/m) | 123.9 | 121.3 | 101.1 | 101.5 | 134.2 |
| Compression set, % | 36.2 | 27 | 27.8 | 26.8 | 21 |
| Bashore rebound, % | 36 | 31 | 23 | 24 | 38 |
| Taber abrasion (mg/1000 cycles) | 58.3 | 37.7 | 52.5 | 48.9 | 87.6 |
| Hardness, Shore-A | 92 | 93 | 89 | 88 | 95 |
| DMA results | | | | | |
| Tan delta at 23 °C (frequency sweep) | 0.076 | 0.094 | 0.122 | 0.121 | 0.08 |
| Tan delta at 25 °C (temp. sweep) | 0.054 | 0.068 | 0.072 | 0.075 | 0.062 |
| Tan delta at 100 °C (temp, sweep) | 0.016 | 0.021 | 0.017 | 0.022 | 0.018 |
| Tg (tan delta peak), °C | – 15.6 | – 14.1 | – 7.7 | – 7.4 | – 19.6 |
| Thermal stability of elastomers (temp. maximum for constant G′) | | | | | |
| Temperature, °C | 170 | 170 | 170 | 170 | 170 |
| G′ (dynes/cm$^2$) $\times 10^8$ | 0.87 | 1.91 | 0.94 | 0.84 | 2.83 |
| DSC analysis results | | | | | |
| First heating endotherm, Peak (°C) | 189 | 189 | 185 | 184 | 188 |
| Energy (J/g) | 19 | 17 | 11 | 16 | 21 |

### 7.9.4.2
### Performance Comparison of HER and HER TG-210 Extenders

TG-210 extender contained about 6 – 10% of higher molecular weight diols produced from the resorcinol and ethylene carbonate reaction. The processability of TG-210 was slightly better than HER extender with MDI prepolymers. To compare the elastomer properties of TG-210 and HER materials, polyether and polyester MDI prepolymers were extended with these extenders and studied [6]. Table 7.77 shows the details on the cast elastomer preparations and properties obtained.

**Table 7.77.** Performance of HER-HP vs HER TG-210 extender in cast elastomers

| Prepolymer type<br>Chain extender used | Baytec MS-242<br>HER-HP | <br>HER-TG-210 | Baytec ME-050<br>HER-HP | <br>HER TG-210 |
|---|---|---|---|---|
| Prepolymer temp. (°C) | 110 | 85 | 110 | 85 |
| Chain extender temp. (°C) | 110 | 85 | 110 | 85 |
| Mix ratio (prepolymer/extender) | 100/14 | 100/14.8 | 100/12.5 | 100/12.1 |
| Stoichiometry, % theory | 90 | 90 | 90 | 90 |
| Cure, h/°C | 16/110 | 16/110 | 16/110 | 16/110 |
| Tensile properties (MPa) | | | | |
| 100% Modulus | 12.4 | 10.3 | 9.3 | 8.2 |
| 200% Modulus | 16.8 | 13.9 | 12.1 | 11.1 |
| 300% Modulus | 21.1 | 18.4 | 15.3 | 14.6 |
| Tensile strength | 31.9 | 34.2 | 23.5 | 26.1 |
| Elongation (%) | 645 | 622 | 605 | 527 |
| Tear strength, Die-C (kN/m) | 134.4 | 137 | 109.6 | 97.7 |
| Die-C (kN/m) | | | | |
| Compression set, % | 30.4 | 32.4 | 14.6 | 17.6 |
| Bashore rebound, % | NA | 34 | NA | 63 |
| Hardness, Shore-D | 45 | 44 | 47 | 43 |
| Dynamic mechanical analysis (DMA) | | | | |
| Tan delta ($\delta$) at 10 Hz/23 °C (frequency) | 0.073 | 0.083 | 0.038 | 0.034 |
| Tan delta ($\delta$) at 25 °C (temp. sweep) | 0.054 | 0.061 | 0.032 | 0.024 |
| Tg (Tan delta peak), °C | – 21.9 | – 24 | – 62.9 | – 59.4 |
| Thermal Stability from DMA (temp. max for constant G″) | | | | |
| Temperature (°C) | 150 | 150 | 160 | 155 |
| Storage Modulus (G′), (dynes/cm$^2$) $\times 10^8$ | 1.48 | 1.47 | 1.55 | 1.42 |

Data taken from [6, 10]

Upon examining the elastomeric properties, it was quite noticeable that HER and TG-210 extender materials performed comparably with respect to tensile, tear, compression set, Tg and hysteresis (tan $\delta$) properties. From the DMA results, it was determined that the thermal stability properties were similar, which suggests that TG-210 could be used in cast elastomer applications in the place of HER without much reduction in elastomer performance.

Similar performance comparison was made with TG-210, HER and HQEE extenders employing polyether prepolymers in the cast urethanes (Table 7.78).

**Table 7.78.** Comparative performance of cast elastomers extended with aromatic diol chain extenders
MDI-polyether prepolymer used: Vibrathane B-625

| Chain extenders | HQEE | HER-HP | HER TG-210 |
|---|---|---|---|
| Casting conditions | | | |
| Prepolymer temperature, °C | 110 | 100 | 85 |
| Chain temperature, °C | 110 | 100 | 85 |
| Mix ratio (prepolymer/extender) | 100/14.6 | 100/14.5 | 100/14.3 |
| Stoichiometry, % theory | 95 | 95 | 95 |
| Cure, h/°C | 16/110 | 16/110 | 16/110 |
| Tensile properties (MPa) | | | |
| 100% Modulus | 11.4 | 10.6 | 9.6 |
| 300% Modulus | 14.6 | 17.3 | 15.9 |
| Tensile strength | 23.6 | 26.3 | 27.6 |
| Elongation (%) | 538 | 559 | 588 |
| Tear strength, Die-C (kN/m) | 126.9 | 115.3 | 119.7 |
| Compression set, % | 21 | 15.4 | 18.8 |
| Bashore rebound, % | 55 | 58 | 58 |
| Hardness, Shore-A | 96 | 95 | 95 |
| DMA Analysis results | | | |
| Tan ($\delta$) at 10 Hz/23 °C (frequency) | 0.055 | 0.044 | 0.049 |
| Tan ($\delta$) at 25 °C (temp. sweep) | 0.039 | 0.038 | 0.043 |
| Tg (Tan ($\delta$) peak, °C) | – 63.3 | – 61.3 | – 57.2 |
| Tg (loss modulus curve peak, G″, °C) | – 70.9 | – 70.8 | – 69.1 |

Data from [6, 10]

From the processing standpoint, TG-210 offers good advantage over HER and HQEE chain extenders because of its lower processing temperature associated with the lower melting point. The Bashore rebound, tensile strength and elongation values of TG-210 based elastomers were higher than HER and HQEE. Low compression set properties could be also achieved with TG-210 extender.

## 7.9.5
## Highly Resilient and Thermally Stable Cast Elastomers

Polyurethane elastomers have been successfully used in dynamic applications such as wheels, roll coverings and tires. In many dynamic applications, polyurethane elastomers are superior elastomeric materials because they can be formulated to high hardness. Urethane elastomers with hardnesses in the range of Shore 90A to Shore 50D have especially shown good load bearing properties. The prepolymers used to develop such materials are predominantly MDI polyethers. Polyurethanes obtained from the poly (tetramethylene ether) glycol (PTMEG) based polyether prepolymers can be engineered to have much resilience than polyester based systems.

MDI polyethers based elastomers are much more resistant to hydrolysis than the TDI polyether based materials. In applications such as the roller skate wheels, sand blast curtains, laundry equipment and tires, MDI polyether prepolymers are preferred due to their high resilience and hydrolysis resistant properties. High quality roller skate wheels are practically made from MDI polyether prepolymers, since they provide good speed performance at low durometer hardness and smoothness for outdoor use. These performances are associated with high resiliency of cast elastomers [16].

Higher resiliency in cast elastomers could be achieved with the use of higher molecular weight PTMEG in the urethane elastomer backbone. The use of higher percentages of PTMEG in the elastomer synthesis can also provide higher resiliency in the urethane system. The low polarity of PTMEG compared to polyester polyol, which keeps the inter-chain forces low, is responsible for the good phase separation of soft and hard segments and high resiliency.

Employing TG-210 aromatic diol chain extender, various blends were made with Terathane polyols (PTMEG) having different molecular weights and highly resilient cast elastomers were developed and studied [17].

### 7.9.5.1
### Highly Resilient Cast Elastomers from TG-210 and Terathane Polyol Blends

Terathane is the trademark for the poly(tetramethylene ether) glycols (PTMEG) available from Dupont Company. By using different molecular weight PTMEG materials, HER TG-210-Terathane blends were made at different weight ratios of TG-210 and PTMEG. The addition of PTMEG into TG-210 extender further lowered the melting point of TG-210, which in turn improved the processability of the blend systems in the cast elastomer synthesis.

Cast elastomers were developed using the Baytec ME-050 prepolymer and various TG-210/PTMEG blend systems. The details are presented in Table 7.79.

In addition to TG-210/PTMEG blends, BD was also used as a chain extender in the elastomer synthesis and cast elastomer properties and performances were compared. The main functions of extenders in the cast elastomer synthesis

**Table 7.79.** Cast elastomers From Baytec ME-50 and HER TG-210/Terathane (PTMEG) blends

| Terathane (PTMEG) M.Wt. | 1000 | 1000 | 1000 | 1000 | 2000 | 2000 | 2000 | 2000 | 2900 | 2900 | 2900 | TG-210 | BD |
|---|---|---|---|---|---|---|---|---|---|---|---|---|---|
| TG-210/PTMEG (Wt. Ratio) | 90/10 | 70/30 | 50/50 | 30/70 | 90/10 | 70/30 | 50/50 | 30/70 | 90/10 | 70/30 | 50/50 | 100% | 100% |
| Blend melting point by DSC (°C) (endothermic peak) | 84.5 | 84.3 | 85.8 | 80.7 | 88.7 | 87.3 | 84.7 | 81.7 | NA | NA | NA | 89.8 | Liquid |
| Casting conditions | | | | | | | | | | | | | |
| Prepolymer temp, °C | 85 | 85 | 85 | 85 | 85 | 85 | 85 | 85 | 85 | 85 | 85 | 85 | 85 |
| Blend temp, °C | 85 | 85 | 85 | 85 | 85 | 85 | 85 | 85 | 85 | 85 | 85 | 85 | RT |
| Mix ratio (prepolymer/blend) | 100/14.7 | 100/17.8 | 100/22.5 | 100/30.5 | 100/14.9 | 100/18.5 | 100/24.5 | 100/36.4 | 100/14.9 | 100/18.8 | 100/25.3 | 100/13.5 | 100/5.9 |
| Stoichiometry, % theory | 95 | 95 | 95 | 95 | 95 | 95 | 95 | 95 | 95 | 95 | 95 | 95 | 95 |
| Cure, h/°C | 16/110 | 16/110 | 16/110 | 16/110 | 16/110 | 16/110 | 16/110 | 16/110 | 16/110 | 16/110 | 16/110 | 16/110 | 16/110 |
| Hard segment content (%) | 33.9 | 32.4 | 30.2 | 26.9 | 34 | 32.6 | 30.5 | 26.9 | 34 | 32.7 | 30.6 | 34.5 | 30.7 |
| Tensile properties (MPa) | | | | | | | | | | | | | |
| 100% Modulus | 8.5 | 7.9 | 6.3 | 5 | 8.4 | 7.9 | 7 | 5.5 | 8.3 | 7.9 | 7 | 8.7 | 6.3 |
| 300% Modulus | 14.8 | 14.1 | 12.4 | 10.8 | 14.8 | 14.5 | 13.7 | 11.8 | 14.7 | 14.4 | 14.1 | 15.4 | 12.8 |
| Tensile strength | 24.4 | 25.7 | 24.7 | 18.8 | 24.1 | 25.5 | 26.3 | 27.3 | 25.6 | 24.8 | 26.6 | 24.8 | 30.7 |
| Elongation (%) | 546 | 519 | 480 | 407 | 544 | 532 | 504 | 467 | 558 | 530 | 504 | 541 | 456 |
| Energy to break (kN/m) (area under the curve) | 76.1 | 69.3 | 57.4 | 35.3 | 81.6 | 79.1 | 68.6 | 51.6 | 85.4 | 77.9 | 69.8 | 84.2 | 52.7 |
| Tear strength, Die-C (kN/m) | 112.9 | 106.1 | 82.6 | 67.4 | 108.9 | 101.9 | 98.9 | 88.6 | 109.2 | 106.6 | 100.5 | 111 | 90.5 |
| Compression set (%) | 16.5 | 16.9 | 18.1 | 25.6 | 16.4 | 15 | 16.3 | 17.7 | 16.8 | 15.8 | 15.1 | 16.5 | 15.3 |
| Bashore rebound (%) | 63 | 66 | 69 | 72 | 65 | 67 | 67 | 72 | 64 | 67 | 67 | 64 | 67 |
| Hardness, Shore-A | 92 | 90 | 87 | 81 | 93 | 92 | 90 | 85 | 92 | 92 | 90 | 94 | 88 |

Terathane is a Trademark of Dupont Company
Data from [13]

is to form the hard segment by reacting with the isocyanate group of the prepolymer. With BD and HER extenders, the hard segments that could be formed have the following structures (Figure 7.143).

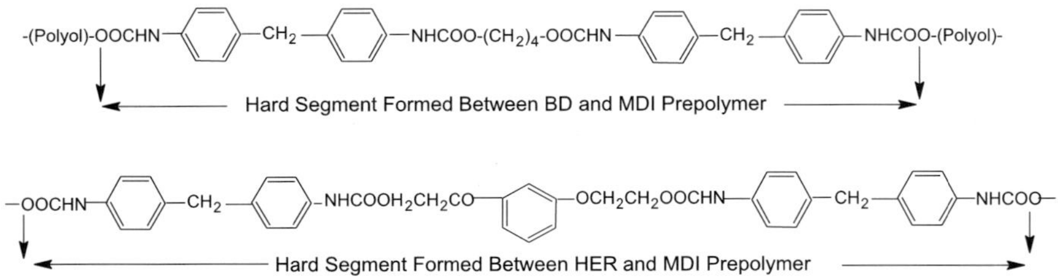


**Figure 7.143.** Hard segments formed from MDI prepolymers and chain extenders

The nature and amount of hard segments present in the urethane elastomers determine the physical and mechanical properties of cast elastomers. By increasing the amount of PTMEG in the blend, the hard segment content of elastomeric materials was decreased, which in turn decreased the 300% tensile modulus, elongation, tear strength and hardness values. On the other-hand, tensile strength and Bashore rebound (resiliency) properties were improved.

Comparing the structures of hard segments (Figure 7.143) formed from the HER-MDI and BD-MDI reaction, it can be seen that *m*-phenylene linkages are present in the HER system. The presence of *m*-phenylene linkage could enhance the rigidity or stiffness of hard segments and also the cohesion of hard segment domains. Due to longer extender chain length of HER compared to BD, the hard segments formed from HER can be expected to be longer. Therefore, it can reasonably be assumed that HER based hard segments are much more crystalline than the BD based hard segments. This can result in greater incompatibility between the soft and hard segments with elastomers having highly and larger crystalline segments. Based on these facts, it can be expected that HER based polyurethane elastomers are more phase separated than the BD based elastomer.

### 7.9.5.2 Dynamic Mechanical Analysis (DMA) of High Resilient Elastomers

DMA is an analytical tool used to determine the dynamic properties of elastomeric materials [18,19]. DMA measures the ability of a material to store and dissipate the mechanical energy. The determination of dynamic properties of elastomeric materials is important, since they influence the performance of certain products such as wheels and tires. From the DMA, the storage modulus ($G'$, elastic behavior), loss modulus ($G''$, energy dissipation), tan $\delta$ ($G''/G'$, hysteresis), loss compliance ($J''$) and glass transition temperature (Tg) properties of elastomers can be determined. The low temperature behavior of elastomers can be determined from the Tg values of soft segments. The Tg property of cast elastomers is not only influenced by the nature of soft segments (polyether or polyester) but also the degree of phase separation between the hard and soft segments.

In order to determine the thermal stability and DMA properties of cast elastomers obtained from the TG-210/Terathane-2000 blends, samples were placed in an air oven and aged for 28 days at 110 °C and 6 days at 150 °C. After aging, all samples were conditioned for at least seven days at room temperature and 50% RH conditions before testing.

The DMA curve obtained for the cast elastomer synthesized from Baytec ME-050 and TG-210/PTMEG system is presented in Figure 7.144.

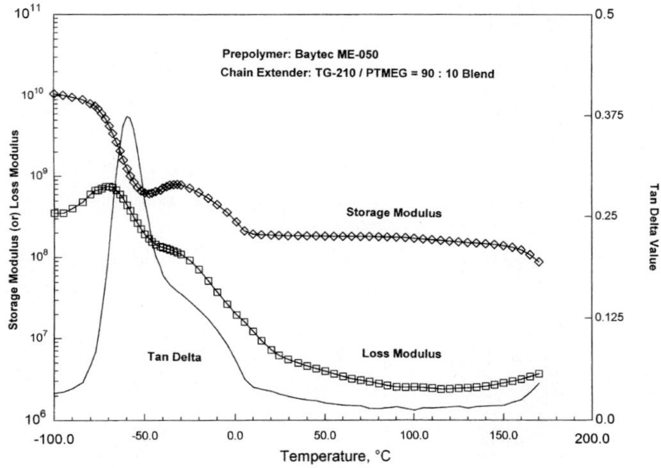


**Figure 7.144.** DMA curves of heigh resilient cast elastomers obtained from Baytec ME-050 and HER TG-210 blend materials

The DMA results obtained on the unaged and heat aged elastomer samples are summarized in Tables 7.80 and 7.81.

**Storage Modulus (G′) Property** The storage modulus (G′) determined from the DMA quantitatively measures the materials elastic property and also qualitatively determines the elastomer's hardness and stiffness. Since the hardness of elastomer is related to its hard segment content, the DMA storage modulus values were found to be decreased as the hardness of elastomers were decreased. The reduction in hardness of elastomers was related to increase in the amount of PTMEG (Terathane 2000 or simply T-2) in the blends. Since the hard segments formed by the HER extender may be longer in length than the BD extender, cast elastomers from the HER extension exhibited higher G′ property.

**Table 7.80.** DMA results of 110 °C/28 days heat aged elastomers of Baytec ME-050-HER TG-210/Terathane 2000 blends

| TG-210/Terathane 2000 ratio (weight) | BD = 100% Before | After | TG-210 = 100% Before | After | 90/10 Before | After | 70/30 Before | After | 50/50 Before | After | 30/70 Before | After |
|---|---|---|---|---|---|---|---|---|---|---|---|---|
| Hardness, Shore-A | 89 | 75 | 91 | 90 | 93 | 85 | 91 | 85 | 90 | 80 | 85 | 70 |
| Storage modulus (G′) (dynes/cm$^2$) ×10$^8$ | | | | | | | | | | | | |
| at 25 °C | 1.18 | 0.61 | 1.85 | 1.31 | 1.82 | 1.16 | 1.76 | 1.1 | 1.34 | 0.74 | 0.79 | 0.39 |
| at 100 °C | 1.12 | 0.24 | 1.79 | 0.66 | 1.67 | 0.56 | 1.67 | 0.55 | 1.29 | 0.32 | 0.76 | 0.11 |
| at 150 °C | 0.94 | 0.08 | 1.41 | 0.23 | 1.37 | ND | 1.29 | 0.18 | 1.03 | ND | 0.62 | ND |
| Thermal stability (temp. max for constant G′) | | | | | | | | | | | | |
| Temperature (°C) | 165 | 130 | 165 | 130 | 160 | 125 | 160 | 125 | 165 | 125 | 165 | 115 |
| G′ (dynes/cm$^2$ ×10$^8$) | 0.75 | 0.17 | 1.13 | 0.48 | 1.24 | 0.43 | 1.18 | 0.44 | 0.78 | 0.25 | 0.49 | 0.08 |
| Glass transition, Tg (°C), at | | | | | | | | | | | | |
| G″ peak | – 69.3 | – 65.5 | – 69.8 | – 67.7 | – 69.2 | – 65.6 | – 70 | – 67.7 | – 71.3 | – 67.7 | – 69 | – 61.7 |
| tan δ peak | – 57.5 | – 57.6 | – 59.9 | – 35.6 | – 59.1 | – 33.5 | – 59.8 | – 31.6 | – 59.6 | – 47.7 | – 57.3 | – 28.3 |
| Tan delta (δ) | | | | | | | | | | | | |
| at 25 °C | 0.025 | 0.103 | 0.034 | 0.074 | 0.029 | 0.067 | 0.027 | 0.07 | 0.022 | 0.077 | 0.016 | 0.072 |
| at 100 °C | 0.016 | 0.119 | 0.017 | 0.076 | 0.014 | 0.079 | 0.013 | 0.077 | 0.015 | 0.085 | 0.014 | 0.082 |
| at 150 °C | 0.02 | 0.154 | 0.022 | 0.114 | 0.017 | ND | 0.017 | 0.123 | 0.022 | ND | 0.027 | ND |
| Loss compilance, J″ (cm$^2$/dynes) ×10$^8$ | | | | | | | | | | | | |
| at 25 °C | 2.1 | 16.6 | 1.81 | 5.6 | 1.63 | 5.7 | 1.57 | 6.3 | 1.66 | 10.3 | 1.98 | 18.4 |
| at 100 °C | 1.4 | 49.3 | 0.96 | 11.3 | 0.88 | 13.9 | 0.78 | 13.9 | 1.13 | 26.5 | 1.87 | 72 |
| at 150 °C | 2.12 | 184.4 | 1.53 | 50 | 1.26 | ND | 1.33 | 67.1 | 2.18 | ND | 4.5 | ND |
| Area under the J″ curve temp. range: – 90 to 110 °C | 5.47 | 31.6 | 3.44 | 8.74 | 2.88 | 9.7 | 3.24 | 10.5 | 3.96 | 17.4 | 7.12 | 40.2 |

**Table 7.81.** DMA results of 150 °C/six days heat aged cast elastomers based on Baytec ME-050-TG-210/Terathane 2000 blends

| TG-210/Terathane 2000 ratio (weight) | BD = 100% Before | After | TG-210 = 100% Before | After | 90/10 Before | After | 70/30 Before | After | 50/50 Before | After | 30/70 Before | After |
|---|---|---|---|---|---|---|---|---|---|---|---|---|
| Hardness, Shore-A | 89 | 70 | 91 | 81 | 93 | 80 | 91 | 80 | 90 | 75 | 85 | 65 |
| Storage modulus (G′) properties (dynes/cm²) × $10^8$ | | | | | | | | | | | | |
| at 25 °C | 1.18 | 0.57 | 1.85 | 0.91 | 1.82 | 0.82 | 1.76 | 0.81 | 1.34 | 0.61 | 0.79 | 0.41 |
| at 100 °C | 1.12 | 0.21 | 1.79 | 0.3 | 1.67 | 0.26 | 1.67 | 0.27 | 1.29 | 0.18 | 0.76 | 0.11 |
| at 150 °C | 0.94 | 0.15 | 1.41 | 0.2 | 1.37 | 0.19 | 1.29 | 0.18 | 1.03 | 0.11 | 0.62 | NA |
| Thermal stability (temp. max for constant G′) | | | | | | | | | | | | |
| Temperarure (°C) | 165 | ND | 165 | ND | 160 | ND | 160 | ND | 165 | ND | 165 | ND |
| G′ (dynes/cm² × $10^8$) | 0.75 | ND | 1.13 | ND | 1.24 | ND | 1.18 | ND | 0.78 | ND | 0.49 | ND |
| Loss modulus (G″) properties | | | | | | | | | | | | |
| Peak temp. (Tg, °C) | − 69.3 | − 64.1 | − 69.8 | − 63.7 | − 69.2 | − 58.9 | − 70 | − 63.5 | − 71.3 | − 65.9 | − 69 | − 63 |
| Tan δ Properties | | | | | | | | | | | | |
| Peak temp, Tg (°C) | − 57.5 | − 33.7 | − 59.9 | − 47.6 | − 59.1 | − 39.1 | − 59.8 | − 47.9 | − 59.6 | − 49.5 | − 57.3 | − 33.3 |
| at 25 °C | 0.025 | 0.165 | 0.034 | 0.116 | 0.029 | 0.127 | 0.027 | 0.123 | 0.022 | 0.113 | 0.016 | 0.119 |
| at 100 °C | 0.016 | 0.145 | 0.017 | 0.109 | 0.014 | 0.104 | 0.013 | 0.123 | 0.015 | 0.113 | 0.014 | 0.139 |
| at 150 °C | 0.02 | 0.162 | 0.022 | 0.103 | 0.017 | 0.094 | 0.017 | 0.128 | 0.022 | 0.12 | 0.027 | NA |
| Loss compliance (J″) properties (cm²/dynes) × $10^{-10}$ | | | | | | | | | | | | |
| Peak temperature, °C | − 49.7 | 20.8 | − 53.7 | 20.3 | − 49.5 | 22 | − 54 | 21 | − 53.4 | 18.9 | − 49.5 | 20.7 |
| Peak compliance | 10.4 | 27.6 | 4.16 | 12.8 | 2.53 | 15 | 4.78 | 15.5 | 7.57 | 17.5 | 1.83 | 28.8 |
| at 25 °C | 2.1 | 28.2 | 1.81 | 12.6 | 1.63 | 15.3 | 1.57 | 15 | 1.66 | 18.2 | 1.98 | 28.6 |
| at 100 °C | 1.4 | 66.6 | 0.96 | 36.2 | 0.88 | 39.6 | 0.78 | 45.3 | 1.13 | 62.9 | 1.87 | 128 |
| at 150 °C | 2.12 | 107.6 | 1.53 | 51.5 | 1.26 | 49.8 | 1.33 | 69.7 | 2.18 | 112.2 | 4.5 | NA |

Data from [13]

**Tan Delta (Tan δ) Property** The measure of tan δ is a qualitative tool to determine the hysteresis or heat build up in an elastomer during the dynamic flex conditions. Tan δ is zero for a perfectly elastic material and infinity for a perfectly inelastic material. Most polyurethane materials are highly elastic and therefore, exhibit tan δ values typically between 0.01 and 1.0. Figures 7.145 and 7.146 show the plots of tan δ values measured at 25 °C and 100 °C temperatures from the temperature sweep DMA analysis performed on the unaged and heat aged elastomer samples.

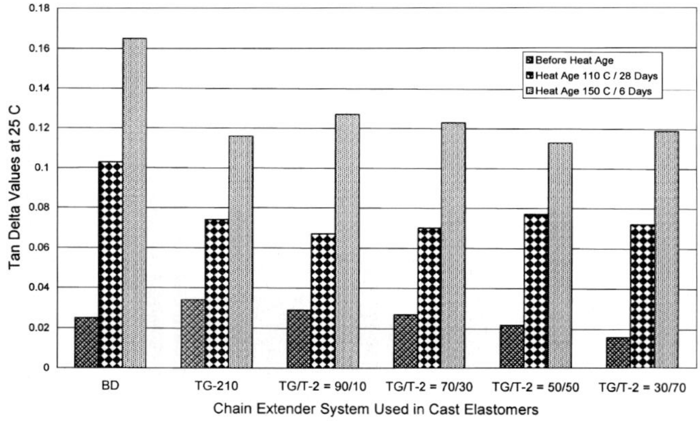


**Figure 7.145.** Tan delta values measured at 25 °C from DMA analyses Baytec ME-050-TG-210/Terathane (T) 2000 blends

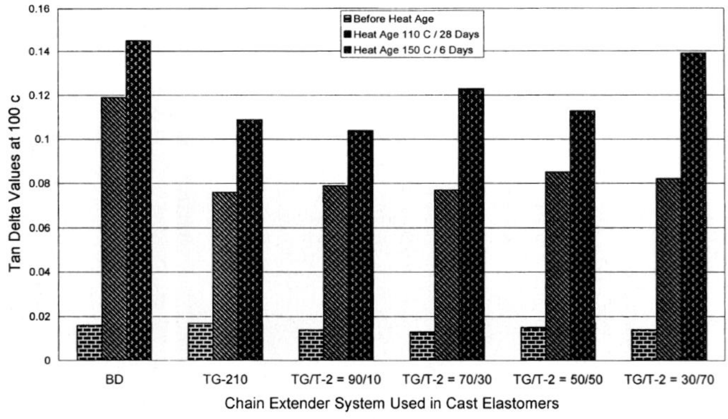


**Figure 7.146.** Tan delta values measured at 100 °C from DMA analyses Baytec ME-050-TG-210/Terathane (T) 2000 blends

The heating of elastomers markedly affected the hysteresis behavior of BD and TG-210/PTMEG extended elastomers. Compared to BD extended elastomer, TG/T-2000 extended elastomers showed lower tan δ values suggesting lower hysteresis for these systems. The Tg associated with the PTMEG soft segments were determined from the tan $\delta$ peak temperatures. By heat aging, the Tg values were shifted to higher temperatures.

**Loss Compliance (J″) Property** Loss compliance measured from the DMA quantifies the heat generated during the dynamic loading in applications like tires, wheels and rollers. Polyurethane elastomers exhibiting lower loss compliance values may experience less heat build up and consequently, may suffer few field failures [19,20]. Figures 7.147 and 7.148 show the plots of loss compliance values determined at 25 °C and 100 °C on the heat aged elastomer samples.

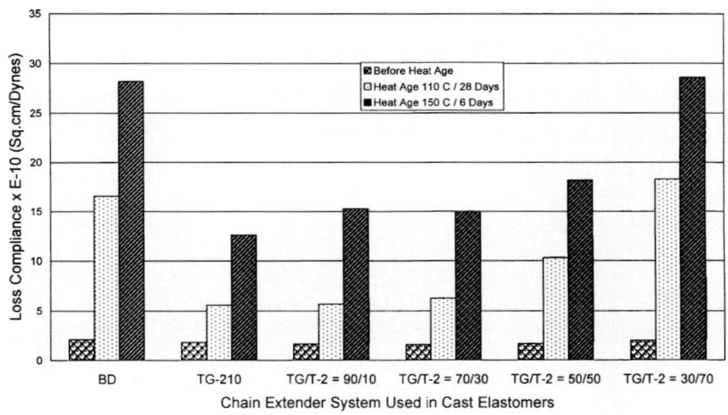


**Figure 7.147.** Loss compliance values measured at 25 °C from DMA analyses Baytec ME-050-TG-210/Terathane (T) 2000 blends

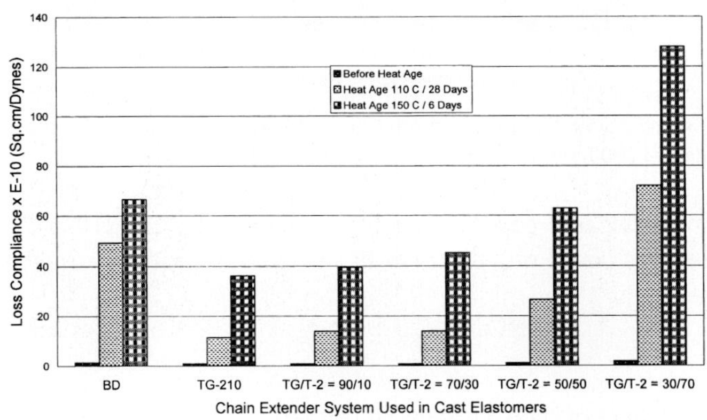


**Figure 7.148.** Loss compliance values measured at 100 °C from DMA analyses Baytec ME-050-TG-210/Terathane (T) 2000 blends

Comparing the results of BD and TG/T-2000 blends extended elastomers, it was observed that TG based elastomers showed lower J″ values even after heat aging for 28 days at 110 °C. Based on the loss compliance data, it can be expected that TG/T-2000 blends containing upto 50% of PTMEG polyol can perform better than BD in elastomer applications.

In addition to loss compliance data, the area under the loss compliance curve from the DMA can also be used to predict the tendency of elastomeric materials to convert mechanical energy into heat energy. A smaller loss compliance curve area means less mechanical energy is converted into heat, which can suggest better dynamic performance of the elastomer over the temperature of interest.

Polyurethane elastomers, in general, are known to withstand continuous use temperature of about 80 °C. At higher temperatures, a reduction in the physical and mechanical properties can be seen, which might be associated with the degradation of materials. With polyether based elastomers, the reduction in properties could be due to hysteresis. Therefore, in order to determine and predict the performance of TG-210/T-2000 based elastomers, the area under the loss compliance curves were determined between – 90 to 110 °C and plots were made (Figure 7.149).

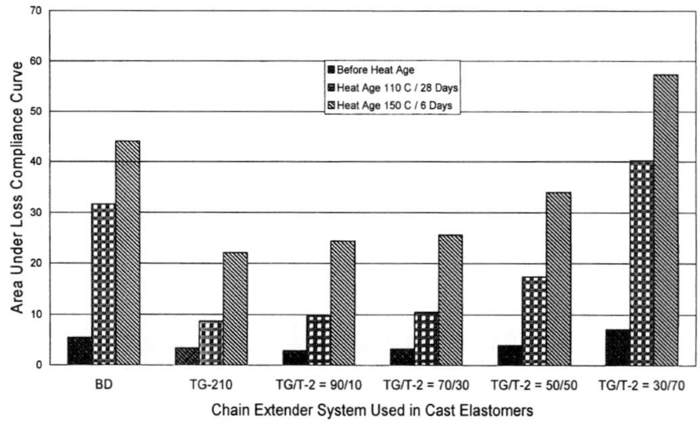


**Figure 7.149.** Area under the loss compilance DMA analysis curves Baytec ME-050-TG-210/Terathane (T) 2000 blends temperature range = – 90 to 110 °C

Again, from the loss compliance curve areas data, TG-210/T-2000 blends based elastomers could out-perform BD extended elastomer in the dynamic applications.

### 7.9.5.3
### Hydrolysis Resistance of High Resilient Elastomers

A hydrolysis resistance study was made on both BD and TG-210/PTMEG based elastomers. Table 7.82 shows the results on the effect of TG-210 concentration and Table 7.83 provides the results on the PTMEG molecular weight effect on the hydrolysis resistance of resilient elastomers synthesized from the MDI prepolymer.

**Table 7.82.** Hydrolysis resistance of cast elastomers from Baytec ME-050-TG-210/Terathane 2000 blends effect of HER TG-210 concentration

| TG-210/Terathane 2000 | BD | | TG-210 | | 90/10 | | 70/30 | | 50/50 | | 30/70 | |
|---|---|---|---|---|---|---|---|---|---|---|---|---|
| ratio (weight) | 100% | 100% | 100% | 100% | | | | | | | | |
| Water immersion | Before | After | Before | After | Before | After | Before | After | Before | After | Before | After |
| Tensile properties (MPa) | | | | | | | | | | | | |
| 100% Modulus | 6.3 | 4.7 (75) | 8.7 | 7.5 (86) | 8.4 | 7.2 (86) | 7.9 | 6.7 (85) | 7 | 5.8 (82) | 5.5 | 4.4 (79) |
| 200% Modulus | 8.8 | 6.9 (79) | 11.9 | 10.2 (86) | 11.4 | 9.9 (87) | 10.9 | 9.3 (86) | 9.9 | 8.2 (83) | 8 | 6.5 (81) |
| 300% Modulus | 12.8 | 9.8 (77) | 15.4 | 13 (84) | 14.8 | 12.8 (86) | 14.5 | 12.2 (84) | 13.7 | 11.2 (82) | 11.8 | 9.4 (82) |
| Tensile strength | 30.2 | 23.3 (76) | 24.8 | 21.5 (87) | 24.1 | 20.8 (86) | 25.5 | 22 (86) | 27 | 23.8 (88) | 27.3 | 24.3 (89) |
| Elongation (%) | 456 | 510 | 541 | 635 | 544 | 611 | 532 | 629 | 504 | 615 | 467 | 572 |
| Energy to break, kN/m (stress-strain curve area) | 52.7 | 49.5 | 84.2 | 88.9 | 81.6 | 83.8 | 79.1 | 85.6 | 68.6 | 79.5 | 51.6 | 58.3 |
| Hardness, Shore-A | 88 | 83 | 94 | 90 | 93 | 90 | 92 | 90 | 90 | 86 | 85 | 80 |

Values in parentheses are percentages of properties retained
Immersed in water at 80 °C for two weeks before testing
Data from [13]

**Table 7.83.** Hydrolysis resistance of cast elastomers from Baytec ME-050-TG-210/Terathane 2000 blends
effect of Terathane molecular weights on properties

| TG-210/Terathane 2000 | BD | | TG-210 | | 1000 | | 2000 | | 2900 | |
|---|---|---|---|---|---|---|---|---|---|---|
| ratio (weight) | 100% | | 100% | | 70/30 | | 70/30 | | 70/30 | |
| Water immersion | Before | After | Before | After | Before | After | Before | After | Before | After |
| Tensile properties (MPa) | | | | | | | | | | |
| 100% Modulus | 6.3 | 4.7 (75) | 8.7 | 7.5 (86) | 7.9 | 6.4 (81) | 7.9 | 6.7 (85) | 7.9 | 6.9 (85) |
| 200% Modulus | 8.8 | 6.9 (79) | 11.9 | 10.2 (86) | 10.7 | 9 (84) | 10.9 | 9.3 (86) | 10.8 | 9.3 (86) |
| 300% Modulus | 12.8 | 9.8 (77) | 15.4 | 13 (84) | 14.1 | 11.9 (84) | 14.5 | 12.2 (84) | 14.4 | 12.2 (84) |
| Tensile strength | 30.2 | 23.3 (76) | 24.8 | 21.5 (87) | 25.7 | 23.8 (92) | 25.5 | 22 (86) | 24.8 | 21.8 (88) |
| Elongation (%) | 456 | 510 | 541 | 635 | 519 | 605 | 532 | 629 | 530 | 628 |
| Energy to break, kN/m (stress-strain curve area) | 52.7 | 49.5 | 84.2 | 88.9 | 69.3 | 73.7 | 79.1 | 85.6 | 77.9 | 85.4 |
| Hardness, Shore-A | 88 | 83 | 94 | 90 | 90 | 89 | 92 | 90 | 92 | 90 |

Values in parentheses are the percentage of property retained
Immersed in water at 80 °C for two weeks before testing
Data from [13]

From the results in Tables 7.82 and 7.83, the cast elastomers developed from the TG-210/PTMEG blend systems showed about 10 to 12% improved hydrolytic stability than the BD extended elastomer.

The cast polyurethane elastomers obtained from the polyether prepolymer and HER TG-210/Terathane polyol blends showed low hysteresis (low tan $\delta$), excellent rebound (Bashore rebound $=>70$), good hydrolytic and thermal stabilities, low compression set and glass transition (Tg) properties. Elastomers exhibiting these properties could find applications in the areas of, for example, recreational wheels, seals and golf balls.

Golf balls made from the HER TG-250 and MDI-PTMEG system exhibited a significant increase in the initial velocity, significantly lower compression (better feel) and retained the same hardness as the control golf balls [21].

### 7.9.6 High Hardness Cast Polyurethanes

High hardness cast polyurethanes have been used in applications such as rolls, rollers and wheels. When the hardness of elastomer has been increased, certain physical and mechanical properties such as the tensile modulus tear strength and compressive strength properties have also been increased. In order to develop materials with the best overall properties, including the hardness, abrasion and heat resistance, MDI-prepolymers with higher NCO content of greater than 9% are often used.

Prepolymers having higher NCO content show very high reactivity towards the curing agents, and therefore mostly result in very short pot or working life. Because of high reactivity, the curing reactions often produce high exotherm, which could take the exothermic temperature as high as 150 °C. Due to these exothermic reactions and conditions involved with high NCO prepolymers, cast elastomer parts could tend to crack, glassy and brittle because of high cross-links formed between the prepolymer and extender.

High hardness polyurethane elastomers have, in general, higher amounts of hard segment content. Since hard segment is hydrophobic and less rapidly attacked, polyurethane elastomers exhibiting high hardness are generally more hydrolytically stable than the soft ones [22]. With high NCO prepolymers, HER chain extension can produce harder and thermally stable elastomers than the BD extender because of longer chain length.

Cast elastomers were prepared using Betathane 23.780 and Betathane 23.782 prepolymers and HER and BD extenders [23]. The prepolymer properties, experimental details on the elastomer castings and mix ratios of prepolymer and extender are presented in Table 7.84.

Cast elastomer properties have been shown in Table 7.85.

In the development of cast elastomers with high NCO prepolymers, HER chain extender processed identically to BD with essentially no loss in work life [4]. Highly resilient and abrasion resistant cast elastomers with ultra hard-

**Table 7.84.** High NCO MDI polyether prepolymers and cure properties

| MDI-PTMEG prepolymer | Betathane 23.780 | | Betathane 23.782 | |
|---|---|---|---|---|
| Physical form | Liquid | | Liquid | |
| Prepolymer % NCO | 13.4 | | 15.3 | |
| Equivalent weight | 313.6 | | 274.6 | |
| Viscosity, Cps at 45 °C | 3700 | | 750 | |
| Specific gravity at 25 °C | 1.15 | | 1.15 | |
| Flash point (°F) | > 200 | | > 200 | |
| Mixing and cure properties | | | | |
| Chain extenders used | BD | HER | BD | HER |
| Mix ratio, prepolymer/extender | 100/13.6 | 100/30 | 100/15.6 | 100/34.3 |
| Stoichiometry, % | 95 | 95 | 95 | 95 |
| Prepolymer temperature, °C | 45±5 | 75±5 | 45±5 | 75±5 |
| Chain extender temperature, °C | 25 | 95±5 | 25 | 95±5 |
| Work life, min | 4–6 | 4–6 | 4–6 | 4–6 |
| Mold temperature, °C | 100–110 | 120 | 100–110 | 120 |
| Demold, min | 15–20 | 10–15 | 15–20 | 15–20 |

BD = 1,4-Butanediol, HER = Resorcinol di($\beta$-hyroxyethyl) ether
Data from [23]

ness were achieved with Betathane/HER systems. In order to determine the retention of hardness (Shore D) at elevated temperatures, hardness measurements were made at higher temperatures (Figure 7.150).

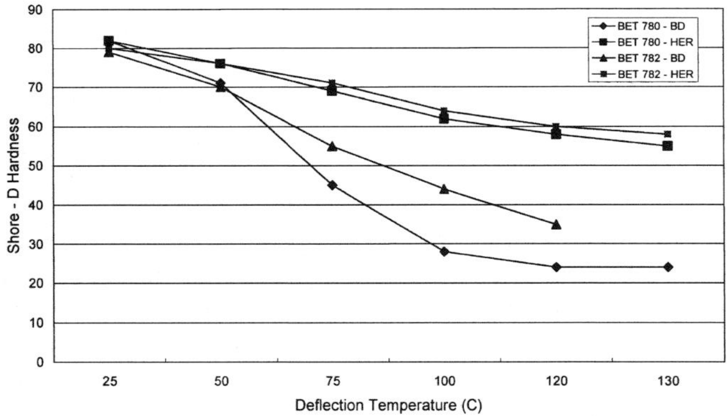


**Figure 7.150.** Ultra hardness cast polyurethanes from HER chain extender

HER extended elastomers retained their hardness better than BD. The marked hardness reduction which was observed in Betathane 23.780/BD at 70 °C was not observed with Betathane 23.780/HER system. Based on this work, it was suggested that these ultra hardness urethane elastomers could be used in most high durometer elastomer applications.

**Table 7.85.** High hardness cast urethane elastomers and properties

| MDI-PTMEG prepolymer | Betathane 23.780 | | Betathane 23.782 | |
|---|---|---|---|---|
| Chain extenders used | BD | HER | BD | HER |
| Hardness, Shore-D | 80–82 | 80–82 | 78–80 | 80–82 |
| Tensile properties (MPa) | | | | |
| 100% Modulus | 32.4 | 41.4 | 37.6 | 42.4 |
| Tensile strength | 48.3 | 51 | 53.1 | 43.4 |
| Tensile at yield | 46.2 | 45.5 | NA | 43.4 |
| Elongation (%) | 135 | 150 | 180 | 125 |
| Tear properties (kN/m) | | | | |
| Die-C Tear | 227.5 | 205.6 | 206.5 | 240 |
| Split tear, Method D 470 | 22.8 | 21 | 21 | 23.6 |
| Split tear, Method D 1935 | 70 | 70 | 75.3 | 70.9 |
| Taber wear index (mg/1000 cycles) (CS-17 Wheel, 1000 g load) | 18 | 11 | 18 | 13 |
| Resiliency, Bashore (%) | 70 | 55–60 | 70 | 55–60 |
| Flexural modulus (GPa) | 2.04 | 1.55 | 1.34 | 1.55 |
| Shore-D deflection at (°C) | | | | |
| 25 | 82 | 82 | 79 | 80 |
| 50 | 71 | 76 | 70 | 76 |
| 75 | 45 | 69 | 55 | 71 |
| 100 | 28 | 62 | 44 | 64 |
| 120 | 24 | 58 | 35 | 60 |
| 130 | 24 | 55 | NA | 58 |

BD = 1,4- Butanediol, HER = Resorcinol di($\beta$-hydroxyethyl) ether
Data from [23]

## 7.9.7 HPR Extended Cast Elastomers

HPR is also a resorcinol based aromatic diol, which could be used as a chain extender in the cast polyurethane applications. Cast polyurethanes made from the HPR extender were studied for their physical and mechanical properties [6]. Tables 7.86 and 7.87 summarize the experimental details on the cast elastomer preparations as well as the properties investigated.

With HPR extender, the presence of methyl groups and secondary hydroxyl groups on the chain showed pronounced effects on the cured elastomer properties compared to HER extender.

Polyester MDI prepolymers and HPR extender produced elastomeric materials, which were softer (Shore A = 70–80) with lower tensile strength, tear and higher compression set properties compared to HER extended elastomers. Interestingly, HPR based polyester materials exhibited very low rebound properties. From the physical appearances of the cast polyurethane films, these elastomers appeared much clearer and nearly transparent. This suggested

**Table 7.86.** HPR extended cast polyurethanes using polyester prepolymers

| Prepolymer type | MS-242 | B-8520 | SMS-85A |
|---|---|---|---|
| Prepolymer manufacturer | Bayer | Uniroyal | Air Products |
| Prepolymer temperature (°C) | 90 | 100 | 90 |
| Chain extender temperature (°C) | 90 | 88 | 90 |
| Mix ratio (prepolymer/extender) | 100/16.3 | 100/17 | 100/16.7 |
| Stoichiometry, % theory | 90 | 90 | 90 |
| Cure, h/°C | 16/110 | 16/110 | 16/110 |
| Tensile properties (MPa) | | | |
| 100% Modulus | 2.6 | 2.5 | 2.8 |
| 300% Modulus | 6.1 | 6.5 | 8.1 |
| Tensile strength | 22 | 11.7 | 21.3 |
| Elongation (%) | 392 | 32.2 | 39.2 |
| Tear strength, Die-C (kN/m) | 37.5 | 32.2 | 39.2 |
| Compression set, % | 57 | 31 | 46 |
| Bashore rebound, % | 6 | 9 | 14 |
| Hardness, Shore-A | 68 | 68 | 80 |
| Dynamic mechanical analysis (DMA) | | | |
| Tan delta ($\delta$) at 10 Hz/23 °C (frequency) | 0.87 | 0.75 | 1.32 |
| Tan delta ($\delta$) at 25 °C (temp. sweep) | 0.3 | 0.21 | 0.63 |
| Tg (tan delta peak), °C | 12 | 8 | 15.6 |
| Thermal stability from DMA (temp. max for constant G″) | | | |
| Temperature (°C) | 150 | 155 | 155 |
| Storage modulus (G′), (dynes/cm$^2$) $\times 10^8$ | 0.28 | 0.29 | 0.28 |

Data from [6, 10]

that, even-though HPR chain extension produced hard segments with MDI prepolymers, they could be more amorphous in character rather than crystalline hard segments. These phenomena might have caused the hard and soft segments mixed with each other.

The DSC analysis made on the elastomers, also, did not show any endothermic transition temperature between room temperature and 200 °C. Published results, similar to this, indicated that polyurethane structures with polyester soft segments containing methyl side chain exhibited lower tensile strength properties compared to linear polyester. This was associated with the effect of the side chain, such as the methyl group, preventing or hindering the crystallization of polyesters. Therefore, it can be assumed that the presence of methyl groups in the HPR extender might have prevented the formation of crystalline hard segments from the reaction with MDI prepolymers.

DMA study made on the HPR extended elastomers showed high hysteresis (high tan $\delta$), higher Tg and lower dynamic stiffness properties at elevated temperatures.

**Table 7.87.** HPR Extended cast polyurethanes using polyether prepolymers

| Prepolymer type | ME-050 | B-625 | SME-90A |
|---|---|---|---|
| Prepolymer manufacturer | Bayer | Uniroyal | Air Products |
| Prepolymer temperature (°C) | 80 | 85 | 90 |
| Chain extender temperature (°C) | 75 | 85 | 90 |
| Mix ratio (prepolymer/extender) | 100/15.2 | 100/15.9 | 100/17.5 |
| Stoichiometry, % theory | 90 | 90 | 90 |
| Cure, h/°C | 16/110 | 16/110 | 16/110 |
| Tensile properties (MPa) | | | |
| 100% Modulus | 2.7 | 2.7 | 2.9 |
| 300% Modulus | 7.5 | 9.3 | 12 |
| Tensile strength | 17.4 | 15 | 21.3 |
| Elongation (%) | 354 | 329 | 326 |
| Tear strength, Die-C (kN/m) | 37.5 | 31.1 | 33.1 |
| Compression set, % | 16.4 | 10.5 | 6.1 |
| Bashore rebound, % | 30 | 19 | 12 |
| Hardness, Shore-A | 70 | 70 | 70 |
| Dynamic mechanical analysis (DMA) | | | |
| Tan delta ($\delta$) at 10 Hz/23 °C (frequency) | 0.053 | 0.131 | 0.187 |
| Tan delta ($\delta$) at 25 °C (temp. sweep) | 0.03 | 0.053 | 0.05 |
| Tg (tan delta peak), °C | – 17.7 | – 10.2 | – 5.9 |
| Thermal stability from DMA (temp. max for constant G″) | | | |
| Temperature (°C) | 150 | 155 | 155 |
| Storage modulus (G′), (dynes/cm$^2$) $\times 10^8$ | 0.33 | 0.31 | 0.35 |

Data from [6, 10]

Cast elastomers obtained with MDI polyether prepolymers behaved similarly like polyester materials. With SME-90A prepolymer, a very low compression set and also low Bashore rebound properties were observed. HPR extension produced very low Bashore rebound properties. Polyurethanes with low Bashore rebound properties (less than 20) are considered as good energy absorbing materials, and therefore HPR elastomers could be valuable in those applications.

## 7.9.8 Cast Elastomers Based on TDI Prepolymers

TDI prepolymers have been widely known in the cast polyurethane industry and used in the manufacture of wide variety of products for numerous applications. With TDI prepolymers, 4,4′-methylene bis(2-chloroaniline) (MBCOA) has been used as the curing agent or extender to produce cast elastomers. For developing low durometer hardness elastomers, for the printing rolls and metal coating industries, TDI prepolymers are often cured with the combination of diol and triol extender systems. In the preparation of low durometer rolls, plasticizers and fillers are often used to modify the properties. It may, sometimes, be difficult to obtain hardness below Shore 50 to 55 A in cast elastomers without the use of plasticizers. Most low durometer elastomers contain varying amounts of plasticizers to reduce the hardness.

In the development of low durometer urethane elastomers, TDI esters are the materials of choice because they can tolerate greater loading of plasticizers without reduction of physical properties. Polyether TDI prepolymers do not provide high physical properties in low durometer and highly plascticized formulations.

The technical grade HER, namely HER TG-300, and HPR extenders were used in the development of cast elastomers using TDI prepolymers. Without adding any plasticizers, cast elastomers have been formulated and tested for their physical and mechanical properties [24]. The results from this work are summarized in Tables 7.88 and 7.89.

**Table 7.88.** Cast elastomers using polyether TDI prepolymer (low free TDI prepolymer Adiprene-L83, NCO content = 3.4%)

| Chain extender | HER-HP | HER TG-300 | HPR |
|---|---|---|---|
| Chain extender/trimethylolpropane | 90/10 | 90/10 | 90/10 |
| Processing temperature (°C) | 90 | 95 | 95 |
| Mix ratio (prepolymer/extender) | 100/7 | 100/8 | 100/7.7 |
| Cure, h/°C | 16/130 | 16/130 | 16/130 |
| Tensile properties (MPa) | | | |
| 50% Modulus | 1.1 | 1.3 | 1.1 |
| 200% Modulus | 1.5 | 1.7 | 1.5 |
| 300% Modulus | 2.2 | 2.8 | 2.6 |
| Tensile strength | 11.7 | 12.9 | 12.3 |
| Elongation (%) | 597 | 529 | 563 |
| Compression set, % | 24.8 | 14.1 | 17.2 |
| Bashore rebound, % | 64 | 68 | 60 |
| Hardness, Shore-A | 58 | 60 | 55 |

Data from [24]

It was observed from the results in Tables 7.88 and 7.89 that low durometer hardness (Shore A 40 to 60) materials could be produced with HPR and TG-300 chain extenders. Avoiding the use of plasticizers in the low durometer elastomers can provide materials with longer service life in applications such as roll covers.

**Table 7.89.** Cast elastomers using low free TDI polyester prepolymer (TDI prepolymer: Adiprene LF 1700 A, NCO content = 2.4%)

| Chain extender | HER-HP | | HER TG-300 | | | |
|---|---|---|---|---|---|---|
| Chain extender/ trimethylolpropane | 100/0 | 90/10 | 100/0 | 90/10 | 80/20 | 70/30 |
| Processing temp. (°C) | 88 | 94 | 90 | 90 | 85 | 85 |
| Mix ratio (prepolymer/extender) | 100/4.9 | 100/6.4 | 100/6.6 | 100/5.7 | 100/4.9 | 100/4.4 |
| Cure, h/°C | 16/130 | 16/130 | 16/130 | 16/130 | 16/130 | 16/130 |
| Tensile Properties (MPa) | | | | | | |
| 50% Modulus | 0.9 | 1 | 0.7 | 0.9 | 0.8 | 1 |
| 200% Modulus | 1.1 | 1.5 | 0.9 | 1.2 | 1.2 | 1.5 |
| 300% Modulus | 1.5 | 2.4 | 1 | 2 | 2.1 | 2.8 |
| Tensile strength | 11.6 | 22.6 | 6.4 | 16.7 | 16.4 | 17.6 |
| Elongation (%) | 728 | 631 | No Break | 674 | 655 | 560 |
| Tear strength, kN/m | 24.2 | 37.6 | 21 | 27.1 | 28.9 | 32.2 |
| Bashore rebound, % | 27 | 37 | NA | 53 | 55 | 56 |
| Compression set, % | 69 | 16.6 | 93 | 10.3 | 4 | 2.7 |
| Hardness, Shore-A | 45 | 48 | 38 | 44 | 46 | 50 |

Data from [24]

## 7.9.9 Cast Polyurethanes Using PPDI Prepolymers

*p*-Phenylene diisocyanate (PPDI) is a specialty diisocyanate developed and marketed by Dupont Company under the trade name Hylene PPDI. Polyurethane elastomers prepared from PPDI based prepolymers have shown superior properties than those obtained from the standard prepolymers based on TDI and MDI [25–27]. The rigid rod-like structure of PPDI produces highly crystalline hard segments with chain extenders, which are responsible for the superior physical and mechanical properties observed. PPDI based cast elastomers exhibited improved hydrolysis resistance, superior dynamic properties, low compression set, high Bashore rebound, low Taber abrasion and good cut and tear resistance at temperature higher than for conventional elastomers. The relatively constant low tan $\delta$ property observed over a wide temperature for the PPDI based elastomers could show the superior performance relative to MDI elastomers.

PPDI prepolymers are commercially available from Uniroyal Chemical Company under the trade name Adiprene. Polyether PPDI prepolymer under the name Adiprene LFPX 950 and polyester prepolymer under the name Adiprene LFPX 1950 are available for cast polyurethanes development. Cast urethanes developed with these two prepolymers and the properties studied are presented in Table 7.90 [28, 29].

**Table 7.90.** Cast elastomers based on PPDI prepolymers

| Prepolymer | Polyether X950 | | Polyester X1950 | |
|---|---|---|---|---|
| Chain extender used | BD | HER-HP | BD | HER-HP |
| Pot-life determination | | | | |
| Temperature (°C) | 100 | 100 | 100 | 100 |
| Pot-life (min) | 2 | 12 | 4.5 | > 20 |
| Pot-life viscosity (Poise) | 7 | 91 | 62 | 60 |
| Casting conditions | | | | |
| Prepolymer temperature, °C | 90 | 105 | 100 | 100 |
| Chain extender temperature, °C | RT | 105 | RT | 100 |
| Mix ratio (prepolymer/extender) | 100/5.6 | 100/12.6 | 100/3.2 | 100/6.9 |
| Stoichiometry, % theory | 95 | 95 | 95 | 95 |
| Cure, h/°C | 16/110 | 16/110 | 16/110 | 16/110 |
| Tensile properties (MPa) | | | | |
| 100% Modulus | 13.2 | 15.1 | 9.04 | 6.6 |
| 300% Modulus | 17.9 | 15.5 | 13 | 10.9 |
| Tensile strength | 33.2 | 31.2 | 43.8 | 36.8 |
| Elongation (%) | 443 | 474 | 642 | 639 |
| Tear strength, Die-C (kN/m) | 114.5 | 121.2 | 136 | 111 |
| Compression set, % | | | | |
| 1. 70°C/22h | 32.4 | 29.1 | 47.2 | 43 |
| 2. 100°C/22h | 43.2 | 35.9 | 58.9 | 50.5 |
| 3. Water Immersion for five weeks at 80°C | 35 | 30 | ND | ND |
| Bashore rebound, % | 59 | 60 | 62 | 58 |
| Hardness, Shore-A (D) | 99 (50) | 99 (55) | 95 (40) | 90 (40) |
| Taber abrasion (mg), | 33 | 22 | 60 | 14 |

(Load = 1000 g, cycles = 1000)
Data from [28, 29]

In the case of polyether PPDI based elastomer, HER extension produced an elastomer with improved tear strength, compression set, Bashore rebound and abrasion resistant properties than the BD extender. With polyester prepolymer, lower compression set and excellent abrasion properties were observed with HER extender.

### 7.9.9.1 Heat Aging of Polyether Elastomers

This study was undertaken to determine the thermal stability of HER extended PPDI polyether elastomer and also to compare its performance against BD. Cast elastomer test samples were cut from a sheet of post-cured material for DMA, DSC and tensile properties determinations. Heating aging of DMA and DSC samples were done in an air oven and aged for 28 days at 110°C and 7 days at 150°C. Samples were conditioned for seven days at 50% RH condition before testing.

### 7.9.9.2 DMA Study Results

The storage modulus (G′), tan δ and loss compliance properties determined from the DMA on the unaged and heat aged elastomer samples are presented in Table 7.91.

**Table 7.91.** DMA results of cast elastomers based on PPDI prepolymers polyether X950-BD and HER chain extender system

| Chain extender | Butanediol | | | HER | | |
|---|---|---|---|---|---|---|
| Heat age temp/days | Unaged | 110 °C/ 28 days | 150 °C/ 7 days | Unaged | 110 °C/ 28 days | 150 °C/ 7 days |
| Storage modulus(G′) (dynes/cm$^2$) $\times 10^8$ | | | | | | |
| at 25 °C | 7.26 | 7.97 | 7.78 | 6.49 | 7.24 | 6.51 |
| at 100 °C | 6.28 | 6.75 | 6.16 | 5.83 | 6.38 | 5.51 |
| at 130 °C | 5.94 | 6.18 | 5.57 | 5.8 | 6.2 | 5.47 |
| at 150 °C | 5.54 | 5.47 | 5.05 | 5.71 | ND | 4.75 |
| Tan δ properties | | | | | | |
| Peak temp, Tg (°C) | – 43.3 | – 45.5 | – 47.2 | – 33.7 | – 35.8 | – 37.5 |
| at 25 °C | 0.009 | 0.016 | 0.025 | 0.015 | 0.021 | 0.055 |
| at 100 °C | 0.013 | 0.026 | 0.051 | 0.009 | 0.018 | 0.055 |
| at 130 °C | 0.013 | 0.032 | 0.058 | 0.008 | 0.019 | 0.051 |
| at 150 °C | 0.013 | 0.039 | 0.064 | 0.009 | ND | 0.051 |
| Loss compliance (J″) (cm$^2$/dynes) $\times 10^{-10}$ | | | | | | |
| Peak temp, °C | – 29.9 | – 29.7 | – 30.7 | – 21.7 | – 19.8 | – 25.8 |
| at 25 °C | 0.12 | 0.2 | 0.32 | 0.23 | 0.3 | 0.84 |
| at 100 °C | 0.21 | 0.39 | 0.83 | 0.17 | 0.28 | 0.99 |
| at 130 °C | 0.22 | 0.52 | 1.12 | 0.13 | 0.31 | 0.94 |
| at 150 °C | 0.24 | 0.72 | 1.25 | 0.16 | ND | 1.06 |
| Peak area, – 90 to 110 °C | 0.65 | 0.73 | 1.04 | 0.76 | 0.77 | 1.58 |

Data from [28, 29]

The temperature sweep DMA curves of BD and HER extended cast elastomers obtained from the polyether PPDI prepolymer are shown in Figures 7.151 and 7.152.

From the data shown in Figures 7.151 and 7.152, the storage modulus of BD and HER extended elastomers have been found to be relatively constant beyond 175 °C temperatures. In addition, the tan δ values were very low and remained relatively unchanged for a wide temperature range. The very low tan δ values observed at 150 °C for these two elastomers suggest these elastomers could perform well in the dynamic applications at this high temperature conditions. Heat aged elastomers also showed very low tan δ and loss compliance

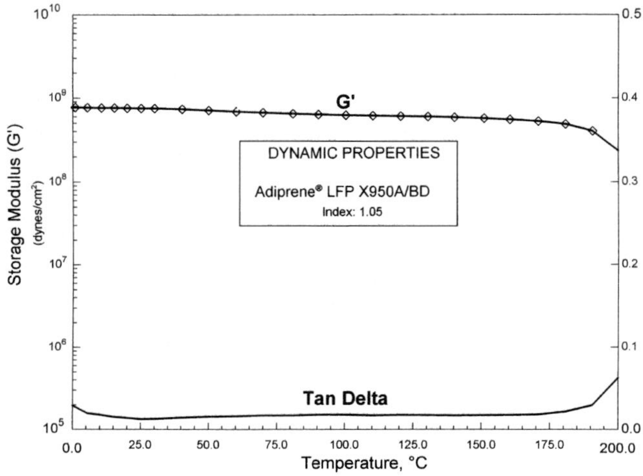


**Figure 7.151.** DMA curves of BD extended polyether-PPDI prepolymer based cast elastomers

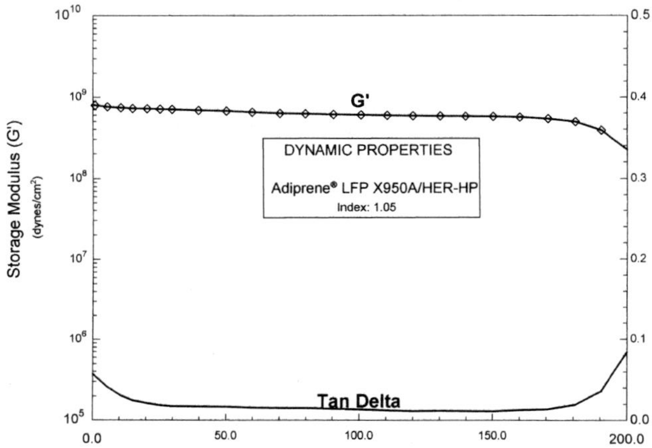


**Figure 7.152.** DMA curves of HER-HP extended polyether-PPDI prepolymer based cast elastomer

properties even at elevated temperatures, which predicts the superior high temperature performance. Due to excellent dynamic properties, HER-PPDI polyether elastomers could be used for high temperature automotive parts, hydraulic seals, packings and gaskets applications where high performance is most desired.

### 7.9.9.3 Hydrolysis Resistance of Elastomers

HER and BD extended cast elastomers were immersed in water conditioned at 80 °C for 14 days and then tested for the retention of tensile, tear and hardness properties (Table 7.92).

**Table 7.92.** Hydrolysis resistance of polyether PPDI based cast elastomers

| Chain extender | Butanediol | | HER-HP | |
|---|---|---|---|---|
| Water immersion | Before | After | Before | After |
| Tensile properties (MPa) | | | | |
| 100% Modulus | 13.2 | 13.1 (99) | 15 | 15 (100) |
| 300% Modulus | 17.9 | 16.8 (93) | 15.5 | 15.2 (98) |
| Elongation (%) | 443 | 498 | 474 | 474 |
| Tear strength, Die-C (kN/m) | 114.5 | 133 | 121.3 | 119.7 |
| Hardness, Shore-A (D) | 99 (49) | 99 (49) | 99 (50–53) | 99 (50–53) |

Values in parentheses are the percentages of property retained
Water immersion conditions: two weeks at 80 °C
Data from [28,29]

Based on Table 7.92 results, PPDI based elastomers exhibited excellent hydrolysis resistant properties for both HER and BD extended systems.

### 7.9.9.4 DSC Analysis

From the DSC analysis, the thermal stability of PPDI polyether cast elastomers was determined from the endothermic transition temperatures associated with the melting of hard segment domains. In the case of HER extended elastomer, due to *m*-phenylene linkage, the hard segments are expected to be more rigid and crystalline compared to BD extender. Heat aged samples were analyzed by the DSC method to determine their stability after the heat aging (Table 7.93).

A DSC thermogram obtained on the HER extended elastomer heat aged for 28 days at 110 °C in an air oven is illustrated in Figure 7.153.

The first heating DSC endothermic peaks for HER extended PPDI elastomer appeared at higher temperatures than the BD based elastomer. Higher endothermic peak energies also suggest that larger size crystalline hard segment domains exist within HER-PPDI elastomer. Considering the molecular structures of BD and HER chain extenders, it could be possible that the higher thermal stability of HER extended elastomer could be associated with the formation of larger hard segments from the PPDI and extender reaction.

On cooling the heated samples (DSC curve shown in Figure 7.153), exothermic peaks associated with the crystallization of hard segment domains were seen. Once again, the higher exothermic peak temperature suggested the existence of larger size hard segments for HER based elastomer.

**Table 7.93.** DSC results of cast polyurethanes prepared from polyether PPDI prepolymer and BD and HER-HP extenders heat aging study

| Chain extender used | First heating Endotherm | | Cooling down Exotherm | | Second heating Endotherm | |
|---|---|---|---|---|---|---|
| | Peak (°C) | Energy, J/g | Peak (°C) | Energy, J/g | Peak (°C) | Energy, J/g |
| Butanediol | | | | | | |
| Unaged | 185.193 | 26 | 168 | 18 | 198 | 19 |
| 110°C/28 days | 178.198 | 25 | 170 | 11 | 195 | 7 |
| 135°C/14 days | 173.199 | 38 | 168 | 11 | 194 | 13 |
| 150°C/7 days | 178.196 | 34 | 171 | 11 | 196 | 11 |
| HER-HP | | | | | | |
| Unaged | 202.213 | 31 | 170 | 25 | 210 | 20 |
| 110°C/28 days | 212 | 39 | 172 | 23 | 208 | 12 |
| 135°C/14 days | 207 | 46 | 170 | 20 | 205 | 23 |
| 150°C/7 days | 208 | 46 | 164, 186 | 18 | 204 | 21 |

Data from [28,29]

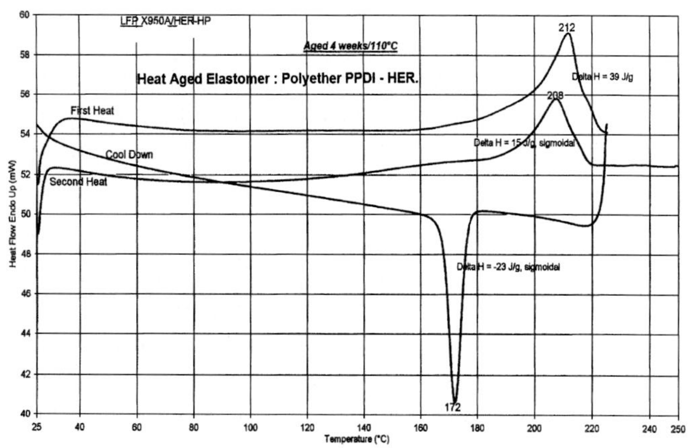


**Figure 7.153.** DSC curves of heat aged cast elastomers made from polyether-PPDI prepolymer and HER-HP materials

During the second heating, DSC curves produced well defined endothermic melting peaks appeared at higher temperatures, which also confirms the higher thermal stability of HER based PPDI elastomer. From the DSC results, HER extension produced elastomers with about 8 to 12°C higher hard segment melting temperatures than the BD.

All the physical, mechanical and DMA data obtained on the HER-PPDI polyether elastomers strongly suggest that they could be very valuable for use in dynamic applications such as wheels, hydraulic seals, drive belts, roll covers, hydrocyclones and power transmission couplings and drive shafts.

### 7.9.10
### Summary and Outlook

Cast polyurethanes obtained from the resorcinol based aromatic diol chain extenders have shown improved performance over the butanediol extender. HER and HPR chain extenders produced urethane elastomers having a wide range of physical and mechanical properties with TDI, MDI and PPDI based polyether and polyester prepolymers. The combination of higher thermal stability, improved dynamic and mechanical properties of HER extended elastomers can make them suitable for a wide range of industrial applications.

## References


1. K. Polamanteer, A. K. Bhowmick and H. L. Stephens, Eds., Handbook of elastomers, Marcel Dekker, New York, p 375 (1988).
2. Z. Wirpsza, Polyurethane chemistry, technology and applications, T.J. Kemp, Ed., Ellis Harwood PTR Prentice Hall, New York (1993).
3. K. C. Frisch and S. L. Reegen, Adv. Urethane Sci. Technol., Vol. 2, p 29 (1973).
4. J. E. Tiedemann, Paper presented at PMA Meeting, Miami, FL, USA (1993).
5. M. A. Mendelsohn, F. W. Navish and D. Kim, Rubber Chem. Technol., Vol. 58, pp 5 and 997 (1986).
6. R. B. Durairaj et al., Paper presented at PMA Meeting, Baltimore, MD, USA (1998).
7. K. G. Reabe and H. Dressler, US Patent, 3,644,534 (1972).
8. H. Dressler, US Patent, 5,059,723 (1991).
9. B. J. Hitch and B. L. Bernard, PCT Appln., WO 91,016,292 (1991).
10. R. B. Durairaj, Paper presented at PMA Meeting, Hilton Head, SC, USA (1999).
11. R. B. Durairaj and J. de Almeida, Adhesives and sealants industry, March (1999).
12. R. B. Durairaj, HER materials for polyurethane applications, In: D. Klempner and K. C. Frisch, Eds., Advances in urethane science and technology, RAPRA Technology, UK (2001).
13. R. B. Durairaj, PCT Appln., WO 00/21913 (2000).
14. G. Oertel, Editor, Polyurethane handbook, 2nd Edn, Hanser, Munich (1993).
15. R. Palinkas, Paper presented at PMA Meeting, October (1990).
16. C. Demarest, Paper presented at PMA Meeting, March (2001).
17. R. B. Durairaj, US Patent, 6,303,732 (2001).
18. J. E. Doyle, Paper presented at PMA Meeting, Milwaukee, WI, USA (1994).
19. S. M. Clift, Paper presented at PMA Meeting, Charlotte, NC, USA (1993).
20. L. F. Nielsen, Mechanical properties of polymers and composites, Marcel Dekker (1974).
21. S. Wu and M. Rajagopalan, US Pat. Appln, 2002 / 0,004,429 (2002).
22. C. Hepburn, Polyurethane elastomers, 2nd Edn, Elsevier Applied Science, London (1992).
23. Data from the Product Brochure from Essex Specialty Products, October (1983).
24. R. B. Durairaj and J. de Almeida, Paper presented at Canadian PMA Meeting (2001).
25. A. Singh, Advan Urethane Sci. Tech, Vol. 13 (1996).
26. B. S. Lambardo et al., Advances in PPDI prepolymer technology, Uniroyal Chemical Company, Middlebury, CT, USA.

27. Technical Data Sheets on Adiprene LFPX950 and LFPX1950 Prepolymers from Uniroyal Chemical Company, Middlebury, CT, USA.
28. R. B. Durairaj and J. de Almeida, Paper presented at PMA Meeting, New Orleans, Louisiana, USA (1999).
29. R. B. Durairaj and J. de Almeida, Poster paper presented at SPI Polyurethanes EXPO 99, Orlando, FL, 12–15 September (1999).

# 8 Polymer Additives

## 8.1
## UV Absorbers

### 8.1.1
### Introduction

Polymeric materials have been widely used in the building industry in the form of thermoplastic and coating products. Thermoplastic compositions have found wide acceptance as useful and most important materials for making roofing materials, plastic sidings, foils, films, fibers and various other molded articles. Articles made from thermoplastic materials are inherently stronger, dimensionally very stable and sufficiently flexible to suit the requirements of their intended use. With proper coating formulations they are easily and variously colored to give them attractive appearances. In outdoor applications these materials are constantly exposed to sunlight during regular use. Therefore, the articles exposed to sunlight or an ultraviolet (UV) radiation source in the atmosphere are unfortunately subject to rapid and severe degradation and deterioration as evidenced by changes in their physical properties. Such photo-degradation can be particularly evidenced by surface cracking, increased brittleness, lost of mechanical strength and discoloration of the polymer [1, 2].

The preliminary effect of UV light or radiation on polymers is the formation of free radicals due to the breaking of molecular bonds. The free radicals thus formed are then reacting with atmospheric oxygen to form peroxy free radicals, and this process is called auto-oxidation. On further attack, these peroxy radicals abstract hydrogen from the polymer backbone, and thus form hydroperoxides and more free radicals. As the process of forming free radicals continues, more and more molecular bonds present in the polymer break, which can lead to deterioration of mechanical properties and discoloration of polymers. In order to prevent or overcome this deterioration of polymers, ultraviolet (UV) inhibitors or light stabilizers have been used, incorporating them with the polymers to stabilize against the deleterious effects of light.

Ultraviolet (UV) light absorbers or stabilizers (in short, UV absorbers) are compounds which, on addition to the polymeric materials, are capable of preventing or retarding the degradation reactions caused by light energy. Short wavelength UV light (UVC) having a wavelength of not more than about 290 nanometer (nm) is absorbed by the ozone layer surrounding the earth. The UV light component of sunlight which influences industrial products, such as plastics, coatings and cosmetics, and human skin is the medium wavelength UV light (UVB) having a wavelength of 290 to 320 nm and long wavelength UV light (UVA) having a wavelength of 320 to 400 nm.

The use of UV absorbers in various thermoplastic polymers, such as polyesters, polyolefins, vinyls, polycarbonates and polystyrene, provides protection against attack by the UV radiation. UV absorbers function by their ability to screen out the damaging UV portion of the light due to very high absorptivity relative to that of polymers. In order to qualify as a UV useful and effective UV absorber for a polymer, there are several requirements which the absorber must fulfil. The absorber must have a high specific absorptivity in the range of wavelength about 290 to 400 nm, which are most deleterious to the polymer and are present in the source of light exposure. In addition, they must be compatible with the polymer, coatings and other compositions and formulations. By their incorporation, they should not degrade either the polymer or coating compositions, which could cause a loss of properties and increase in color. Most importantly, the absorber should not significantly absorb in the visible region of the spectrum or a color may be imparted to the polymer to which the absorber has been added.

If the absorbers are highly volatile, then they will be volatilized out of the plastic or polymer composition during the processing, which may result in leaving the polymer unprotected against UV light sooner or later. Therefore the absorber must have a sufficiently low volatility to permit its continued presence in the polymer. In addition to above requirements, the absorbers must be stable to UV light, heat and moisture.

In general, UV absorbers vary in their ability to absorb UV light, in resistance to removal by extraction or volatilization, and in stability against the effects of light, oxygen and water. High UV absorber or high UV stability means that polymer or coating compositions containing the absorbers are not damaged or only very slightly damaged by sunlight or other UV radiation, which could make them suitable for outdoor and critical indoor applications. When used outdoors for a number of years, the plastics or coatings should not yellow and not exhibit embrittlement or surface cracking, nor show any impairment of their physical and mechanical properties.

The use of plastic materials in the outdoor applications is feasible only because of the use of efficient and effective photo stabilizers or UV absorbers designed to slow down the degradation process. The UV wavelength sensitivity for the degradation processes in common and commercial polymers have been reviewed and published [3, 4]. A number of well known polymers and their UV wavelength sensitivity are presented in Table 8.1.

**Table 8.1.** Photochemical sensitivity of polymers

| Polymers | Wavelength, nm |
|---|---|
| Polyester | 325 |
| Polystyrene | 318 |
| Polyethylene | 300 |
| Polypropylene | 310 |
| Polyvinyl chloride | 310–320 |
| Polymethyl methacrylate | 315 |
| Polycarbonate | 295, 345 |
| Styrene acrylonitrile | 290, 325 |

Data from [4]

The UV absorbers protect these polymers against photo-degradation by competing with the polymer for the absorption of UV light. UV absorbers may act as optical filters or deactivate the substrate molecules that have been excited by light. The protection mechanism of UV absorbers is based on essentially absorbing the harmful UV radiation and dissipation as heat. An ideal UV absorber (UVA) to protect the thermoplastic polymers showed in Table 8.1 should be very light stable and have high absorption over the UV range from 290 to 400 nm. The performance of UV absorbers in plastics and coatings have been reviewed and well documented in the literature [5–10]. The most commercially important UV absorbers are based on benzophenone, benzotriazole and triazine chemistries derived from resorcinol and its derivatives.

In this chapter, a detailed overview of different UV absorbers technologies developed and applied based on the resorcinol chemistry is presented. Due to the unique structure of resorcinol, the products developed using this chemical exhibited outstanding performance in the UV stabilization applications. Though numerous products and derivatives have already been developed and successfully used in the protection of industrially and commercially important polymeric materials against UV degradation, greater potential still exists in the development and applications of new UV absorbers from the resorcinol chemistry.

### 8.1.2
### Resorcinol Based UV Absorbers for Polymers

Hydroxyphenyl benzophenones, hydroxyphenyl benzotriazoles and hydroxyphenyl triazines derived from the resorcinol chemistry, are the most extensively studied resorcinol based UV absorbers and have the following generalized chemical structures (Figures 8.1–8.3).

Resorcinol based chemical compounds exhibiting chemical structures, shown in Figures 8.1–8.3 contain both a functional group capable of absorb-

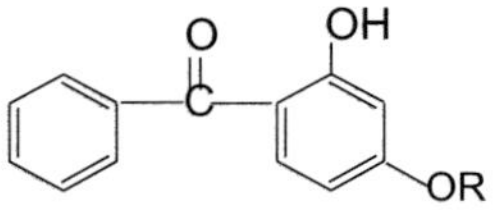


"Hydoxyphenyl Benzophenone"

**Figure 8.1.**

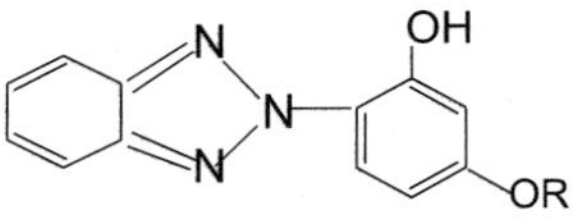


"Hydroxyphenyl Benzotriazole"

**Figure 8.2.**

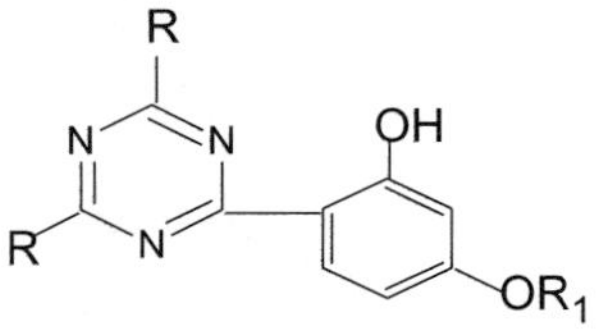


"Hydroxyphenyl Triazine"

**Figure 8.3.**

ing UV light and labile hydrogen closely positioned to this group. The hydrogen bonding, which could be formed between the UV absorbing functional group and the *ortho*-hydroxyl group present in the phenolic ring, exerts a decisive influence on the spectroscopic and photochemical properties of UV stabilizer [11, 12]. The UV stabilizing effectiveness of *ortho*-hydroxyphenyl substituted derivatives increases with the strength of hydrogen bonds. Therefore, in order to function as effective and efficient UV stabilizer, the organic compound should possess a functional group capable of absorbing the UV light and an *ortho*-phenolic hydroxyl group in the same molecule.

Resorcinol and its derivatives have the unique advantages of forming various organic compounds containing both a UV absorbing functional group and *ortho*-phenolic hydroxyl group, which could be used as UV absorbers. In the case of resorcinol, the two most reactive sites on the benzene ring for the substitution reactions are 4 and 6 and are identical with respect to the hydroxyl groups. Therefore, the reaction at 4 or 6 position of resorcinol will lead to primarily the *ortho*-phenolic hydroxyl containing derivatives. With this unique structure and properties of resorcinol, various possibilities existed on the development of numerous organic compounds that exhibited UV absorber properties.

## 8.1.3 Benzophenones from Resorcinol

It is well known that benzophenone derivatives, for an example 2,4-dihydroxy-benzophenone or 4-benzoylresorcinol, can be synthesized from resorcinol by several different procedures and methods as illustrated in Figure 8.4.

In the Hoesch synthesis process, resorcinol was reacted with benzonitrile in the presence of zinc chloride catalyst and the resulting 2,4-dihydroxy benzophenoneoxime produced in the reaction was hydrolyzed to obtain 4-benzoylresorcinol [13]. With Friedel-Crafts acylation reaction, resorcinol was reacted with benzoic acid or benzoyl chloride in the presence of Lewis acid catalysts, such as aluminum chloride, zinc chloride or boron trifluoride [14–16]. From the Friedel-Crafts reaction, benzoylresorcinol derivatives containing one, two or three benzoyl groups could be produced depending upon the reaction conditions and procedures.

An alternate and, also a low cost method of producing 4-benzoylresorcinol was the use of benzotrichloride ($\alpha$,$\alpha$,$\alpha$-trichloro toluene) in the resorcinol reaction [17–21]. In this process, benzotrichloride was partially hydrolyzed into benzoic acid, and also produced a xanthenone byproduct capable of producing reddish brown color in the benzoylresorcinol product. The formation of xanthenone compound was the result of a secondary polyhydroxyphenol produced in the reaction. In spite of these side reactions, benzotrichloride process produces 4-benzoylresorcinol in higher yields. A preparative procedure for the synthesis of 4-benzoylresorcinol using the benzotrichloride is outlined below.

**Procedure for the Synthesis of 2,4-Dihydroxy benzophenone from Resorcinol and Benzotrichloride [20]** Water (1000 ml) is placed in a vessel having a capacity of 2 l, and 220 g (2.0 mol) of resorcinol added with stirring. When this has dissolved, 2.12 ml (2.5 g) of *cis*-9-octadecen-1-ol (oleyl alcohol) are added. Then 294 ml (2.0 mol) of benzotrichloride are added drop-wise over 1.5 h. A strongly exothermic reaction takes place and this addition must be carried out at such a rate that the temperature remains in the range of 40 – 50 °C. When all of the benzotrichloride has been added, the reaction mixture is heated at 70 °C for 1 h, after which the heater is switched off and the product is filtered off in vacuo and washed with hot water. After drying in vacuo at 80 °C, there are obtained 420 g (98% of theory) of crude 2,4-dihydroxybenzophenone as a yellowish orange powder melting at 139 – 140 °C.

As mentioned previously, the benzoylation of resorcinol using benzoyl chloride in the presence of Friedel-Crafts catalysts produces various benzoyl derivatives as illustrated in Figure 8.5.

Melting points and UV absorption data obtained on the benzoylresorcinol derivatives are presented in Table 8.2 and Figure 8.6 [22].

From the UV spectra of benzoylresorcinol derivatives, peak maxima have been observed in the wavelength regions of 245 to 255 nm, 280 to 290 nm and

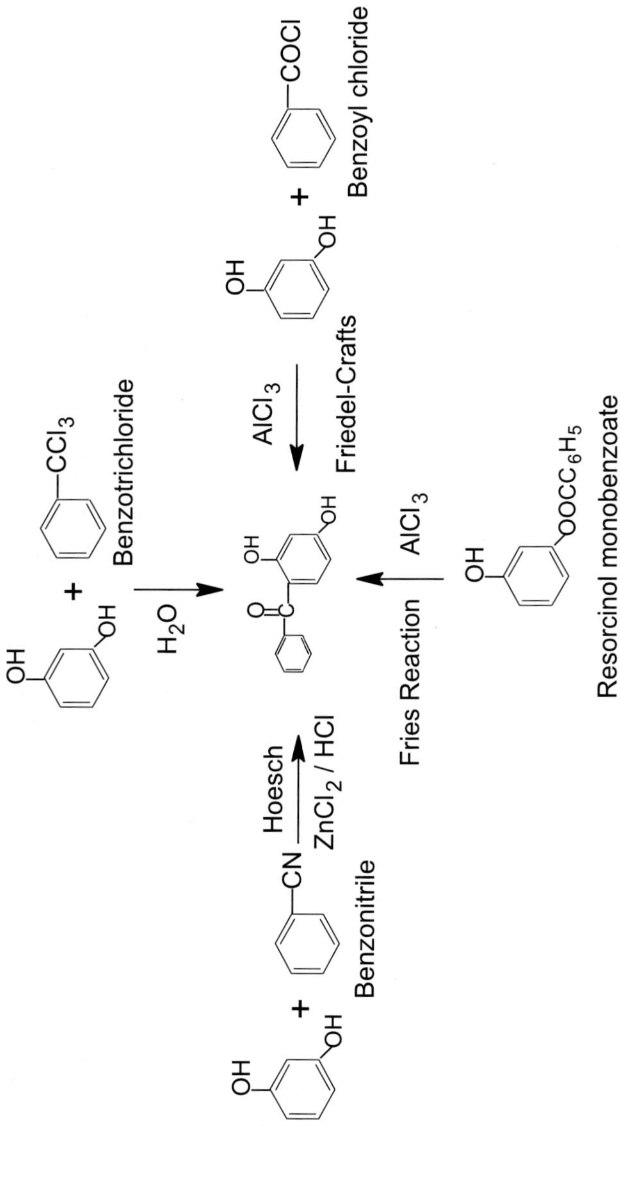


**Figure 8.4.** Synthesis schemes for the preparation of 4-benzoylresorcinol (2,4-dihydroxybenzophenone)

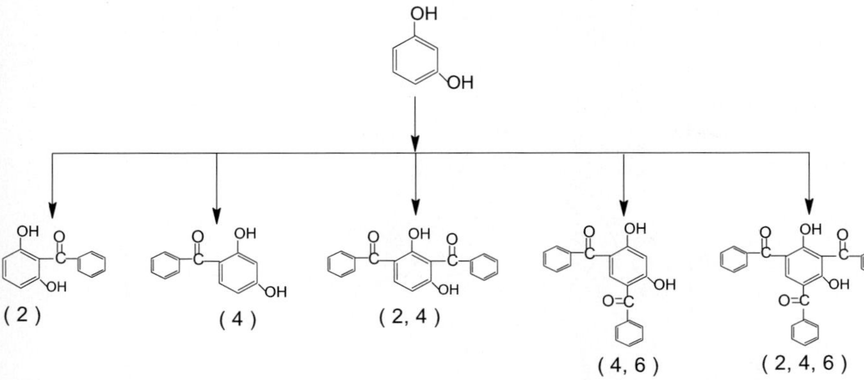


**Figure 8.5.** Benzophenone derivatives of resorcinol

**Table 8.2.** Structure and UV absorption properties of benzoyl substituted resorcinols

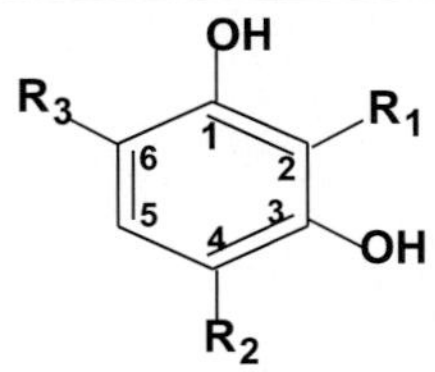


| Substitutions | | | Melting | UV absorption data | | | | | |
|---|---|---|---|---|---|---|---|---|---|
| $R_1$ | $R_2$ | $R_3$ | Point (°C) | $\lambda_{max}$ | $\log \varepsilon$ | $\lambda_{max}$ | $\log \varepsilon$ | $\lambda_{max}$ | $\log \varepsilon$ |
| $C_6H_5CO$ | H | H | 135 | 250 | 4.07 | 280 | 3.74 | ND | ND |
| H | $C_6H_5CO$ | H | 145 | 242 | 3.94 | 290 | 3.96 | 338 | 4.12 |
| $C_6H_5CO$ | $C_6H_5CO$ | H | 105 | 248 | 4.3 | 290 | 3.9 | 348 | 4.34 |
| H | $C_6H_5CO$ | $C_6H_5CO$ | 145 | 258 | 4.27 | 275 | 4.26 | 338 | 4.15 |
| $C_6H_5CO$ | $C_6H_5CO$ | $C_6H_5CO$ | 185 | 258 | 4.3 | 280 | 4.53 | 330 | 3.91 |

ND = None detected
Data from [22]

335 to 350 nm [22]. In the case of 2-benzoylresorcinol, the peak maximum in the 335 to 350 nm range had disappeared, which may be associated with the steric hindrance in the molecule. UV spectra of 2,4- and 4,6-dibenzoylresorcinols appeared similar. 2,4,6-Tribenzoylresorcinol produced a unique UV spectrum as a result of distortion in the molecule due to the presence of three bulkier substituents. By comparing the UV spectral data of 2,4- and 4,6-derivatives, the peak maximum had shifted to higher wavelength region and higher absorption coefficient for the 2,4-dibenzoylresorcinol. The UV absorption property in the longer wavelength region, namely 335 to 350 nm, of 4-benzoylresorcinol and its derivatives appeared to be associated with the conjugate chelation, which could occur between the *ortho*-hydroxy and carbonyl group of the molecules.

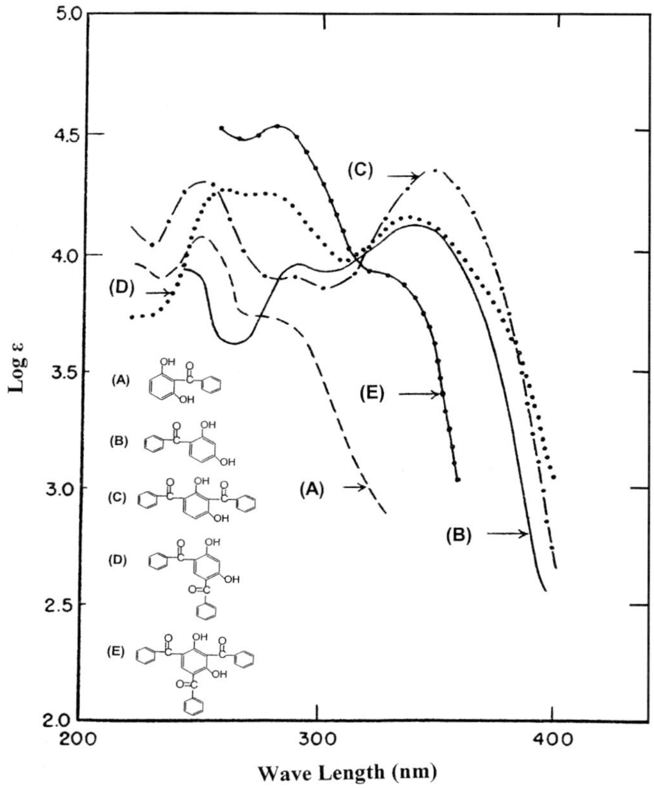


**Figure 8.6.** UV spectra of benzoyl derivates of resorcinol [22]

This conjugation could be reinforced by suitably placed electron sources or suppressed by steric effects [22].

### 8.1.3.1 UV Absorption Mechanism for Benzophenones Compounds

Resorcinol based benzophenone derivatives have been employed as UV absorbing compounds in numerous applications including the thermoplastic polymers, coatings and cosmetic formulations. Due to the fact that resorcinol derivatives have high absorption coefficients in the wavelength regions of 280 to 350 nm, they have been very effective in protecting the performances of various materials in their applications. UV absorbers have the ability to convert the energy they absorbed from the UV light in to thermal energy in the form of heat, through a keto-enol tautomerism mechanism (Figure 8.7).

From the published spectral data, it was concluded that the transformation of energy involved fast radiationless transitions. The heat released from the

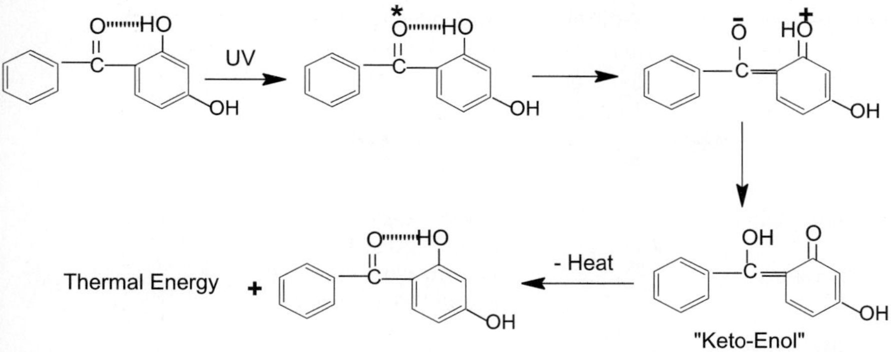


**Figure 8.7.** Possible UV stabilization mechanism for resorcinolbenzophenone absorbers

UV absorber could be easily dissipated through the substrate. The amount of heat released from this UV absorption mechanistic process is considered very small compared to the total heat load on the samples or polymers, and therefore does not contribute to the thermal degradation of polymers.

### 8.1.3.2
### 2,4-Dihydroxy Benzophenone and Derivatives

**2,4-Dihydroxy Benzophenone** Benzophenone UV absorbers based on resorcinol chemistry have been commercially manufactured by various chemical companies and marketed under different trade or commercial names. Table 8.3 presents some company names and their trade marks for resorcinolic benzophenone UV absorbers.

**Table 8.3.** Companies marketing UV absorbers

| Name of company | Trade name |
|---|---|
| BASF | Uvinul |
| Lycus Ltd. | Maxgard |
| Cytec | Cyasorb |
| Great Lakes Chemical | Lowilite |
| Shipro Kasei Kaisha Ltd | Seesorb |

2,4-Dihydroxybenzophenone, also known as benzophenone-1, is compatible with a wide range of polymers. This compound's physical and solubility properties are presented in Table 8.4 and UV spectrum is shown in Figure 8.8.

This UV absorber is particularly recommended for using in polystyrene, epoxy resins and also effective in protecting PMMA and coatings.

**Table 8.4.** Properties of 2,4-dihydroxy benzophenone

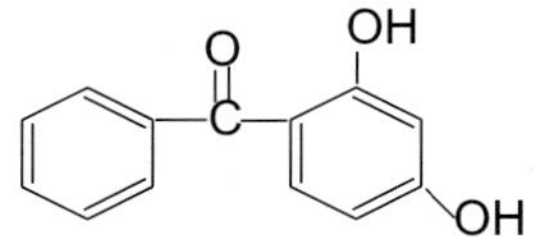

| | |
|---|---|
| Chemical name | 2,4-Dihydroxy benzophenone, benzophenone-1 |
| Trade Name | Maxgard 400, Uvinul 3000, Seesorb 100, Lowilite 24 |
| Physical properties | |
| Physical form | Off-white powder |
| Melting point (°C | 140 – 143 |
| Density (g/cc, 25 °C) | 1.27 |
| Absorptivity (at 288 nm) | 62 |
| Absorptivity (at 323 nm) | 47 |
| Toxicity, Rat LD50 (g/kg) | 8.6 |
| Solubility (g/100 ml at 30 °C) | |
| Acetone | 10 |
| Ethyl acetate | 25 |
| Methanol | 40 |
| Ethanol | 40 |
| Methyl methacrylate (MMA) | 5 |
| Methyl ethyl ketone (MEK) | 40 |
| Water | $< 0.001$ |

Applications: polystyrene, PMMA, polyisoprene,
Pressure sensitive acrylic adhesives, varnishes, lacquers
Data source: technical information from the manufacturers

**Derivatives of 2,4-Dihydroxybenzophenone** In addition to functioning as a UV stabilizer, a successful absorber additive must have both excellent compatibility and solubility in numerous polymer substrates, along with superior resistance to loss from the stabilized composition during processing and end-use applications. Many UV absorbers exhibit limited compatibility in certain substrates, and excessive tendency to exude, sublime and/or volatilize during weathering or processing of stabilized compositions at elevated temperatures. Because of this problem, various structural modifications have been made to increase the compatibility and, also to reduce the volatility of 2,4-dihydroxybenzophenone.

**Alkyl Ether Derivatives of 2,4-Dihydroxybenzophenone** Alkyl ether derivatives, namely 4-alkoxy-2-hydroxybenzophenone, have been prepared by various synthetic routes as illustrated in Figure 8.9.

Depending upon the raw materials used and synthetic conditions employed, various alkyl ether derivatives have been synthesized in high purity. The use of alkyl halides [23–27], alkyl sulfates [26, 28], alkyl carbonates [29] and alkyl

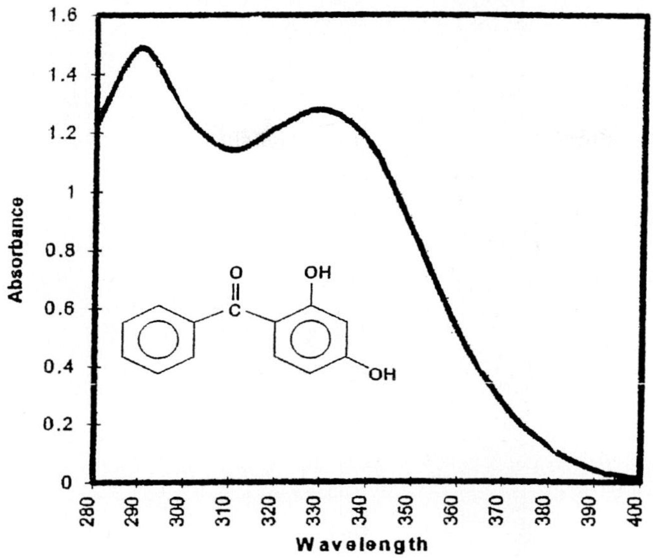


**Figure 8.8.** UV spectrum of 2,4-dihydroxybenzophenone from Lycus

esters of aromatic sulfonates [18,30] has been reported to produce compounds with improved yields.

**4-Methoxybenzophenone** 2-Hydroxy-4-methoxybenzophenone, a methyl ether derivative, has the following chemical structure (Figure 8.10).

This compound could be prepared either by using dimethyl sulfate or methyl benzene sulfonates in the reaction with 4-benzoylresorcinol, and based on the preparative procedures given below.

**Procedure for the Preparation of 2-Hydroxy-4-methoxybenzophenone from 2,4-Dihydroxybenzophenone and Dimethyl Sulfate [28]** 2,4-Dihydroxybenzophenone (214 g = 1.0 mol) was added to a mixture of 500 ml of 2 N sodium hydroxide solution and 150 ml of water at room temperature and stirring was carried out until everything had dissolved (20 min). Thereafter, 31.6 ml (= 0.33 mol) of dimethyl sulfate were added and stirring was continued for 0.5 h, after which a further 31.6 ml (= 0.33 mol) of dimethyl sulfate were added while stirring. After 0.5 h, a further 31.6 ml (= 0.33 mol) of dimethyl sulfate were added while stirring and, after 0.5 h, 50 ml of 2 N sodium hydroxide solution and 9.5 ml of dimethyl sulfate were added. The mixture was stirred for 1 h at room temperature, 30 ml of concentrated ammonia solution were added and stirring was then continued for a further 0.5 h. The precipitate which had separated out was filtered off under suction and washed with a little water to give 349 g (202.4 g of dry product = 89% of theory) of moist 2-hydroxy-4-methoxybenzophenone, which contains 1.07% of 2,4-dimethoxybenzophenone. The moist filtration

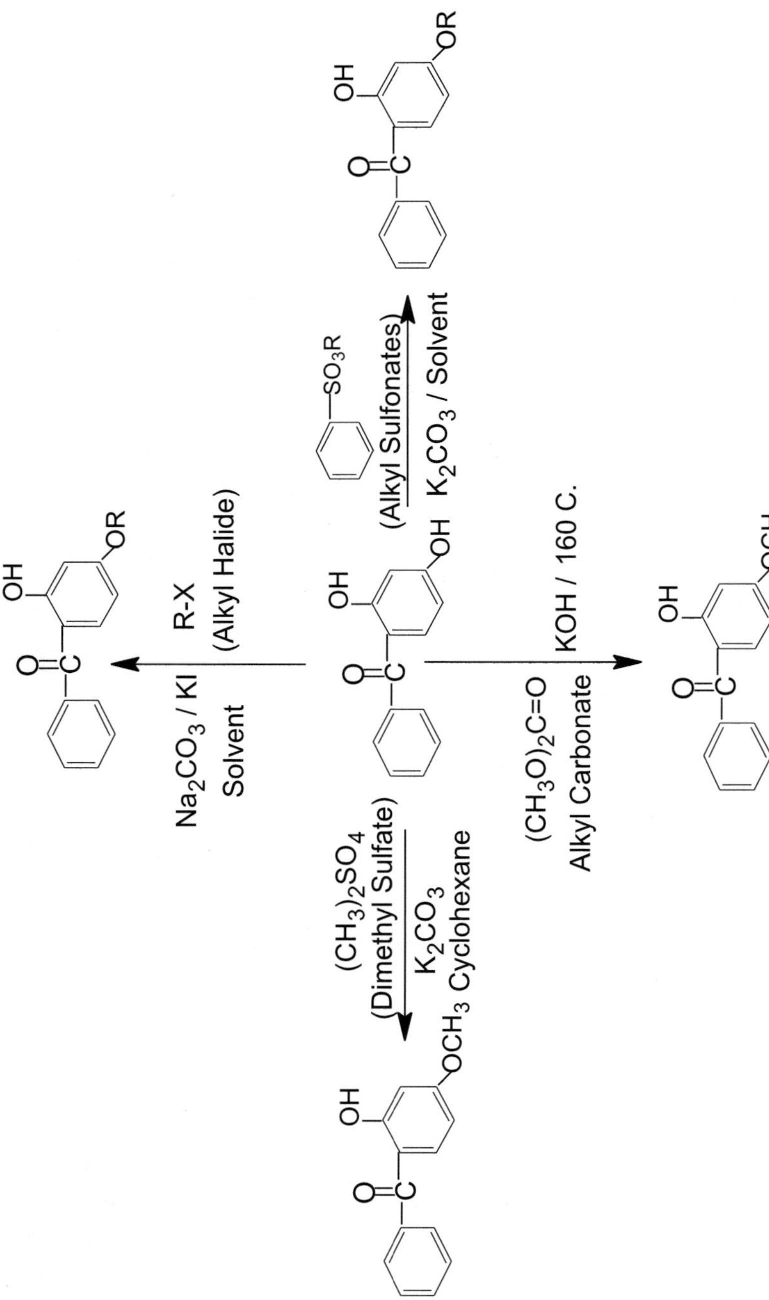


**Figure 8.9.** Synthesis schemes for the preparation of alkyl ether derivates of 2,4-dihydroxybenzophnenone

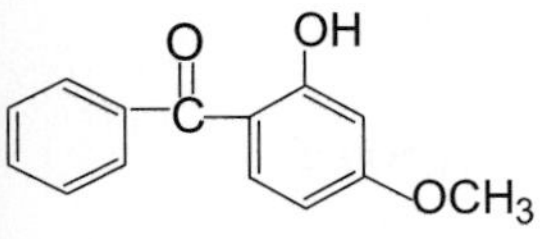


**Figure 8.10.**

residue was recrystallized from 1170 ml of methanol with the addition of 5 g of active carbon. After filtration under suction and drying under reduced pressure from a water pump at 50 °C, 171 g (= 75% of theory) of pure 2-hydroxy-4-methoxybenzophenone of melting point 62 – 63 °C were obtained. According to gas chromatography, the product did not contain any dimethoxy derivative.

**Procedure for the Preparation of 2-Hydroxy-4-methoxybenzophenone from 2,4-Dihydroxybenzophenone and Methylbenzene Sulphonate [30]** 2,4-Dihydroxybenzophenone (21.4 g of (0.1 mol), 19.0 g (0.11 mol) of methylbenzene sulphonate, 12.7 g (0.12 mol) of anhydrous potassium carbonate were mixed in a sixfold quantity of cyclohexanone at a temperature of 95 to 100 °C over the course of 5 h. The reaction mixture was cooled, filtered and the solvent was removed from the filtrate by distillation under reduced pressure. The residue was submitted to crystallization from a $C_1$–$C_4$ aliphatic alcohol to obtain white 2-hydroxy-4-methoxybenzophenone of melting point of 62.4 to 63 °C and of a yield of approximately 90 percent.

Table 8.5 presents the properties and Figure 8.11 shows the UV spectrum of 2-hydroxy-4-methoxybenzophenone.

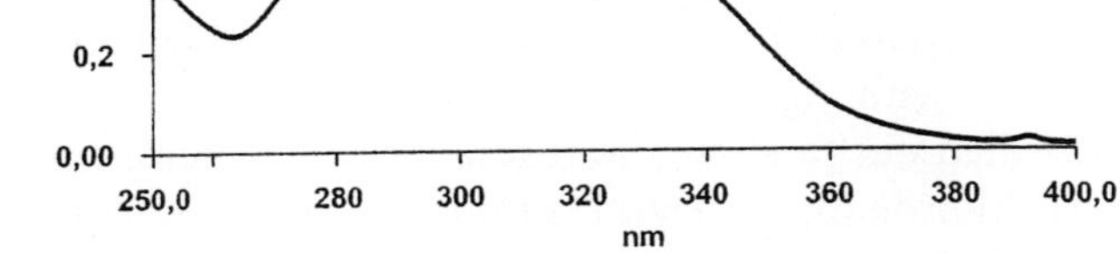


**Figure 8.11.** UV spectrum of 2-hydroxy-4-methoxybenzophenone from BASF

This UV absorber compound has been proven to be effective in polyesters, polystyrene, PVC, PMMA, coatings and oil based cosmetic materials.

**2-Hydroxy-4-Octoxybenzophenone** 2-Hydroxy-4-octoxybenzophenone, shown in Figure 8.12, could be synthesized employing either the octyl halides or octyl ester of *p*-toluene sulfonates.

**Table 8.5.** Properties of 2-hydroxy-4-methoxy benzophenone

| | |
|---|---|
| Chemical name | 2-Hydroxy-4-methoxy benzophenone benzophenone-3 |
| Trade name | Maxgard 500, Uvinul M40, Seesorb 101, Lowilite 20 |
| Physical properties | |
| Physical form | Slightly yellowish powder |
| Melting point (°C) | 62 |
| Density (g/cc, 25 °C) | 1.33 |
| Absorptivity (at 287 nm) | 63 |
| Toxicity, LD50 (g/kg) | 7.4 |
| Solubility (g/100 ml at 25 °C) | |
| Isopropyl alcohol (IPA) | 5 |
| Ethyl alcohol (95%) | 5 |
| Ethylene glycol | 0.7 |
| Propylene glycol | 1.8 |

Soluble in usual cosmetic oils
Applications: oil based cosmetics, PVC, polyacrylates, unsaturated polyesters, coatings, varnishes, lacquers, alkyds, sunscreen for skin protection oils.
Data source: technical information from the manufacturers

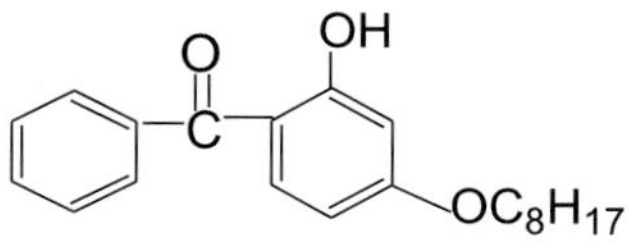


**Figure 8.12.**

**Synthesis of 2-Hydroxy-4-*n*-octoxybenzophenone from 2,4-Dihydroxybenzophenone and Octyl Chloride [27]** To a 250 ml three-neck round bottom flask is charged 21.4 g of 2,4-dihydroxybenzophenone, 16.3 g of octyl chloride, 5.6 g of sodium carbonate, 0.664 g of potassium iodide, and 21.4 g of Igepal CO-630. The mixture is heated to 155 °C with vigorous stirring. Sufficient water, generated by the reaction, is distilled out to allow the reaction to reach temperature; any co-distilling octyl chloride (upper layer) is returned to the reactor. The reaction is then kept at 155 °C for 2 h. Thereafter excess octyl chloride and water are recovered at 11 mm and 155 °C; which about 15 min is required to recover 0.7 g of a cloudy yellow liquid. The reactor then is cooled to 45 °C and 30 ml of acetone is added to precipitate salts and facilitate filtration. The mixture is filtered through a warm fitted glass funnel and the salt cake is washed once with 20 ml of acetone. The acetone in the filtrate is recovered in vacuo for recycle, initially at a pot $T = 80$ °C at 18–23" of Hg and finally at 110 °C and 11 mm. The filtrate, containing surfactant and product, a volume of 40 ml is diluted with 85 ml of methanol and cooled to – 8 °C to precipitate the product, which is filtered and washed twice with 40 ml of – 8 °C methanol. The product is sucked dry for 5 min and then used in the next step.

From the filtrate, methanol and surfactant are recovered for recycle. A vacuum dried sample is used for yield. The yield = 30.4 g, 93.3%.

**Synthesis of 2-Hydroxy-4-*n*-octoxybenzophenone from 2,4-Dihydroxybenzophenone and *n*-Octyl-*p*-toluenesulfonate [31]** *n*-Octyl-*p*-toluenesulfonate (58.7 g), 57 g of potassium carbonate, 44.2 g of 2,4-dihydroxybenzophenone and 103 g of water were mixed and the resulting mixture was boiled under reflux for 5 h with stirring. The mixture was cooled to separate out a precipitate, which was filtered, washed with water and dried to give 60.7 g of yellow acicular crystals having a melting point of 42 to 44 °C. These crystals were recrystallized from methanol, giving pale yellow acicular crystals having a melting point of 47 to 48 °C. The yield was 74.3 mol% based on resorcinol.

Solubility data and UV spectrum of 2-hydroxy-4-*n*-octoxybenzophenone are presented in Table 8.6 and Figure 8.13.

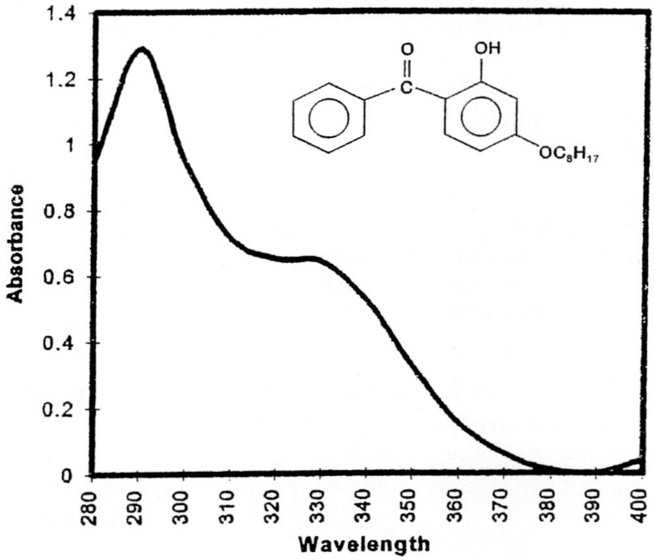


**Figure 8.13.** UV spectrum of 2-hydroxy-4-octoxybenzophenone from Lycus

This compound is an extremely effective UV absorber that has been recommended for polypropylene, polyethylene, plasticized PVC, acrylic polymers, polycarbonate and polyester. It is also recommended to be used in adhesives and sealants.

**2-Hydroxy-4-Dodecyloxybenzophenone** 2-Hydroxy-4-dodecyloxybenzophenone, sold under the trade name Seesorb 103, could be synthesized from

**Table 8.6.** Properties of 2-hydroxy-4-*n*-octoxybenzophenone

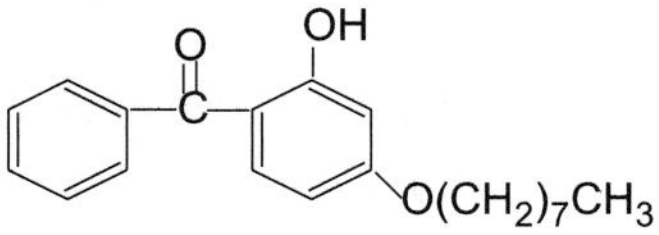


| | |
|---|---|
| Chemical name | 2-Hydroxy-4-*n*-octoxybenzophenone, benzophenone-12 |
| Trade name | Maxgard 700, Cyasorb UV 531 Lowilite 22, Uvinul 408, Seesorb 102 |
| Physical properties | |
| Physical form | Slightly yellowish crystalline powder |
| Melting point (°C) | 46.5 – 49 |
| Density (g/cc, 25 °C) | 1.16 |
| Absorptivity (in toluene) | 31 |
| $\lambda_{max}$ | 329 |
| Toxicity (LD50, g/kg) | > 10 |
| Solubility (g/100 ml at 25 °C) | |
| Methanol | 1 |
| Ethanol | 3 |
| Benzene | 72 |
| Acetone | 74 |
| Methyl ethyl ketone (MEK) | 65 |
| Dioctyl phthalate (DOP) | 25 |
| Water | < 0.01 |

Applications: PP, low and high density PE, PVC (plastisols), architectural coatings, acrylics, varnishes, lacquers and alkyds, polyesters, polystyrene, ABS and synthetic resins.
Data source: technical information from the manufacturers

the reaction of 4-benzoylresorcinol with 1-bomododecane in the presence of anhydrous potassium carbonate, Figure 8.14 [32].

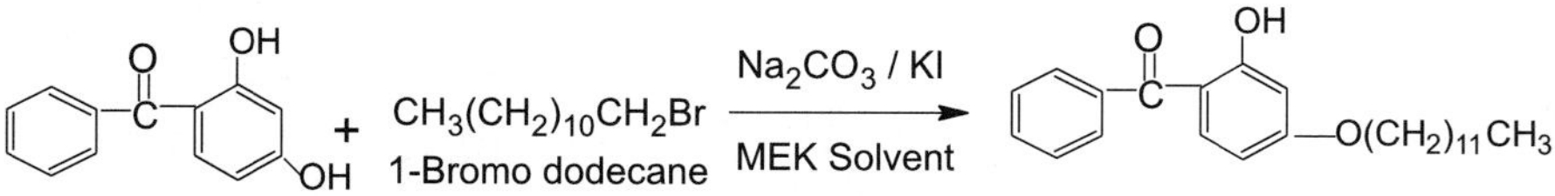


**Figure 8.14.**

A procedure for the preparation of this compound is shown below.

**Preparation of 2-Hydroxy-4-dodecyloxy benzophenone [32]** Into a 1-l flask, there were charged 107 g of 2,4-dihydroxybenzophenone (0.5 mol), 322 g of methyl ethyl ketone (MEK), 8.3 g of potassium iodide, 137 g of 1-bromo dodecane and 70 g of anhydrous potassium carbonate. The mixture was refluxed for 19 h and then about 300 ml of MEK was distilled out (about 74% of charge). The residue in the flask was then drowned in 3 l of water and the pH of the mixture was adjusted to 9.5 with a 40% by weight aqueous solution of sodium hydroxide. The charge was then filtered and the filter cake was reslurried in 3 l of water, heated to 85 °C at a pH $= 9.5$ and thereafter cooled to 10 °C. The mixture was then filtered and the cake was washed with water. The cake was then recrystallized from 1.5 l of isopropanol containing 3 g of activated carbon. The product was dried in a desiccator and then air dried at 40 °C. The yield was 166 g. The product had a $K_{max} = 27.5$ at 325 nm. The product had a melting point of 54.1 to 55.8 °C.

Due to the presence of lengthy alkyl chain in the molecule, the compatibility with hydrocarbon polymers such as polyethylene, polypropylene and polystyrene are expected to be improved with this absorber. In addition, the low melting point of this material could enhance its processability during the polymer compounding process.

**2-Hydroxy-4-Benzyloxybenzophenone** 4-Benzyloxy-2-hydroxybenzophenone, also commercially known as Seesorb 105 UV absorber, is a white crystalline powder with a melting point of > 115 °C. Most of the benzophenone UV absorbers appeared to be yellowish in color, and therefore, with the use of either colorless or white colored materials, the preparation of white or clear articles could be possible. Compatibility with polymers such as polyethylene and polypropylene was reported with this UV absorber. Based on the synthetic scheme presented in Figure 8.15, this compound could be prepared using the following procedure.

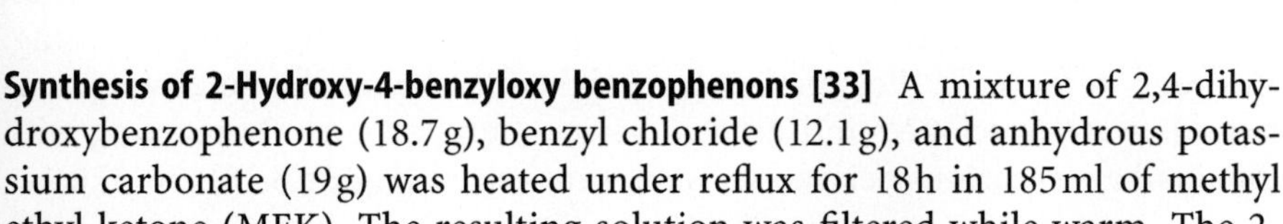


**Figure 8.15.**

**Synthesis of 2-Hydroxy-4-benzyloxy benzophenons [33]** A mixture of 2,4-dihydroxybenzophenone (18.7 g), benzyl chloride (12.1 g), and anhydrous potassium carbonate (19 g) was heated under reflux for 18 h in 185 ml of methyl ethyl ketone (MEK). The resulting solution was filtered while warm. The 2-hydroxy-4-benzyloxy-benzophenone crystallized from the filtrate as light tan plate shaped crystals of about one millimeter diameter and an additional quantity was recovered by the evaporation of mother liquor. The 2-hydroxy-4-benzyloxy-benzophenone had a melting point range between 119 and 122 °C. 65% of theoretical yield was obtained.

**4-Allyloxy-2-hydroxybenzophenone** This compound was synthesized from the reaction of 4-benzoylresorcinol with allyl bromide in the presence of potassium carbonate (Figure 8.16) [34].

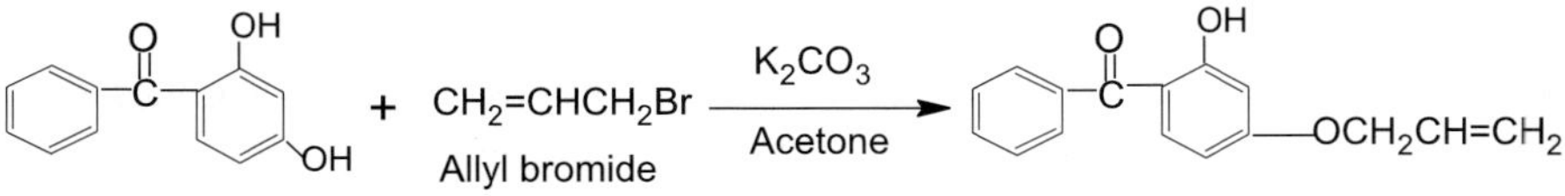


**Figure 8.16.**

**Procedure for the Preparation of 4-Allyloxy-2-hydroxybenzophenone [34]** A mixture of 21.4 g (0.1 mol) of 2,4-dihydroxybenzophenone, 13.2 g (0.11 mol) of freshly distilled allyl bromide, and 14 g (0.1 mol) of potassium carbonate in 100 ml of dry acetone is refluxed under nitrogen for 10 h. After cooling, the inorganic salts are filtered off and the organic solution is washed with water and extracted with chloroform. After drying and evaporating the solvent, the light yellow oil is crystallized from ether to give light yellow crystals, 18 g (73% yields), and m.p. 68 – 70 °C.

This allyloxy compound was then converted into silane derivatives (Figure 8.17) [34, 35].

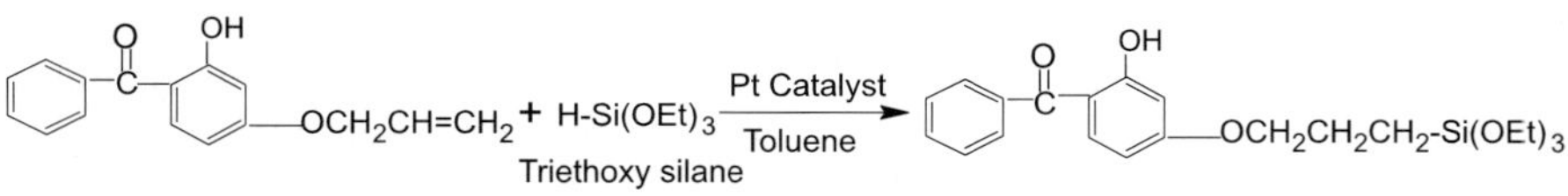


**Figure 8.17.**

Hard coat formulations were made using the silane modified benzophenone and applied on the polycarbonates to obtain scratch resistant coatings with UV protection.

**1,4-Bis-(4-benzoyl-3-hydroxyphenoxy) Alkanes** Although the *o*-hydroxybenzophenones and their derivatives have been effective in UV absorbing properties, due to lower molecular weights they tend to volatilize during their processing. In order to increase their molecular weights, bis-benzophenone type compounds were developed and used in various applications. By reacting with dihaloalkanes, various derivatives were synthesized (Figure 8.18) [36, 37].

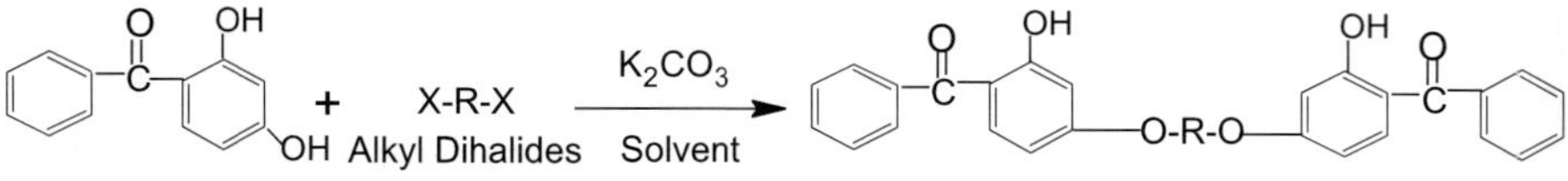


**Figure 8.18.**

Table 8.7 shows the derivatives obtained based on the above reaction.

**Table 8.7.** Melting points of di-(3-hydroxy-4-benzoylphenoxy) alkanes

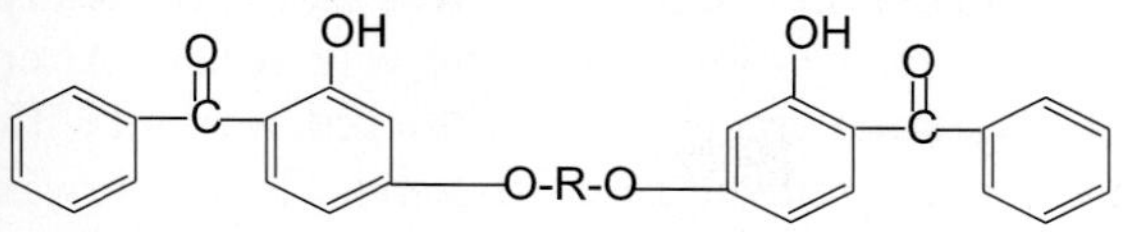


| R Group | Melting point (°C) |
|---|---|
| $CH_2$ | 114 – 117 |
| $C_2H_4$ | 211 – 213 |
| $C_3H_6$ | 128.8 – 130 |
| $C_4H_8$ | 159 – 163 |
| $C_5H_{10}$ | 145 – 147 |
| $C_{10}H_{20}$ | 118 – 121 |
| $C_2H_4OC_2H_4$ | 128 – 129.5 |
| $C_2H_4OC_2H_4OC_2H_4$ | 104.5 – 106 |
| $C_4H_8OC_4H_8$ | 77 – 82 |

Data from [37]

The effect of chain length on the melting points of di-(3-hydroxy-4-benzoylphenoxy)-alkanes is presented in Figure 8.19. By employing the dihalo compounds of varying structures, it could be possible to develop bis-benzophenone derivatives having different melting points and solubility properties suited for various application needs.

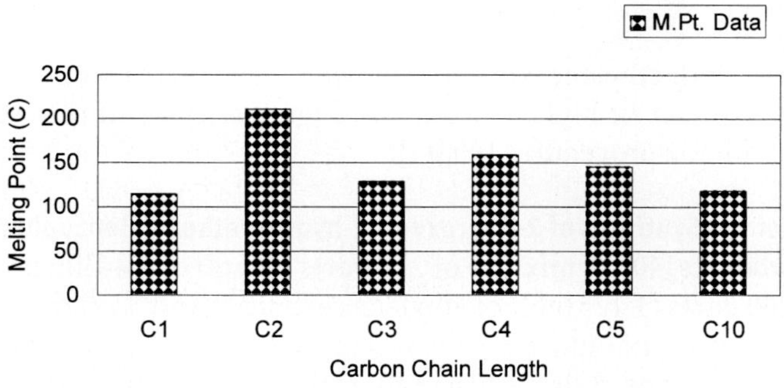


**Figure 8.19.** Effect of chain length on melting points of di-(3-hydroxy-4-benzoylphenoxy) alkanes

From the TGA determinations, for some derivatives, a 5% weight loss occurred above 400 °C, which suggests higher thermal stability and lower volatility for these materials [37].

1,4-Bis-(4-benzoyl-3-hydroxyphenyl)-butane is available commercially under the trade name Seesorb 151, which may be prepared by employing the following procedure [36].

**Synthesis of 1,4-Di-(3-hydroxy-4-benzoylphenoxy)butane [36]** 2,4-Dihydroxybenzophenone (0.25 mol) and 0.25 mol potassium carbonate were added gradually into 300 ml acetone, and then 0.1 mol of dibromobutane was added. After boiling under reflux for 24 h, the resulting precipitate was filtered, washed with 50 ml acetone and then dried and recrystallized from benzene. The result was a 62% yield of a crystalline substance with a melting point of 165 to 166 °C.

**2-Hydroxy-4-(2-hydroxyethoxy) Benzophenone and Derivatives** UV absorbers having functional groups capable of reacting with other organic compounds and polymers could be more valuable in their applications. By incorporating the hydroxyethyl ether group at the 4-position of 2,4-dihydroxybenzophenone, the volatility, solubility, reactivity and UV absorbing properties could be changed. 2-Hydroxy-4-(2-hydroxyethoxy) benzophenone compound was prepared based on the reaction schemes presented in Figure 8.20 [38–42].

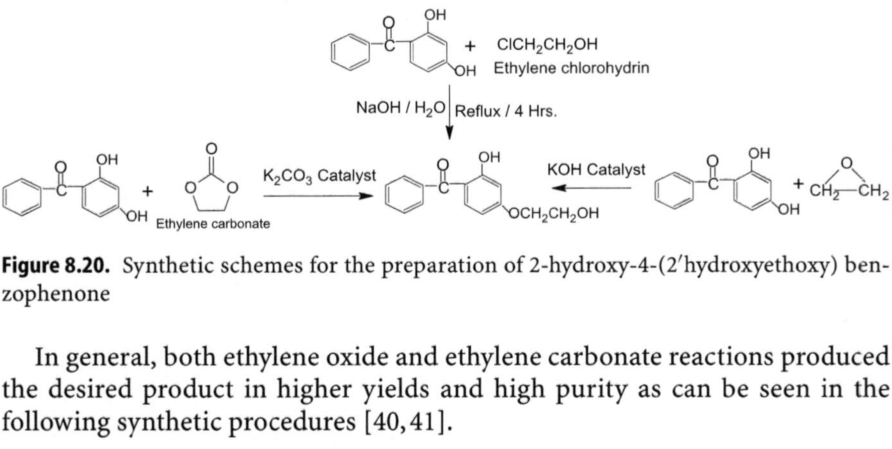


**Figure 8.20.** Synthetic schemes for the preparation of 2-hydroxy-4-(2′hydroxyethoxy) benzophenone

In general, both ethylene oxide and ethylene carbonate reactions produced the desired product in higher yields and high purity as can be seen in the following synthetic procedures [40,41].

**Procedure for the Synthesis of 2-Hydroxy-4-(2-hydroxyethoxy) Benzophenone Using Ethylene Carbonate [40]** A mixture of 214 parts (1 mol) of 2,4-dihydroxybenzophenone, 100 parts (1.125 mol) of ethylene carbonate and 5.7 parts (0.025 mol) of benzyltriethylammonium chloride was heated while stirring to 140 to 150 °C at which point carbon dioxide began to evolve slowly. After the evolution of carbon dioxide stopped (11 h) the reaction mixture was cooled to about 100 °C and then within 5 min 750 ml of water was added while stirring. The mixture was cooled to room temperature while stirring, at which point the oily reaction product crystallized. After suctioning off the solid, washing with water, and drying; 2-hydroxy-4-(2′-hydroxyethoxy) benzophenone was obtained having purity of 97% (HPLC). Yield: 248 parts (93 percent of the theoretical).

**Procedure for the Synthesis of 2-hydroxy-4-(2-hydroxyethoxy) Benzophenone Using Ethylene Oxide [41]** To a prepared one gallon autoclave, 1000 g of 2,4-dihydroxybenzophenone (4.5 mol, 2.9% $H_2O$) 16.2 g of 45% KOH and 70.3 g of distilled water were added. The autoclave was then sealed, purged, and pressure checked. The reactor was heated slowly to 105 to 110 °C under 2 to 3 psig nitrogen while allowing the 2,4-dihydroxybenzophenone to melt. The autoclave was pressurized to 34 psig with nitrogen with agitation, and 210 g of ethylene oxide (4.77 mol) were added at 150 to 175 g/h and $<$ 90 psig. The reaction mixture was reacted out to constant pressure and then evacuated slowly to 70 mm Hg to strip low volatiles. The autoclave was vented to 0 psig with nitrogen and the product was discharged. Then 15.7 g of the molten reaction mixture were stirred with about 200 g of water. The solidified product was filtered off and the wet product cake was dried under vacuum. 14.6 g of light tan colored product was recovered in a yield of 99%, having a melting point of 91.5 to 97.5 °C. Analysis: 98.9% 2-hydroxy-4-($\beta$-hydroxyethoxy)-benzophenone, 1.0% 2,4-dihydroxybenzophenone.

This compound is also available commercially from Cytec Industries (Cyasorb 198) and from BASF (Uvinul X-19), and some of the physical and solubility properties of this compound have been shown in Table 8.8.

**Table 8.8.** Properties of 2-hydroxy-4-(2-hydroxy ethoxy) benzophenone

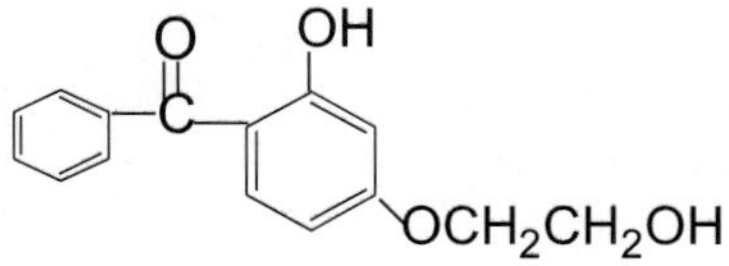


| | |
|---|---|
| Trade name | Cyasorb 198<br>Uvinul x-19 |
| Physical properties | |
| Physical form | Pale yellow powder |
| Melting point (°C) | 89–92 |
| UV absorption data | |
| Absorptivity at 242 nm | 40.02 |
| Absorptivity at 288 nm | 59.11 |
| Absorptivity at 325 nm | 38.27 |
| Solubility (wt% at RT) | |
| Water | 0.4 |
| Methanol | 16 |
| Acetone | 37 |
| Toluene | 3 |
| *n*-Hexane | 0.1 |
| Ethyl acetate | 19 |
| Ethyl alcohol | 13 |
| Methyl ethyl ketone (MEK) | 30 |

Data Source: Technical information from the manufacturers

In order to enhance the processability, thermal stability and also polymer compatibility, this UV absorber had been reacted with longer chain carboxylic acids, aliphatic and aromatic dicarboxylic acid anhydrides and dimethyl carbonates (Figure 8.21) [39,43,44].

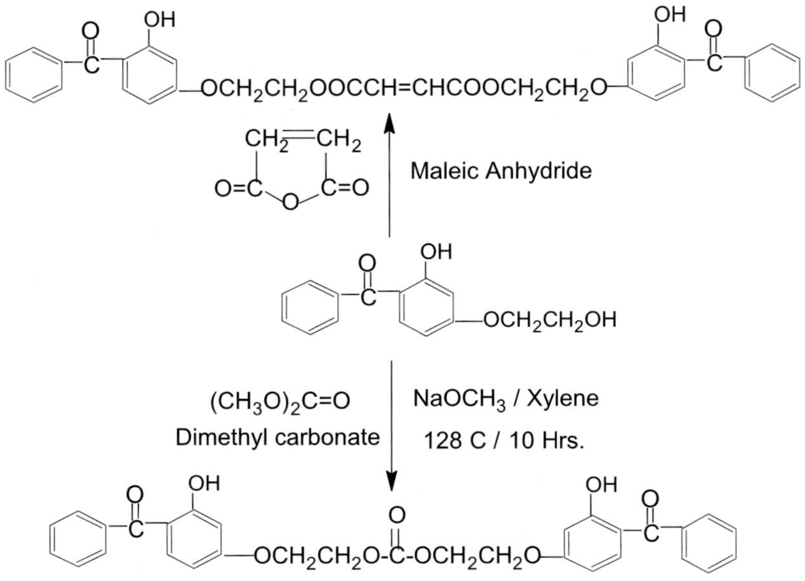


**Figure 8.21.**

Applications of 2-hydroxy-4-(2-hydroxyethoxy) benzophenone and its derivatives include photo stabilizing organic compounds, oils, fats, cosmetics, dyes, photographic materials, plastics and coatings.

**2-Hydroxy-4-methoxybenzophenone-5-sulfonic Acid** UV absorber materials utilized in the cosmetic solutions or lotions, hairsprays or sun tan lotions should be of highest degree of purity, completely soluble in water and also stable upon storage. The water solubility of resorcinol based benzophenone compounds could be enhanced by sulfonating these derivatives. Employing chlorosulfonic acid as the sulfonating agent, sulfonic acid derivative of 2-hydroxy-4-methoxybenzophenone was prepared (Figure 8.22) [45–47].

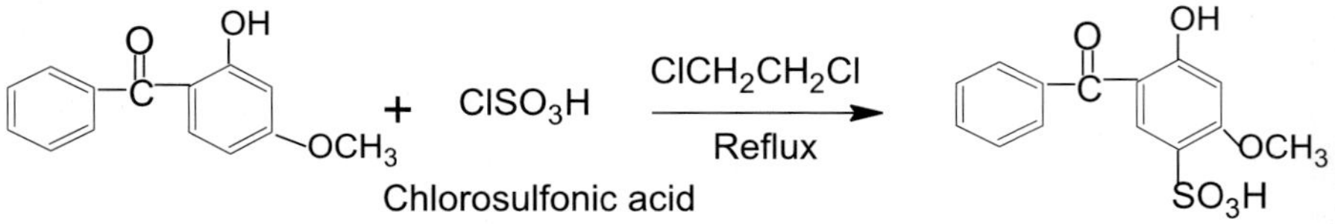


**Figure 8.22.**

Two preparative procedures for the synthesis of 2-hydroxy-4-methoxybenzophenone-5-sulfonic acid are presented below.

**Procedure for the Preparation of 2-Hydroxy-4-methoxybenzophenone-5-sulfonic acid Trihydrate [46]** 2-Hydroxy-4-methoxybenzophenone (40 g) was dissolved in 200 ml of ethylene dichloride. To this solution, 215 g of chlorosulfonic acid were added. This charge was then refluxed for 1 h and cooled to about 65 °C. Air was blown through the charge to remove the hydrochloric acid. The charge was then treated with 80 cc of water, cooled to 45 °C, allowed to stand for 2 h and the ethylene dichloride layer was separated from the aqueous solution.

The water content was adjusted to about 28% solution which was then sprayed into spray drier at an ingress temperature of about 225 °F, the egression temperature being about 135 °F, and the time for pass being from about 0.5 to 2 s. The product, 2-hydroxy-4-methoxybenzophenone-5-sulfonic acid, was collected from the drier as a fluffy white powder and had a melting point of 106 to 111 °C and a K value of 40 (the K value being defined as the absorption of 1 g of material per liter of solution measured in a 1 cm cell), and was of high purity and exhibited good storage stability.

**Preparation Procedure for 2-Hydroxy-4-methoxybenzophenone-5-sulfonic Acid [45]** Dichloroethane (663 g) and 74.6 g of 2-hydroxy-4-methoxybenzophenone were charged into a three-neck flask equipped with stirrer, thermometer, reflux condenser and dropping funnel and heating mantle. The solution was heated to the reflux temperature (85–86 °C) and was dehydrated by distilling of 66.5 g 1,2-dichloroethane. While maintaining at reflux, 30 g of chlorosulfonic acid was added slowly over a period of about 2 h. The rate of addition was regulated by the speed of evolution of the HCl. After all the chlorosulfonic acid was added, the charge was still maintained at reflux for an additional 15 min to remove traces of HCl. It was then cooled to 5 °C and filtered. The filter cake was washed with 500 g of cold 1,2-dichloroethane and dried. 98 g of product were obtained.

When the product was tested for ultraviolet light absorbing properties at 348 nm, a K value of 47 to 48 was found. This was in comparison to a K value of 42 for a commercial hydrated form of 3-benzoyl-4-hydroxy-6-methoxybenzene sulfonic acid.

Based on the products obtained from the above procedures, it was observed that the K values, a measure of the absorptivity of a compound, depended upon the presence of water in the material. Excluding the water resulted in a product showed unusually and also unexpected superior UV absorber property.

Table 8.9 outlines some of the physical properties and Figure 8.23 shows the UV spectrum of this benzophenone sulfonic acid derivative.

This benzophenone compound is considered as UVA and UVB absorber and especially suitable for formulations requiring water solubility.

**Bis-(2,4-dihydroxybenzophenone) Compounds** The thermal stability and absorptivity of UV absorbers could be enhanced by making the benzophenone com-

**Table 8.9.** Properties of 2-hydroxy-4-methoxybenzophenone-5-sulfonic acid

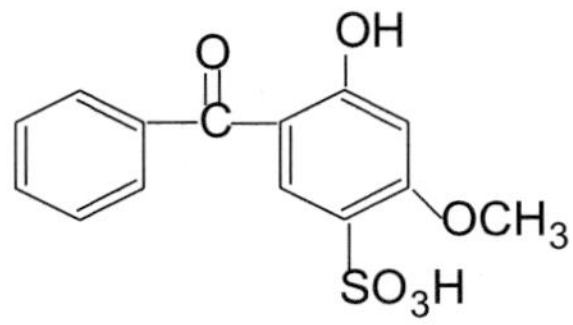


| | |
|---|---|
| Chemical name | 2-Hydroxy-4-methoxybenzophenone-5-sulfonic acid, benzophenone-4 |
| Trade name | Maxgard 1900, Uvinul MS40, Seesorb 101S, Lowilite 20S |
| Physical properties | |
| Physical form | Off-white fine to coarse powder |
| Odor | Practically odorless |
| Melting point (°C) | 108–110 |
| K value | 46–49 |
| Toxicity, rat LD50 (g/kg) | 2.7 |
| Solubility (g/100 ml at 20 °C) | |
| Acetone | 8.5 |
| *n*-Hexane | 0 |
| Isopropyl alcohol | 25 |
| Toluene | < 0.5 |
| Ethyl acetate | 0 |
| Water | 20 |

Applications: water based cosmetics such as sun tan lotions,
shampoos and hair sprays, coatings, photographic films, lithographic plates Data source: technical data sheets from BASF, May 2000
Technical information from Great Lakes Chemical

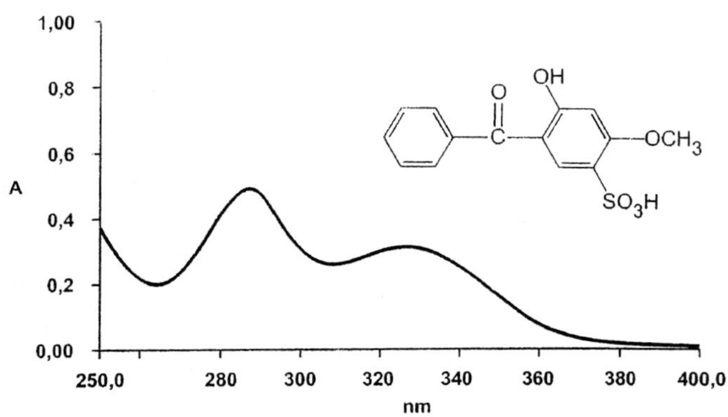


**Figure 8.23.** UV spectrum of 2-hydroxy-4-methoxybenzophenone-5-sulfonic acid from BASF

pounds having multiple *o*-hydroxyphenyl groups in the molecule, which were developed using 2,4-dihydroxybenzophenone using the following chemical reaction scheme (Figure 8.24) [48–51].

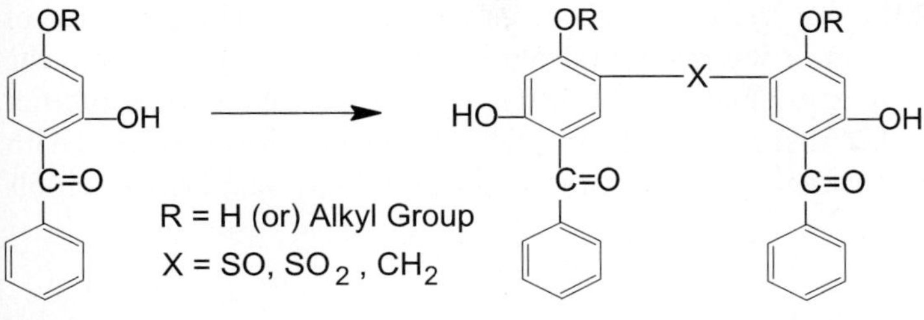


**Figure 8.24.**

**Benzophenone Derivatives Having Antioxidant Properties** Unstabilized polymers, in general, absorb UV radiation and can undergo decomposition by the oxidation process. To stop or reduce this decomposition process in polymers such as polypropylene, both UV absorbers and antioxidants have been used separately in the compositions. The presence of UV absorber can help reduce the UV catalyzed carbonyl group formation in the polymer. Similarly, the embrittlement caused by the heat could be reduced or retarded by the incorporation of an antioxidant in the polymer system. Therefore, it is always advantageous to develop and use compounds having UV absorber and antioxidant properties together in the same molecule. The condensation of 2,4-dihydroxybenzophenone and its derivatives with hindered phenols produced compounds possessing UV absorber and antioxidant properties (Figures 8.25 and 8.26) [52, 53].

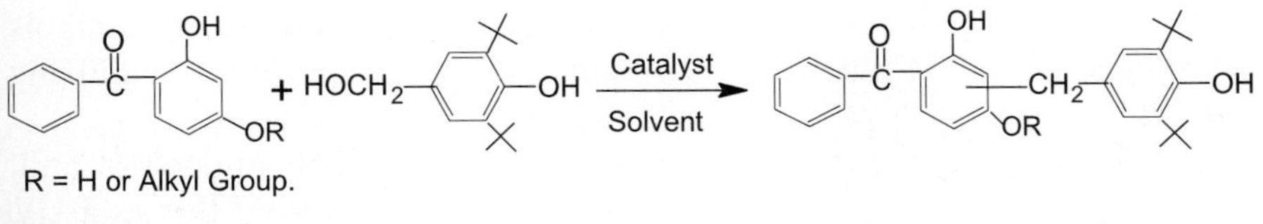


**Figure 8.25.**

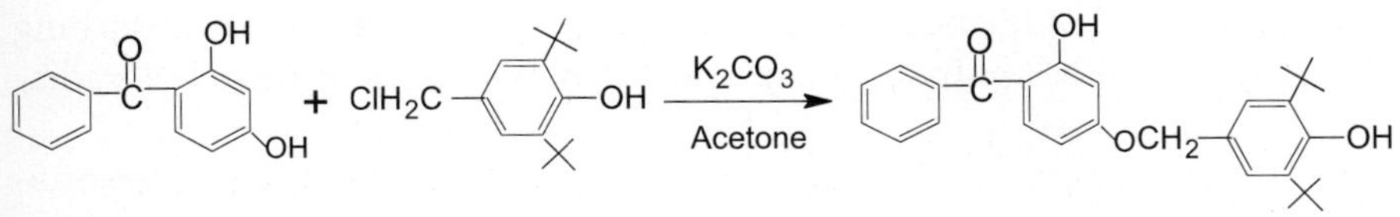


**Figure 8.26.**

### 8.1.3.3
### Dibenzoyl Resorcinol and Derivatives

As mentioned previously in this chapter, Friedel-Crafts acylation of resorcinol could produce different benzoyl derivatives of resorcinol. The presence of second benzoyl group in the resorcinol molecule considerably enhances the UV absorbing property. Therefore, in order to develop dibenzoylresorcinol derivatives for the UV absorber applications, benzoyl chloride was reacted with resorcinol, which resulted in a mixture of 2,4- and 4,6-dibenzoylresorcinols (Figure 8.27) [54].

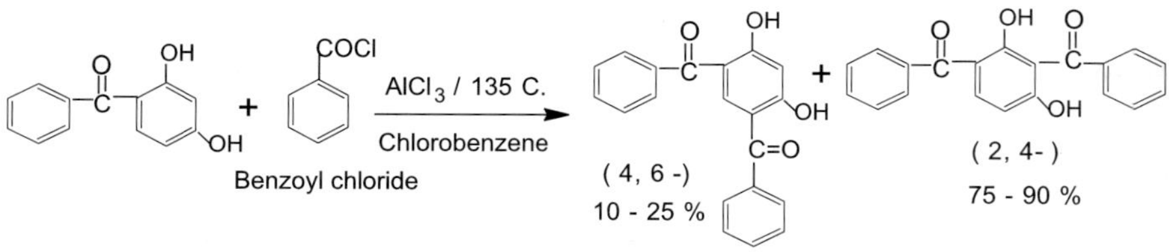


**Figure 8.27.**

From this reaction, the first benzoyl group normally enters into the 4-position of resorcinol due to high reactivity (4 and 6 positions in resorcinol are the same for the first group substitution). After the first benzoyl group reaction, the second benzoyl group can enter either at the 2- or 6-positions for the substitution. Due to the steric hindrance, theoretically, the 2-position is not the preferred position for the second benzoyl group. However, the carbonyl (benzoyl) and hydroxyl interactions, which could lead to the formation of a chelate structure, make the 2-position more receptive for the second benzoyl group substitution. This might be the reason why the 2,4-dibenzoylresorcinol yields were higher.

Based on the UV stabilization effects of 2,4- and 4,6-dibenzoylresorcinols on cellulose esters, 2,4-dibenzoylresorcinol has shown superior UV stabilizing properties [55].

2,4-Dibenzoylresorcinol has been reported as a yellow compound and, consequently, white formulations may not be produced using this as a UV stabilizer. On the other hand, 4,6-dibenzoylresorcinol is known as a white crystalline material, which could allow the production of white formulations [56]. Therefore, considerable efforts were made in the development and applications of 4,6-dibenzoylresorcinol and its derivatives as UV absorbing materials to protect polycarbonate and other polymers against UV radiation [57–65].

4,6-Dibenzoylresorcinol was conveniently prepared by employing resorcinol dimethyl ether or resorcinol according to the following reactions (Figure 8.28) [56,57].

Several derivatives based on the reaction schemes (shown in Figures 8.29 and 8.30) have been developed and utilized in the coating formulations, which provided resistance to scratch and photo degradation to polycarbonate materials [58–65].

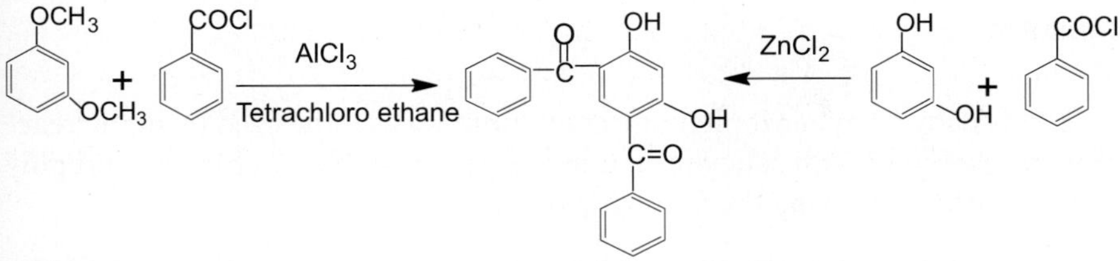


**Figure 8.28.**

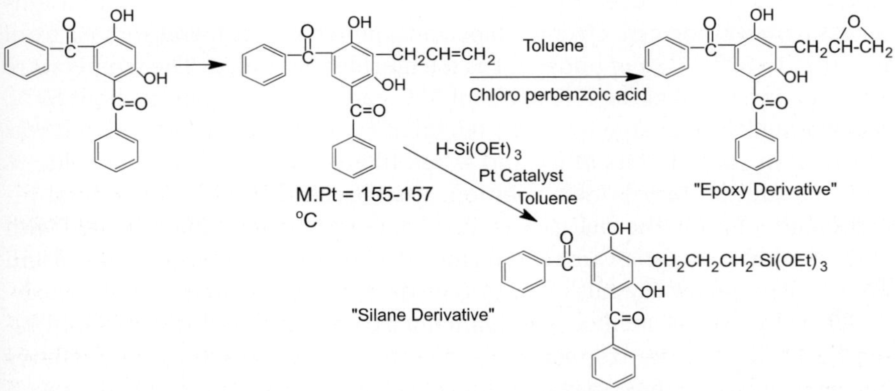


**Figure 8.29.**

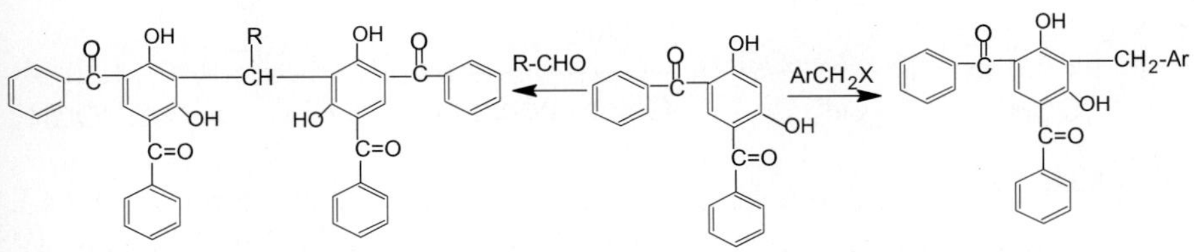


**Figure 8.30.**

## 8.1.3.4 Trihydroxybenzophenone Derivatives

The introduction of functional groups, such as hydroxyl and alkyl, in the phenyl ring of the benzoyl portion of the molecule may be expected to markedly alter the UV absorption properties of 2,4-dihydroxybenzophenone. The presence of hydroxyl group at the 2′-position of the 4-benzoylresorcinol was observed to improve the photo stability of the resulting molecule (Figure 8.31) [66].

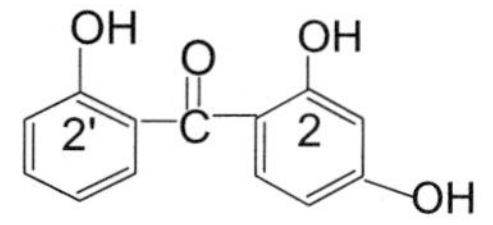


2, 2', 4 - Trihydroxybenzophenone

**Figure 8.31.**

2,2′,4-Trihydroxybenzophenone compound was synthesized from the reaction of resorcinol with salicylic acid in the presence of zinc chloride and poly (phosphoric acid) using the following procedure [67].

**Synthesis of 2,2′,4-Trihydroxybenzophenone [67]** Polyphosphoric acid (317 g, 103% $H_3PO_4$ content), 30 g of salicylic acid, 25.2 g of resorcinol and 67 g of zinc chloride were stirred together at room temperature (27 °C) until a uniform slurry was formed. Over a 2-h period wherein the temperature rose from initial 27 to 37 °C, 38 g of phosphorus trichloride was added. The temperature was then increased at the rate of about 5 °C per hour for 4 h, until about 60 °C was attained. The charge was maintained at 60 °C for 16 h, after which it was drowned in several liters of ice and water, filtered, and washed with cold 5% aqueous sodium bicarbonate solution. After treatment with the sodium bicarbonate solution, the charge was dissolved in hot isopropanol, treated with Nuchar with a very small amount of zinc dust, filtered, cooled, crystallized and filtered. Pale yellow crystals of 2,2′,4-trihydroxybenzophenone was obtained.

2,2′-Dihydroxy-4-methoxybenzophenone compound could be prepared using 2,2′4-trihydroxybenzophenone or directly from the reaction of dimethoxy benzene with *o*-methoxy benzoyl chloride (Figure 8.32) [68].

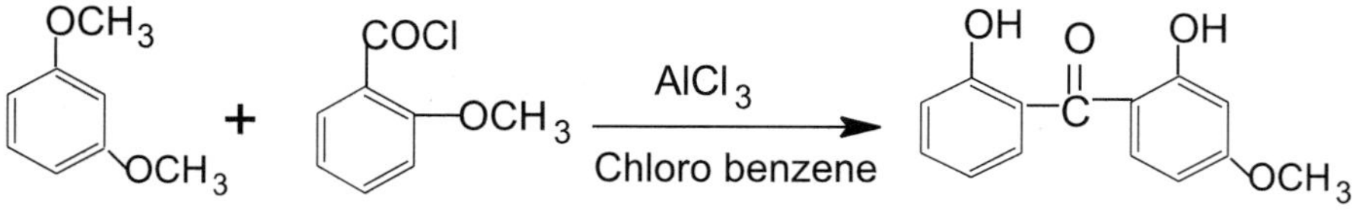


**Figure 8.32.**

Table 8.10 and Figure 8.33 show the properties and UV spectrum of 2,2′-dihydroxy-4-methoxybenzophenone absorber.

### 8.1.3.5 Tetrahydroxybenzophenone and Derivatives

The most important and well known tetrahydroxybenzophenone UV absorber obtained from resorcinol is 2,2′,4,4′-tetrahydroxybenzophenone having the following structure (Figure 8.34).

This compound was prepared from resorcinol and $\beta$-resorcylic acid in the presence of zinc chloride and $PCl_3$ [67,69,70].

**Table 8.10.** Properties of 2,2′-dihydroxy-4-methoxybenzophenone

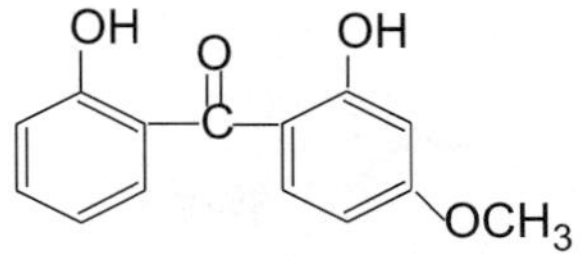


| | |
|---|---|
| Chemical name | 2,2′-Dihydroxy-4-methoxybenzophenone, benzophenone-8 |
| Trade name | Maxgard 900, Cyasorb UV 24 |
| Physical properties | |
| Physical form | Yellow powder |
| Melting point (°C) | 68 |
| Density (g/cc, 25 °C) | 1.38 |
| Absorptivity (in toluene) | 54 |
| $\lambda_{max}$ | 356 |
| Solubility (g/100 ml at 25 °C) | |
| Methyl isobutyl ketone (MIBK) | 100 |
| Methyl ethyl ketone (MEK) | > 50 |
| *n*-Butyl acetate | 47 |
| Xylene | 27.1 |
| Mineral spirit | 3.1 |
| Ethanol (95%) | 21.4 |
| *n*-Hexane | 2.3 |
| Benzene | 46.6 |
| Xylene | 30.6 |
| Di-(2-ethyl hexyl) phthalate | 14.8 |
| Tricresyl phosphate | 20.7 |
| Tri-(2-ethyl hexyl) phosphate | 31.1 |
| Di-(2-ethyl hexyl) sebacate | 14.8 |

Applications: architectural coatings, wood coatings, acrylics, varnishes, lacquers
Data source: Light Stabilizers for Coatings, Product Overview, Cytec Industries
Reference [68]

**Procedure for the Preparation of 2,2′,4,4′-Tetrahydroxy Benzophenone from Resorcinol and β-Resorcylic Acid [69]** Into a flask was charged 317 g of a mixture consisting of 103% $H_3PO_4$ prepared by mixing 190 g of commercial polyphosphoric acid and 127 g of 85% phosphoric acid. To this was added, with constant stirring to uniform slurry, at 25–30 °C, 33 g of β-resorcylic acid, 29.5 g of resorcinol and 67 g of zinc chloride. This was heated to 60 °C in 1 h. 38 g of phosphorus trichloride was added in 1½ h. After the phosphorus trichloride was added, the charge was held at 60 °C for 8.5 h. It was then submerged in several liters of ice and water, filtered and washed with cold 5% aqueous sodium bicarbonate solution, the charge was dissolved in hot water made slightly acidic with HCl, treated with Nuchar and a very small amount of zinc dust, filtered,

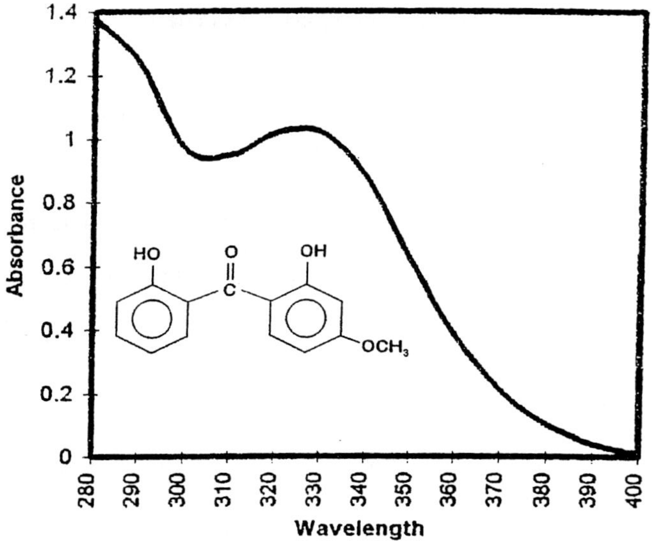


**Figure 8.33.** UV spectrum of 2,2′-dihydroxy-4-methoxybenzophenone from Lycus

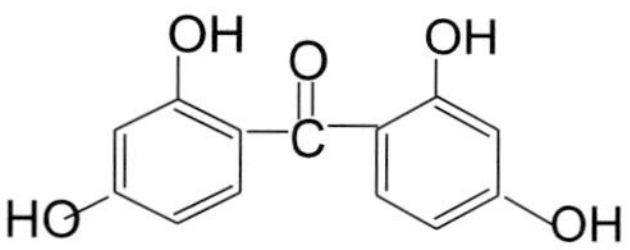


**Figure 8.34.**

cooled, crystallized and filtered. 45 g of 2,2′,4,4′-tetrahydroxybenzophenone, m.p. 200–202 °C was obtained. This corresponds to a yield of 85% of theory.

Physical properties and UV spectrum are shown in Table 8.11 and Figure 8.35.

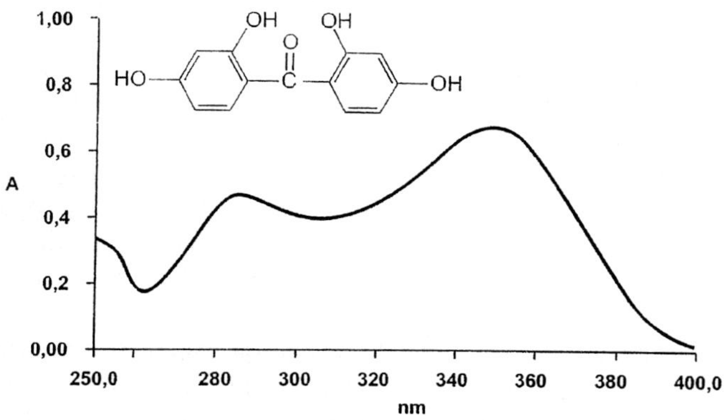


**Figure 8.35.** UV spectrum of 2,2′,4,4′-tetrahydroxybenzophenone from BASF

**Table 8.11.** Properties of 2,2′,4,4′-tetrahydroxy benzophenone

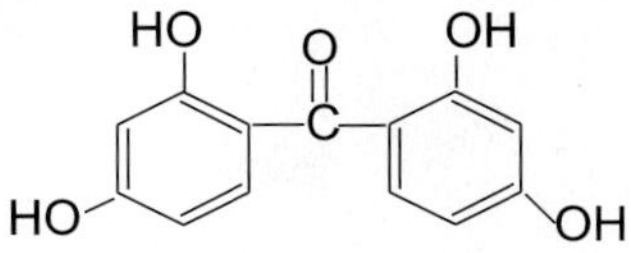


| | |
|---|---|
| Chemical name | 2,2′,4,4′-Tetrahydroxy benzophenone, benzophenone-2 |
| Trade name | Maxgard 1000, Uvinul D50, Seesorb 106 |
| Physical properties | |
| Physical form | Light yellow powder |
| Melting point (°C) | 195 – 197 |
| Density (g/cc, 25 °C) | 1.21 |
| Absorptivity (at 344 nm) | 60 |
| Solubility (g/100 ml at 30 °C) | |
| Water | < 1 |
| Ethyl acetate | 10 |
| Methanol | 50 |
| Methyl ethyl ketone (MEK) | 22 |
| Toluene | < 1 |

Applications: plastic optical lenses, polyurethane coatings, oil paints, acrylic adhesives
Data: technical information on Maxgard 1000 from Lycus Ltd

UV spectrum of this compound shows broad absorption range, which may be ideal for using as a UVA filter for protecting various materials. This compound has been frequently used to improve the stability of perfumes against oxidation.

2,2′-Dihydroxy-4,4′-dialkoxybenzophenone compounds are not only compatible with various types of substantially colorless film-forming plastics, but also their efficiency of absorption at 400 nm is much greater [71]. This compound could be synthesized directly from the dimethoxy benzene derivatives as illustrated in Figure 8.36 [71–74].

Physical properties and UV spectrum of this compound are presented in Table 8.12 and Figure 8.37.

Another important and widely used UV absorber derivative to protect colorants and fragrances in cosmetics is 2,2′-dihydroxy-4,4′-dimethoxybenzophenone-5,5′-disodium sulfonate, for which the properties and UV spectrum are presented in Table 8.13 and in Figure 8.38 [75].

**Tetrahydroxybenzophenone Compounds From Benzene-dicarboxylic Acid Chlorides**
Reacting terephthaloyl chloride in the presence of Friedel-Crafts catalyst with resorcinol produced a tetrahydroxy-diketo compound based on the following reaction (Figure 8.39) [76,77].

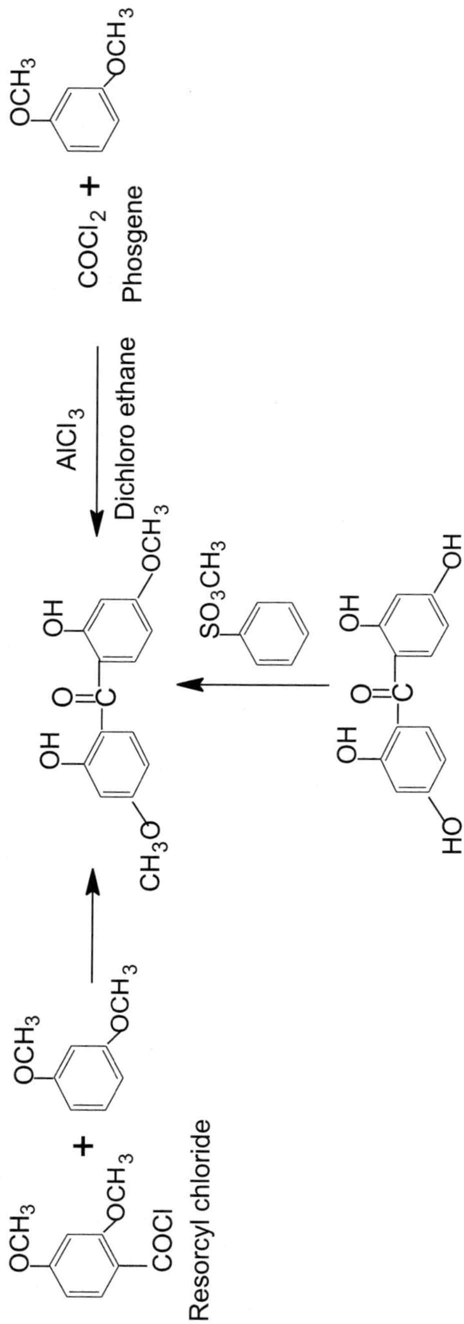
COCl2 +
Phosgene
AlCl3
Dichloro ethane
Resorcyl chloride
OCH3
COCl
OH
CH3O
SO3CH3
HO

**Figure 8.36.**

**Table 8.12.** Properties of 2,2′-dihydroxy-4,4′-dimethoxybenzophenone

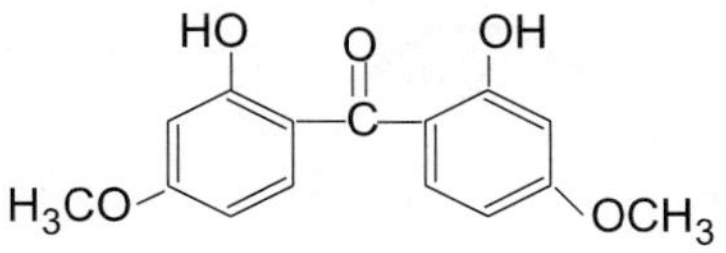


| | |
|---|---|
| Chemical name | 2,2′-Dihydroxy-4,4′-dimethoxy-benzophenone, benzophenone-6 |
| Trade name | Maxgard 300, Uvinul D49, Seesorb 107 |
| Physical properties | |
| Physical form | Light yellow crystalline powder |
| Melting point (°C) | 130 – 132 |
| Density (g/cc, 25 °C) | 1.34 |
| Absorptivity (at 339 nm) | 48 |
| Solubility (g/100 ml at 25 °C)[a] | |
| Ethanol (95%) | 0.5 |
| *n*-Hexane | 0.1 |
| Benzene | 5.2 |
| Xylene | 2.9 |
| Di-(2-ethyl hexyl) phthalate | 1.3 |
| Tricresyl phosphate | 1 |
| Tri-(2-ethyl hexyl) phosphate | 2.8 |
| Di-(2-ethyl hexyl) sebacate | 1.1 |

Applications: polyester film, rubber, polyurethane coatings, lacquers, fluorescent pigments, photographic emulsions, adhesives
[a]Data from [68]

The above compound was recrystallized using a mixture of water and acetone, which showed that it could be soluble in a standard solvent system. Also, the UV spectrum of this compound showed absorption values at $\lambda_{max} = 255\,nm$ ($\varepsilon = 16\,800$), $\lambda_{max} = 298\,nm$ ($\varepsilon = 22\,000$) and $\lambda_{max} = 333\,nm$ ($\varepsilon = 21\,750$). These absorption values suggest that it is capable of absorbing UV rays in a wide range of wavelengths. In addition, this compound exhibited substantially no absorption above 400 nm.

When isophthaloyl chloride was employed in the place of terephthaloyl chloride, a tetrahydroxy compound having isomeric structure was obtained [78]. Various dialkyl ether derivatives were prepared using this tetrahydroxy-diketo product (Figure 8.40).

Table 8.14 summarizes the details on the derivatives, melting points and UV absorptivity data obtained from these compounds.

Plot of melting points and absorptivity as a function of alkyl chain length carbons, Figure 8.41, showed that both properties decreased as the number of carbon atoms in the alkyl ether chain was increased.

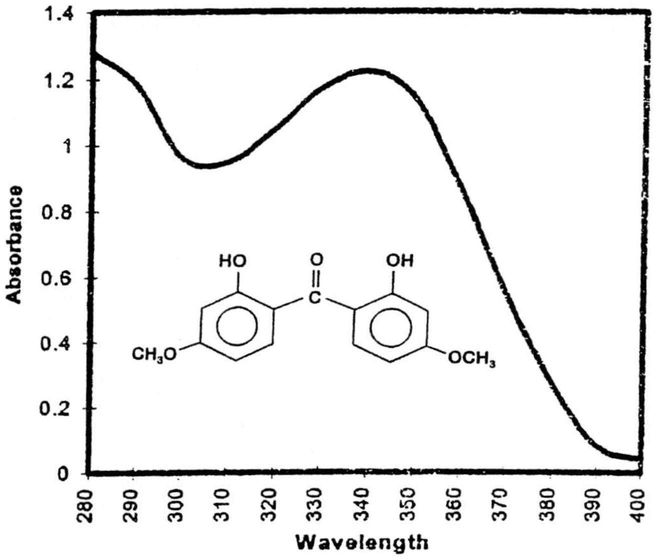


**Figure 8.37.** UV spectrum of 2,2′-dihydroxy-4.4′-dimethoxybenzophenone from Lycus

**Table 8.13.** Properties of disodium salt of 2,2′-dihydroxy-4,4′-dimethoxy-5,5′-disulfobenzophenone

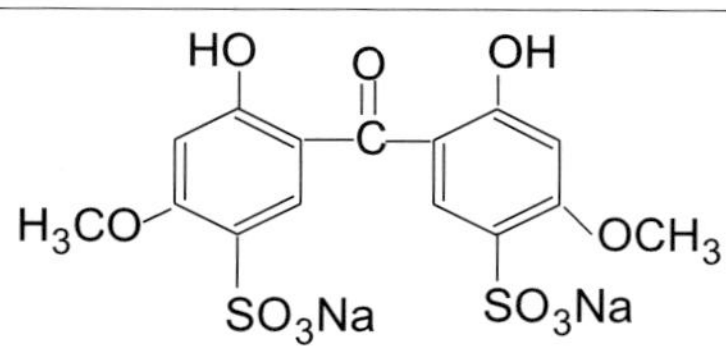


| | |
|---|---|
| Chemical name | Disodium-2,2′-dihydroxy-4,4-dimethoxy-5,5′-disulfobenzophenone, benzophenone-9 |
| Trade name | Maxgard 1800, Uvinul DS49 |
| Physical properties | |
| Physical form | Light yellow powder |
| K value (water), | 16.5 |
| (Highest absorbance near 330 nm) | |
| Solubility (g/100 ml at 25 °C) | |
| Water | 5 |
| Ethyl acetate | 0 |
| Methyl ethyl ketone (MEK) | 0 |
| Ethanol | < 1 |

Applications: protects colorants, active ingredients and Fragrances in cosmetics
Data source: technical information on Maxgard 1800, Lycus Ltd

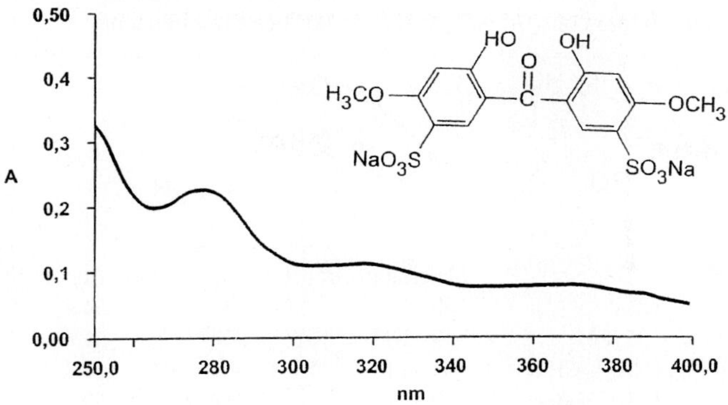


**Figure 8.38.** UV spectrum of 2,2′-dihydroxy-4,4′-dimethoxy-5,5′-disulfonic acid sodium salt from BASF

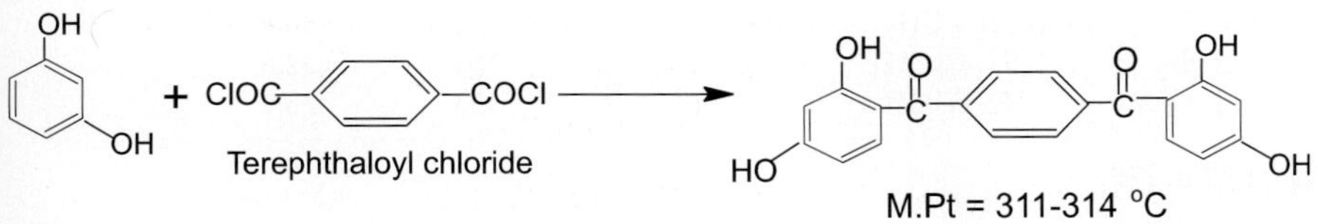


**Figure 8.39.**

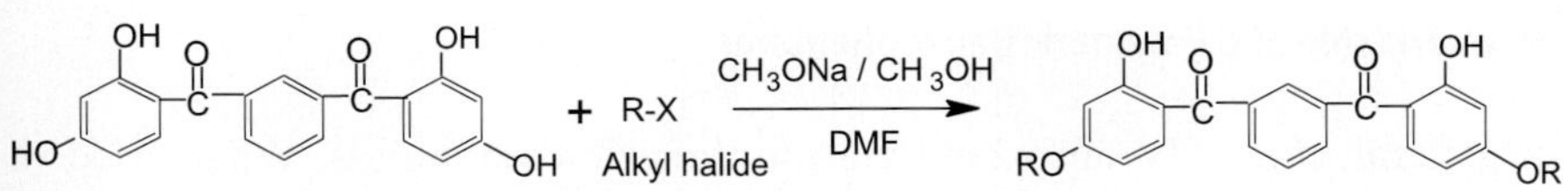


**Figure 8.40.**

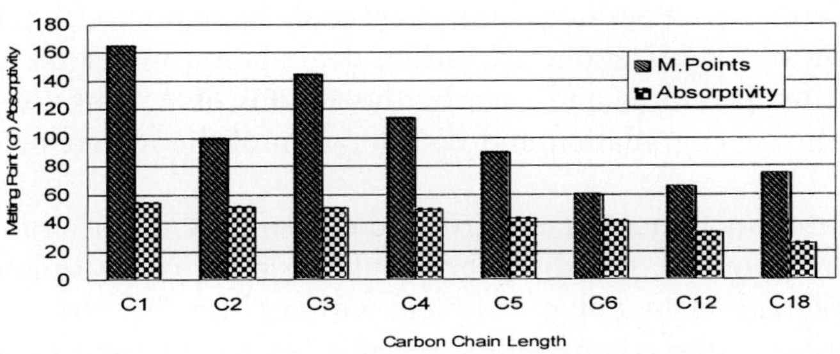


**Figure 8.41.** Effect of alkyl chain length on melting point and UV absorption properties

Based on the absorptivity and melting point data, these derivatives could be utilized as good UV absorbers in varieties of applications.

**Table 8.14.** Dialkyl ethers of 1,3-bis(2′,4′-dihydroxy-benzoyl) benzene

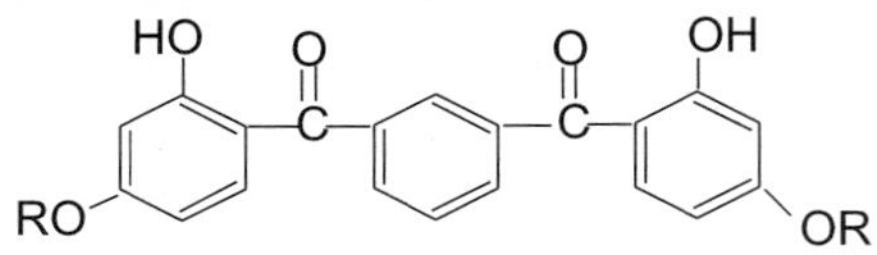


| R Group | | Melting point (°C) | $\lambda_{max}$ (nm) | Absorptivity (a) |
|---|---|---|---|---|
| Methyl | $CH_3$ | 165–166 | 332 | 54.4 |
| Ethyl | $CH_2CH_3$ | 99–100 | 330 | 51.6 |
| Propyl | $(CH_2)_2CH_3$ | 145–146 | 333 | 50.4 |
| Isopropyl | $CH(CH_3)_2$ | 120–121 | 332 | 50.3 |
| Butyl | $(CH_2)_3CH_3$ | 114–115 | 332 | 49.2 |
| Pentyl | $(CH_2)_4CH_3$ | 89–90 | 332 | 43.3 |
| Hexyl | $(CH_2)_5CH_3$ | 60 | 332 | 40.9 |
| Octyl | $(CH_2)_7CH_3$ | Resin | 331 | 36.2 |
| Dodecyl | $(CH_2)_{11}CH_3$ | 64–65 | 333 | 32.8 |
| Octadecyl | $(CH_2)_{17}CH_3$ | 75–76 | 333 | 25.2 |

Data from [78]

### 8.1.3.6 Polymerizable and Polymeric Benzophenones

In general, most UV absorbers are monomeric compounds. When used in coatings and plastic formulations, due to lower molecular weight of monomeric materials, they could be dispersed well and the initial performance may be good. However, over a period of time, they could be removed from the matrix composition by vaporization, migration, degradation under the conditions of elevated temperatures and also, by the actions of various solvents. This will result in the degradation and discoloration of the matrix or polymeric materials.

In order to function as an effective and efficient UV absorber in polymers or plastic compounds, the absorber should be more compatible or soluble and also, remain in the polymer system without lose. UV absorbers having lower volatility, more compatibility and less leachability and for use in the thermoplastic and coatings applications is most desired.

Various approaches have been used to overcome the volatility and compatibility problems of UV absorbers. One solution was to synthesize UV absorbers containing polymerizable functional groups. UV absorbers with polymerizable groups, capable of undergoing either an addition or condensation polymerization, can be homo-polymerized or copolymerized with other monomers [79–84]. Considerable efforts were made to develop and use UV absorbers having polymerizable groups. The polymerizable stabilizer was incorporated into the polymer as a co-monomer in the synthesis or copolymerized with another monomer to develop a stabilizing polymer, which was then added to the desired polymer system. Since UV absorber has been attached to the polymer backbone, it can remain in the polymer composition and the stabilizing benefits are retained for a long period of time.

The polymerizable UV absorbers possessing functional groups, such as vinylic or allylic, capable of undergoing addition polymerization in the presence of various initiator systems are presented below (Figures 8.42–8.46) [85–90]. These monomers could be used to produce homo-polymers or co-polymers with other vinyl monomers, such as acrylates and methacrylates. The polymers obtained from these UV monomers are expected to improve the compatibility, stability and decreased UV absorber migration from the matrix polymer.

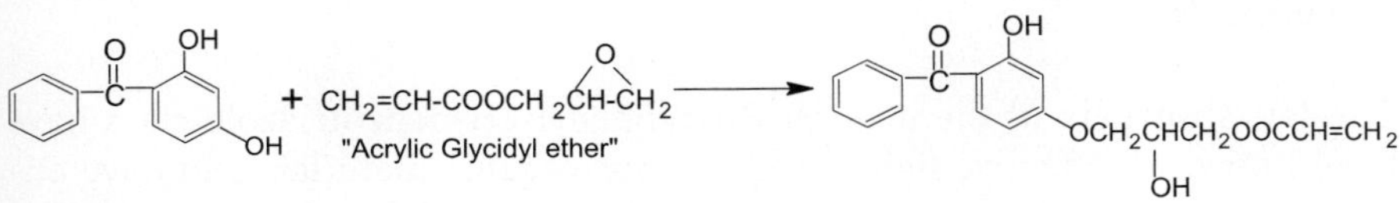


**Figure 8.42.**

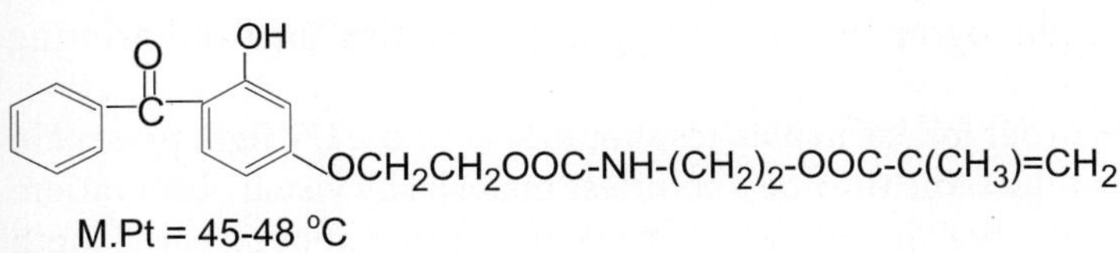


**Figure 8.43.**

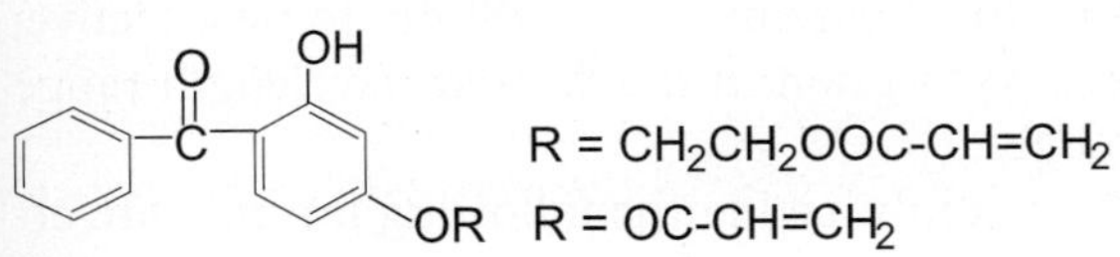

**Figure 8.44.**

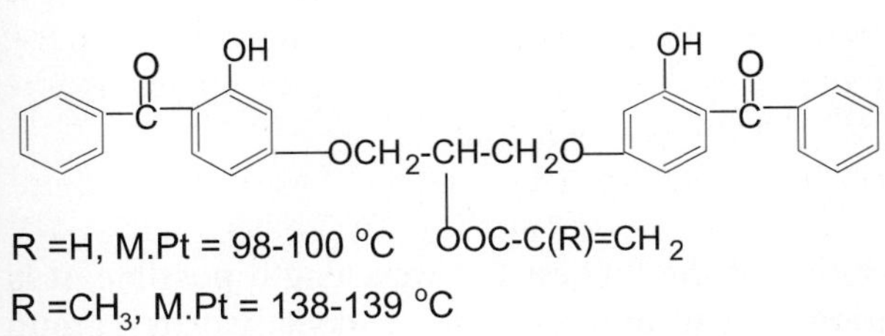


**Figure 8.45.**

In addition to 2,4-dihydroxybenzophenone, 2-hydroxy-4-(2-hydroxyethoxy)-benzophenone UV absorber compound has also been utilized in the

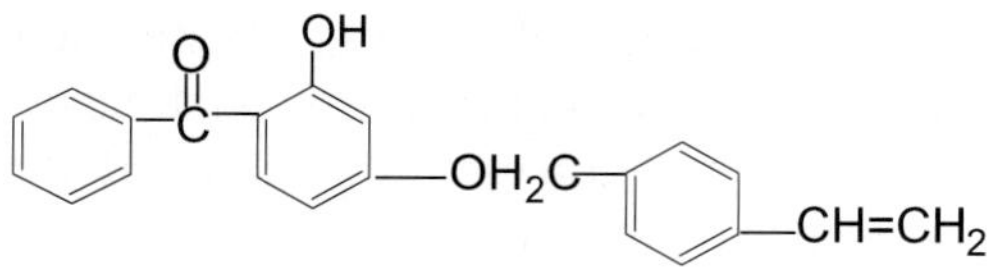


**Figure 8.46.**

development of different vinylic monomers, polymers and copolymers. These materials have been used in polymers and cosmetic compositions to protect them against UV radiations [91–94].

## 8.1.4 Benzotriazoles

The UV absorbers of the *ortho*-hydroxyphenyl-2H-benzotriazole class have been known as effective light stabilizers for organic materials and have enjoyed considerable commercial success. They are widely used in practice as light stabilizers for a large number of substrates, for example for stabilizing thermoplastic and coating materials, and also in various recording materials which include photographic layers and papers, printing inks and printing papers [95].

A good UV absorber for use in plastics should absorb the UV light present in the sunlight and at the same time be a colorless material by visual observation. An effective UV absorber should have its peak absorption above a wavelength of 320 nm. The absorption peak may be at higher wavelength, as long as absorption drops off significantly as it approaches the visual range so that no visible color could be shown by the compound. In addition, to be effective, it should show a high degree of absorption in the desired wavelength range sufficiently below the visual range.

It is known that organic compounds having the following formula are effective UV absorbers (Figure 8.47).

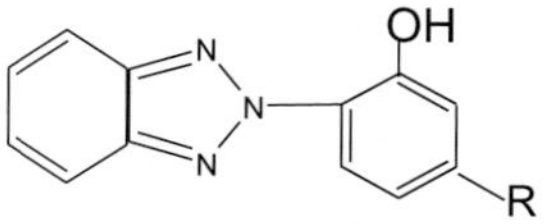


**Figure 8.47.**

In order to show absorption at the longest UV wavelength possible, it is advantageous to have additional substitutions like lower alkyls, hydroxyl group or alkoxy groups present in the *ortho*-hydroxy substituted benzene ring. The presence of hydroxy and alkylene ether groups at the 4-position was found to enhance the UV absorption of benzotriazole derivatives, particularly in the longer wavelength region.

An overview of resorcinol based benzotriazoles useful in the UV absorber applications, which have been published in the literature are presented here.

### 8.1.4.1
### Benzotriazoles from Resorcinol

The synthetic scheme by which the benzotriazole substituted resorcinol derivatives could be prepared is illustrated in Figure 8.48.

In addition to *o*-nitro aniline, substituted nitro aniline and also resorcinol derivatives could be utilized to develop various benzotriazole derivatives.

A typical procedure for the synthesis of benzotriazole compound from resorcinol is illustrated below [96].

**Procedure for the Synthesis of 5-Chloro-(2′,4′-dihydroxyphenyl)-benzotriazole [96]**

**Step 1: Preparation of 4-chloro-2-nitro-2′,4′-dihydroxyazobenzene**
In 616 g of a 17.5% hydrochloric acid were dissolved at 70 to 80 °C 121 g of 4-chloro-2-nitroaniline and the resulting solution was added to 300 g of ice at a time while thoroughly stirring. Diazotization was carried out by adding a solution of 51.0 g of sodium nitrite in 70 g of water to the above solution at below 0 °C. Separately, 81.1 g of resorcinol were dissolved in a mixture of 510 g of a 35% hydrochloric acid and 1750 g of ice water, and then the resulting solution was kept at about 0 °C. Then, the diazo liquor obtained above was added drop-wise to the resorcinol solution at about 0 °C over about 1.5 h. After completion of the addition, the solution was stirred at the same temperature for further 1 h and then kept at 50 °C for 1 h. Thus the coupling reaction was completed. Separated red 4-chloro-2-nitro-2′,4′-dihydroxyazobenzene was filtered, thoroughly washed with water and dried. The amount obtained was 197.1 g (yield 96.0%).

**Step 2: Preparation of 5-chloro-2-(2′,4′-dihydroxyphenyl)-benzotriazole**
To a suspension of 29.3 g of 4-chloro-2-nitro-2′,4′-dihydroxy-azobenzene obtained in 100 g of methanol were added 50 g of a 40% aqueous caustic soda solution and then 25.5 g of zinc powder at room temperature. On heating the reaction solution under reflux for about 1 h, the solution changed in color. Thereafter, the solution was further heated for 30 min and filtered hot. The filtrate was neutralized with acetic acid at about 10 °C to separate a solid matter. The matter was filtered, dried and recrystallized from toluene repeatedly to obtain pale yellow 5-chloro-2-(2′,4′-dihydroxyphenyl)-benzotriazole in a yield of 60% (m.p. 188–89 °C).

Table 8.15 presents some details on different benzotriazole derivatives obtained from resorcinol and their UV absorption data [97].

Employing $\beta$-resorcylic acid in the place of resorcinol, mono-benzotriazole derivative containing carboxylic acid functional group was synthesized based on the procedure given below (Figure 8.49) [98].

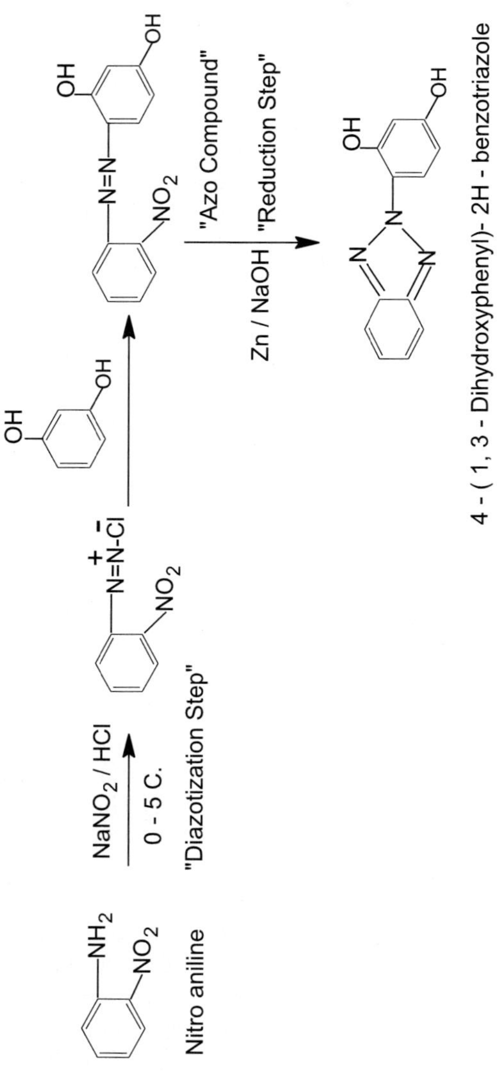
Nitro aniline
NaNO2 / HCl
0 - 5 C.
"Diazotization Step"
"Azo Compound"
Zn / NaOH
"Reduction Step"
4 - ( 1, 3 - Dihydroxyphenyl)- 2H - benzotriazole

**Figure 8.48.**

**Table 8.15.** UV Absorption properties of resorcinol based benzotriazoles

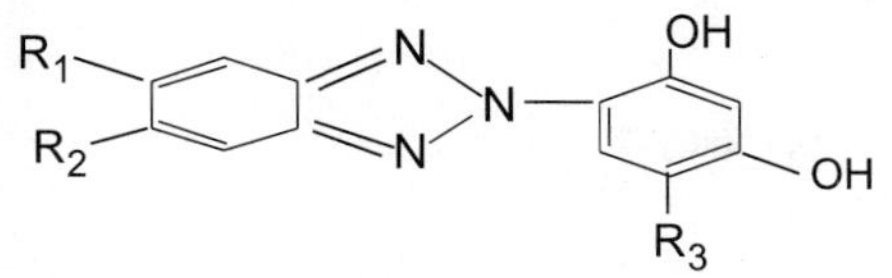

| Substitution group | | | UV absorption (in $CH_3OH$) | |
|---|---|---|---|---|
| $R_1$ | $R_2$ | $R_3$ | $\lambda_{max}$ (nm) | $\varepsilon \times 10^4$ |
| H | H | H | 339 | 2.18 |
| H | $CH_3$ | H | 341 | 2.27 |
| H | $OCH_3$ | H | 346 | 2.46 |
| H | Cl | H | 348 | 1.98 |
| $OCH_3$ | $OCH_3$ | H | 346 | 2.94 |
| H | H | $CH_2CH_3$ | 348 | 2.12 |
| H | H | Cl | 345 | 2.02 |

Data obtained from [97]

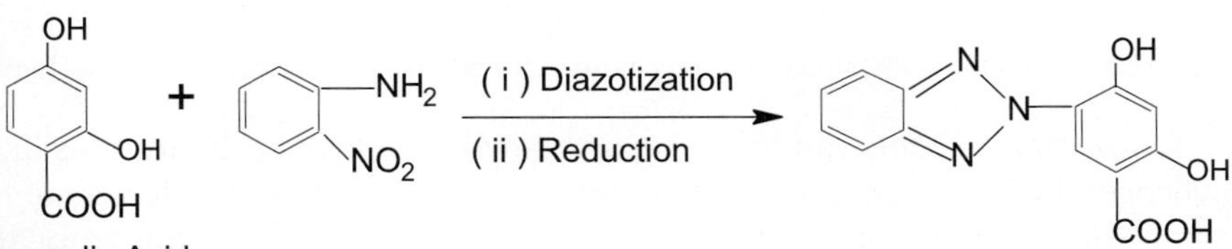


Resocylic Acid

**Figure 8.49.**

**Preparation of 2-(5-Carboxy-2,4-dihydroxyphenyl) Benzotriazole from β-Resorcylic Acid [98]** *o*-Nitroaniline (40.5 g) was diazotized in the normal manner with concentrated hydrochloric acid (104 cc) and sodium nitrite (22.1 g). The excess sodium nitrite was destroyed with sulfamic acid (3.0 g) and the clear diazonium chloride solution was then added, below the surface, to a cooled solution (0 to 5 °C) of β-resorcylic acid (46.3 g) prepared using 1.5 l of water containing 60 g of sodium hydroxide. The solution was stirred for 2 h after the addition was completed and then acidified with dilute hydrochloric acid.

The crude 2-(5-carboxy-2,4-dihydroxy) phenylazo-nitrobenzene thus obtained (30.3 g) was treated with zinc dust (39.2 g) and sodium hydroxide (240 cc, 5N). The product thus obtained was recrystallized from toluene-hexane, followed by another recrystallization from ethanol-benzene yielded 2-(5-carboxy-2,4-dihydroxyphenyl) benzotriazole, m.p. about 250–251 °C.

If 2 mol of *o*-nitroaniline derivatives, instead of 1 mol, were used in the synthesis then, bis-(benzotriazole) resorcinol derivatives could be produced, for which a generalized chemical structure is shown below (Figure 8.50) [99–103].

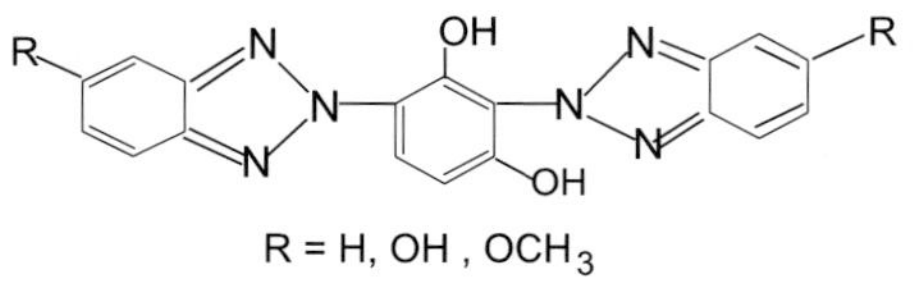


**Figure 8.50.**

## 8.1.4.2
## UV Absorbing Mechanism for Benzotriazole Compounds

A proton transfer mechanism proposed for 2-hydroxybenzophenone compounds for the absorption of UV light energy and dissipates it as heat energy can also be proposed for the benzotriazole based UV absorbers as shown in Figure 8.51.

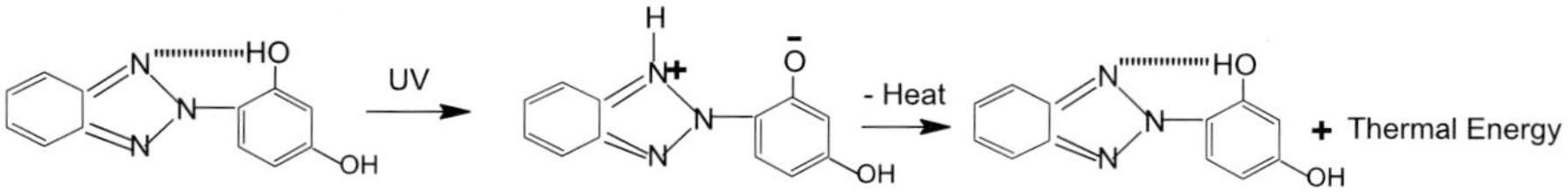


**Figure 8.51.** Possible UV stabilization mechanism for resorcinol benzotriazole absorbers

Benzotriazole compounds having *ortho*-hydroxyphenyl substitutions have hydrogen bonded structures. On absorbing the UV energy, an electronic rearrangement takes place, which results in the transfer of proton from the hydroxyl group to the nitrogen atom of the benzotriazole ring. This rearranged molecule is very unstable, and therefore rapidly returns to the original ground state through the dissipation of heat energy. This process is repeated indefinitely and thus providing prolonged UV stability to substrate where the UV stabilizers have been employed.

## 8.1.4.3
## Effect of Substitution on the UV Performance

Studies have indicated that substitution on the benzotriazole ring has greater influence on the UV absorption power as well as the absorption wavelength of resorcinol based benzotriazole UV absorbers [101]. The influence of electron donating and electron withdrawing groups in the 4-position of benzotriazole on the UV absorption molar extinction coefficient values can be seen in Table 8.16.

Obviously, from Table 8.16 results, the introduction of hydroxyl, methoxy and chloro groups at the 4′-position of the benzotriazole ring resulted in the increase of UV absorption coefficients and also, the wavelength of absorption. These results suggest that the enhancement in UV properties may be associated with the formation or presence of efficient intra-molecular hydrogen bonding in the absorbers.

**Table 8.16.** UV absorption data for benzotriazole derivatives

Mono-benzotriazole derivatives

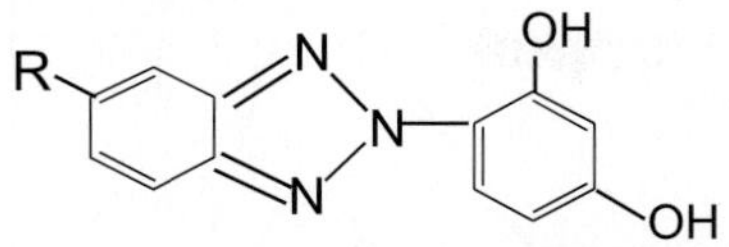


| R-Group | | UV spectrum (in dichloroethane) | | | |
|---|---|---|---|---|---|
| | | $\lambda_{max}$ (nm) | $\varepsilon \times 10^4$ | $\lambda_{max}$ (nm) | $\varepsilon \times 10^4$ |
| H[a] | Proton | ND | ND | 339 | 2.18 |
| OH | Hydroxyl | 300 | 0.96 | 350 | 2.9 |
| $OCH_3$ | Methoxy | 300 | 1.4 | 350 | 3.37 |
| Cl | Chloro | NA | NA | 354 | NA |

Di-benzotriazole derivatives

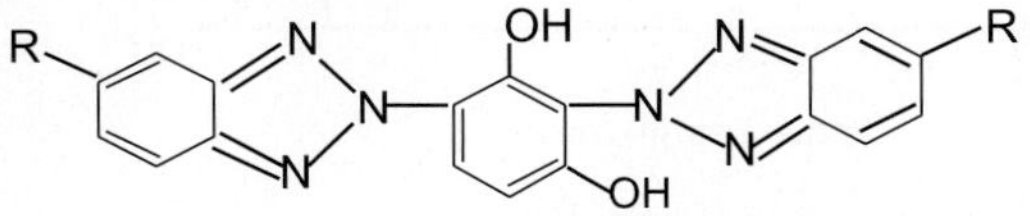


| R-Group | | UV spectrum (in dichloroethane) | | | |
|---|---|---|---|---|---|
| | | $\lambda_{max}$ (nm) | $\varepsilon \times 10^4$ | $\lambda_{max}$ (nm) | $\varepsilon \times 10^4$ |
| H | Proton | ND | ND | 330 | 3.4 |
| OH | Hydroxyl | 300 | 2.2 | 346 | 3.45 |
| $OCH_3$ | Methoxy | 300 | 1.7 | 344 | 3.6 |

[a]From [97]
Data obtained in methanol Data obtained from [96, 101]
ND = Not determined or none detected. NA = Not available

By comparing the extinction coefficient results of mono and di-benzotriazole substituted resorcinol derivatives (Figure 8.52 and Table 8.16), higher values were observed for the di-benzotriazole resorcinol derivatives. The two benzotriazole groups present in the structure might have exhibited stronger hydrogen bonding in the di-benzotriazole molecule [99–102].

This result was similar to the results observed for 4,6 and 2,4-dibenzoyl substituted resorcinol derivatives, as discussed and disclosed in the benzophenone derivatives section.

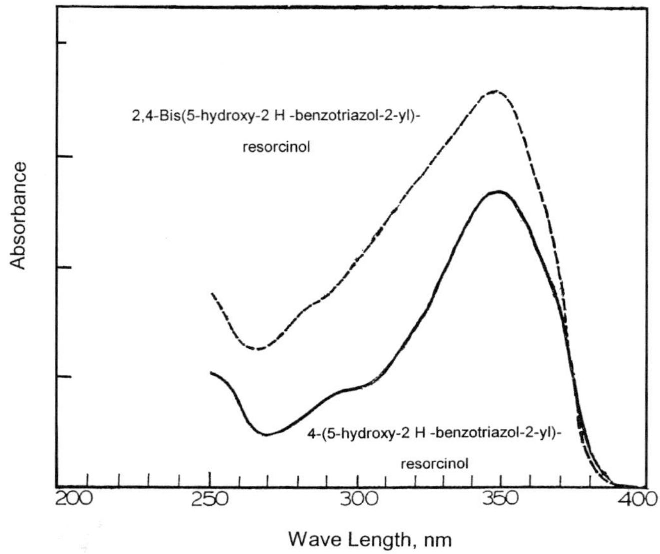


**Figure 8.52.** UV spectrum of benzotriazole compounds [99–102]

### 8.1.4.4
### Benzotriazole Derivatives

Mono and di-benzotriazole substituted resorcinolic UV absorbers have two phenolic hydroxyl groups in the resorcinol portion of the molecule. The hydroxyl group present at the 4-position of the resorcinolic ring can be utilized for the modification of benzotriazole derivatives to enhance their solubility, thermal stability and substrate compatibility. Reactions typical for the phenolic hydroxyl group can be done with these resorcinolic derivatives. Some of the modifications and products obtained are presented in Figure 8.53 [103, 104].

### 8.1.4.5
### Polymerizable and Polymeric Benzotriazoles

A number of polymerizable UV stabilizers based on resorcinol-benzotriazole compounds have been developed and polymerized [105–112]. In these monomers, vinylic groups were introduced by the chemical reaction of resorcinolic hydroxyl group with either the acrylic (or methacrylic) compounds or glycidyl ethers having vinylic groups. The synthesis of acrylate and methacrylate groups containing resorcinol based benzotriazole monomers and their homo as well as copolymers obtained from these UV absorber materials are schematically presented in Figure 8.54.

The exited state intra-molecular proton transfer (ESIPT) mechanism operates in the monomeric benzotriazole compounds for the deactivation energy

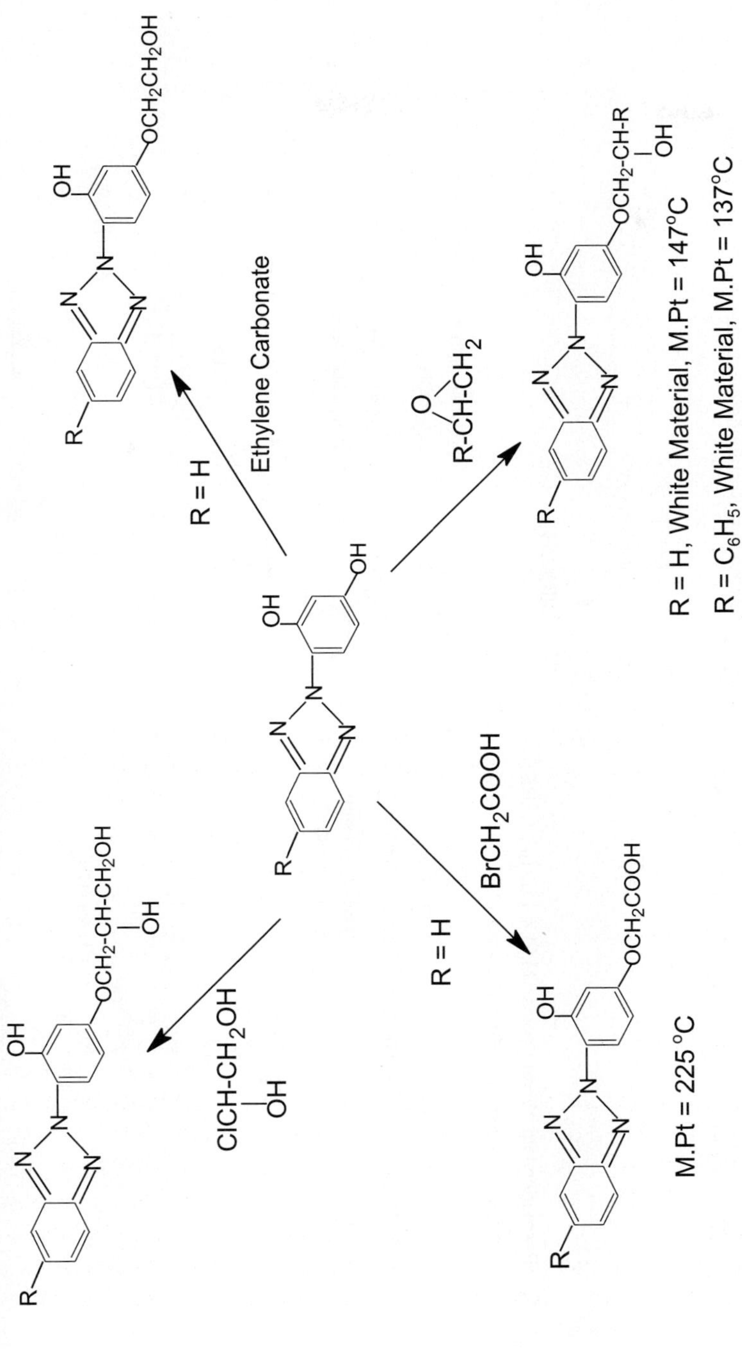
R = H
Ethylene Carbonate
OCH2CH2OH
R-CH-CH2
OCH2-CH-R
OH
R = H, White Material, M.Pt = 147°C
R = C6H5, White Material, M.Pt = 137°C
BrCH2COOH
R = H
OCH2COOH
M.Pt = 225 °C
ClCH-CH2OH
OCH2-CH-CH2OH

**Figure 8.53.**

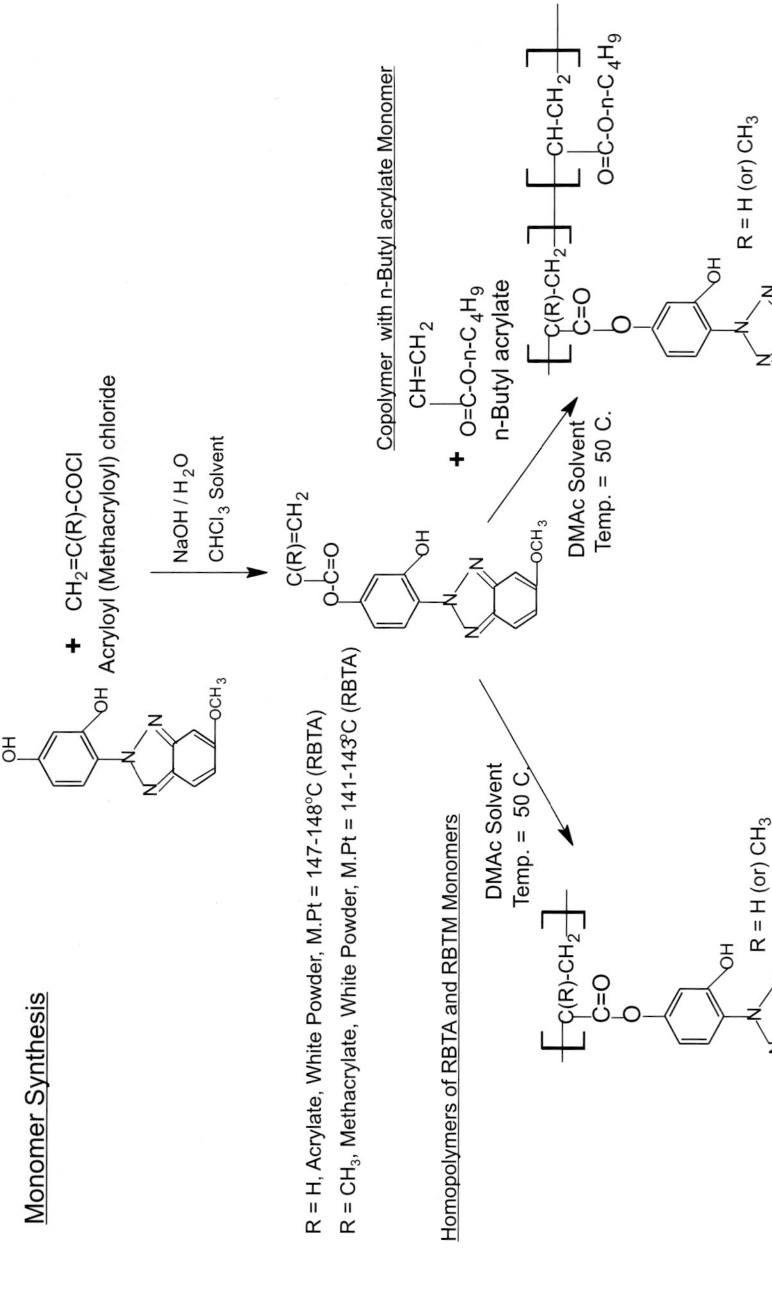


**Figure 8.54.** Monomeric and polymerized resorcinol-benzotriazole based UV absorbers

may be applicable to the polymerized benzotriazole system also. Polymerized stabilizers have been found to be less sensitive to strong hydrogen bonding solvents like DMSO. It was theorized that, though the intra molecular hydrogen bonds present in the benzotriazole monomers could be disrupted by the strong interacting solvent such as DMSO, the polymer chain could shield the benzotriazole chromophores from the solvent interactions in the polymerized absorbers [95]. This can be seen from Table 8.17 results, which shows that the absorption maximum and molar extinction coefficient of monomeric and polymeric benzotriazole systems were observed to be similar.

**Table 8.17.** UV absorption data of resorcinol based benzotriazole monomers, homopolymers and copolymers (measured in chloroform)

| Monomer/polymer | $\lambda_{max}$ (nm) | $\varepsilon \times 10^4$ (l/mol cm) |
|---|---|---|
| Acrylate system | | |
| RBTA Monomer | 344 | 2.83 |
| Homopolymer-poly(RBTA) | 338 | 2.62 |
| Copolymer-poly(MMA-*co*-RBTA) | 344 | 3.27 |
| Copolymer-poly(*n*-BA-*co*-RBTA) | 344 | 2.84 |
| Methacrylate system | | |
| RBTM monomer | 344 | 2.78 |
| Homopolymer-poly(RBTM) | 340 | 3.18 |
| Copolymer-poly(MMA-*co*-RBTM) | 344 | 2.62 |
| Copolymer-poly(*n*-BA-*co*-RBTM) | 344 | 3.09 |

MMA = Methyl methacrylate
*n*-BA = *n*-Butyl acrylate
RBTA = Resorcinol based methoxybenzotriazole acrylate
RBTM = Resorcinol based methoxybenzotriazole methacrylate
Data from [109]

Some other polymerizable benzotriazole based UV absorbers developed are presented below (Figures 8.55 through 8.59) [105–112].

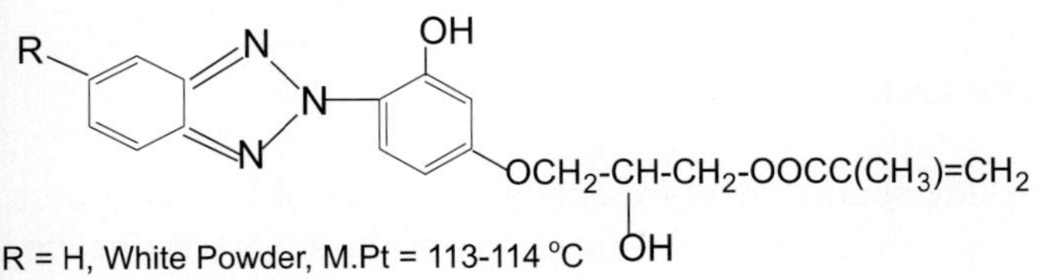


R = H, White Powder, M.Pt = 113-114 °C

R = $OCH_3$, White Powder, M.Pt = 100-102°C

**Figure 8.55.**

As an example, the UV spectrum of one polymerizable benzotriazole monomer is shown in Figure 8.60 [110].

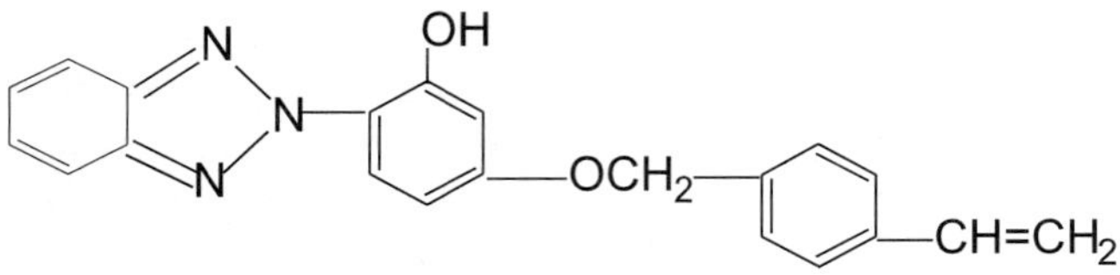

**Figure 8.56.**

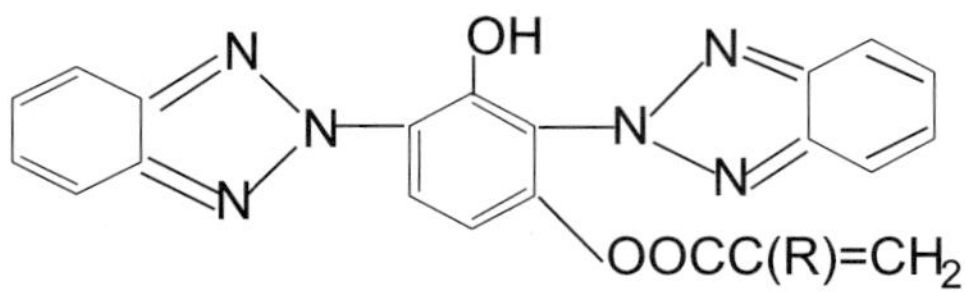

R = H, White Crystals, M.Pt = 230 - 232°C

R = $CH_3$, White Crystals, M.Pt = 216 - 218°C

**Figure 8.57.**

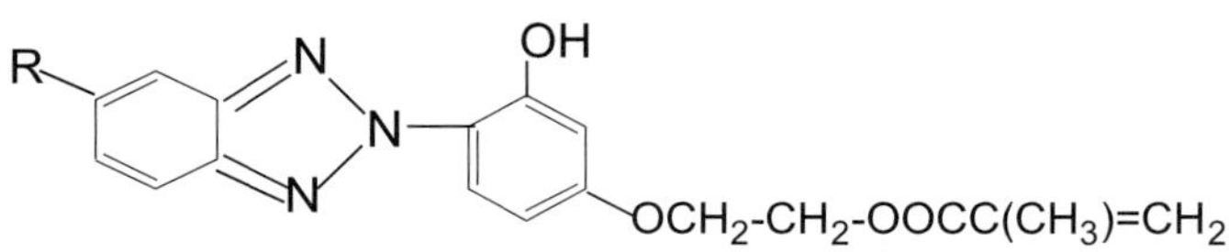

R = H (or) Cl, White Solid.

**Figure 8.58.**

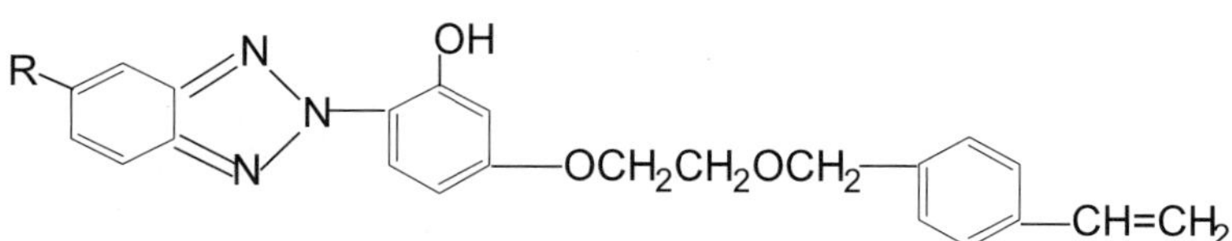

R = H, Cl or $OCH_3$

**Figure 8.59.**

## 8.1.5
## Resorcinol Based Triazine Chemistry

Resorcinol based triazine chemistry represents a new class of UV absorbers. In this class of materials, the 2-, 4- and 6-positions of the 1,3,5-triazine ring are substituted with aromatic groups, which have a hydroxyl group at the *ortho*-position to the point of attachment to the triazine ring as shown in Figure 8.61.

Based on the published reports, that compounds representing the structure shown in Figure 8.61 exhibited high degree of absorbance in the ultraviolet light wavelength region, which could effectively protect the plastics, coat-

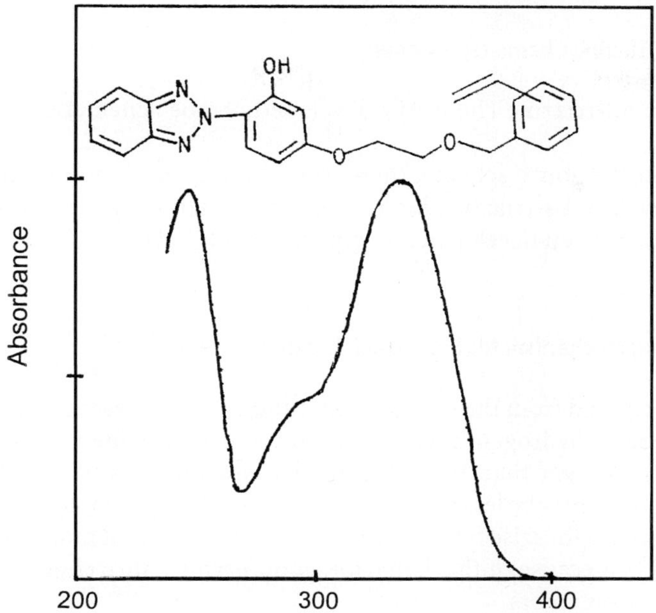


**Figure 8.60.** UV spectrum of polymerizable benzotriazole monomer [110]

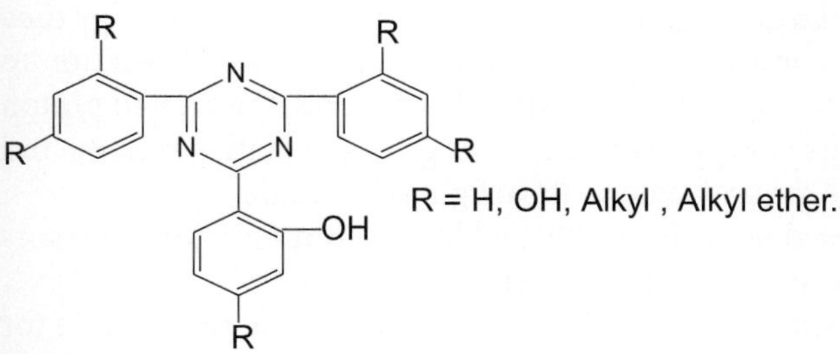


**Figure 8.61.**

ings and other organic materials, and also transmitting visible light almost completely. High thermal stability could be expected with compounds having triazine bound structures due to an extended aromaticity presents in this type of structures. Compared to benzophenone and benzotriazole based UV absorbers, triazine based compound exhibited high molar extinction coefficients [5, 113, 114].

Triazine-resorcinol chemistry for the UV absorber applications was developed in the 1960s. Various synthetic methods utilized to develop triazine-resorcinol have been reviewed [115–117]. In this chapter, the high performance UV absorbers developed based on the resorcinol-triazine chemistry and its application towards the protection of various polymers and coating materials have been discussed.

### 8.1.5.1 Resorcinol-Triazine Chemistry – General

The resorcinol-triazine chemistry developed can be schematically shown in Figure 8.62.

Based on the above scheme, there were three types of derivatives namely, the tris-resorcinol-*s*-triazine, bis-resorcinol-*s*-triazine and mono-resorcinol-*s*-triazine have been developed and reported [118–121].

### 8.1.5.2 UV Absorbing Mechanism for Resorcinol-Triazine System

The data obtained from the spectroscopic and X-ray analyses revealed that the intra molecular hydrogen bonds in the resorcinol-triazine compounds have been much stronger than benzotriazole based compounds. Therefore, these hydrogen bonds may be less likely to be broken by the inter-molecular hydrogen bonds, which might arise from the interactions with solvent molecules or polar polymers [5]. Because of this behavior, good performance could be expected even in polar polymers.

The triazine-resorcinol compounds have at least one hydroxyl group in the 2′-position and an alkyl ether group in the 4′-position. Similar to benzophenone and benzotriazole compounds, after absorbing the UV energy they undergo an excited state intra molecular proton transfer (ESIPT) reaction to create a proton transferred intermediate. After this process, the excited proton excited molecule loses its energy as thermal energy to the matrix system, and thereby returning to the ground state as the original molecule.

A mechanism proposed, and also published in the literature, for the resorcinol-triazine based UV absorber can be seen in Figure 8.63.

The ESIPT mechanism operates in these UV absorbers is responsible for the inherent photo stability and absorption properties of these materials. The UV absorbing capacity with triazine based compounds remains intact as long as the intra molecular proton transfer reaction is not broken.

### 8.1.5.3 Tris-Resorcinol-Triazine and Derivatives

Tris-resorcinol-triazine compound can be obtained from the reaction of reaction of resorcinol with cyanuric chloride in the presence of Friedel-Crafts catalyst such as aluminum chloride [118, 119].

This compound could be obtained by the procedure shown below.

**Synthesis of 2,4,6-Tris-(2,4-dihydroxyphenyl)-s-Triazin [118]** 100 Parts of cyanuric chloride, 180 parts of resorcinol, 100 parts of aluminum chloride and 1500 parts of tetrachloroethane were refluxed, with stirring, under anhydrous conditions

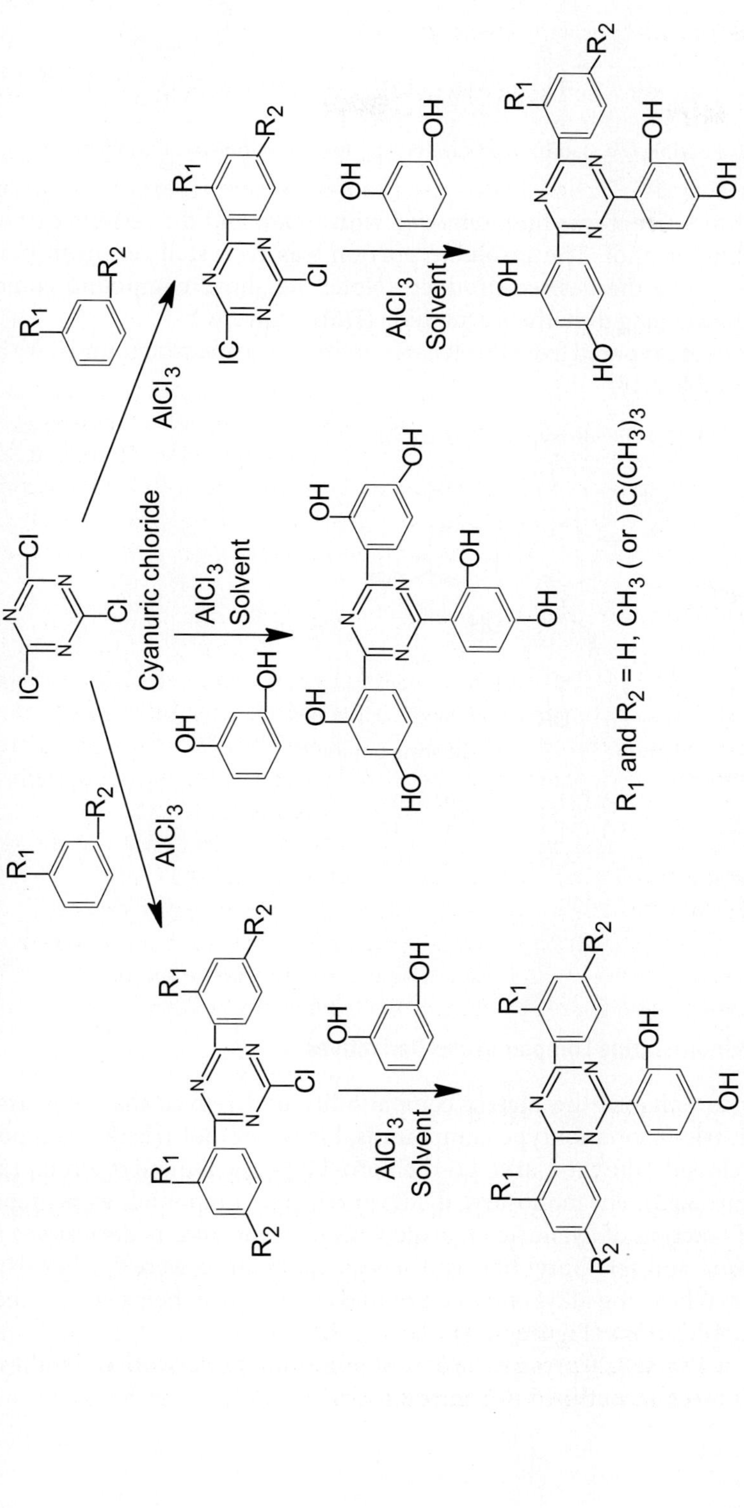


**Figure 8.62.** Resorcinol-triazine based UV absorber chemistry

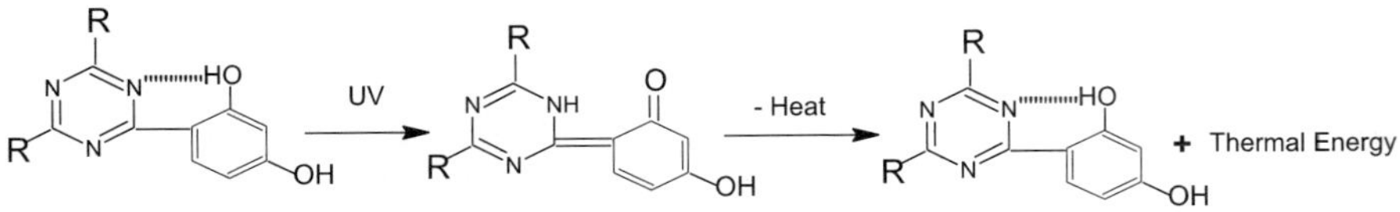


**Figure 8.63.** Possible UV stabilization mechanism for resorcinol-triazine absorbers

for 18 h. The solvent was then removed with steam and the residue extracted with boiling alcohol. The insoluble portion was recrystallized from boiling pyridine to give the desired product. (Note: the above compound could be recrystallized using dimethylformamide (DMF) solvent.)

Derivatives prepared from this tris-resorcinol-triazine compound have been shown in Table 8.18.

**Table 8.18.** Tris-(2,4-dihydroxyphenyl)-*s*-triazine

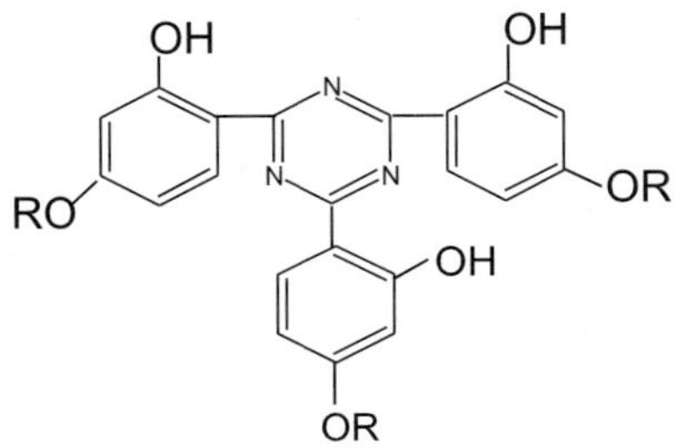


| R Group | | Melting point (°C) |
|---|---|---|
| *n*-Propyl | $(CH_2)_2CH_3$ | 202 – 203 |
| *n*-Octyl | $(CH_2)_7CH_3$ | 145 – 147 |
| Allyl | $CH_2CH{=}CH_2$ | 168 – 169 |

Data from [118–120]

### 8.1.5.4
### Bis-Resorcinol-Triazine Compound and Derivatives

In order to enhance the matrix compatibility and reduce the color associated with tris-resorcinol type compounds, bis-resorcinol-triazine compound were developed from the 2-aryl-4,6-dichloro-1,3,5-triazine and resorcinol reaction (Figure 8.62). The mono-aryl-dichloro-triazine compounds were prepared from the reaction of cyanuric chloride with benzene and its derivatives such as *m*-xylene and *tert*-butyl benzene. These compounds were further derivatized by synthesizing alkyl ether compounds to enhance their performance in various applications (Figure 8.64).

Tables 8.19 and 8.20 present the data obtained on the derivatives synthesized using the reaction outlined in Figure 8.64.

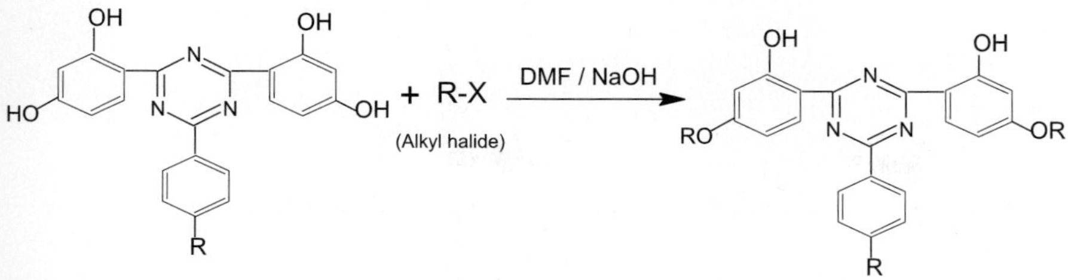


**Figure 8.64.**

**Table 8.19.** Bis-(2,4-dihydroxyphenyl)-*s*-triazine derivatives and properties

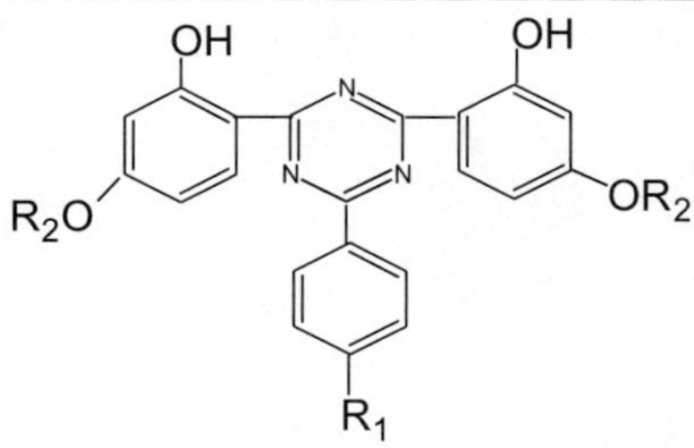


| $R_1$ Group | Reactant used Name | Formula | $R_2$ Group | Melting point (°C) |
|---|---|---|---|---|
| Hydrogen, H | | | H | > 300 |
| | Diethyl sulphate | $(CH_3CH_2)_2SO_4$ | $CH_3CH_2$ | 233–234 |
| | *n*-Propyl bromide | $CH_3(CH_2)_2Br$ | $CH_3(CH_2)_2$ | 174–175 |
| | *n*-Butyl bromide | $CH_3(CH_2)_3Br$ | $CH_3(CH_2)_3$ | 169–170 |
| | *n*-Octyl bromide | $CH_3(CH_2)_7Br$ | $CH_3(CH_2)_7$ | 112–113 |
| | *n*-Dodecyl bromide | $CH_3(CH_2)_{11}Br$ | $CH_3(CH_2)_{11}$ | 105.5–106 |
| *tert*-Butyl, $C(CH_3)_3$ | | | H | > 300 |
| | Diethyl sulphate | $(CH_3CH_2)_2SO_4$ | $CH_3CH_2$ | 183.5–184 |
| | *n*-Propyl bromide | $CH_3(CH_2)_2Br$ | $CH_3(CH_2)_2$ | 178–179 |
| | *n*-Octyl bromide | $CH_3(CH_2)_7Br$ | $CH_3(CH_2)_7$ | 132.5–133 |

Data from [118]

In addition to ether derivatives, sulfonate and hydroxyethyl ether compounds have been synthesized (Figure 8.65) [122–124].

### 8.1.5.5 Mono-resorcinol-Triazine Compounds and Their Derivatives

A more common class of triaryl-triazine UV absorbers is based on 2-(2,4-dihydroxyphenyl)-4,6-diaryl-1,3,5-triazine chemistry outlined in Figure 8.62. In these compounds, two non-phenolic aromatic groups and one resorcinol based aromatic groups are attached to the 1,3,5-triazine ring. Most of the

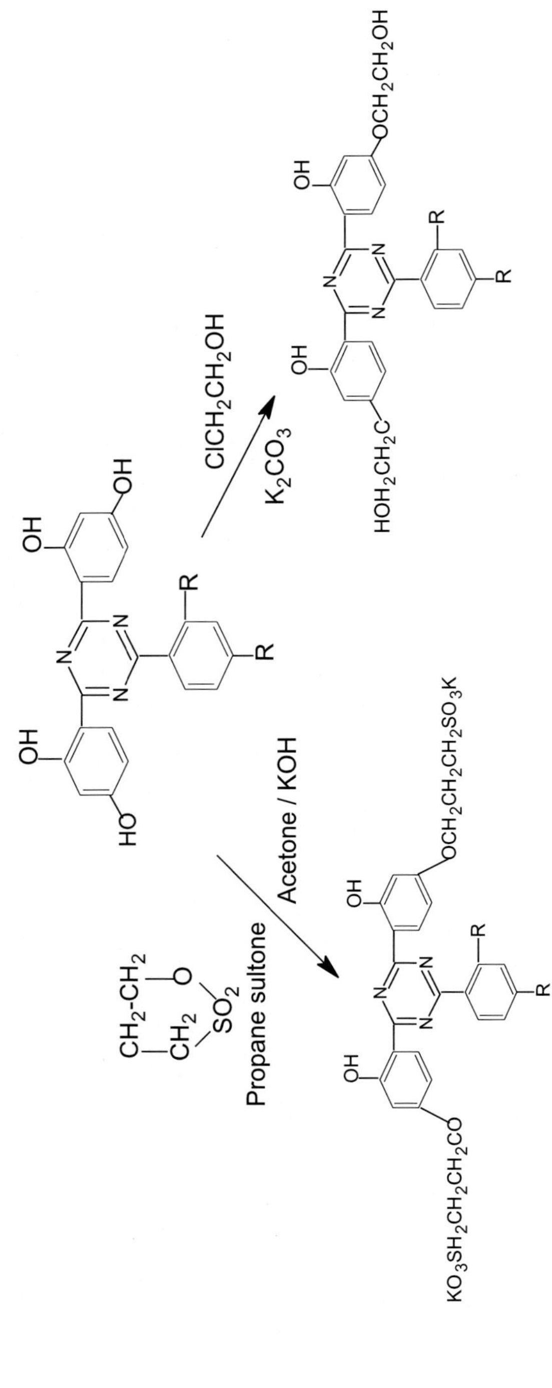

OH
HO
OH
OH
R
R
N
N
N
ClCH2CH2OH
K2CO3
HOH2CH2C
OCH2CH2OH
CH2-CH2
CH2
O
SO2
Propane sultone
Acetone / KOH
KO3SH2CH2CH2CO
OCH2CH2CH2SO3K


**Figure 8.65.**

**Table 8.20.** Dialkyl ether derivatives of bis-(2,4-dihydroxyphenyl)-*s*-triazine and properties

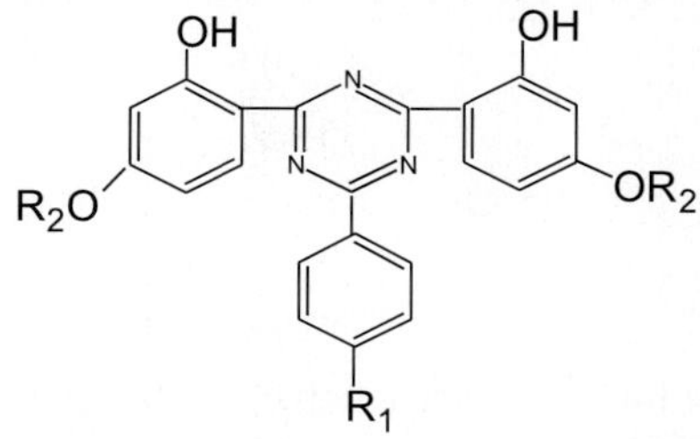


| $R_1$ Group | Reactant used Name | Formula | $R_2$ Group | Melting point (°C) |
|---|---|---|---|---|
| *tert*-Butyl, $C(CH_3)_3$ | | | H | > 300 |
| | Allyl bromide | $CH_2{=}CHCH_2Br$ | $CH_2{=}CHCH_2$ | 146–147 |
| | Benzyl bromide | $C_6H_5\text{-}CH_2Br$ | $C_6H_5\text{-}CH_2$ | 212–214 |
| | Ethylene chlorohydrin | $HOCH_2CH_2Cl$ | $HOCH_2CH_2$ | 181–188 |
| | Chloro acetic acid ester | $C_2H_5OOCCH_2Br$ | $C_2H_5OOCCH_2$ | 169–170 |
| Methoxy, $OCH_3$ | | | H | 341–342.5 |
| | Allyl bromide | $CH_2{=}CHCH_2Br$ | $CH_2{=}CHCH_2$ | 135–137.5 |
| | Benzyl bromide | $C_6H_5\text{-}CH_2Br$ | $C_6H_5\text{-}CH_2$ | 87–88 |
| | Chloro acetic acid ester | $C_2H_5OOCCH_2Br$ | $C_2H_5OOCCH_2$ | 152–153 |
| Chloro, Cl | | | H | 350 |
| | Allyl bromide | $CH_2{=}CHCH_2Br$ | $CH_2{=}CHCH_2$ | 143–144 |
| | Benzyl chloride | $C_6H_5\text{-}CH_2Cl$ | $C_6H_5\text{-}CH_2$ | 177–178 |

Data from [121]

compounds are obtained from the use of either the bis-phenyl or bis-(2,4-dimethyl phenyl)-mono-chloro-triazine compound in the synthesis, which has the following, structures (Figures 8.66 and 8.67) [121,125–133].

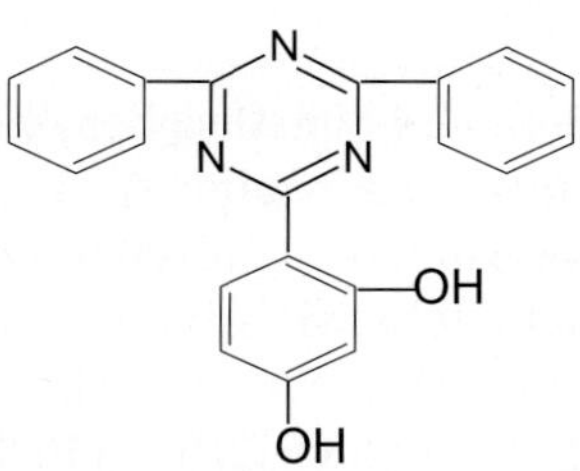


2-(2,4-Dihydroxyphenyl)-4,6-diaryl-1,3,5-triazine

**Figure 8.66.**

A procedure for the preparation of compound shown in Figure 8.67 is presented below [121].

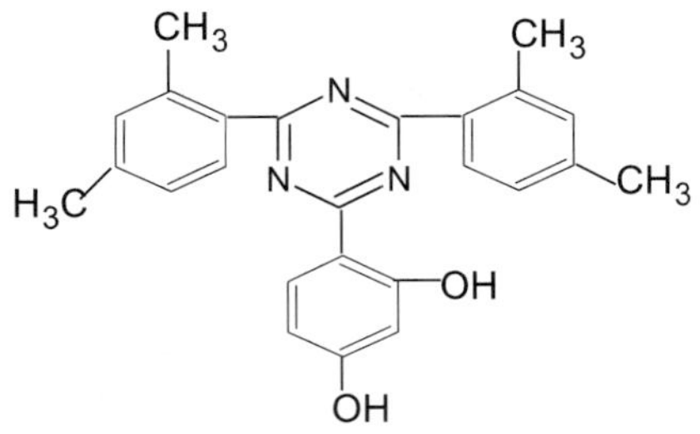

2-(2,4-Dihydroxyphenyl)-4,6-(2,4-dimethylphenyl)-1,3,5-triazine.

**Figure 8.67.**

**Synthesis of 2,4-Bis-(2,4-dimethylphenyl)-6-(2,4-dihydroxyphenyl)-1,3,5-triazin [121]** Here 6.6 parts of 2-chloro-4,6-bis-(2,4-dimethylphenyl)-1,3,5-triazine, prepared by condensing 1 mol of cyanuric chloride with 2 mol of 1,3-dimethylbenzene in the presence of aluminum chloride in benzene having a melting point of 138 °C, and 3 parts of resorcinol were dissolved in nitrobenzene, and at 15 to 20 °C 3 parts of anhydrous aluminum chloride were added. The temperature was raised within 30 min at 110 °C the batch was stirred for 3 h longer at this temperature and then poured into a mixture of 250 parts of water, 200 parts of ice and 50 parts of concentrated hydrochloric acid. This was repeatedly decanted from the nitrobenzene layer and then the mixture was subjected to steam distillation. The precipitated product was suctioned off. After drying, about 8 parts of product was obtained in the form of pale yellow crystals melting at 197.5 to 198.5 °C from DMF.

Since nitrobenzene is considered a toxic solvent to be used in organic synthesis, alternative solvents such as chlorobenzene and tetrachloroethane have been employed in the synthesis of triazine compounds. Using *n*-hexylresorcinol, a triazine compound has been developed based on the following synthetic procedure.

**Synthesis of 2-(2,4-Dihydroxy-5-hexylphenyl)-4,6-bis-(2,4-dimethylphenyl)-*s*-triazin [126]** To a 500-ml three-necked, round-bottomed flask equipped with a magnetic stirrer, condenser, and a nitrogen atmosphere were charged 5.10 g (16.0 mmol) of 2-chloro-4,6-bis-(2,4-dimethylphenyl)-*s*-triazine, 3.10 g (16.0 mmol) of hexylresorcinol, and 40 ml of tetrachloroethane. The mixture became homogeneous with gentle warming after which 2.10 g (16.0 mmol) of aluminum chloride were added quickly in small portions. The mixture was heated for 5 h with an oil bath maintained at 148 °C and was allowed to sit overnight at room temperature. Following the addition of 200 ml of 2 N hydrochloric acid, the mixture was refluxed for 2 h. After cooling to room temperature, a portion of ethyl acetate was added and the layers were separated. The aqueous layer was extracted twice more with ethyl acetate and the combined organic layers were dried over anhydrous magnesium sulfate. Following filtration, the solvent was removed under reduced pressure to afford 7.35 g of yellow solid. The crude product was recrystallized from ethyl acetate/heptane to yield 4.15 g of the title compound as a yellow solid; mp 165 – 167 °C. UV $\lambda_{max}$ (ethyl acetate) 288; 348 nm ($\varepsilon$ = 38 000; 16 200).

Various derivatives have been developed using the 2-(2,4-dihydroxy phenyl)-4,6-bis-phenyl-1,3,5-triazine and 2-(2,4-dihydroxy phenyl)-4,6-bis-(2,4-dimethylphenyl)-*s*-triazine compounds, which are presented in Tables 8.21 and 8.22.

**Table 8.21.** Diether derivatives of bis-(2,4-dihydroxyphenyl)-*s*-triazine and properties

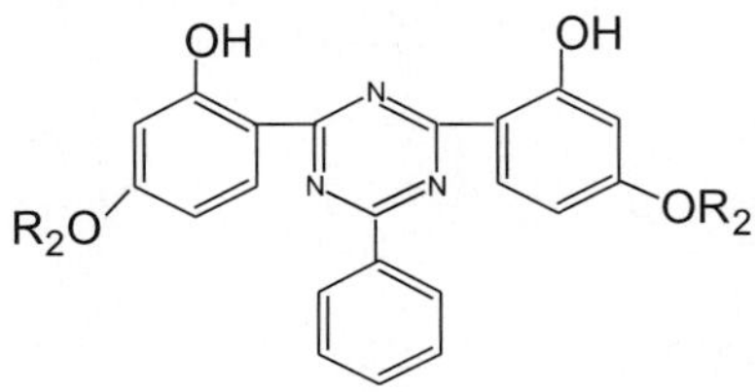


| Reactant used Name | Formula | $R_2$ Group | Melting point (°C) |
|---|---|---|---|
| | | H | > 300 |
| Allyl bromide | $CH_2$=$CHCH_2Br$ | $CH_2$=$CHCH_2$ | 178 |
| Benzyl chloride | $C_6H_5$-$CH_2Cl$ | $C_6H_5$-$CH_2$ | 171 – 172.5 |
| *p*-Chloro benzyl chloride | *p*-Cl-$C_6H_4$-$CH_2Cl$ | *p*-Cl-$C_6H_4$-$CH_2$ | 265 – 267 |
| Ethylene chlorohydrin | $HOCH_2CH_2Cl$ | $HOCH_2CH_2$ | 252 – 253 |
| Crotyl bromide | $CH_3CH$=CH-$CH_2Br$ | $CH_3CH$=CH-$CH_2$ | 212 – 213 |
| Bromo butyronitrile | $NCCH_2CH_2CH_2Br$ | $NCCH_2CH_2CH_2$ | 223 – 224 |
| 1-Chloro-3-bromopropane | $BrCH_2CH_2CH_2Cl$ | $CH_2CH_2CH_2Cl$ | 181 – 183 |
| Phenacyl bromide | $C_6H_5$-C(O)$CH_2Br$ | $C_6H_5$-C(O)$CH_2$ | 280 – 281 |

Data from [121]

Polymerizable vinyl groups containing triazine derivatives were also developed and reported (Figures 8.68 and 8.69) [134, 135].

Two commercially well known UV absorbers based on resorcinol-triazine chemistry are available from Ciba Specialty Chemicals and Cytec Industries under their trade names Tinuvin 1577 and Cyasorb UV-1164, for which the properties, UV spectra and applications are presented in Tables 8.23 and 8.24 and in Figures 8.70 and 8.71.

The resorcinol-triazine based UV absorber compounds have shown excellent thermal stability and outstanding UV performance, and also proved to be good UV absorbers in the applications of polymer films, coatings, molded thermoplastic articles, photographic recording materials and sunscreen formulations.

The triazines exhibited superior photo stability in coating applications. The coating compositions containing Ciba Specialty Chemical's Tinuvin 400 UV

**Table 8.22.** 2,4-Bis-(2,4-dimethylphenyl)-6-(2,4-didhydroxyphenyl)-*s*-1,3,5-triazine

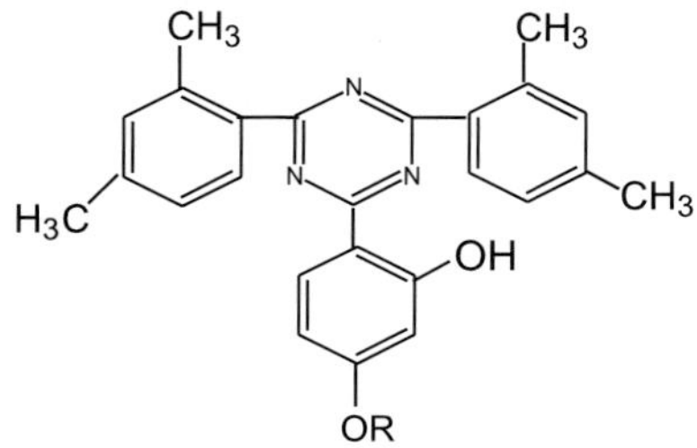


| Name | Reactant used Formula | $R_2$ Group | Melting point (°C) |
|---|---|---|---|
| | | H | 197.5 – 198.5 |
| Allyl bromide | $CH_2=CHCH_2Br$ | $CH_2=CHCH_2$ | 141 – 142 |
| Benzyl chloride | $C_6H_5$-$CH_2Cl$ | $C_6H_5$-$CH_2$ | 164 |
| *p*-Chloro benzyl chloride | *p*-Cl-$C_6H_4$-$CH_2Cl$ | *p*-Cl-$C_6H_4$-$CH_2$ | 153 – 155 |
| Ethylene chloro-hydrin | $HOCH_2CH_2Cl$ | $HOCH_2CH_2$ | 176 |
| Sodium chloro-acetate | $ClCH_2COONa$ | $CH_2COOH$ | 215 – 216 |
| Bromo butyric acid ester | $BrCH_2CH_2CH_2COOC_2H_5$ | $CH_2CH_2CH_2COOC_2H_5$ | 125 – 126 |
| Chloro acetic acid ester | $ClCH_2COOC_2H_5$ | $CH_2COOC_2H_5$ | 94 – 97 |
| Bromo acetic acid glycol ester | $ClCH_2COOCH_2CH_2OH$ | $CH_2COOCH_2CH_2OH$ | 93 – 94 |

Data from [121]

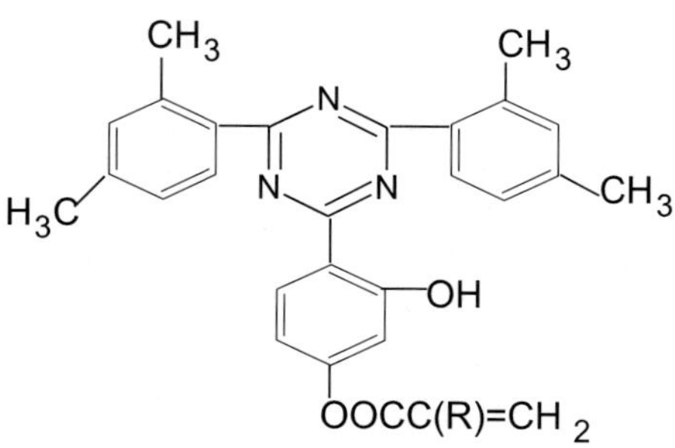


R = H, M.Pt = 154 - 155 °C

**Figure 8.68.**

absorber (shown in Figure 8.72) was reported to be an innovative UV light stabilizer class for coatings that sets new performance for automotive and refinish coatings, as well as high performance industrial coating system.

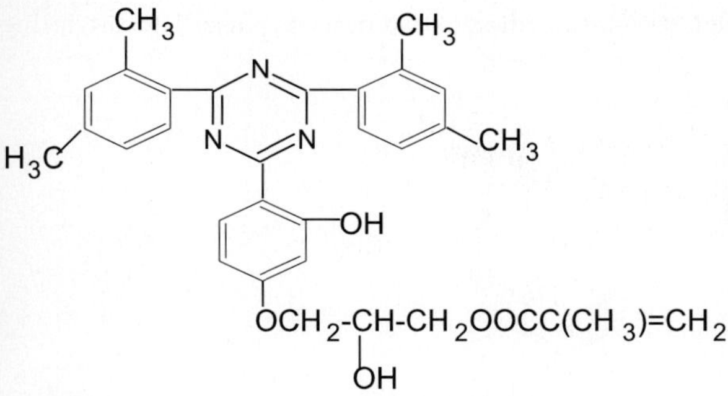


**Figure 8.69.**

**Table 8.23.** 2-(4,6-Diphenyl-1,3,5-triazin-2-yl)-5-hexyloxy phenol)

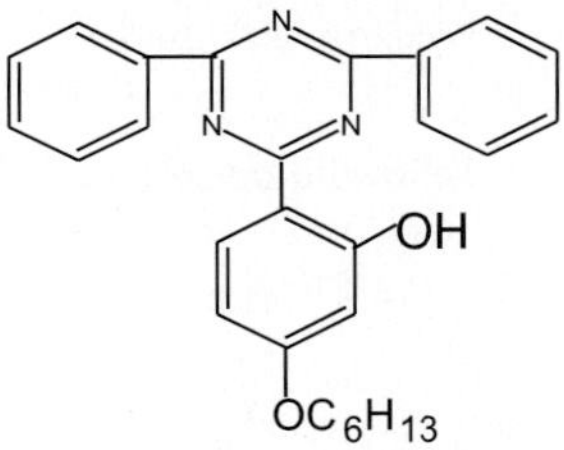


| | |
|---|---|
| Chemical name | 2-(4,6-Diphenyl-1,3,5-triazin-2-yl)-5-hexyloxy-phenol |
| Trade Name | TINUVIN 1577 |
| Physical properties | |
| Melting point (°C), neat | 148 |
| UV absorption, $\varepsilon$ at $\lambda_{max}$ 341 | 22 000 |
| TGA (heating rate 20 °C/min, air) | |
| 1% Weight loss, °C | 300 |
| 5% Weight loss, °C | 335 |
| 10% Weight loss, °C | 353 |
| Solubility (at 20 °C, g/l) | |
| *n*-Hexane | 0.7 |
| Acetone | 3.1 |
| Methylene chloride | 169 |
| Chloroform | 254 |
| Toluene | 51 |
| Ethanol | 0.2 |
| Ethyl acetate | 6.6 |

Applications: plastics such as PC, PET, PEN and thermoplastic polymers, polymer blends and alloys such as PC/ABS, PC/PBT, PPE/IPS and PPE/PA
Data source: technical information on Tinuvin 1577 from Ciba Specialty Chemicals

**Table 8.24.** 2,4-Bis(2,4-dimethylphenyl)-6-(2-hydroxy-4-iso-octyloxyphenyl)-1,3,5-triazine properties

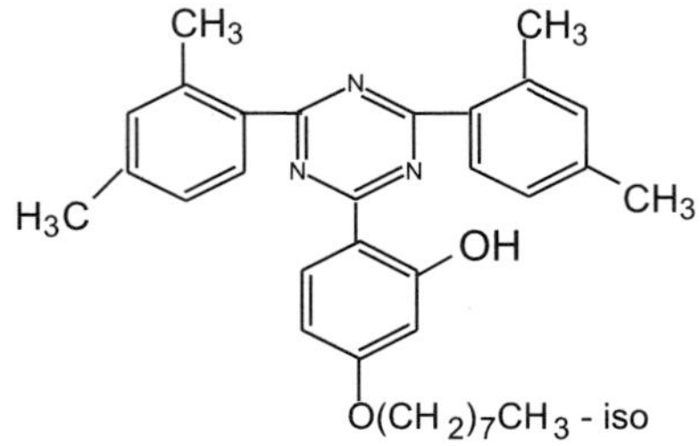


| | |
|---|---|
| Chemical name | 2,4-Bis(2,4-dimethylphenyl)-6-(2-hydroxy-4-iso-octyloxyphenyl)-1,3,5-triazine (iso-octyl group is a mixture of octyl groups) |
| Trade name | Cyasorb UV-1164L |
| Physical properties | |
| Physical form | Yellow liquid (65% active in xylenes) |
| Melting point (°C), neat | 60 |
| Specific gravity (g/cc, 25 °C), neat | 0.978 |
| Absorptivity (at $\lambda_{max}$ 341 nm), l/g cm | 44 |
| TGA (heating rate 10 °C/min, air) | |
| 10% Weight loss, °C | 381 |
| 20% Weight loss, °C | 402 |
| Solubility (wt % at 24 °C) | |
| Water | Insoluble |
| *n*-Heptane | 63 |
| Methyl isobutyl ketone (MIBK) | 69 |
| Ethyl acetate | 68 |
| *N*-Methylpyrrolidone (NMP) | 65 |
| 2-Butoxyethanol | 13 |
| 1-Butoxy-2-propanol | 6 |

Applications: high solids coatings, water-borne coatings, coil coatings, acrylic melamines, acrylic and polyester urethanes, silane-modified acrylics
Data source: Light stabilizers for coatings, Product Overview, Cytec Industries

All the spectral and physical properties of the well known resorcinol based UV absorber compounds reported in this chapter have been obtained from the manufacturers' published information and data sheets [136–141].

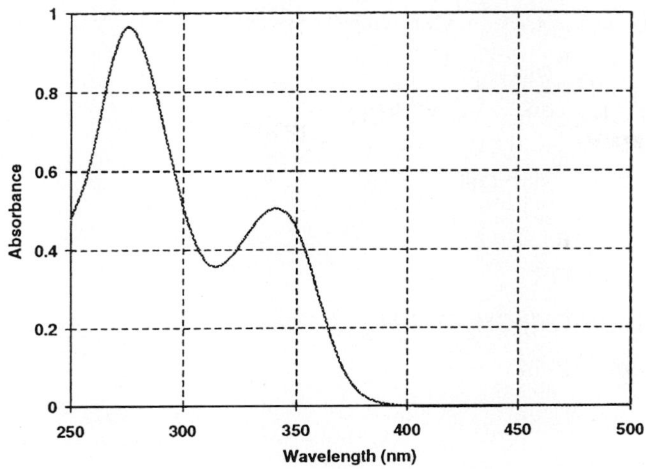


**Figure 8.70.** UV spectrum of 2-(4,6-diphenyl-1,3,5-triazin-2-yl)-5-hexyloxyphenol from Ciba Speciality Chemicals

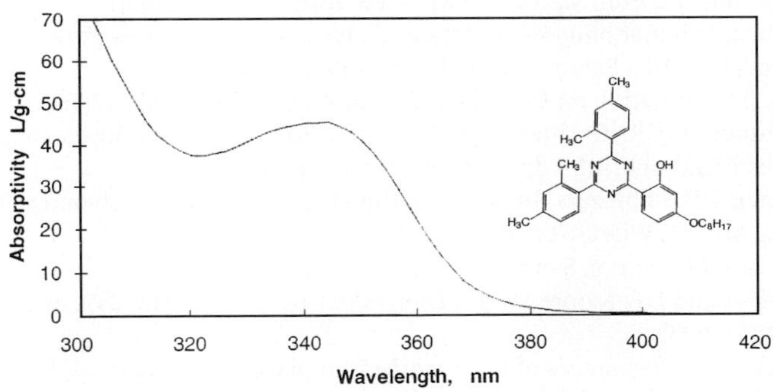


**Figure 8.71.** UV spectrum of 2,4-bis-(2,4-dimethylphenyl)-6-(2-hydroxy-iso-octyloxyphenyl)-1,3,5-triazine from Cytec Industries

## 8.1.6 Summary and Outlook

Research on UV absorbers based on resorcinol chemistry is still very active, which involves the synthesis of new compounds and determination of their properties. The advancement in the new UV absorber technology could enable industrial people to develop better protection methods and techniques for the protection of plastics, coatings and cosmetics material against the harmful UV radiation, and also to extend their service life.

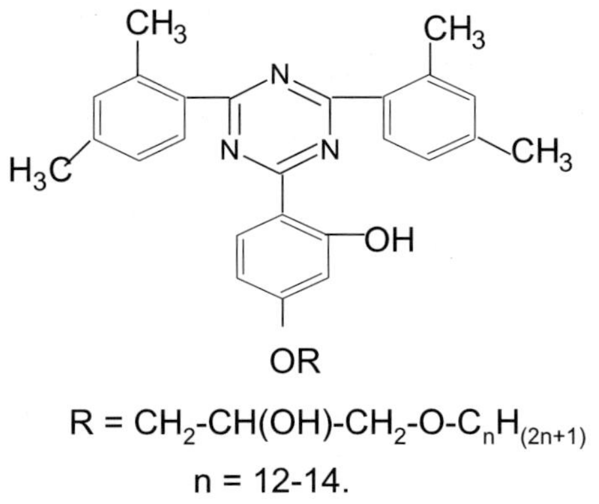

**Figure 8.72.**

## References


1. A. Torikai, Photodegrdation of polymers, In: S. H. Hamid, Ed., Handbook of polymer degradation, 2nd Edn, Marcel Dekker, New York, p 573 (2000).
2. J. F. Rabek, Polymer photodegradation, Chapman and Hall, New York (1995).
3. A. L. Andrady, Adv. Polym. Sci., Vol. 128, p 49 (1997).
4. Technical Information on Uvinul Grades from BASF, November (2002).
5. F. Gugumus, In: R. Gachter and H. Muller, Eds., Light stabilizers in plastics and additives handbook, 4th Edn, Hanser, Munich, p 175 (1996).
6. M. Dexter, UV Stabilizers. In: Kirk – Othmer Encyclopedia of Chemical Technology, 3rd Edn, Vol. 23, Wiley, New York (1983).
7. J. E. Pickett, Macromol. Symp, Vol. 115, p 127 (1997).
8. J. E. Pickett and J. E. Moore, ACS Advances in Chemistry Series 249, ACs, Washington DC, p 287 (1995).
9. J. E. Pickett, Performances of UV absorbers in plastics and coatings, Technical Information Series, General Electric Company R&D Center, December (1997).
10. J. F. Rabek, Photostabilization of polymers, Elsevier Applied Sci, p 202 (1996).
11. F. Gugumus, Light stabilizers, In: H. Zweiful, Ed., Plastics Additives Handbook, Hanser, Munich,(2001).
12. B. Ranby and J. F. Rabek, Photodegradation, photoxidation and photostabilization of polymers, Wiley Interscience, London (1975).
13. M. Whelen, US Patent, 3,371,119 (1968).
14. H. Nufer, US Patent, 3,407,235 (1968).
15. American Cyanamid Company, British Patent, 947,686 (1962).
16. E. Dobratz and A. Kolka, US Patent, 3,403,183 (1968).
17. F. Ponder, US Patent, 3,526,666 (1970).
18. H. Shioda et al., US Patent, 3,639,483 (1972).
19. M. Yukutomi et al., US Patent, 3,769,349 (1973).
20. P. Neumann, US Patent, 4,990,680 (1991).
21. J. Beau, US Patent, 5,629,453 (1997).
22. J. van Allen and J. Tinker, J. Org. Chem., Vol. 19, p 1243 (1954).
23. W. Hady et al., US Patent, 2,853,521 (1958).
24. I. Heckenbleikner, J. Hussar and R. Besser, US Patent, 3,632,650 (1972).
25. T. Dalbey, US Patent, 3,697,599 (1972).

26. B. Lachmann and H. Jurgen, US Patent, 3,843,729 (1974).
27. A. Wexler, P. Chakraborti and M. Brown, US Patent, 4,323,710 (1982).
28. A. Aumueller and P. Neumann, US Patent, 4,980,512 (1990).
29. M. Notari, F. Mizia and F. Rivetti, US Patent, 5,849,955 (1998).
30. E. Boboli et al., US Patent, 3,584,053 (1971).
31. H. Shioda et al., British Patent, 1,246,958 (1971).
32. A. F. Strobel and S.C. Catino, US Patent, 3,448,133 (1969).
33. O. Kander and A. Rosenbaum, US Patent, 3,322,705 (1967).
34. B. Ashby, US Patent, 4,278,804 (1981).
35. B. Anthony, US Patent, 4,495,360 (1985).
36. M. Karvas et al., British Patent, 1,212,466 (1969).
37. R. Wear, US Patent, 3,580,927 (1971).
38. R. Coleman, US Patent, 3,391,110 (1968).
39. H. Eggensperger, F. Franzen and W. Kloss, US Patent, 3,676,471 (1972).
40. E. Hahn and P. Neumann, US Patent, 4,885,396 (1989).
41. B. Kim and H. Kaack, US Patent, 4,978,797 (1990).
42. A. Aumueller and W. Goetze, US Patent, 5,118,859 (1992).
43. G. Lappin and W. Mc Connell, US Patent, 3,644,485 (1972).
44. T. Sassi, PCT WO 02/46136 A2 (2002).
45. GAF Corporation, British Patent, 1,136,525 (1967).
46. A. Cofrancesco and R. Ferrigan, US Patent, 3,468,938 (1969).
47. D. Stauffer, US Patent, 3,636,077 (1972).
48. H. Dressler and K. Reabe, US Patent, 3,399,237 (1968).
49. J. Milionis, US Patent, 3,649,695 (1972).
50. N. Kubota, T. Shibata and K. Sugibuchi, US Patent, 4,186,151 (1980).
51. M. Minagawa, N. Kubota and T. Shibata, US Patent, 4,189,409 (1980).
52. H. Seki et al., US Patent, 3,632,651 (1972).
53. J. Milionis and F. Furman, US Patent, 3,632,858 (1972).
54. C. Havens, US Patent, 2,900,361 (1959).
55. C. Kibler, M. Knowles and J. Tamblyn, US Patent, 3,075,850 (1963).
56. D. Gordon and R. Hudson, US Patent, 2,794,052 (1957).
57. J. Pickett, US Patent, 5,567,850 (1996).
58. J. Pickett and G. Gillette, US Patent, 5,679,820 (1997).
59. J. Pickett and A. Simonian, US Patent, 5,905,172 (1999).
60. J. Pickett and A. Simonian, US Patent, 5,763,674 (1998).
61. J. Pickett and A. Simonian, US Patent, 6,037,059 (2000).
62. J. Pickett, US Patent, 5,712,419 (1998).
63. A. Simonian, G. Gillette and J. Pickett, US Patent, 5,959,012 (1999).
64. J. Pickett et al., US Patent, 5,981,073 (1999).
65. G. Gillette and G. Patel, US Patent, 6,352,778 (2002).
66. J. E. Pickett and J. Moore, Angew. Makromol. Chem., Vol. 232, p 229 (1995).
67. L. Staley, US Patent, 3,073,866 (1963).
68. W. Hardy and F. Forster, US Patent, 2,853,521 (1958).
69. K. Shinichi, H. Yasuhiko and K. Osamu, Jap. Pat. Publ. 06-172253 (1994).
70. S. Jae-han, Korean Patent, 9,205,953 (1992).
71. P. Hock, US Patent, 2,693,492 (1954).
72. J. Dayan and S. Roberts, US Patent, 2,853,522 (1958).
73. American Cyanamid Company, British Patent, 889,366 (1959).
74. W. Hardy and W. Forster, US Patent, 2,773,903 (1956).
75. P. Neumann, H. Eilingsfeld and A. Aumueller, US Patent, 5,072,034 (1991).

76. Ciba Ltd, British Patent, 893,236 (1962).
77. Ciba Ltd, German Patent, DE 1,109,172 (1961).
78. H. Brunetti et al., US Patent, 2,693,492 (1954).
79. J. Fertig, A. I. Goldberg and M. Shoultchi, J. Appl. Polym. Sci., Vol. 10, p 663 (1966).
80. D. Bailey and O. Vogl, J. Macromol. Sci, Rev. Macromol. Chem., Vol. C14, p 267, (1976).
81. J. Fertig et al., US Patent, 3,173,893 (1965).
82. J. Fertig et al., US Patent, 3,340,231 (1967).
83. J. Fertig et al., US Patent, 3,202,716 (1965).
84. J. Fertig et al., US Patent, 3,341,493 (1967).
85. M. Johnson, US Patent, 4,504,625 (1985).
86. R. Horton and H. Brooks, US Patent, 3,313,866 (1967).
87. K. Eiglmeier and J. Schulz, US Patent, 4,824,892 (1989).
88. J. Milionis and F. Arthen, US Patent, 3,049,503 (1962).
89. S. Tocker, Makromol. Chem., Vol. 101, p 23 (1967).
90. S. Tocker, US Patent, 3,107,199 (1963).
91. H. Keller et al., US Patent, 6,001,337 (1999).
92. D. Olson and D. Burns, US Patent, 6,235,375 (2001).
93. D. Olson and D. Burns, US Patent, 6,464,817 (2002).
94. D. Olson and D. Burns, US Patent, 6,582,790 (2003).
95. J. C. Crawford, Prog. Polym. Sci., Vol. 24, p 7 (1999).
96. S. Hotta et al., US Patent, 4,028,331 (1977).
97. L. Vishwakarma, US Patent, 5,455,152 (1995).
98. R. Boyle and J. Milionis, US Patent, 3,214,436 (1965).
99. S. Li and O. Vogl, Polym. Bull., Vol. 12, p 375 (1984).
100. S. Li et al., Polym. Bull., Vol. 12, p 237 (1984).
101. Y. Jiang et al., Polym. Bull., Vol. 20, p 169 (1988).
102. F. Xi, W. Bassett and O. Vogl, Makromol. Chem., Vol. 185, p 2497 (1984).
103. S. Li, A. Gupta and O. Vogl, Monatsh. Chem., Vol. 114, p 937 (1983).
104. H. Saito and M. Makino, US Patent, 5,803,929 (1998).
105. O. Vogl and C. Zhang, US Patent, 5,099,027 (1992).
106. M. Skoultchi, J. Fertig and L. Nass, US Patent, 3,493,539 (1970).
107. J. Milionis and W. Hardy, US Patent, 3,072,585 (1963).
108. R. Liu et al., Polym. Bull., Vol. 20, p 59 (1988).
109. F. Xi, W. Bassett and O. Vogl, Polym. Bull., Vol. 11, p 329 (1984).
110. L. Vishwakarma, US Patent, 5,585,228 (1996).
111. L. Vishwakarma, US Patent, 5,670,654 (1997).
112. L. Vishwakarma, H. Yau and T. Chen, US Patent, 5,674,670 (1997).
113. G. Rytz, R. Hilfiker, R. Schmidt and E. Schmitter, Angew. Makromol. Chem., Vol. 247, p 213 (1997).
114. G. Stuber et al., J. Phys. Chem., Vol. 99, p 10097 (1995).
115. H. Brunetti and C. Luethi, Helv. Chim. Acta, Vol. 55, p 1566 (1972).
116. S. Tanimoyo and M. Yamagata, Senryo to Yakahin, Vol. 40, p 325 (1995).
117. R. Gupta, H. Singh and R. Cappadona, US Patent, 6,632,944 (2003).
118. W. Hardy, J. Milionis and F. Pinto, US Patent, 3,118,887 (1964).
119. W. Hardy, J. Milionis and F. Pinto, US Patent, 3,268,474 (1966).
120. M. Deunnenberger, H. Bailand and C. Luethi, US Patent, 3,242,175 (1966).
121. M. Deunnenberger, H. Bailand and C. Luethi, US Patent, 3,244,708 (1966).
122. C. Luethi and K. Eschle, US Patent, 3,444,164 (1969).
123. C. Midgal, J. Hines and E. Klugar, US Patent, 4,826,978 (1989).
124. C. Midgal, J. Hines and E. Klugar, US Patent, 4,962,142 (1990).

125. P. Waterman, US Patent, 5,461,151 (1995).
126. T. Stevenson et al., US Patent, 5,556,973 (1996).
127. T. Stevenson et al., US Patent, 5,637,706 (1997).
128. R. Gupta et al., US Patent, 6,486,316 (2002).
129. P. Susi, US Patent, 4,619,956 (1986).
130. P. Susi, US Patent, 4,740,542 (1988).
131. R. Gupta et al., US Patent, 6,225,468 (2001).
132. R. Gupta and D. Jakiela, US Patent, 6,239,276 (2001).
133. R. Gupta and D. Jakiela, US Patent Appln, 2002/0013462 A1 (2002).
134. H. Huber et al., US Patent, 3,423,360 (1969).
135. A. Vallet, US Patent, 5,106,891 (1992).
136. UV Stabilizers for Plastics and Coatings, Lycus Ltd, El Dorado, AR, USA.
137. Ultraviolet Light Absorbers (UVA), MAYZO, Norcross, GA, USA.
138. Additives: Ultraviolet Absorbers, Shipro Kasei Kaisha, Ltd, Japan.
139. Polymer Additives, Great Lakes Chemical Corporation, West Lafayette, IN, USA.
140. Technical Information on Polymer Additives, Ciba Specialty Chemicals, USA.
141. Light Absorbers Product Information, Polymer Additives, Cytec Industries, Inc, West Paterson, NJ.

## 8.2 Flame Retardants

### 8.2.1 Introduction

Thermoplastic materials have been widely used in building construction, home furnishings, electric and electronic parts, automobiles and other industrial applications because they are lightweight materials with chemical and water resistant properties as well as electrical insulating and good mechanical properties. In general, thermoplastic resins used for the manufacture of housing materials such as home appliances and computers must be flame resistant for the sake of fire prevention and safety. Flame retarded thermoplastic resin compositions and their applications have been steadily increasing worldwide. Therefore, the demand for flame retardants is strong globally, especially for engineering thermoplastics, because of their increased use in electrical equipments, consumer products and transportation applications. Several published articles and books have outlined and discussed the types and nature of commercially available flame retardants, flame retarding mechanisms, and their advantages in the thermoplastic and thermoset polymers [1–10].

Though thermoplastic polymers have excellent physical and mechanical properties, and are well suited for manufacturing household and industrial parts and products, they are basically organic materials, and therefore combustible in nature. The only way to protect these organic polymers from initiating fires is to use flame retardants in their compositions. Preventing fires or delaying them makes escape possible, and over a long period of time this will result in the protection of life, health and property.

The most important families of flame retardant materials are based on the halogens, phosphorus, nitrogen, antimony and metal salts. These flame retardant materials, when used in polymers at various loading levels, offer higher fire safety levels to these combustible materials. The presence of flame retardant in polymers can be expected to inhibit or suppress the combustion process in the event of a fire. Depending upon the nature of flame retardants used, they can act chemically and/or physically in the solid, liquid or gas phase to minimize or retard the fire. Flame retardants can interfere with the combustion during a particular stage of this process, for example during heating, decomposition, ignition or flame spread [10].

A well known and also widely adapted method for producing flame retardancy in polymeric materials is the use of halogen-containing compounds along with antimony trioxide. In the case of halogenated compounds, brominated flame retardants are widely used due to their good flame retardation capacity and low cost. With halogenated flame retardants in thermoplastic resins, evolution of harmful toxic gases are observed when these resins burn. Because of concern over environmental contamination, the use of halogen-containing flame retardants is restricted in many countries. Because of the release of toxic gases such as dioxin and difuran during the combustion of thermoplastic resins flame retarded with polybrominated diphenyl ethers, these types of materials are prohibited in certain European countries.

Due to the concerns of toxic gases and fumes and their environmental impacts on the use of halogenated flame retardants in the most widely used thermoplastics for the manufacture of electronic and electric components and house hold appliances, non-halogenated flame retardants were required in these applications. Therefore, phosphorus compounds, particularly phosphate esters, have attracted much attention because of their good flame resistance properties in the solid phase and very little or no liberation of toxic gases during combustion. Organo phosphorus flame retardants work very effectively in the solid or condensed phase by inhibiting ignition and promoting the char formation. The formation of good char layer is not only difficult to ignite but also insulates the under lying virgin polymer by minimizing thermal degradation. These organo phosphorus compounds were found to be very effective in oxygen and nitrogen groups containing polymers.

In the case of organophosphorus compounds, organic phosphoric acid ester compounds such as triphenyl phosphate (TPP), cresyl diphenyl phosphate and tricresyl phosphate compounds have been widely used industrially. Though these flame retardant materials show good flame retarding properties in thermoplastic resins, due to lower molecular weights, they tend to migrate to the polymer surface when they are injection molded, and this phenomenon is called "juicing". In order to obtain a good flame retardant rating, namely the V-0 in UL-94 test, higher amounts of monophosphate esters should be used in thermoplastic compositions, which ultimately could affect the heat resistance property of polymers. Because of lower molecular weights, TPP type flame retardants have the drawbacks in causing smoking and volatilization during the molding process.

To overcome the difficulties associated with the use of monophosphate esters as flame retardants, new flame retardant products and technologies have been introduced based on resorcinol chemistries, which offered high temperature stability and resistance and also, enhanced the performances of commercially important thermoplastic materials.

In this chapter, resorcinol based flame retardant chemistries and technologies developed and utilized in improving the flame retardant and processing performances of various thermoplastic polymers used in the manufacture of home furnishings and electronic materials are discussed.

### 8.2.2 Thermoplastic Polymers Need Flame Retardants

Some of the important thermoplastic resins and their physical characteristics are described below.

### 8.2.2.1
### Polycarbonate (PC) Resins

These thermoplastic materials exhibit high impact strength and heat resistant properties, and have shown a limited oxygen index (LOI) value of about 25. They are widely used in automobiles and electrical appliances because of their excellent transparency, good flame resistance and high mechanical strength properties. But, they have shown poor processability during the molding process. To improve the processability, PC resins are often blended or alloyed with ABS, HIPS and PBT thermoplastic polymers. Molding compositions containing a PC and styrene co-polymer resins have good processability and high notched impact strength properties. PC has been widely used in the production of CDs, CD-ROMs and digital video disks (DVDs). Non-halogenated flame retardants are primarily used in the PC, PC/ABS and PC/HIPS alloys, and the alloys or blends are employed in the manufacture of housings of desk top computers and lap tops.

### 8.2.2.2
### Polyphenylene Ether (PPE or PPO) Resins

PPO resins are known to exhibit a desirable combination of chemical, physical and electrical properties over a wide range of temperatures extending from their brittle points to heat distortion temperatures. This combination of properties renders PPO resins suitable for a broad range of applications. But, the usefulness of PPO resins is limited in some applications due to their high temperature processability, low impact resistance and poor chemical resistance. In order to improve the processability and other desirable properties, PPO resins are often alloyed or blended with other thermoplastic resins such as homo polystyrene (PS), acrylonitrile-butadiene-styrene (ABS) copolymers and rubber modified polystyrene (HIPS) resins. Commercial interests have focused primarily on PPO and HIPS blends because of their good physical properties for molding applications. Modified PPO (with HIPS) resins have wide range of applications, which include computer and business machine housings, components of electrical/electronic equipment, building and construction, and appliances and house wares.

### 8.2.2.3
### Acrylonitrile-Butadiene-Styrene (ABS) Resins

ABS resins are copolymers made from the use of acrylonitrile, butadiene and styrene monomers in the polymerization process. They are tough and glassy materials, and also possess easier processing, chemical resistance, increased surface hardness and good impact strength properties. ABS resins are used in various electrical and electronic components, automobile parts, toys, cameras and pipes. In general, when ABS resin is burned, a char is not produced due to extensive decompositions and volatilization of decomposed products. Halogenated and non-halogenated flame retardant materials are often used to enhance the flame retardant properties. In applications where the toxic fumes from the halogenated compounds are a concern, non-halogen flame retardants are employed.

#### 8.2.2.4 High Impact Polystyrene (HIPS) Resins

HIPS has been made by blending the general purpose polystyrene (GPPS) with styrene-butadiene rubber (SBR) to enhance the impact strength. SBR is more flammable than polystyrene due to the presence of butadiene rubber in the composition. Therefore, higher impact strength HIPS materials, which contain more SBR, normally require slightly more flame retardant additive to achieve a given flame retardant rating. When the HIPS material is hit with a shock load, the polybutadiene rubber present in the composition absorbs the impact energy. In some HIPS formulations, the impact strength may be increased up to five times that of GPPS. A rubber modified styrene resin has a good processability, high impact strength and good appearance. Therefore, HIPS polymers have been utilized in television, computer monitor casings, facsimiles, electric appliances and office supplies applications.

#### 8.2.2.5 Poly (Butylene Terephthalate) (PBT) Resin

This thermoplastic polyester material has good dimensional stability, high heat resistance and chemical resistance properties. Due to good flow in thin-wall sections, rapid crystallization, low mold temperatures and fast cycle times, PBT material is more economical to mold. PBT has been used in the production of electronic and electrical connectors for the computer and business machine industries.

### 8.2.3 Flammability Testing of Plastic Materials

The determination on the effectiveness of flame retardants in the thermoplastic materials has been made by using the following test methods.

#### 8.2.3.1 Limiting Oxygen Index (LOI) Method [11]

A limited oxygen index (LOI) value is assigned to specimens when tested in accordance with an ASTM Standard D-2863, "Measuring the Minimum Oxygen

Concentration to Support Candle-Like Combustion of Plastics". The oxygen index (OI) test measures the relative flammability by determining the amount of oxygen required to sustain burning of the polymer using a standardized specimen under controlled environmental conditions. Specimens are burned in a laboratory controlled mixtures of nitrogen and oxygen. The OI represents the minimum amount of oxygen that must be present to sustain flame on a thermoplastic material. The higher the LOI value, the less likely the thermoplastic will burn.

#### 8.2.3.2 Cone Calorimeter Method

The cone calorimeter is standardized as ASTM E-1354 and ISO-5660. This equipment can be used to study the pyrolysis and burning behavior of flame retarded thermoplastic compounds. It is an analytical tool to perform fire tests, and the results obtained can be fairly comparable with real fires. In a typical cone calorimeter experiment, the polymer sample, measuring 100 mm × 100 mm and no more than 50 mm thick, placed in an aluminum dish is subjected to radiant heat flux of up to 100 $kW/m^2$ in the presence of a spark ignition source. In a single test, the properties such as heat release rate, peak of heat release, time to ignition, total heat released, mass loss rate as a function of time and the density of smoke produced (called extinction coefficient) can be measured simultaneously.

From the cone calorimeter test results, if the ignition time for the flame retarded material is longer then the material is expected to show low ignitability properties, which in turn could reduce the possibility of fire starting immediately. By determining the maximum heat release rate, one can find out the escape time from the fire associated with the burning of polymers. Overall, the cone calorimeter results can be used in fire hazard assessment and also in predictions of full scale fire performance.

#### 8.2.3.3 Underwriters Laboratory Vertical Burning Test (UL-94)

UL-94 Test is one of the most widely accepted flammability performance standards for thermoplastic polymers. For this test, the resin to be tested is molded into bars having standard size of 5″ (12.7 cm) ×0. 5″ (1.3 cm) width and thicknesses in the range of one-eighth and one-sixteenth of an inch (0.125 and 0.0625 inch). Normally five bars are tested. In a controlled environment protected from drafts and under specified temperature and humidity conditions, the bars are held in a vertical position while a flame having an inner cone of height 0.75 inch (20 mm) from a Bunsen burner is applied to the center of the lower end of bar for 10 s and then removed. Typically, the bar initially ignites and is then timed until it self extinguishes ($T_1$). As soon as the burning part (bar) extinguishes, the flame is reapplied for a second 10 s time period. Again, the time is recorded until the part (or bar) self extinguishes ($T_2$). If the thermoplastic material (bar) does not ignite at the first and/or second flame application, then $T_1$ and/or $T_2$ are 0 s. This is repeated for all five bars and the $T_1$ and $T_2$ are recorded for each bar. The criteria for V-0, V-1 and V-2 classification ratings are based on the test results outlined in Table 8.25.

**Table 8.25.** Material fire retardant classifications from UL vertical burning test method

| Test conditions and criteria for classifications | UL-94 Classifiaction V-0 | V-1 | V-2 |
|---|---|---|---|
| 1. Total flaming combustion for each specimene, $T_1$ and $T_2$ (s) | < 10 | < 30 | < 30 |
| 2. Total flaming combustion for all five specimen of any set, $5 \times (T_1 + T_2)$, (s) | < 50 | < 250 | < 250 |
| 3. Flaming and glowing combustion for each specimen after second burner application (s) | < 30 | < 60 | < 60 |
| 4. Cotton ignited by flaming drips from any specimen | No | No | Yes |
| 5. Glowing or flaming combustion of any specimen to holding clamp | No | No | No |

Reference: Underwriters Laboratories Inc

Various flame retardant materials have been used as additives in thermoplastic polymer compositions, which enabled these polymers to obtain different flame retarded classifications in the UL-94 test. Each thermoplastic composition requires a specified UL-94 rating for a given application. The effectiveness of a flame retardant additive in a particular thermoplastic polymer depends on its chemical structure, dispersability and compatibility with the polymer. In addition to these characteristics, the additive should not adversely affect the physical and mechanical properties of the matrix polymer.

### 8.2.4 Flame Retardants from Resorcinol Chemistries

In order to overcome the volatility of TPP, aryl diphosphate esters of resorcinol were developed, which exhibited lower vapor pressure, high thermal stability and very unusual plasticizing properties in the thermoplastic compositions. Resorcinol diphosphates having different chemical structures and physical properties were synthesized and successfully commercialized for the demanding flame retarded thermoplastic polymer applications.

In this section, the different resorcinol based phosphate ester compounds synthesized are detailed.

### 8.2.4.1 Resorcinol Bis-(Diphenyl Phosphate) Ester (RDP)

RDP was synthesized by reacting resorcinol with phosphorus oxychloride ($POCl_3$) first to form the *m*-phenylene diphosphotetrachloridate intermediate, and then the resultant compound (without isolation) was reacted with a known amount of phenol (Figure 8.73) [12–19, 21].

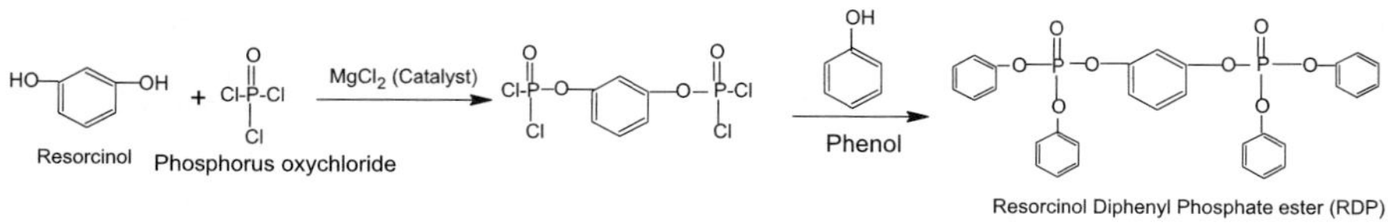


**Figure 8.73.** Synthesis scheme for the preparation of resorcinol bis-(diphenyl phosphate)

In the above synthesis, in order to obtain RDP with high purity (without higher molecular weight products), resorcinol should be reacted first with large excess of $POCl_3$ in the presence of Friedel-Crafts or Lewis acid type catalyst, such as $AlCl_3$ or $MgCl_2$, to obtain exclusively the tetrachloridate intermediate. After the formation of this intermediate compound, the excess $POCl_3$ is normally distilled out before reacting with phenol to obtain the diphosphate ester compound.

A synthesis procedure for the preparation of RDP is illustrated below.

**Procedure for the Synthesis of Resorcinol Bis-(Diphenyl Phosphate) (RDP [18])** $POCl_3$ (406 g, 2.65 mol), 400 g of xylene and 0.8 g of aluminum chloride were charged to a 1-l flask equipped with a stirrer, thermometer, reflux condenser and scrubber. The mixture was heated to 113 °C and 60 g (0.545 mol) of resorcinol were added. Over a 2-h period, 41.0 g (1.12 mol) of HCl gas were trapped in the caustic scrubber.

At the end of the HCl evolution, 200 ml of xylene were added to the pot which was then heated and the $POCl_3$ was distilled at atmospheric pressure over and chased out by the xylene solvent. The distillation was continued until the overhead temperature reached 142 °C. The pot temperature reached 150 °C at the end. GC analysis indicated the pot contents to be free of $POCl_3$.

Phenol (185 g, 1.96 mol) was then added to the pot which was heated for 2 h at 150 °C. HCl (66 g) was collected. GC analysis of the pot contents after the phenol reaction showed 1.5% of phenol remaining in the mixture. The pot contents were then washed twice with 300 ml portions of a 5% solution of sodium hydroxide in water to remove the excess phenol. This was followed by a 300 ml water wash. The washed material was then vacuum stripped to remove xylene and water. The dried product weighed 246.0 g. HPLC analysis showed 1.87% triphenyl phosphate, 0.1% diphenyl hydroxy phenyl phosphate and 97.7% tetra phenyl resorcinol diphosphate and oligomers

**Oligomeric Mixtures of Resorcinol Diphosphate Ester** RDP can also be synthesized from the reaction of diphenyl chlorophosphate (DPCP) with resorcinol. When the phenol is reacted with $POCl_3$, the DPCP is obtained. From this reaction, in addition to DPCP, monophenyl dichlorophosphate (MPCP) is also produced, and the composition of DPCP and MPCP depends on the reaction conditions and molar ratios of the reactants, namely phenol and $POCl_3$. By employing DPCP and MPCP monomers in the synthesis, oligomeric *m*-phenylene poly (diaryl phosphate) compounds having the following generalized structures were synthesized (Figure 8.74) [22].

The following synthesis procedure could be used to develop oligomeric resorcinol diphosphate ester compounds.

**Procedure for the Synthesis of Oligomeric Resorcinol Diphosphate Este [22]** A 500 ml four-necked flask equipped with mechanical stirrer, nitrogen inlet and outlet, thermometer, reflux condenser was charged with 180 g (0.67 mol) of diphenyl chlorophosphate (DPCP) and 20.0 g (0.095 mol) of monophenyl dichlorophosphate (MPCP), weight ratio of DPCP to MPCP 90 : 10, 47.3 g of resorcinol (0.43 mol), and 250 mg of magnesium chloride. The reaction mixture was heated to 150 °C. At 130 °C, the hydrogen chloride evolution was quite fast. The reaction was complete (as measured by the amount of hydrogen chloride evolved) after 4 h. The reaction mixture was washed at 60 °C with 2 × 200 ml of 2% caustic, followed by 3 × 200 ml of water. After removal of traces of water, under vacuum and at 80 °C, there was left 195.2 g of light yellow oil (90% yields). Compositional analysis showed that this oligomeric mixture contained 2.7% TPP, 66.1% RDP (called dimmer, $P_2$), 20.4% trimer (called $P_3$), 6.7% tetramer ($P_4$), 2.2% pentamer ($P_5$) and the rest being higher molecular weight materials. RDP material synthesized from this procedure is called RDP (oligomer).

### 8.2.4.2
### Mono-Hydroxy Terminated Resorcinol Diphosphate (RDP-OH)

In the presence of Lewis acid catalyst, such as $MgCl_2$, a lower molecular weight mono-hydroxy terminated diphosphate ester compound was obtained from the reaction of resorcinol with DPCP (Figure 8.75) [23].

The procedure used in the synthesis of RDP-OH containing resorcinolic phosphate ester is presented below.

**Procedure for the Preparation Monohydroxy-Terminated Resorcinol Diphosphate Ester (RDP-OH [23]** Diphenyl chlorophosphate (134.3 g, 0.50 mol), resorcinol (55.0 g, 0.5 mol, and magnesium chloride (500 mg) were heated to 110 °C for 4 h, until cessation of hydrogen halide evolution. The reaction mixture was washed with 2 × 150 ml of water at 40 °C and was dried under vacuum. The product had the following composition by liquid chromatography (LC) resorcinol bis-(diphenyl phosphate (49.1%); resorcinol diphenyl phosphate (41.7%); resorcinol (3.7%); and triphenyl phosphate (1.0%).

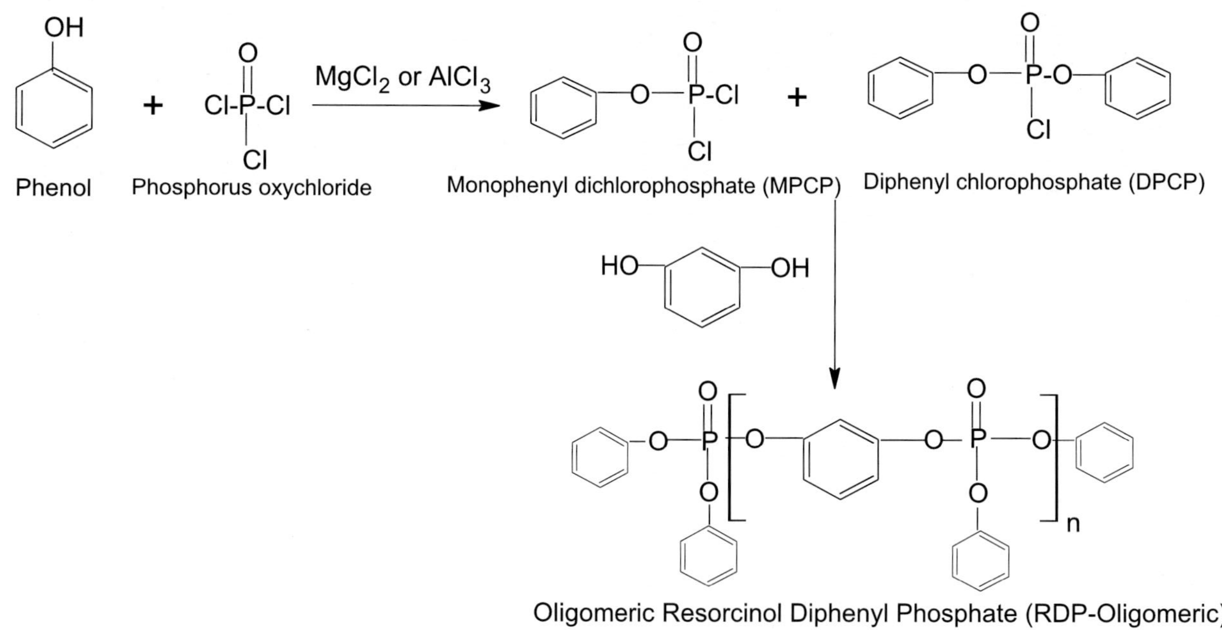


**Figure 8.74.** Synthesis of oligomeric mixtures of resorcinol bis-(diphenyl phosphate)

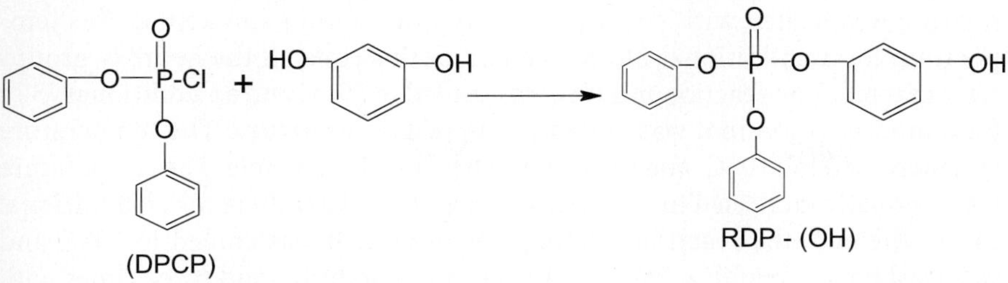


**Figure 8.75.** Synthesis scheme for monohydroxy-terminated resorcinol diphospahte (RDP-(OH)

### 8.2.4.3
### Dihydroxy Terminated Resorcinol Diphosphate [RDP-(OH)$_2$]

The use of MPCP and resorcinol in the phosphate ester synthesis could produce either a polymer or lower molecular weight oligomers depending upon the molar ratios of these two reactants in the reaction. By employing stoichiometric excess of resorcinol, oligomeric diphosphates containing terminal hydroxyl groups could be produced based on the reaction shown below (Figure 8.76).

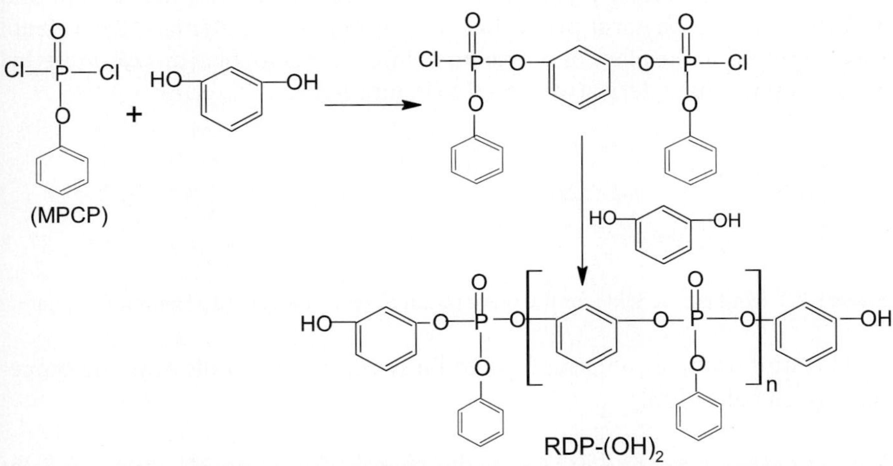


**Figure 8.76.** Synthesis of dihydroxy-terminated resorcinol diphosphate (RDP-(OH)$_2$

Dihydroxy terminated oligomeric phosphate mixtures obtained from the resorcinol and MPCP monomers was based on the following preparative procedure [24].

**Procedure for the Preparation of Dihydroxy-Terminated Resorcinol Diphosphate Ester (RDP-(OH)$_2$ [24]** To 1403 g (6.65 miles) of monophenyl dichlorophosphate was added 3.5 g of magnesium chloride and 366 g of resorcinol. The reaction mixture was heated with stirring to 110 °C over 30 min and kept at this temperature for an additional 2 h. At the end of this time all the hydroxy groups were reacted. The reaction mixture was cooled to 50 °C and an additional 750 g (6.82 mol) of resorcinol was added to the reaction mixture. The temperature was increased to 110 °C, and the mixture became dark purple. The temperature was gradually increased to 150 °C over 6 h and was kept there for an additional 2 h. At the end, the reaction mixture was orange. It was cooled to 70 °C and was washed twice with a 5% sodium carbonate solution and three times with water.

After removal of traces of water at 80 °C under vacuum, there was left 1380.5 g of an extremely viscous orange oil that did not flow at room temperature. The percent yield was 82%. The material had the following composition: $P_1$–$(OH)_3$: 2%; $P_1$–$(OH)_2$: 22%; $P_2$–$(OH)_2$: 29%; $P_3$–$(OH)_2$: 14%; $P_4$–$(OH)_2$: 9%; $P_5$–$(OH)_2$: 5% and higher molecular weights.

### 8.2.4.4 Resorcinol Bis-(dixylenyl) Phosphate

The RDP obtained from the resorcinol exhibited liquid property. In order to enhance the handling performance of resorcinolic diphosphate compound and improve the physical properties of thermoplastic polymers, 2,6-xylenol was employed in the place of phenol to obtain resorcinol bis-[di-(2,6-dimethyl phenyl)] phosphate derivative [25–30] (Figure 8.77).

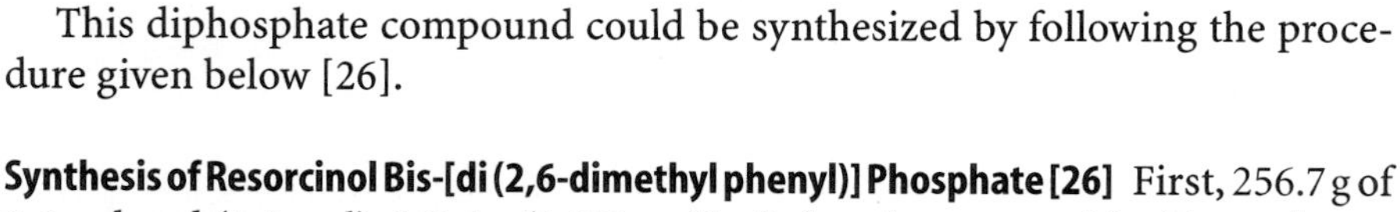


**Figure 8.77.** Synthesis scheme for the preparation of resorcinol bis (dixylenyl) phosphate

This diphosphate compound could be synthesized by following the procedure given below [26].

**Synthesis of Resorcinol Bis-[di (2,6-dimethyl phenyl)] Phosphate [26]** First, 256.7 g of 2,6-xylenol (2.1 mol), 161.1 g (1.051 mol) of phosphorus oxychloride, and 3.7 g of aluminum chloride were slowly heated to 150 °C and kept at this temperature for 3 h. At the end of this time, hydrogen chloride evolution had subsided. Then, 57.8 g (0.525 mol) of resorcinol were added to the reaction mixture. The hydrogen chloride temperature was then raised to and kept at 180 °C until evolution ceased. The reaction mixture was cooled to 120 °C, and then methanol (500 ml) was added. The mixture was allowed to cool slowly with stirring to induce crystallization. A white solid was filtered from the reaction mixture and was rinsed with cold methanol. The product was then dried. There was obtained 280.1 g of a white solid (78% yield) melting at 77–79 °C. HPLC analysis showed 85.0 area % of the desired product.

The diphosphate compound obtained from the 2,6-xylenol was a white crystalline compound and exhibited 95–96 °C melting points [27].

### 8.2.4.5
### Bis-(3-Hydroxyphenyl) Phenyl Phosphate (BHPP)

By adjusting the molar ratios of resorcinol and MPCP monomers and, also carefully controlling the reaction conditions to avoid the formation of high molecular weight products, bis-(3-hydroxyphenyl) phenyl phosphate was synthesized based on the following reaction scheme and procedure [31] (Figure 8.78).

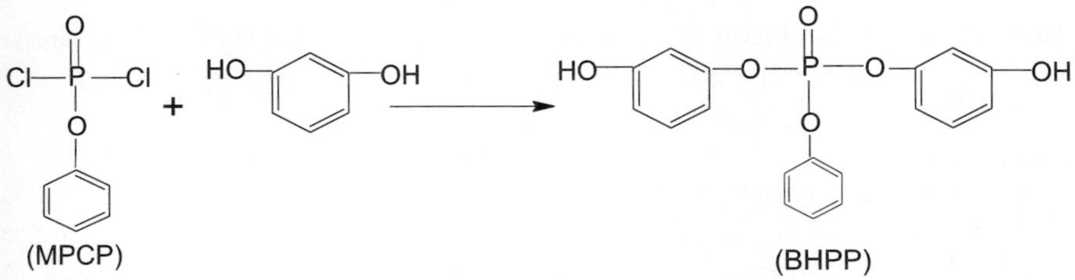


**Figure 8.78.** Synthesis of bis(3-hydroxyphenyl) phenyl phosphate (BHPP)

**Synthesis of Bis-(3-hydroxy phenyl) Phenyl Phosphate (BHPP) [31]** To a 1-l reaction vessel equipped with a temperature controller, overhead stirrer and a reflux condenser were added 220 g of resorcinol and 360 g of xylene. The mixture was first heated slowly to 70 °C to dissolve resorcinol and then heated to 90 °C. To the resultant solution, 195 g of phenyl dichloro phosphate was added continuously at a constant rate over a period of 2 h. The evolution of HCl gas was detected immediately. The reaction mixture was further heated at reflux for 2.5 h until HCl evolution has subsided. The reaction was assumed to be complete when no more HCl evolution was detected at 128 °C. After the removal of solvent, the resultant residue was recrystallized from tetrahydrofuran to give BHPP (Yield = 95%).

**Commercial Sources of Resorcinol Phosphate Ester Compounds** Resorcinolic phosphate ester compounds have been commercially produced and marketed under different trade names by several companies. The two most widely known resorcinol phosphate esters are resorcinol bis-(diphenyl) phosphate (RDP) and resorcinol bis-(dixylenyl) phosphate. Table 8.26 shows the names of companies producing these compounds and their commercial names.

The physical properties of widely known RDP flame retardant compound, available from Akzo Nobel under the trade name Fyroflex RDP, are presented in Table 8.27.

**Table 8.26.** Commercial sources of resorcinol based flame retardants

| Resorcinol bis-(diphosphate) (RDP)<br>Company source | Trade names |
|---|---|
| AKZO Nobel | Fyroflex RDP |
| Great Lakes Chemical | Reofos RDP |
| Daihachi Chemical | CR-733 S |
| Asahi Denka | Phosphate PFR |
| **Resorcinol bis-(dixylenyl phosphate)<br>Company Source** | **Trade Names** |
| Daihachi Chemical | PX-200 |
| Asahi Denka | Phosphate FP-500 |

**Table 8.27.** Properties of resorcinol bis-(diphenyl phosphate), RDP (oligomeric mixtures)

| Properties | |
|---|---|
| Physical appearance | Clear and transparent liquid |
| Phosphorus content (weight %) | 10.9 |
| Specific gravity at 25 °C | 1.3 |
| Density at 25 °C (kg/m$^3$) | 1300 |
| Viscosity at 25 °C, MPa/s | 600 |
| Color, APHA | $< 100$ |
| Assay, wt.% | 60 – 75 |
| Boiling point (°C) | $> 300$ |
| Solubility in water | Insoluble |

Data source: tech data sheet on Fyroflex RDP from Akzo Nobel

### 8.2.5 Toxicity, Environmental and Health Aspects of RDP

For achieving greater success in the flame retardant applications, the additives should exhibit not only good flame retarding properties to the thermoplastic polymers but also, show less toxicity, environmentally friendly and minimal health risks to humans. In addition to acute and chronic toxicities, the flame retardant materials' leachability from their substrates with time and eco-toxicity in streams, soil and ground water are also important. Extensive studies have been performed on Akzo Nobel's Fyroflex RDP by the European Flame Retardant Association (EFRA) on the environmental and health aspects of RDP, and the details are given below.

#### Health

- RDP was determined to be non-toxic in acute oral, dermal or inhalation studies, not irritating to skin and eyes and also, not causing allergic reaction to skin.

- RDP is not mutagenic in a series of in vitro and in vivo mutagenic tests.
- RDP was tested for its nuero-toxicity and found not to be toxic to the nervous system. In a two generation reprotoxicity study no effects on reproductive performance or off spring were observed [32].
- RDP is not harmful to the unborn (not teratogenic) and not immuno-toxic [33,34].
- In sub-chronic (28-day exposure) studies, no significant findings were reported for RDP.
- A comprehensive study in the rat, mouse and monkey confirmed that RDP is metabolized in an identical manner [35].

These studies have shown that RDP has low impact on health.

### Environmental

- RDP is ultimately biodegradable. Its half life in water at 20 °C and pH = 7 is 17 days. RDP is therefore not anticipated to bio-accumulative. RDP is not toxic to fish, algae and bacteria.

## 8.2.6 Thermal Stabilities of Resorcinol Phosphate Esters

Detailed investigations have been made on the thermal stabilities of different phosphate ester compounds synthesized from resorcinol [36,37]. TGA results are presented in Table 8.28.

**Table 8.28.** Resorcinol based phosphate esters and thermal properties

| Flame retardants | Phosphorus (%) | Physical appearance | TGA weight loss (%) at | | | Residue (%) remaining at 560 °C |
|---|---|---|---|---|---|---|
| | | | 1 | 5 | 10 | |
| RDP (oligomer) | 10.8 | Colorless liquid | 251 | 317 | 340 | 5.7 |
| RDP (purity = 95%) | 10.8 | White solid | 267 | 321 | 344 | 0.8 |
| RDP-OH | 9.8 | Yellow liquid | 195 | 277 | 298 | 1 |
| RDP-$(OH)_2$ | 10.9 | Viscous liquid | 99 | 306 | 332 | 60 |

TGA was determined in air at a heating rate of 10 °C/min
Data from [36]

From the TGA weight loss data (Table 8.28), RDP–OH exhibited the lowest thermal stability compared to the other derivatives, which may be associated with its lower molecular weight. The weight percent residue, determined at 560 °C, showed considerable char yield for the RDP–$(OH)_2$, which indicated that this compound might be a good char former in applications where high char formation is most required. Based on the TGA data, resorcinol based diphosphate esters have shown good thermal stabilities and therefore, can be expected to be non-volatile or less volatile during the processing conditions of thermoplastic polymers.

The thermal stability comparison of RDP and TPP was made and the results are shown in Table 8.29 [38].

**Table 8.29.** Thermal properties of RDP and TPP TGA weight loss data in air (heating rate = 10 °C/min)

| Flame retardant | Weight loss (%) temperature (°C) | | | | |
|---|---|---|---|---|---|
| | 1 | 5 | 10 | 50 | 99 |
| RDP | 242 | 299 | 340 | 404 | 724 |
| TPP | 187 | 219 | 235 | 281 | 398 |

RDP = Resorcinol diphenyl phosphate, TPP = Triphenyl phosphate
Data from [38]

The initial, 5 and 10 wt. % loss data clearly show that TPP is more volatile than RDP, and therefore considerable loss of TPP material could be expected from the thermoplastic compositions during the extrusion and injection molding processes.

## 8.2.7 Applications of Resorcinol Phosphate Esters

Handling and blending of RDP in various thermoplastic polymers could be much easier due to liquid. Since RDP is low in color, it may not impart any color to the blending materials. RDP is expected to be more efficient flame retardant than TPP or alkylated TPP (ATPP) in part because of its higher phosphorus content. The phosphorus contents of ATTP, TPP and RDP are about 8.3, 9.5 and 10.9 wt. % respectively [39]. RDP has higher thermal stability and molecular weight than TPP, and therefore could show lower volatility at elevated temperatures.

RDP is expected to be more compatible with engineering polymers containing polar functional groups in their back bone structure, partly due to the presence of ether groups in RDP which could make physical interactions with polymers' polar groups. Because of this anticipated interactions, RDP could provide additional benefit of improving the processing (or plasticizing) characteristics of thermoplastic polymers. The plasticizing character of RDP can significantly improve the melt flow properties of thermoplastic compositions, which might be ultimately beneficial in the compounding and molding operations. RDP can be used as processing aid in various high performance thermoplastic resin systems where improvements in processing, flame retardancy and thermal stabilities are required.

### 8.2.7.1
### Flow Modifier in Thermoplastic Resins

Melt flow modifications of thermoplastic resin compositions is becoming increasingly important in the design of thin wall moldings. Since RDP exhibited liquid and lower viscosity properties at room temperature, and also expected to be compatible with thermoplastics, it was evaluated as a flow modifier in several resin systems [38,40].

Table 8.30 presents the results obtained on the melt flow indices, heat distortion temperatures (HDT) and UL-94 ratings for the PC, PPO, PC/ABS and PPO/HIPS polymer systems.

In the case of unmodified HIPS, ABS and PC resins, considerable increase in melt flow rates was seen even with very low levels of RDP in the composition. The use of RDP in these polymers can be multi-functional, since it acts both as flame retardant and flow enhancers. The addition of RDP into these polymers appeared to affect the HDT, but with the appropriate selection of blend compositions, this could be adjusted for suitable applications.

### 8.2.7.2
### Flame Retardant Properties of ABS Resin Systems

ABS resins have excellent mechanical, heat resistant and impact resistance properties but they are highly flammable. The LOI value for the unmodified ABS resin is about 17.6. To enhance the flame retardant property of ABS resin, RDP was added up to 30 wt. % level, which showed only marginal increase in the LOI value as shown in Figure 8.79 [37,41].

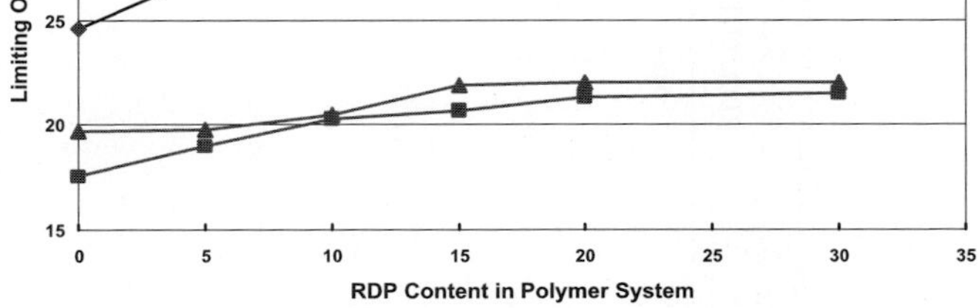


**Figure 8.79.** LOI values of PC and PC/ABS systems flame retarded with RDP

Flame retarded compositions of ABS resins were prepared using resorcinol diphosphate and the phenolic novolac resin, and tested for properties. The results are presented in Table 8.31 [42–44].

**Table 8.30.** Melt flow and flame retardant properties of plastics containing RDP

| | PPO/HIPS blend | | PC/ABS blend | | HIPS | | ABS | | PC | |
|---|---|---|---|---|---|---|---|---|---|---|
| Properties | Mod. PPO (Noryl) | Mod. PPO/ 20% RDP | PC/ABS (5 : 1) | PC/ABS/ RDP | HIPS | HIPS/ 10% RDP | ABS | ABS/ 2% RDP | PC | PC/ 2% RDP |
| MFI (g/10 min) (Temp, °C/kg) | 1.5 (300/1.2) | 8.7 (300/1.2) | 13 (300/1.2) | 19 (300/1.2) | 6 (200/5) | 18 (200/5) | 7 (200/5) | 10 (200/5) | 26 (300/1.2) | 32 (300/1.2) |
| HDT (°C) (at 1.82 MPa) | 125 | 81 | 110 | 84 | 72 | 58 | 74 | 72 | 121 | 111 |
| UL-94 (Sample: 1.6/3.2 mm) | Fail/fail | V-0/V-0 | Fail/fail | V-0/V-0 | Fail/fail | ND/V-2 | ND | ND | ND | ND |

MFI = Measured according to ASTM-D 1238, HDT = Measured according to ASTM-D 648, ND = Not determined or not reported
Data from [38]

**Table 8.31.** Flame retarded ABS resin compositions and properties

| Composition (by weight) | 1 | 2 | 3 | 4 | 5 | 6 | 7 | 8 |
|---|---|---|---|---|---|---|---|---|
| 1. ABS resin | 100 | 100 | 100 | 100 | 100 | 100 | 100 | 100 |
| Flame retardants used | | | | | | | | |
| 1. Resorcinol bis-(2,6-dimethyl-phenyl) phosphate | 10 | 13.3 | | 7.5 | | | | |
| 2. Tris-(2,6-dimethylphenyl) phosphate | | | 15 | 7.5 | | | | |
| 3. Triphenyl phosphate | | | | | 10 | 13.3 | | |
| 4. Resorcinol (diphenyl phosphate) oligomer | | | | | | | 10 | 13.3 |
| 5. Phenolic novolac resin | 10 | 6.7 | 5 | 5 | 10 | 6.7 | 10 | 6.7 |
| Properties measured | | | | | | | | |
| LOI (%) | 28.9 | 33.3 | 32.5 | 35.5 | 31.1 | 32.9 | 31.6 | 34.6 |
| Izod impact strength | 16.7 | 16.6 | 15.9 | 13.7 | 13.8 | 12.8 | 13.2 | 11.1 |
| Deflection temperature under load (°C) | 72 | 70.5 | 72.6 | 70.3 | 61.1 | 59.3 | 65.2 | 62.1 |
| Vicat softening point (°C) | 84 | 82 | 84.8 | 81.5 | 72.2 | 69.7 | 75.4 | 72.8 |

ABS resin (Starlight) from Cheil Industries,
Resorcinol bis-(2,6-dimethylphenyl) phosphate (PX-200) from Daihachi Chemical,
Resorcinol (diphenyl phosphate) oligomer (CR-733) from Daihachi Chemical,
Tris-(2,6-dimethylphenyl) phosphate (PX-130) from Daihachi Chemical,
Phenolic novolac resin (Tamanol-759) from Arakawa.
Data from [42–44]

The addition of phenolic novolac resin in the ABS resin compositions improved the LOI values mainly due to its char forming characteristics. In-spite of higher loading of additives, the impact strength, HDT and softening points were not severely affected.

Table 8.32 shows another flame retarded ABS compositions and properties [45].

The ABS resin composition contained SAN and PPO polymers, and their addition was observed to show good impact strength, heat resistance and flame retardant properties.

#### 8.2.7.3 Flame Retardant Properties in HIPS Polymers

HIPS is also a very combustible thermoplastic resin and shows LOI value of 17.6, and also not classified in UL-94 test [37, 46]. With RDP flame retardant only a slight increase in LOI, to 19.3, and no changes in UL-94 in UL-94 classifications were observed even up to 15 wt. % of RDP loadings (Figure 8.80).

**Table 8.32.** Flame retarded ABS composition and properties

| Composition (by weight) | 1 | 2 | 3 |
|---|---|---|---|
| 1. ABS resin | 80 | 80 | 80 |
| 2. Poly(phenylene ether), PPO | 20 | 20 | 20 |
| 3. SAN (AN = 13 wt. %) | 10 | 10 | 10 |
| Flame retardants used | | | |
| 1. Tris-(2,6-dimethylphenyl) phosphate | 16 | 10 | |
| 2. Resorcinol bis-(2,6-dimethylphenyl) phosphate | | 6 | |
| 3. Resorcinol (diphenyl phosphate) oligomer | | | 16 |
| 4. Phenolic novolac resin | 6 | 6 | 6 |
| Properties measured | | | |
| Impact strength (kg cm $cm^{-1}$; 1/8″) | 18 | 16 | 10 |
| Vicat softening point (°C) | 87 | 86 | 79 |
| Flammability, UL-94 (1/10″) | V-1 | V-1 | V-1 |

SAN = Styrene-acrylonitrile copolymer
Data from [45]

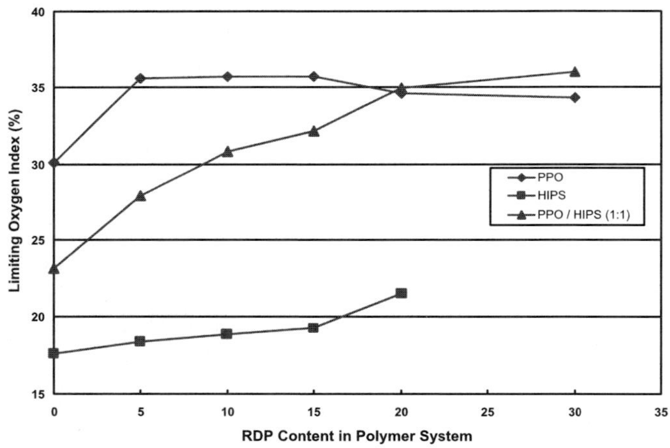


**Figure 8.80.** Liniting Oxygen values of PPO and PPO/HIPS systems containing RDP

When RDP loading was increased to 20 wt. %, then a V-2 rating in UL-94 and LOI value of 21.5 were obtained for HIPS polymer [46].

The melt flow, flame retardant, HDT and impact strength properties of flame retarded HIPS materials were evaluated, in which RDP compounds having different structural characteristics were used as flame retardants (Table 8.33) [36].

Melt flow properties were increased by the addition of RDP flame retardants, which indicated that HIPS compounds could be processed much easier.

**Table 8.33.** Physical, mechanical and flame retardant properties of high impact polystyrene (HIPS) containing different resorcinol diphosphate ester (RDP) compounds

| RDP compounds: | HIPS (control) | RDP (oligomer) | RDP (95% purity) | RDP-(OH) |
|---|---|---|---|---|
| Phosphorus content (%) | 0 | 1.1 | 1 | 1 |
| Melt flow index (g/10 min) | 2.4 | 12 | 14.4 | 16.6 |
| Heat distortion temp. (°C) | 91 | 67 | 66 | 63 |
| Izod impact strength (kJ/m$^2$) | 8.2 | 6 | 5.5 | 6 |
| UL-94 rating (1.6 mm) | Fail | V-2 | V-2 | V-2 |
| After flame time (AFT, s) | > 30 | 15.9 | 17.5 | 14 |

HIPS from Dow Chemical, Trade Name: Styron 404
Data from [36]

### 8.2.7.4 Flame Retarded Properties in PPO/HIPS Systems

PPO resin is less combustible than either HIPS or ABS polymer with an LOI value of 30 [37]. On the addition of 5 wt. % RDP in PPO, the LOI was increased to 35.7 and this value remained almost constant with a further increase in the loading of RDP (Figure 8.80).

HIPS/PPO blends, most commonly known as modified HIPS resins, can be very effectively flame retarded with RDP. Noryl resin is a blend of HIPS and PPO polymer and exhibits LOI value of 23.2, which is an intermediate value between HIPS (17.6) and PPO (30.1). The addition of RDP into Noryl resin, even at 1.5 wt. % level, dramatically improved its flame retardant property. The formulation did not drip and exhibited V-2 rating in UL-94 testing [37]. RDP showed much greater flame retarding effect on Noryl resin than in PPO at higher loading. Because of this, a V-0 rating could be easily achieved with RDP content 5 wt. % and higher in PPO/HIPS resins (Table 8.34) [39, 47–53].

Resorcinol phosphates in PPO/HIPS blends provided good molding quality, heat dimensional stability, and greater improvements in dripping characteristics, shorter flame out time and UL-94 ratings.

### 8.2.7.5 Flame Retarding Properties in PC and PC/ABS Resin Systems

As can be seen in Figure 8.79, the flame retardant property of neat PC resin was improved by the addition of RDP. The LOI value of PC resin was steadily increasing up to 20 wt. % RDP loading and then remained constant beyond this level, which shows that RDP is a good flame retardant for PC. Even with 5 wt. % RDP loading, V-2 classification can be obtained for the PC resin. If the loading of RDP could be increased to 15 wt. %, then a V-0 rating with non flaming drips classification can be easily achieved [46]. Complete dripping suppression of

**Table 8.34.** Flame retarded modified poly(phenylene oxide), PPO (performance of RDP vs alkylated-TPP)

| | | |
|---|---|---|
| Composition | | |
| Modified PPO (Noryl PX-888) | 87 | 90 |
| (Blend of PPO and HIPS) | | |
| Alkylated triphenyl phosphate (ATPP) | 13 | |
| Resorcinol diphenyl phosphate (RDP) | | 10 |
| Phosphorus content (wt.%) | 1.08 | 1.08 |
| Properties | | |
| Oxygen Index | 36.5 | 37.2 |
| UL-94 rating (1.6 mm) | V-0 | V-0 |
| HDT at 1.82 MPa, °C | 86 | 94 |
| Mechanical properties | | |
| Flexural strength, MPa | 89 | 102 |
| Flexural modulus, GPa | 3.5 | 3.6 |
| Tensile at yield, MPa | 48 | 49 |
| Elongation (%) | 19 | 23 |

Noryl PX-888 from GE Plastics,
Triphenyl phosphate (Reofos 50) from FMC Corporation,
RDP (Reofos RDP) from FMC Corporation.
Data from [39]

PC resin could be seen with 20 wt. % RDP in the thermoplastic composition.

The addition of ABS into PC resin makes the blend more combustible than PC itself. From the results in the figure, it can be observed that the addition of RDP up to 30 wt. % in PC/ABS blend, only improved the flame retardant property marginally, which was very comparable to unmodified ABS resin. In order to achieve a V-0 rating in PC/ABS resin and also to attain the maximum performances in these blends, extensive developmental efforts and formulations have been made [54–62].

Triphenyl phosphate (TPP) is an effective flame retardant for PC/ABS blends due to its high phosphorus content, and also its ability to act both in the solid and gas phase. TPP also shows good plasticizing behavior and enhances the flow properties of PC/ABS blends with minimal reduction in impact properties. The major drawback with TPP is its high volatility, which translates into juicing and subsequent stress cracking in thermoplastic polymers. Stress cracking is the appearance of crazes that ultimately crack the thermoplastic polymer surface. RDP was developed to overcome the stress cracking phenomena occurring with the use of TPP in PC/ABS and other thermoplastic polymers. The combination of low volatility, high phosphorus content and outstanding charring ability property of RDP made an excellent alternative flame retardant to TPP in the thermoplastic polymers.

Performance comparison of RDP was made against TPP in the PC/ABS blends systems and the results are presented in Table 8.35 [63].

**Table 8.35.** Flame retardant performance of oligomeric RDP against TPP in PC/ABS blend systems

| Composition (parts by weight) | 1 | 2 | 3 | 4 |
|---|---|---|---|---|
| 1. Polycarbonate | 75 | 64.5 | 64.5 | 64.5 |
| 2. ABS resin | 25 | 25 | 25 | 25 |
| 3. Oligomeric RDP | | | 10 | 10 |
| 4. Triphenyl phosphate (TPP) | | 10 | | |
| 5. Teflon | | 0.5 | 0.5 | 0.5 |
| 6. STB | | 0.02 | | 0.02 |
| Properties | | | | |
| 1. Flame retardation, UL-94 (1.6 mm) | HB | V-1 | V-2 | V-0 |
| 2. Vicat softening, °C | 133 | 89 | 95 | 95 |
| 3. Juicing behavior | Yes | Yes | No | No |

STB = Sodium salt of trichlorobenzene sulfonate
Data from [63]

The softening temperature of PC/ABS resin was considerably lowered by the addition of phosphate ester flame retardants. However, compared to TPP, RDP showed considerable improvements in the juicing behavior and heat distortion temperature of PC/ABS resin.

Resorcinol diphosphate compounds having different molecular structure were also evaluated for their performances in PC/ABS system, and the results are presented in Table 8.36.

**Table 8.36.** Physical, mechanical and flame retardant properties of PC/ABS blends containing different resorcinol diphosphate ester (RDP) compounds

| RDP compounds: | PC/ABS (control) | RDP (oligomer) | RDP (95% purity) | RDP-(OH) | RDP-$(OH)_2$ |
|---|---|---|---|---|---|
| Phosphorus content (%) | 0 | 0.9 | 1 | 1 | 1 |
| Melt Flow Index (g/10 min) | 2.2 | 7.7 | 10.1 | ND | ND |
| Heat distortion temp (°C) | 116 | 84 | 86 | 71 | 104 |
| Tensile yield (MPa) | 53.8 | 53.8 | 58.6 | 14.8 | ND |
| UL-94 Rating (1.6 mm) | Fail | V-0 | V-0 | V-0 | V-2 |
| After flame time (AFT, s) | > 30 | 0.7 | 1.7 | 1.1 | 5.8 |

PC/ABS blend ratio = 5 : 1. Lexan 141 (PC from GE)/Magnum 9010 (ABS from Dow)
Data from [36]

As expected, the addition of RDP flame retardant additives improved the melt flow properties and flame retardant characteristics of PC/ABS resins.

The performance of RDP in PC/ABS resins having different blend ratios were evaluated, and the results obtained and reported are presented in Table 8.37 [64].

When compared to ATTP, RDP exhibited superior flame retarding properties with V-0 ratings for 1/16″ samples, high heat deflection temperatures and increased melt flow properties.

RDP was also evaluated as the flame retardant additive in various PC/ABS blend ratios with loadings less than 9 wt. % in the thermoplastic composition (Table 8.38).

From the Table 8.38 results, with an increase in the PC content of PC/ABS blend, less RDP was needed to achieve the V-0 rating since PC itself can show good LOI and flame retardant properties. The heat distortion temperature was also observed to increase with an increase in PC concentration. By the addition of an impact modifier, the HDT and notched Izod impact properties were considerably improved.

In the development of thermoplastic molding compositions, particularly in PC/ABS blends, fluorinated polyolefins, such as Teflon, are most often used as anti-dripping agents. One of the major disadvantages of Teflon present in these polymeric materials is the formation of hydrogen fluoride gas arising from the Teflon in the event of fire. The specific use of Teflon in the PC/ABS molding compositions is the only way to achieve self extinguishing property with V-0 ratings in 1.6 mm thick test samples [65–68]. To avoid Teflon in the PC/ABS blends, PPO and other materials have been combined with RDP to achieve the self extinguishing property without dripping in accordance with the UL-94 test [69–71].

Table 8.39 presents the results obtained on the use of PPO in the PC/ABS resins to achieve the V-0 ratings.

The molding compositions free of Teflon in the formulations showed flame retardant properties, which were highly effective with respect to non dripping and self-extinguishing properties. Compositions containing RDP and SAN polymer exhibited a good range of properties, which included notched impact strength, heat resistance, surface quality and improved stress cracking resistance.

### 8.2.8
### Flame Retardant Mechanisms in PC/ABS and PPO/HIPS Resins

In general, phosphorus based flame retardants mainly influence the reactions occurring in the condensed phase. They are highly effective, particularly in polymers having high oxygen content or oxygen containing polymers. In the event of a fire, phosphorus containing flame retardants are converted by thermal decomposition into phosphoric acid, which in the condensed phase

**Table 8.37.** Flame retardant compositions of PC/ABS blends and properties

| PC/ABS blend ratio | PC/ABS = 3/1 | | | | | PC/ABS = 4/1 | | | | | PC/ABS = 5/1 | | | | |
|---|---|---|---|---|---|---|---|---|---|---|---|---|---|---|---|
| Flame retardant (FR) used | FR | ATPP | RDP | | | FR | ATPP | RDP | | | FR | ATPP | RDP | | |
| FR loading (%) | 0 | 12.9 | 8 | 10 | 12 | 0 | 12.9 | 8 | 10 | 12 | 0 | 12.9 | 8 | 10 | 12 |
| Mechanical properties | | | | | | | | | | | | | | | |
| Tensile strength (MPa) | 48.1 | 48.1 | 50.9 | 47.4 | 46.9 | 48.4 | 45.7 | 47 | 46.9 | 47.4 | 48.8 | 46.2 | 53.3 | 49.2 | 50.1 |
| Tensile modulus (GPa) | 0.9 | 1 | 1.07 | 1.02 | 1.06 | 0.86 | 1.02 | 1 | 1.01 | 1.03 | 0.86 | 1.11 | 0.97 | 0.94 | 1.05 |
| Flexural strength (MPa) | 59.6 | 73.5 | 57.6 | 71.7 | 70.9 | 58.6 | 71.3 | 65.6 | 71.8 | 71.5 | 58.9 | 74.7 | 64.9 | 72.3 | 73.7 |
| Flexural modulus (GPa) | 2.18 | 2.64 | 2.07 | 2.46 | 2.54 | 2.16 | 2.53 | 2.31 | 2.56 | 2.51 | 2.16 | 2.66 | 2.33 | 2.58 | 2.63 |
| HDT, °C (1.82 MPa) | 111.7 | 72.5 | 88.5 | 80.2 | 75.4 | 114.1 | 72.5 | 90.7 | 82.4 | 78.8 | 115.9 | 73 | 91.2 | 83.1 | 80.3 |
| Flammability (UL-94) | | | | | | | | | | | | | | | |
| Sample: 1/8″ | NR | V-2 | V-0 | V-0 | V-0 | NR | V-2 | V-0 | V-0 | V-0 | NR | V-2 | V-0 | V-0 | V-0 |
| Sample: 1/16″ | NR | V-2 | V-0 | V-0 | V-0 | NR | V-2 | V-0 | V-0 | V-0 | NR | V-2 | V-0 | V-0 | V-0 |

NR = Not reported or no rating. All formulations contained 0.5% Teflon
ATTP = Alkylated triphenyl phosphate. RDP = RDP from Akzo Chemicals,
The "phosphorus" contents corresponding to 12.9% ATTP and 10% RDP are about equal
Data from [64]

**Table 8.38.** Flame retardation of polycarbonate/ABS blends with RDP and their properties

| | | | | |
|---|---|---|---|---|
| Composition | | | | |
| Polycarbonate/ABS ratio | 3 : 1 | 4 : 1 | 8 : 1 | 8 : 1 |
| RDP used (weight) | 9 | 8 | 4 | 6 |
| Impact modifier | No | No | No | 4 |
| Teflon 6C | 0.5 | 0.5 | 0.5 | 0.5 |
| Properties | | | | |
| Oxygen index | 29.3 | 28.5 | 28.5 | 28.1 |
| UL-94 rating (1.6 mm) | V-0 | V-0 | V-0 | V-0 |
| HDT at 1.82 MPa, °C | 88 | 94 | 104 | 123 |
| Notched Izod, J/m | 360 | 600 | 150 | 760 |
| Mechanical properties | | | | |
| Flexural strength, MPa | 114 | 108 | 133 | 109 |
| Flexural modulus, GPa | 4.1 | 3.9 | 4.5 | 3.4 |
| Tensile at yield, MPa | 44 | 46 | 60 | 47 |
| Elongation (%) | 37 | 39 | 22 | 20 |

Polycarbonate: Lexan 141 from GE Plastics
ABS: Magnum 213 from Dow Chemical
RDP: Reofos from FMC Corporation
Impact modifier: Paraloid EXL-3607 from Rohm and Haas
Data from [39]

**Table 8.39.** Flame retardant properties of PC/ABS blends in the absence of Teflon as an anti-drip agent

| Composition (wt%) | 1 | 2 | 3 | 4 | 5 |
|---|---|---|---|---|---|
| 1. Polycarbonate (PC) | 67 | 66 | 65 | 68 | 69.7 |
| 2. SAN copolymer | 5 | 5 | 5 | 5 | 5 |
| 3. ABS resin | 10 | 10 | 10 | 10 | 10 |
| 4. RDP | 15 | 15 | 15 | 12 | 15 |
| 5. Polyphenylene ether (PPO) | 3 | 4 | 5 | 5 | |
| 6. Teflon | | | | | 0.3 |
| Properties | | | | | |
| 1. UL-94 (2.1 mm) | V-0 | V-0 | V-0 | V-0 | V-0 |
| 2. Burn time (s) | 13 | 14 | 12 | 11 | 10 |
| 3. Vicat softening point (°C) | 88 | 89 | 89 | 98 | 92 |
| 4. LOI (% $O_2$) | 30.5 | ND | 32.5 | 33.5 | ND |

ND = Not determined or not reported
Data from [69]

extracts water from the pyrolyzing substrates and causing them to char. The phosphoric acid thus produced, can be converted into pyro phosphoric acid, which can act as heat transfer barrier in the condensed phase [1].

In the case of PC/ABS polymers, PC is a highly char forming material whereas the ABS is not. It has been reported that PC was observed to retain more phosphorus at elevated temperature conditions than ABS polymer [37]. Aryl phosphate esters, in general, are not very effective flame retardants for non-char forming polymers such as ABS. This is because, in the condensed phase, they cannot be involved in the charring process, and also cannot act as radical scavengers in the vapor phase due to their low volatility at elevated temperatures.

Based on the combined analytical results, such as TGA, chemical analysis for phosphorus and IR analysis on solid residues, it was determined that PC/ABS blend can effectively retain the RDP in the course of combustion or decomposition [72]. This may be due to higher molecular weight of RDP as well as its interaction with PC at elevated temperatures. PC material could under go Fries type rearrangement under various conditions including the high temperature and in the presence of catalysts, and produces phenolic type reaction products as shown in Figure 8.81 [73].

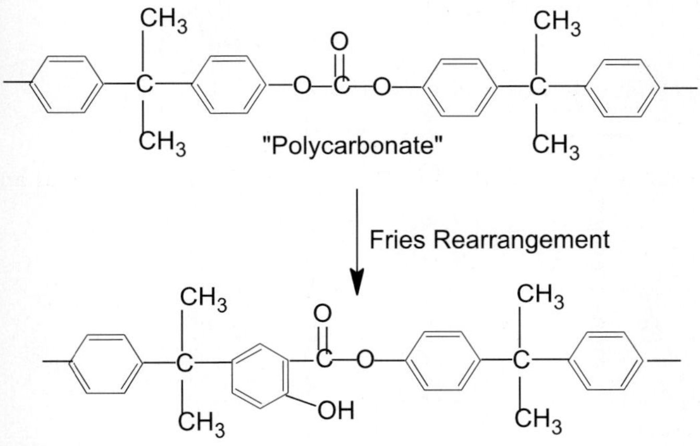


**Figure 8.81.**

Published information shows that a *trans*-esterification reaction of triphenyl phosphate with phenolic hydroxyl group present in resorcinol could be possible [74]. Therefore, the phenolic hydroxyl groups thus formed from the PC decomposition can undergo similar *trans*-esterification type reaction with RDP compounds, and can produce phenol as the volatile by-product (Figure 8.82).

This *trans*-esterification process might continue, with the liberation of more phenolic molecules from the reactions of polycarbonate hydroxyl groups and the partially condensed RDP structure, and form a cross-linked network type structure. Because of this reaction, the accumulation of phosphorus in the solid phase can occur, which can show that PC acts as a good char former [75]. On the oxidation of resorcinol phosphate ester during the combustion, phosphoric

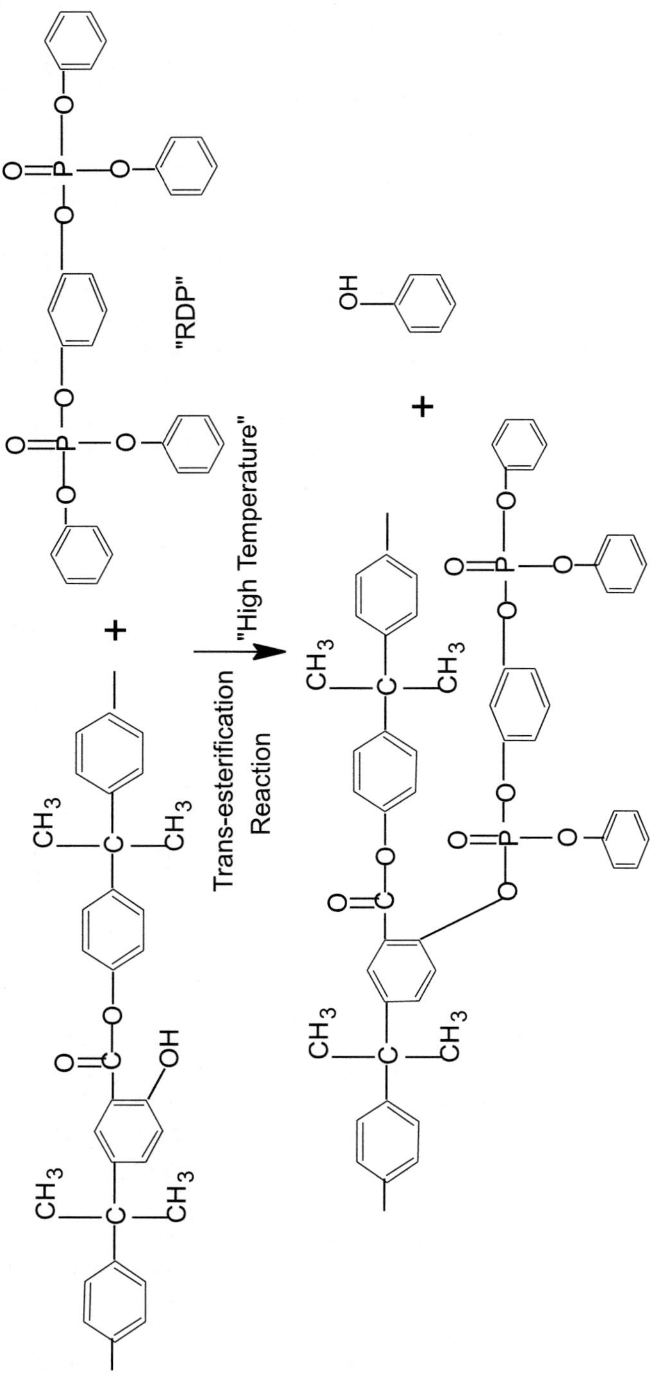


**Figure 8.82.**

acid is expected to be produced, which can provide additional surface barrier against the diffusion of combustible gases to the flame.

The reaction between phenolic hydroxyl groups and aryl phosphate ester compounds could take place at elevated temperatures, which suggests that the incorporation of phenolic novolac resins and resorcinol phosphates could enhance the thermal stability of highly flammable ABS resin [76]. Phenolic novolac resins having different molecular weights were used in combination with resorcinol bis-(dixylenyl) phosphate in the ABS resin and their synergistic flame retardant properties were studied [77]. Results from the study showed that the incorporation of resorcinol phosphate and novolac resin together in the ABS resin was very effective in enhancing the flame retardant properties. LOI values as high as 53 and V-1 rating in UL-94 test were obtained for certain ABS formulations. The enhancement in flame retardant property of ABS resin was related to the formation of chars from the cross-linking reaction of resorcinol phosphate with novolac compounds, which might have taken place during the thermal decomposition process.

Aryl phosphate ester compounds, particularly resorcinol phosphates, have been successfully employed as flame retardant additives for PPO and HIPS blends [78–81]. In these blends, PPO is a good char former. With the presence of resorcinol phosphates in PPO resins, the char forming ability of PPO could be enhanced further due to possible interactions between these two compounds. Similar to PC resin, PPO resin was also reported to undergo Fries type rearrangement upon thermal decomposition, as shown in Figure 8.83 [46,82–85].

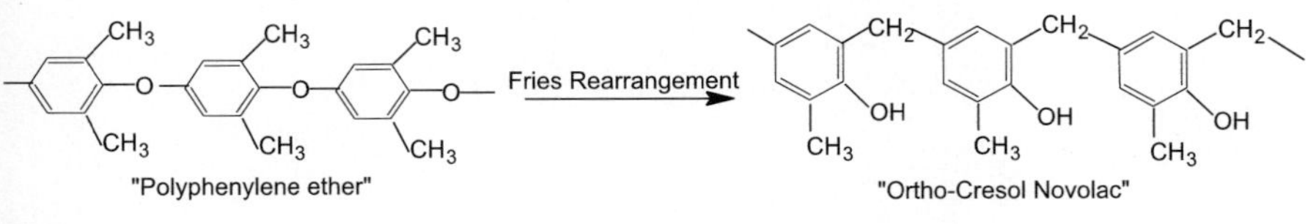


**Figure 8.83.**

Upon rearrangement, polyphenylene ether polymer is converted into phenolic novolac type compounds, and therefore contains more phenolic hydroxyl groups in the polymer chain. This could now enhance the interaction between hydroxyl groups and phosphate compounds. The *trans*-esterification reaction, which could take place with RDP additive, can cross-link the polymer chains and thereby, increase in the accumulation of phosphorus in the condensed phase (Figure 8.84) [46].

The possible reaction shown in Figure 8.84 strongly suggests that the presence of RDP in PPO/HIPS blends increases the char yield as well as accelerates the PPO char forming process. In this way, phosphorus containing compounds tend to accumulate in the char on the surface of burning polymer. Based on the mechanistic study made on RDP additive, it was concluded that the flame retarding action of RDP in the PPO/HIPS polymers was primarily a condensed phase flame retardant mechanism.

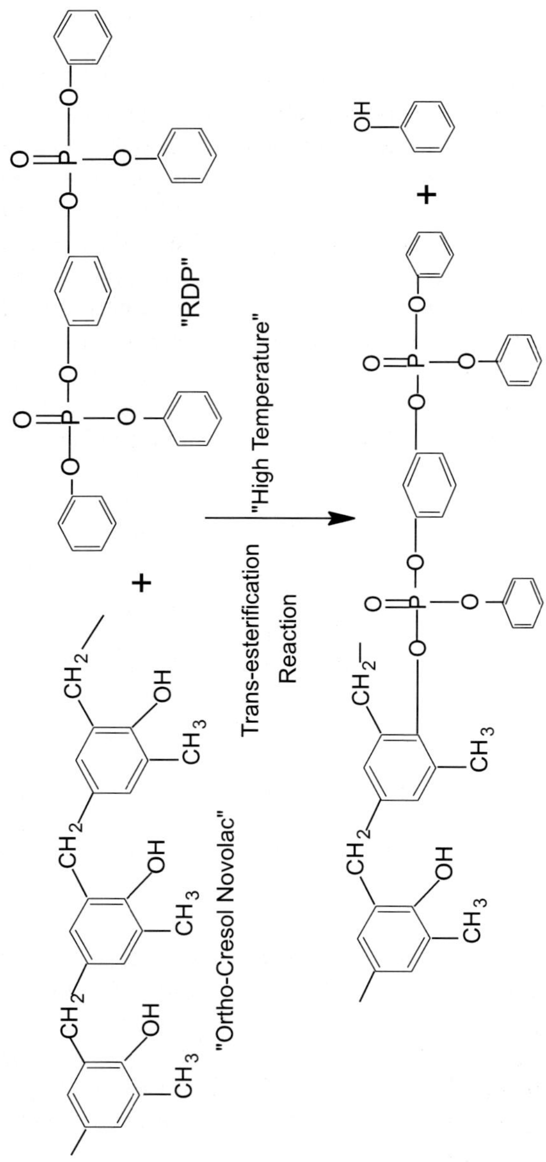
"RDP"
"Ortho-Cresol Novolac"
"High Temperature"
Trans-esterification
Reaction
CH2
OH
CH3
O
P
OH

**Figure 8.84.**

## 8.2.9 Flame Retardant Additives in Polyester Materials

Polyester materials, namely, poly(ethylene terephthalate), PET, and poly(butylene terephthalate), PBT, are commercially important polymers and have been used in the applications of films, fibers and molding resins. These polymers cannot resist ignition upon application of fire and may not self-extinguish, and therefore unsuitable in the applications of children's clothing, sleepwear, carpets, sheets and draperies. In an effort to find effective flame retardant additives for polyesters, resorcinol phosphate esters were found to be more suitable, particular for applications in textile fibers and films [86,87]. Table 8.40 presents the LOI results obtained on the use of resorcinol diphosphate in polyester compositions.

**Table 8.40.** Limiting oxygen index values of polyester systems containing resorcinol diphosphate ester

| Polyester/flame retardant | Flame retardant level (wt.%) | Limiting Oxygen Index (%) |
|---|---|---|
| Poly(ethylene terephthalate), PET | | |
| 1. PET (control) | | 17 |
| 2. PET/RDP | 5 | 17.7 |
| 3. PET/RDP | 10 | 19.4 |
| 4. PET/RDP | 15 | 21.9 |
| Poly(butylene terephthalate), PBT | | |
| 1. PBT (control) | | 16.5 |
| 2. PBT/RDP | 5 | 18.5 |
| 3. PBT/RDP | 10 | 18.8 |
| 4. PBT/RDP | 15 | 19.1 |

Data from [87]

The flammability properties of PET and PBT show that the increase in LOI values was directly related to the phosphorus content in the polyester composition.

Flame retardant mechanism of phosphate esters in polyester materials was examined from the results obtained by TGA and mass spectroscopic studies [88]. From the TGA analysis, a 36 wt. % residue obtained at 700 °C for the polyester and phosphate ester additive, corresponded to the composition of formula $(PO_3)_x$, which indicated that phosphate ester additive was operating strictly in the solid phase. The mechanism of flame retardant activity involved both a thermal stabilization of the base polyester and also, the formation of a barrier film on the burning surface which prevented the polyester oxidation and subsequent fuel formation [88].

PBT polyesters are thermoplastic polymers and used in various insulating parts in the electric and electronic industries, since they have good dimensional stability and processability properties. In order to improve the flame retarding properties, environmentally friendlier phosphate esters were the logical substitute for the toxic gases liberating halogenated flame retardants. Phosphate esters, in general, have limited solubility in crystalline polyesters, and therefore tend to migrate to polymer surface and drastically affect the polymer surface, and combustion performance. To overcome this problem, a high charring phenolic novolac resin was used in combination with resorcinol diphosphate compound in the polyester composition, and the results obtained are presented in Table 8.41 [89].

**Table 8.41.** Flame retardant properties of poly(butylene terephthalate), PBT

| Composition (wt.%) | 1 | 2 | 3 | 4 | 5 | 6 |
|---|---|---|---|---|---|---|
| 1. PBT (Valox 325) | 85 | 74.5 | 74.7 | 74.7 | 74.7 | 74.9 |
| 2. Fyroflex RDP | 15 | 15 | | 12.5 | 12.5 | |
| 3. Oligomeric RDP | | | 15 | | | |
| 4. Fyroflex BDP | | | | | | 15 |
| 5. Phenolic novolac (HRJ 12 700) | | 10 | 10 | | | 10 |
| 6. PC (Lexan 101) | | | | 12.5 | | |
| 7. PPO (BHPP 820) | | | | | 12.5 | |
| 8. PTFE (Teflon 6C) | | 0.5 | 0.3 | 0.3 | 0.3 | 0.1 |
| Flame retardant properties | | | | | | |
| UL-94, sample: 1/8″ | Fail | V-0 | V-0 | Fail | Fail | V-0 |
| UL-94, sample: 1/16″ | | V-0 | V-1 | Fail | Fail | V-2 |
| Plate out behavior (“juicing”) after molding | Yes | No | No | No | No | No |

Data from [89]

The addition of novolac resin was observed to help the RDP more compatible with the PBT and produced synergistic flame retardant effect. The combustion behavior of PBT was drastically changed in the presence of novolac. UL-94 test rating of V-0 can be achieved for 3.2 and 1.6 mm samples of PBT containing novolac and RDP additives, respectively, at 10 and 15 wt. % levels. In addition, the presence of novolac resin prevented the exudation of RDP to the polyester surface.

Due to limited compatibility, the use of phosphate esters in polyesters is limited. By adding an effective amount of a liquid crystalline polymer (LCP) additive, the compatibility of RDP with polyesters was improved, as shown in Table 8.42 [90].

The addition of LCP additive not only enhanced the polyester and RDP compatibility, but also highly improved the PC and RDP compatibility. From the scanning electron microscopic (SEM) micrographs, the PBT composition containing RDP and LCP binary additives showed a smoother morphology without any phase separation. It appeared that the LCP was acting as a surface modifier by improving the interfacial adhesion between the PBT and RDP.

**Table 8.42.** Melt Flow Index data of flame retarded thermoplastic polymers modified with liquid crystalline polymer (LCP)

| Thermoplastic blend system | Composition (PPH) | Melt Flow Index (g/10 min) | Increase, % |
|---|---|---|---|
| Poly(butylene terephthalate), PBT | | | |
| 1. PBT (control) | | 22 | |
| 2. PBT/RDP | 100/15 | 39 | 77 |
| 3. PBT/LCP | 100/5 | 30 | 36 |
| 4. PBT/RDP/LCP | 100/15/5 | 47 | 114 |
| 5. PBT/RDP/LCP | 100/3/3 | 32 | 45 |
| Poly(ethylene terephthalate), PET | | | |
| 1. PET (control) | | 22 | |
| 2. PET/RDP | 100/15 | 95 | 239 |
| 3. PET/RDP/LCP | 100/15/5 | 120 | 330 |
| Poly(carbonate), PC | | | |
| 1. PC (control) | | 7 | |
| 2. PC/RDP | 100/15 | 26 | 270 |
| 3. PC/RDP/LCP | 100/15/5 | 31 | 343 |
| 4. PC/RDP/LCP | 100/3/3 | 11 | 57 |

MFI for PBT and PET systems were measured at 260 °C at a load of 2.16 kg
MFI for PC systems were measured at 300 °C at a load of 1.2 kg
Data from [90]

### 8.2.10 Phosphorus–Nitrogen Synergism

In flame retardant systems, synergism is often defined as an effect greater than the additive effect, produced by a combination of two or more ingredients [1]. Studies have shown that phosphorus containing and nitrogen containing flame retardant materials, when used together, exhibited synergistic effects on the flame retarding properties. Though the phosphate esters have been enhancing the flame retardant properties of HIPS polymers, they tend to affect the impact resistance of HIPS. The addition of melamine cyanurate (MC) along with hydroxy-terminated resorcinol phosphate ester (shown in Figure 8.75) worked synergistically in the PPO/HIPS polymers by enhancing their flame retardant properties without much sacrifice in the impact resistance properties [91].

The results are presented in Table 8.43.

FT-IR studies showed that hydrogen bonding between the hydroxyl group of the resorcinol phosphate and amino group of MC was occurring, which suggested a strong interaction between these compounds in the polymer composition. From the cone calorimeter results, it was concluded that the presence of MC in the PPO/HIPS promoted the char formation, lowered the burning surface temperature and also, the heat release rate [91]. By combining the

**Table 8.43.** Flame retardant properties of mono(3-hdroxyphenyl) diphenyl phosphate and MC in poly(phenylene ether)/high impact poly(styrene) system (PPO/HIPS = 74/26 weight ratio)

| Properties | Flame retardants | | |
|---|---|---|---|
| | Control | MHDP | MHDP + MC |
| Flame retardant loading (parts) | 0 | 19 | 19 + 8 |
| Flame retardance, UL-94 | HB | V-1 | V-0 |
| Izod impact strength, kg cm cm$^{-1}$ | 14.4 | 14.3 | 11.2 |
| Melt flow rate, g/10 min | 0.4 | 5.5 | 5 |
| Vicat softening temp., °C | 124.4 | 102 | 102.8 |
| Average heat release rate (kW m$^{-2}$ (cone calorimeter) | 686 | 288 | 273 |
| Maximum surface temp., °C | | 465 | 335 |
| Residue at 550 °C, % | 1 | 3 | 9 |

MHDP = Mono(3-hydroxyphenyl) diphenyl phosphate
MC = Melamine cyanurate
Data from [91]

hydroxy-terminated resorcinol phosphate and MC, a completely halogen free flame retardant system, exhibiting outstanding flame retardancy with excellent melt flow and heat resistance, for the PPO/HIPS polymers was achieved.

A new class of flame retardants, namely polyphosphate ester morpholide, was synthesized using phenol, morpholine, phosphorus oxychloride and resorcinol, and was based on the following reaction scheme (Figure 8.85) [92–94].

The compound appeared as a liquid and contained 11.2 wt. % of phosphorus. Since the flame retardant property is directly related to the phosphorus content, for a given loading of flame retardant in the thermoplastic polymer, this morpholide compound can be expected to show better flammability properties.

Table 8.44 presents the flame retardant, melt flow and HDT properties of PC/ABS and PPO/HIPS resins blended with resorcinol based polyphosphate ester morpholide and diphosphate (RDP).

As can be seen in Table 8.44 results, for PC/ABS blend, improved melt flow, better heat resistance and shorter combustion time properties were observed for morpholide compound compared to RDP. With PPO/HIPS blend, on the other hand, in addition to the above improved properties, good flame retardancy was achieved with polyphosphate ester morpholide.

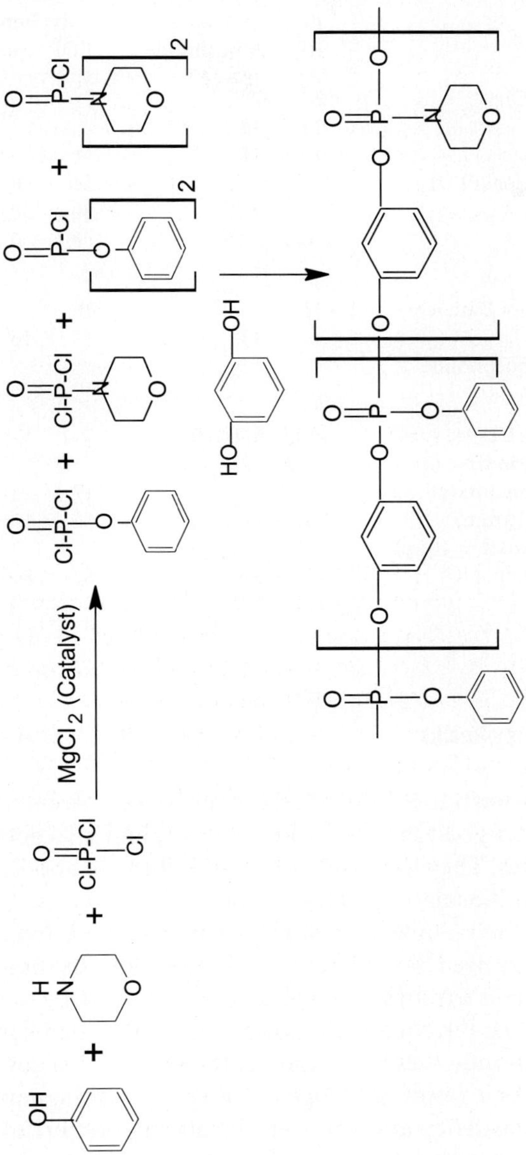


**Figure 8.85.** Reaction scheme for the preparation of polyphosphate ester morpholide

**Table 8.44.** Flame retardant properties of resorcinol based polyphosphate ester morpholide

| Composition | Polycarbonate system RDP | Polycarbonate system Morpholide | Polyphenylene ether RDP | Polyphenylene ether Morpholide |
|---|---|---|---|---|
| 1. Polycarbonate (PC) | 75 | 75 | | |
| 2. ABS resin | 13 | 13 | | |
| 3. SAN polymer | 12 | 12 | | |
| 4. Polyphenylene ether (PPO) | | | 50 | 50 |
| 5. HIPS | | | 50 | 50 |
| 6. Teflon 7AJ | 0.5 | 0.5 | 0.1 | 0.1 |
| Flame retardants | | | | |
| 1. RDP (CR-733S from Daihachi) | 13 | | 16 | |
| 2. Resorcinol based phosphate ester morpholide | | 13 | | 16 |
| Properties | | | | |
| 1. Flame retardancy, UL-94, 1/16″ | V-0 | V-0 | V-1 | V-0 |
| 2. Average combustion time (s) | 3.2 | 2.3 | 5.5 | 2.8 |
| 3. Max. single combustion (s) | 7 | 3 | 17 | 5 |
| 4. Melt flow rate (g/10 min) (temp. = 220 °C, load = 10 kg) | 22 | 27 | 13 | 17 |
| 5. Heat distortion temp. (°C) | 85 | 91 | 82 | 86 |

Data from [94]

## 8.2.11 Flame Retarded Epoxy Resins

Epoxy resins are known to exhibit excellent moisture, solvent, chemical, electrical resistance and good mechanical properties, and also good adhesion to variety of substrates. They have wide range of industrial applications, such as the composites, and insulating materials for electric devices. A major use for epoxy resins is in the manufacture of glass reinforced printed circuit boards. The most commonly used flame retardants for the epoxy resins are brominated epoxy resins based on tetrabromo bisphenol-A (TBBA). Due to the generation of toxic fumes during the combustion from the halogenated flame retardants, the trend to discontinue these materials in the epoxy resin compositions has been intensified. As a result, a halogen free epoxy resin composition showing good flame retardancy and sufficient insulating properties in the circuit boards and other encapsulation applications has been required. Resorcinol based phosphate ester compound (shown in Figure 8.78) having terminal hydroxyl groups was evaluated as reactive flame retardant for diglycidyl ether of bisphenol-A (DGEBA) [95].

The flame retardant properties of BHPP compound were compared against TBBA in the DGEBA resin system (Table 8.45).

**Table 8.45.** Flame retardant properties of cured bisphenol-A epoxy resin with bis(3-hydoxyhenyl) phenyl phosphate and tetrabromo-bisphenol-A

| Curing system | P/Br (%) | Tg (°C) | Char yield, % (Air) | | Flammability results | | | | |
|---|---|---|---|---|---|---|---|---|---|
| | | | 600 °C | 700 °C | Burn time (s) | Drip | Visible smoke | UL-94 rating | LOI (%) |
| 1. BPA/PN | 0 | 121 | 21 | 2 | 87 | No | Slight | V-2 | 26 |
| 2. BHPP/PN (25/75) | 0.75 | 111 | 30 | 23 | 42 | No | Slight | V-2 | 28 |
| 3. BHPP/PN (50/50) | 1.51 | 102 | 32 | 27 | 7 | No | No | V-0 | 31 |
| 4. BHPP/PN (75/25) | 2.25 | 96 | 36 | 32 | 0 | No | No | V-0 | 33 |
| 5. TBBA/PN (25/75) | 5.1 | 128 | 23 | 0 | 38 | No | Heavy | V-2 | 28 |
| 6. TBBA/PN (50/50) | 10.2 | 121 | 25 | 2 | 18 | No | Slight | V-1 | 30 |
| 7. TBBA/PN (75/25) | 15.3 | 113 | 27 | 6 | 0 | No | No | V-0 | 35 |

Bisphenol-A epoxy = EEW = 189, BPA = Bisphenol-A
PN = Phenolic novolac resin from Schenectady Chemical, HRJ-2210 (OH Equivalet = 104)
BHHP = Bis(3-hydroxyphenyl) phenyl phosphate
TBBA = Tetrabromo bisphenol-A
Data from [95]

From Table 8.45 results, the curing system containing 2.25 wt. % phosphorus (BHPP/PN = 75/25) or 15.3 wt. % bromine (TBBA/PN = 75/25) showed V-0 ratings in UL-94 test, which suggests that phosphate ester is a better flame retardant than the brominated material. In the cured epoxy resin, BHPP produced no fumes and toxic gas emission during the combustion when compared to TBBA. Also, BHPP based epoxy system produced higher char yields at high temperature than TBBA, which indicated that phosphorus containing flame retardants are better for the encapsulation applications.

A halogen free flame retarded insulating epoxy resin composition, containing resorcinol diphosphate, displayed flame retardant and insulation properties, which after the formation of an insulating layer in the circuit board, withstood a plating process and did not cause any bleed out in the insulation layer [96].

### 8.2.12 Flame Retardancy in Polyurethanes

Fire retardant materials have often been used to improve the flammability characteristics of rigid and flexible polyurethane materials. The use of RDP in a flexible polyurethane foam not only improved the flame retardant property, but also reduced the fogging phenomena associated with the liberation of volatile products from the foams used in the car interiors [97, 98].

### 8.2.13 Summary and Outlook

The application of flame retardants in the thermoplastic and other polymeric products are beneficial in protecting these flammable materials against fire, and also by preventing or delaying fires, which in turn could make life, health and property are effectively protected. Greater potential exists for organo phosphorus compounds to play increasingly greater role in the flame retardant applications. The success for the new flame retardants depends on their cost-performance effectiveness as well as environmentally friendlier materials. Since, unique molecular structures and properties could be obtained with resorcinolic flame retardants; new compound synthesized from the resorcinol chemistry could offer valuable processing and flame retardant properties to thermoplastic polymers.

## References


1. A. F. Grand and C. A. Wilkie, Fire retardancy of polymeric materials, Marcel Dekker, New York (2000).
2. W. Lyons, The chemistry and uses of fire retardants, Wiley, New York (1970).
3. F. Cullis and M. Hirschler, The combustion of organic polymers, Clarendon, Oxford (1981).
4. M. Aseeva and G. Zaikov, Combustion of polymer materials, Carl Hanser, New York (1985).
5. R. Gann, R. Dipert and M. Drews, Flammability, In: Encyl. Polym. Sci. Eng. Vol. 7, p 154, Wiley, New York (1987).
6. G. Bertelli et al., Performance and mechanisms of fire retardants, In: Polymers – Review, Polym. Degrad. Stab. Vol. 20, p 271 (1988).
7. J. Troitzsch, J. Prog. Org. Coatings, Vol. 11, p 41 (1983).
8. E. Flick, Plastics Additives, Section 1X, Fire and Flame Retardants, p 213, Noyes Publications, Park Ridge, NJ (1986).
9. J. Troitzsch, Chemistry today, Vol. 16, Jan (1998).
10. J. Troitzsch, Int. Plast. Flammability Handbook, Hanser, Munich (1990).
11. C. Fennimore, Candle-like test for flammability of polymers, In: M. Lewin, S. Atlas and E. Pearce, Eds., Flame retardant polymeric materials, Vol. 1, Plenum, New York (1975).
12. G. Barrett, US Patent, 2,520,090 (1947).
13. I. Schumacher, US Patent, 3,965,220 (1976).
14. A. Morgan, I. Schumacher and W. Vaderlinde, US Patent, 4,062,909 (1977).
15. A. Morgan, I. Schumacher and W. Vaderlinde, US Patent, 4,086,302 (1978).
16. A. Morgan, I. Schumacher and W. Vaderlinde, US Patent, 4,144,394 (1979).
17. L. Gunkel and H. Barda, Euro. Pat. Appln, 0,485,807 A2 (1992).
18. L. Gunkel and H. Barda, US Patent, 5,281,741 (1994).
19. S. Kwata et al., US Patent, 5,616,768 (1997).
20. J. Stults, US Patent, 5,756,798 (1998).
21. N. Tokuyasu and S. Nakamura, US Patent, 6,204,404 (2001).
22. B. Brady, D. Bright and F. Schafer, US Patent, 5,457,221(1995).
23. D. Bright and R. Pirelli, US Patent, 5,728,859 (1998).
24. D. Bright and R. Pirelli, US Patent, 5,508,462 (1996).
25. P. Horner and R. Rashbrook, US Patent, 4,134,876 (1979).
26. D. Bright and R. Pirelli, US Patent, 5,420,327 (1995).
27. T. Matsumura and Y. Tanaka, Euro. Patent, 0,509,506 B1 (1997).
28. K. Hirao, S. Oda and Y. Omada, US Patent, 5,769,824 (1997).
29. S. Kwata, K. Hirao and K. Noguchi, US Patent, 5,880,308 (1999).
30. S. Kwata, K. Hirao and K. Noguchi, Euro. Patent, 0,822,197 B1 (2003).
31. C. Wang and J. Shieh, Eur. Polym. J., Vol. 36, p 443 (2000).
32. R. Henrich et al., Int. J. Toxicol., Vol. 19, p 243 (2000).
33. B. Ryan et al., Int. J. Toxicol., Vol. 19, p 257 (2000).
34. R. Sherwood et al., Int. J. Toxicol., Vol. 19, p 265 (2000).
35. R. Frendenthal et al., Int. J. Toxicol., Vol. 19, p 233 (2000).
36. D. Bright, S. Dashevsky, P. Moy and K. Tu, ANTEC Conference Proceedings, Atlanta, GA (1998).
37. S. Levchik et al., Applications and mode of fire retardant action of aromatic phosphates, In: S. Al-Malaika, A. Golovoy and C. Wilkie, Eds., Specialty polymer additives, Blackwell Publishing, USA (2001).

38. D. Bright, S. Dashevsky, P. Moy and B. Williams, J. Vinyl & Additive Technol., Vol. 3, p 170 (1997).
39. J. Green, ANTEC Conference Proceedings, p 3544 (1995).
40. P. Moy, J. Vinyl & Additive Technol., Vol. 4, p 216 (1988).
41. E. Murasko et al., J. Fire Sci., Vol. 16, p 278 (1998).
42. K. Otsuki, Euro. Patent Appln, 0,947,547 A1 (1999).
43. K. Otsuki, US Patent Publn, 2001 /0,049,407 A1 (2001).
44. K. Otsuki, US Patent, 6,462,112 (2002).
45. B. Jang et al., US Patent, 6,646,032 (2003).
46. E. Murasko et al., J. Fire Sci., Vol. 16, p 233 (1998).
47. T. Matsumura and Y. Tanaka, Euro. Patent Appln, 0,509,506 A2 (1992).
48. H. Nishihara et al., US Patent, 5,278,212 (1994).
49. E. Ueda et al., US Patent, 6,083,428 (2000).
50. A. Adedeji et al., US Patent, 6,486,244 (2002).
51. A. Adedeji et al., US Patent, 6,258,879 (2001).
52. M. Weber et al., US Patent, 6,197,869 (2001).
53. A. Landa and J. Keulen, US Patent, 6,583,205 (2003).
54. S. Vilasagar and H. Rawlings, US Patent, 5,302,646 (1994).
55. S. Ogoe, N. Schrock and J. Spoon, US Patent, 5,663,280 (1997).
56. T. Eckel, D. Wittmann, M. Oller and H. Alberts, US Patent, 5,672,645 (1997).
57. M. Kobayashi, US Patent, 5,994,433 (1999).
58. J. Barrer et al., US Patent, 6,084,054 (2000).
59. J. Bastiaens et al., US Patent, 6,174,942 (2001).
60. J. Bastiaens, J. Campbell and Govaerts, US Patent, 6,204,313 (2001).
61. J. Campbell et al., US Patent, 6,228,912 (2001).
62. A. Nodera, US Patent, 6,384,114 (2002).
63. J. Gosens, C. Pratt, H. Savenije and C. Classen, US Patent, 5,204,394 (1993).
64. P. Moy, Paper Presented at Annual BCC Conference on Flame Retardancy, May (1993).
65. T. Eckel, D. Whitmann, M. Oller and H. Alberts, US Patent, 5,672,645 (1997).
66. S. Kim, Y. Choi and J. Song, US Patent, 5,864,004 (1999).
67. T. Eckel, D. Whitmann, J. Stolting and M. Zobel, US Patent, 6,583,204 (2003).
68. T. Eckel, M. Zobel, B. Keller and D. Whitmann, US Patent, 6,596,794 (2003).
69. K. Fuhr, T. Eckel, D. Whitmann and H. Alberts, US Patent, 5,741,838 (1998).
70. K. Fuhr, T. Eckel, D. Whitmann and H. Alberts, US Patent, 5,658,974 (1997).
71. K. Fuhr, T. Eckel, D. Whitmann and H. Alberts, US Patent, 5,621,029 (1997).
72. E. Murasko, G. Levchik, S. Levchik, D. Bright and S. Dashevsky, J. Appl. Polym. Sci., Vol. 71, p 1863 (1999).
73. J. Green, J. Fire Sci., Vol. 14, p 426 (1996).
74. J. Hill, US Patent, 6,075,158 (2000).
75. R. Vaia, H. Ishii and E. Giannelis, Chem. Mater. Vol. 5, p 1164 (1993).
76. E. Weil, W. Zhu, N. Patel and S. Mukhopadhyay, Polym. Degrad. Stab., Vol. 54, p 125 (1996).
77. K. Lee et al., Polym. Degrad. Stab., Vol. 81, p 173 (2003).
78. E. Weil, Conference Proceedings on Recent Advances in Flame Retardancy of Polymeric Materials, Vol. 1, p 1, Stamford, CT (1990).
79. J. Green, Conference Proceedings on Recent Advances in Flame Retardancy of Polymeric Materials, Vol. 2, p 1, Stamford, CT (1991).
80. J. Green, Conference Proceedings on Recent Advances in Flame Retardancy of Polymeric Materials, Vol. 3, p 162, Stamford, CT (1992).
81. K. Takeda et al., J. Appl. Polym. Sci., Vol. 64, p 1175 (1997).

82. A. Boscolletto et al., J. Appl. Polym. Sci., Vol. 67, p 2231 (1998).
83. A. Factor, J. Polym. Sci., Part A-1, Vol. 7, p 363 (1969).
84. J. Jachowicz, M. Kryszewski and P. Kowalski, J. Appl. Polym. Sci., Vol. 22, p 2891 (1979).
85. J. Jachowicz, M. Kryszewski and M. Mucha, Macromolecules, Vol. 17, p 1315 (1980).
86. R. Rashbrook, US Patent, 4,203,888 (1980).
87. R. Stackman, US Patent, 3,987,008 (1976).
88. R. Stackman, Ind. Eng. Chem. Prod. Res. Dev. Vol. 21, p 332 (1982).
89. S. Levchik, D. Bright and G. Alessio, PCT Appln, WO 00/49077 (2000).
90. K. Kim, S. Dashevsky, A. Aaronson and J. Burk, US Patent, 5,679,288 (1997).
91. H. Nishihara, S. Tanji and R. Kanatani, Polym. J., Vol. 30, p 163 (1998).
92. J. Lim, K. Seo and S. Yang, Euro. Patent Appln, 1,209,163 (2001).
93. J. Lim, K. Seo and S. Yang, US Patent Appln, 2002/0,077,392 (2002).
94. J. Lim, K. Seo and S. Yang, US Patent Appln. 2003/0,040,560 (2003).
95. C. Wang and Y. Shieh, Euro. Polym. J., Vol. 36, p 443 (2000).
96. N. Cooray, K. Tsukamoto and T. Ishitsuka, US Patent Publn, 2003/0,211,328 (2003).
97. C. Blundell and J. Wuestenenk, US Patent, 5,958,993 (1999).
98. G. Fesman and A. Aaronson, US Patent, 5,547,614 (1996).

# 9 Resorcinol Chemistry in Pharmaceuticals Applications

## 9.1 Introduction

Pharmaceutical companies have been actively involved in the research and development of drug compounds for the treatment and prevention of various human illnesses. With an increasing elderly population consuming many drugs, drug companies are speeding up their innovations, developments and introduction of new products to meet the demands. In the drug discovery efforts, new technologies offer both challenges and opportunities. In the recent years, research and developmental efforts made in the pharmaceutical chemistry have led to the development of many new innovative technologies. Innovations in medicines often provide many new therapeutic choices and opportunities in terms of prevention, diagnosis and treatment of diseases. Driven by the increase in knowledge in the areas of biotechnology, genomics and improved research and development tools and techniques, the role of natural sciences in the drug development has been constantly changing.

Resorcinol possesses unique structural characteristics for developing drug compounds exhibiting chemical structures and properties suitable for treating various human diseases. Chemical compounds derived from the resorcinolic derivatives can be potent and effective medicines. Historically, the active ingredients of various plant extracts contained resorcinolic derivatives or structures, and were primarily used in the treatment of severe diseases such as cancer, osteoporosis, and gastric ulcers and so on. For example, the daidzein present in the soya extract was found to be effective in the treatment of osteoporosis and symptoms of postmenopausal women. Similarly, isoliquiritigenin isolated from the tonka bean and licorice root was effective in inhibiting cancer formation and the treatment of gastric ulcers. Now, due to advancement in chemical technology, synthetic procedures have been developed to synthesis and produce commercially both daidzein and isoliquiritigenin compounds directly from resorcinol and its derivatives.

Resorcinol chemistry has the potential to introduce new and improved active drug compounds, which can lead to quicker development of effective medicines and faster solutions for the emerging diseases. The new technologies and therapies based on resorcinol chemistry may have good potential to bring significant benefits to patients. Considerable efforts are now underway in the major pharmaceutical companies' R&D to use resorcinolic derivatives in developing new drugs to cure and treat severe and life threatening diseases.

Resorcinol derivatives have been used by the major drug companies to develop new drugs, and also for the manufacture of existing therapeutic compounds. Some of the most important derivatives used in the drug developments are schematically illustrated in the following figure (Figure 9.1).

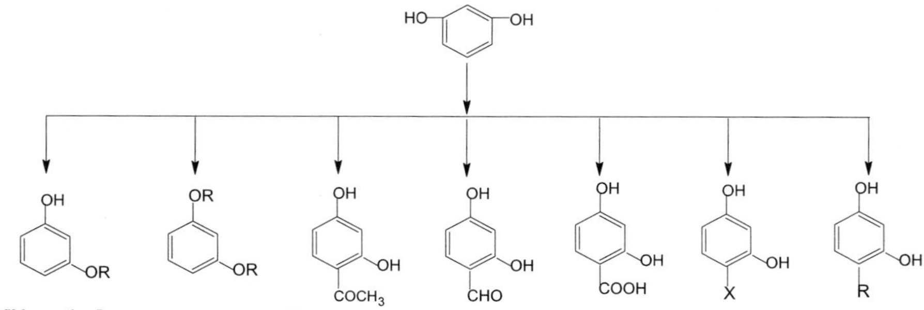


**Figure 9.1.** Resorcinol derivates used in the development of pharmaceutical drug compounds

In this chapter, various drug compounds developed based on the above resorcinolic derivatives have been discussed. In addition, the synthetic procedures used in the preparation of some important and most useful resorcinol derivatives and drug compounds, obtained from the literature, have been presented at the end of this chapter.

## 9.2 Alcoholism

Alcoholism is a chronic disease and can lead to serious physical, mental and social illnesses. When alcohol (ethanol) is consumed, the metabolism of alcohol takes place in the liver in two stages. In the first step, the alcohol dehydrogenase (ADH) enzyme converts ethanol into acetaldehyde, which is a highly toxic substance. The second step of ethanol metabolism is catalyzed by acetaldehyde dehydrogenase-2 (ALDH 2) enzyme, which converts acetaldehyde into acetic acid. Acetic acid is non-toxic and can be readily used by human body to produce energy. These two enzyme process occur gastrally as well as hepatically, with the liver supplying the co-enzyme nicotinamide-adenine-dinucleotide (NAD+) required in the process. In the absence of NAD+ co-enzyme, the ethanol cannot be metabolized. A large quantity of ethanol is converted into acetaldehyde in the presence of NAD+ co-enzyme and ADH enzyme. The excess acetaldehyde thus produced enters the blood stream and brain. Since acetaldehyde is a toxic compound, this can cause headaches, nausea, vomiting, diarrhea, upset-stomach, dizziness and confused consciousness.

Alcohol addicts have increased gastric and hepatic dehydrogenase and NAD+, and the constant alcohol consumption forces the liver to convert NADH to NAD+ on a larger scale. Higher the production of NAD+ results in more conversion of ethanol into acetaldehyde. This can be avoided or reduced if ethanol is retained in the stomach for a longer period of time, so that this can be converted into $CO_2$ and $H_2O$ by gastric metabolism. Daidzein and daidzin, having the following structural formulae, exhibited an inhibition action on alcohol dehydrogenase enzyme to slow the conversion of ethanol into acetaldehyde [1] (Figure 9.2).

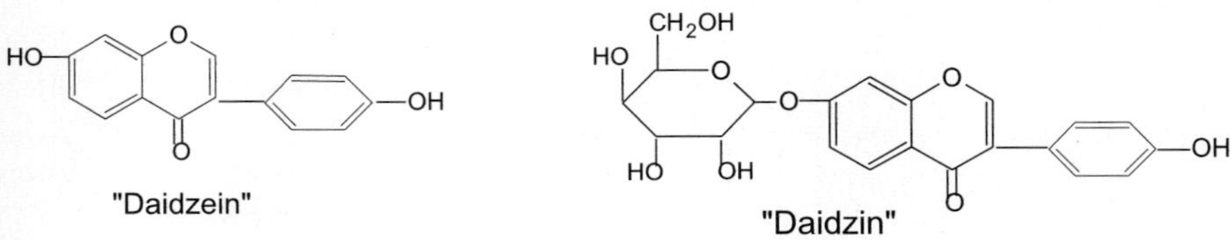


**Figure 9.2.**

Historically, daidzein and daidzin have been obtained from the "kudzu" plant as an extract of kudzu blossom, and used to inhibit alcohol and acetaldehyde dehydrogenase enzymes. Daidzin is the major active component in extracts of "radix puerariae", a traditional Chinese medication that suppresses the ethanol intake of Syrian golden hamsters. Now daidzein can be synthesized directly from resorcinol, and daidzin from daidzein from the following reactions [2,3] (Figures 9.3 and 9.4).

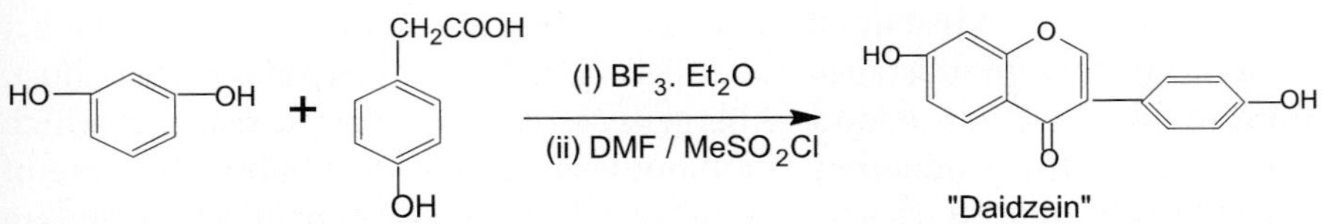


**Figure 9.3.**

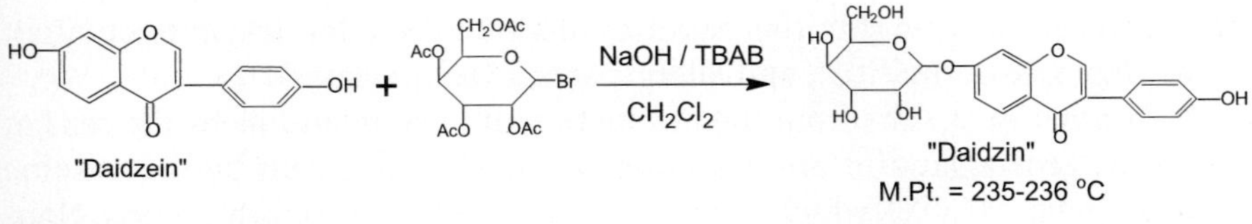


**Figure 9.4.**

Daidzin's antidipsotropic activity has been studied, and the results showed that this compound was effective an agent to reduce alcohol consumption [4–6]. The mechanism of action of daidzin's antidipsotropic property was associated with the inhibition of ALDH-2 enzyme, and its interaction with neurotransmitter enzyme systems. Daidzein and daidzin analogs were effective in increasing the concentration of aldehydes formed during the catabolism of certain neurotransmitters, and this increase in aldehyde amount was effective in reducing the alcohol consumption [7–9].

Daidzin analogs containing the carboxylic acid groups were synthesized from the reaction of daidzein and alkyl-halide carboxylic acid derivatives [7,8]. These compounds were used to determine their structure property relationship on the inhibiting of ALDH-1 and ALDH-2 enzymes (Figure 9.5).

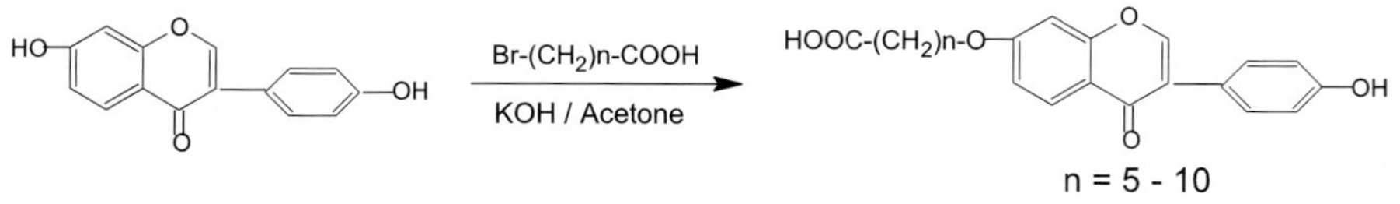


**Figure 9.5.**

Antidipsotropic activity of the carboxyl derivatives of 7-O-substituted daidzein showed that these derivatives were better ALDH-2 inhibitors than daidzein and daidzin [7].

## 9.3 Allergy and Inflammation

Allergy refers to the reaction by immune system in response to bodily contact with certain foreign substances called "allergens", which includes pollens, dust, mite, molds, dander, and foods. Allergens can cause an allergic reaction, which may result in the production of antibodies or immunoglobulins. Because of this, the cells, due to irritation, tend to release various chemicals including histamine. These chemicals, in turn, cause inflammation and exhibit typical allergic symptoms. The parts of body that are prone to react to allergies include eyes, nose, lungs, skin and stomach. Some of the common allergic disorders are hay fever (allergic rhinitis), asthma, allergic eyes (allergic conjunctivitis), eczema (atopic dermatitis), and allergic shock (anaphylaxis).

Inflammation results from the production of prostaglandins by the cells of the body. Prostaglandins are produced within the body cells by the enzyme cyclooxygenase (COX), which can promote inflammation, pain and fever. Non-steroidal anti-inflammatory drugs (NSAIDs) are used primarily to treat the inflammation. These non-steroidal drugs are expected to block the cyclooxygenase enzymes and reduce prostaglandins production throughout the body. NSAIDs can reduce both pain and inflammation.

Resorcinol derivatives have been employed in the synthesis of different non-steroidal type organic compounds having anti-allergic and anti-inflammatory properties. 1,3-Dimethoxyaniline was used as the starting material in the synthesis of benzobisoxazinetetrones which showed anti-allergic activity when administered topically or intra nasally [10]. These compounds were acting as anti-allergic agents by blocking the histamine release from passively sensitized rat mat cells (Figure 9.6).

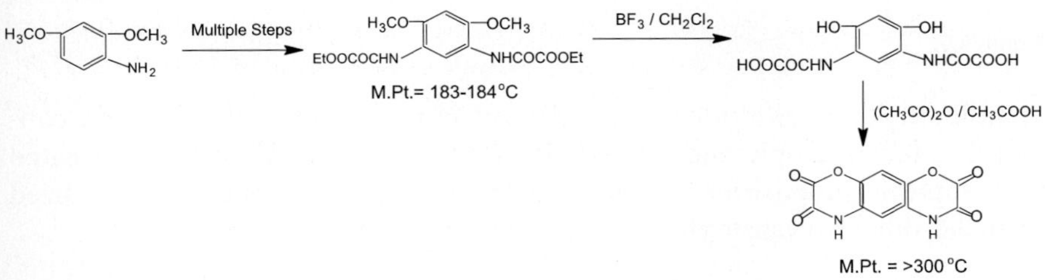


**Figure 9.6.**

Compounds having the following general structures were synthesized using 2,4-dihydroxy-3-propylacetophenone as the raw material in a multi-step reaction [11–16]. These drug compounds were found to be useful in the treatment of allergies and inflammatory conditions (Figures 9.7 and 9.8).

Isoflavone derivatives possessing anti-inflammatory and provitaminic properties in regard to vitamin P (or C2), and suited for the treatment of vascular

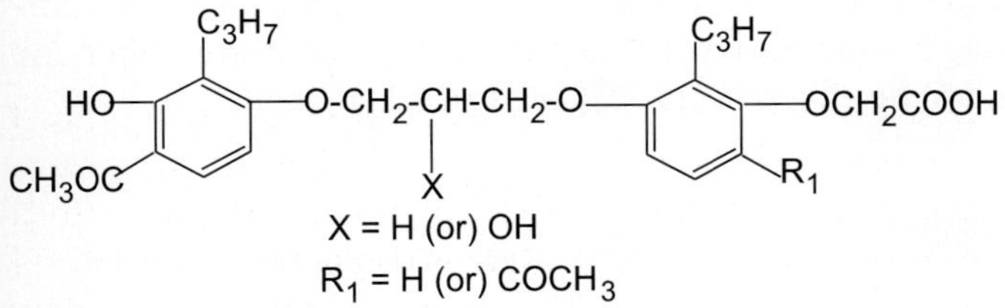


**Figure 9.7.**

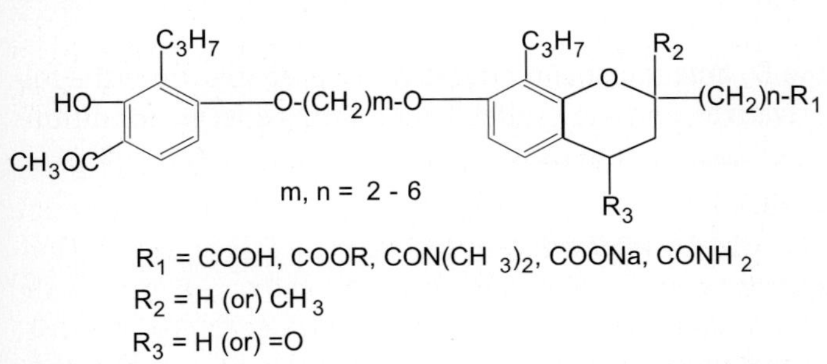


**Figure 9.8.**

disorders and P-hypovitaminosis conditions such as blood vessel rupture, and heightened blood pressure, were prepared based on the following reaction scheme [17] (Figure 9.9).

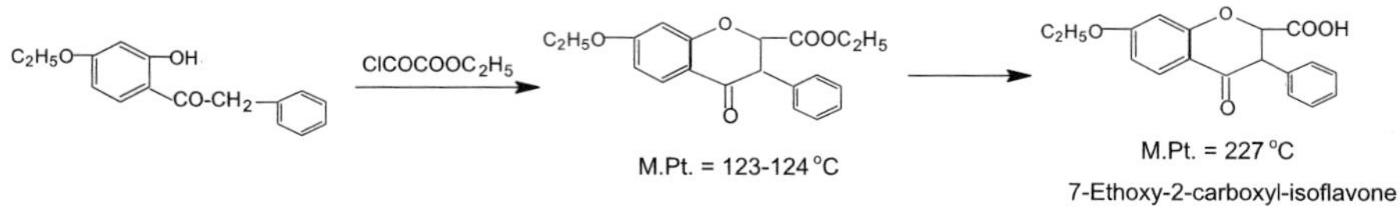


**Figure 9.9.**

Inflammatory conditions associated with chronic dermatitis, psoriasis, contact dermatitis, atopic and seborrheic dermatitis, and acne could be treated with topical compositions containing the chalcone compound synthesized from 2,6-dihydroxyacetophenone [18] (Figure 9.10).

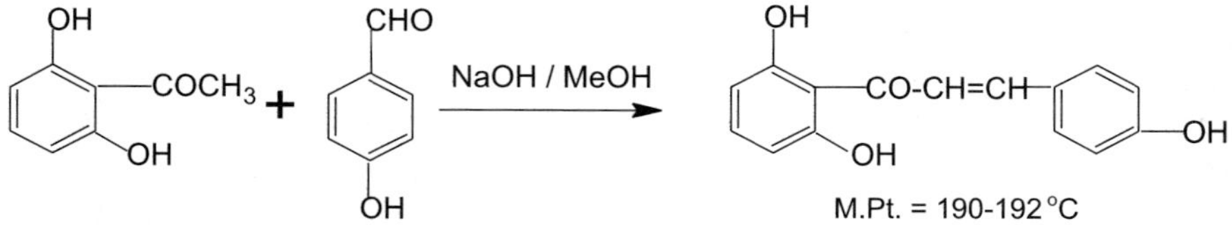


**Figure 9.10.**

For treating allergic conditions such as asthma, hay-fever, rhinitis, and skin allergies, and inflammatory conditions such as contact dermatitis, burns, mucosa of the eyes, lips, mouth, and genital or anal regions, the drug compounds prepared using 2,4-dihydroxy-3-(3,3,3-trifluoropropyl)-acetophenone derivative were found to be effective [19–21] (Figure 9.11).

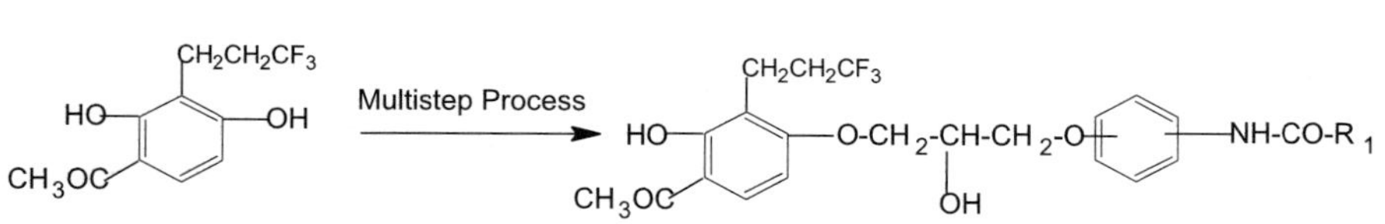


**Figure 9.11.**

Several leukotriene $B_4$ antagonist compounds were prepared from the following resorcinol derivatives, and were suitable for treating allergic conditions such as psoriasis, bowel diseases and asthma [22–24] (Figure 9.12).

## 9.4 Alzheimer's Disease

Alzheimer's disease (AD) is a progressive disease of the human brain that affects memory, thinking, behavior and emotions. This disease may be caused

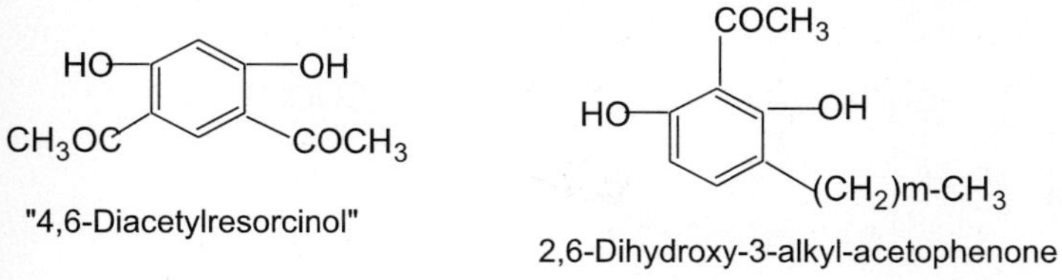


**Figure 9.12.**

by the abnormal deposition of $\beta$-amyloid protein or peptide in the brain which leads to the loss of nerve or neuron cells. The loss of nerve cells in the brain area causes deficits in the neurotransmitters, which are the brain's chemical messengers. Studies have shown that a major component of the pathology of Alzheimer's disease is chronic inflammation [25, 26]. The inflammatory component of Alzheimer's disease is the action of cytosolic phospholipase $A_2$ ($PLA_2$) enzyme in the production of phospholipids metabolites called eicosanoids. By inhibiting phospholipase $A_2$ activity, the inflammation, and in turn Alzheimer's disease, can be treated or prevented. Administration of non-steroidal anti-inflammatory drugs appeared to slow the advancement of Alzheimer's disease.

Chemical compounds (shown below) that appeared to inhibit the actions of cytosolic phospholipase $A_2$ enzyme were prepared using resorcinol dimethyl ether and 2,4-dihydroxyacetophenone derivatives and found useful to treat Alzheimer's disease [27,28] (Figures 9.13, 9.14 and 9.15).

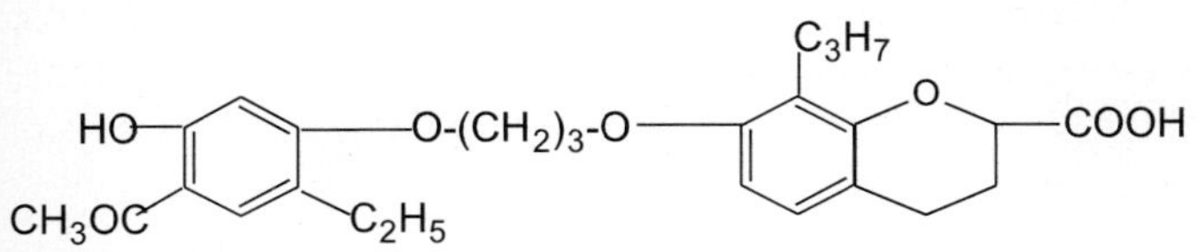


**Figure 9.13.**

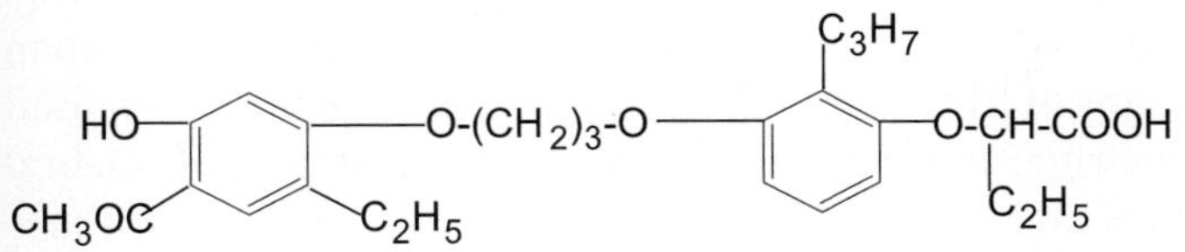


**Figure 9.14.**

Isoflavanoid compounds have shown significant estrogenic activity, and may act in the brain to enhance neurotransmission and restore synaptic density. Alzheimer's disease may be strongly associated with decreased choline acetyl transferase activity and the loss of cholinergic neurons, which appears to be essential for learning and memory processes. Phytoestrogen isoflavone compounds such as daidzein can be useful for up-regulating choline acetyl

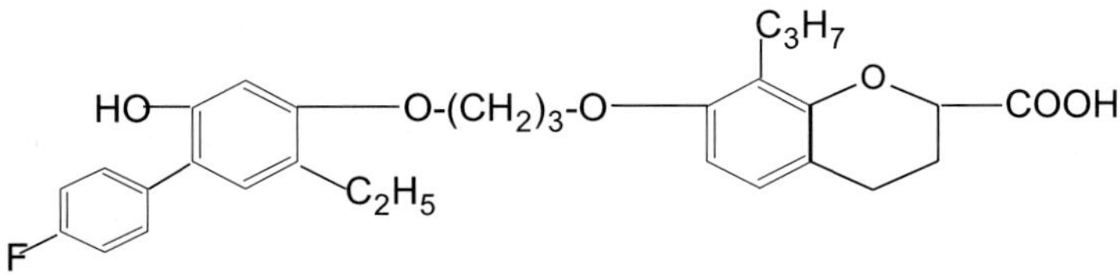


**Figure 9.15.**

transferase and nerve growth factor in the brain, and therefore may be effective to relieve the symptoms of Alzheimer's disease or related dementia in human. Resorcinol based isoflavanoid compounds such as daidzein, formononetin and equol were suggested for the treatment or prevention of Alzheimer's dementia and reduced cognitive functions associated with advancing age [29, 30].

Centchroman is a resorcinol derived compound and has been marketed in India under the trade names Centron and Saheli (Figure 9.16).

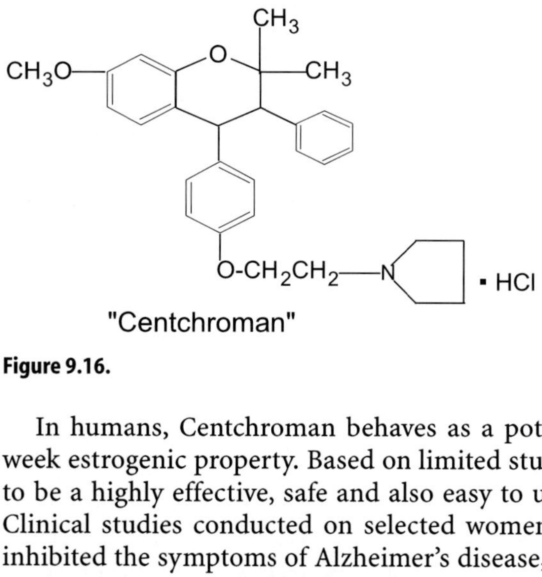


**Figure 9.16.**

In humans, Centchroman behaves as a potent antiestrogen and also has week estrogenic property. Based on limited studies, this compound appeared to be a highly effective, safe and also easy to use as oral contraceptive drug. Clinical studies conducted on selected women indicated that Centchroman inhibited the symptoms of Alzheimer's disease, and therefore this compound can be used in the pharmaceutical compositions for the treatment of cerebral degenerative disorders [31].

## 9.5 Anaemia

Sickle-cell anaemia is an inherited blood disease of childhood which can cause lots of pain, damage vital organs, stroke, proliferative retinopathy and cerebral thrombosis. This disease affects a protein inside the red blood cells called hemoglobin, which carries oxygen from the lungs and takes it to every part of the body. The red blood cells of sickle cell disease appeared to be shaped like sickles instead of the normal round shape. These cells tend to become trapped and destroyed in the liver and spleen, which may result in a shortage of red blood cells and is called anaemia [32,33].

For the treatment of sickle-cell anaemia, resorcinol derivative based compounds have been developed and tested [34,35] (Figure 9.17).

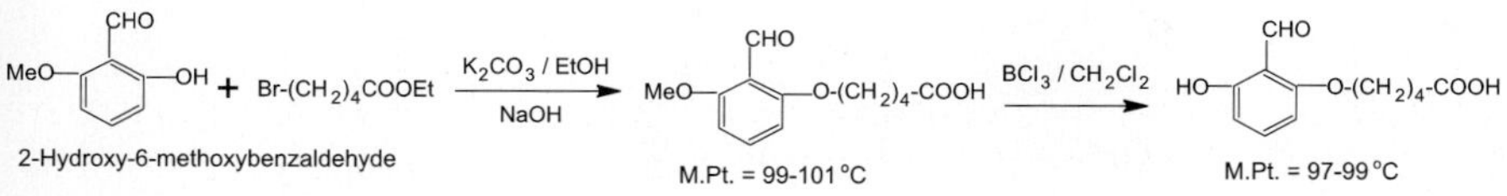


**Figure 9.17.**

These compounds were found to be useful medicines in the palliation of hemoglobinopathies in sickle cell anaemia and pulmonary dysfunction.

## 9.6 Antiseptic, Bactericide and Germicide

An antiseptic is a substance which destroys microorganisms that carry the disease and does not harm the body tissue. Similarly, bactericide is a drug or chemical agent that destroys the bacteria or inhibits bacterial growth.

Alkyl and aralkyl substituted resorcinol have shown remarkable antiseptic and germicidal properties [36]. Resorcinol derivatives possessing these properties, in general, have the following chemical structures (Figure 9.18).

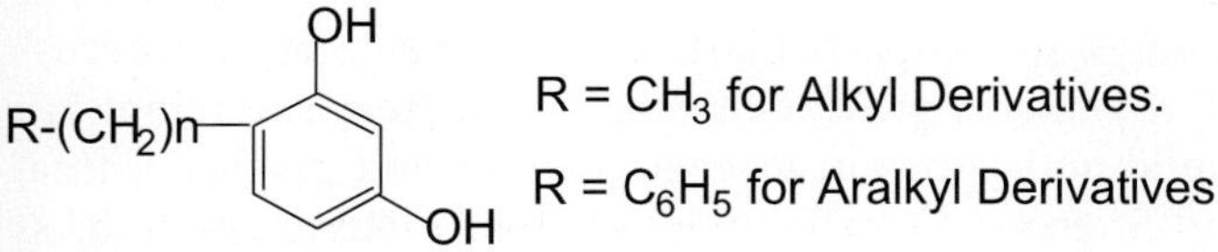


**Figure 9.18.**

In the case of aralkyl compounds, the phenol coefficient, which is an indicator of germicidal activity of a compound, indicated the values of 23, 41 and 31 respectively for $n = 1$, 2 and 3, suggesting that 2,4-dihydroxydiphenyl ethane derivative has the maximum germicidal strength. Hexylresorcinol had shown the maximal germicidal action (phenol coefficient = 45) among the mono-alkyl derivative of resorcinol. The di-alkyl substituted resorcinol, on the other hand, showed phenol coefficients several times higher than that of a mono-alkyl derivative indicating stronger germicidal property [36].

The antiseptic power of mono-alkyl substituted resorcinol derivatives may be associated with their ability to lower surface tension, which can make these derivatives penetrate more easily through the membranes. In fact, the excellent germicidal action exhibited by the resorcinol derivatives are due to a combination of their physical ability to lower surface tension and chemical reactivity associated with the two hydroxyl groups [36, 37].

Various mono-alkyl resorcinol derivatives having useful antiseptic and germicidal properties were prepared by the reaction of resorcinol with alkyl alcohols or aliphatic carboxylic acids, and are based on the following reaction schemes [38–44] (Figures 9.19 and 9.20).

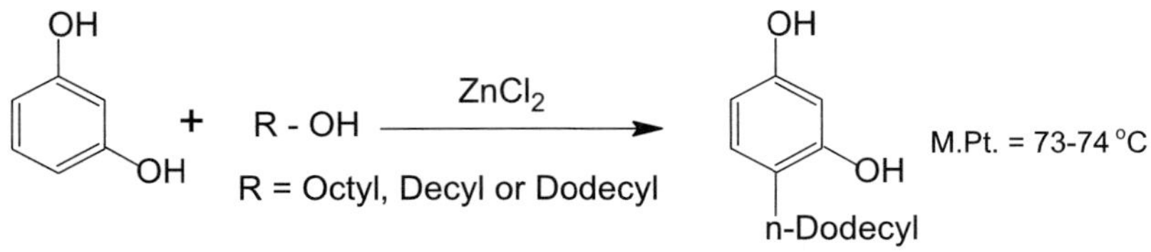


**Figure 9.19.**

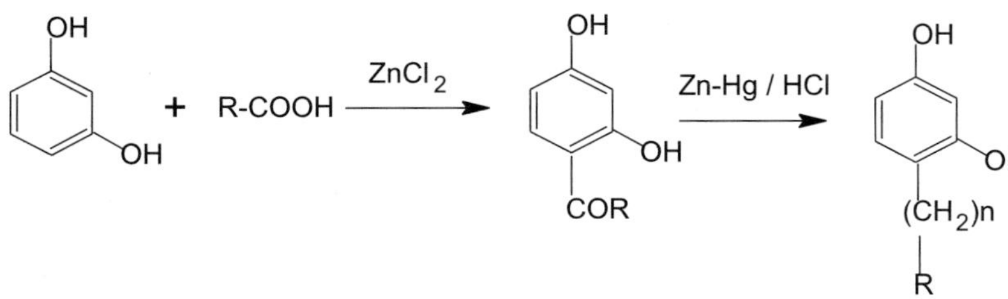


**Figure 9.20.**

4-*n*-Dodecylresorcinol exhibited highly antiseptic and germicidal actions in dilute aqueous solutions, and particularly effective on "Staphylococcus" and "Streptococcus" organisms present in oral cavity. Hexylresorcinol has been employed as mild antiseptics in lozenges, sprays and gargles. Throat Plus (Beechams) and Strepsils Extra lozenges contain *n*-hexylresorcinol as the antiseptic component. The bacterial effect of hexylresorcinol was assessed using bacterial aggregates, and showed highly effective in the destruction of bacteria [45]. Drugs containing hexylresorcinol may be effective against dental plaque. With hexylresorcinol as the antiseptic component, improved dental floss having highly effective antibacterial, analgesic and germicidal properties were developed [46–48].

High bacterial value against both cocci and bacilli and low toxicity for the local and general tissue cells were observed for the chloro-substituted alkyl resorcinol derivatives shown below [43, 49] (Figure 9.21).

Chalcone compounds synthesized from the reactions of 2,4-dihydroxyacetophenone and hydroxybenzaldehydes, and substituted resorcinaldehyde derivative with hydroxyacetophenones, as shown below, were observed to show bacteriostatic activities [50–52] (Figures 9.22 and 9.23).

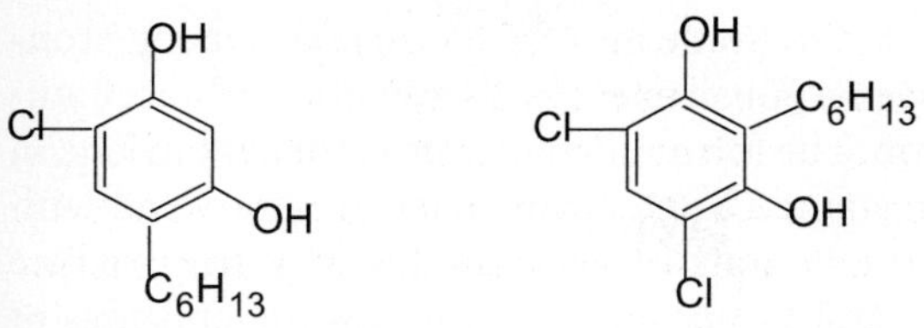


**Figure 9.21.**

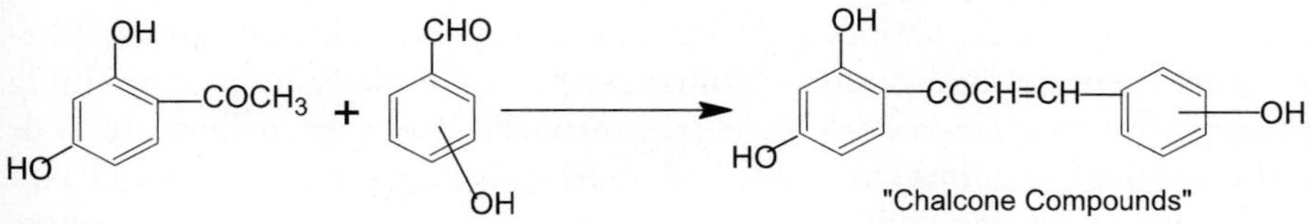


**Figure 9.22.**

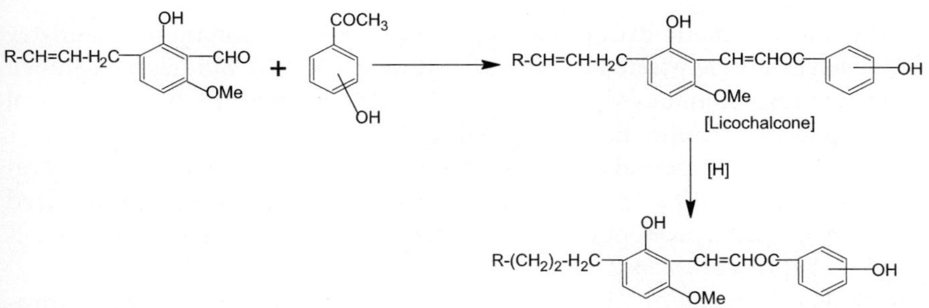


**Figure 9.23.**

The antimicrobial activity of chalcone compounds against "Staphylococcus" was enhanced when the electron donating hydroxyl group was also present in the aldehyde portion of the aromatic group. Licochalcones were very effective on "Mycobacterium" species and therefore, they were useful in the treatment or prophylaxis of diseases caused by microorganisms or parasites.

## 9.7 Asthma

Asthma is a chronic inflammation of the bronchial tubes (airways) that causes swelling and narrowing (constriction) of the airways. Leukotrienes, particularly cysteinyl-LTs, have major effects on the smaller peripheral airways of the lungs and also, on the larger central passages which include the trachea and bronchi [53–56]. In the presence of an allergy trigger, like pollen or dust, leukotrienes are manufactured from the fatty substances trapped in the membrane of the triggered cell. A series of reactions occur in the cell, which results in the generation of a set of different leukotrienes. These leukotrienes are then transported through the cell membrane into the blood and causing bronchoconstriction, bronchial hyper responsiveness, inflammatory cell accumulation, and mucus hyper secretion. The formation and release of leukotriene in the lung tissues contribute to sustained bronchoconstriction associated with asthma. Therefore, inhibiting the actions of leukotrienes may help reduce the inflammatory processes, control symptoms, and prevent the episodes of asthma [55].

In the treatment of asthma, chemical compounds acting as antagonists to prevent leukotrienes action or as inhibitors to prevent leukotriene synthesis, have been developed and administered. Asthma medications containing the organic compounds work by relaxing bronchospasm (bronchodilators) or reducing the inflammation. Since leukotrienes have proven to be important mediators of asthma conditions, anti-leukotriene drugs, acting as receptor antagonists and biosynthesis inhibitors, have shown improved effects on controlling asthma.

The anti-asthmatic drugs developed based on the resorcinolic chemistry have been extensively tested in asthma treatments [57–69]. Studies have shown that some resorcinol based organic compounds can provide excellent control and long term reliability in treating asthma.

Bis-chromyl compound useful as inhibitors of certain antigen-antibody reactions, particularly for the relief and prophylaxis of asthma, were synthesized using 2,6-dihydroxyacetophenone derivative, and based on the following reaction [57–59] (Figure 9.24).

The disodium salt of 1,3-bis-(2-carboxy chromon-5-yloxy)-2-hydroxypropane has been commercially produced and marketed under the brand name "Cromolyn Sodium" (Intal), and used to prevent the release of asthma causing chemicals such as histamine in the lungs. Cromolyn was approved by the Food and Drug Administration (FDA) in 1973. In 1997, the FDA approved an over the counter status for the nasal solution of this product. Cromolyn can help prevent asthma triggered by exercise, cold air, and allergic substances such as cat dander.

Chromonealdehyde compound obtained from 2-hydroxy-4-acetoxyacetophenone can be useful not only for treating asthma, dermatitis and allergic diseases, but also as intermediate in the organic synthesis [60] (Figure 9.25).

In the development of asthma treating drug compounds, 2,4-dihydroxy-3-*n*-propylacetophenone derivative was also used. Compounds containing the following chemical structures synthesized from the propylacetophenone derivative were useful in the prophylaxis and treatment of bronchial asthma, rhinitis, hay fever, and allergic eczema [61,64,66,67] (Figures 9.26 and 9.27).

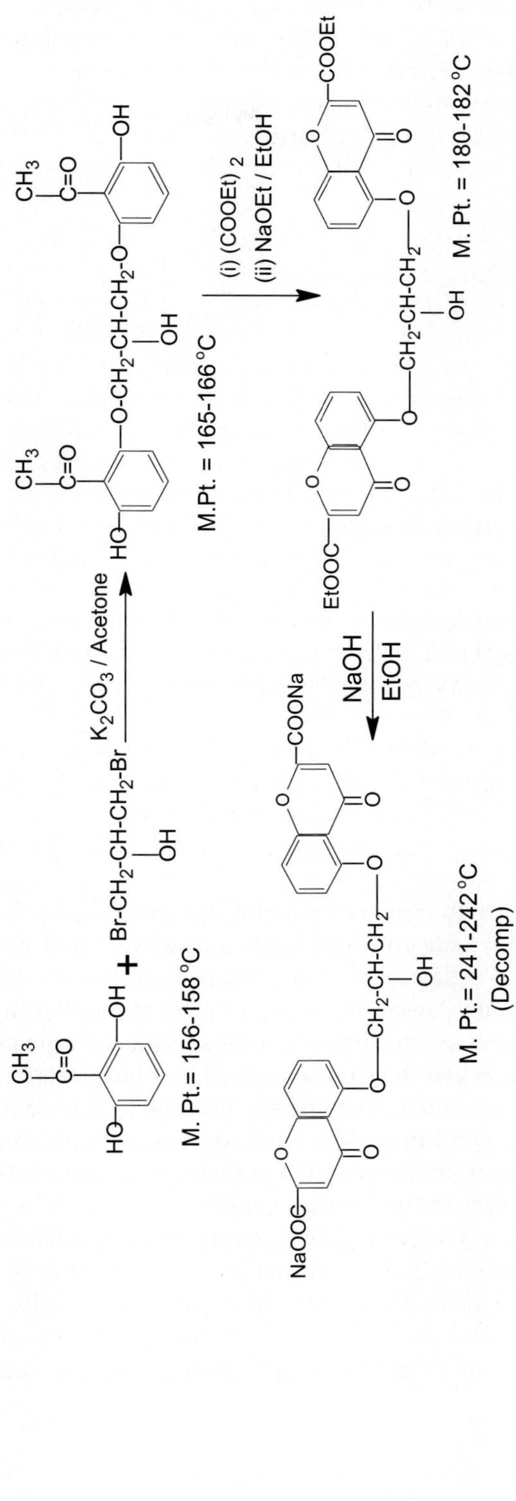
M. Pt.= 156-158 °C
K2CO3 / Acetone
M.Pt. = 165-166 °C
(i) (COOEt) 2
(ii) NaOEt / EtOH
M. Pt. = 180-182 °C
NaOH
EtOH
M. Pt.= 241-242 °C
(Decomp)

**Figure 9.24.**

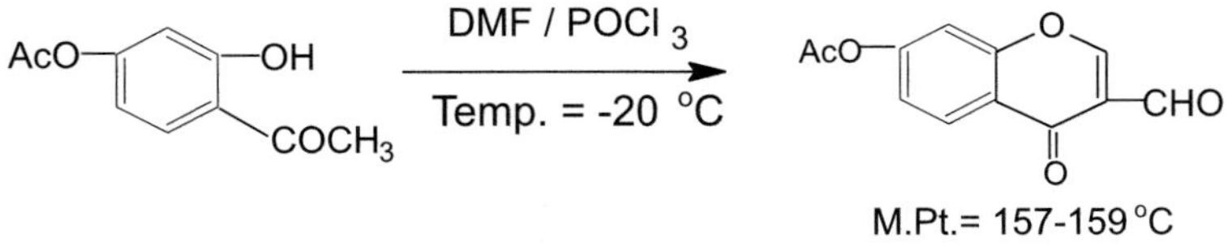


**Figure 9.25.**

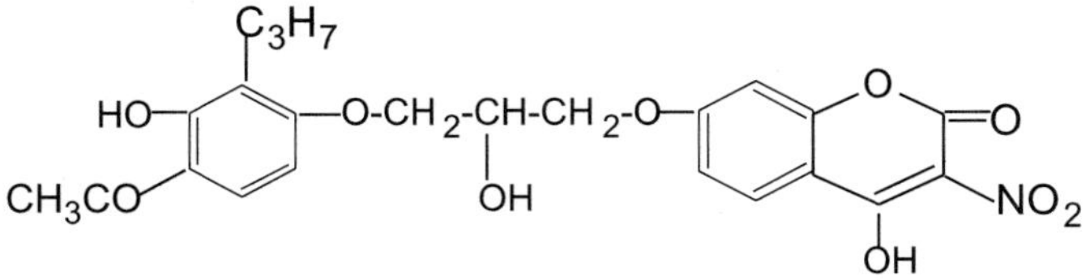


**Figure 9.26.**

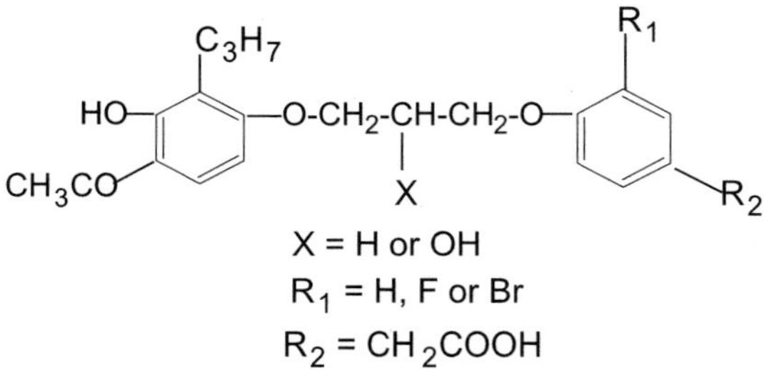


**Figure 9.27.**

## 9.8 Cancer

Cancer refers to a group of diseases that affect the body's cell structures. In general, normal body cells grow and die in a controlled way. But in cancer, an abnormal growth of cells occurs which tends to proliferate without control. Cancer can involve any tissue of the body, and has many different forms in each of the body areas. Most cancers are named for the type of cell or organ in which they start. When the cancer spreads to other parts of the body, it is then called metastatic cancer. Most deaths associated with cancer disease are due to its metastatic conditions. The most common types of cancer occurring in men and women are skin cancer, lung cancer, brain cancer, bladder cancer, breast cancer, liver cancer and prostate cancer.

In order to treat cancerous tumors, chemotherapy, radiation and surgery procedures or treatments have been made. Chemotherapy is the treatment of cancer with anti-cancer drugs that can destroy cancer cells, thus stopping them from growing and multiplying. Chemotherapy treatment is performed in cancer patients alone or in addition to surgery, radiation and/or biological therapy.

In chemotherapy treatments, the anti-cancer drugs are administered either by direct injection into a vein (IV) or by means of a catheter. Some drugs are given to the patients in the form of a pill. Several types of cancer are now considered to be curable by chemotherapy, and include Hodgkin's disease, large cell lymphoma, testicular cancer, and the early stages of breast cancer. Chemotherapy continues to be an effective treatment for many cancers along with surgery and radiation therapy.

In cancer treatments, multidrug resistance, the mechanism by which many cancers develop resistance to chemotherapy drugs, is a major problem. Drug resistance includes both the intrinsic resistance at the time of chemotherapy treatment and acquired drug resistance. Chemotherapy normally kills drug sensitive cells, but leaves behind a larger proportion of drug-resistance cells. When the tumor cells begin to grow again, chemotherapy may fail because the remaining tumor or cancer cells are now drug resistance. Because of this problem, combination chemotherapy by which two or more drugs are given at the same time to avoid the emergence of resistance cells, and also to kill pre-existing drug resistance cells. Multidrug resistance affects patients having breast, blood, ovarian, lung, and low gastro intestinal tract cancers [70,71].

Currently, considerable effort has been made by various drug and pharmaceutical companies to develop ideal anti-cancer drugs that would kill the malignant cells without affecting normal cells. The anti-cancer drugs that are currently in use by the cancer patients may not be targeted against the specific features of the cancer cells. These drugs usually cause severe toxicity to the non-malignant rapidly growing tissues, and resulting in bone-marrow, suppression, and diarrhea. The mode of action of cancer drugs involves either directly damage the DNA in the nucleus of the cell or inducing irreversible metabolic changes which affect the normal cellular physiology and results in cell deaths.

The search for anti-cancer drugs which can increase the chances of cancer cells to apoptosis may be a primary goal in cancer chemotherapy. Resorcinol based drug compounds have been developed and were found to be more useful in the treatment and prevention of various types of cancers.

Researchers have found that isoliquiritigenin, a compound isolated from the tonka bean, might be capable of inhibiting cancer formation [72]. Isoliquiritigenin is a chalcone compound and was prepared in higher yields from the reaction of 2,4-dihydroxyacetophenone with *p*-hydroxybenzaldehydes [73] (Figure 9.28).

This chalcone compound exhibited anti-ulcer activity, and it was suggested that it can also be used as an active ingredients in anti-tumor drugs. Isoliquiritigenin was found to be effective in the treatment of kidney diseases caused by anti-tumor drugs.

The hydroxylated chalcone, namely 1-propanone-1-(2,4-dihydroxyphenyl)-3-hydroxy-3-(4-hydroxyphenyl), having the following chemical structure

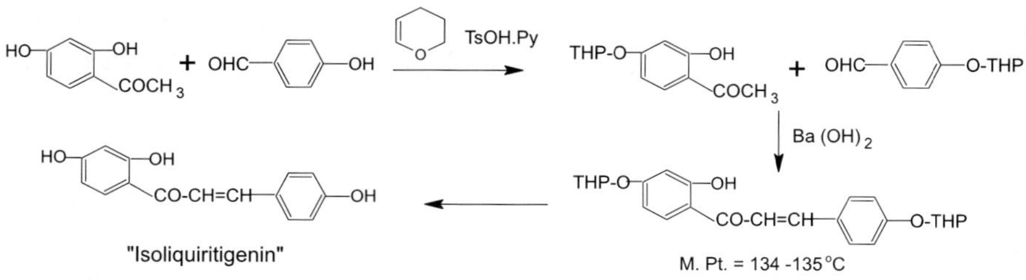


**Figure 9.28.**

showed activity consistent with an induction of apoptosis and potential activity as anti-tumorigenic and anti-carcinogenic agent [74] (Figure 9.29).

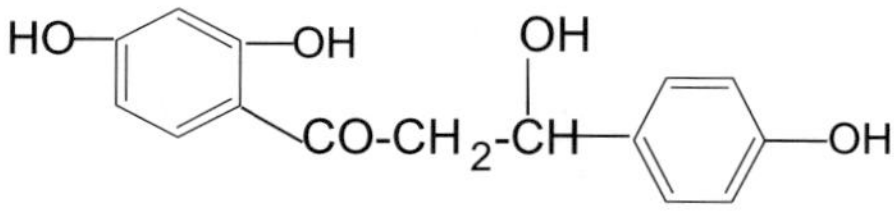


**Figure 9.29.**

Flavones and isoflavones have shown excellent antioxidant activities and are potent candidates for use in medicaments for the treatment of cancer. Isoflavones are considered as phytoestrogens and could exhibit anti-carcinogenic effects due to their antioxidant properties, and also their ability to suppress proto-oncogene expression, or inhibiting enzymes that are associated with cell proliferation, such as protein tyrosine kinase [75].

Isoflavones, like daidzein and genistein, are converted into hormone-like compounds, which may help block the growth of hormone dependent cancers. Daidzein can act as an inhibitor of protein tyrosine kinase enzyme and may block the signals from receptors which control the cells to grow [76].

Phytoestrogen compounds such as tetrahydrodaidzein and dehydroequol were synthesized from daidzein based on the following reaction scheme [77] (Figure 9.30).

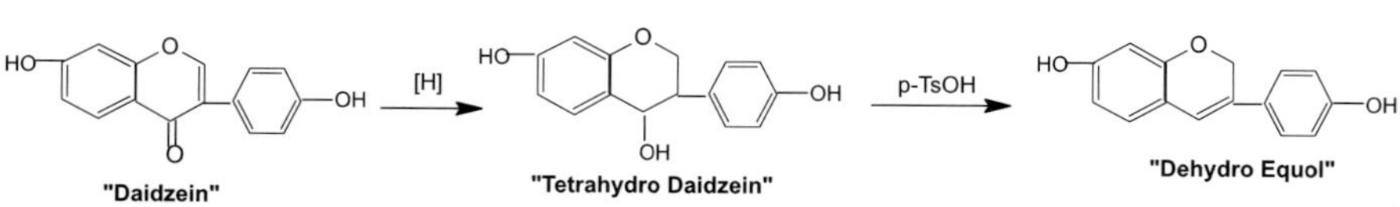


**Figure 9.30.**

Antioxidant studies relevant to the treatment of cancer on the tetrahydrodaidzein and dehydroequol have shown that these compounds were highly effective antioxidants. Tetrahydrodaidzein showed strong inhibiting effect on the growth of hormone responsive cancer cells, such as HL60 and K562 cell lines. These results suggest that they can show high therapeutic benefits in treating cancer diseases.

Novel flavanoids for use as adjuvants in chemotherapy, to protect against the side effects of cytostatics have been developed and tested [78] (Figure 9.31).

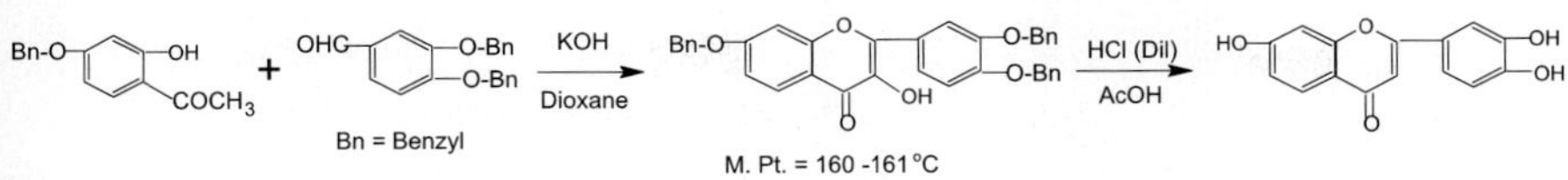


**Figure 9.31.**

These flavanoids compounds exhibited more potent than quercetin (a known cytostatics compound) in inhibiting the growth of OVCAR-3 cells.

Using 2,4-dihydroxyacetophenone derivative, a flavone compound was synthesized from the following reaction [79] (Figure 9.32).

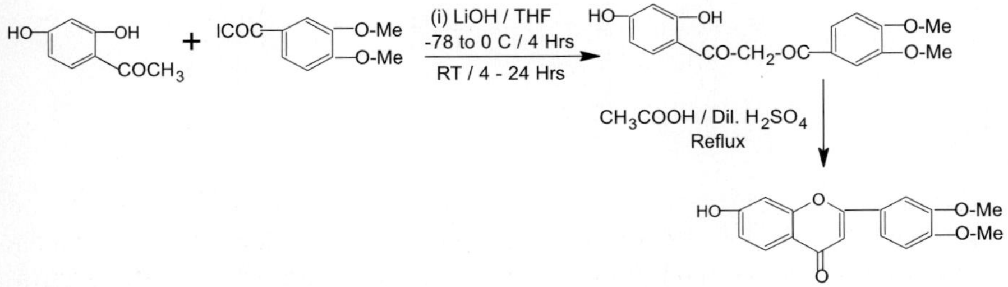


**Figure 9.32.**

Chemotherapy continues to be an effective therapy for many cancers. For the treatment of breast cancer, Adriamycin is a drug of choice and was very effective in the combination therapy. The phenoxy type compound, structure shown below, synthesized from the 2-propylresorcinol was very effective to reverse the resistance of HL60/ADR cells, human acute myeloblastic leukemia cell line, with Adriamycin anti-cancer drug [80,81]. In the absence of resorcinolic compound, the anti-cancer Adriamycin drug did not inhibit the growth of HL60/ADR cell lines at the tested concentrations (Figure 9.33).

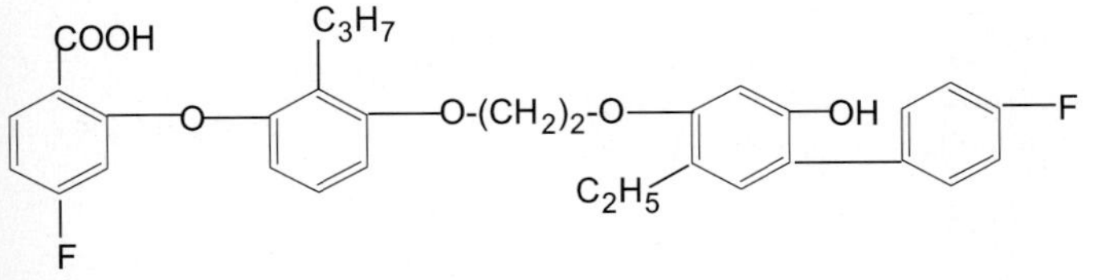


**Figure 9.33.**

Compounds possessing anti-proliferative activities act as modulators in multiple drug resistance and are highly desirable in chemotherapy. Synthetic drug compounds having structures similar to the naturally occurring flavanoids were synthesized as follows [82] (Figure 9.34).

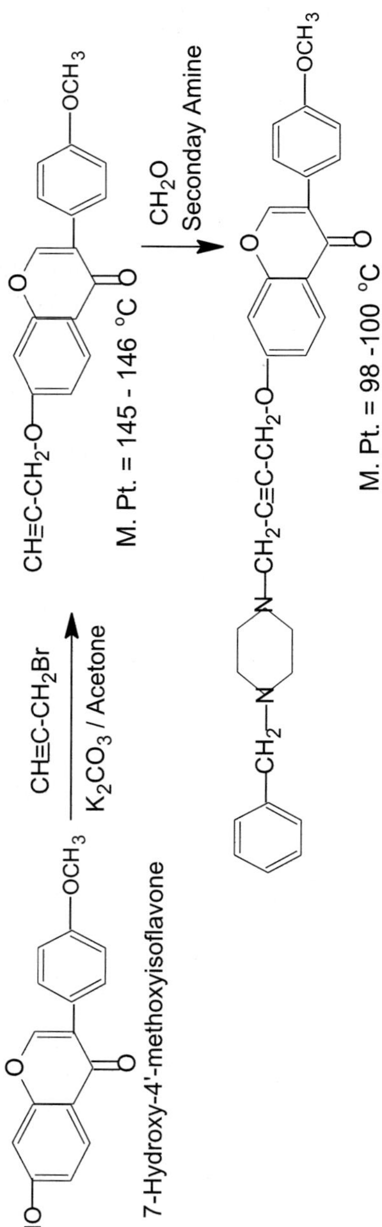
HO
O
O
OCH3
7-Hydroxy-4'-methoxyisoflavone
CH≡C-CH2Br
K2CO3 / Acetone
CH≡C-CH2-O
OCH3
M. Pt. = 145 - 146 °C
CH2O
Seconday Amine
CH2
N
N
CH2-C≡C-CH2-O
OCH3
M. Pt. = 98 -100 °C

**Figure 9.34.**

This isoflavanoid compound was tested for its cytotoxicity against drug resistant cancer cells, such as MDA 435/LCC6-MDR, both alone and in combination with the well known cancer drug Paclitaxel. In the presence of isoflavone, Paclitaxel can recover its excellent inhibitory activity against the drug resistance of cancer cells.

Benzopyran compounds for the manufacture of medicaments having anti-estrogenic character and for the treatment of breast cancer were developed from resorcinol [83] (Figure 9.35).

4-Methylbenzopyran compounds synthesized were determined to be useful as in the preparation of therapeutic compounds and compositions for the treatment of estrogen related breast cancer, uterine cancer and ovarian cancer [84] (Figure 9.36).

The compound EM-800 is an orally active pure antiestrogen and extremely potent against breast cancer cells in a culture [85,86]. Generally, an orally active and pure estrogen drug is extremely valuable as a second-line therapy after tamoxifen treatment has failed.

A spirolactone derivative, which could be an intermediate for the preparation of compounds possessing possible cytotoxic activity towards various human cancer cells, was prepared using a 2,4-dihydroxybenzaldehyde derivative [87] (Figure 9.37).

## 9.9 Cholesterol

Cholesterol is a fatty substance found in the blood of humans. The two main sources for cholesterol in the body are dietary intake and the liver. Liver produces cholesterol and it is subsequently carried into the bloodstream by lipoproteins. Depending upon whether the cholesterol is combined with low-density or high-density lipoproteins, it is called LDL cholesterol (bad) or HDL cholesterol (good). Higher amounts of LDL cholesterol in the blood stream increase the risk of coronary heart disease, which is associated with the formation of a hard, thick substance known as cholesterol plaque. The build up of cholesterol plaque in the heart arteries causes the thickening of the artery walls and narrowing arteries, which is called atherosclerosis. Because of this, coronary arteries can not supply enough blood and oxygen to the heart muscle during exertion. The lack of oxygen to the heart muscle results in chest pain and formation of blood clot in the artery, which can lead to heart muscle death and heart attack. HDL cholesterol, on the other-hand, prevents atherosclerosis by extracting cholesterol from the artery walls and disposing it through the liver.

For the prevention and treatment of atherosclerosis, the LDL cholesterol should be lowered or HDL cholesterol should be increased in the blood stream. Various drug compounds have been found to be effective in decreasing the LDL cholesterol levels. Resorcinolic compounds were developed and determined to be useful to treat atherosclerosis.

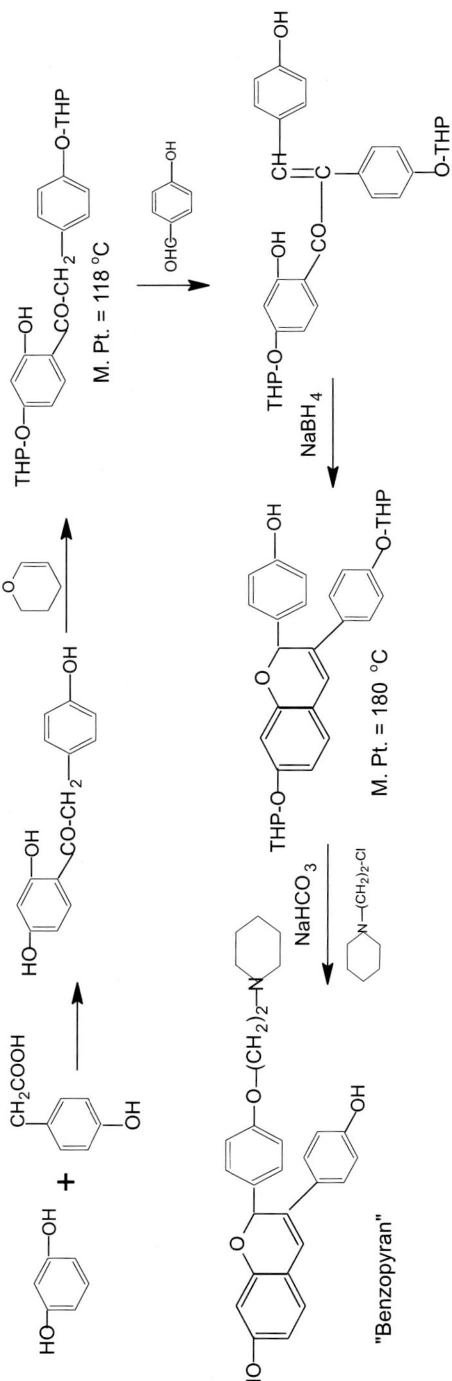
HO
OH
CH2COOH
OH
HO
OH
CO-CH2
OH
THP-O
OH
CO-CH2
O-THP
M. Pt. = 118 °C
OHC
OH
THP-O
OH
CO
C
CH
OH
O-THP
NaBH4
THP-O
O
OH
O-THP
M. Pt. = 180 °C
NaHCO3
N-(CH2)2-Cl
HO
O
O-(CH2)2-N
OH
"Benzopyran"

Figure 9.35.

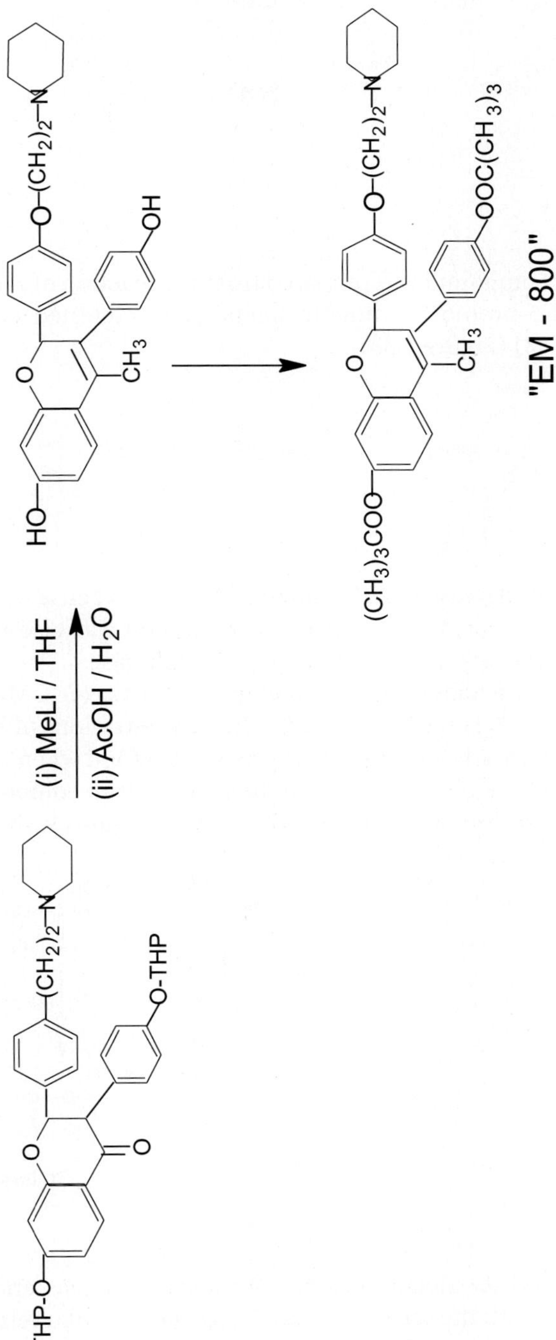
THP-O
O-THP
$(CH_2)_2$–N
(i) MeLi / THF
(ii) AcOH / $H_2O$
HO
OH
$CH_3$
O–$(CH_2)_2$–N
$(CH_3)_3COO$
$OOC(CH_3)_3$
"EM - 800"

Figure 9.36.

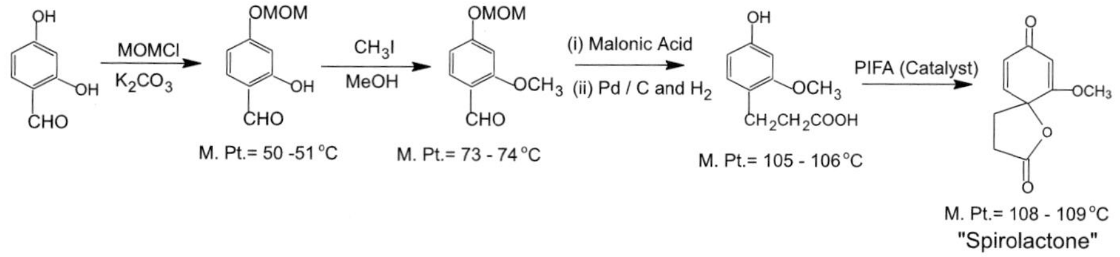


**Figure 9.37.**

The following compound was prepared from the reaction of 2-methylresorcinol and methyl-5-bromo-2,2-dimethylpentanoate and tested as hypocholesterolemic agent [88] (Figure 9.38).

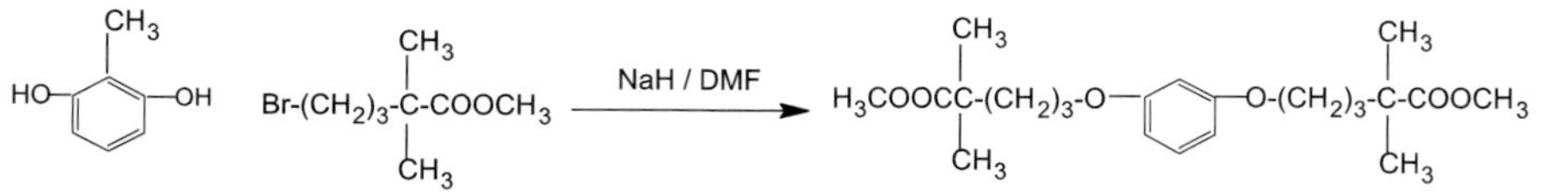


**Figure 9.38.**

This compound increased HDL cholesterol at the expense of LDL cholesterol. In addition to this, the total plasma cholesterol was also reduced, and thereby reducing the risk factor of coronary heart disease.

Daidzein and its metabolites, *o*-desmethylangolensin (O-DMA) and dihydrodaidzein, were determined to increase the concentrations of HDL-cholesterol, and also decreased the LDL-cholesterol in the blood when administered to humans [89]. These pharmaceutically important drug compounds can be synthesized from the following reactions [2, 90, 91] (Figures 9.39 and 9.40).

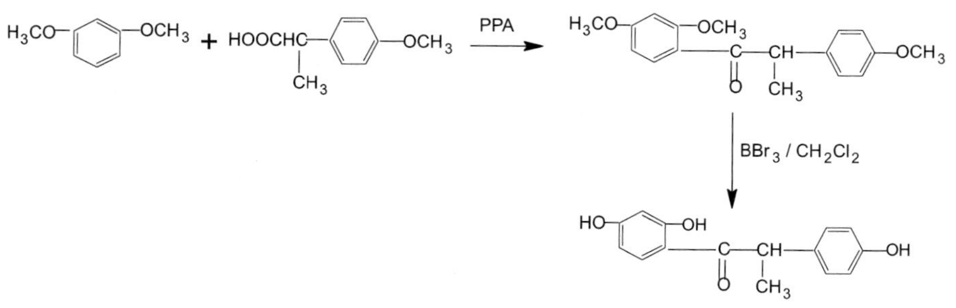


**Figure 9.39.**

The daidzein and its metabolites are the weak estrogens and have been shown to decrease both the serum LDL and total cholesterol levels. It has been theorized that daidzein was functioning like oral estrogens without the side effects associated with other synthetic pharmaceutical chemicals.

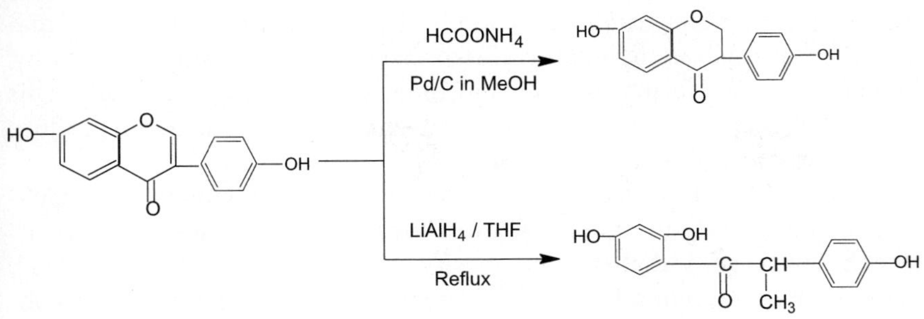


**Figure 9.40.**

## 9.10 Diabetes and Obesity

Diabetes is a disorder of metabolism and is characterized by high blood sugar (glucose) levels, which may result from defects in insulin secretion in the pancreas. In general, blood sugar levels are highly controlled by insulin. After food digestion, glucose enters into the bloodstream, where it is used by cells for growth and energy. When the blood glucose level is increased, insulin is released from the pancreas to normalize the glucose level. In people with diabetes, the pancreas either produces little or no insulin, and therefore the glucose level in the blood is high, which is also known as hyperglycemia. Diabetes can lead to blindness, kidney failure and nerve damage. This disease is also an important factor in accelerating the hardening and narrowing of the arteries, known as atherosclerosis, which can lead to strokes and coronary heart diseases.

Hypoglycemia, on the other hand, means abnormally low blood sugar, which is the result of excessive use of insulin. Since blood glucose level is essential for the proper functioning of nerve cells in the brain, low blood sugar condition can lead to dizziness, confusion, weakness, and tremors.

There are two main types of diabetes, namely type 1 and type 2. Type 1 diabetes is also called insulin-dependent diabetes. People with type 1 diabetes do not make insulin at all. Type 2 diabetes is called non-insulin dependent mellitus (NIDM). In type 2 diabetes, patients may not make enough insulin or cannot use the insulin that body produces. The chronic complications of diabetes mellitus are retinopathy (eye), nephropathy (kidney damage), and neuropathy (nerve damage).

Obesity means having excess of body fat. This is a chronic medical disease that can lead to diabetes, high blood pressure, heart disease, gallstones, and other chronic illnesses.

In order to overcome the diseases associated with diabetes and obesity, a number of oral anti-diabetic agents or drugs have been developed and used by the diabetic patients. These drugs served in several different ways, including stimulating insulin secretion, improving insulin sensitivity, and enhancing insulin actions, to control the blood glucose level, and thereby to prevent the acute and long term diabetic disease complications. Hyperglycemia related diabetic conditions such as angiopathy, atherosclerosis, nephropathy, neuropathy, retinopathy, cataract formation and glaucoma are increasing worldwide. This results in the need for an oral anti-diabetic drugs or agents that do not produce side effects or more severe complications.

Resorcinol based derivatives were employed in the development of pharmaceutical compounds, and have been tested for their effectiveness in the treatment of human chronic diabetes and obesity conditions [79,89,92–96,99–102].

Chronic hyperglycemia leads to a decrease in insulin secretion and contributes to increased insulin resistance. These results in increased blood glucose concentration, and therefore the diabetic conditions are self exacerbated. By treating hyperglycemia, the self exacerbating cycle of diabetes is interrupted, and thus the prophylaxis or treatment of diabetes mellitus could be possible. Compounds containing the propiophenone groups showed hypoglycemia activity, and inhibited renal glucose reabsorption by which it showed a urine glucose increasing behavior [93, 94]. 2,6-Dihydroxyacetophenone was used to synthesize propiophenone compounds for the treatment of diabetes [93] (Figure 9.41).

Antidiabetic compounds were developed from the following reaction scheme using 2,4-dihydroxy-3-propylpropiophenone derivative [95, 97] (Figure 9.42).

Chemical compounds having the above structure could be effective in the treatment of various diseases that includes diabetes, atherosclerosis, hyperglycemia, hyperlipidemia, and/or obesity because they could lower one or more of the biological entities such as glucose, insulin, triglycerides, fatty acids, and cholesterol.

Glucocorticoids have been shown to exhibit major influence on the glucose production. The excess glucocorticoid aggravates established diabetes, and similarly its deficiency reduces the blood sugar level. Dibenzopyranyl sulfonamide compounds synthesized from the resorcinol dimethylether derivative can exhibit glucocorticoid receptor antagonist activities [98, 99] (Figure 9.43).

These sulfonamide compounds were non-steroidal types and found to be useful for treating the type 2 diabetes and also the symptoms of hyperglycemia, obesity, hyperinsulinemia, and hypertriglyceridemia.

Aryloxyacetic acid compounds were known to be potent agonists of peroxisome proliferator activated receptor (PPAR). Resorcinol was used as the starting material for the synthesis of active drug compounds containing the aryloxyacetic acid group from the following of reactions [100] (Figure 9.44).

Drug compounds possessing the above chemical structures were shown to be useful in the treatment, control or prevention of non-insulin dependent diabetic conditions.

In addition to the above chemicals, compounds having the following generalized structure have been developed and tested for treating diabetic disease [101, 102] (Figure 9.45).

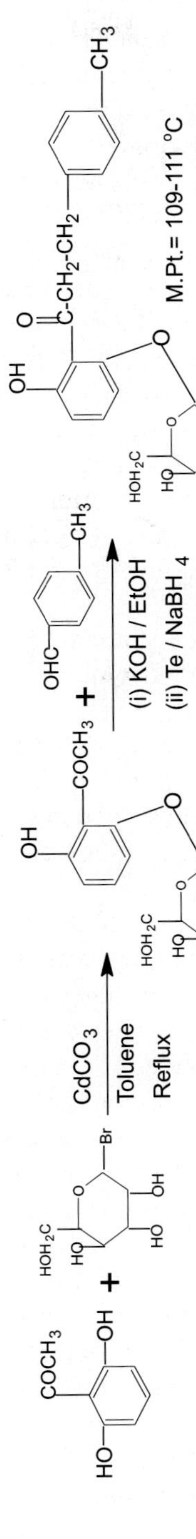


**Figure 9.41.**

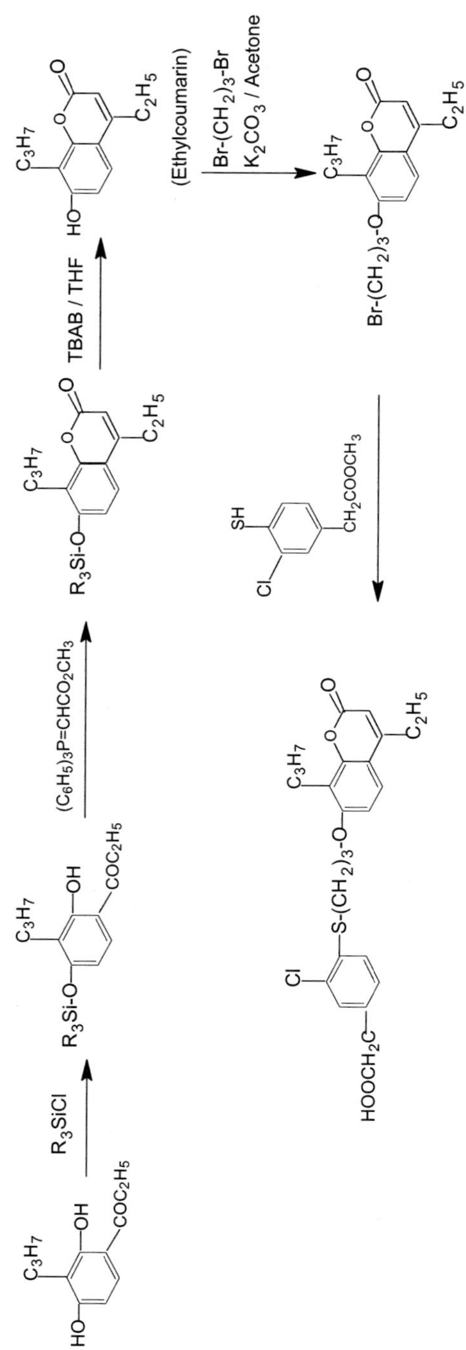

R3SiCl
R3Si-O
(C6H5)3P=CHCO2CH3
TBAB / THF
(Ethylcoumarin)
Br-(CH2)3-Br
K2CO3 / Acetone
Br-(CH2)3-O
CH2COOCH3
SH
Cl
S-(CH2)3-O
HOOCH2C
C3H7
C2H5
COC2H5
OH
HO


Figure 9.42.

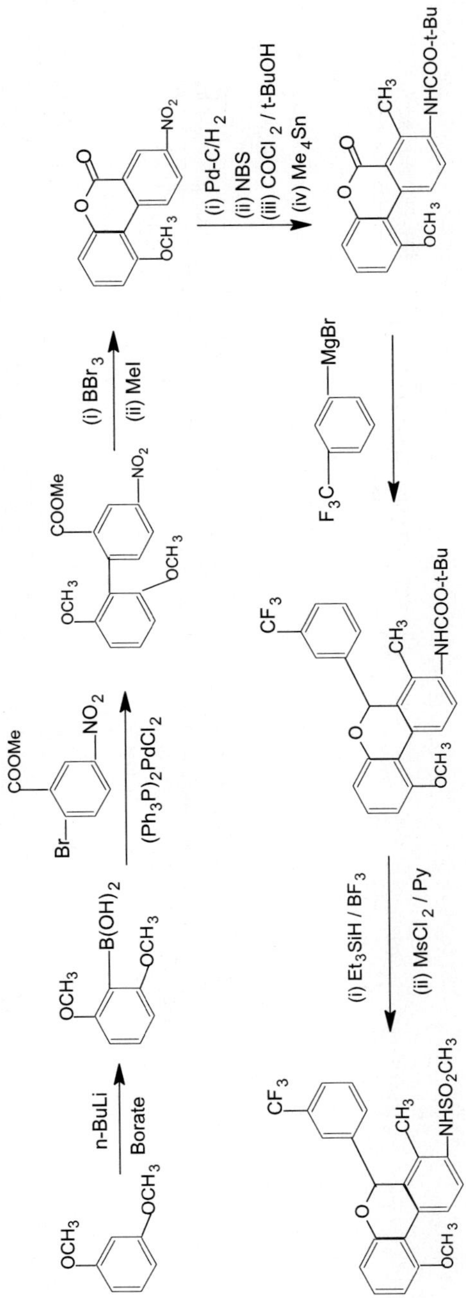

OCH3
OCH3
n-BuLi
Borate
B(OH)2
Br
COOMe
NO2
(Ph3P)2PdCl2
(i) BBr3
(ii) MeI
(i) Pd-C/H2
(ii) NBS
(iii) COCl2 / t-BuOH
(iv) Me4Sn
CH3
NHCOO-t-Bu
F3C
MgBr
CF3
(i) Et3SiH / BF3
(ii) MsCl2 / Py
NHSO2CH3


**Figure 9.43.**

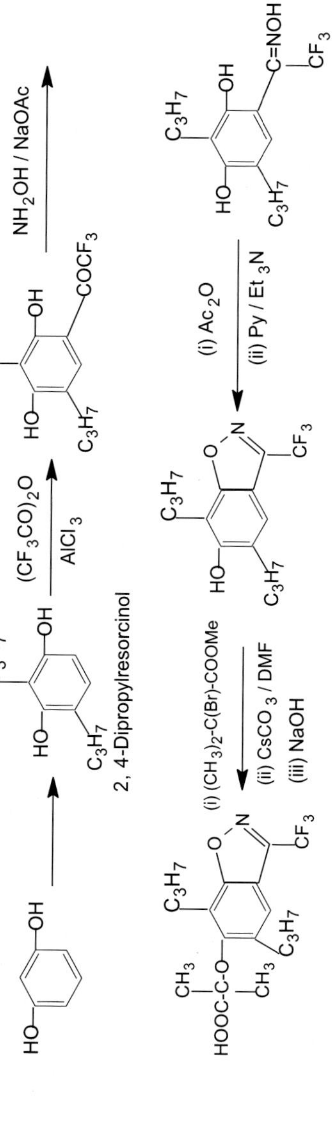
HO
OH
C3H7
2, 4-Dipropylresorcinol
(CF3CO)2O
AlCl3
COCF3
NH2OH / NaOAc
C=NOH
CF3
(i) Ac2O
(ii) Py / Et3N
O
N
(i) (CH3)2-C(Br)-COOMe
(ii) CsCO3 / DMF
(iii) NaOH
HOOC-C-O
CH3

**Figure 9.44.**

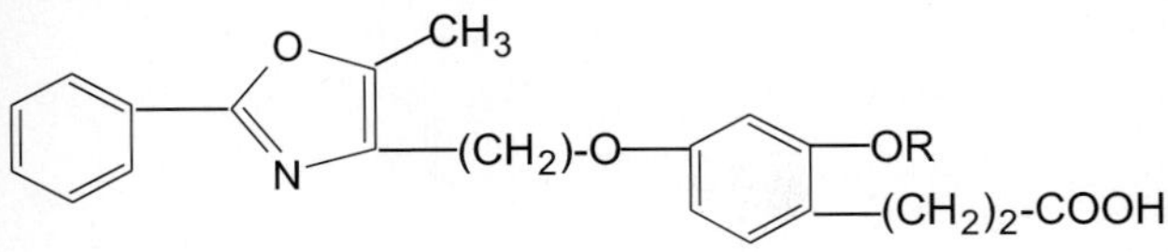


**Figure 9.45.**

## 9.11 Diagnosis and Biological Assays

Many diagnostic and analytical techniques require chemical compounds used to study the mechanism of actions of drugs. Resorcinol and its derivatives have been utilized to synthesize different kinds of active and effective organic compounds suitable in these applications.

In order to diagnose and study the drug actions, $\gamma$-emitting pharmaceutical compounds are being administered to patients. For detecting the blood flow in the brain, a radio tracer compound has been used. The radio labeled amphetamine compound developed from the resorcinol chemistry may be useful as radio tracer for positron or single photon imaging of the blood flow, and also for locating the sites of brain strokes [103] (Figure 9.46).

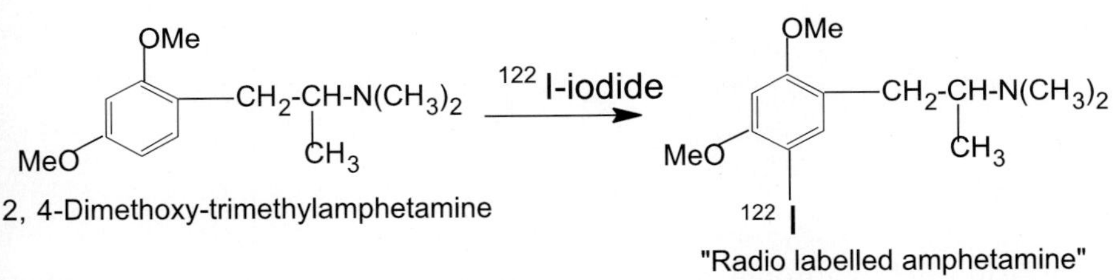


**Figure 9.46.**

Fluorescent compounds are widely used as labels in a variety of diagnostic procedures. These compounds are normally linked to other substances to form probes, and then used to assay a variety of materials of interest. Electrophoretic probes may be used in the detection and quantitation of polypeptides, polynucleotides, hormones and drugs. Fluorescent compounds were developed from the alkyl-substituted resorcinol derivatives from the following reaction [104] (Figure 9.47).

These compounds can be used in the simultaneous determination of multiple analytes in a sample.

Long wavelength and narrow emission bandwidth fluorescent dyes were prepared from resorcinol and 3,6-dichloro trimellitic acid were useful in the methods to determine DNA sequence analysis [105].

The presence and activity of enzymes can be used to determine the health and metabolic state of a cell or to distinguish one cell from another. Synthetic

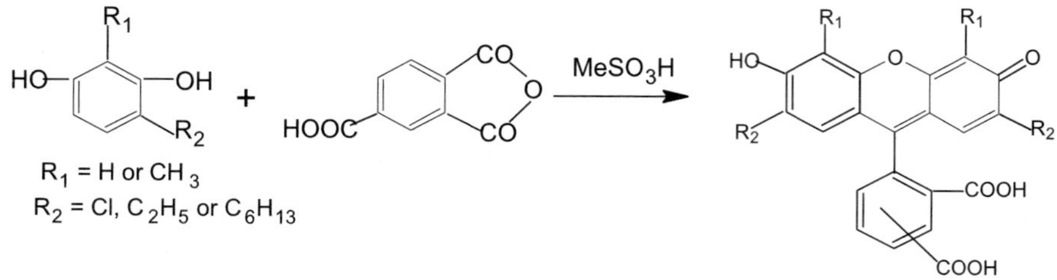


**Figure 9.47.**

organic compounds are normally employed to detect or assay isolated enzymes, enzyme in body fluids and in cell lysates. Assessing enzymatic activity in cells are important for the evaluation of drugs and diseases. Coumarin compounds were known to show fluorescent properties, which could be used in assaying the enzyme activities, and may be conjugated to antibodies for the determination of immunoassays [106, 107] (Figure 9.48).

**Figure 9.48.**

## 9.12 Depression

Depression is a syndrome and characterized not only by negative thoughts, moods and behaviors, but also by specific body functions such as eating and sleeping. Scientists have been making progress in understanding the causes for the depressive disorders, and also in the development of various drug compounds to treat depression [108]. Depression in humans is the result of various complicated process involving genetic, psychological and environmental factors. The biological basis of depression originates in the brain. The four most important brain chemicals, also called neurotransmitters, are norepinephrine, serotonin, dopamine and aceteylcholine. Different neuropsychiatric illnesses are associated with an excess or lack of some of the neurotransmitters in certain parts of the brain. A lack of dopamine at the base of brain may cause Parkinson's disease. Similarly, Alzheimer's dementia appears to be related to lower acetylcholine levels in the brain. The depressive disorders may be associated with altered brain serotonin and norepinephrine systems.

Modern brain imaging techniques have revealed that neural circuits responsible for the regulation of moods, thinking, sleep and an appetite appeared to fail to function properly in people with depression disorders. The two neurotransmitters, namely norepinephrine and serotonin, are not produced in sufficient quantities in a depressed person's brain, and are responsible for the transmission of messages from one neuron to another.

Several drug compounds have been successfully used to treat the depression. In these drugs, selective serotonin reuptake inhibitors (SSRIs) drugs are the medications that can increase the levels of serotonin chemical in the brain. Newer antidepressant drugs have been under constant development in order to bring robust effects on both the norepinephrine and serotonin systems. Resorcinol based compounds showing antidepressant properties were developed and evaluated [109–118].

Substituted phenyl chroman derivatives possessing the following chemical structure showed anorexic or mood modifying effects in mammals depending upon the dosage [109] (Figure 9.49).

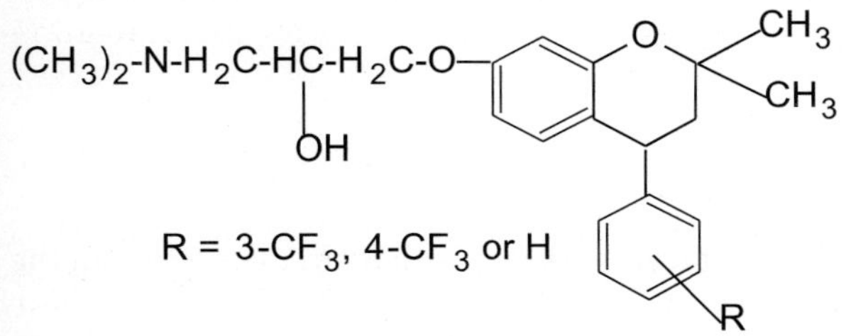


**Figure 9.49.**

Compounds useful for the treatment of central nervous system (CNS) disorders, in particular neurodegenerative disorders, and Parkinson's disease were prepared from the 7-hydroxy-3,4-dimethylcoumarin derivative [113] (Figure 9.50).

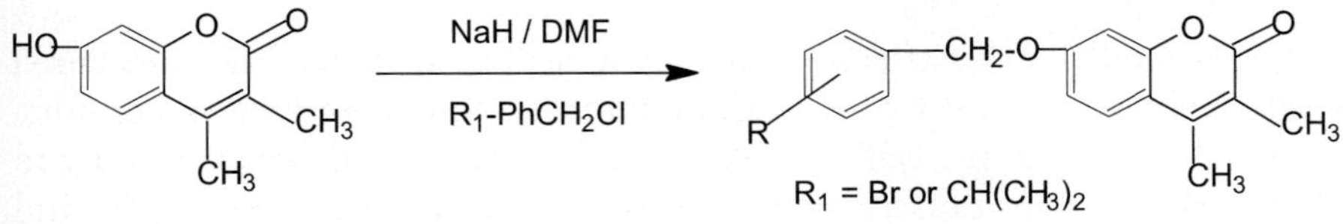


**Figure 9.50.**

These coumarin compounds showed inhibiting activity on manoaminooxidase (MAO) enzyme. MAO is the main enzyme that breaks down norepinephrine neurochemical. When MAO is inhibited, the amount of norepinephrine in the brain is increased.

Serotonin or 5-hydroxy tryptamine (5-HT) neurotransmitter is responsible for various central nervous system (CNS) disorders such as anxiety and depression. This chemical is known to affect many types of conditions including psychiatric disorders, feeling behavior and sexual activity, and the effects of serotonin are regulated by various 5-HT receptors. The 5-HT receptor appears to be exclusively present in the brain. Therefore, the 5-HT6 receptor ligands are believed to be potential use in the treatment of certain CNS disorders such as anxiety, depression, epilepsy, Alzheimer's disease, sleep disorders and anorexia. For the treatment of CNS disorders and also to act as 5-HT6 receptor, chroman and benzofuran compounds were developed using 2,4-dihydroxyacetophenone and 2,4-dihydroxybenzaldehyde derivatives [114,115] (Figures 9.51 and 9.52).

Antipsychotic agents that could provide functional modulations of dopamine system of the brain and without the excessive blockade of the postsynaptic dopamine receptors were synthesized and tested [116–118] (Figure 9.53).

The chromyl methylamino derivatives showed high affinity for dopamine D-2 receptor. These compounds improved the synthesis of dopamine neurotransmitter, and therefore could be useful in the treatment of dopaminergic disorders such as Parkinson's disease, schizophrenia, alcohol addiction and cocaine addition.

## 9.13 Estrogens

In humans, hormones are the chemical messengers that circulate through the blood stream and control numerous activities of the body functions. These hormones are produced by different glands and organs present in the human body. The major classes of human steroidal hormones are estrogens, progesterones and androgens. Estrogens and progesterone hormones are produced by the female ovaries and their production is controlled by the hormones secreted in the brain. The three primary steroids present in the estrogens are estradiol, estrone and estriol. Estrogens are traveling through the blood stream and interacting with cells in a variety of tissues such as breast, uterus, brain, bone, liver and heart [119].

Estrogens can exert effects on tissues in several ways, and the most well characterized mechanism of action is their interaction with estrogen receptors. They are located in the cell's nucleus along with DNA molecules, leading to alterations in gene transcription. Two estrogen receptors, namely $ER_\alpha$ and $ER_\beta$ have been discovered and are known to bind estrogens as well as other agonists and antagonists [120, 121]. The estrogens are binding to estrogen receptors ($ER_\alpha$ and $ER_\beta$) present in the cells of breast and uterus, and trigger the synthesis of growth promoting factors, which lead to cell growth and divisions.

Estrogens have been shown to exert positive as well as negative effects on various tissues. With liver cells, the estrogens alter the production of proteins that influence the cholesterol level in the blood. On the other hand, in breast and uterus tissues, the major effect of estrogen is on the cell proliferation, which could increase the cancer risk. Estrogen deprivation is also an important cause for postmenopausal osteoporosis, which affects about one-third to one-half of all postmenopausal women.

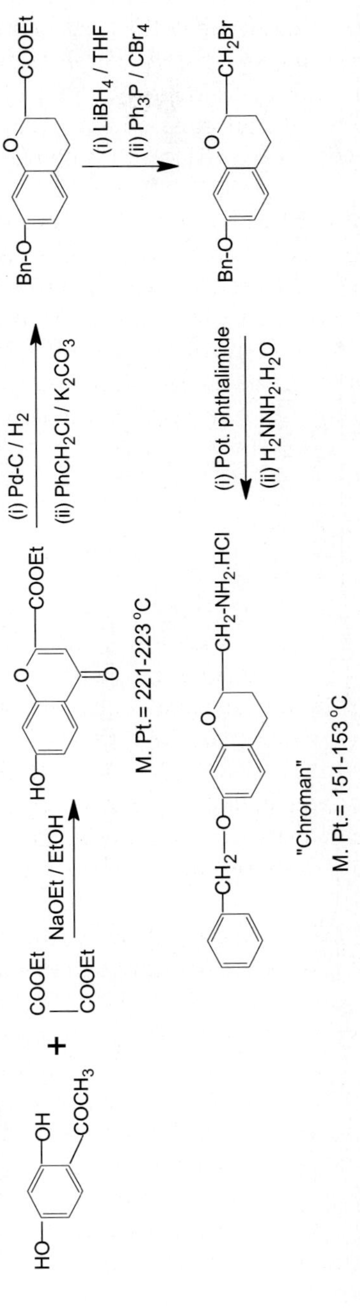
HO
OH
COCH3
+
COOEt
COOEt
NaOEt / EtOH
HO
O
COOEt
O
M. Pt.= 221-223 °C
(i) Pd-C / H2
(ii) PhCH2Cl / K2CO3
Bn-O
O
COOEt
(i) LiBH4 / THF
(ii) Ph3P / CBr4
Bn-O
O
CH2Br
(i) Pot. phthalimide
(ii) H2NNH2.H2O
CH2
O
O
CH2-NH2.HCl
"Chroman"
M. Pt.= 151-153 °C

**Figure 9.51.**

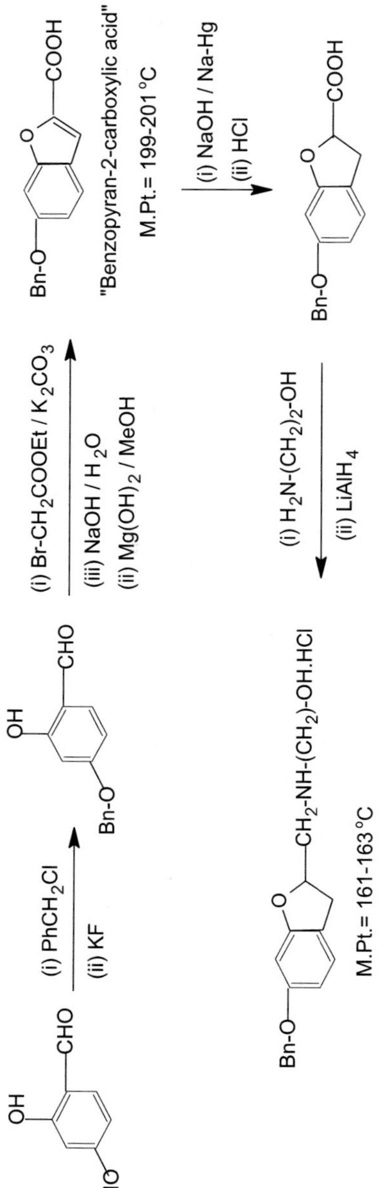

OH
HO
CHO
(i) PhCH2Cl
(ii) KF
Bn-O
(i) Br-CH2COOEt / K2CO3
(iii) NaOH / H2O
(ii) Mg(OH)2 / MeOH
COOH
"Benzopyran-2-carboxylic acid"
M.Pt.= 199-201 °C
(i) NaOH / Na-Hg
(ii) HCl
(i) H2N-(CH2)2-OH
(ii) LiAlH4
CH2-NH-(CH2)-OH.HCl
M.Pt.= 161-163 °C


Figure 9.52.

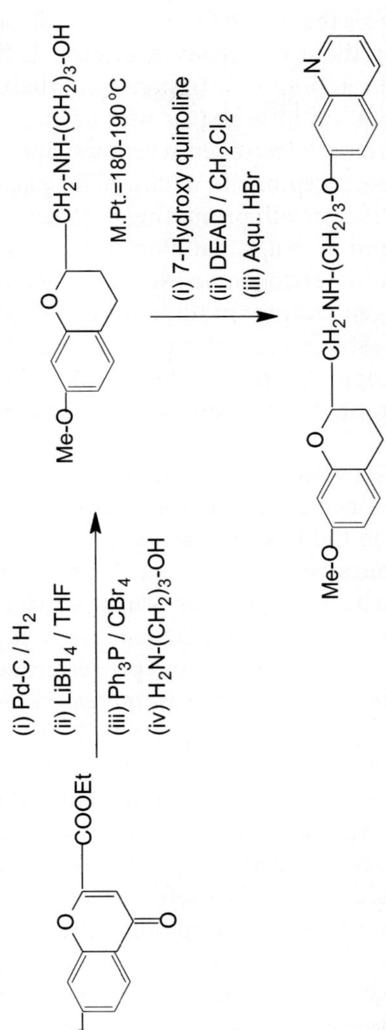
Me-O
COOEt
(i) Pd-C / H2
(ii) LiBH4 / THF
(iii) Ph3P / CBr4
(iv) H2N-(CH2)3-OH
Me-O
CH2-NH-(CH2)3-OH
M.Pt.=180-190 °C
(i) 7-Hydroxy quinoline
(ii) DEAD / CH2Cl2
(iii) Aqu. HBr
Me-O
CH2-NH-(CH2)3-O
N

**Figure 9.53.**

Since estrogen can promote the development of cancer in the breast and uterus, chemical substances that block the actions of estrogens might be useful in preventing or treating these two types of cancer. Antiestrogens are compounds that can block the actions of estrogens, and thereby interfere with or even prevent the proliferation of breast or uterine cancer cells [122]. These compounds work by binding to estrogen receptors, and thus blocking estrogens from binding to these receptors. In addition, they also block the estrogen activating genes for specific growth promoting proteins.

Antiestrogen compound or drugs that block the actions of estrogen in certain tissues can also mimic estrogen's action in other tissues. These drugs are called selective estrogen receptor modulators or SERMs. SERM drug compound might inhibit the estrogen receptor present in the breast cells but could activate the estrogen receptor present in the uterine cells. In this way, SERM drug could inhibit the proliferation of breast cells, but enhances the proliferation of uterine cells.

Estrogens have positive effects on blood vessels and on bones. After the menopause, women produce little estrogens, which could lead to several health problems including the risk of heart diseases and osteoporosis. In order to strengthen the bones and control other menopausal symptoms, postmenopausal women take hormone pills or drugs, which can increase the risk of breast and uterus cancer. For treating menopausal symptoms, hormone replacement therapy (HRT) using estrogen and progesterone are normally used. This treatment, most often, could lead to an increase in heart attack, strokes and blood clot problems.

SERM drugs inhibiting the positive effects of estrogens on bones, heart and blood vessels, and without inhibiting the potential harmful effects of estrogen on the breast and uterus are most needed. Resorcinolic derivatives were utilized in the synthesis of various drug compounds that showed estrogenic and antiestrogenic properties, and could be useful for the treatment of estrogen related diseases. Some of the important compounds obtained from the resorcinol based chemistry are illustrated below.

Coumestrol and coumarone derivative exhibiting estrogenic properties were synthesized from the following reactions [123–126] (Figure 9.54).

2-Phenylbenzofuran compounds having strong and selective antiestrogenic properties to treat estrogen related illnesses were prepared from the 2,4-dimethoxyacetophenone derivative [127]. Since benzofuran compounds have shown marked affinity to the estradiol receptor, they could inhibit the growth of hormone dependent tumor cells. These compounds can be used in applications of the therapeutic treatment of estrogen dependent illnesses (Figure 9.55).

Benzofluorenone compounds showing SERM type activities were synthesized from another acetophenone derivative [128, 129] (Figure 9.56).

These compounds were antiestrogenic in the uterus and could completely antagonize the tropic effects of estrogen agonists in the uterine tissue. Also, they can act as estrogen agonists in the bone, cardiovascular and central nervous

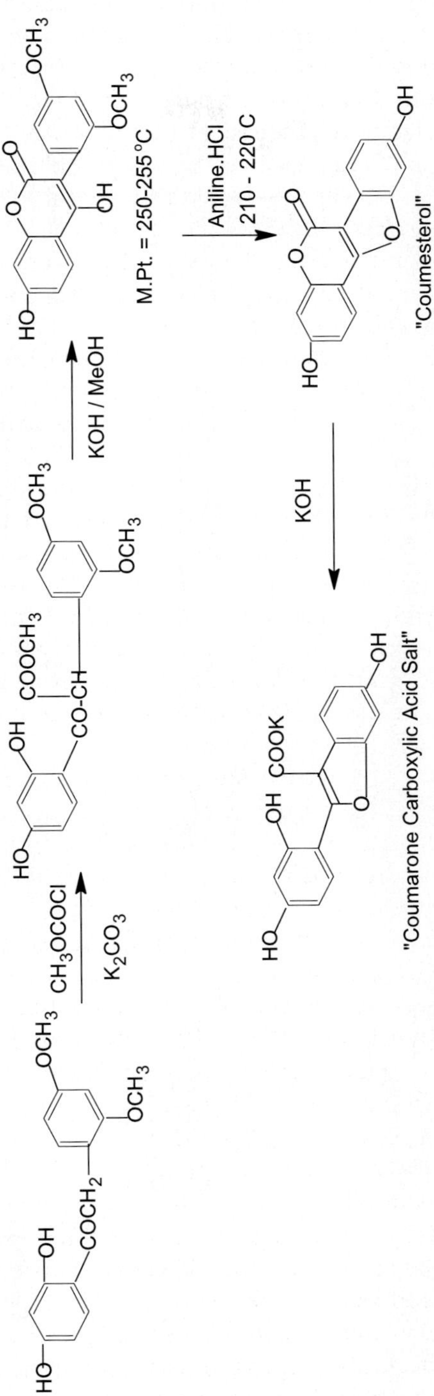
CH3OCOCl
K2CO3
KOH / MeOH
M.Pt. = 250-255 °C
Aniline.HCl
210 - 220 C
"Coumesterol"
KOH
"Coumarone Carboxylic Acid Salt"

**Figure 9.54.**

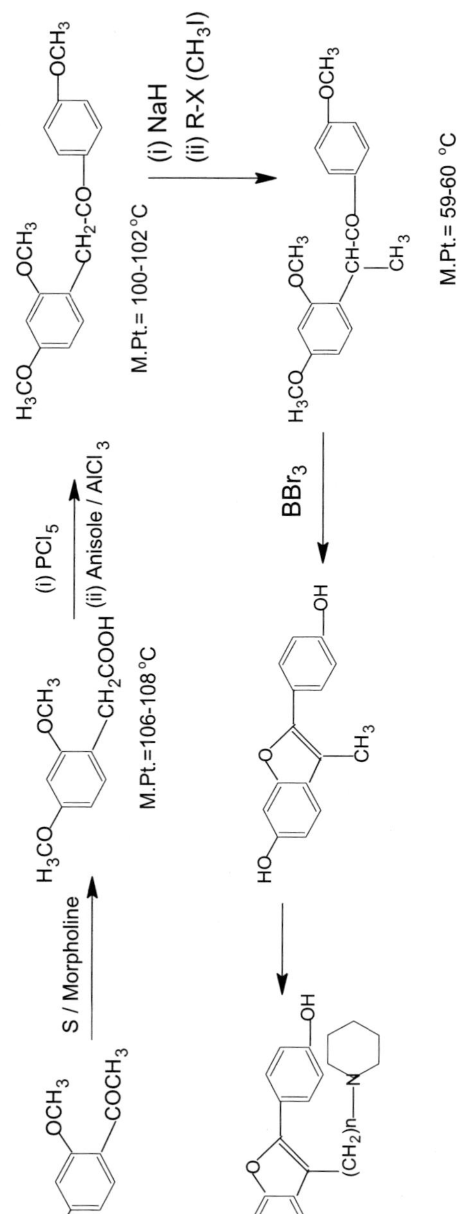
H3CO
OCH3
COCH3
S / Morpholine
H3CO
OCH3
CH2COOH
M.Pt.=106-108 °C
(i) PCl5
(ii) Anisole / AlCl 3
H3CO
OCH3
CH2-CO
OCH3
M.Pt.= 100-102 °C
(i) NaH
(ii) R-X (CH3I)
H3CO
OCH3
CH-CO
CH3
OCH3
M.Pt.= 59-60 °C
BBr3
HO
OH
CH3
HO
OH
(CH2)n
N

**Figure 9.55.**

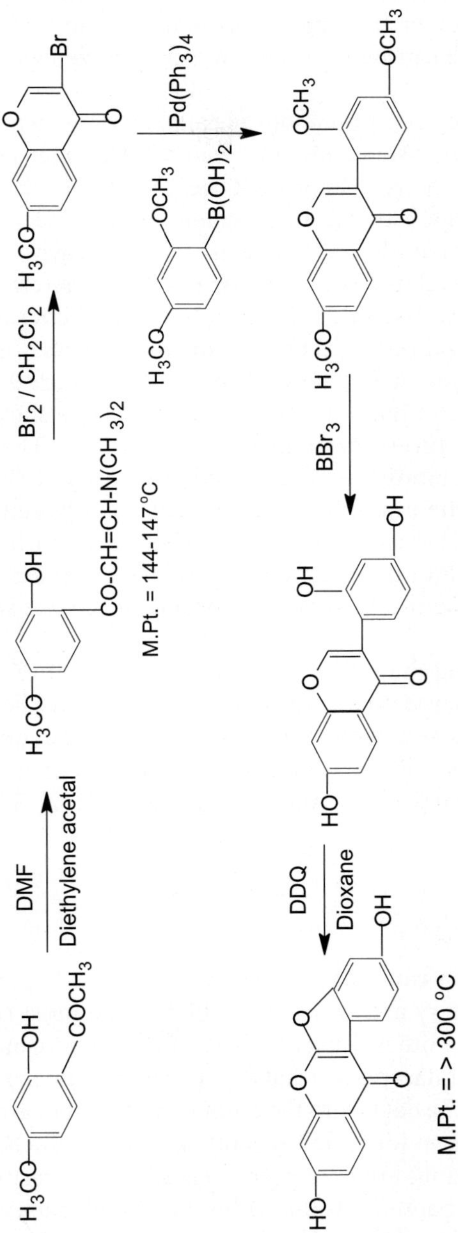
DMF
Diethylene acetal
CO-CH=CH-N(CH 3)2
M.Pt. = 144-147°C
Br2 / CH2Cl2
B(OH)2
Pd(Ph3)4
BBr3
DDQ
Dioxane
M.Pt. = > 300 °C

**Figure 9.56.**

systems. Because of these actions, benzofluorenone compounds could function as SERMs and useful in treating or preventing mammal disease states or syndromes which are caused by either an estrogen deficiency or an excess of estrogens.

Estrogen deficiency is also one of the causes for the degenerative changes in the central nervous system, and may lead to Alzheimer's disease (AD) and decline of cognition. In order to prevent the Alzheimer's disease and improve the cognitive functions, estrogen replacement therapy (ERT) has shown some benefits. To avoid the undesirable effects associated with ERT, constant efforts are made to develop SERM type drugs. Novel hetero atom containing tetracyclic derivatives developed from the resorcinol chemistry have shown applications in the treatment and prevention of disorders caused by the depletion of estrogens in postmenopausal women [130] (Figure 9.57).

A comparison of two human estrogen receptors has shown that $ER_\beta$ possessed physical structure shorter in length than $ER_\alpha$. These two receptors have shown different binding activities in various tissues. The $ER_\beta$ has shown express strongly in the brain, bone and vascular epithelium, but weakly expressed in uterus and breast relative to $ER_\alpha$. Therefore, SERM type compounds selectively targeting $ER_\beta$ over $ER_\alpha$ could produce beneficial effects in several diseases such as Alzheimer's, anxiety disorders, osteoporosis and cardiovascular [131].

Compounds having the following generalized structure synthesized from 2-chloro-4,6-dihydroxyacetophenone was shown to have selectivity for $ER_\beta$ over $ER_\alpha$, and may possess agonist activity on $ER_\beta$ without undesirable uterine effects. These compounds could be useful for the treatment of CNS diseases related to $ER_\beta$, and particularly Alzheimer's disease [132] (Figure 9.58).

## 9.14 Heart Failure

Heart failure has been caused by coronary artery and hypertension. Besides atherosclerotic coronary artery disease, which is the most common cause of heart failure, a host of other things such as valvular abnormalities, hypertension, cardiac arrhythmias, pericardial diseases and acute myocarditis may result in heart failure. The data from the animal models of heart failure and from the clinical studies have identified several substances, such as endothelin-1, and other heart functions are strongly correlated with severity of heart failure.

Endothelin-1 is a peptide produced by the vascular endothelial cells and causes vasoconstriction. Endothelin has a number of other actions besides vasoconstriction, namely the stimulation of aldosterone secretion, positive inotropy and chronotropy in heart, decreasing renal blood flow and release of atrial natriuretic peptide. This compound has been implicated in the pathogenesis of hypertension, vasospasm and heart failure [133].

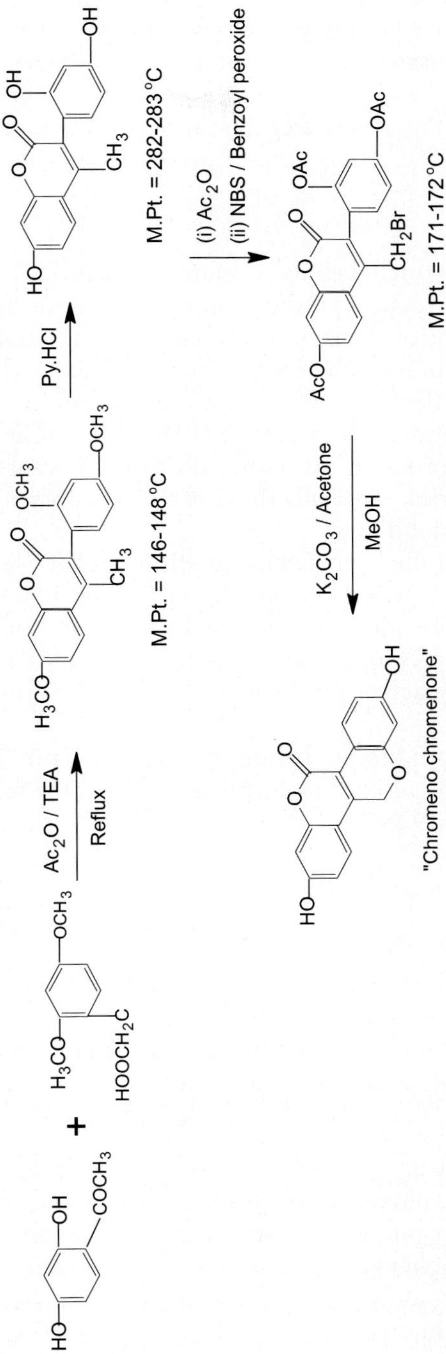
Ac2O / TEA
Reflux
M.Pt. = 146-148 °C
Py.HCl
M.Pt. = 282-283 °C
(i) Ac2O
(ii) NBS / Benzoyl peroxide
M.Pt. = 171-172 °C
K2CO3 / Acetone
MeOH
"Chromeno chromenone"

**Figure 9.57.**

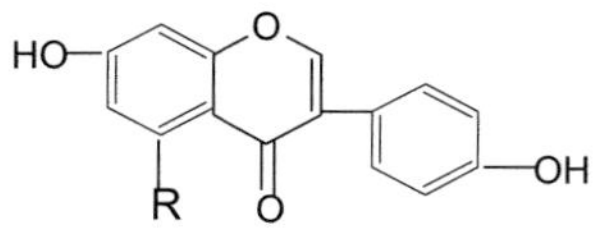


R = F, Cl, Br, I, CN , SMe
"Benzopyran Compounds"

**Figure 9.58.**

Vasopressin and oxytocin belong to a family of peptide hormone from the neurohypophysis. Oxytocin is produced in neurons in the hypothalamus and has involved in the control of uterine contraction during labor and milk release afterwards. Vasopressin is synthesized in the hypothalamus like oxytocin, and the main biological effects of vasopressin are the antidiuresis and vascular contraction [134]. Antidiuretic hormone (ADH) even at low doses regulates the osmotic content of blood. At higher doses, ADH causes contraction of arterioles and capillaries, especially those of the coronary vessels, to produce localized increase in blood flow.

In response to treat the heart failure conditions, effective drugs have been developed and clinically tested. Current therapeutic options to treat heart failure include vasodilators, positive inotropic therapy, diuretics and Endothelin antagonists [135]. Resorcinol derivatives were used in the development of various pharmaceutically active drug compounds for the cure of heart failure diseases [136–142].

A pharmaceutical product sold under the generic name "Carbochromene" was produced from the resorcinol derivative and has the following chemical structure [137] (Figure 9.59).

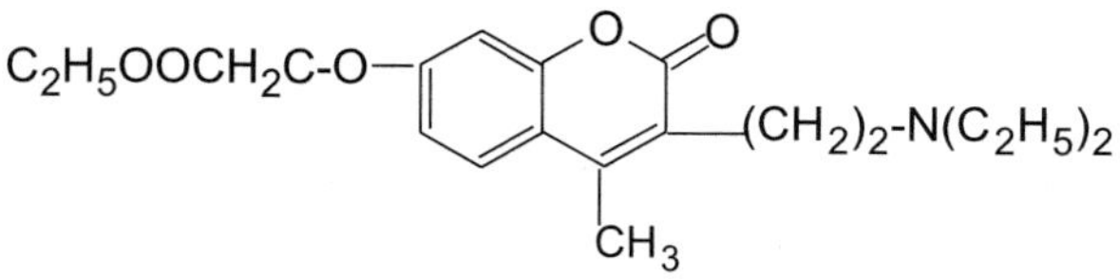


**Figure 9.59.**

This drug has been used as a vasodilator, which could open the blood vessel by relaxing the muscular wall of blood vessel.

Antihypertension drugs are used to reduce the blood pressure in humans. Chromone compounds obtained from resorcinol derivatives have exhibited superior antihypertensive properties at relatively low doses and were synthesized from 7-hydroxy-2,3-diphenylchromone [138] (Figure 9.60).

Inotropic agents are drug compounds that make the heart beat stronger. Phenylimadazole compounds synthesized from the reaction of $\alpha$-bromo-2,4-dimthoxyacetone with amino pyrimidine and amino pyrazine were effective as positive inotropic agents. These compounds possessed vasodilating activ-

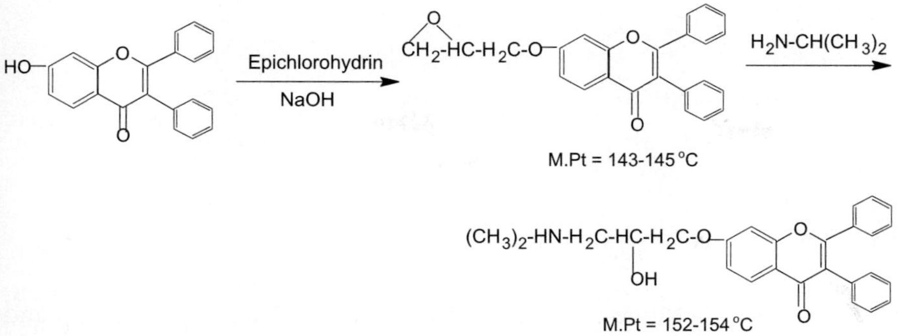


**Figure 9.60.**

ity and exhibited minimal effects on blood pressure and heart rates [139] (Figure 9.61).

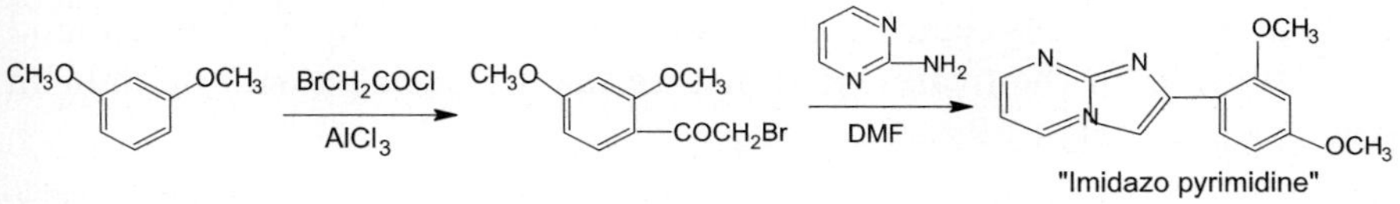


**Figure 9.61.**

Positive inotropic agents such as imidazopyrimidines are useful for patients presenting with pulmonary congestion and a systemic low out put rate.

Anti-platelet agents can reduce the tendency of platelets in the blood to clump or clot. If the blood platelet aggregation is prevented, then the glutinous properties of platelets are reduced, and thereby the formation of blood clots (thrombi) is prevented. Thrombolytic agents are blood clot dissolving medicines which can open the acutely occluded coronary arteries. The resorcinol based compounds exhibited antithrombic properties and were effective in combating cardiac infarctions, thromboembolic illnesses and arteriosclerosis [140] (Figure 9.62).

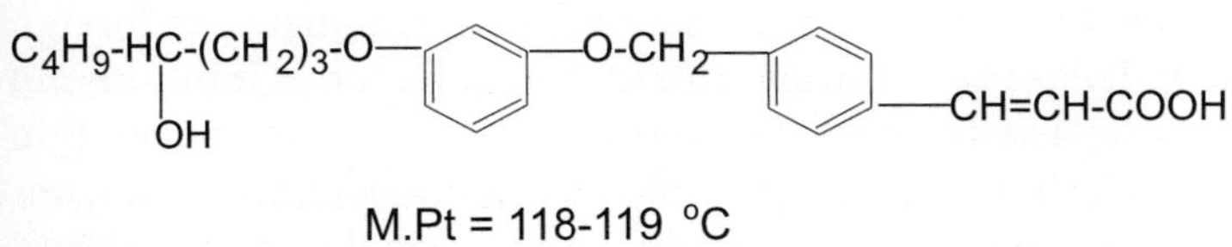


**Figure 9.62.**

When 2,4-dimethoxyphenyl acetic acid was reacted with spiro[1H]indene-1,4′-piperidine hydrochloride, 1′-(2,4-dimethoxyphenyl acetyl)-spiro-[1H]-inden-1,4′-piperidine was obtained [141] (Figure 9.63).

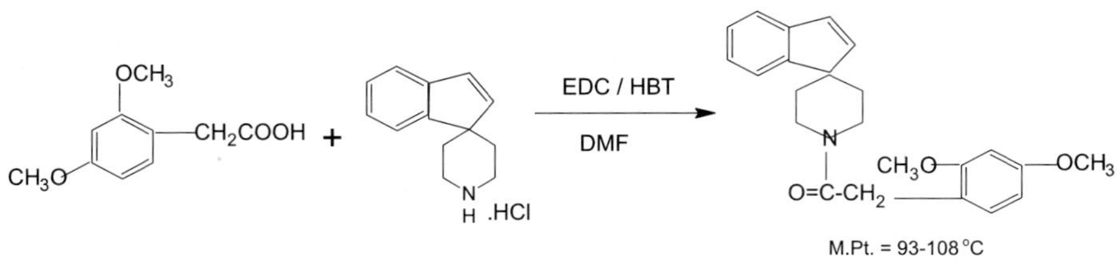


**Figure 9.63.**

The above piperidine compounds exhibited inhibition of vasopressin and observed to bind the vasopressin receptor. They were found to be useful in inducing vasodilation, treating hypertension, inducing diuresis and inhibiting agglutination.

Endothelin is an important potent vasoconstrictor producing long-lasting effects in arteries and veins. This causes profound actions on the cardiovascular systems and in particular the coronary, renal and cerebral circulation. Pharmaceutical compositions containing the resorcinol based compound were found to be suitable for the treatment of disease states associated with Endothelin peptides [142] (Figure 9.64).

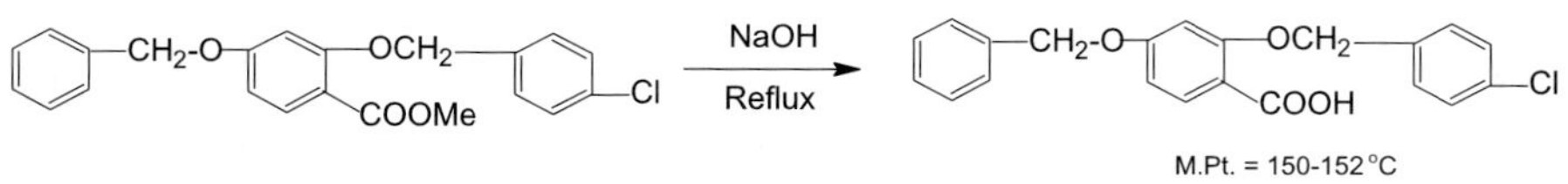


**Figure 9.64.**

## 9.15 Immune Diseases

Compounds synthesized from the resorcinol derivatives were found to be suitable for treating diseases caused by two types of immune related disorders.

The complement system found in the human blood is composed of heart labile serum proteins that combine with antibodies to eliminate the foreign cell. Complement (C) plays an important role in the mediation of immune and allergic reactions. The activation of C-complements leads to the generation of chemotactic peptides that mediate the inflammation associated with complement dependent diseases. Complement activation not only causes cell lysis but also involves in contraction of smooth muscles, release of histamine from mast cells and platelets, enhanced phagocytosis, chemotaxis of phagocytes, and activation of lymphocytes and macrophages.

The complement system can be activated via two different pathways, namely the classical pathway and alternate pathway. Once initiated, a cascade of events ensues, which results in the release of biologically active peptide fragments, the anaphylatoxins C3a, C4a and C5a. The main activity of C3a and C5a is anaphylaxis. These cause histamine release from the mast cells and basophiles, which can affect the activity of smooth muscle. Some of the clinical implications of C5a release are rheumatoid arthritis and pulmonary disorders.

Benzofuran compounds prepared from the scheme of reactions illustrated below were shown to suppress immune responses and selectively inhibit complement. In addition, these compounds interrupted the proteolytic processing of C5 to bioactive compounds by blocking the release of C5a. They also exhibited immunosuppressive activities to inhibit natural killer activity, lymphocyte proliferation, and T cell activation [143–146] (Figure 9.65).

Glucocorticoids are the most powerful anti-inflammatory agents known. In human health, glucocorticoid receptor (GR) has a major role in regulating human physiology and immune response. Steroids have been shown to interact with GR and exhibit potent anti-inflammatory properties. These steroids have shown a number of side effects such as osteoporosis, diabetes, skin atrophy (thinning of skin), muscle wasting and psychotic manifestations. Drug companies are developing potent anti-inflammatory glucocorticoids devoid of most serious side effects. Due to the detailed understanding of glucocorticoid receptor structure and its functions, new drug compounds that could activate the GR's anti-inflammatory properties without side effects are now under development.

Glucocorticoid receptor selective benzopyrano-[3,4-*f*]-quinolines and chromeno-[3,4-*f*]-quinolines, which were shown to be useful in the treatment of various immune or autoimmune diseases have been synthesized using resorcinol dimethyl ether derivative [147–149] (Figure 9.66).

## 9.16 Leukotrienes

Leukotrienes (LTs) are lipid mediators derived from arachidonic acid metabolism [150–154]. The following figure shows the pathways for the production of arachidonic acid metabolites (Figure 9.67).

Leukotrienes, thromboxanes and prostaglandins are all classified as members of the prostaglandin or eicosanoids class. Prostaglandins are synthesized from arachidonate in the cell membrane by the action of phospholipase $A_2$ enzyme. The arachidonic acid produced can be metabolized to oxygenated products by several distinct enzymic pathways. The two most important pathways are lipoxygenase and cyclooxygenase enzyme acted pathways. From the cyclooxygenase pathway, prostaglandins and thromboxanes (collectively known as prostanoids) are formed. The chemically different leukotrienes (LTs) are produced by the lipoxygenase pathway.

When arachidonic acid is metabolized by the lipoxygenase enzyme, hydroperoxy eicosatetraenoic acid (HPETE) is first produced, which is then converted into an unstable $LTA_4$. Leukotriene $A_4$ is then converted into either $LTB_4$

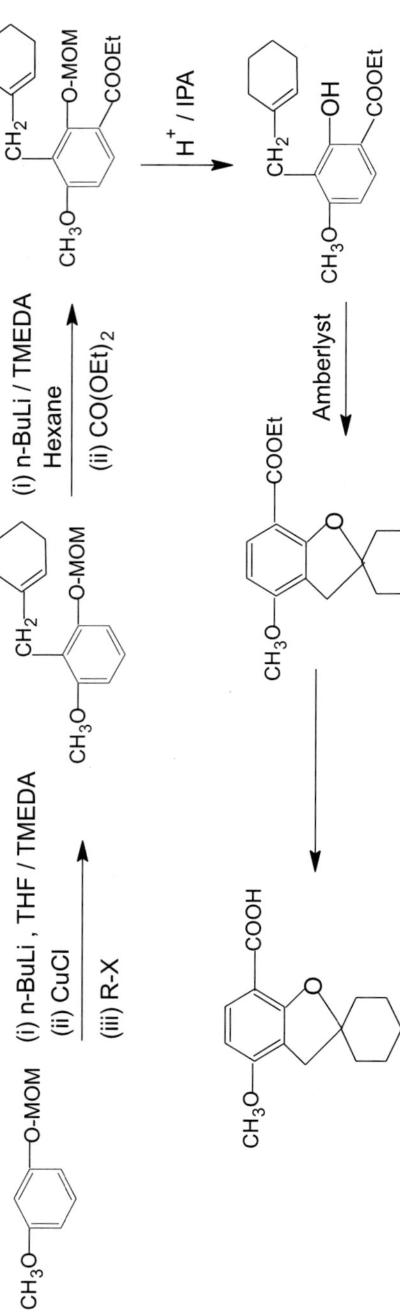


7-Carboxyl-4-methoxyspiro-(benzofuran-2[ 3H]-cyclohexane"

**Figure 9.65.**

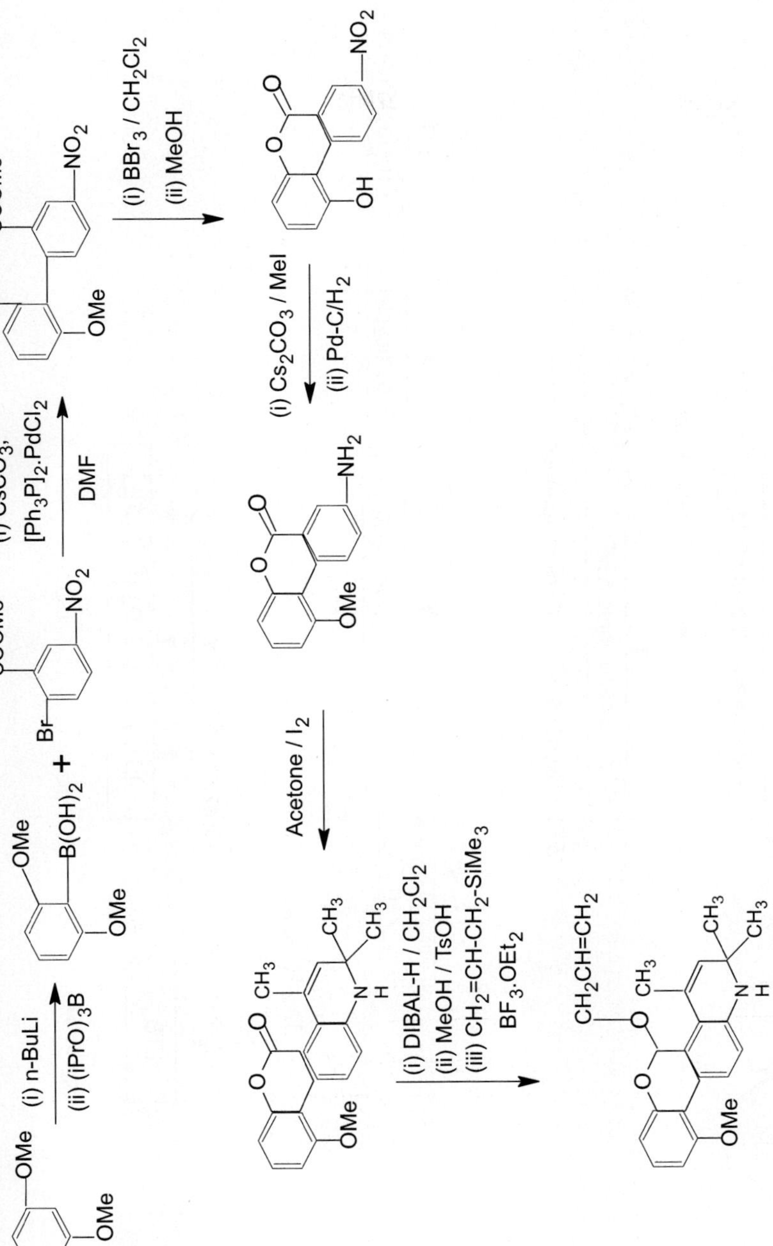


"Chromeno[ 3, 4-f]-quinoline"

**Figure 9.66.**

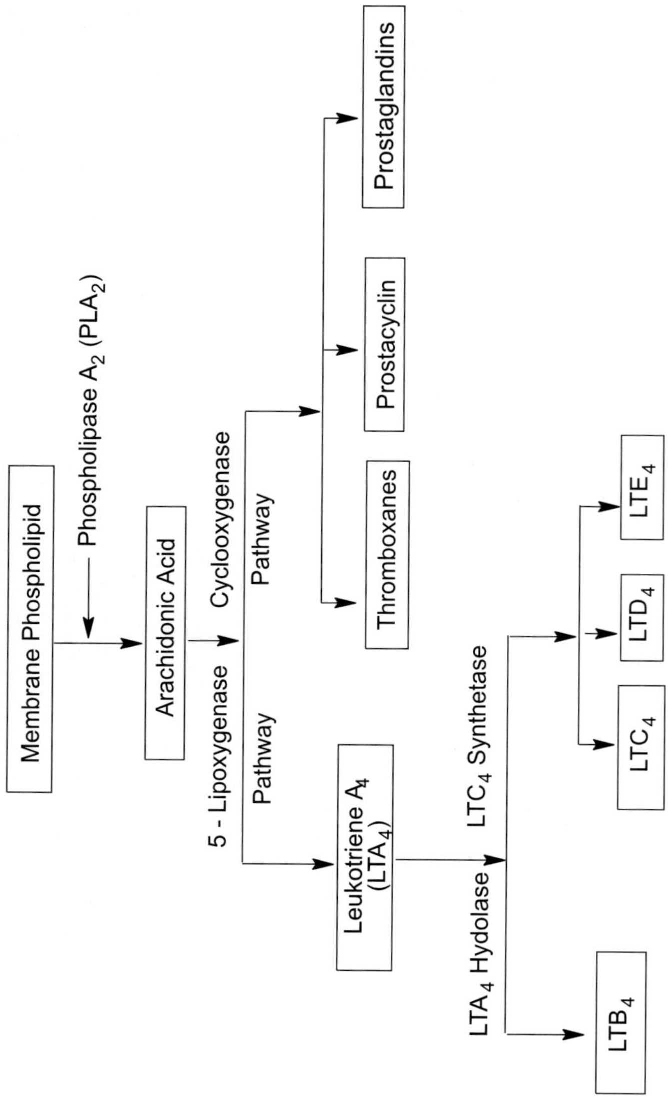
Membrane Phospholipid
Phospholipase $A_2$ ($PLA_2$)
Arachidonic Acid
Cyclooxygenase
Pathway
5 - Lipoxygenase
Pathway
Thromboxanes
Prostacyclin
Prostaglandins
Leukotriene $A_4$
($LTA_4$)
$LTC_4$ Synthetase
$LTA_4$ Hydolase
$LTB_4$
$LTC_4$
$LTD_4$
$LTE_4$

**Figure 9.67.**

or $LTC_4$ depending upon whether the reaction is metabolized by hydolase or synthetase enzymes. $LTC_4$ is actively transported out of the cells, and rapidly metabolized first into $LTD_4$ and then to $LTE_4$. Leukotrienes $LTC_4$, $LTD_4$ and $LTE_4$ are collectively referred to as the cysteinyl leukotrienes because of their chemical structures.

The leukotrienes are transported through the cell membranes into the blood and are responsible for the inflammatory and allergic reactions in lungs, spleen, brain and heart. It has been well known that the slow reacting substance of anaphylaxis (SRSA) is actually a mixture of leukotrienes. Anti-inflammatory compounds such as aspirin and ibuprofen inhibit the enzyme cyclooxygenase, and therefore decrease prostaglandin synthesis. On the other hand, anti-inflammatory corticosteroids inhibit the activation of phospholipase $A_2$ enzyme, and thus restrict prostaglandin production.

Leukotrienes are known to cause slow contractions of smooth muscles, and have been implicated as mediators of numerous other disease states, including inflammations and skin diseases. In order to prevent leukotrienes action or their synthesis various antagonist compounds were developed and used. These compounds were administered orally, intravenously, rectally, topically, intramuscularly or nasally to prevent or reverse leukotriene actions.

Resorcinol derivatives were used in the synthesis of leukotriene antagonist compounds [155]. In this section, compounds exhibiting antagonist behavior towards various leukotrienes which were primarily developed from the resorcinol chemistry are discussed.

2,4-Dihydroxy-3-propylacetophenone was used as the starting material for the synthesis of a tetrazole type compound based on the following reaction [156]. These compounds exhibited leukotriene $D_4$ antagonism, and therefore can be therapeutically used in the treatment of allergic disorders (Figure 9.68).

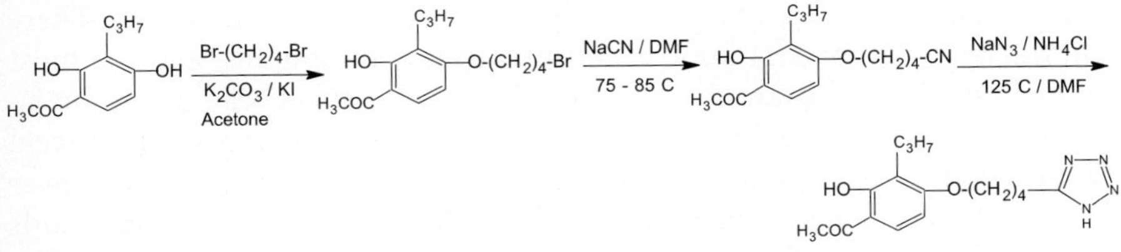


**Figure 9.68.**

To prevent or alleviate the symptoms or conditions associated with $LTD_4$, the following compounds were also developed using 2,4-dihydroxy-3-propylacetophenone derivative [157] (Figures 9.69 and 9.70).

Compounds exhibiting selective antagonistic behavior against $LTB_4$ with little or no antagonism of $LTD_4$ can be used as anti-inflammatory agents for treating bowel diseases, rheumatoid arthritis, gout and psoriasis. Though $LTB_4$, $LTC_4$ and $LTD_4$ are the products obtained from arachidonic acid, $LTC_4$

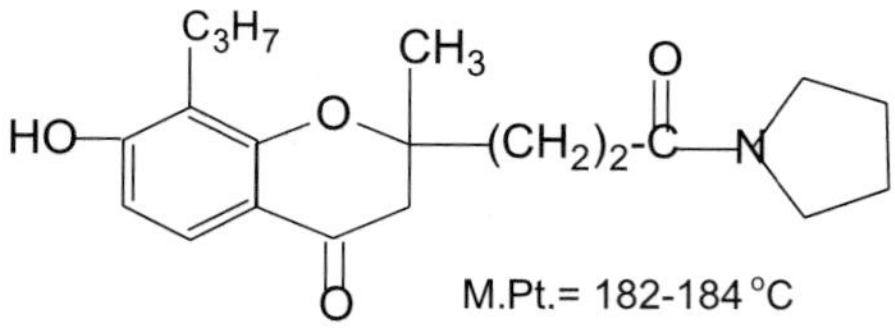


**Figure 9.69.**

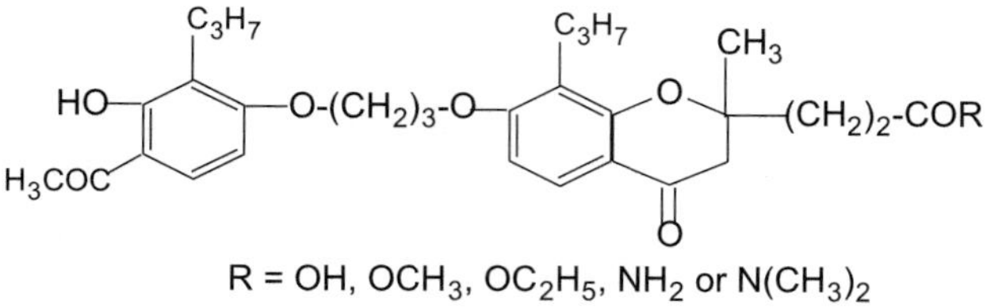


**Figure 9.70.**

and $LTD_4$ can cause the contraction of smooth muscle, human bronchi, pulmonary artery and vein. $LTB_4$ is associated with the cause of arthritis and bowel diseases. Resorcinol based compound capable of acting against the actions of $LTB_4$ was synthesized from the following reaction [158] (Figure 9.71).

Benzopyran type compounds, obtained from the resorcinol, have been shown to inhibit the actions of $LTB_4$, and therefore can be used in the treatment of $LTB_4$ induced illnesses such as rheumatoid arthritis, osteoarthritis, bowel diseases and skin disorders [159] (Figure 9.72).

## 9.17 Malaria

Malaria is a disease caused by a parasite that infects the red blood cells. There are four types of malaria known to cause disease in humans, and the most serious type is falciparum malaria. Life threatening effects can occur due to the restriction of blood flow to the vital organs as a result of malarial attack. Persons with falciparum malaria may have bleeding problems, kidney or liver failure, CNS problems, coma and die. New family of anti-malarial compounds developed based on the following resorcinol chemistry were active against falciparum malaria and have low toxicity [160] (Figure 9.73).

The thiazole carboxylic acid compounds have good chelating properties towards iron. Therefore, by forming iron complexes these compounds deprive malaria parasite of much needed iron for its metabolic process.

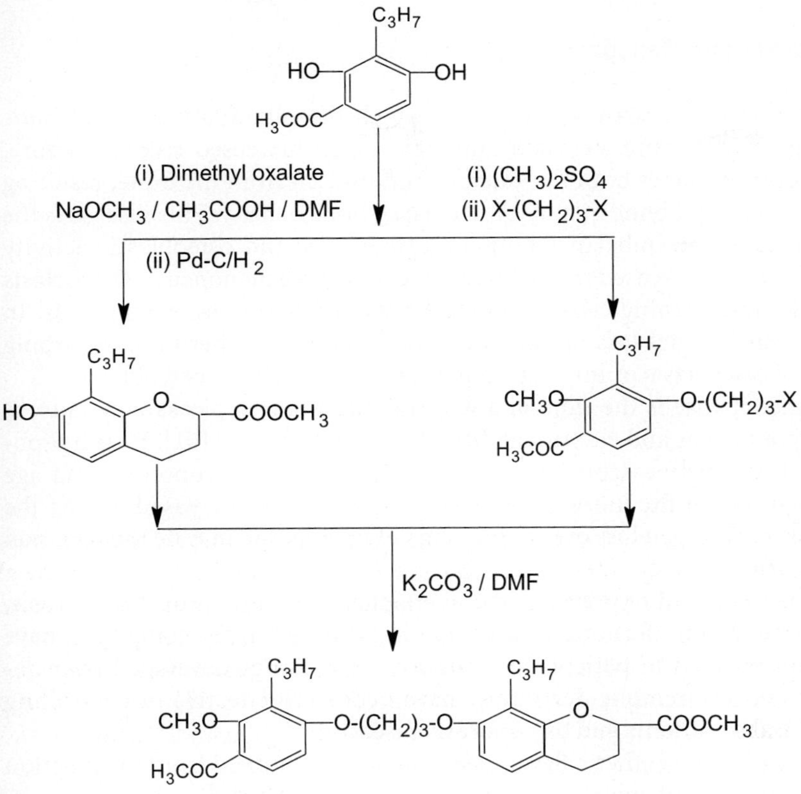


**Figure 9.71.**

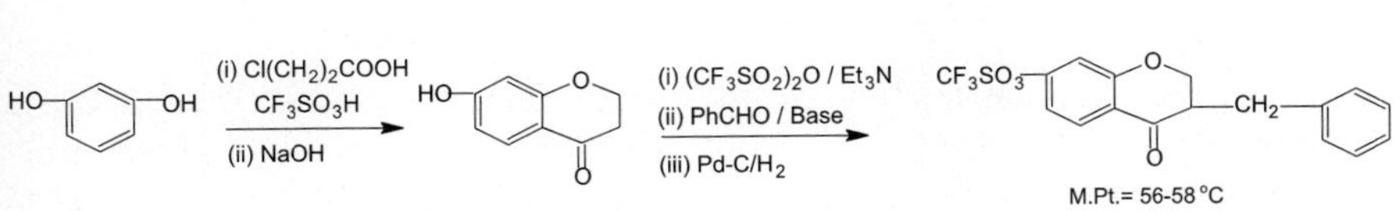


**Figure 9.72.**

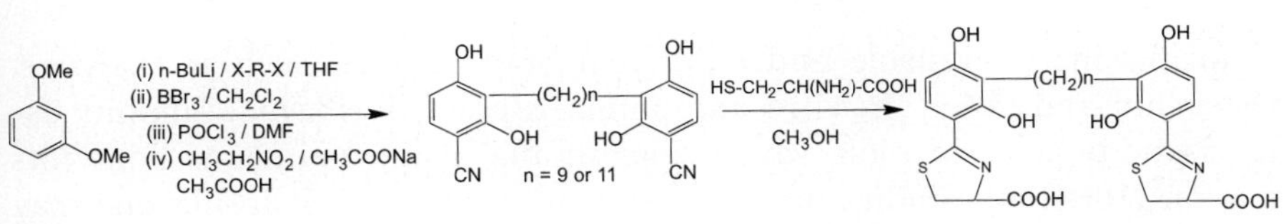


**Figure 9.73.**

## 9.18 Osteoporosis and Menopause

Osteoporosis is a disease condition in which the quantitative loss of bone has progressed beyond a certain limit causing an increased risk of fracture. Osteoporosis depletes both the calcium and protein from the bone, resulting in either abnormal bone quality or decreased bone density. This disease is the consequence of an imbalance between osteoclastic and osteoblastic activity coupled with an increased rate of bone loss occurs with menopause. Osteoclasts are the bone-resorbing cells, and osteoblasts are the bone-forming cells. In osteoporosis, the net loss of bone mass occurs due to either bone-resorbing activity of osteoclasts or impaired bone-forming activity of osteoblasts.

The menopause is the time in a woman's life when unpleasant symptoms such as hot flushes and emotional disturbances take place [161]. Health problems such as cardiovascular disease, high cholesterol, osteoporosis and age related cancers of the uterus, breast and colon may be increased during the menopause. Estrogen loss over many years of menopause may be the common cause of osteoporosis.

For the treatment or prevention of menopausal symptoms or osteoporosis, various therapeutic formulations containing different drug compound have been administered to patients at a specific time. Drugs developed from resorcinol and resorcinolic derivatives have been very effective in controlling menopausal symptoms and osteoporosis disease.

Ipriflavone is a synthetic flavanoids compound obtained from the reaction of resorcinol with phenylacetic acid [162–165] (Figure 9.74).

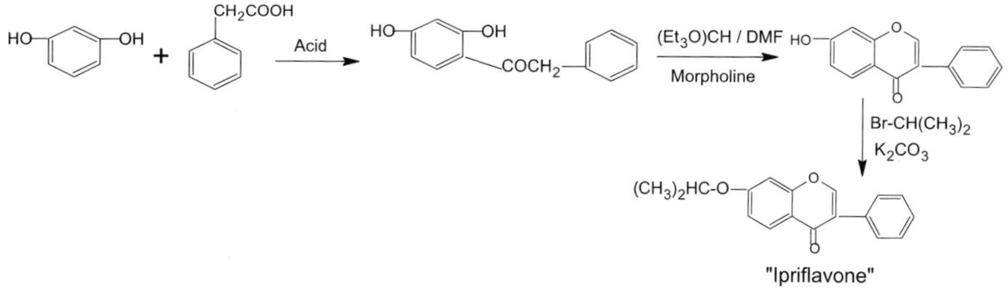


**Figure 9.74.**

Ipriflavone is available under different brand names, such as Ostivone, Osteochin and Osten. In vitro and animal studies, Ipriflavone inhibits osteoclastic bone resorption, which suggests that it may stimulate bone formation [166–168]. Studies have shown some evidence that Ipriflavone may stimulate osteoblast activity by down-regulating of endothelin receptors. Ipriflavone may also reduce the ability of endothelin-1 mineralization. Ipriflavone appeared to the incorporation of calcium into bone, and thus stopping bone loss and increasing the bone density and strength. This compound is sometimes classified as phytoestrogen, but does not activate any estrogen receptors. Ipriflavone has approved for the treatment of osteoporosis in some European countries and Japan. Ipriflavone is very effective in preventing and treating postmenopausal and senile osteoporosis.

Isoflavone compounds such as formononetin, daidzein, dihydrodaidzein and tetrahydrodaidzein were also effective in the treatment or prevention of menopausal symptoms or osteoporosis [77, 169]. These compounds were synthesized from resorcinol as can be seen in the following reaction scheme (Figure 9.75).

Dihydrodaidzein, tetrahydrodaidzein and equol compounds have shown inhibiting vasoconstrictor responses in the vascular reactivity studies on the inhibiting effect on restrictor responses to noradrenaline in the rat aortic rings. Clinical studies made on these compounds have shown that they were useful in treating menopausal syndrome, hot flushes, hypertension and atherosclerosis [2, 77].

Fluoroalkyl benzopyran compounds obtained from the following reaction showed antiestrogenic activities, and therefore can be used in the production of pharmaceutical agents for treating estrogen dependent diseases and tumors and osteoporosis [170] (Figure 9.76).

In general, isoflavone compounds, such as daidzein, have been shown to stimulate osteoblastic bone formation [171, 172]. They may have the potential to prevent bone loss with increasing age. Isoflavones may be important in maintaining healthy bones, and therefore these compounds could play a major role in the prevention of osteoporosis in healthy people.

## 9.19 Pathological Conditions – Iron Overload Diseases

Iron is an essential factor in many important cell functions, such as growth, immunological response and energy production. Iron imbalance is one of the main reasons for numerous diseases, including Alzheimer's, Parkinson and some cancers [173–175]. Too much iron is known to damage vital organs, particularly the liver, heart and pancreas. Iron overload is basically the accumulation of excess iron in body tissues. Significant iron overload results in hemochromatosis disease. The advanced stages of this disease include liver cancer, liver cirrhosis, arthritis, diabetes and heart failure. Therapeutic phlebotomy is the preferred treatment for reducing iron overload in hemochromatosis patients. In these treatments, patients use a chelator drug that attaches to iron, and in this way the iron can escape body cells and be excreted in stool and urine.

For treating pathological conditions associated with an excess of iron in humans, bioactive chelator compounds were synthesized from 2,4-dihydroxybenzaldehyde [176–181] (Figure 9.77).

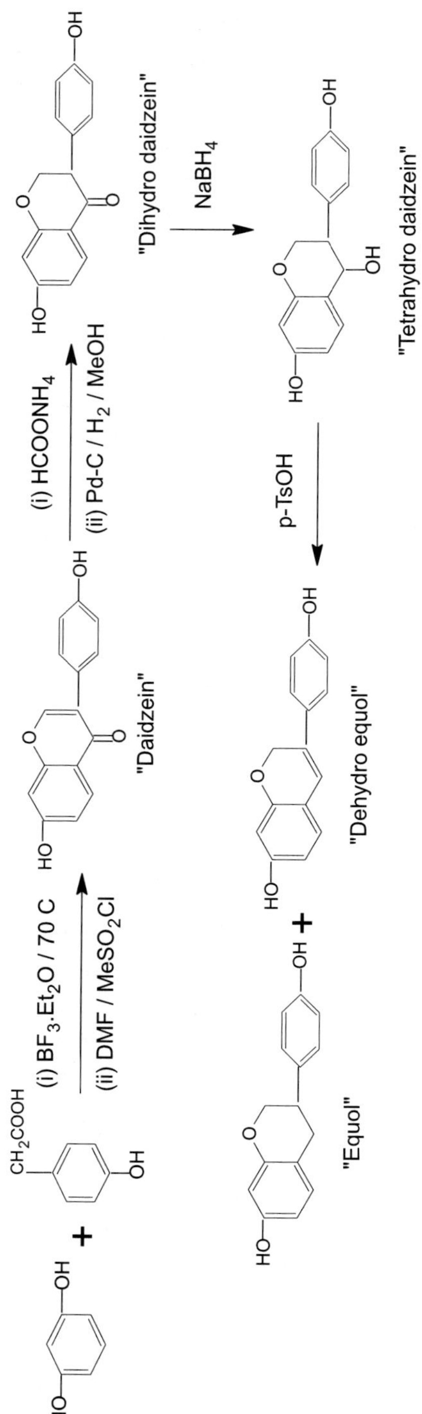
(i) BF3.Et2O / 70 C
(ii) DMF / MeSO2Cl
CH2COOH
"Daidzein"
(i) HCOONH4
(ii) Pd-C / H2 / MeOH
"Dihydro daidzein"
NaBH4
"Tetrahydro daidzein"
p-TsOH
"Dehydro equol"
"Equol"

**Figure 9.75.**

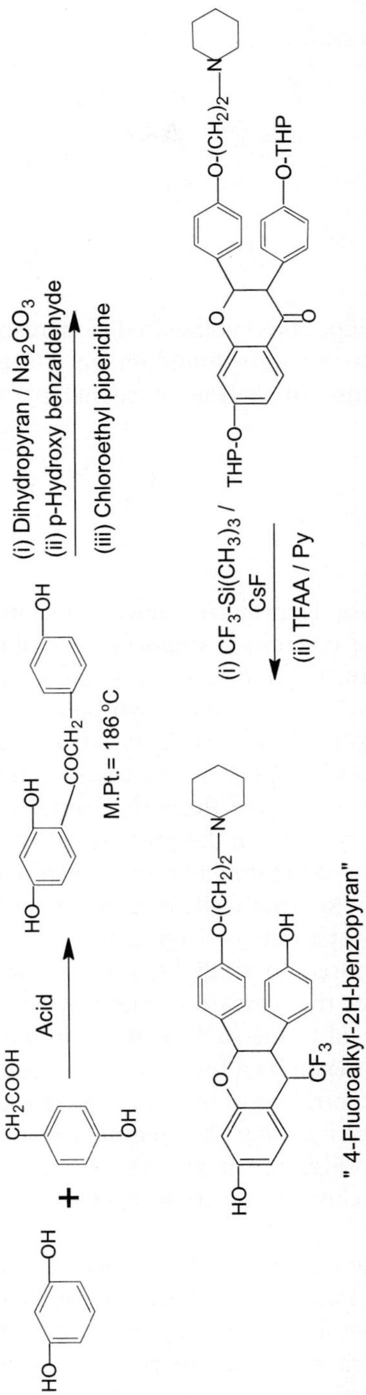
Acid
M.Pt.= 186 °C
(i) Dihydropyran / $Na_2CO_3$
(ii) p-Hydroxy benzaldehyde
(iii) Chloroethyl piperidine
(i) $CF_3$-$Si(CH_3)_3$ / CsF
(ii) TFAA / Py
" 4-Fluoroalkyl-2H-benzopyran"

**Figure 9.76.**

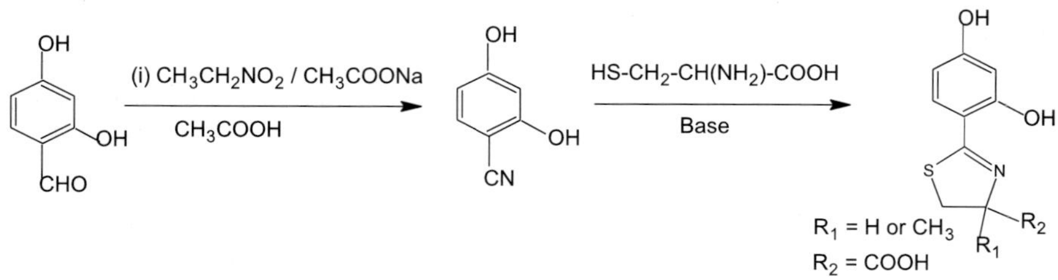


**Figure 9.77.**

The thiazoline acid compounds synthesized from resorcinol derivative and D-cysteine or 2-alkylcysteine were found to be effective in binding iron, and could find applications in dialysis, encephalopathy, osteomalacia and Alzheimer's disease.

## 9.20 Skin Lightening Agents

Normal human skin color is primarily due to melanin, hemoglobin and carotenoids. Out of these melanin is important for skin pigmentation and cosmetology. Skin pigment, melanin, is produced in melanocyte cells and consists of the orange pheomelanin and the brown-black eumelanin. Melanocyte cells are located in the lower part of dermis, and their main function is to synthesize melanin pigment, which protects the body from the damaging effects of UV radiation. Melanin is formed through a series of oxidative reactions involving the amino acid tyrosine in the presence of tyrosinase enzyme. The mechanism of formation of skin pigmentation, melanin, is a complex process, and the main stages can be schematically represented as below.

Tyrosine → Dopa → Dopamine → Dopachrome → Melanins

When the skin is exposed to sunlight, melanocyte cells increase their melanin production. After the formation, melanin is first deposited in melanosomes and then carried to the surface of the skin by keratinocytes. The end result is that the skin exhibits a brown or tan color, and the color darkness depends on the concentration of melanin. There are a number of factors that can cause uneven and irregular skin pigmentation including inflammation, excessive sunlight, pregnancy, hormone treatments, inherited genetics, aging process, toxic lever, chemical reactions from perfume and bacteria (acne vulgaris).

The damaging effects of sunlight on skin are well documented, and include erythema (sunburn) and malignant changes on the skin surface. Numerous epidemiologic studies have shown a strong relationship between sunlight exposure and human skin cancer. Acute or persistent UVB exposure can not only result in the formation of hyperpigmental lesions or regions of skin but also malignant melanoma skin cancer [182]. Malignant melanoma starts in melanocyte cells normally as a mole and then turns cancerous.

Abnormalities of human skin pigmentation occur as a result of both genetic and environmental factors (UV radiation). The excess production of melanin, known as hyper pigmentation, is closely related to melanoma and may induce melanistic skin freckles. Freckles are areas where the melanocyte cells are more active and responsive to UV radiation than in neighboring skin.

A variety of cosmetics and medicaments have been actively developed for the prevention of excess production of melanins. There is a strong demand for compounds or agents that enable the brown spots on skin (called chloasma or melasma) to be restored to a normal skin color. Compositions containing skin lightening agents, when topically applied, make the skin appear lighter or inhibit pigmentation. The effects of depigmentation are achieved by the mechanisms of either reducing pigmentation by decolorizing the melanin already present in the skin or preventing the new melanin formation. The actual mechanism of action is believed to involve the inhibitions of the enzyme tyrosinase that is involved in the metabolism of tyrosine to melanin in the melanocyte cells. For the effective function of skin lightening agents, they should inhibit tyrosinase enzyme effectively, should be permeable through both the cell and melanocyte membranes and safe and non-toxic to skin cells at the desired concentrations.

In order to deliver effectively into the skin, low molecular weight depigmenting or lightening agents are often required. With phenolic agents, their depigmenting effect is closely related to the antioxidant properties of the phenolic hydroxyl groups and their cytotoxicity within melanocyte cells [183]. 4-Substituted resorcinol compounds were observed to show tyrosinase inhibitory activities [184, 185]. From the structure-activity relationship, it has been demonstrated that 4-substituted resorcinol group is important in expressing the tyrosinase enzyme activity of the polyphenolic compounds [186–189].

Resorcinolic compounds were developed and employed in the treatment of various skin conditions including the diseases associated with tyrosinase enzymes [190–205].

Alkyl substituted resorcinols having propyl, isopropyl, hexyl and octyl groups at the 4-position have showed very good tyrosinase activity-inhibiting effect and were less toxic compared to hydroquinone [190].

Compounds synthesized from the reaction of 4-alkyl substituted resorcinols with trifluoro acetic anhydride were active against bacteria, dermatophytes, yeasts, molds and phytopathogenic fungi [191] (Figure 9.78).

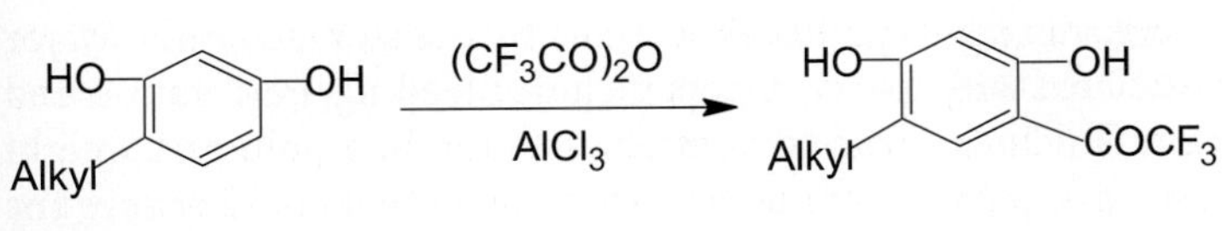


**Figure 9.78.**

These fluorinated derivatives could be useful for the therapy and treatment of mycoses, acne, dandruff, bacterial skin infections and hyperkeratotic states of the skin.

A large number of proliferative skin disorders such as mycosis fungoids, psoriasis and eczema are treated by the application of photosensitizing chemicals and UV light. Psoralens or furocoumarins are the well known photosensitizing chemicals and have been utilized in the photo-chemotherapy of psoriasis. Psoralen compounds display phototherapy effects against several kinds of malignancies and ability to halt proliferation of cells of epidermal origin. 4,5′,8′-trimethyl psoralen useful for the treatment of certain cancers (cutaneous T-cell lymphoma) and auto immune diseases (Scleroderma and phemphigus vulgaris) was synthesized from the following reaction [192, 193] (Figure 9.79).

4-Substituted resorcinol derivatives exhibiting depigmental action and show effective in inhibiting tyrosinase enzyme activity were synthesized as follows [195, 198, 201]. (Figures 9.80, 9.81 and 9.82)

These derivatives could be used in cosmetic compositions as skin lightening agents or treating pigmental blemishes.

1,3-Bis-(2,4-dihydroxyphenyl)-propane compounds synthesized from the 4-substituted resorcinol derivatives were highly effective in suppressing pigment deposition in the skin caused by the UV radiation. They had an efficacy in readily discoloring the pigment deposited on the skin, caused no skin irritation and were sufficiently stable during the storage [196, 197] (Figure 9.83).

Strong demand for compound acting as skin lightening agents which could restore the normal skin color by removing the acquired spots or freckles from the skin is continuing. The 4-(2,4-dihydroxyphenyl)-cyclohexanol, obtained from the following scheme of reactions, and its salts were found to be useful for treating inflammatory disorders such as psoriasis and acne, and also dandruff conditions [199] (Figure 9.84).

Compounds useful as skin lightening agents and also for the treatment of psoriasis, dermatitis, acne and dandruff were synthesized as shown below [202, 203] (Figure 9.85).

## 9.21 Ulcers

A peptic ulcer is a hole in the gut lining of the stomach, duodenum or esophagus. Depending upon the location of the ulcer, it can be called gastric ulcer (stomach), duodenal ulcer or esophageal ulcer. Ulcers occur in the stomach due to corrosion by the acidic digestive juices secreted by the stomach cells. Major complications associated with gastric ulcers include bleeding, perforation and gastric obstruction which can lead to vomiting, abdominal pain and weight loss. For treating ulcers, patients are using medications that could relieve the pain and prevent ulcer complications. These medications work by neutralizing

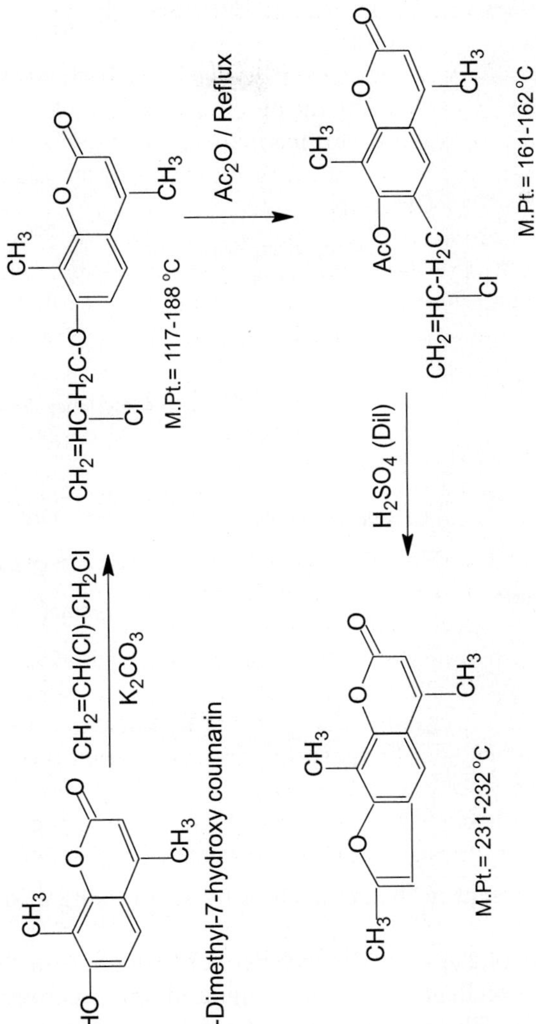
4, 8-Dimethyl-7-hydroxy coumarin
CH2=CH(Cl)-CH2Cl
K2CO3
CH2=HC-H2C-O
M.Pt.= 117-188 °C
Ac2O / Reflux
AcO
M.Pt.= 161-162 °C
H2SO4 (Dil)
M.Pt.= 231-232 °C

**Figure 9.79.**

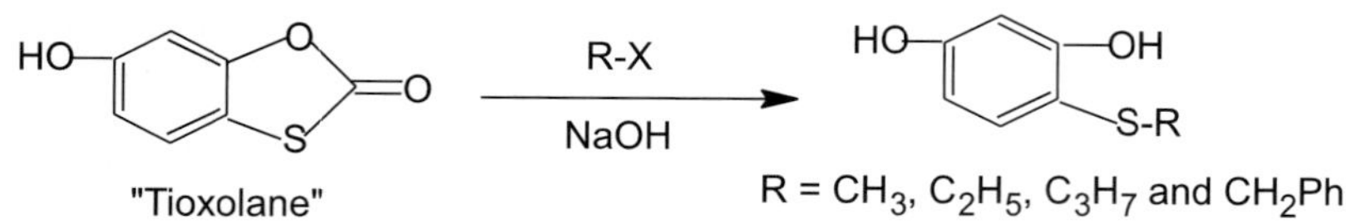


**Figure 9.80.**

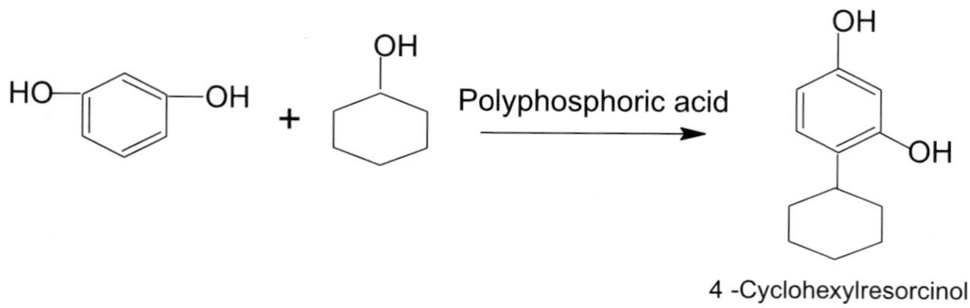


**Figure 9.81.**

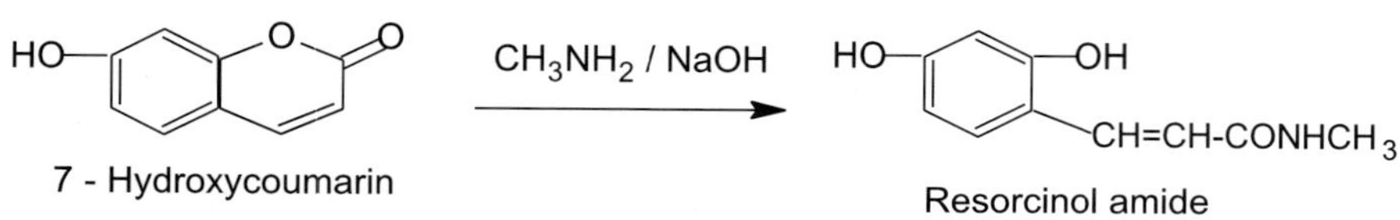


**Figure 9.82.**

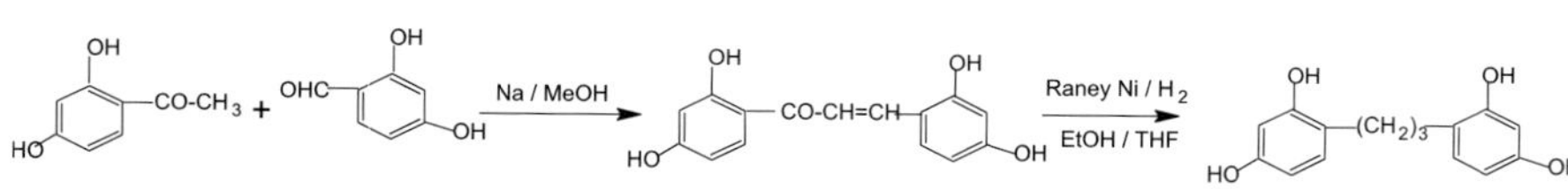


**Figure 9.83.**

the acids already present in the stomach or by suppressing acid secretion by the acid cells.

Isoliquiritigenin (4,2′,4′-trihydroxychalcone) contained in the extract of licorice root is an excellent anti-ulcer compound and has been used in the gastric ulcer therapy. This compound now can be synthesized directly from the 2,4-dihydroxyacetophenone derivative [206].

Chalcone compounds having excellent anti-gastric and anti-duodenal ulcer activities together with high absorptive ratio in the living body, and also exhibiting low acute and chronic toxicity were synthesized from 2,4-dihydroxyacetophenone [207,208] (Figure 9.86).

After administering these chalcone compounds, no unfavorable side effects were observed for CNS or autoimmune nervous system. In addition to anti-ulcer properties, they were found suitable for treating inflammatory and allergic conditions with good analgesic activity.

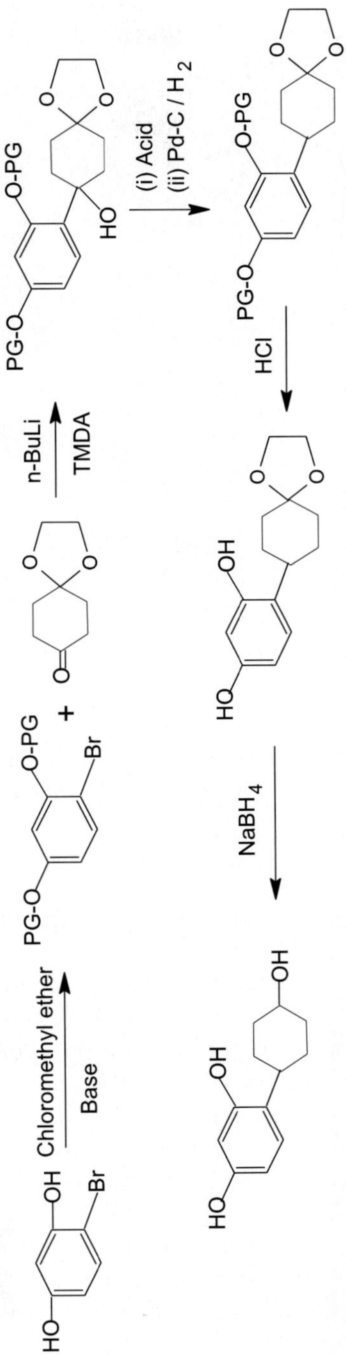
Chloromethyl ether
Base
n-BuLi
TMDA
(i) Acid
(ii) Pd-C / H 2
HCl
NaBH 4

**Figure 9.84.**

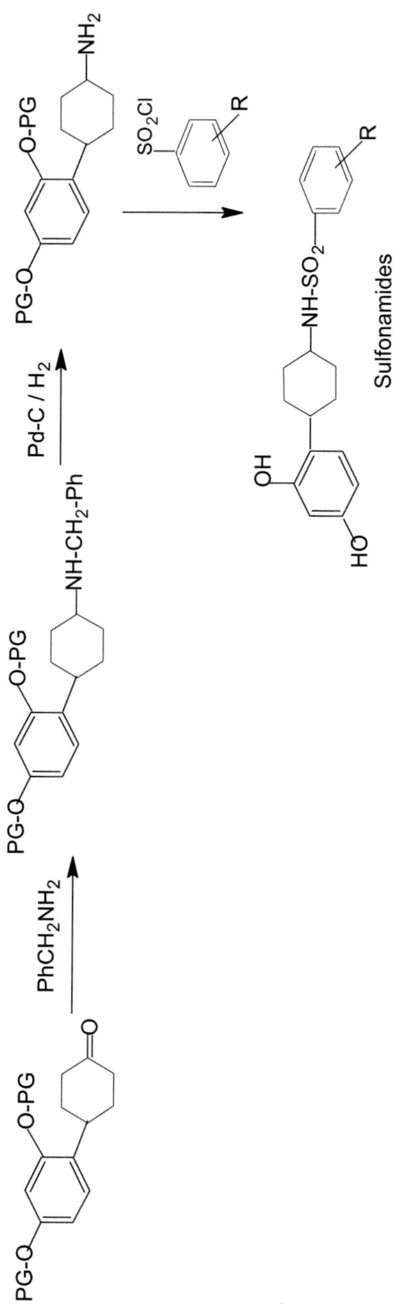

PG-O
O-PG
PhCH2NH2
NH-CH2-Ph
Pd-C / H2
NH2
SO2Cl
R
OH
HO
NH-SO2
Sulfonamides


**Figure 9.85.**

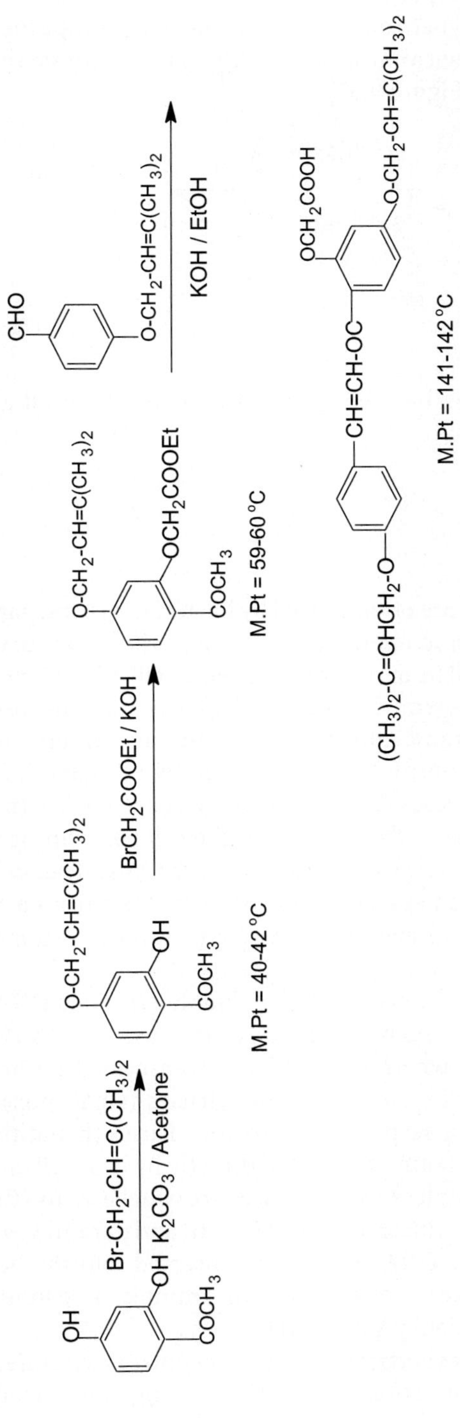
Br-CH2-CH=C(CH3)2
K2CO3 / Acetone
M.Pt = 40-42 °C
BrCH2COOEt / KOH
M.Pt = 59-60 °C
KOH / EtOH
M.Pt = 141-142 °C

**Figure 9.86.**

2-(2-Methoxy-4-propargyloxy)-1,3,4,7-tetraazaindene synthesized from methyl-2,4-dihydroxybenzoate exhibited anti-ulcer properties as well as other valuable pharmacological properties such as blood pressure and positive inotropic effect [209] (Figure 9.87).

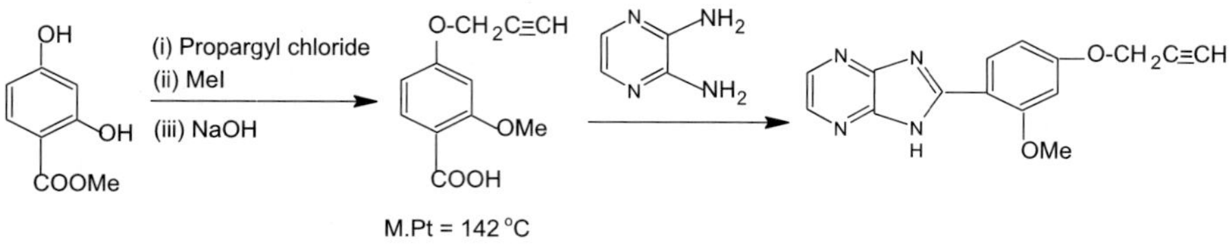


**Figure 9.87.**

This compound can be used as an active medical ingredient in human and veterinary medicines.

## 9.22 Viral Infections

Infections in humans are often caused by bacteria, viruses, parasites and fungi. Viruses are microscopic infectious organisms and their survival entirely depends on using the DNA of other living cells, called host cells, to develop and replicate. In their proliferation process, the virus first attaches to a receptor on the host cell's membrane, and then penetrates and enters the cell. Inside the cell, the virus takes control of the living cell by releasing its DNA or RNA to replicate and then releases the newly made virus particles. In this process, the infected cell dies because the virus stops the cell performing its normal functions. Now, the newly released viruses from the host dead cell go on to infect other cells. Viruses can spread by airborne droplets from a sneeze or cough, sexual contact, hand-to-eye or mouth contact, mouth to mouth or through contaminated blood.

Human viruses can be classified into double-stranded DNA viruses, single-stranded RNA viruses and retroviruses. In the case of DNA viruses, they enter the host cell nucleus where their DNA is transcribed by host polymerases. Some examples of DNA viruses are poxviruses (small pox), herpes viruses (chicken pox, shingles, herpes), and hepadnaviruses (hepatitis B). The genetic code of RNA viruses is often translated directly by host cell enzymes in the cytoplasm, and the examples of such viruses are rubella virus (German measles), rhabdoviruses (rabies), picornaviruses (polio) and paramyxoviruses (mumps). With retroviruses, the DNA copy is first inserted into the host genome, and then copied to produce a new virus. An example of a human retrovirus is Human Immunodeficiency Virus (HIV).

Antiviral drugs have been used to cure or control virus infections. The most general mechanism of action with antiviral drugs involves the inhibition of viral nucleic acid synthesis. These drugs may be active against both DNA and RNA viruses, and depend on the viral enzymes affected by the drug. Antiviral drugs or compounds have been developed using resorcinolic derivatives and tested for their effectiveness.

In humans, picornaviruses cause a number of syndromes of different degree of severity which include life threatening diseases such as poliomyelitis, hepatitis A and influenza. Hexamethylene-bis-(7-coumarin) synthesized from 7-hydroxycoumarin was found to be effective in the treatment of infections caused by picornaviruses, and particularly rhinoviruses 2 [210] (Figure 9.88).

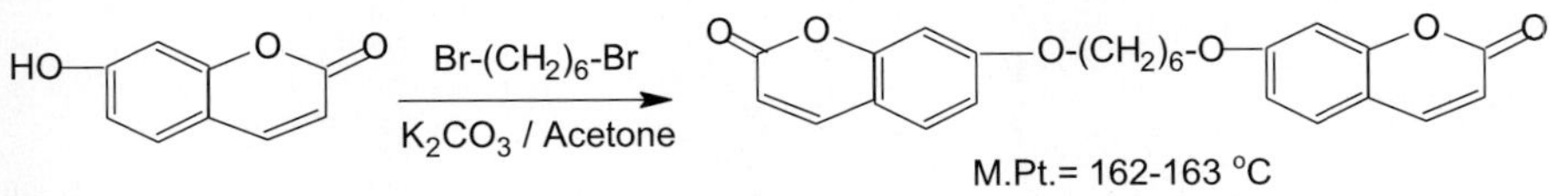


**Figure 9.88.**

Flaviviruses can cause a range of acute febrile illnesses and, encephalitic and hemorrhagic diseases. Thioxodihydroxy pyrimidine made from 2,4-dihydroxybenzaldehyde was useful in the prevention and treating of viral diseases caused by viruses in the flaviviridae family [211] (Figure 9.89).

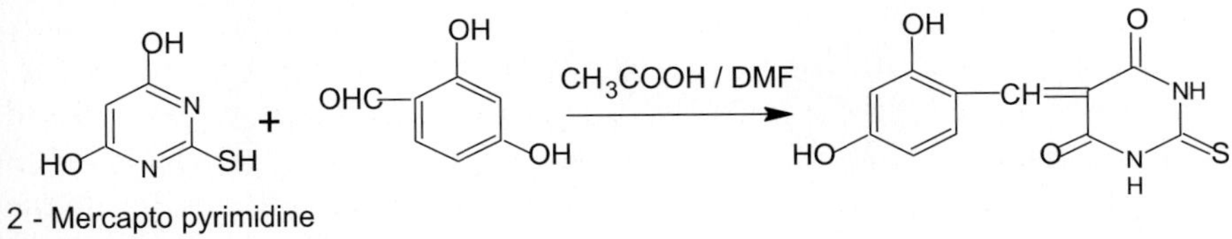


**Figure 9.89.**

Methyl-2,4-dihydroxybenzoate was used to synthesize drug compounds, as shown below, for the treatment of various disorders including viral infections, cancer, bone diseases, angiogenesis, inflammations and retenosis [212] (Figure 9.90).

4,5-Dihydro-2-(2,4-dihydroxyphenyl)-4-methylthiazole-4-(*S*)-carboxylic acid is a pharmaceutical compound synthesized from the reaction of 2,4-dihydroxybenzonitrile with 2-methylcysteine. This chemical was useful in different drug applications including antipicornaviral agents [213,214].

Retroviruses are small, single-stranded positive-sense RNA viruses. HIV is also a retrovirus, and is responsible for the Acquired Immune Deficiency Syndrome (AIDS) in human and several very unusual diseases of the immune systems of other animals. In addition, many retroviruses are well known carcinogens. Intensive efforts are currently underway to develop therapies to prevent or intervene in the development of clinical symptoms of HIV-infected patients. Suksdorfin compound was effective in inhibiting retroviral growth, replication, binding and other metabolism. Suksdorfin analogs were synthesized from the resorcinolic derivative to treat HIV type diseases [215–217] (Figure 9.91).

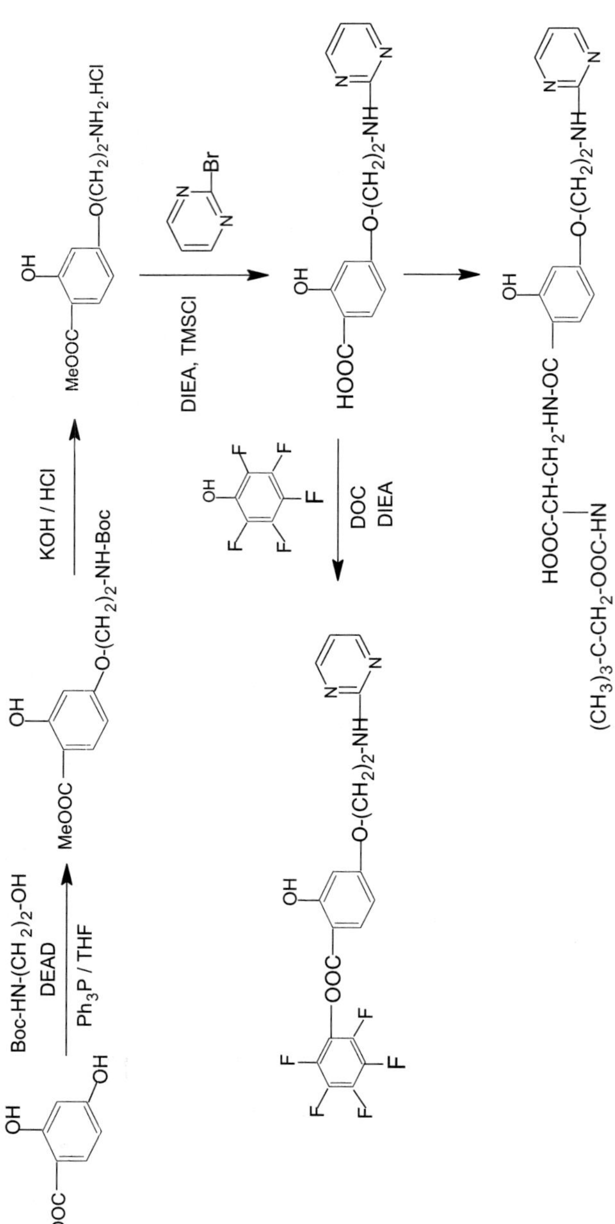

MeOOC
OH
Boc-HN-(CH 2)2-OH
DEAD
Ph3P / THF
O-(CH2)2-NH-Boc
KOH / HCl
O(CH2)2-NH2.HCl
Br
DIEA, TMSCl
HOOC
O-(CH2)2-NH
DOC
DIEA
OOC
F
HOOC-CH-CH2-HN-OC
(CH3)3-C-CH2-OOC-HN


Figure 9.90.

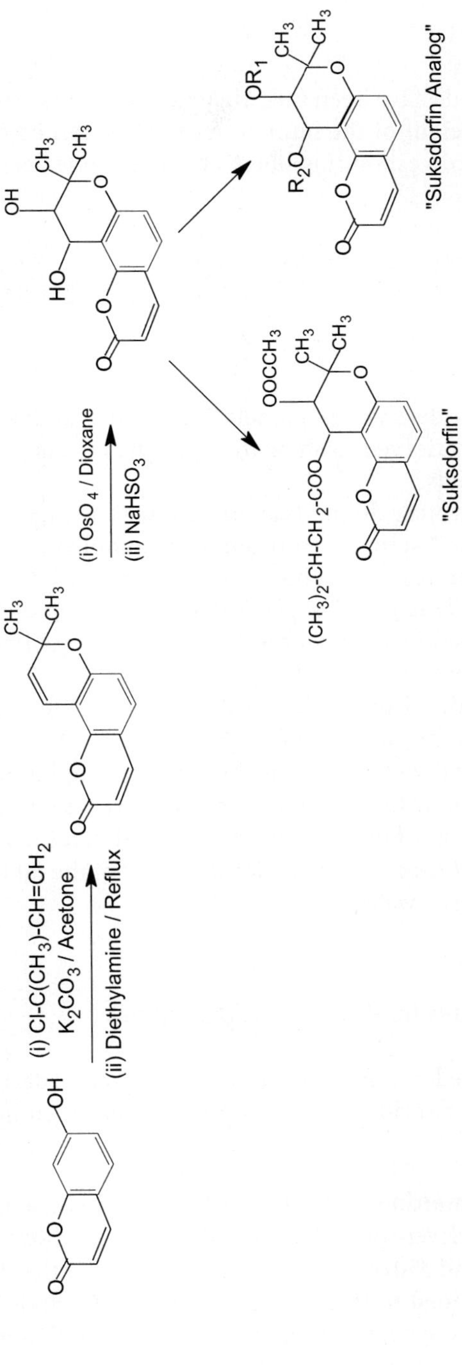
(i) Cl-C(CH3)-CH=CH2
K2CO3 / Acetone
(ii) Diethylamine / Reflux
CH3
CH3
OH
(i) OsO4 / Dioxane
(ii) NaHSO3
HO
OH
(CH3)2-CH-CH2-COO
OOCCH3
"Suksdorfin"
R2O
OR1
"Suksdorfin Analog"

**Figure 9.91.**

## 9.23 Weight Management

Isoflavone compounds have been investigated to determine their effectiveness in controlling the weight of the human body. They can be conveniently synthesized from their corresponding phenyl benzylketone derivatives [218–220] (Figure 9.92).

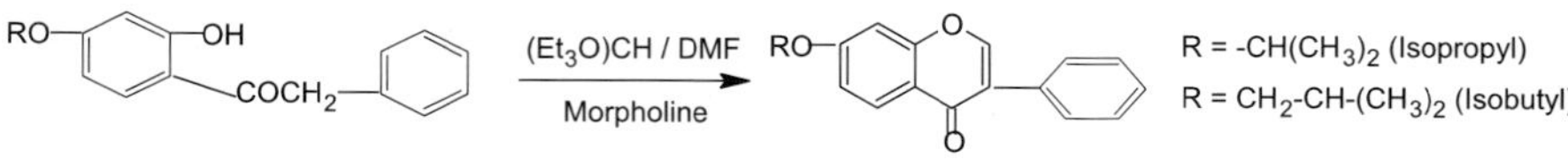


**Figure 9.92.**

The two isoflavone derivatives, namely 7-isobutoxyisoflavone and 7-isopropylisoflavone, were observed to show the opposite effects when administered to humans and animals.

7-Isobutoxyisoflavone exhibited catabolic and anorexigenic properties suitable for use in animal subjects to promote metabolism and reduce weight. Based on biochemical tests, 7-isobutoxyflavone increased the oxygenation of liver mitochondria. This could suggest that anorexigenic effect occurred at the cell level, and the biochemical mechanism of energy liberation occurred towards increased combustion [218].

On the other hand, 7-isopropyloxyisoflavone was determined to be an anabolic compound having weight gain promoting effectiveness. The anabolic effect of this chemical was tested on thinned, dystrophic patients suffering from pathological conditions, and as a result of treatment, the patients have gained weight. Analysis of the body weight showed that the weight increase of the muscle tissue was specifically greater than that of the fat tissue, and the fat content of muscle tissue was decreased [219, 220].

## 9.24 Preparative Procedures for Resorcinolic Derivatives

The preparative procedures of some important resorcinol derivatives employed in the development of various drug compounds discussed in this chapter are given below.

**Synthesis of 4-Bromoresorcinol [221]** In a 1-l flask fitted with a mechanical stirrer and a dropping funnel were placed 46.2 g (0.3 mol) of 2,4-dihydroxybenzoic acid ($\beta$-resorcylic acid) and 350 cc of glacial acetic acid. After the stirrer was started the mixture was warmed until solution results (45 °C) and then was allowed to cool to 35 °C. Through the dropping funnel was added a solution of 48 g (15 cc, 0.3 mol) of bromine in 240 cc of glacial acetic acid with vigorous stirring over a period of about 1 h. The temperature of the reaction mixture remained at 30–35 °C. When all the bromine was added the solution was poured into 5 l of water, and the mixture was cooled to 0–5 °C and allowed to stand for several hours. The fine, white crystals of 2,4-dihydroxy-5-bromobenzoic acid were collected on a 10-cm. Büchner funnel and washed with about 500 cc of cold water. The crude product, after air drying at room temperature, melted at 194–200 °C and weighed 55–60 g. For purification it was dissolved in 1.5 l of boiling water, and the solution was refluxed for 1 h, filtered while hot, and cooled in an ice bath. The material which crystallized was collected, washed with 100 cc of cold water, and air dried. The yield of colorless 2,4-dihydroxy-5-bromobenzoic acid, melting at 206.5–208.5 °C, was 40–44 g (57–63% of the theoretical amount).

Thirty grams of purified 2,4-dihydroxy-5-bromobenzoic acid was refluxed for 24 h with 375 cc of water, and the resulting solution was filtered, cooled, and extracted with a 400-cc and a 200-cc portion of ether. The ether was removed by evaporation, and the 4-bromoresorcinol was dried on a steam bath. The yield of product melted at 100–102 °C was 22–22.5 g (90–92% of the theoretical amount).

**Preparation of 1-Chloro-2,4-dimethoxybenzene [222]** A 30 l, multi-neck jacketed flask, equipped with a thermocouple, mechanical stirrer, nitrogen purge and peristaltic pump was charged with 4-chlororesorcinol (1.4 kg) and water (4.94 kg). The jacket temperature was set at 20 °C and the reactor was charged with 50% NaOH (1.8 kg) keeping the temperature below 35 °C. The reactor was then charged with *tert*-butyl methyl ether ("TMBE") (5.85 kg) and a nitrogen purge was started through the reactor. The reactor was charged with dimethylsulfate (2.66 kg) and the mixture stirred for approximately 18 h at ambient temperature. Stirring was stopped and the layers allowed to separate and the aqueous layer was drained from the reactor. Stirring was started and the reactor charged with water (6.68 kg) and 2% ammonium hydroxide (0.76 kg) and the stirring was then stopped and layers allowed to separate. The aqueous layer was drained from the reactor. The TMBE was removed by rotary evaporation, the residue dissolved in heptane (2.7 kg) and the vacuum filtered through silica gel and solvent removed by rotary evaporation. The residue was distilled to remove heptane to give 1-chloro-2,4-dimethoxybenzene as pale yellow oil.

**1-Bromo-2,4-dibenzyloxybenzene [223]** A mixture of 75.0 g (0.397 mol) of 4-bromoresorcinol, 95.1 ml (0.80 mol) of benzylbromide and 331 g (2.4 mol) of anhydrous potassium carbonate in 400 ml of *N*, *N*-dimethylformamide was stirred for 12 h at 25 °C and for 4 h at 85 °C. The reaction mixture was cooled and added to 1 l of ice-200 ml pentane-100 ml ether. The organic phase was washed with three 500 ml portions of water, dried over magnesium sulfate and evaporated to oil. The oil was rapidly chromatographed on 400 g of silica gel eluted with 20% ether-pentane to yield 80 g of oil. The chromatographed oil was crystallized from pentane at 0 °C to yield 45.0 g (30%) of the title compound, m.p. 37–38 °C.

**3-Benzyloxy-4-bromophenol [223]** To a 0 °C slurry of 1.7 g (42.5 mmol) of potassium hydride in 35 ml of *N*, *N*-dimethylformamide was slowly added a solution of 7.22 g (38.2 mmol) of 4-bromoresorcinol. The resultant mixture was stirred for 30 min and then 4.54 ml (38.2 mmol) of benzyl bromide was slowly added. The reaction mixture was stirred 3 h longer at 0 °C and then added to 200 ml of cold water and 200 ml of ether. The ether extract was washed twice with 200 ml portions of water, dried over magnesium sulfate and evaporated to an oil. The crude oil was purified via column chromatography on 400 g of silica gel eluted with 25% ether-pentane to yield (in order of elution) 2.2 g (16%) of 2,4-dibenzyloxybromobenzene, 0.21 g (2%) of 5-benzyloxy-2-bromophenol and 3.52 g (33%) of 3-benzyloxy-4-bromophenol.

**Synthesis of 2,4-Dihydroxybenzaldehyde [224]** A 1-l three neck flask equipped with a temperature thermocouple and an overhead stirrer was charged with DMF (49.3 g, 0.675 mol) and acetonitrile (150 ml). The flask was treated with $POCl_3$ (88.16 g, 0.575 mol) in acetonitrile dropwise over 20 min so that the temperature was maintained at 22 – 28 °C with a water bath. This was stirred at ambient temperature for 1 h to insure complete conversion to the Vilsmeier reagent. The solution remained clear throughout. The reaction was cooled in a dry-ice bath to – 14 to – 17 °C and a solution of resorcinol (55.06 g, 0.5 mol) in acetonitrile (150 ml) was slowly added to maintain – 10 to – 17 °C during the addition. Precipitation of the Vilsmeier formamidinium-phosphoro-dichloridate occurred during this addition. The reaction was stirred for an additional 2 h at (– 15±2) °C and then at 28 – 32 °C for 1 h The HPLC of the reaction solution showed < 5% of the starting material. (A small amount of an unknown with similar retention time to the 2,4-dihyroxybenzaldehye was present.) The reaction was cooled to 5 °C and after 1 h the product was isolated by filtration and rinsed with cold acetonitrile. The product was dried at 30 °C at 10 mm of Hg to constant weight. The light yellow solid, 123 g had a melting point of 158 – 159 °C; HPLC wt/wt assay of 98.3% by hydrolysis to the title compound in the HPLC mobile phase. The corrected yield was 80%. The product can be stored in the dark at 5 °C for several weeks without significant coloration.

**Synthesis of 2,4-dibenzyloxybenzaldehyde [225]** A mixture of 2,4-dihydroxybenzaldehyde (17.5 g), potassium carbonate (42 g) and benzyl bromide (35 ml) in dimethyl formamide (100 ml) was stirred at room temperature for 18 h. The reaction mixture was poured into vigorously stirred ice-water (300 ml) and stirring was continued for 30 min. The resulting brown solid was filtered and dried to give 2,4-dibenzyloxybenzaldehyde (28.5 g), m.p. 80 – 81 °C.

**Synthesis of 4-Benzyloxy-2-hydroxybenzaldehyde [224]** A 2-l three neck flask equipped with a temperature thermocouple and an overhead stirrer was charged with 2,4-dihydroxybenzaldehyde (103.59 g, 0.75 mol, 1 eq), sodium bicarbonate powder (71.83 g, 0.855 mol, 1.14 eq), potassium iodide (12.45 g, 0.075 mol, 0.1 eq), and acetonitrile (700 ml). The reaction was heated to 60 °C and benzyl chloride (123.43 g, 0.975 mol, 1.30 eq) was added over 1 min. The reaction was stirred and heated at reflux until the PAR (% area response) of 2,4-dihydroxybenzaldehyde was 4–6% by HPLC assay. The solvent was removed in vacuo, and the solution was treated with water (260 ml), concentrated hydrochloric acid (12 ml), and ethyl acetate (500 ml, at 30 °C). The layers were separated, and the aqueous layer extracted with additional ethyl acetate (150 ml). The organic layers were combined and washed successively with 3% aqueous potassium carbonate saturated with ethyl acetate (twice), water, 5% aqueous citric acid, and brine. The organic layer was evaporated in vacuo (rotovap). The residue was dissolved by heating in TBME (500 ml) at 40–50 °C and on cooling to 30 °C; hexane (620 ml) was added. The resulting precipitate was stirred for 2 h at 5–10 °C, and the product collected by filtration. The flask was rinsed with cold TBME-hexane (200 ml of a 1:1 (v/v) mixture). After air drying for 1.5 h, the product was dried at <35 °C for 16 h (10 mm of Hg). The product, 126.2 g, produced an HPLC wt/wt assay of 93.1% compared to a purified standard; the corrected yield was 69%; mp 76–78 °C.

**Synthesis of 2,4-Dibenzyloxybenzyl Alcohol [225]** A solution of 2,4-dibenzyloxybenzaldehyde (2.26 g) in methanol (20 ml) was treated portionwise with sodium borohydride (0.268 g) over 5 min. Water (5 ml) was added dropwise to the reaction mixture. The resulting white solid was filtered, dried and crystallized from a mixture of ether and petroleum ether to give 2,4-dibenzyloxybenzyl alcohol as a white solid, m.p. 87 °C.

**Synthesis of 2,4-Dibenzyloxybenzonitrile [225]** A mixture of 2,4-dibenzyloxybenzaldehyde (1.59 g), diammonium hydrogen phosphate (3.5 g) and *n*-propyl nitrate (15 ml) in glacial acetic acid (5 ml) was refluxed for 15 h. The reaction mixture was evaporated to dryness in vacuo and the residue was stirred with water (100 ml). The insoluble material was filtered and dried at 50 °C to yield 2,4-dibenzyloxybenzonitrile as a tan colored solid (0.6 g, 40%), m.p. 100 °C.

**Preparation of 2,4-Dihydroxybenzonitrile [176]** A mixture of 2,4-dihydroxybenzaldehyde (5.0 g, 36.2 mmol), sodium acetate (5.94 g, 72.4 mmol), nitroethane (5.44 g, 72.4 mmol) and glacial acetic acid (10 ml) was refluxed for 6 h. After cooling, the mixture was poured onto ice (100 g) and extracted with ethyl acetate (four times 50 ml). The combined organic layers were washed with saturated $NaHCO_3$ until the pH of the aqueous layer remained at 8, dried ($Na_2SO_4$) and the solvent removed in vacuo. Flash chromatography ($SiO_2$, cyclohexane: ethyl acetate = 1 : 1) afforded 2,4-dihydroxybenzonitrile (2.87 g, 59%) as a pale yellow solid.

**Process for Producing 2,4-Dihydroxyacetophenone [226]** Resorcinol (110 g, 1 mol), 440 g (7.3 mol) of acetic acid, 16 g of diisopropyl ether and 5.5 g of Amberlyst 15 (strongly acidic ion-exchange resin) were charged into a 500 ml three-necked flask equipped with a thermometer and also with a distillation column having a Dean-Stark trap (water separator) and a reflux condenser, followed by heating to an inner temperature of 124 °C. While discharging water distilled off from the Dean-Stark trap, the reaction of resorcinol and acetic acid was continued at the temperature for 10 h. The reaction mixture was cooled down to room temperature, and then the strongly acidic ion-exchange resin was removed by filtration. The filtrate was subjected to distillation under reduced pressure to remove about 380 g of the acetic acid and diisopropyl ether. 640 g of water containing 1.3 g (0.013 mol) of sulfuric acid was added to the residue, and the mixture was refluxed for 6 h to give a homogeneous solution. The thus obtained solution was cooled down to room temperature and the precipitates were collected to give 109 g (0.72 mol) of 2,4-dihydroxyacetophenone (melting point: 144.8 to 145.5 °C, yield: 72%). The analysis of the 2,4-dihydroxyacetophenone through gas chromatography revealed that the purity of the product was 99.3%.

**Synthesis of 2,4-Dibenzyloxyacetophenone [225]** A mixture of 2,4-dihydroxyacetophenone (15.2 g), potassium carbonate (30.4 g), potassium iodide (0.5 g) and benzyl bromide (37.6 g) in dimethyl formamide (200 ml) was stirred at room temperature for 18 h. The reaction mixture was filtered and the insoluble material washed with a little dimethyl formamide. The combined filtrate plus washings were evaporated to dryness and the residue partitioned between ether and water. The organic phase was dried over magnesium sulphate, evaporated and the residual pink-red colored solid crystallized from cyclohexane to give 2,4-dibenzyloxyacetophenone (28.4 g) as a pale pink colored powder, m.p. 64–65 °C.

**Synthesis of 5-Acetyl-2,4-dihydroxybenzaldehyde [22]** To a mixture of 49.8 g of 2,4-dimethoxybenzaldehyde, 23.6 ml of acetyl chloride, and 1000 ml of methylene chloride cooled to approximately 0 °C were added 119.7 g of aluminum chloride with stirring under a nitrogen atmosphere. The reaction was allowed to warm to room temperature while stirring overnight. The mixture was added to slurry of ice and concentrated hydrochloric acid. The organic layer was separated and washed with a saturated sodium chloride solution, dried over sodium sulfate, and concentrated in vacuo. The residue was purified by high pressure liquid chromatography over silica gel eluting with a 0–40% ethyl acetate in hexane gradient. The appropriate fractions were pooled and concentrated to provide 5.85 g of the desired title intermediate, m.p. = 142–143 °C.

**Synthesis of 5-Acetyl-2,4-dihydroxybenzyl alcohol [22]** One gram of 5-acetyl-2,4-dihydroxybenzaldehyde was hydrogenated in the presence of 1 g of 10% palladium on carbon in 150 ml of ethanol. After about 1 h, hydrogen uptake ceased and hydrogenation was terminated. The reaction mixture was filtered and concentrated in vacuo. The residue was dissolved in approximately 30 ml of ethyl acetate and cooled in the refrigerator. The resulting solid was recovered by filtration affording 0.66 g of the desired title intermediate, m.p. = 151–152 °C.

**Preparation of 4-Benzyloxy-2-hydroxyacetophenone [28]** In a dry round-bottom flask under nitrogen, 2,4-dihydroxyacetophenone (15.2 g, 100 mmol) was dissolved in methyl ethyl ketone (400 ml) and dimethylsulfoxide (100 ml). To this solution were added benzyl bromide (17.0 g, 100 mmol) and potassium carbonate (27.6 g, 200 mmol). The reaction was heated to reflux and stirred for 15 h. The methyl ethyl ketone was removed in vacuo, and the dimethylsulfoxide solution was diluted with ethyl acetate and washed several times with brine. The organic material was collected, dried (magnesium sulfate), filtered, and concentrated to provide a dark solid. The solid was recrystallized from hexane/toluene to provide the title benzyl ether as a tan solid (12.8 g, 55.7%); m.p. 143–144.5 °C.

**Synthesis of 3-Propyl-2,4-dihydroxyacetophenone [63]**

**Step A: 4-Allyloxy-2-hydroxyacetophenone**

Mix 100 g of 2,4-dihydroxyacetophenone, 88 g of allyl bromide and 120 g of potassium carbonate in 250 ml of acetone. Reflux while stirring overnight. Filter and strip the residue to dryness. Dissolve the residue in 700 ml of ether and add 250 ml of 4 N aqueous sodium hydroxide. Separate the solids by filtration and wash with ether. Suspend the solids in a mixture of water and ether (1:1) and acidify with 10% aqueous hydrochloric acid. Separate the ether layer and extract the aqueous layer with ether. Combine the ether extracts, wash with water, dry and strip to an oil (yield 100.2 g) which is employed in the next step without further purification.

**Step B: 3-Allyl-2,4-dihydroxyacetophenone**

Heat 50 g of the 4-allyloxy-2-hydroxyacetophenone of Step A at 210–220 °C under a nitrogen atmosphere with stirring for 8 h. Cool and boil the residue in 250 ml of carbon tetrachloride. Cool and allow the residue to stand overnight. Separate the solids by filtration to obtain the title product (m.p. 132–134 °C).

**Step C: 2,4-Dihydroxy-3-propylacetophenone**

Dissolve 9.6 g of 3-allyl-2,4-dihydroxyacetophenone in 100 ml of ethanol; flush with nitrogen and add 0.5 g of 5% palladium/charcoal catalyst. Hydrogenate in a Parr apparatus at room temperature until 45 p.s.i. of hydrogen is taken up. Filter and strip the filtrate to dryness to obtain the title product. (m.p. 125–127 °C).

**Synthesis of 4,6-Diacetoresorcinol [22]** A mixture of 3 g of 4,6-diacetoresorcinol dimethyl ether, 50 ml of acetic acid, and 50 ml of 48% hydrobromic acid was refluxed for 2 h. After concentration in vacuo, the residue was partitioned between ethyl acetate and a saturated sodium chloride solution. The layers were separated and the organic solution was washed with a saturated sodium bicarbonate solution. Evaporation of the solvent provided crude product which was purified by high pressure liquid chromatography over silica gel eluting with a 0–50% ethyl acetate in hexane gradient. The appropriate fractions were combined and concentrated in vacuo to provide 0.9 g of the desired title intermediate, m.p. = 182–184 °C.

**Synthesis of 2-Allyl-4,6-diacetylresorcinol [227]** 4,6-Diacetylresorcinol (19.4 g; 0.1 mol), sodium hydroxide (8.0 g; 0.2 mol) and allyl bromide (12.1 g; 9 ml; 0.1 mol) in water (160 ml) and 1,2-dichloroethane (150 ml) were stirred at reflux for 3 h. The mixture was cooled to room temperature, the organic layer was separated and the aqueous layer was extracted with 1,2-dichloroethane (3 × 50 ml). The combined organic layers were dried (anhydrous sodium sulphate) and evaporated to give 2-allyl-4,6-diacetylresorcinol as a pink oil (9 g; 38%) which rapidly crystallized, m.p. 84–86 °C.

**Synthesis of 4,6-Diacetyl-2-*n*-propylresorcinol [227]** 4,6-Diacetylresorcinol (600 g; 3.09 mol), sodium hydroxide (247 g; 6.18 mol) and *n*-propyl bromide (645 g; 477 ml; 5.25 mol) in water (2.47 l) were stirred at reflux for 22 h. The mixture was cooled to room temperature, diluted with water (2 l), and acidified with concentrated hydrochloric acid. Filtration gave a light brown solid which was washed with water (2 l), sucked dry and recrystallized from ethanol (three times) to give a white crystalline solid (408 g), m.p. 84–92 °C. A portion (300 g) of the solid was recrystallized from chloroform (600 ml)/petroleum-ether (60–80 °C; 1200 ml) to give 4,6-diacetyl-2-*n*-propylresorcinol (225 g; 30%) as a white crystalline solid, m.p. 97–99 °C.

**Preparation of 2-Propyl-1,3-dimethoxybenzene [228]** 1,3-Dimethoxybenzene (20 g, 145 mmol) in 200 ml of dry tetrahydrofuran (THF) was cooled to – 10 °C. To this solution at – 10 °C added *n*-BuLi (100 ml of a 1.6 mol/l solution in hexane, 160 mmol) over 20 min. The reaction was then stirred for 2.5 h at 0 °C. At 0 °C, propyliodide (24.65 g, 145 mmol) was added slowly over 15 min. When the addition was complete, the reaction was allowed to warm to room temperature and stirred overnight. After stirring overnight, the reaction was refluxed for 1.5 h then cooled to room temperature and quenched with ice. The THF was removed under vacuum, and the resulting aqueous layer was extracted several times with $Et_2O$. The organic extract was dried over $MgSO_4$ and filtered to give clear oil after solvent removal (26.11 g). The oil was purified by vacuum distillation (24.0 g, 92%). Bpt. 80–82 °C at 10 mm Hg.

**Preparation of 2-Propyl-1,3-dihydroxybenzene [228]** A mixture of solid 1,3-dimethoxy-2-propylbenzene (33.70 g, 190 mmol) and solid pyridine hyrochloride (150 g, 1.30 mol) were warmed to 180 °C. After 7.5 h, the reaction was cooled to 110 °C and 50 ml of $H_2O$ was added slowly. After the reaction cooled to room temperature, it was diluted with 100 ml of water and extracted several times with EtOAc. The EtOAc extract was washed once with 2 N HCl and then dried over $MgSO_4$. Filtration and solvent removal gave 38.5 g of an orange solid. The product was purified by recrystallization from dichloromethane providing 11.86 g (41%) of yellow crystals

**Preparation of 2,4-dimethoxyphenylacetic Acid [127]** In a 250 ml round-bottomed flask with a reflux condenser, 20.0 g (0.11 mol) 2,4-dimethoxyacetophenone, 19.32 g (0.22 mol) morpholine (19.3 ml) and 7.12 g (0.22 mol) sulphur were heated for ca. 20 h at 135 °C. Following this, any remaining morpholine was removed in a vacuum. The remaining thiomorpholide was saponified without further purification.

**Saponification:** The brown oil was mixed with 90 g 50% KOH in 160 ml of ethanol and refluxed for 6 h. Following this a large part of the alcohol was distilled off, it was diluted with water and solid constituents filtered out. It was then ice-cooled, acidified with concentrated hydrochloric acid, extracted three times with dichloromethane, dried over $MgSO_4$, filtered and concentrated in a water jet vacuum. The raw product was recrystallized from water as beige crystals: m.p.: 106–108 °C; yield: 51%.

**Synthesis of Daidzein [2]** Resorcinol (29 mmol) and 4-hydroxyphenyl acetic acid (29 mmol) were dissolved in freshly distilled borontrifluoride etherate (20 mol eq) under nitrogen. The resulting mixture was stirred and heated at 70 °C overnight. The reaction was monitored by TLC (80% $Et_2O$/Hexane). The resulting mixture was cooled down at room temperature, and then *N,N*-dimethylformamide (46.2 ml) was added dropwise. The mixture was again heated up at 50 °C for 30 min, then methanesulphonyl chloride (7 ml in 10 ml DMF) was added dropwise and the resulting mixture was heated at 60–70 °C until LC (80% $Et_2O$/Hexane) shown that the reaction was almost finished, about 10 h. After cooling down at room temperature, the mixture was poured into 400 ml ice-cold water. The precipitate was filtered. The filtrate was collected and dried. The cure product was recrystallized in 94% ethanol (aq) and gave a quite pure daidzein (3 g) in 44% yield.

**Preparation of Isoflavone Formononetin [229]** In a 1 l three-necked RB flask fitted with a reflux condenser, a dropping funnel and stopper, *p*-methoxyphenylacetic acid (50.0 g, 0.3 mol), resorcinol (33.2 g, 0.3 mol) and borontrifluoride etherate (194 ml, 1.53 mol) were stirred with heating (90 °C) for 1 h and cooled to 10 °C. To this cooled solution, DMF (230 ml) was added drop-wise with stirring. In another 500 ml RB flask fitted with a drying tube, DMF (390 ml) was cooled to 10 °C. To this cooled DMF, *p*-toluenesulfonyl chloride (170.0 g, 0.9 mol) was added in small portions and the mixture was allowed to stand at room temperature for 20 min. This mixture containing $N,N'$-dimethyl (chloromethylene) ammonium chloride was then added to the above reaction mixture and the mixture was stirred at room temperature for 1 h. The dark orange-yellow solution was then poured slowly into boiling dilute HCl (0.1 N) with vigorous stirring and allowed to stand for 30 min. During this process, the yellow precipitate slowly became white with the formation of formononetin. The product was filtered off and washed with water. The product was further purified by recrystallization from aqueous methanol giving a yield of 80%. The melting point was 256–257 °C.

**Synthesis of 2,4,4′-Trimethoxy-α-methyldesoxybenzoin [2]** 2-(*p*-Methoxyphenyl) propionic acid (0.39 g, 4 mmol) and 1,3-dimethoxybenzene (0.5 g, 0.5 ml, 4 mmol) were mixed in polyphosphoric acid (PPA) (10 g) and the reaction mixture was mechanically stirred at 75 °C for 5 h. The reaction mixture was then allowed to cool to room temperature and mechanically stirred for a further 12 h. The reaction was then quenched with ice water and the product extracted with $CHCl_3$ (3 × 25 ml). The $CHCl_3$ layer was dried ($Na_2SO_4$) and the solvent removed under reduced pressure. The residual crude product was purified by silica gel column chromatography (eluent 7 : 2 $CH_2Cl_2$ : EtOAc) to give the pure 2,4,4′-trimethoxy-α-methyldesoxybenzoin (0.68 g, 58%).

**Synthesis of 2,4,4′-Trihydroxyphenyl-α-methyldesoxybenzoin (*O*-Desmethylangolensin or O-DMA) [2]** 2,4,4′-Trimethoxy-α-methyldesoxybenzoin (0.312 g 1.04 mmol) was dissolved in dry $CH_2Cl_2$ (10 ml). To this solution five equivalents of 10 mol/l $BBr_3$ in hexane (1.3 g, 5.2 ml, 5.2 mmol) was added slowly and the reaction mixture allowed to stir under $N_2$ at room temperature for 6 days. Reaction was quenched with ice/water and after stirring for 1 h the product was extracted with diethylether (3 × 25 ml). The ether layer was dried ($Na_2SO_4$) and the solvent removed under reduced pressure. The residual crude product was purified by silica gel column chromatography (eluent 7 : 1 $CH_2Cl_2$ : EtOAc) to give the pure 2,4,4′-trihydroxyphenyl-α-methyldesoxybenzoin.

**Synthesis of 2,4-Dimethoxyphenol [230]** In a 500 ml flask were placed 5.0 g (30.1 mmol) of 2,4-dimethoxybenzaldehyde and 150.4 ml of methylene chloride (0.2 mol/l). This mixture was vigorously stirred with a magnetic stir bar. To the homogeneous solution were added 7.7 ml of 30% aqueous hydrogen peroxide (75.0 mmol 2.5 equivalents) and 4.5 ml of formic acid (120.0 mmol, 4.0 equivalents). The flask was fitted with a reflux condenser and heated to reflux for 22 h with stirring. After cooling, 97.5 ml of 1.5 N sodium hydroxide (146 mmol, 4.86 equivalents) was added to the flask. The mixture was stirred for 15 min. The organic layer was separated and concentrated to a residue using a rotary evaporator. The residue was combined with the aqueous solution and 65.0 ml of methanol was added. The solution was stirred for 30 min. The methanol was removed using a rotary evaporator.

The neutral materials were removed from the aqueous residue by extracting with two 100 ml portions of methylene chloride. The solution was adjusted to a pH of 1 to 2 with concentrated hydrochloric acid. The 2,4-dimethoxyphenol was extracted with three 100 ml portions of methylene chloride. The organic solution containing the neutrals as well as the one containing the product were separately dried over anhydrous magnesium sulfate and filtered into tared round-bottom flasks. The methylene chloride was removed using a rotary evaporator. A total of 0.267 g of neutrals was recovered. A total of 3.52 g of the crude 2,4-dimethoxyphenol was obtained (22.9 mmol, 76.0% yield). The 2,4-dimethoxyphenol was purified utilizing bulb-to-bulb distillation. A mass of 2.94 g of 2,4-dimethoxyphenol as brown liquid was obtained (19.1 mmol, 63.4% yield). The purity was determined by GC (97.9%) as well as HPLC (>99.9%).

**Preparation of 2,6-Dihydroxybenzoic acid [231]** Fifty parts of resorcinol was dissolved in 160 parts of ethanol, and 62.8 parts of anhydrous potassium carbonate was added thereto. After heating to 140 °C, the mixture was kept at 140 °C for 4 h under a carbon dioxide gas pressure of 14 kg/cm$^2$ while allowing to absorb 14 parts of carbon dioxide. To the obtained reaction mixture containing 47 parts of a resorcylic acid mixture (63.1% of 2,6-dihydroxybenzoic acid and 36.9% of 2,4-dihydroxybenzoic acid), 300 parts of water was added and sulfuric acid was further added thereto until the pH value of the mixture reached 6. After distilling off 150 parts of a mixture of ethanol and water, the residue was refluxed at 98 to 100 °C for 3 h to decompose the 2,4-dihydroxybenzoic acid. During the decomposition reaction, the pH value of the reaction mixture was controlled at 6 by appropriately adding sulfuric acid. The undecomposed 2,4-dihydroxybenzoic acid was monitored liquid-chromatographically and thus the end point of the decomposition was determined.

After completion of the decomposition reaction, sulfuric acid was added to the mixture until the pH value reached 3 and the insoluble matters thus formed were separated by filtration. Sulfuric acid was added to the filtrate until the pH value thereof reached 1. After cooling to 5 °C, crystals of 2,6-dihydroxybenzoic acid thus formed were separated on a filter, washed with water and dried at 70 °C. Thus 25 parts of a dry cake of 2,6-dihydroxybenzoic acid was obtained. The result of the composition analysis on this product by liquid chromatography indicated that the purity of the 2,6-dihydroxybenzoic acid was 99.0%.

**Preparation of 2,4-Dihydroxybenzoic acid [232]** A 35-1 solids mixer was charged with 10 kg of sodium bicarbonate, 5 kg of potassium bicarbonate and 3.3 kg of resorcinol and heated with mixing to 120 °C in a stream of carbon dioxide. After 3 h the contents were cooled down with mixing, and 281 of water were added. The hot suspension at 60 °C was then discharged and brought to pH 3 at room temperature with 20 kg of concentrated hydrochloric acid. The suspension was cooled to 10–15 °C and filtered on a filter press, and the filter residue was washed with water and dried, leaving 4.1 kg (81%) of 2,4-dihydroxybenzoic acid; purity (HPLC): 99.5%.

**Preparation of Resorcinol Diacetate [233]** 110 g (1.0 mol) of resorcinol and 255 g (2.5 mol) of acetic anhydride were added to a reaction vessel fitted with a distillation head, condenser, thermometer, and receiving vessel and heated to a temperature of 100 °C for a period of about 1.0 h under an atmosphere of nitrogen. The temperature was then raised to 130 °C to distil acetic acid and finally to 140 °C to distil the excess acetic anhydride. The resulting product was purified by vacuum distillation under a pressure of about 10 mm Hg. The yield of purified resorcinol diacetate was about 85 to 90% by weight.

## 9.25 Summary and Outlook

As can be seen from this chapter, resorcinol and its derivatives have been successfully used in the development and manufacturing of pharmaceutically most important drug compounds to treat the devastating diseases that affect the whole world. Drug discovery to cure certain disease is remarkably a robust field, and has become an increasingly time consuming and expensive process. Modern medications to treat the old and new diseases are constantly emerging. Resorcinol based chemistry could help the pharmaceutical companies to synthesize advanced intermediates and active drug compounds for developing and introducing promising drugs for the treatment of life-threatening illnesses quickly into the market place.

## References


1. P. Duthinh, US Patent, 5,547,671 (1996).
2. G. Kelly et al., US Patent, 6,649,648 (2003).
3. P. Lewis, S. Kaltia and K. Wahala, J. Chem. Soc., Perkin Trans. 1, p 2481 (1998).
4. W. Keung, O. Lazo, L. Kunze and B. Valle, Proc. Natl. Acad. Sci. USA, Vol. 93, p 4284 (1996).
5. W. Keung, A. Klyosov and B. Valle, Proc. Natl. Acad. Sci. USA, Vol. 94, p 1675 (1997).
6. W. Keung and B. Valle, Proc. Natl. Acad. Sci. USA, Vol. 95, p 2198 (1998).
7. B. Valle and W. Keung, US Patent, 6,121,010 (2000).
8. B. Valle and W. Keung, US Patent, 6,255,497 (2001).
9. G. Gao, D. Li and W. Keung, Bioorg. Medicin. Chem., Vol. 11, p 4069 (2003).
10. B. Love and F. Huang, US Patent, 4,476,125 (1984).
11. M. Carson, R. LeMahieu and W. Nason, US Patent, 4,507,498 (1985).
12. J. Musser and K. Kees, US Patent, 4,528,392 (1985).
13. M. Miyano and R. Shone, US Patent, 4,565,882 (1986).
14. M. Miyano, R. Shone and D. Sohn, US Patent, 4,778,903 (1988).
15. M. Ohashi et al., US Patent, 5,023,364 (1991).
16. H. Takagaki et al., US Patent, 5,428,059 (1995).
17. J. Gazave, US Patent, 3,352,754 (1967).
18. C. Hall and E. Glenn, US Patent, 4,279,930 (1981).
19. A. Beck, A. Sallmann, R. Lang and P. Wenk, US Patent, 4,670,604 (1987).
20. A. Beck, A. Sallmann, R. Lang and P. Wenk, US Patent, 4,808,604 (1989).
21. A. Beck, A. Sallmann, R. Lang and P. Wenk, US Patent, 4,820,726 (1989).
22. N. Bollinger, T. Goodson and D. Herron, US Patent, 4,945,099 (1990).

23. N. Bollinger, T. Goodson and D. Herron, US Patent, 5,098,613 (1992).
24. N. Bollinger, T. Goodson and D. Herron, US Patent, 5,294,613(1994).
25. J. Schnabel, Science, Vol. 260, p 1719 (1993).
26. Chemical & Engineering News, p 8, March 22 (2004).
27. J. Clemens, M. Sofia and D. Stephenson, US Patent, 5,478,857 (1995).
28. J. Clemens, M. Sofia and D. Stephenson, US Patent, 5,563,164 (1996).
29. S. Gorbach, US Patent, 5,733,926 (1998).
30. T. Clakson et al., US Patent, 5,952,374 (1999).
31. K. Korsgaard, M. Shalmi and B. Guldhammer, US Patent, 5,827,873 (1998).
32. J. Fixler and L. Styles, Pediatr. Clin. North Am., Vol. 49, p 1193 (2002).
33. E. Vichinsky, Lancet, Vol. 360, p 629 (2002).
34. G. Kneen, US Patent, 4,410,537 (1983).
35. G. Kneen, US Patent, 4,535,183 (1985).
36. E. Klarmann, J. Am. Chem. Soc., Vol. 48, p 2358 (1926).
37. Leonard and Wood, J. Am. Med. Assoc., Vol. 85, p 24 and 1855 (1925).
38. H. Hirzel, US Patent, 1,717,105 (1929).
39. L. Kyrides, US Patent, 1,974,821 (1934).
40. W. Austin, US Patent, 2,030,423 (1936).
41. E. Bousquet, US Patent, 2,146,007 (1939).
42. W. Austin, US Patent, 2,060,654 (1936).
43. L. Kyrides, US Patent, 2,171,494 (1939).
44. C. Starks, US Patent, 4,093,667 (1978).
45. J. Kraal, A. Hussain, S. Gregorio and E. Akaho, J. Dent. Res., Vol. 58, p 2125 (1979).
46. E. Rosenberger, US Patent, 5,174,313 (1992).
47. E. Rosenberger, US Patent, 5,280,796 (1997).
48. E. Rosenberger, US Patent, RE 35,439 (1997).
49. F. Hartman and V. Schelling, US Patent, 2,318,388 (1943).
50. C. Devia et al., Rev. Microbiol., Vol. 29, No. 4 (1998).
51. A. Kharazmi et al., US Patent, 5,985,935 (1999).
52. A. Kharazmi et al., US Patent, 6,609,046 (2003).
53. S. Nicosia, V. Capra and G. Rovati, Pulm. Pharm. Thera., Vol. 14, p 3 (2001).
54. W. Busse, Ann. Allergy Asthma Immunol., Vol. 81, p 17 (1998).
55. J. Drazen, Leukotrienes, Asthma and Rhinitis, Blackwell Scientific (1995).
56. P. Korenblat, Hospital Physician, p 50, December (2000).
57. C. Fitzmaurice and T. Lee, US Patent, 3,419,578(1968).
58. R. Altounyan, US Patent, 3,671,625 (1972).
59. B. Warren and J. Spicer, US Patent, 3,899,513 (1975).
60. A. Nohara, T. Umetani, Y. Miyata and Y. Sanno, US Patent, 3,984,441 (1976).
61. D. Buckle and H. Smith, US Patent, 4,211,791 (1980).
62. H. Cairns, K. Brown and R. Altounyan, US Patent, 4,239,748 (1980).
63. J. Rokach, P. Hamel and R. Hirschmann, US Patent, 4,252,818 (1981).
64. A. Nohara and Y. Maki, US Patent, 4,672,073 (1987).
65. R. Young and J. Rokach, US Patent, 4,666,928 (1987).
66. A. Nohara and Y. Maki, US Patent, 4,714,774 (1987).
67. P. Belanger et al., US Patent, 4,820,867 (1989).
68. S. Baker et al., US Patent, 5,462,954 (1995).
69. M. Dombroski, K. Koch and A. Piscopio, US Patent, 6,117,874 (2000).
70. J. Holdie and A. Coldman, Drug Resistance in Cancer, Cambridge University Press, (1998).
71. R. Ozols, Drug Resistance in Cancer Therapy, Kluwer Academic, Boston (1989).

72. Today's Chemist at Work, September 2003, p 8 (2003).
73. H. Takagaki et al., US Patent, 5,068,364 (1991).
74. R. Rosen et al., US Patent, 6,498,195 (2002).
75. J. Pavitan, J. Indones. Med. Assoc., November 2001, pp 1–6 (2001).
76. M. Messina, The Simple Soybean and Your Health, Avery Publishing Group, Garden City Park, New York (1994).
77. G. Kelly and G. Jonnou, US Patent Appln., 2003/0,018,060 (2003).
78. W. Van Der Vijgh et al., US Patent Appln., 2002/0,147,353 (2002).
79. H. Buchholz and S. Perruchon, US Patent Appln., 2004/0,059,100 (2004).
80. J. Sawyer, S. Spaethe and J. Starling, US Patent, 5,543,428 (1996).
81. G. Jedlitschky, I. Leier and D. Keppler, US Patent, 6,673,813 (2004).
82. E. Bombardelli and P. Valenti, US Patent, 6,608,089 (2003).
83. R. Kapil, S. Durani, J. Dhar and B. Setty, US Patent, 5,254,568 (1993).
84. F. Labrie, Y. Merand and S. Gauthier, US Patent, 6,060,503 (2002).
85. S. Gauthier et al., J. Med. Chem., Vol. 40, p 2117 (1997).
86. J. Simard et al., Int. J. Cancer, Vol. 73, p 104 (1997).
87. G. Pluorde and B. Fisher, Molecules, Vol. 7, p 315 (2002).
88. I. Sircar, US Patent, 4,351,950 (1982).
89. S. Potter, E. Henley and D. Waggle, US Patent, 5,855,892 (1999).
90. K. Wahala et al., J. Chem. Soc, Perkin Trans. 1, p 642 (2001).
91. A. Salakka and Wahala, 2nd Int. Elect. Conf. Symp. Org. Chem., Sept. 1–30 (1998).
92. M. Shlyankevich, US Patent, 5,523,087 (1996).
93. K. Tsujihara et al., US Patent, 5,830,873 (1998).
94. K. Tsujihara et al., US Patent, 5,731,292 (1998).
95. A. Adams et al., US Patent, 5,859,051 (1999).
96. N. Korsgarrd, et al., US Patent, 6,008,242 (1999).
97. A. Adams et al., US Patent, 6,090,836 (2000).
98. P. Kym et al., US Patent, 6,593,480 (2003).
99. P. Kym et al., US Patent, 6,329,534 (2001).
100. K. Liu, L. Xu and A. Jones, US Patent, 6,569,879 (2003).
101. T. Fort et al., US Patent, 6,610,890 (2003).
102. S. Fakhoury et al., US Pat. Appln., 2003/0,225,083 (2003).
103. T. Sargent, A. Shulgin and C. Mathis, US Patent, 4,647,446 (1987).
104. T. Matray, V. Hernandez and S. Singh, US Patent, 6,673,550 (2004).
105. S. Menchen et al., US Patent, 6,649,598 (2003).
106. J. Bloomer and C. Leah, US Pat. Appln., 2004/0,014,160 (2004).
107. J. Celebuski, US Patent, 5,247,099 (1993).
108. S. Rovner, Depression, C&E News, Feb. 9, 2004, p 33 (2004).
109. D. Gardner, US Patent, 4,093,631 (1978).
110. H. Hausberg et al., US Patent, 4,508,732 (1985).
111. H. Hausberg et al., US Patent, 4,780,478 (1988).
112. J. Freedman, US Patent, 3,859,306 (1975).
113. B. Mueller et al., US Patent, 5,100,914 (1992).
114. M. Kelly et al., US Patent, 2003/0,153,599 (2003).
115. M. Kelly et al., US Patent, 6,638,972 (2003).
116. R. Mewshaw, US Patent, 5,541,199 (1996).
117. R. Mewshaw, US Patent, 5,670,667 (1997).
118. R. Mewshaw, US Patent, 5,684,039 (1997).
119. C. Gruber et al., N. Engl. J. Med., Vol. 346, p 340 (2002).
120. J. Gustafsson, J. Endocrinol., Vol. 163, p 379 (1999).

121. K. Pettersson and J. Gustafsson, Ann. Rev. Physiol., Vol. 63, p 165 (2001).
122. A. Barron-Gonzalez at al., Salud. Publ. Mexico, Vol. 43, No. 6, p 1 (2001).
123. O. Emerson and E. Bickoff, US Patent, 2,884,427 (1959).
124. E. Bickoff and A. Booth, US Patent, 2,890,116 (1959).
125. E. Bickoff, A. Livingston and A. Booth, US Patent, 3,027,382 (1962).
126. K. Horoya et al., J. Chem. Soc., Perkin Trans. 1, p 4339 (2000).
127. E. Angerer, S. Erber and M. Schneider, US Patent, 5,470,854 (1995).
128. P. Miller et al., US Pat. Appln., 2003/0,087,955 (2003).
129. P. Miller et al., US Patent, 6,589,980 (2003).
130. R. Konojia et al., US Pat. Appln., 2003/0,216,463 (2003).
131. E. McInerney et al., Endocrinol., Vol. 139, p 4513 (1998).
132. B. Bariaam, J. Folmer and T. Piser, US Pat. Appln., 2004/0,039,015 (2004).
133. P. Cowburn and J. Cleland, J. Clin. Basic Cardiol., Vol. 2, p 41 (1999).
134. C. Barberis, B. Mouillac and T. Durroux, J. Endocrin., Vol. 156, p 223 (1998).
135. G. Haas and J. Young, Acute Heart Failure Management, In: E. Topol, Ed., Textbook of Cardiovascular Medicine, Lippincott William and Wilkins, Philadelphia (2002).
136. H. Havera, J. Johnson and H. Vidrio, US Patent, 3,995,041 (1976).
137. F. Valle, US Patent, 4,452,811 (1984).
138. E. Wu, US Patent, 4,495,198 (1985).
139. W. Spitzer, US Patent, 4,537,889 (1985).
140. H. Boshagen et al., US Patent, 4,590,291 (1986).
141. M. Bock et al., US Patent, 5,464,788 (1995).
142. P. Astles et al., US Patent, 6,211,234 (2001).
143. R. Sindelar et al., US Patent, 5,173,499 (1992).
144. R. Sindelar et al., US Patent, 5,366,986 (1994).
145. R. Sindelar et al., US Patent, 5,401,767 (1995).
146. R. Sindelar et al., US Patent, 5,808,109 (1998).
147. Y. Ku et al., US Patent, 6,518,430 (2003).
148. Y. Ku et al., US Patent, 6,673,930 (2004).
149. M. Coghlan, US Patent, 6,506,766 (2003).
150. B. Samuelson, A new Class of Mediators of Immediate Hypersensitivity Reactions and Inflammations, In: B. Samuelson, R. Paoletti and P. Ramell, Eds., Advances in Prostaglandins, Thromboxane and Leukotriene Research, Raven Press, New York (1983).
151. R. Lewis and K. Austen, J. Clin. Invest., Vol. 73, p 889 (1984).
152. P. Loftus and R. Bernstein, J. Org. Chem., Vol. 48, p 40 (1983).
153. R. Bernstein et al., J. Med. Chem., Vol. 31, p 692 (1988).
154. P. Bernstein, Am. J. Respir. Crit. Care Med., Vol. 157, p S220 (1998).
155. W. Marshall, S. Sigmund and C. White Sitt, US Patent, 4,777,299 (1998).
156. W. Marshall and J. Verge, US Patent, 4,661,505 (1987).
157. M. Miyano, R. Shone and D. Sohn, US Patent, 4,665,203 (1987).
158. S. Djuric et al., US Patent, 4,889,871 (1989).
159. K. Koch et al., US Patent, 5,550,152 (1996).
160. R. Bergeron, US Pat. Appln., 2003/0,083,349 (2003).
161. J. Greene and D. Cooke, Brit. J. Psychiatry, Vol. 136, p 486 (1980).
162. M. Tsuda et al., US Patent, 4,644,012 (1987).
163. T. Kelly et al., US Patent, 5,247,102 (1993).
164. P. Chiesi and L. Pavesi, US Patent, 5,504,105 (1996).
165. M. Ferrari, US Patent, 5,973,169 (1999).
166. C. Albanese et al., Biochem. Biophys. Res. Commn., Vol. 199, p 930 (1994).

167. K. Head, Altern. Med. Rev., Vol. 4, p 10 (1999).
168. K. Kitatani and H. Morii, Nippon Rinsho, Vol 56, p 1537 (1998).
169. G. Kelly, US Patent, 6,340,703 (2002).
170. H. Kuenzer et al., US Pat. Appln., 2004/0,034,017 (2004).
171. Y. Gao and M. Yamaguchi, Mol. Cell Biochem., Vol. 194, p 93 (1999).
172. E. Sugimoto and M. Yamaguchi, Biochem. Pharmacol., Vol. 59, p 471 (2000).
173. M. Smith et al., Proc. Natl. Acad. Sci., Vol. 94, p 9866 (1997).
174. D. Ben-Shachar et al., Ann. Neurol., Vol. 32, S105 (1992).
175. S. Levine, Brain Res., Vol. 760, p 298 (1997).
176. R. Bergeron, US Patent, 6,083,966 (2000).
177. R. Bergeron, US Patent, 6,525,080 (2003).
178. R. Bergeron, US Patent, 6,559,315 (2003).
179. R. Bergeron, US Pat. Appln., 2003/0,236,417 (2003).
180. M. Chorghade et al., US Pat. Appln., 2003/0,225,287 (2003).
181. M. Chorghade et al., US Pat. Appln., 2003/0,229,231 (2003).
182. T. Dooley, Oncol. Res., Vol. 6, p 50 (1998).
183. T. Dooley et al., Skin Pharmacol., Vol. 7, p 188 (1994).
184. A. McEviley and R. Iyengar, Crit. Rev. Food Sci., Vol. 32, p 253 (1992).
185. K. Shimizu et al., Planta. Med., Vol. 66, p 11 (2000).
186. K. Shimizu et al., Planta. Med., Vol. 64, p 408 (1998).
187. T. Yokota et al., Pigment Cell Res., Vol. 11, p 355 (1998).
188. T. Ha et al., Bull. Korean Chem. Soc., Vol. 22, p 97 (2001).
189. K. Likhitwitayawvid et al., J. Nat. Prod., Vol. 64, p 1457 (2001).
190. M. Torihara et al., Euro. Pat., 0,341,664 (1992).
191. R. Bicki et al., US Patent, 4,225,619 (1980).
192. N. Heindel et al., US Patent, 5,216,176 (1993).
193. D. Bender et al., US Patent, 4,398,031 (1983).
194. Y. Miura et al., US Patent, 5,399,785 (1995).
195. A. LaGrange et al., US Patent, 5,468,472 (1995).
196. T. Shinomiya et al., US Patent, 5,880,314 (1999).
197. Y. Rhee et al., US Patent, 6,093,836 (2000).
198. L. Hu, US Patent, 6,132,740 (2000).
199. E. Collington, US Pat. Appln., 2002/0,155,075 (2002).
200. S. Haremza et al., US Pat. Appln., 2002/0,025,301 (2002).
201. B. Harichian et al., US Pat. Appln., 2004/0,042,983 (2004).
202. S. Bradley et al., US Patent, 6,541,473 (2003).
203. S. Bradley et al., US Patent, 6,590,105 (2003).
204. H. Richard et al., US Patent, 5,976,558 (1999).
205. H. Pyo et al., US Patent, 6,503,941 (2003).
206. T. Takagaki et al., US Patent, 5,068,364 (1991).
207. K. Kyogaku et al., US Patent, 4,085,135 (1978).
208. A. Vastone et al., US Patent, 4,656,305 (1987).
209. R. Jones et al., US Patent, 4,477,454 (1984).
210. N. Beuscher et al., US Patent, 5,489,585 (1996).
211. T. Bailey and D. Young, US Patent, 6,440,985 (2002).
212. K. Kees et al., US Pat. Appln., 2003/0,186,967 (2003).
213. M. Chorghade et al., US Pat. Appln., 2003/0,220,504 (2003).
214. M. Chorghade et al., US Pat. Appln., 2004/0,002,613 (2004).
215. K. Lee et al., US Patent, 5,637,589 (1997).
216. K. Lee et al., US Patent, 5,726,204 (1998).

217. K. Lee et al., US Patent, 5,847,165 (1998).
218. L. Feuer et al., US Patent, 3,864,362 (1975).
219. L. Feuer, et al., US Patent, 3,907,830 (1975).
220. L. Feuer et al., US Patent, 4,166,862 (1979).
221. R. Sandin et al., Organic Syntheses, Coll. Vol. 2, p 100 (1966).
222. W. Bauta et al., US Pat. Appln., 2003/0,105,340 (2003).
223. T. Althuis et al., US Patent, 4,283,563 (1981).
224. W. Mendelson et al., US Patent, 5,599,988 (1997).
225. P. Astles et al., US Patent, 6,211,234 (2001).
226. T. Fukumoto et al., US Patent, 5,621,146 (1997).
227. R. Rainey et al., US Patent, 4,400,542 (1983).
228. R. Dillard et al., US Patent, 5,552,441 (1996).
229. B. Sreenivasan et al., US Patent, 5,981,775 (1999).
230. P. Pansegrau et al., US Patent, 5,840,997 (1998).
231. T. Nakamatsu et al., US Patent, 5,304,677 (1994).
232. P. Neumann et al., US Patent, 4,996,354 (1991).
233. C. Kramer, US Patent, 4,127,560 (1978).

# 10 Resorcinol Chemistry in Photoresist Technology

## 10.1 Introduction

In the semiconductor industry, the manufacture of an integrated circuit (commonly known as a chip or microchip) starts with a semiconductor silicon wafer on which thousands or millions of tiny transistors, capacitors and resistors are fabricated. The integrated circuit (simply called an "IC") can function as an amplifier, oscillator, timer, counter, computer memory or microprocessor. Nowadays, IC is at the heart of all the electronic equipments and, used in the manufacture of computers, navigational systems, pocket calculators, industrial monitoring and control systems, digital watches, word processors and communication networks.

Fabrication of integrated circuits relies heavily on photolithography, which is the process of transferring geometric shapes and patterns from a mask to the surface of a silicon wafer [1]. The performance of microprocessors and DRAM chips are directly coupled to the dimensions of the transistors and wiring interconnects that comprise the circuits. The patterns of conductors on circuit boards and the tiny transistors on microchips are printed by a lithographic process on a layer of polymeric materials, called photoresists. Photoresist technology has been refined to the point where millions of micron sized devices can be fabricated simultaneously and reliably on silicon substrates.

Photolithography is the process of making precise designs on thin films of photoresist materials, which consists of the following steps [1–4].

- First, the wafer surface is cleaned to improve the resist adhesion and then, the photoresist material is spin coated on the wafer.
- In the next step, the resist layer is selectively exposed to a form of radiation, such as the ultraviolet (UV) light, electron beam or X-rays, by means of an exposure tool to transfer the image of a mask or reticle to the surface of resist coated wafer.
- When the resist is developed by immersing the wafer in an alkaline solution, the patterns in the resist are formed. Depending upon the nature of resist materials applied, positive or negative resist, on the surface of silicon wafer, positive or negative patterns can be obtained. With positive photoresist

materials, the exposed regions are rendered soluble, since they are degraded by the exposure. On the other hand, due to the cross linking reactions during the radiation, the exposed regions become insoluble with negative resists.
- After developing the wafer, it is then subjected to etching by treatment with buffered HF or by using dry plasma etching methods. In either of etching process, the photoresist material must protect the regions of silicon wafer it still covers, while allowing the exposed areas of the substrate to be removed.
- Finally, after the pattern is transferred onto the substrate (silicon wafer) surface, the photoresist is stripped off and discarded.

A schematic representation of positive and negative photoresist processes employed in photolithography is illustrated in Figure 10.1 [4].

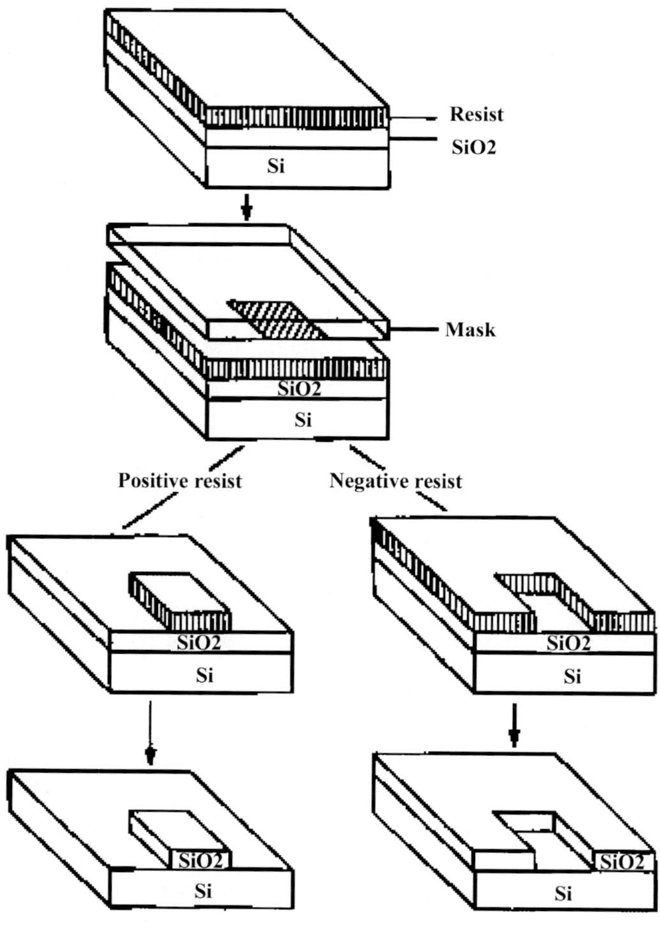


**Figure 10.1.** Positive and negative photoresist processes employed in photolithography

## 10.2 Positive and Negative Photoresist Materials

A photoresist composition consists of a polymer (resist), photoactive compound (PAC) and solvents. In the case of resist polymers, they are classified in to two groups, namely the positive and negative photoresists.

### 10.2.1 Positive Photoresists

Positive photoresist compositions normally comprise a photoactive compound (Figure 10.2), a phenolic novolac resin (Figure 10.3) and solvents.

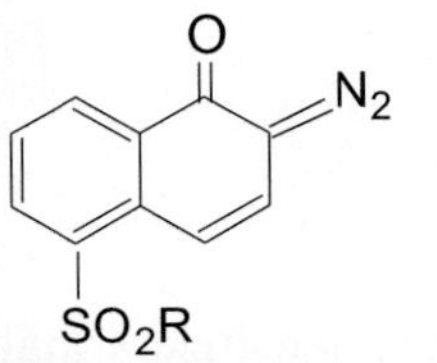


"Naphthoquinone Diazide"

**Figure 10.2.**

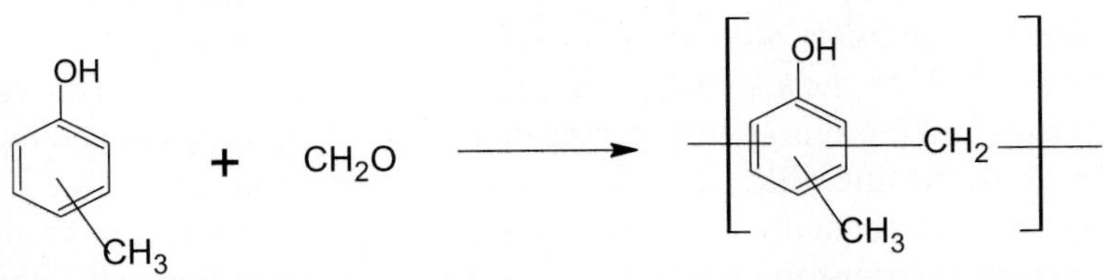


o, p and m - Cresols Formaldehyde "Phenolic Novolac Resin"

**Figure 10.3.**

The photoactive compound, which is usually naphthoquinone diazide, is insoluble in basic solvents and also inhibits the dissolution of base soluble resist polymer. The novolac resin (resist) is relatively a low molecular weight (M.Wts. $=<$ 10 000) compound and prepared from the reaction of alkyl substituted phenols with formaldehyde.

Upon exposure to a light source, the photoactive or sensitizer compound loses 1 mol of nitrogen followed by a Wolff rearrangement, after which it is converted in to an acid derivative by reacting with water present in the resist composition. The photoactive compound decomposition mechanism is outlined in Figure 10.4.

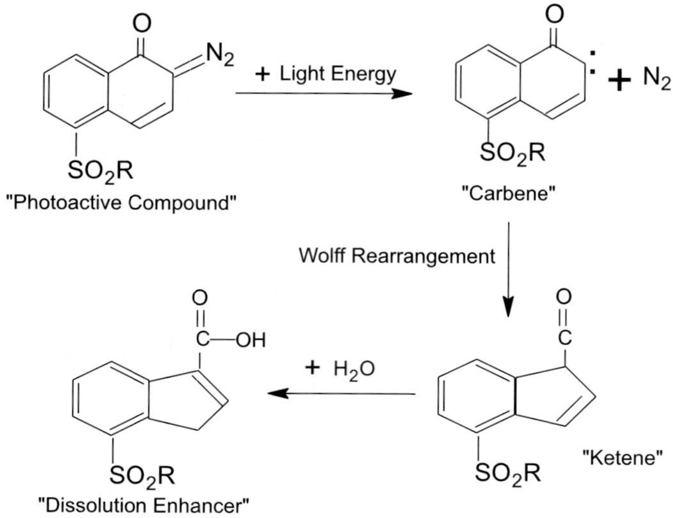


**Figure 10.4.**

The formation of carboxylic acid derivatives from the sensitizers makes the exposed area of photoresists to be more soluble in aqueous solutions of metallic or quaternary ammonium hydroxides, called the developers. Unexposed regions remain much less soluble in developer (about 100 times less soluble), due to the presence of hydrophobic naphthoquinone diazide, which is the primary mechanism of image formation in positive resists.

Positive resists have the advantages of producing higher resolutions because the unexposed film regions are not permeated or essentially unchanged by the developer. Thus, the line width and shape of a pattern is precisely retained. Novolac resins contain aromatic rings connected through the methylene bridges, which are fairly resistant to chemical attack and show high thermal stability for plasma etching.

## 10.2.2 Negative Photoresists

Negative resist compositions are based on a cyclized or partially cyclized rubber with azide type photosensitizers. On exposure to UV light, azide compounds decompose to produce nitrenes, which crosslink the polymer systems. Negative photoresists have been the workhorse of the microelectronics industry, and was replaced when the integrated circuits resolution requirements exceeded 3 μm. The advantages of negative resists over positive resists include better adhesion to substrate without pre-treatment and have faster photospeed. With negative resists, the developer permeates both the exposed and unexposed regions of the film, which can lead to swelling or dissolution of film and also distort the pattern.

## 10.3 Radiation Sensitivity of Photoresists

Integrated circuits with increasing sophistication and higher integration require the need for the fabrication of ultra-fine patterns with line widths of less than one micron in the manufacturing process of semiconductor substrate. Photolithography normally involves the use of actinic light with wavelengths in the near to mid UV-region (300 to 450 nm) or deep UV (DUV) for exposing photoresists. In the case of Hg-Xe lamp light source, the out put in the spectral region 300 to 450 nm has several strong peaks (g-, h- and i-lines) which may be isolated and exposed to the resists. Exposure of the positive resist containing novolac resin and naphthoquinone diazide compounds with light source of 436 nm (g-line) could produce resist line width of less than one micron. Similarly, using i-line (365 nm) light exposure, photoresist line width in the range of 0.35 to 0.5 μm could be achieved.

The vast majority of i-line processes use positive single layer resists based on phenolic novolac and diazidonaphthoquinone or naphthoquinone diazide (DNQ) technology. Though i-line chemistry is robust and mature, incremental improvements in resolution, exposure latitude and depth of focus are achieved by the formulation changes in the novolac and DNQ systems. In order to achieve these improved properties, photoresists with good thermal stability, good adhesion to substrates, high purity, aqueous base development and compatibility with aggressive plasma etching processes are required.

Most integrated circuits produced are printed with i-line (365 nm) exposure technology. For producing future generation integrated circuits, photoresist materials should be exposed with deep UV (DUV, 248 nm) to obtain higher resolution. With i-line technology, a resolution of approximately 0.5 μm can be achieved, whereas a 0.25 μm resolution can be obtained with 248 nm (DUV) photoresists. The novolac-DNQ based resists have not been utilized for DUV integrated circuit applications, mainly due to low sensitivity and the absorbance of DNQ photoactive compound (PAC) at 248 nm. Novolac resin is too strongly absorbing for use in 248 nm exposure formulations, but shows an absorbance maximum at 257 nm that provides transparency similar to poly (hydroxy styrene) [5].

## 10.4 Resists with Good Lithographic Properties

Semiconductor devices and electronic components with increased density or compactness in sizes are the recent trends in the electronic technology. In order to manufacture these devices, great improvements in the photoresist compositions possessing the following lithographic properties are needed [6].

- Good resolution capabilities in both the micron and sub micron ranges without incomplete development in the exposed areas (no scumming).

- Higher heat resistant resist images (above 120 °C).
- Relatively fast photospeeds.
- Good adhesion to substrate (silicon wafer).
- Good developer dissolution rates.
- Near to absolute vertical profiles between the exposed and unexposed photoresist after development.
- Resistance to etching solutions and plasma etching techniques.

In an effort to improve the lithographic properties of positive photoresist compositions, resorcinol chemistry has been utilized, and the details are presented below.

## 10.5 Resorcinol Chemistry in Photoresist Applications

### 10.5.1 Heat Resistant Novolac Resins

For the production of highly integrated ICs, positive photoresist compositions capable of producing excellent resolutions have been used. The resolution of photoresists can be enhanced not only by improving the resist properties, but also by changing the methods of producing ICs, for example, changing the wet etching to a dry etching method. In the dry etching process, the substrate is treated with reactive ions, which results in the increase of substrate temperature. Because of this temperature rise, the resist patterns may undergo thermal deformation and reduce dimension accuracy. Also, the highly reactive chloride and fluoride ions can impair the resist patterns.

A positive photoresist compositions having enhanced dry etching resistance properties, and also improved resolution and thermal resistance were developed from the resorcinolic novolac resins synthesized using resorcinol, 2-methyl resorcinol, *m*-cresol and 3,5-dimethyl phenol (Figure 10.5) [7].

When the resorcinol based novolac resins were employed in the positive resist applications, resist films exhibited high heat resistance, dry etching resistance and resolution properties, and the results are presented in Table 10.1.

After the photoresist film has been selectively exposed and developed, the film must be able to withstand the rigors of subsequent processing conditions without significant physical changes. For example, with the plasma etching process, the temperatures of the resist films may exceed their glass transition temperatures (Tg), which is around 100 °C for the conventional phenolic novolac resins. The Tgs of phenolic novolac resins can be increased by increasing their molecular weights, which could drastically reduce their dissolution rates in the developer solutions, and also film resolutions.

The dissolution rate of resist films is also decreased when the concentrations of multi-substituted phenol is increased in the novolac preparations.

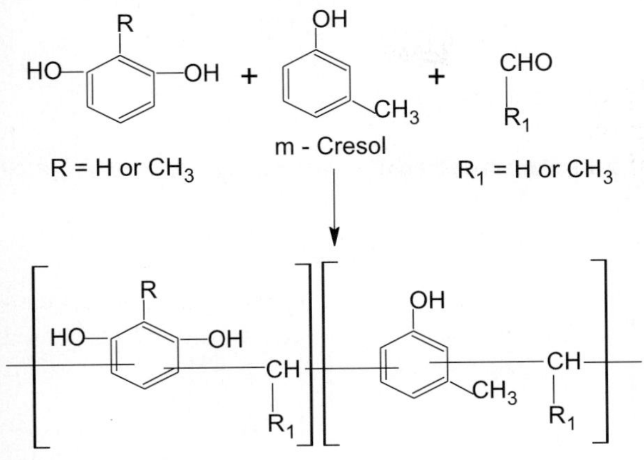


**Figure 10.5.**

**Table 10.1.** Heat resistance novolac resins and photoresist properties

| Molar ratios of phenols | | | | Aldehyde used | Heat resistance tempera-ture | Dry etch resistance (selection ratio) | Reso-lution |
|---|---|---|---|---|---|---|---|
| Resor-cinol | 2-Methyl resorcinol | *m*-Cresol | 3,5-Dimethyl phenol | | (°C) | | (μm) |
| 100 | | | | Acetaldehyde | 170 | 4 | 0.9 |
| 30 | | 70 | | Acetaldehyde | 160 | 4.2 | 0.8 |
| 20 | | 64 | 16 | Acetaldehyde | 170 | 4.3 | 0.7 |
| | 100 | | | Formaldehyde | 190 | 3.9 | 0.9 |
| 20 | 80 | | | Acetaldehyde | 160 | 4 | 0.8 |
| Control resin: *m*-cresol/*p*-cresol/formaldehyde resin | | | | | 110 | 2.1 | 1.2 |

Selection ratio = speed at which $SiO_2$ film is etched/speed at which resist pattern is etched
Data from [7]

When 2-methyl resorcinol was employed, novolac resins possessing high glass transition temperatures with improved dissolution rate in the developer solutions were obtained (Figure 10.6) [8].

Resin compositions, their glass transition temperatures and dissolution rates are presented in Table 10.2.

The resolution of the resist film, obtained from the 2-methyl resorcinol based novolac, was measured by SEM, and was found to be 0.7 μm.

The use of mono-substituted resorcinol can provide novolac resins or polymers with linear structures, and their presence could improve the solution solubility and enhance developer solubility. 4-Chloro resorcinol was allowed to react with 2,6-bis-(hydroxy methyl)-*p*-cresol to synthesize polymeric type products as shown in Figure 10.7 [9–11].

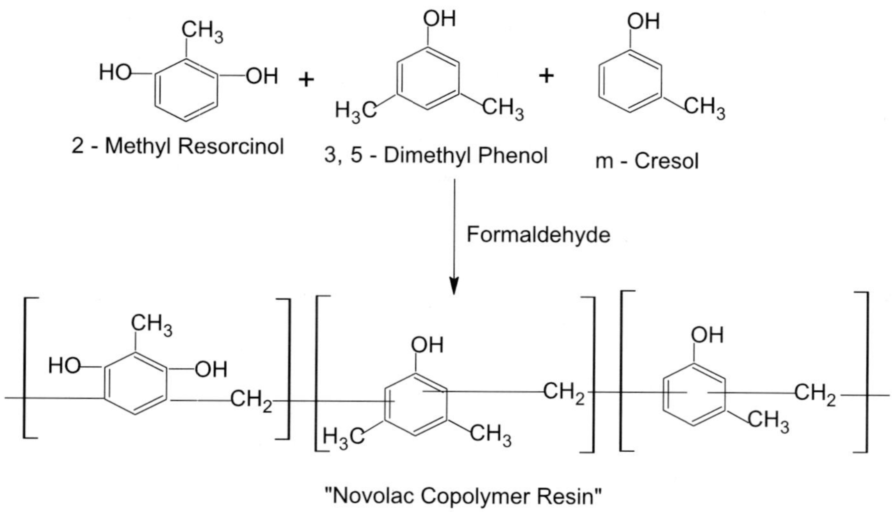


**Figure 10.6.**

**Table 10.2.** Novolac resins from 2-methyl resorcinol (2-MR), *m*-cresol (MC) and 3,5-dimethyl phenol (3,5-DMP) and properties

| Novolac resin composition (wt%) | | | Mol. weight (m.w., by GPC) | Tg (°C) (by DSC) | Dissolution rate |
|---|---|---|---|---|---|
| MC | 3,5-DMP | 2-MR | | | |
| 95 | 5 | 0 | 8990 | 89 | NA |
| 82 | 18 | 0 | 22 100 | 80 | NA |
| 59 | 41 | 0 | 35 500 | 127 | NA |
| 39 | 34 | 27 | 6520 | 142 | 2670[a] |
| 38 | 31 | 31 | 4550 | 129 | NA |
| 55 | 36 | 9 | 19 300 | 136 | 401[a] |
| 59 | 35 | 6 | 54 900 | 138 | 12[b] |
| 54 | 39 | 7 | 27 200 | 138 | 208[b] |
| 39 | 50 | 11 | NA | 144 | 33[b] |

NA = Not available
[a] A°/Sec. in 0.135 N Tetramethyl ammonium hydroxide (TMAH) solution
[b] A°/Sec in 0.27 N TMAH solution.
Data from [8]

Due to possible higher molecular weights, the novolac polymer obtained from 4-chloro resorcinol showed high glass transition temperatures compared to the novolac resins synthesized using *m*- and *p*-cresols (Table 10.3). Photoresist compositions containing 4-chloro resorcinol novolac showed greater thermal stability.

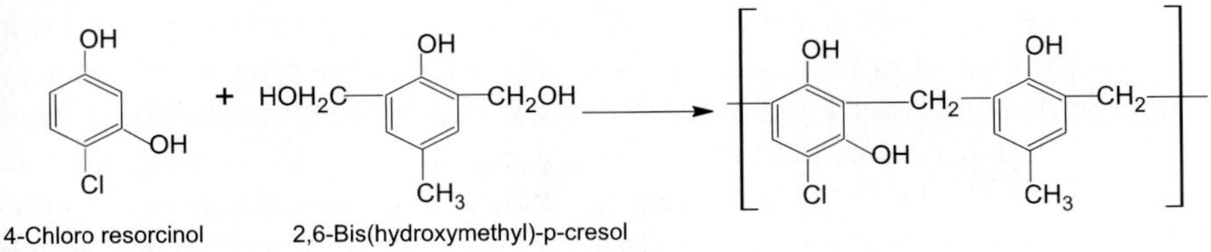


**Figure 10.7.**

**Table 10.3.** Novolac resins from 2,6-bis(hydroxymethyl)-*p*-cresol (BHMC) and 4-chloro resorcinol (4-CR) and dissolution properties

| Novolac resin composition (mol) BHMC | 4-CR | Tg (°C) | Dissolution Time (s)[a] NaOH (0.123 N) | TMAH (0.3 N) |
|---|---|---|---|---|
| 1 | 1.05 | 176 | 13 | 6 |
| 1 | 1.02 | 186 | NA | NA |
| 1 | 0.85 | > 200 | 37 | 6 |
| 1 | 0.65 | > 200 | 200 | 17 |
| Control Resin: *m*-, *p*-cresol novolac | | 107 | 63 | 50 |

[a] Approximate film thickness = 1 μm
TMAH = Tetramethyl ammonium hydroxide
Data from [10]

## 10.5.2 Photoactive or Photosensitizer Compounds

The performance of photoresists with respect to their photospeed, line and space resolutions, scum and image profiles are important when the resist patterns are formed on silicon wafers.

Photospeed or sensitivity is the incident exposure energy required to effect enough chemical change in the resist to obtain the desired image corresponding to that of the mask. Photospeed increases as the energy required to produce the image decreases. Greater sensitivity or lower exposure energy of a photoresist could increase the wafer throughput. Resolution, for positive resists, is a function of both the rate of degradation of photoactive compounds (PAC) and the rate of solubility of the resist on exposure. Resolution is the minimum line and space dimensions measured in microns in which, the smaller the value shows better resolution of the resist.

Scum is associated with the presence of small quantities of residual photoresist material remaining in exposed and developed areas of the wafer substrate. Its presence can present problems in the subsequent processing steps. Profile is the slope of the side wall of the resist lines at the smallest resolutions and determined by the scanning electron microscopic (SEM) techniques.

The properties and performance of photoresist compositions depend not only on the type of radiation source, but primarily on the type of photoactive compounds employed. The photoactive compounds or photosensitizers used in the positive resist formulations are predominantly naphthoquinone diazide derivatives as shown in Figure 10.2. The nature of R group influences the solubility characteristics of the sensitizers and their absorption properties, which is usually 1,2,3-trihydroxy benzophenone. Photoactive compounds containing polyhydroxy benzophenone groups show high absorption at 365 nm (i-line) region, and therefore resist compositions with these sensitizers provide poor pattern form from i-line exposure. For these reasons, benzophenone type photoactive compounds have been employed for exposure to g-line (436 nm) lithography.

Resorcinol based benzophenone compounds have been synthesized and utilized in the development of photoactive compounds. 2,2′,4,4′-Tetrahydroxy-3,3′-dimethyl benzophenone and 4,2,4′-trihydroxy-3′-benzophenone compounds (Figure 10.8), prepared from 2-methyl resorcinol, were further reacted with naphthoquinone-(1,2)-diazide-(2)-5-sulfonyl chloride to obtain DNQ photoactive compounds [12].

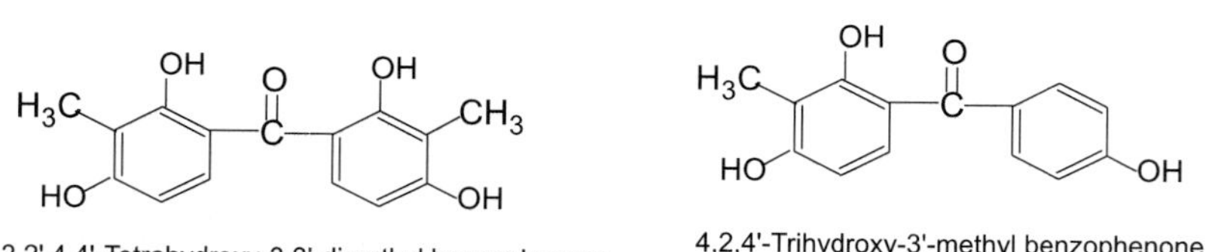


2,2',4,4'-Tetrahydroxy-3,3'-dimethyl benzophenone

4,2,4'-Trihydroxy-3'-methyl benzophenone

**Figure 10.8.**

Positive resist compositions containing phenolic novolac resin and the above benzophenone sensitizers were coated on silicon wafers, exposed to light having a wavelength of 436 nm (g-line) and developed, and showed excellent resolution as evidenced from the high $\gamma$-values determined.

To increase the sensitivity of photoresist compositions, DNQ derivatives obtained from the polyhydroxy flavans synthesized from resorcinol, 2-methyl resorcinol and ketones, such as acetone and cyclohexanone, were employed (Figure 10.9) [13–16].

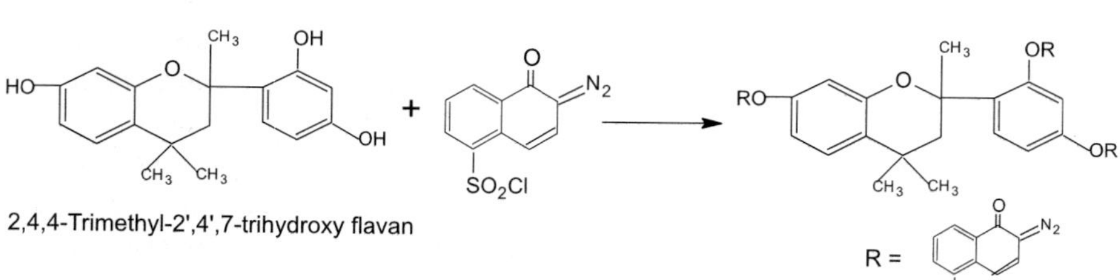


2,4,4-Trimethyl-2',4',7-trihydroxy flavan

**Figure 10.9.**

These photoactive compounds were employed in g-line and i-line exposure applications. When used in formulations containing m and *p*-cresol novolac and exposed to i-line lithography, a resist film with a resolution of 0.5 μm, $\gamma$-value of 5.2, heat resistance of 130 °C, depth of focus 3.5 μm and narrow wall profile properties was obtained [15].

Photoresists employed for i-line (365 nm) generally contain lower molecular weight novolac resins. The use of poly(naphthoquinone diazide ester) derivatives as photoactive compounds could enhance the photospeed or sensitivity and resolution of the resulting resists. With resorcinol, cyclic polyhydroxy compounds, called resorcinarenes, were synthesized and utilized to develop poly(DNQ) type photosensitizers [17]. A reaction scheme illustrating the synthesis is shown in Figure 10.10.

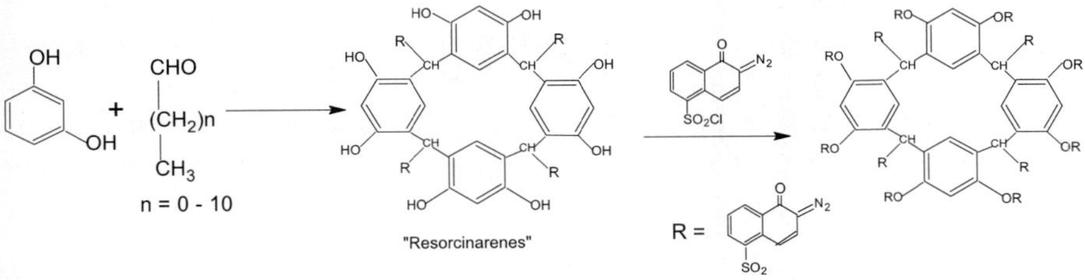


**Figure 10.10.**

The photoactive compounds thus prepared from resorcinarenes exhibited good solubility in photoresist solvents such as ethyl acetate and propylene glycol methyl ether acetate, and resist compositions showed excellent storage stability and high sensitivity for exposure to i-line lithography.

Photoactive compounds containing higher number of sensitizer groups are beneficial for fast photospeeds, higher resolutions and enhanced developer dissolutions. Sensitizer compounds of higher molecular weights and having greater number of naphthoquinone diazide groups substituted onto these compounds were synthesized from trinuclear novolac oligomers. These compounds were prepared from the reaction of resorcinol with 2,6-bis(hydroxy methyl)-*p*-cresol as illustrated in the reaction scheme shown in Figure 10.11 [18–20].

The details on the composition of photoactive compounds, their solution solubility and photoresist performance with respect to photospeed, resolution and wall angle are presented in Table 10.4.

From Table 10.4 results, the positive photoresist compositions containing photoactive compounds having penta and tetra esters of naphthoquinone diazide groups produced films with good resolution and wall profiles.

Novel radiation sensitive photoresist compositions suitable as a positive resist, which showed good developability, sensitivity, resolution and depth of focus in the i-line lithographic process, were obtained by employing resorcinol based naphthoquinone diazide photosensitizer compounds (Figure 10.12) [21,22].

**Table 10.4.** Photoactive compounds (PAC) made from the reaction product of resorcinol derivatives and 2,6-bis(hydroxymethyl)-*p*-cresol and their positive resist properties

| Resorcinol derivative | Photoactive compound | | Solution stability in hours[a] | | | Photoresist performance[b,c] | | |
|---|---|---|---|---|---|---|---|---|
| | Tetra-ester (%) | Penta-ester (%) | RT | 35 °C | 45 °C | Photospeed ($mJ/cm^2$ at 0.35 μm) | Resolution (μm) | Wall angle (degrees) |
| Resorcinol | 56.7 | 28.7 | 138 | 65 | 65 | 128 | 0.33 | 87.4 |
| 2-Methyl resorcinol | 23.6 | 68.7 | > 1558 | > 1558 | 129 | 129 | 0.345 | 88 |
| 4-Ethyl resorcinol | 27.7 | 66.3 | > 3119 | > 3119 | 528 | NA | NA | NA |
| 4-Hexyl resorcinol | 32.6 | 57.8 | > 2416 | > 2416 | > 2416 | NA | NA | NA |

[a] Solvent used was a mixture containing 90% ethyl acetate, 5% butyl acetate and 5% xylene
[b] Novolac resin used was made from *m*-cresol, *p*-cresol, *o*-cresol and formaldehyde monomers
[c] Solvent mixture comprised of ethyl acetate, anisole and amyl acetate
Data from [20]

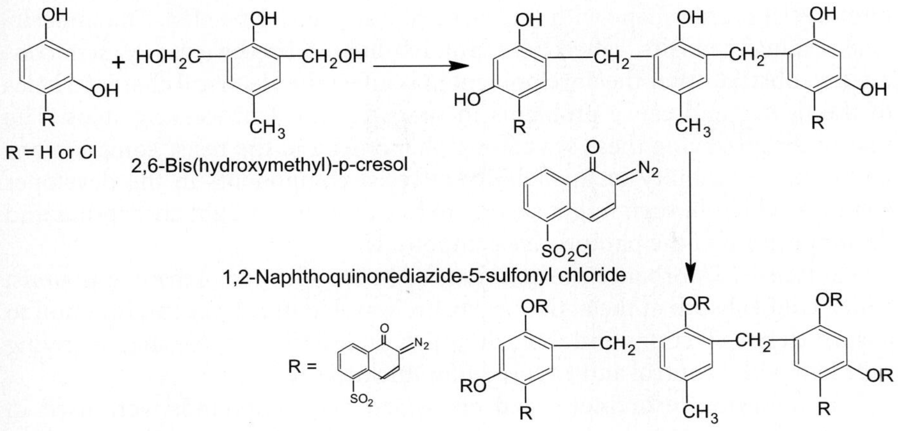


**Figure 10.11.**

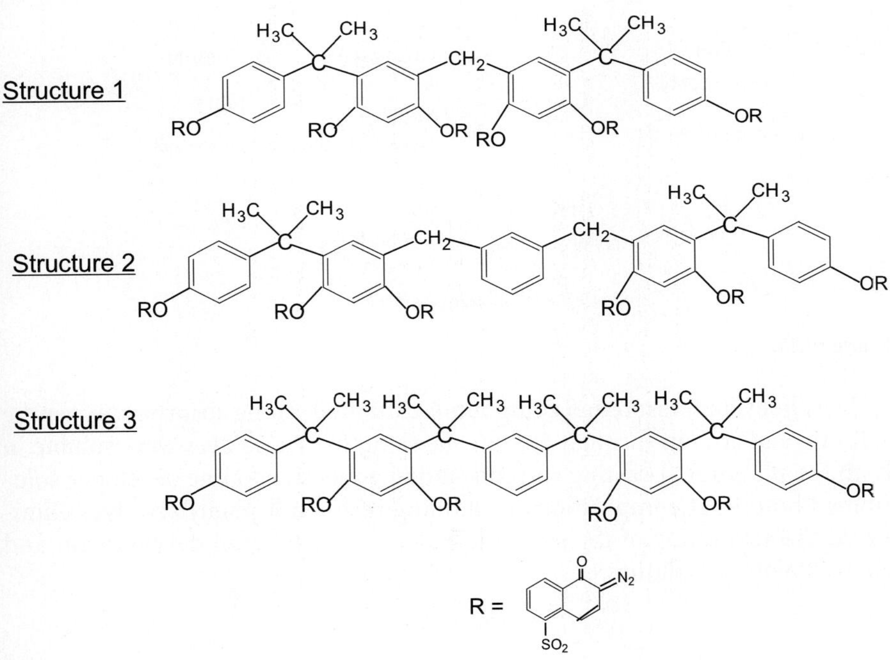


**Figure 10.12.**

## 10.5.3 Positive Dye Photoresist Compositions

For the fabrication of very large scale integrated (VLSI) electronic circuits on silicon chips, photoresist materials providing high degree of resolution are required. In the manufacture of miniaturized integrated circuits, a scum free photoresist profile shape with improved photospeeds is needed. The undesirable residual deposits, called the scum, left behind on the exposed semiconductor substrate after the development may affect the electrical characteristics of the device and cause problems in the subsequent processing steps. The reason for observing these scumming phenomena in the resist compositions is the incompatibility or insolubility of resist components in the developer solution, which in turn might be due to low or reduced light absorption and decomposition of the photoactive compounds.

Photoresist absorbance can be enhanced by the addition of dye compounds, which could absorb at the actinic exposure wavelengths. Dyes can function to absorb stray reflections and thus, diminish undesirable profile shapes, giving excellent width control and exposed film thickness retention.

The following resorcinol based phenylazo dye compounds were used in positive photoresist compositions (Figure 10.13) [22–24].

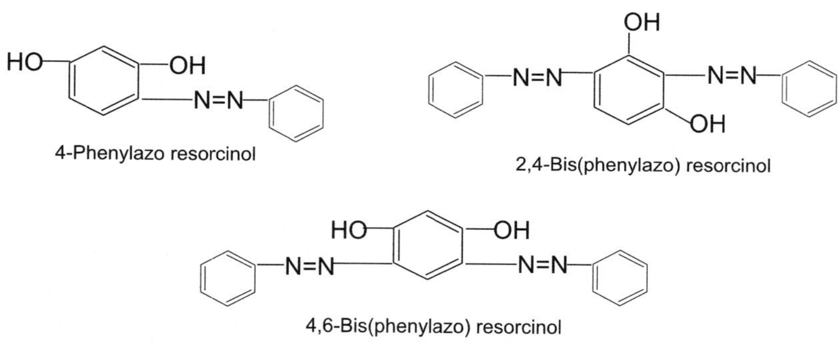


**Figure 10.13.**

Bis(phenylazo) resorcinol compounds exhibited strong absorbance over the full range of 365 nm through 436 nm wavelengths. These dyes were soluble in both the photoresist casting solvents and in aqueous alkaline developer solutions. Photoresist compositions containing resorcinol-phenylazo dyes eliminated the scumming of the semiconductor substrate upon development and showed better resolution.

## 10.5.4
## Dissolution Inhibitors in Photoresist Compositions

Semiconductor industry is in need of constant improvements in the photolithographic processes to reduce the microelectronic feature sizes. Lithography involving shorter wavelengths, namely deep UV radiation (less than 300 nm), can provide resolutions of the order of less than 0.15 μm with chemically amplified photoresist (CAR) materials. Deep ultraviolet (DUV) resists are typically poly(hydroxy styrene) (PHOST) based polymers with a photoacid generator (PAG) providing the solubility change. The chemically amplified (CA) resists based on PHOST used at 248 nm wavelength represent a radical departure from i-line's (356 nm) novolac based resist chemistry. The PHOST has protecting groups that becomes de-protected when a PAG compound decomposes upon exposure to 248 nm wavelength light, and the polymer undergoes cleavage. The de-protection mechanism causes a polarity change in the resist polymer from hydrophobic to hydrophilic, making exposed regions soluble in the developer. The differential in solubility between the exposed polymer and the acid cleaved polymeric material serves as the basis for developing the image.

The use of dissolution inhibitors in the chemically amplified photoresist compositions can inhibit the dissolution of the resist under basic conditions but facilitate dissolution under acidic conditions after exposure to radiation. In addition, the dissolution inhibitors can reduce unexposed film loss and also, improve other resist properties such as the distribution of resist additives, resolution and etch resistance.

Calix [4] resorcinarenes based compounds were developed based on the resorcinol chemistry outlined in Figure 10.14 and employed as dissolution inhibitors [25].

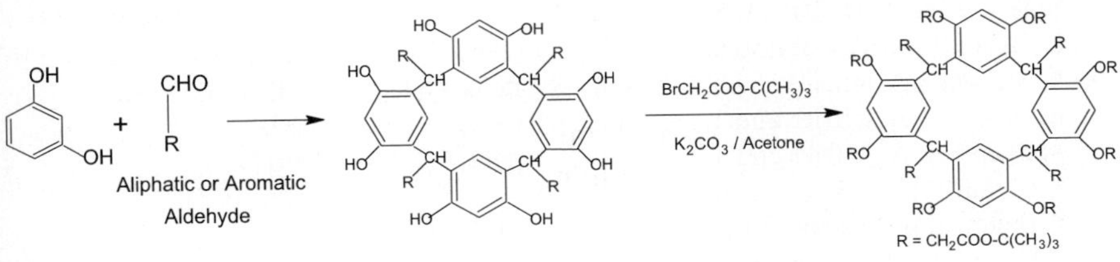


**Figure 10.14.**

The calixarene or resorcinarene esters were particularly impressive in their dissolution inhibition capability and developer selectivity was greatly increased. These compounds were expected to improve the lithographic performance of chemically amplified photoresists.

## 10.6 Summary and Future Directions

Advanced photoresist chemistries and materials have been the driving forces for the success of the electronic industry. Photoresists are the primary reasons for observing modern electronic revolution. With phenolic novolac/DNQ photoresists, i-line lithography has been very successful in developing 0.35 μm resolution patterns on the silicon wafers. The resorcinol based advanced sensitizers and high heat resistant novolacs were observed to enhance the performance of positive photoresists in i-line exposure technology. Electronic industry is moving to shorter irradiation wavelengths, such as 248 nm (DUV) and 193 nm, to obtain finer features. Chemical amplification resists, for example PHOST with PAG, have been successful in the DUV region lithography. Developing and employing unique resorcinol chemistries could aid in the improvements and performances needed in photoresist compositions, which could provide viable manufacturing technology for the production of sub-0.20 μm geometries.

## References

1. W. Moreau, Semiconductor Lithography: Principles, Practices and Materials, Plenum Press, New York (1988).
2. L. Thompson, C. Wilson and M. Bowden, Introduction to Microlithography, ACS Symposium Series (1983).
3. Photolithography, Texas Engineering Extension Services (2000).
4. www.ee.gatech.edu/research/labs/vc/graphics/photolith/plithp1.gif. Cited 21 July (2004).
5. Wilson Research Group, Photomask Lithography – University of Texas at Austin, TX (2000).
6. A. Jeffries, A. Blakeney and M. Toukhy, US Patent, 4,992,596 (1991).
7. Y. Hosaka et al., US Patent, 5,087,548 (1992).
8. L. Bogan, Euro. Patent Appln., 0,435,502 A1 (1990).
9. A. Blakeney, A. Jeffries and T. Sarubbi, US Patent, 4,837,121 (1989).
10. A. Blakeney, A. Jeffries and T. Sarubbi, US Patent, 4,970,287 (1990).
11. A. Blakeney, A. Jeffries and T. Sarubbi, US Patent, 5,024,921 (1991).
12. Y. Uetani et al., U.S Patent, 5,124,228 (1992).
13. Y. Uetani and H. Nakanishi, US Patent, 5,290,657 (1994).
14. Y. Uetani, H. Nakanishi and Y. Doi, US Patent, 5,283,155 (1994).
15. Y. Doi et al., Euro. Patent, 0,510,671 B1 (1992).
16. Y. Uetani, J. Tomioka and S. Lee, US Patent, 6,383,709 (2002).
17. H. Iida et al., US Patent Appln., 2003/0,175,613 (2003).
18. A. Jeffries, A. Blakeney and M. Toukhy, US Patent, 4,957,846 (1990).
19. A. Jeffries, A. Blakeney and M. Toukhy, US Patent, 4,992,356 (1991).
20. A. Zampini and H. Sanford, US Patent, 6,051,358 (2000).
21. K. Inomata, M. Akiyama, S. Iwanaga and A. Tsuji, US Patent, 6,007,961 (1999).
22. K. Inomata, K. Douki, T. Mizumachi and I. Iwanaga. US Patent, 6,171,750 (2001).
23. P. Trefonas, A. Zampini and D. Madoux, US Patent, 4,983,492 (1991).
24. P. Trefonas, A. Zampini and D. Madoux, US Patent, 5,164,279 (1992).
25. H. Ito, T. Nakayama and M. Ueda, US Patent, 6,093,517 (2000).

# Subject Index

Printing: Krips bv, Meppel
Binding: Stürtz, Würzburg